EVOLUTION SINCE DARWIN

The First 150 Years

EVOLUTION SINCE DARWIN

The First 150 Years

Edited by

Michael A. Bell

Douglas J. Futuyma

Walter F. Eanes

Jeffrey S. Levinton

Department of Ecology and Evolution
Stony Brook University

Sinauer Associates, Inc. • Publishers
Sunderland, Massachusetts U.S.A.

About the Cover

The portrait of Charles Darwin is by George Richmond; the sketch is Darwin's first phylogenetic tree, from his *Notebook B on Transmutation of Species* (1837–1838); the major branches of the Tree of Life image is courtesy of David M. Hillis.

EVOLUTION SINCE DARWIN: The First 150 Years

For information, address:
Sinauer Associates, Inc., 23 Plumtree Road, Sunderland, MA 01375 USA
Fax: 413-549-1118
E-mail: orders@sinauer.com; publish@sinauer.com
Internet: www.sinauer.com

Library of Congress Cataloging-in-Publication Data

Evolution since Darwin : the first 150 years / edited by Michael A. Bell ... [et al.].
p. cm.
Includes bibliographical references and index.
Proceedings of a workshop held Nov. 4-7, 2009 at Stony Brook University to mark the bicentennial anniversary of Darwin's birth and the sesquicentennial of the publication of On the Origin of Species
ISBN 978-0-87893-413-3
1. Evolution (Biology)--History--Congresses. 2. Darwin, Charles, 1809-1882--Congresses. I. Bell, Michael A.
QH361.E925 2010
576.8--dc22

2010019644

Printed in China
5 4 3 2 1

For our teachers

Contents

Preface

In 2009, worldwide celebrations marked the bicentennial anniversary of Darwin's birth and the sesquicentennial of the publication of *The Origin of Species*. But the year also celebrated the flowering of evolutionary biology. The many celebratory meetings and lectures were perhaps the most prolific and visible acknowledgments of a scientist and a science in recent history. They dwarfed the Darwin centennial events of 1959, which transpired only twelve years after the first publication of the journal *Evolution*, and six years after Watson and Crick had proposed the structure of DNA. In the interim, the whole of biology has been transformed, but so has evolutionary science. It has become a self-sustaining field of inquiry, a central focus of at least 15 new periodicals and of numerous university departments. The field has retained allegiance to the big questions posed in the *Origin*, even while incorporating new concepts, biological knowledge, and technology to address questions and explain phenomena that could not be imagined in 1859 or often even in 1959. Ironically—and possibly accounting in part for the extraordinary attention to the subject—in 2009, opposition to the idea and to teaching of evolution (and rejection of the science itself) was at an all-time frenzy in the United States and continues to spread widely, even as evolution has become increasingly integrated into the framework of modern biology and its applications.

As 2009 drew near, the Department of Ecology and Evolution at Stony Brook University worked with Samuel M. Scheiner of the National Science Foundation to develop funding for "Darwin 2009," a symposium that would conspicuously mark the occasion and include public education and outreach. The result was a meeting on November 4–7, 2009 at Stony Brook University, whose format evolved into a workshop with 22 invited speakers, additional invited discussants, and an audience of Stony Brook faculty and students. Informal discussion was encouraged and greatly amplified the value of the formal presentations. We judged that the impact of the public educational effort could be maximized by engaging the experience and outstanding capabilities of the Gottesman Center for Science Teaching and Learning at the American Museum of Natural History. The result was a separate, well attended, day-long teacher-training workshop on evolutionary biology at the Museum.

The celebrations of Darwin's work in 1959 were commemorated in several books that stand as portraits of evolutionary science at that time. Our planning committee conceived the Stony Brook workshop and this volume as a similar record of our time, which 50 years hence may well be viewed much as we now view the inspiring but less advanced ideas and understanding of 1959. It was challenging to organize a program that could fully encompass a field that has grown so enormously in scope since 1959, but we assembled a slate of speakers that would represent diverse subdisciplines within evolutionary biology, as well as perspectives from the history and philosophy of science and the applications and implications of evolution for society. We explicitly invited a diversity of speakers and discussants, with respect to nationality, age, gender, and sexual orientation.

We are gratified that all the speakers, including two who were prevented at the last minute from attending the Stony Brook workshop, enthusiastically contributed manuscripts for this volume. We did not ask the speakers to use a particular format or prescribed content for their chapters, but we requested that they discuss Darwin and try to trace the history of their subjects from his work. Thus, the chapters of this book are as varied as those in the 1959 symposium records. Therefore, this volume provides an overview of the current state of major research areas in evolutionary biology for experienced investigators, graduate students, and advanced undergraduates. It also places many of the major themes of evolutionary biology in their historical context and attempts to look to the future. We hope that this volume will serve as a milepost by which our intellectual successors can measure how far evolutionary science has come in another 50 years, when they celebrate the 200th anniversary of the publication of *The Origin of Species*.

Because this volume is the product of a workshop, we are indebted both to those who contributed to the organization and execution of the workshop and to production of the book. Massimo Pigliucci was crucial to organizing and bringing the Darwin 2009 workshop to Stony Brook. Ann Brody of the Office of Conferences and Special Events at Stony Brook University provided valuable advice and then executed all the arrangements for housing, meals, local transportation, and facilities. M. Caitlin Fisher-Reid served as the graduate assistant and managed a huge array of tasks for the workshop with truly exceptional commitment, skill, and efficiency. Jessica Gurevitch, Chair, Department of Ecology and Evolution, provided departmental staff support, funding, and resources. James D. Staros, the former Dean of the College of Arts and Sciences and Eric W. Kaler, Provost of Stony Brook University, provided financial support for the workshop, and Lawrence B. Martin, the Dean of the Graduate School, contributed support for graduate student participation in the workshop. Many graduate students generously served as personal contacts for the visiting speakers and discussants or executed numerous tasks that arose during the workshop. An external advisory board, which included May R. Berenbaum, Sean B. Carroll, Scott F. Gilbert, David M. Hillis, Andrew H. Knoll, Jay Labov, Antonio Lazcano, Michael Lynch, Nancy Moran, Michael Ruse, Johanna Schmitt, Eugenie Scott, George N. Somero, Kim Sterelny, Tim D. White, and David Sloan Wilson, contributed valuable advice in the early stages of the workshop. The workshop would not have been possible without these contributions, and we are grateful for them all.

The public outreach workshop at the American Museum of Natural History was organized and executed by Lisa J. Gugenheim (Senior Vice-President for Education, Government Relations and Strategic Planning) and James Short (Director of the Center) with the aid of curators Joel L. Cracraft, Rob DeSalle, and Susan L. Perkins.

We would like to thank Samuel M. Scheiner for managing our proposal for the workshop and for encouraging production of this volume, and George W. Gilchrist for managing the grant for Darwn 2009 at the National Science Foundation. Ivar Strand and Celeste Radgowski in The Office of Sponsored Programs at Stony Brook University helped revise the grant for Darwin 2009 when it was converted from a large meeting to a more focused workshop. This volume and the workshop on which it is based were supported by National Science Foundation Grant No. 0838101. Any opinions, findings, conclusions, or recommendations expressed in this volume are those of the authors and do not necessarily reflect the views of the National Science Foundation.

Twenty-two speakers and eight additional participants in the workshop quickly produced the articles that constitute this volume. Several participants in Darwin 2009 offered dissenting opinions or amplified elements of the presentations in the workshop, and they were asked to write commentaries that expressed their views. M. Caitlin Fisher-Reid continued to assist us by finding or scanning illustrations for the book, and Peter Park and Fumio Aoki helped convert or transmit large image files. Several external reviewers commented on specific chapters from the perspective of their specialties to help enhance and clarify the contents of this volume. We sincerely thank May R. Berenbaum, Peter J. Bowler, Sean B. Carroll, H. James Cleaves, Michael R. Dietrich, Michael Foote, Charles Fox, David M. Hillis, David Houle, Corbin D. Jones, William L. Jungers, Junhyong Kim, Antonio Lazcano, Scott Lidgard, David P. Mindell, Mohamed A. F. Noor, Robert Richards, Vasiliki Betty Smocovitis, Stephen C. Stearns, Bruce E. Tabashnik, Kevin R. Thornton, John J. Wiens, and Marlene Zuk for their advice. We are especially grateful to Peter J. Bowler and David M. Hillis, who generously reviewed multiple manuscripts. We thank Andy Sinauer for his thoughtful guidance in structuring and assembling the contents of this book. As copy editor, Lucy Anderson created a consistent and graceful style for the text, and Christopher Small, Joan Gemme, and Sydney Carroll at Sinauer Associates managed the project, turned image files in diverse formats and styles into a consistent set of figures, and assembled the chapters and commentaries into a coherent unit. We thank them all for their roles in the production of this volume.

MICHAEL A. BELL
DOUGLAS J. FUTUYMA
WALTER F. EANES
JEFFREY S. LEVINTON

Stony Brook, New York
May 2010

Contributors

ANURAG A. AGRAWAL[1]
Department of Ecology and Evolutionary Biology
Cornell University
Ithaca, NY 14853-2701, USA
aa337@cornell.edu

MAY R. BERENBAUM[2]
Department of Entomology
University of Illinois at Urbana-Champaign
Urbana, IL 61801, USA
maybe@illinois.edu

PETER J. BOWLER[1]
School of History and Anthropology
Queens University Belfast
Belfast BT7 1NN, Northern Ireland, UK
p.bowler@qub.ac.uk

ROBERT BOYD[3]
Department of Anthropology
University of California, Los Angeles
Los Angeles, California 90024, USA
rboyd@anthro.ucla.edu

JEFFREY K. CONNER[3]
Kellogg Biological Station *and*
Department of Plant Biology
Michigan State University
Hickory Corners, MI 49060, USA
connerj@msu.edu

JOEL L. CRACRAFT[4]
Division of Vertebrate Zoology—Ornithology
American Museum of Natural History
Central Park West at 79th St.
New York, NY 10024-5192, USA
jlc@amnh.org

CHARLES C. DAVIS[4]
Department of Organismic and Evolutionary
Biology *and* Harvard University Herbaria
Harvard University
Cambridge, MA 02138, USA
cdavis@oeb.harvard.edu

MICHAEL J. DONOGHUE[3]
Department of Ecology and Evolutionary Biology
Yale University
New Haven, CT 06405-8106, USA
michael.donoghue@yale.edu

DANIEL E. DYKHUIZEN[4]
Department of Ecology and Evolution
Stony Brook University
Stony Brook, NY 11794-5245, USA
dandyk@life.bio.sunysb.edu

ERIKA J. EDWARDS[3]
Department of Ecology and Evolutionary Biology
Brown University
Providence, RI 02912, USA
erika_edwards@brown.edu

[1] Speaker in the Darwin 2009 workshop.

[2] Invited speaker who could not attend the Darwin 2009 workshop.

[3] Co-author who did not attend the Darwin 2009 workshop.

[4] Discussant in the Darwin 2009 workshop.

MICHAEL FOOTE[1]
Department of the Geophysical Sciences
University of Chicago
Chicago, IL 60637, USA
mfoote@uchicago.edu

DOUGLAS J. FUTUYMA[1]
Department of Ecology and Evolution
Stony Brook University
Stony Brook, NY 11794-5245, USA
futuyma@life.bio.sunysb.edu

FRED GOULD[1]
Department of Entomology
North Carolina State University
Raleigh, NC 27695-7634, USA
fred_gould@ncsu.edu

RICHARD G. HARRISON[1]
Department of Ecology and Evolutionary Biology
Cornell University
Ithaca, NY 14853, USA
rgh4@cornell.edu

DAVID M. HILLIS[1]
Section of Integrative Biology *and* Center for Computational Biology and Bioinformatics
University of Texas at Austin
Austin, TX 78712, USA
dhillis@mail.utexas.edu

HOPI E. HOEKSTRA[1]
Department of Organismic and Evolutionary Biology *and* Museum of Comparative Zoology
Harvard University
Cambridge, MA 02138, USA
hoekstra@oeb.harvard.edu

MICHAEL D. JENNIONS[3]
Evolution, Ecology and Genetics
Research School of Biology
The Australian National University
Canberra ACT 0200, Australia
michael.jennions@anu.edu.au

ANDREW D. KERN[1]
Department of Biological Sciences
Dartmouth College
Hanover, NH 03755, USA
andrew.d.kern@dartmouth.edu

MARK KIRKPATRICK[1]
Section of Integrative Biology
University of Texas at Austin
Austin, TX 78712, USA
kirkp@mail.utexas.edu

HANNA KOKKO[1]
Department of Biological and Environmental Science
University of Helsinki
00014 Helsinki, Finland
and Evolution, Ecology and Genetics Research
School of Biology
The Australian National University
Canberra ACT 0200, Australia
hanna.kokko@helsinki.fi

BRYAN KOLACZKOWSKI[3]
Department of Biological Sciences
Dartmouth College
Hanover, NH 03755, USA
bryan.kolaczkowski@dartmouth.edu

CHRISTOPHER E. LANE[4]
Department of Biological Sciences
University of Rhode Island
Kingston, RI 02881, USA
clane@mail.uri.edu

ANTONIO LAZCANO[1]
Facultad de Ciencias
Universidad Nacional Autónoma de México
Apdo. Postal 70-407,
04510 Cd. Universitaria
México D.F., México
alar@correo.unam.mx

JONATHAN B. LOSOS[1]
Department of Organismic and Evolutionary Biology *and* Museum of Comparative Zoology
Harvard University
Cambridge, MA 02138, USA
jlosos@oeb.harvard.edu

D. LUKE MAHLER[3]
Department of Organismic and Evolutionary Biology *and* Museum of Comparative Zoology
Harvard University
Cambridge, MA 02138, USA
lmahler@oeb.harvard.edu

CHARLES R. MARSHALL[4]
Department of Organismic and Evolutionary Biology, Earth Planetary Sciences, *and* Museum of Comparative Zoology
Harvard University
Cambridge, MA 02138, USA
Present Address: University of California Museum of Paleontology
and Department of Integrative Biology
University of California, Berkeley
Berkeley, CA 94720, USA
crmarshall@berkeley.edu

MARK A. MCPEEK[1,4]
Department of Biological Sciences
Dartmouth College
Hanover, NH 03755, USA
mark.mcpeek@dartmouth.edu

ROBERTA L. MILLSTEIN[2]
Department of Philosophy
University of California, Davis
Davis, CA 95616-8673, USA
RLMillstein@ucdavis.edu

SERGIO RASMANN[3]
Department of Ecology and Evolutionary Biology
Cornell University
Ithaca, NY 14853-2701, USA
sgr37@cornell.edu

JOSHUA S. REST[4]
Department of Ecology and Evolution
Stony Brook University
Stony Brook, NY 11794-5245, USA
joshua.rest@stonybrook.edu

PETER J. RICHERSON[1]
Department of Environmental Science and Policy
University of California, Davis
Davis, CA 95616, USA
pjricherson@ucdavis.edu

MARY A. SCHULER[3]
Department of Cell and Developmental Biology
University of Illinois at Urbana-Champaign
Urbana, IL 61801, USA
cmcdonne@illinois.edu

VASSILIKI BETTY SMOCOVITIS[4]
Departments of Biology and History
University of Florida
Gainesville, FL 32611, USA
bsmocovi@ufl.edu

GÜNTER P. WAGNER[1]
Department of Ecology and Evolutionary Biology
Yale University
New Haven, CT 06405-8106, USA
gunter.wagner@yale.edu

PETER J. WAGNER[1]
National Museum of Natural History
Smithsonian Institution
Washington, DC 20013-7012, USA
wagnerpj@si.edu

JOHN WAKELEY[1]
Department of Organsimic and Evolutionary Biology
Harvard University
Cambridge, MA 02138, USA
wakeley@fas.harvard.edu

TIM D. WHITE[1]
Department of Integrative Biology
University of California, Berkeley
Berkeley, CA 94720-3140, USA
timwhite@berkeley.edu

GREGORY A. WRAY[1]
Department of Biology
Duke University
Durham, NC 27708, USA
gwray@duke.edu

JIANZHI ZHANG[1]
Department of Ecology and Evolutionary Biology
University of Michigan
Ann Arbor, MI 48109-1048, USA
jianzhi@umich.edu

Part I

EVOLUTION SINCE DARWIN

Chapter 1

Evolutionary Biology: 150 Years of Progress

Douglas J. Futuyma

Two hundred years after the birth of Darwin and 150 years after the publication of *The Origin of Species*, it seems appropriate to celebrate the progress in the field of evolutionary biology, paralleling advances in biology generally. I will focus on the progress of evolutionary biology since the centennial celebrations in 1959. Although a proper review of this history would require a trained historian, I can at least offer some personal perspective on the more recent period. It was during that centennial year that I entered college and experienced my first formal exposure to some elements of evolutionary thought. So, only three career generations have elapsed since Darwin transformed biology, yet evolutionary biology itself has been transformed during these 50 years.

In the proceedings of major symposia in 1959, such as those held at the University of Chicago (Tax 1960) and the Cold Spring Harbor Laboratory (1959), one can find the accomplishments of the evolutionary synthesis that transpired in the 1930s and 1940s and guided research in the 1950s. The major themes of that synthesis were the randomness of mutations with respect to adaptive need; the surprising and puzzling abundance of genetic variation within populations; population genetics theory formulated by Fisher, Wright, and Haldane (Figures 1.1–1.3), which especially emphasized the power and ubiquity of natural selection; affirmation of Darwin's gradualist position that most phenotypic evolution occurs by a succession of small changes; definition of species as sets of reproductively isolated populations that arise gradually via divergence of allopatric populations as a consequence of ecological selection (and founder effect, according to Mayr 1954); and the continuity between microevolution and macroevolution. The synthetic theory, it was asserted, is sufficient to explain patterns of macroevolution described by morphologists and paleontologists, so justifying repudiation of anti-Darwinian orthogenetic and neo-Lamarckian ideas. These conclusions were well grounded. For example, abundant evidence of the action of natural selection, even on such surprising features as

FIGURE 1.1 Ronald A. Fisher (Courtesy of Joan Fisher Box.)

FIGURE 1.2 Sewall Wright (left) and John Maynard Smith, in 1980 (Courtesy of J. J. Bull.)

chromosome inversions, was provided by Dobzhansky, Ford, and Clausen, Keck and Hiesey, as well as many others. Rensch and Simpson (Figure 1.4) showed how patterns of variation among both living and fossil taxa comported with Darwinian explanations. Evidence and theory from genetics, systematics, and paleontology converged on the conclusion that evolution should generally be gradual rather than saltational.

It is edifying to reflect on the state of research on evolution and of biology, generally, in 1959. Watson and Crick had proposed the structure of DNA only 6 years before, and Meselson and Stahl had demonstrated the semiconservative replication of DNA only 1 year before. Miller (1953) had started work on abiotic organic syntheses, under what were thought to be conditions on the early Earth (see Lazcano, Chapter 14). The first steps in deciphering the genetic code, the identification of messenger RNA, and the operon model of Jacob and Monod did not occur until 1961. Amino acid sequences of some few proteins had been determined, but DNA sequencing lay far in the future. The theory of population genetics could seldom be applied, because there existed almost no molecular markers and the frequencies of alleles and genotypes could be estimated for only a few discrete polymorphisms in morphology or chromosome structure. Although evidence for natural selection had been mounting, there were neither many estimates of the strength of selection, nor much convincing evidence of genetic drift. Computers were primitive; the first use of computer simulation in evolutionary biology may have been Lewontin and White's analysis of inversion polymorphisms in the grasshopper *Moraba scurra* in 1960.

FIGURE 1.3 John Burton Sanderson Haldane (Courtesy of Dr. K. Patau.)

FIGURE 1.4 G. Ledyard Stebbins, George Gaylord Simpson, and Theodosius Dobzhansky (Courtesy of G. L. Stebbins.)

Although systematists had continued to publish phylogenetic hypotheses ever since Haeckel, there existed in 1959 neither the DNA sequencing nor the computational power for anything like modern phylogenetic analyses; furthermore, there was no broad understanding of how phylogenies might be reliably inferred. In the 1930s, the German botanist Walter Zimmermann anticipated many of Hennig's ideas (Donoghue and Kadereit 1992), but neither his work nor Hennig's *Grundzüge einer Theorie der phylogenetischer Systematik* (1950) had broad impact until the English translation of this book appeared in 1966. Moreover, phylogenetic relationships that systematists proposed for their study organisms generally drew little interest from evolutionary biologists; phylogenetics played almost no role in the evolutionary synthesis (see Hillis, Chapter 16).

Against this background, I will sketch my impression of some of the major developments since 1959 in several areas of evolutionary biology. It will be apparent that the progress in these areas has built on several foundations, including conceptual and theoretical development, attention to neglected questions, technical advances (especially in molecular biology and information processing), integration with other areas of biology (especially molecular biology, developmental biology, and ecology), and increasing synthesis among the several fields of evolutionary biology. I suspect that a broader education of students in evolutionary biology may have played a role.

Evolutionary Genetics

The salient feature of evolutionary genetics since 1959 has obviously been its continual incorporation of the spectacular advances in molecular

FIGURE 1.5 Richard Lewontin (Courtesy of Richard Lewontin.)

genetics and now genomics. Two instances of this incorporation have had particularly great consequences. When Lewontin and Hubby (1966; Lewontin 1974) revealed abundant allozyme polymorphisms, they described the difficulty of accounting for such rich variation, raised the possibility that drift rather than selection might be involved, and revived the controversy between classical and balance interpretations of genetic variation in a new form (Figure 1.5).

Perhaps more importantly, it became possible for the first time to estimate the allele and genotype frequencies of population genetics theory, so that the theory might be applied more easily and broadly. Moreover, genetic variation in any species could be studied, no matter how intractable it might be for classical genetic analysis, and a new window into the genetics of species differences was opened.

The other great event in the 1960s was Kimura's (1968) introduction of the neutral theory of molecular evolution (Figure 1.6). Given the conviction of the universality of selection that had developed in the aftermath of the evolutionary synthesis, the neutral theory (even as modified by Ohta 1973) engendered a passionate neutralist/selectionist debate, stimulated and received support from the study of molecular variation within and among species, and impelled the development of both theory and tests for selection. Among these, the Hudson-Kreitman-Aguadé (HKA) test (Hudson et al. 1987) stood out as the first test for selection at particular genes, as opposed to inferring selection from average patterns among sites or genes. This and other advances, of course, were enabled by the shift from allozyme polymorphism to DNA sequence variation, first studied at the population level by Kreitman (1983).

FIGURE 1.6 Motoo Kimura (Courtesy of William Provine.)

The ability to study variation at the molecular level has given birth to many new kinds of studies. Reasonably constant rates of sequence evolution (molecular clock), first noted for proteins by Zuckerkandl and Pauling (1965) and that provided the impetus for Kimura's neutral theory, offered an unprecedented basis for estimating the absolute time of many evolutionary events whose dating would otherwise have been impossible, especially for taxa or characters that have little fossil record. Together with the development of phylogenetic methods, molecular data enabled the growth of the new field of phylogeography (Avise et al. 1987), giving insight into population histories. Sequence data enabled the application of a new retrospective population genetics in the form of coalescent theory that continues to provide insights into demographic history, gene flow, and selection (see Wakeley, Chapter 5). As the field of molecular evolution grew, it expanded into evolutionary genomics, which today includes several major kinds of studies. Evolutionary biologists can explore traditional questions, such as those concerning the role of selection versus drift, for large ensembles of genes (see Kolaczkowski and Kern, Chapter 6); they can study the evolution of genome structure, such as the birth and death of genes; and they are embarked on a path, only dimly foreseeable, toward understanding the evolution of functional complexes of genes and the phenotypes they affect (see Hoekstra, Chapter 22). And, a most familiar yet profoundly important fact should not be overlooked: that all of molecular biology and genomics triumphantly affirms the unity of life and its common ancestry. How satisfying it is that we can hope to understand the origin of life and the last universal common ancestor (see Lazcano, Chapter 14)... how wonderful, that we can apply yeast genomics to mammals!

Some major advances in evolutionary genetics depended only slightly, if at all, on molecular data. Most phenotypic characters are quantitative traits. Although developed for plant and animal breeding, quantitative genetics had also played some role in evolutionary genetics, ever since Sewall Wright. It became a more prominent topic in evolutionary biology in the 1970s, as Lande (1976) and others developed explicit evolutionary models for quantitative traits and ways to describe selection on them in nature. The genetic variance–covariance matrix (**G**) is now familiar to most students of evolution (see Kirkpatrick, Chapter 7), and there are now hundreds of estimates of selection gradients for natural populations (Kingsolver et al. 2001; Endler 1986). This work has revealed that selection is often surprisingly strong—an observation related to the many instances in which very rapid adaptive evolutionary change has been documented, especially in populations subjected to anthropogenic changes in their environment (Hendry et al. 2008). We have long left behind the need to cite industrial melanism in *Biston betularia* for evidence of evolution in action.

However, mounting evidence that paucity of genetic variation or strong genetic correlations may restrain adaptive evolution has sparked increasing theoretical and empirical research on genetic constraints (Blows and Hoffmann 2005; Walsh and Blows 2009). Several contemporary trends bear on this shift in emphasis. Quantitative trait locus (QTL) mapping, made possible by abundant molecular markers, now provides estimates of the numbers and magnitude of phenotypic effects of loci (chromosome regions) that influence character variation and sometimes indicates causal candidate genes. During and after the synthesis, attention focused more on the fate of genetic and phenotypic variation than its origin, perhaps because the response to selection was thought assured by an almost unlimited supply of standing variation. Today, the origin of variation is increasingly an object of study, partly because of the perception that the mutational process may limit adaptation in some instances (Hartl and Taubes 1998). The origin of phenotypic variation cannot be understood without reference to developmental processes, but evolutionary developmental biology (evo-devo) has now emerged as the phoenix of evolutionary biology.

The spectacular advances in the understanding of evolution at a genetic level could suggest that no vestige might remain of the genetic views held in 1959. But, this is not at all the case. The theory of population genetics developed by Fisher, Haldane, and Wright was based on pre-molecular concepts of alleles, mutation, and recombination that are still the central mathematical theory of evolutionary change and have been extended to many phenomena in terms that a time-traveler from 1959 would fully recognize. The dynamics of mutations and alleles in this theory usually do not depend on their molecular specification, and the theory of selection, which could readily accommodate phenomena such as meiotic drive and canalization, specifies only fitness differences among genotypes, with no reference to what their phenotypes might be.

Adaptation

The population genetic theory of selection was very general, possessed an abstract quality, and had been tailored, in only a few instances before 1959, to explain classes of presumably adaptive traits. To be sure, Fisher in 1930 had addressed phenomena such as sex ratio, sexual selection, and the evolution of self-fertilization (although mostly not with equations!), and had offered a theory of the evolution of dominance (Fisher 1930). The life history theory of evolution had been adumbrated by Lack (1954) and Williams (1957), while the theory of what would later be called "kin selection for altruism" had been sketched by Haldane (1932) and then developed by George Williams and Doris Williams (1957). However, the publication of *Animal Dispersion in Relation to Social Behaviour* by Wynne-Edwards (1962) explicitly invoked group selection to explain the self-sacrificial traits he posited, leading to a response by George Williams in *Adaptation and Natural Selection* (1966) that clarified for many biologists the nature of levels of selection and the perceived difficulties of group selection (Figure 1.7). Moreover, Williams presented hypotheses on the evolution of sex and modes of reproduction, life history evolution, and social adaptations. Shortly before then, Hamilton (1964) had developed his seminal theory of inclusive fitness (Figure 1.8).

FIGURE 1.7 George C. Williams (Courtesy of Doris Williams.)

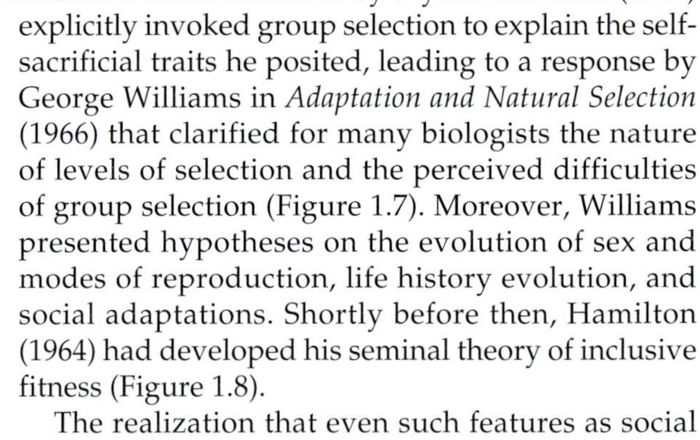

The realization that even such features as social behaviors could be explained by selection at the level of individuals and genes seems to have contributed to an explosive growth of models and tests of adaptation in organisms' characteristics. (Curiously, this efflorescence did not extend to morphological and physiological features.) A phenotypic approach to modeling, as introduced by Levins (1968), Maynard

Smith (1972, 1978; see Figure 1.2), Parker (1974), and others, introduced optimality and evolutionarily stable strategies as important new ways of developing adaptive hypotheses. This adaptationist program doubtless benefited from some severe criticism (Gould and Lewontin 1979), but probably no one today questions whether there is a far deeper understanding of the evolution of many traits than was possible in 1959. A very incomplete list might include senescence, litter size, semelparity and iteroparity, sex allocation, sex change, and phenotypic plasticity. Many models have been offered to explain the prevalence of sexual reproduction, although empirical tests of the models have unfortunately not kept pace. Among the most striking developments has been recognition and exploration of many kinds of genetic conflicts: between parents and offspring (Trivers 1974; Haig 2010), between sexes (Partridge and Hurst 1998; Rice and Holland 1997), and between genes within genomes (Hurst et al.1996). Attention to sexual selection by female choice, which had been in disrepute almost since Darwin introduced the concept in 1871, grew in the 1980s, giving rise to a huge theoretical and empirical literature. The understanding of mating systems, parental care, habitat utilization, and numerous aspects of social behavior has been greatly advanced by behavioral ecology (see Kokko and Jennions, Chapter 12).

FIGURE 1.8 William D. Hamilton (Courtesy of the University of Oxford.)

Although ecology and evolution have always been related (Collins 1986), the relationship became somewhat more intimate in the 1960s, when MacArthur, Wilson, Levins, and others developed an evolutionary approach to topics such as interspecific competition and its impact on species diversity within communities. Ehrlich and Raven's (1964) scenario of coevolution between plants and herbivorous insects, likewise, was proposed to explain the growth of species diversity (see Berenbaum and Schuler, Chapter 11). At its apogee of optimism, evolutionary ecology supposed that the structure of communities that evolved in similar environments would converge almost deterministically—a supposition that has since found little support (Orians and Paine 1983; Schluter and Ricklefs 1993; Ricklefs 2006). Today, species diversity and other aspects of community structure are increasingly thought to have been influenced more by long-term evolutionary history and historical contingency (see following) than by law-like processes of multispecies coevolution. Nonetheless, interspecific interactions have shaped many features of populations and species (Thompson 2004; see Agrawal et al., Chapter 10; Berenbaum and Schuler, Chapter 11) and may have affected adaptive radiation and species diversity (see Losos and Mahler, Chapter 15). Evolutionary models show that coevolution need not stabilize predator–prey interactions, parasite–host interactions, or mutualism (Doebeli and Knowlton 1998; Lipsitch et al. 1995).

Speciation

In the early 1960s, Mayr's (Figure 1.9) biological species concept (BSC) and his views on speciation (Mayr 1963) held sway among zoologists and some botanists (Stebbins 1950; Grant 1971; see Figure 1.4). Since then, diverse species concepts have been proposed. Today, practicing systematists are increasingly applying one or another phylogenetic species concept in place of the BSC. Speciation then would consist of the evolution of a distinguishing character, not reproductive isolation. The use of two meanings for the word "species" may cause confusion, but because definitions are conventions, the choice of a definition or concept of species is not a matter of scientific truth or falsehood.

Dobzhansky, Mayr, and Stebbins held that speciation (i.e., the evolution of reproductive isolation) was caused by divergent ecological selection in allopatric populations, sometimes reinforced in areas of secondary contact. A special form of allopatric speciation, Mayr's founder-effect (peripatric) speciation, became widely accepted for animals, and Lewis (1966) suggested a similar hypothesis for plants. Dobzhansky and others had adduced evidence for a polygenic basis of reproductive isolation, as well as for epistatic differences between species that have since been termed "Dobzhansky–Muller incompatibilities." Data bearing on the causes of the genetic divergence underlying reproductive isolation were surprisingly few. Except for obvious examples, such as ecological isolation in different habitats or pollination of plants by different pollinators, there was little evidence that reproductive isolation was caused by ecological selection, because in almost no instances had a causal connection between the ecological differentiation of sister species and their reproductive isolation been shown. Moreover, sexual selection, which had been utterly neglected since 1871, played no part in anyone's thinking about speciation. Almost no mathematical population genetic models of the geography, genetics, or possible causes of speciation had been advanced by the early 1960s.

FIGURE 1.9 Ernst Mayr (Courtesy of Harvard News Service and E. Mayr.)

Since then, the understanding of speciation (using a BSC-related definition) has increased appreciably (Coyne and Orr 2004; Gavrilets 2004). Starting with models of assortative mating and sympatric speciation by O'Donald in 1960 and Maynard Smith in 1966, population genetic models of speciation have proliferated (Gavrilets 2004). Speciation in the face of gene flow is possible, but the conditions are highly restrictive if gene flow is high (as in sympatric or nearly sympatric speciation). Sexual selection can be a potent agent of speciation, as promoted by Lande (1981), Kirkpatrick (1982a), and West-Eberhard (1983). Founder-effect speciation is unlikely if it requires a drift-induced shift across a fitness valley between peaks on an adaptive landscape, but similar effects may arise from drift along ridges on "holey" multidimensional landscapes (Gavrilets 2004). And from coalescent theory, it is known that sister species may inherit polymorphisms from their common ancestor and become genetically monophyletic at a rate determined by their effective population size. (The consequence for phylogenetic analysis is that gene trees may not match the phylogeny of three or more closely related species.)

On the empirical side (Coyne and Orr 2004), genetic analysis of species differences, including isolating barriers, has shown that they are generally polygenic, although a few substitutions of large effect can contribute a large fraction of the character difference. Some such loci and their functions have been identified, especially those contributing to Dobzhansky–Muller incompatibility, and they appear generally to show evidence of selection (Maheshwari et al. 2008; Tang and Presgraves 2009). Whether these loci were instrumental in speciation or followed it is not certain. Founder-effect speciation has found little support in studies of bottlenecked experimental populations (Galiana et al. 1993; Rundle et al. 1998), and coalescence-based estimates of effective population sizes suggest that they have been larger than founder-effect speciation would require (Walsh and Friesen 2005). Although molecular evidence indicates that some speciation events have been completed in the presence of gene flow between the incipient species (Noor et al. 2001), only a few of the proposed cases of sympatric speciation, chiefly those in which an allopatric alternative seems very unlikely, have convinced skeptics (Coyne and Orr 2004; Futuyma 2008). Perhaps the greatest change, although still limited, has been the development of evidence on selection as the cause of speciation. In addition to signatures of selection on specific genes, there is now considerably more evidence of a role for ecological selection in speciation (Rundle and Nosil 2005). Although the evidence is still largely correlative, there is good reason to think that sexual selection has caused speciation in some groups. The role of polyploidy in plant speciation has been affirmed, and cases of rapid origin of species from diploid hybrids have been documented (Rieseberg 1997). A more pluralistic view of speciation appears to be emerging (see Harrison, Chapter 13).

Evolutionary Developmental Biology

I will venture only a few comments on a field in which I am especially unqualified and that Wray treats elsewhere in this volume (see Chapter 9). Descriptive embryology was a major basis for phylogenetically oriented research in the decades following the publication of *The Origin of Species* (Bowler 1996, Chapter 2). It provided important evidence for some patterns of morphological transformation and identified phenomena such as allometry and heterochrony. In the absence of an understanding of developmental mechanisms, some authors offered phenomenological, somewhat metaphorical ideas on the evolution of developmental systems. Schmalhausen (1949) emphasized internal selection and homeostasis; Olson and Miller (1958) discussed functional and developmental integration of characters; and Waddington (1953) developed concepts of canalization and genetic assimilation. Some of these ideas may be viewed as predecessors of current thinking on concepts such as biological homology, modularity, and evolvability (Wagner 1989, Chapter 8; Wagner and Altenberg 1996).

A renaissance of interest in phenotypic macroevolution and its basis in developmental phenomena (e.g., heterochrony), urged by Gould (1977) and others, created an intensely receptive audience for the breakthroughs in geneticists' understanding of gene regulation and the genetic basis of developmental pathways. The discovery in the 1980s that *Hox* genes and other elements in the so-called genetic toolkit are conserved among phyla was a stunning revelation, which seemed to promise that the mechanisms of macroevolutionary divergence in body plans might soon be discovered. Modern evo-devo, with its mechanistic perspective, is a new, exciting field. Together with studies of genome evolution, it is beginning to shed light on the paths by which major evolutionary changes can occur. For example, the enclosure of a turtle's pectoral girdle by the rib cage is a long-standing puzzle, but it appears that this evolutionary change can be rather easily explained by slight differences in relative growth that is in part mediated by fibroblast growth factor (Cebra-Thomas et al. 2005; Nagashima et al. 2009). In this and other cases, macroevolution is becoming more comprehensible. Great attention is now focused on questions of how developmental changes evolve within populations (Colosimo et al. 2004; Hoekstra 2006). Because the mapping of genetic variation onto phenotypic variation is not a simple one-to-one relationship but rather is shaped by complex developmental pathways, we cannot fully understand the origin of and limitations on the phenotypic variation on which selection acts until we know the mechanisms of development and how they evolve. Progress toward this goal will be one of the most rewarding changes in the field of evolutionary biology.

Evolutionary History

Following Darwin, the chief research programs in evolutionary biology were centered in paleontology and systematics and were concerned with determining histories of evolutionary change (Bowler 1996). This goal remained the basis for the continuing inferences of relationships by systematists, mostly in the form of classifications. Systematics entered the evolutionary synthesis insofar as it provided information on species and populations, while phylogenetic speculation played almost no part (see Hillis, Chapter 16). The training and culture of most systematists differed greatly from that of population geneticists and students of evolutionary process well into the

1970s and beyond. In paleontology, the synthesis largely swept away the commonly espoused neo-Lamarckian, orthogenetic, and saltational theories, but the training, culture, and social interactions of paleontologists (who were and are often in geology departments) were still largely divorced from those of biologists. Consequently, the study of long-term evolutionary history and macroevolution, by paleontologists and systematists was almost entirely disconnected from the study of evolutionary processes in populations and species. This situation prevailed well into the 1980s and to some extent persists today. Indeed, during the 1980s, what may be described as mutual indifference became transformed into a certain amount of active conflict.

Paleobiology

As a neontologist, I can offer only a very limited perspective on changes in paleontology. First, there have been two immensely important developments in geology. It was only in about 1960 that geology was transformed by unquestionable evidence of continental drift and plate tectonics—thus, all previous suppositions about the history of climates and the distributions of organisms, such as Darlington's (1957) zoogeography, had to be completely reexamined. Then in 1980, Alvarez et al. stunned science with evidence of a bolide impact that may have caused the Cretaceous–Tertiary mass extinction. Within paleontology itself, at least in the United States, changes in emphasis and approach seem to have started in the 1970s, when Raup, Valentine, Schopf, Sepkoski, Gould and others championed a broad, question-oriented approach to what some called paleobiology—an attempt to build a nomothetic discipline (Raup and Gould 1974). For example, they tested data on taxonomic diversity and morphological diversity (or "disparity," as it was later dubbed) against random models (Gould et al. 1977; Sepkoski 1978; Foote 1993). Conceptual and empirical understanding of evolutionary rates and trends, factors affecting historical changes in diversity, and paleoecology has grown so greatly (see Wagner, Chapter 17; Foote, Chapter 18) that a contemporary textbook (Foote and Miller 2007) differs profoundly from its predecessor two decades earlier.

Perhaps paleobiology and population genetics would have continued on their separate self-absorbed paths longer, had it not been for Eldredge and Gould's explosive essay on "punctuated equilibria" in 1972 (Figure 1.10), and Gould's subsequent campaign for biologists to support this idea. Population geneticists reacted by arguing that the pattern of punctuated equilibria (i.e., rapid

FIGURE 1.10 Stephen Jay Gould (left) and Niles Eldredge (right) with their mentor Norman Newell, a leading paleontologist at Columbia University (Courtesy of Gillian Newell.)

shifts between more static morphologies) was readily explicable by traditional natural selection, and they rejected Eldredge and Gould's hypothesis that such a shift marks escape from epistatic genetic constraint in the course of founder-effect speciation. Eldredge and Gould claimed they were merely applying Mayr's broadly accepted hypothesis on speciation to the fossil record, but they and Stanley (1975, 1979) drew further inferences that population geneticists could not accept: that phenotypic evolution *requires* speciation, that trends must be viewed as the product of species selection rather than anagenesis within species lineages, and that macroevolution is decoupled from microevolution. The intensity of this dispute is reflected in papers (Stanley 1982; Charlesworth et al. 1982) by participants in a memorably contentious conference on macroevolution in 1980 (Lewin 1980; see also Macroevolution Conference [Letters]).

The dispute led population geneticists to model and generally to question founder-effect speciation. In the light of this and other arguments, Eldredge (1989) and Gould (2002) eventually abandoned the position that speciation enables phenotypic evolution by reducing genetic constraint. Gould (2002) embraced a hypothesis I offered to explain the possible (but not well documented) association between speciation and morphological change, namely that many adaptive changes in populations will be geologically ephemeral (and often invisible), unless they are protected by reproductive isolation from inevitable interbreeding, in the fullness of time, with genetically different populations (Futuyma 1987, 2010; Eldredge et al. 2005). The debates eventually had salutary consequences. Stasis is now broadly understood to pose an important problem and perhaps has helped to focus attention on evolutionary constraints. Many characters have been proposed to affect differential rates of speciation and extinction rates, and selection, or sorting, at the levels of species and clades has clearly shaped the history of life (Jablonski 2008; see Wagner, Chapter 17; Foote, Chapter 18), although it is less clear that species selection is commonly the basis of evolutionary trends in morphological or other phenotypic traits (Jablonski 2008). Perhaps most importantly, paleobiologists, on the basis of increasingly diverse and comprehensive studies (and with some help from phylogenetics), have impressed on other evolutionary biologists the importance of a historical perspective in explaining contemporary patterns (Jablonski 2000). Conversely, it appears that many paleobiologists today have a deeper appreciation of microevolutionary processes than did their predecessors and that mutual understanding is greater than before.

Phylogenetics

For many evolutionary biologists, the beginning of modern phylogenetics is considered to have occurred with the introduction of numerical taxonomy by Michener and Sokal (1957; Felsenstein 2004) and the publication of Willi Hennig's book in English translation (1966) (Figure 1.11). The first phylogenetic methods, based on parsimony, were introduced by the population geneticists Edwards and Cavalli-Sforza (1963) and by Camin and Sokal (1965). These approaches were soon followed by phylogenetic inferences from cytochrome c sequences, based on a distance method by Fitch and Margoliash (1967), and by Farris's (1970) development of parsimony methods. The systematics wars that followed (Hull 1988) probably did little to hasten interest in phylogeny among other biologists, but these hostilities slowly subsided as younger systematists distanced themselves from ideology and as increasingly abundant molecular data and increasingly powerful computers provided the basis for greater confidence in phylogenetic inferences. Important advances included the adoption of statistical methods of phylogenetic inference, as introduced by Felsenstein and others.

FIGURE 1.11 Willi Hennig (Courtesy of Gerd Hennig.)

In the broader community of evolutionary biologists, interest in phylogeny lagged behind progress in phylogenetic methods. In 1988, I found it difficult to muster enough speakers for a symposium on "Phylogeny and Evolutionary Processes," and in one of the papers that issued from that symposium, Donoghue (1989: 1137) wrote that "phylogenetic trees appear only rarely in *Evolution*, *The American Naturalist*, or *Paleobiology*, and have seldom been integrated into any general approach to evolution." But soon afterward, perhaps spurred in part by Felsenstein's (1985) introduction of a phylogenetically rigorous comparative method for testing hypotheses on adaptation, interest mounted rapidly in using a phylogenetic framework in studies of evolutionary process. This trend has resulted in nothing less than a transfiguration of much of evolutionary biology. There is now high confidence in many parts of the tree of life—no one doubts that *Homo* and *Pan* are sister groups—and often fair knowledge of divergence times can be claimed, based on admittedly imprecise molecular clocks. Phylogenies are used to test and sometimes to generate hypotheses about evolutionary processes. Historical biogeography has been reborn on a far firmer (and pluralistic) basis. Evidence of phylogenetic conservatism of many characters (long obvious to every systematist, but largely neglected in evolutionary thought) raises questions about genetic or developmental constraints and design limitations (Wake 1991; Gould 2002). Historical biogeography and recognition of "phylogenetic niche conservatism" provide new interpretations of patterns of species richness (Wiens and Graham 2005; Mittelbach et al. 2007); community ecology is increasingly being revised using a historical perspective (Cavender-Bares et al. 2009). As phylogenies are inferred for both species and genes, population genetics and phylogenetics have become increasingly indistinct. Phylogeny is the lens through which W.-H. Li (1997) views almost all of molecular evolution in his authoritative textbook on the subject, and it is indispensable in genomics. The phylogenetic perspective, like paleobiology, brings with it an awareness of the explanatory importance of history and macroevolution. The *Sturm und Drang* (i.e., the storm and stress—to borrow from an eighteenth century German movement), which long beset the relationship between historical and process-oriented evolutionary biology (Futuyma 1988), seems to be waning and a more unified science has been emerging.

Challenges to the Synthesis

The evolutionary synthesis has invited many challenges, and it has frequently been asserted that it is a failed paradigm. The notion that science progresses by "paradigm shifts," as postulated by the philosopher Thomas Kuhn (1962), has been enthusiastically adopted by challengers of evolutionary synthesis, such as Ho and Saunders (1984), who entitled an edited volume *Beyond Neo-Darwinism: An Introduction to the New Evolutionary Paradigm* and wrote that "all the signs are that evolutionary theory is in crisis."

As almost all the foregoing discussion makes clear, evolutionary theory and information have been almost continually extended since the evolutionary synthesis. Evolutionary theory stretched comfortably to accommodate major discoveries, such as the genetic code, codon bias, and mobile genetic elements. Fisher, Wright, and Haldane provided the theoretical tools that later workers used, with some additions, to develop new models of biased gene conversion, evolution of sex, life history evolution, and reproductive isolation. But, have any elements of the synthetic theory been rejected, completely replaced, or altered beyond recognition? Has the profession of evolutionary biology been forced by "knowledge of another, incompatible sort," as Kuhn put it,[1] to change its view of the field? Note that some very significant changes in evolutionary biology, such as the spectacular rise in importance of phylogenetic thinking, do not challenge traditional evolutionary theory. However, there has been no lack of claims that the theory is not only incomplete, but also wrong in critical respects.

Challenges from Systematics

Some systematists claimed to challenge Darwinian or neo-Darwinian theory on what they considered to be logical consequences of cladistic methodology or principles of classification. Nelson and Platnick (1984), for example, argued (if I correctly interpret them) that ancestral taxa do not exist, because they would be paraphyletic; thus, ancestral taxa and areas are artifacts,

---

[1] Kuhn's portrait of paradigm shifts and scientific revolutions, which was controversial from the outset, has fallen into disfavor among historians of science (V. B. Smocovitis, personal communication). I do not adopt it as a framework here, but mention it only because of its popularity among challengers of the synthetic theory.

created only because Darwinian theory required them. These authors rejected the concept of a center or area of origin and adopted the mystifying panbiogeography of Croizat (1958), which states that ancestral distributions were originally broadly inclusive and subsequently underwent subdivision by new barriers. Their insistence that only such a vicariant history is falsifiable, justified their rejection of dispersal—an argument that would be laughable, if it had not shackled biogeography for years to come. Such ideas never spread beyond cladist circles and may be viewed as historical oddities.

Physicalism

Over the course of history, several developmental biologists have argued that the morphology of organisms should be attributed to physical processes, with the action of natural selection on hereditary variation playing a minor role. Brian C. Goodwin, for instance, illustrating his points with physical descriptions of cleavage and of the role of cytoskeletons in the dynamics of groups of cells (e.g., epithelia), argued that "development and evolution may…be deducible from the properties of the living state, involving both molecules and fields, which together provide the temporal and spatial order revealed in ontogenesis and phylogenesis" (Goodwin 1984: 240). A "field" here refers to "a spatial domain in which every part has a state determined by the state of neighboring parts so that the whole has a specific relational structure" (Goodwin 1984: 228). Goodwin downplayed the roles of genes and natural selection. It is certainly true that maternally inherited membranes, cytoplasmic organization, and RNA are necessary for the development of an egg into an embryo. Yet, contemporary developmental biology strongly affirms the role of genes in both species-specific and more phylogenetically global aspects of ontogeny. Of course, there must exist physical explanations of molecular and cellular events, for how else could proteins and other biomolecules exert their effects? The physicalistic argument seems to confuse proximate causes (the action of molecules and cells) with more ultimate causes. The gene products that act to produce, via physical principles, the observed cell and tissue events are the products of historical evolution and represent a series of transformations, not merely at the level of the individual organism in which an alteration first arises, but also at the level of populations and species—and these alterations require change in frequencies of relevant alleles, resulting from selection or drift. To include physical, mechanistic explanations as part of a hierarchy of explanations (Newman and Bhat 2009) is to enrich our understanding and may well be necessary for a proper ultimate explanation. However, physicalistic explanations are complementary to and compatible with traditional evolutionary explanation.

Levels of Selection

The major figures in the evolutionary synthesis followed Darwin's thinking in considering natural selection to consist of differences in fitness among individual organisms, in almost all contexts. Wynne-Edwards (1962) challenged this position by explicitly invoking selection among groups. The result of his book was surely the opposite of what he had hoped for, because Maynard Smith (1964), Williams (1966), and others developed strong arguments against the efficacy of group selection and, together with Hamilton's (1964) formulation of kin selection, showed that many phenomena could more parsimoniously be explained by selection at the level of individuals or genes. Since 1975, D. S. Wilson has argued for the importance of group selection, though in a very different form (i.e., trait-group selection) than Wynne-Edwards described (Wilson 1975). The argument has enriched evolutionary biology by stimulating focus on levels of selection and attendant phenomena, such as the evolution of multicellularity, individuality, cancer, and social groups (Frank 1998; Michod 1999, 2007). Multilevel selection involves difficult conceptual problems, which were largely clarified by the mathematician George Price (see Frank 1995), but has continued to engage philosophers of science (Sober 1984; Okasha 2006). One approach is to recognize two kinds of multilevel selection of genes and groups: (1) one that focuses on the frequency of an allele in the entire population, with fitness differences among groups measured by how many genes they contribute to the population (e.g., Wilson's trait-group model), and (2) another that focuses on the number of genetically differentiated groups, with fitness differences among groups measured by the number of like groups they produce (i.e., the Wynne-Edwards scenario; Okasha 2006). If I understand the current state of play rightly, most researchers in the area agree that the second form of multilevel selection is generally unimportant, but that the first form can play an important role in the evolution of some altruistic and other traits. However, most researchers appear to agree that in this case, the trait groups are almost always kin groups, so that kin selection and group selection are two ways of describing the same process (Lehmann and Keller 2006; West et al. 2007).

Meanwhile, Eldredge and Gould (1972), Stanley (1975), and other paleontologists postulated that evolutionary trends are ascribable to species selection, rather than the individual selection that operates within species and alters characters (but, they claimed at random with respect to the long-term trend). The scope of gene-level selection, which had always been recognized in phenomena such as meiotic drive, expanded when the abundance and behavior of transposable genetic elements came to light. As I noted above, the importance of clade selection in the history of life is now widely appreciated (Williams 1992), but the role of species selection in long-term trends is probably much less important than its advocates claimed, considering how few cases have been documented. However, I judge genic selection to be an important but modest expansion of the conceptual framework of natural selection—one in which individual selection still holds pride of place.

Punctuated Equilibria

As earlier noted, punctuated equilibria and species selection, advanced by Eldredge, Stanley, and Gould (the latter most insistently), was a major challenge to neo-Darwinian orthodoxy, a conscious effort at revolution. Population geneticists held that the punctuated pattern of rapid morphological shifts, usually in one or a few characters (e.g., the number of rows of eye lenses in the trilobite *Phacops rana*, described by Eldredge 1971), is easily accommodated by neo-Darwinian theory (Charlesworth et al. 1982; Kirkpatrick 1982b). As to Eldredge and Gould's suggestion that punctuational changes mark peripatric speciation events, many authors noted that intraspecific geographic variation shows that phenotypic evolution does not require speciation. Moreover, mathematical models of Mayr's peripatric speciation hypothesis (and closely related ideas put forward by Carson and Templeton) led some population geneticists to cast considerable doubt on these proposed mechanisms of speciation (see the exchange between Carson and Templeton [1984] and Barton and Charlesworth [1984]).

As I have noted earlier, the role of species selection in macroevolutionary trends, cited by advocates as a major consequence of punctuated equilibria, remains equivocal. However, its role in shaping the frequency distribution of characters within biotas is not doubted: clades have differed in diversification rate, and in some instances, the difference can be ascribed to certain characters (Jablonski 2008). On its two major, specific claims, then, Eldredge and Gould's challenge to the synthetic theory failed, and no paradigms have fallen. But, in a broader sense, they were highly successful. The punctuated equilibrium debate was heuristically highly productive, and the wider issues that in part grew out of it, such as the significance of tiers of sorting (e.g., effects of mass extinctions) and of historical contingency, became part of the fabric of evolutionary science. Paleobiology and the importance of history have gained renewed appreciation, and I think that despite his hyperbole and some downright errors, much of the credit must go to Gould's tireless promotion of his field.

Saltation and Discontinuity

Darwin's gradualism—his postulate that phenotypic evolution has proceeded by slight "insensible" steps—was challenged during his day (Huxley 1860) and has been ever since. In the face of mutationism and the extreme saltationist views of some paleontologists and of the geneticist Richard Goldschmidt (1940), major figures in the evolutionary synthesis, especially Fisher, Mayr, and Simpson, defended gradualism on both theoretical and empirical grounds. Nonetheless, it is impossible to demonstrate that all phenotypic gaps among Recent or extinct taxa represent missing information, and challenges to thoroughgoing gradualism have been frequent. Perhaps the most extreme approach, in recent decades, was Gould's flirtation with what might be called neo-Goldschmidtism (Gould

1980), which was effectively rebutted in reviews of Goldschmidt's book by Charlesworth (1982), Templeton (1982), and others. Gottlieb (1984) and Orr and Coyne (1992), advanced a much softer argument for discontinuity by compiling evidence for the quite large effect of some single-gene substitutions on certain phenotypic differences between closely related species and populations. This conclusion is supported by some subsequent studies; for example, single-allelic differences account for most of the presence versus absence of the pelvic girdle or of lateral plates in the stickleback *Gasterosteus aculeatus* (Colosimo et al. 2004, 2005; Shapiro et al. 2004; Chan et al. 2010). These and other instances of apparently simple genetic or developmental bases for major morphological differences, some marking higher taxa, warrant cautious interpretation. For example, the loss of a structure like the pelvic girdle by mutation of a regulatory gene at the start of a complex developmental pathway does not mean that the structure originated by mutation of that or any other single gene. Nonetheless, this is a subject that will surely be clarified by further research, and a modest shift away from thoroughgoing gradualism should not be ruled out.

The Neutral Theory

In 1959, there existed few examples of phenomena that could most plausibly be explained by genetic drift, and adherents to the synthetic theory certainly placed "nearly exclusive reliance on selection" (Gould 1980). Kimura's (1968) neutral theory, echoed by King and Jukes's (1969) announcement of "non-Darwinian evolution," extended Wright's and Fisher's models of genetic drift and met great resistance at first. Yet now, the neutral theory has become the null model in analyses of molecular evolution—a radical change in perspective that I think has the best claim to being labeled a "paradigm shift" in evolutionary biology in the last 50 years. That the neutrality of some features of sequence differences and genomes (e.g., introns) is now being questioned does not vitiate the significance and success of the neutralist challenge.

Recent Challenges

As molecular and developmental features and processes have been revealed, there has been no lack of proclamations of major challenges to the synthetic theory. At this time, it appears that some of these challenges are stronger than others.

Some claims seem simply to be wrong, based on current evidence. For instance, evidence for adaptively directed mutation has been called into serious question (Sniegowski et al. 2000). Likewise, although transposable elements and biased gene conversion may have important evolutionary consequences, there is no evidence that they can drive the evolution of phenotypic (e.g., morphological) traits independently of individual selection, as Dover (2002) proposed in his hypothesis of molecular drive.

Some real phenomena may be important, yet not constitute the challenge claimed. Almost since transposable elements were found to be abundant and to be responsible for many of the classical mutations studied by geneticists, they have been proposed as a major basis for evolutionary change, perhaps by inserting themselves into regulatory regions. Convincing evidence for this hypothesis has been scarce, but may be mounting (G. P. Wagner, personal communication). If the hypothesis should be supported, it will be interesting and may have important consequences for understanding rates and effects of different kinds of mutations. However, the discovery that mobile sequences are a source of evolutionarily significant mutations does not challenge traditional theory. As Bodmer (1983: 203) remarked, the rich variety of mechanisms that produce genetic variants does not explain why they should increase in frequency in a population, which "still needs to be explained in terms of the fundamental ideas of population genetics."

In some cases, we simply do not know enough to judge if a mechanism or phenomenon is common enough in natural populations to be considered important in evolution. Two phenomena that are claimed to warrant a "new expanded synthesis" (Pigliucci 2007) are examples. One is "genetic accommodation" (West-Eberhard 2003), a concept closely related to Waddington's (1953) genetic assimilation, whereby phenotypic changes of individual organisms that are evoked by stimuli in a novel environment become genetically determined by selection of alleles that narrow and fix the norm of reaction. Thus, phenotypic adaptation precedes genetic adaptation (West-Eberhard 2003). The traditional alternative scenario is simply that selection builds adaptive norms of reaction by increasing the frequency of those mutations that are expressed in the environments in which they enhance fitness. If I correctly understand genetic accommodation, the difference is that the initial phenotypic response to a novel environment has not already been shaped by a history of selection of the kind envisioned in the traditional view. Whether or not this is too subtle a distinction to constitute a major rewriting of evolutionary theory may be debated, but the major question now is empirical: how commonly does this process occur? So far, genetic assimilation is known mostly from artificial selection experiments, and only a few fairly convincing examples from natural populations have been described (Aubret and Shine 2009).

Another example of a possible challenge to traditional theory is epigenetic modification of gene expression, which may be inherited for a considerable number of generations (Jablonka and Lamb 2005; Gilbert and Epel 2009). Perhaps such modifications may be viewed simply as another kind of mutation, playing the same role in evolutionary theory as traditional gene or chromosome mutations. Determining their rate of back mutation would be important for judging their long-term effects. If there should be convincing evidence of "adaptively directed" epigenetic modification, it will be important to determine how it comes about. It is easy to conceive of a natural selection of epigenetic effects, but the precise basis of such selection would be interesting. Again, whether transgenerational epigenetic effects will prove to be widespread and important, or just curiosities of genomic natural history, remains to be seen: decades of research have provided innumerable examples of a traditional genetic basis for adaptive traits, including differences between populations and species. Yet, the possibility that cell and molecular biology will reveal phenomena that call for substantial revision of evolutionary theory cannot be ruled out, even if any such proclamation at this time is premature.

Conclusions

Since 1959, evolutionary biology has experienced episodic but generally accelerating progress. Conceptual and theoretical advances, deeper biological understanding arising from molecular biology, computational and other technological development, and the sheer growth in the number of researchers (many of whom have had a broader training in quantitative methods and across the evolutionary disciplines than earlier biologists) have all contributed to deeper understanding and more robust knowledge, at the same time revealing, as all scientific progress does, the inadequacy of our understanding of a great many problems. The growth of evolutionary biology has been marked by expansion of theory and growing synthesis among its several disciplines, as well as synthesis with other areas of biology, from molecular genetics to ecology. Evolutionary biology may enjoy greater respect across the biological sciences now than ever before. The success of the neutral theory of molecular evolution comes closest to counting as a paradigm shift within the field. An expanded synthesis of evolutionary biology may well be underway, but the synthesis has expanded frequently and in several dimensions in the last few decades.

When we glance further back, to the publication of *The Origin of Species* 150 years ago, we see that evolutionary biology, like biology generally, has traversed an immense distance. Darwin did not know about particulate inheritance or the genetical theory of natural selection, much less molecular biology. But even a partial list of Darwin's insights that hold strong today testifies to the breadth and depth of his knowledge and thinking. Granted, we would not entirely agree with his portrait of the Tree of Life as a strictly branching figure (although his pattern indeed dominates over reticulation) or with his dictum that *natura non facit saltum* (although the saltations currently recognized are very modest indeed), and we would supplement dispersal with some vicariance and moving land masses to explain biogeographic patterns. We would tell Darwin that evolution by natural selection can occur much faster than he suspected and that he should expunge inheritance of acquired characters almost altogether. However, we should congratulate him on getting so much right: the fundamental notion of common ancestry and phylogeny; the origin of all known life "from some one primordial form;" variable rates of evolution; hereditary variation that arises without reference to need; evolution as a population process of frequency change; natural selection at the individual level, as the chief cause of adaptation; sexual selection; kin and colony selection; coadaptation of characters; the importance of the mysterious laws of developmental growth; speciation as a gradual process (though some of us might urge him to give greater emphasis to spatial separation); the importance of interspecific interactions as major agents of selection; competition and "divergence of character;" and the incompleteness of the fossil record. Perhaps, this year, we are compensating for the earlier "eclipse of Darwinism" (Bowler 1983) with excessive adulation, but consider: that we can cite so much support for so many of his ideas after 150 years and the utter transformation of biology is a remarkable measure of the man.

Acknowledgments

I thank Mike Bell, Jerry Coyne, Michael Foote, Betty Smocovitis, and Stephen Stearns for valuable comments on the manuscript, Jeffrey Levinton, Mark Kirkpatrick, and Mark Siegal for enlightening conversation on certain points, and all the speakers at the Darwin 2009 Workshop for perspectives or reminders that helped to shape my text. I have dared to address this range of issues only because of the innumerable biologists from whom I have learned and among whom I wish to credit in particular my colleagues, present and past, in the Department of Ecology and Evolution at Stony Brook for generously answering questions and for maintaining an intellectually stimulating environment. I owe a special debt of gratitude to Robert Sokal, who set standards of professional dedication and rigor that were inspiring, even if unattainable by a younger colleague, and who has honored me with his friendship.

Literature Cited

Alvarez, L. W., W. Alvarez, F. Asaro, and 1 other. 1980. Extraterrestrial cause for the Cretaceous-Tertiary extinction. *Science* 208: 1095–1108.

Aubret, F. and R. Shine. 2009. Genetic assimilation and the postcolonization erosion of phenotypic plasticity in island tiger snakes. *Cur. Biol.* 19: 1932–1936.

Avise, J. C., J. Arnold, R. M. Ball, and 8 others. 1987. Intraspecific phylogeography: The mitochondrial bridge between population genetics and systematics. *Ann. Rev. Ecol. Syst.* 18: 489–522.

Barton, N. H. and B. Charlesworth. 1984. Genetic revolutions, founder effects, and speciation. *Ann. Rev. Ecol. Syst.* 15: 133–164.

Blows, M. W. and A. A. Hoffmann. 2005. A reassessment of genetic limits to evolutionary change. *Ecology* 86: 1371–1384.

Bodmer, W. 1983. Gene clusters and genome evolution. In D. S. Bendall (ed.), *Evolution from Molecules to Men*, pp. 197–208. Cambridge University Press, Cambridge, UK.

Bowler, P. J. 1983. *The Eclipse of Darwinism: Anti-Darwinian Evolution Theories in the Decades Around 1900.* Johns Hopkins University Press, Baltimore.

Bowler, P. J. 1996. *Life's Splendid Drama: Evolutionary Biology and the Reconstruction of Life's Ancestry 1860–1940*. University of Chicago Press, Chicago.
Camin, J. H. and R. R. Sokal. 1965. A method for deducing branching sequences in phylogeny. *Evolution* 19: 311–326.
Carson, H. L. and A. R. Templeton. 1984. Genetic revolutions in relation to speciation phenomena: The founding of new populations. *Ann. Rev. Ecol. Syst.* 15: 97–131.
Cavender-Bares, J., K. H. Kozak, P. V. A. Fine, and 1 other. 2009. The merging of community ecology and phylogenetic biology. *Ecol. Lett.* 12: 693–715.
Cebra-Thomas, J., F. Tan, S. Sistla, and 5 others. 2005. How the turtle forms its shell: A paracrine hypothesis of carapace development. *J. Exp. Zool. (Mol. Devel. Evol.)* 304B: 558–569.
Chan, Y. F., M. E. Marks, F. C. Jones, and 13 others. 2010. Adaptive evolution of pelvic reduction in sticklebacks by recurrent deletion of a *Pitx1* enhancer. *Science* 327: 302–305.
Charlesworth, B. 1982. Review: Hopeful monsters cannot fly. *Paleobiology* 8: 469–474.
Charlesworth, B., R. Lande, and M. Slatkin. 1982. A neo-Darwinian commentary on macroevolution. *Evolution* 36: 474–498.
Cold Spring Harbor Symp. Quant. Biol. XXIV. 1959. *Genetics and Twentieth Century Darwinism*. The Biological Laboratory, Cold Spring Harbor, New York.
Collins, J. P. 1986. Evolutionary ecology and the use of natural selection in ecological theory. *J. Hist. Biol.* 19: 257–288.
Colosimo, P. F., K. E. Hosemann, S. Balahadra, and 7 others. 2005. Widespread parallel evolution in sticklebacks by repeated fixation of *Ectodysplasin* alleles. *Science* 307: 1928–1933.
Colosimo, P. F., C. L. Peichel, K. Nereng, and 4 others. 2004. The genetic architecture of parallel armor plate reduction in threespine sticklebacks. *PLoS Biology* 2: 635–641.
Coyne, J. A. and H. A. Orr. 2004. *Speciation*. Sinauer Associates, Sunderland, MA.
Croizat, L. 1958. *Panbiogeography*. Published by the author, Caracas, Venezuela.
Darlington, P. J. Jr. 1957. *Zoogeography: The Geographical Distribution of Animals*. Wiley, New York.
Doebeli, M. and N. Knowlton. 1998. The evolution of interspecific mutualisms. *Proc. Natl. Acad. Sci. USA* 95: 8676–8680.
Donoghue, M. J. 1989. Phylogenies and the analysis of evolutionary sequences with examples from seed plants. *Evolution* 43: 1137–1156.
Donoghue, M. J. and J. W. Kadereit. 1992. Walter Zimmermann and the growth of phylogenetic theory. *Syst. Biol.* 41: 74–85.
Dover, G. 2002. Molecular drive. *Trends Genet.* 18: 587–589.
Edwards, A. W. F. and L. L. Cavalli-Sforza. 1963. The reconstruction of evolution. *Ann. Human. Genet.* 27: 105–106.
Ehrlich. P. R. and P. H. Raven. 1964. Butterflies and plants: A study in coevolution. *Evolution* 18: 586–608.
Eldredge, N. 1971. The allopatric model and phylogeny in Paleozoic invertebrates. *Evolution* 25: 156–167.
Eldredge, N. 1989. *Macroevolutionary Dynamics*. McGraw-Hill, New York.
Eldredge, N. and S. J. Gould. 1972. Punctuated equilibria: An alternative to phyletic gradualism. In T. J. M. Schopf (ed.), *Models in Paleobiology*, pp. 82–115. Freeman, Cooper, San Francisco.

Eldredge, N., J. N. Thompson, P. M. Brakefield, and 7 others. 2005. The dynamics of evolutionary stasis. *Paleobiology* 31: 133–145.
Endler, J. A. 1986. *Natural Selection in the Wild*. Princeton University Press, Princeton.
Farris, J. S. 1970. Methods for computing Wagner trees. *Syst. Zool.* 19: 83–92.
Felsenstein, J. 1985. Phylogenies and the comparative method. *Am. Nat.* 125: 1–15.
Felsenstein, J. 2004. *Inferring Phylogenies*. Sinauer Associates, Sunderland, MA.
Fisher, R. A. 1930. *The Genetical Theory of Natural Selection*. Clarendon, Oxford.
Fitch, W. M. and E. Margoliash. 1967. Construction of phylogenetic trees. *Science* 155: 279–284.
Foote, M. 1993. Discordance and concordance between morphological and taxonomic diversity. *Paleobiology* 19: 184–205.
Foote, M. and A. I. Miller. 2007. *Principles of Paleontology*, 3rd ed. W. H. Freeman, New York.
Frank, S. A. 1995. George Price's contributions to evolutionary genetics. *J. Theor. Biol.* 175: 373–388.
Frank, S. A. 1998. *Foundations of Social Evolution*. Princeton University Press, Princeton.
Futuyma, D. J. 1987. On the role of species in anagenesis. *Am. Nat.* 130: 465–473.
Futuyma, D. J. 1988. *Sturm und Drang* and the evolutionary synthesis. *Evolution* 42: 217–226.
Futuyma, D. J. 2008. Sympatric speciation: Norm or exception? In K. J. Tilmon (ed.), *Specialization, Speciation, and Radiation: The Evolutionary Biology of Herbivorous Insects*, pp. 136–148. University of California Press, Berkeley.
Futuyma, D. J. 2010. Evolutionary constraint and ecological consequences. *Evolution* 64, in press.
Galiana, A., A. Moya, and F. J. Ayala. 1993. Founder-flush speciation in *Drosophila pseudoobscura*: A large-scale experiment. *Evolution* 47: 432–444.
Gavrilets, S. 2004. *Fitness Landscapes and the Origin of Species*. Princeton University Press, Princeton.
Gilbert, S. F. and D. Epel. 2009. *Ecological Developmental Biology: Integrating Epigenetics, Medicine, and Evolution*. Sinauer Associates, Sunderland, MA.
Goldschmidt, R. 1940. *The Material Basis of Evolution*. Yale University Press, New Haven.
Goodwin, B. C. 1984. A relational or field theory of reproduction and its evolutionary implications. In M.-W. Ho and P. T. Saunders (eds.), *Beyond Neo-Darwinism: An Introduction to the New Evolutionary Paradigm*, pp. 219–241. Academic Press, London.
Gottlieb, L. D. 1984. Genetics and morphological evolution in plants. *Am. Nat.* 123: 681–709.
Gould, S. J. 1977. *Ontogeny and Phylogeny*. Harvard University Press, Cambridge, MA.
Gould, S. J. 1980. Is a new and general theory of evolution emerging? *Paleobiology* 6: 119–130.
Gould, S. J. 2002. *The Structure of Evolutionary Theory*. Harvard University Press, Cambridge, MA.
Gould, S. J. and R. C. Lewontin, 1979. The spandrels of San Marco and the panglossian paradigm. *Proc. Roy. Soc. Lond.* B 205: 581–598.
Gould, S. J., D. M. Raup, J. J. Sepkoski Jr., and 2 others. 1977. The shape of evolution: A comparison of real and random clades. *Paleobiology* 3: 23–40.
Grant, V. 1971. *Plant Speciation*, 2nd ed. Columbia University Press, New York.

Haig, D. 2004. Genomic imprinting and kinship: How good is the evidence? *Ann. Rev. Genet.* 38: 553–585.

Haldane, J. B. S. 1932. *The Causes of Evolution*. Longman, Green, New York.

Hamilton, W. F. 1964. The genetical evolution of social behavior, I and II. *J. Theor. Bio.* 7: 1–52.

Hartl, D. L. and C. H. Taubes. 1998. Towards a theory of evolutionary adaptation. *Genetica* 102–103 (special issue S1): 525–533.

Hendry, A. P., T. J. Farrugia, and M. T. Kinnison. 2008. Human influences on rates of phenotypic change in wild populations. *Mol. Ecol.* 17: 20–29.

Hennig, W. 1950. *Grundzüge einer Theorie der phylogenetischen Systematik*. Deutscher Zentralverlag, Berlin.

Ho, M.-W. and P. T. Saunders (eds.). 1984. *Beyond Neo-Darwinism: An Introduction to the New Evolutionary Paradigm*. Academic Press, London.

Hoekstra, H. E. 2006. Genetics, development and evolution of adaptive pigmentation in vertebrates. *Heredity* 97: 222–234.

Hudson, R. R., M. Kreitman, and M. Aguadé. 1987. A test of neutral molecular evolution based on nucleotide data. *Genetics* 116: 153–159.

Hull, D. L. 1988. *Science as a Process: An Evolutionary Account of the Social and Conceptual Development of Science*. University of Chicago Press, Chicago.

Hurst, L. D., A. Atlan, and B. D. Bengtsson. 1996. Genetic conflicts. *Quart. Rev. Biol.* 7: 317–364.

Huxley, T. H. 1860. *Collected Essays*. MacMillan, London.

Jablonka, E. and M. J. Lamb. 2005. *Evolution in Four Dimensions: Genetic, Epigenetic, Behavioral, and Symbolic Variation in the History of Life*. MIT Press, Cambridge, MA.

Jablonski, D. 2000. Micro- and macroevolution: Scale and hierarchy in evolutionary biology and paleobiology. *Paleobiology* 26: 15–52.

Jablonski, D. 2008. Species selection: Theory and data. *Ann. Rev. Ecol. Syst.* 39: 501–524.

Kimura, M. 1968. Evolutionary rate at the molecular level. *Nature* 217: 624–626.

King, J. L. and T. H. Jukes. 1969. Non-Darwinian evolution. *Science* 164: 788–798.

Kingsolver, J. G., H. E. Hoekstra, J. M. Hoekstra, and 6 others. 2001. The strength of phenotypic selection in natural populations. *Am. Nat.* 157: 245–261.

Kirkpatrick, M. 1982a. Sexual selection and the evolution of female choice. *Evolution* 36: 1–12.

Kirkpatrick, M. 1982b. Quantum evolution and punctuated equilibria in continuous genetic characters. *Am. Nat.* 119: 833–848.

Kreitman, M. 1983. Nucelotide polymorphism at the alcohol dehydrogenase locus of *Drosophila melanogaster*. *Nature* 304: 412–417.

Kuhn, T. S. 1962. *The Structure of Scientific Revolutions*. University of Chicago Press, Chicago.

Lack, D. 1954. *The Natural Regulation of Animal Numbers*. Oxford University Press, Oxford.

Lande, R. 1976. Natural selection and random genetic drift in phenotypic evolution. *Evolution* 30: 314–334.

Lande, R. 1981. Models of speciation by sexual selection on polygenic characters. *Proc. Natl. Acad. Sci. USA* 78: 3721–3725.

Lehmann, L. and L. Keller. 2006. The evolution of cooperation and altruism—a general framework and a classification of models. *J. Evol. Biol.* 19: 1365–1376.

Levins, R. 1968. *Evolution in Changing Environments*. Princeton University Press, Princeton.

Lewin, R. 1980. Evolutionary theory under fire. *Science* 210: 883–887.
Lewis, H. 1966. Speciation in flowering plants. *Science* 152: 167–172.
Lewontin, R. C. 1974. *The Genetic Basis of Evolutionary Change*. Columbia University Press, New York.
Lewontin, R. C. and J. L. Hubby. 1966. A molecular approach to the study of genic heterozygosity in natural populations. II. Amount of variation and degree of heterozygosity in natural populations of *Drosophila pseudoobscura*. *Genetics* 54: 595–609.
Lewontin, R. C. and M. J. D. White. 1960. Interaction between inversion polymorphisms of two chromosome pairs in the grasshopper, *Moraba scurra*. *Evolution* 14: 116–129.
Li, W.-H. 1997. *Molecular Evolution*. Sinauer Associates, Sunderland, MA.
Lipsitch, M., M. Nowak, D. Ebert, and 1 other. 1995. The population dynamics of vertically and horizontally transmitted parasites. *Proc. Roy. Soc. Lond. B* 260: 321–327.
Macroevolution conference [letters]. Futuyma, D. J., R. C. Lewontin, G. C. Mayer, J. Seger, and J. W. Stubblefield III. 1981. *Science* 211: 770.
Maheshwari, S., J. Wang, and D. A. Barbash. 2008. Recurrent positive selection of the *Drosophila* hybrid incompatibility gene *Hmr*. *Mol. Biol. Evol.* 25: 2421–2430.
Maynard Smith, J. 1964. Group selection and kin selection. *Nature* 200: 1145–1147.
Maynard Smith, J. 1966. Sympatric speciation. *Am. Nat.* 100: 637–650.
Maynard Smith, J. 1972. *On Evolution*. Edinburgh University Press, Edinburgh.
Maynard Smith, J. 1978. Optimization theory in evolution. *Ann. Rev. Ecol. Syst.* 9: 31–56.
Mayr, E. 1954. Change of genetic environment and evolution. In J. Huxley, A. C. Hardy, and E. B. Ford (eds.), *Evolution as a Process*, pp. 157–180. MacMillan, New York.
Mayr, E. 1963. *Animal Species and Evolution*. Harvard University Press, Cambridge, MA.
Michener, C. D. and R. R. Sokal. 1957. A quantitative approach to a problem in classification. *Evolution* 11: 130–162.
Michod, R. E. 1999. *Darwinian Dynamics: Evolutionary Transitions in Fitness and Individuality*. Princeton University Press, Princeton.
Michod, R. E. 2007. Evolution of individuality during the transition from unicellular to multicellular life. *Proc. Natl. Acad. Sci. USA* 104: 8613–8618.
Miller, S. J. 1953. A production of amino acids under possible primitive earth conditions. *Science* 117: 528.
Mittelbach, G. G., D. W. Schemske, H. V. Cornell, and 19 others. 2007. Evolution and the biodiversity gradient: Speciation, extinction and biogeography. *Ecol. Lett.* 10: 315–331.
Nagashima, H., F. Sugahara, M. Takechi, and 4 others. 2009. Evolution of the turtle body plan by the folding and creation of new muscle connections. *Science* 325: 193–196.
Nelson, G. and N. Platnick. 1984. Systematics and evolution. In M.-W. Ho and P. T. Saunders (eds.), *Beyond Neo-Darwinism: An Introduction to the New Evolutionary Paradigm*, pp. 143–158. Academic Press, London.
Newman, S. A. and R. Bhat. 2009. Dynamical patterning modules: a "pattern language" for development and evolution of multicellular form. *Int. J. Devel. Biol.* 53 (special issue S1): 693–705.

Noor, M. A. F., K. L. Grams, L. A. Bertucci, and 1 other. 2001. Chromosomal inversions and the reproductive isolation of species. *Proc. Natl. Acad. Sci. USA* 98: 12084–12088.

O'Donald, P. 1960. Assortative mating in a population in which two alleles are segregating. *Heredity* 15: 389–396.

Ohta, T. 1973. Slightly deleterious substitutions in evolution. *Nature* 246: 96–98.

Okasha, S. 2006. *Evolution and the Levels of Selection*. Oxford University Press, Oxford.

Olson, E. C. and R. L. Miller. 1958. *Morphological Integration*. University of Chicago Press, Chicago.

Orians, G. H. and R. T. Paine. 1983. Convergent evolution at the community level. In D. J. Futuyma and M. Slatkin (eds.), *Coevolution*, pp. 431–458. Sinauer Associates, Sunderland, MA.

Orr, H. A. and J. A. Coyne. 1992. The genetics of adaptation revisited. *Am. Nat.* 140: 725–742.

Parker, G. A. 1974. The reproductive behavior and the nature of sexual selection in *Scatophaga stercoraria* L. (Diptera: Scatophagidae). VII. The origin and evolution of the passive phase. *Evolution* 24: 774–788.

Partridge, L. and L. D. Hurst. 1998. Sex and conflict. *Science* 281: 2003–2008.

Pigliucci, M. 2007. Do we need an extended evolutionary synthesis? *Evolution* 61: 2743–2749.

Raup, D. M. and S. J. Gould. 1974. Stochastic simulation and evolution of morphology—towards a nomothetic paleontology. *Syst. Zool.* 23: 305–322.

Rice, W. R. and B. Holland. 1997. The enemies within: Intergenomic conflict, interlocus contest evolution (ICE), and the intraspecific Red Queen. *Behav. Ecol. Sociobiol.* 41: 1–10.

Ricklefs, R. E. 2006. Evolutionary diversification and the origin of the diversity-environment relationship. *Ecology* 87 (suppl.): S3–S13.

Rieseberg, L. H. 1997. Hybrid origins of plant species. *Ann. Rev. Ecol. Evol. Syst.* 28: 359–389.

Rundle, H. D. and P. Nosil. 2005. Ecological speciation. *Ecol. Lett.* 8: 336–352.

Rundle, H. D., A. O. Mooers, and M. C. Whitlock. 1998. Single founder-flush events and the evolution of reproductive isolation. *Evolution* 52: 1850–1855.

Schluter, D. and R. E. Ricklefs. 1993. Convergence and the regional component of species diversity. In R. E. Ricklefs and D. Schluter (eds.), *Species Diversity in Ecological Communities: Historical and Geographical Perspectives*, pp. 230–240. University of Chicago Press, Chicago.

Schmalhausen, I. I. 1949. *Factors of Evolution*. Blakiston, Philadelphia.

Sepkoski, J. J. Jr. 1978. A kinetic model of Phanerozoic taxonomic diversity. I. Analysis of marine orders. *Paleobiology* 5: 337–352.

Shapiro, M. D., M. E. Marks, C. L. Peichel, and 5 others. 2004. Genetic and developmental basis of evolutionary pelvic reduction in threespine sticklebacks. *Nature* 428: 717–723.

Sniegowski, P. D., P. J. Gerrish, T. Johnson, and 1 other. 2000. The evolution of mutation rates: Separating causes from consequences. *BioEssays* 22: 1057–1066.

Sober, E. 1984. *The Nature of Selection: Evolutionary Theory in Philosophical Focus*. MIT Press, Cambridge, MA.

Stanley, S. M. 1975. A theory of evolution above the species level. *Proc. Natl. Acad. Sci. USA* 72: 646–650.

Stanley, S. M. 1979. *Macroevolution: Pattern and Process*. Freeman, San Francisco.

Stanley, S. M. 1982. Macroevolution and the fossil record. *Evolution* 36: 460–473.

Stebbins, G. L. 1950. *Variation and Evolution in Plants*. Columbia University Press, New York.
Tang, S. W. and D. A. Presgraves. 2009. Evolution in the *Drosophila* nuclear pore complex results in multiple hybrid incompatibilities. *Science* 323: 779–782.
Tax, S. 1960. *Evolution after Darwin: The University of Chicago Centennial*. University of Chicago Press, Chicago.
Templeton, A. R. 1982. Review: Why read Goldschmidt? *Paleobiology* 8: 474–481.
Thompson, J. N. 2004. *The Geographic Mosaic of Coevolution*. University of Chicago Press, Chicago.
Trivers, R. L. 1974. Parent-offspring conflict. *Am. Zool.* 11: 249–264.
Waddington, C. H. 1953. Genetic assimilation of an acquired character. *Evolution* 7: 118–126.
Wagner, G. P. 1989. The biological homology concept. *Ann. Rev. Ecol. Syst.* 20: 51–69.
Wagner, G. P. and L. Altenberg. 1996. Perspective: Complex adaptations and the evolution of evolvability. *Evolution* 50: 967–976.
Wake, D. B. 1991. Homoplasy: The result of natural selection, or evidence of design limitations? *Am. Nat.* 138: 543–567.
Walsh, B. and M. W. Blows. 2009. Abundant genetic variation + strong selection = multivariate genetic constraints: A geometric view of adaptation. *Ann. Rev. Ecol. Syst.* 40: 41–59.
Walsh, H. E., I. L. Jones, and V. L. Friesen. 2005. A test of founder speciation using multiple loci in the auklets (*Aethia* spp.). *Genetics* 171: 1885–1894.
West, S. A., A. S. Griffin, and A. Gardner. 2007. Social semantics: Altruism, cooperation, mutualism, strong reciprocity and group selection. *J. Evol. Biol.* 20: 415–432.
West-Eberhard, M. J. 1983. Sexual selection, social competition, and speciation. *Quart. Rev. Biol.* 58: 155–183.
West-Eberhard, M. J. 2003. *Developmental Plasticity and Evolution*. Oxford University Press, Oxford.
Wiens, J. J. and C. H. Graham. 2005. Niche conservatism: Integrating evolution, ecology, and conservation biology. *Ann. Rev. Ecol. Syst.* 36: 519–539.
Williams, G. C. 1957. Pleiotropy, natural selection, and the evolution of senescence. *Evolution* 11: 398–411.
Williams, G. C. 1966. *Adaptation and Natural Selection: A Critique of Some Current Evolutionary Thought*. Princeton University Press, Princeton.
Williams, G. C. 1992. *Natural Selection: Domains, Levels, and Challenges*. Oxford University Press, Oxford.
Williams, G. C. and D. C. Williams. 1957. Natural selection of individually harmful social adaptations among sibs with special reference to social insects. *Evolution* 11: 32–39.
Wilson, D. S. 1975. A theory of group selection. *Proc. Natl. Acad. Sci. USA* 72: 143–146.
Wynne-Edwards, V. C. 1962. *Animal Dispersion in Relation to Social Behaviour*. Oliver and Boyd, Edinburgh.
Zuckerkandl, E. and L. Pauling. 1965. Evolutionary divergence and convergence in proteins. In V. Bryson and H. J. Vogel (eds.), *Evolving Genes and Proteins*, pp. 97–116. Academic Press, New York.

Chapter 2

Rethinking Darwin's Position in the History of Science

Peter J. Bowler

The sheer number of celebrations taking place to mark the bicentenary of Darwin's birth is a clear indication of his iconic status as the discoverer of a major scientific theory and as a symbol of the power of science to challenge established beliefs. But, there may be a danger that this cult of Darwin as a celebrity may blind us to the complexity of the effects his work had both on science and on Western culture. We speak of a Darwinian revolution as though his books were the sole source of the new ideas that transformed science and society in the late nineteenth century and that still have a controversial impact today. However, it has long been recognized by historians that Darwin's work affected his contemporaries in several different ways. His book *The Origin of Species* of 1859 certainly triggered the debate that finally converted the scientific world to evolutionism and accelerated (although it did not initiate) the conversion of public opinion on the topic. It also introduced a radical new explanation of how evolution works: natural selection. Although it is now recognized that the selection theory is a major innovation that serves as the basis of much modern science, historians are aware that the idea proved hard for most nineteenth-century thinkers to accept. The first generation of evolutionary biologists was not very Darwinian, by modern standards, and searched actively for alternatives to the selection theory. Thus, it is valuable to think more carefully about what Darwin achieved in his own time, avoiding the temptation to read our modern enthusiasm for the selection theory into the reactions of his contemporaries (Bowler 1983, 1988, 2009a).

Such a reassessment of Darwin's position is of more than academic interest. In the modern world, Darwinism enjoys huge support within the sciences, but finds itself under massive attacks from forces outside science, most notably religious fundamentalism. The opponents frequently claim that Darwin's theory has had a malign influence on modern culture as the principal source of atheism, materialism, and a series of evil ideologies that are often known as social Darwinism. For critics, such as Richard

Weikart (2004), who is a supporter of intelligent design, Darwinism was actually responsible for the emergence of Nazism and the horrors of the holocaust. I am anxious to challenge Weikart's interpretation of history, and I have recently begun to explore the technique of counterfactual history to help uncover the extent to which the theory of natural selection can be held responsible for its alleged consequences (Bowler 2008). In counterfactual history, one selects a crucial turning point at which history might have been channeled in another direction and tries to explore the consequences. The technique has been hitherto used mostly in science fiction, but it has also begun to attract the attention of serious historians, especially those with an interest in military history and its wider implications (Cowley 1999, 2001; Ferguson 1997). What if the Confederates had won Gettysburg or the Luftwaffe at the Battle of Britain? The consequences for the subsequent development of America and Britain would be enormous. A parallel example from the field of economic history is Robert Fogel's classic study (1964) of how America would have developed without the advent of the railroads.

I intend to apply the same technique to the history of science by asking what would have happened if Darwin had drowned on the voyage of the *Beagle* or had died at an early age from the illness that affected him in subsequent years. There would obviously have been no *The Origin of Species*, and I will argue that without that book, there might not have been any serious discussion of natural selection until the early twentieth century, which was when evolutionary theory actually began to take hold in science in our world, built on the foundation of Darwinian thought. There would have been an evolutionary movement, but it would have been based almost exclusively on the non-Darwinian mechanisms that are known as the alternatives to selection and that were developed during the so-called eclipse of Darwinism. The potential value of the counterfactual history technique comes when it is applied to the wider debates centered on the theory and its implications. I will argue that most of the ideological positions, known collectively as social Darwinism, would have emerged, even if the theory of natural selection had not been available. Most social Darwinism was, in fact, social evolutionism with only superficial links to the core theory that Darwin had proposed. It can be shown from examples that it is perfectly possible to generate support for unrestrained free enterprise, militarism, imperialism, and racism based on non-Darwinian theories, such as Lamarckism. As Desmond and Moore (2009) have recently argued, Darwin's passionate opposition to slavery was fuelled by his rejection of the efforts made by creationist and anti-Darwinian evolutionists to defend a polygenist view of the origin of human races. The alleged link between Darwin and Hitler is thus rendered implausible by showing that most of the alleged consequences would have occurred even if the selection theory had not been part of the debates.

Having suggested some of the potential advantages of the counterfactual technique for the wider debate, in this paper I want to concentrate on the more limited objective of defending and exploring it in the history of science itself. First, I will address the key assumption of the counterfactual scenario outlined above, namely that without Darwin there would have been no debate over the theory of natural selection in the late nineteenth century. There is an obvious objection to this position arising from the widely held belief that the theory was somehow in the air at the time—either as a product of the logic of scientific discovery or because it reflected prevailing cultural developments—and that if Darwin hadn't promoted the theory, then someone else (probably Alfred Russel Wallace) would have, and the Darwinian Revolution would have been precipitated under another name. I will argue against this position by showing that there was, in fact, no one who could plausibly have taken Darwin's place. I will then try to reconstruct how science might have developed in a world without the selection theory, which is much less speculative than most versions of counterfactual history, since other explanations of evolution were actually developed in our world as alternatives to Darwinism. I will, however, address the questions of when and how natural selection would have entered into science in a world without Darwin.

Before getting to details, I want to make it clear that my use of counterfactual history is neither intended to undermine the credibility of the Darwinian theory nor the objectivity of science. As will be seen, the outcome is a world with a biology composed of more or less the same factors as our world—it is just that they were developed in a different sequence. Darwin's theory of natural selection, far from being a product of its own time, was so radical that hardly anyone could accept it as the primary mechanism of evolution, including most scientists. These scientists wanted less materialistic theories of evolution and explored various alternative theories that, in hindsight, might have been blind alleys. But, those alternatives were proposed as serious scientific theories, and some of the issues they addressed have reemerged in modern evolutionary developmental biology (see Wray, Chapter 9). Evolutionary biology today has moved beyond the neo-Darwinism of the mid-twentieth century, because we no longer accept the model of transmission genetics that ignored the whole issue of individual development. In this chapter, I will imagine a world in which development never ceased to be seen as a central theme for evolutionary biology and genetics and the selection theory emerged as modifications and extensions of the developmental view of evolution. Currently, advocates of the more radical interpretations of evolutionary developmental (evo-devo) biology have the viewpoint of needing to fight a battle for acceptance against a narrowly defined neo-Darwinian paradigm. But, in the alternative, counterfactual history universe, the non-Darwinian aspects of evo-devo would have been part of the original foundation on which the theory of natural selection would have been grafted, once a clearer idea of the nature of heredity had been established. Thus, the transition to a balanced synthesis might have been achieved much more smoothly.

Darwin's Originality

The claim that the theory of natural selection was somehow *in the air* in the mid-nineteenth century rests on the assumption that since all of the components of the theory had become widely available, it was only a matter of time before someone put them together and promoted the idea. It is argued that others besides Darwin did indeed articulate it, most obviously Patrick Matthew and Alfred Russel Wallace; there are minor literary industries devoted to arguing that these people should be recognized as the true discoverers of the theory (Brackman 1980; Brooks 1983; Dempster 1996). Darwin, it is claimed, only achieved his pre-eminent status because he and his followers had suppressed the claims of the rivals. However, some of these claims are based on a huge oversimplification of what Darwin's theory entailed. It was not just a theory of natural selection (i.e., of the selection of random variations that are due to their differential abilities to reproduce in an environment with limited resources). Darwin discovered natural selection after he had already recognized that evolution must be represented as a branching tree in which populations diverge from a common ancestry when they are exposed to different environments, most obviously as a result of migration and consequent separation by geographical barriers. He also realized that this model (which was very original when he came to it in the late 1830s, after his experiences on the *Beagle* voyage) could explain the way in which organisms are classified into groups, with common ancestry being the basis of the underlying similarities that unite superficially divergent forms. In addition, he looked to artificial selection, both as a means to help understand how the natural equivalent worked and as a source of information about variability and heredity (Bowler 2009b).

It can thus be argued that selection, by itself, is not sufficient foundation for a theory that has the ability to transform our understanding of the natural world—which is where the case for Patrick Matthew's theory breaks down. Although he appreciated that selection could produce newly adapted forms, he conceived the idea within a catastrophist geological framework, which encouraged him to see most evolution as taking place in sudden bursts following mass extinctions. McKinney (1971) provides a collection of relevant texts by Matthew and other precursors of Darwin. Matthew did appreciate that artificial selection provided a scientific model (his comments came in the appendix to an 1831 book on forestry); but, beyond offering a few suggestions as to the topics that might be investigated, he went no further with the idea. He also seems to have had no inclination to follow up his discovery by showing in detail how it could explain the facts of the geographical and geological distribution of forms or the morphological and embryological relationships that underpin our ability to recognize the tree of life (Figure 2.1).

Alfred Russel Wallace is in a different category altogether. He came to his theory by a route that is in many ways was very similar to Darwin's.

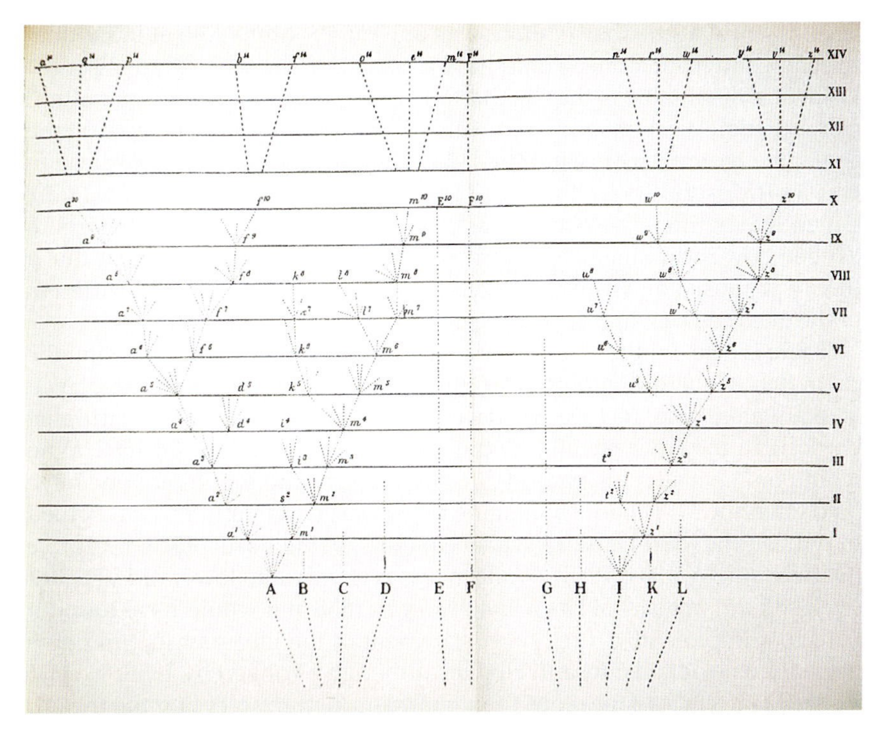

FIGURE 2.1 Hypothetical Phylogenetic Tree (From Darwin 1859.)

He was inspired by Charles Lyell's uniformitarian geology, he worked on biogeography, and he independently articulated the model of the tree of life in an 1855 paper. Moreover, his inspiration for the discovery of natural selection in 1858 was based, again like Darwin, on recognizing the implications of T. R. Malthus's principle of population. Darwin himself thought that Wallace had independently discovered the whole theory, and this assumption has underpinned most subsequent interpretations of Wallace's 1858 paper, allowing him to be hailed as the co-discoverer of the theory.

However, several historians have pointed out that there were, in fact, significant differences between how the two discoverers articulated the idea (Bowler 1976; Nicholson 1960). Kottler (1985) offers an assessment of these interpretations. Wallace does not seem to have conceived the process of selection in terms of a ruthless struggle between individual variants within a population (perhaps not surprisingly given his very different ideological position compared to Darwin's). He thought in terms of variants being measured against a standard that was set by the environment, not against each other. He also seems unclear on the nature of the varieties being selected. The traditional assumption is that he meant individual variants, but that would be a curious use of the term variety, which, even in his day, was normally applied to distinct subpopulations within a species. At the very least, I would argue that Wallace was unclear about the role of individual selection and if left alone, would have focused on a kind of group selection. What is indisputable is that he did not see a relationship between natural and artificial selection, and later warned Darwin that by emphasizing the analogy, Darwin encouraged the misunderstanding that natural selection was a purposeful force (Burkhardt et al. 2004: 227, letter of 2 July 1862). Given that artificial selection was a key resource in Darwin's efforts to explain how individualistic natural selection worked, it seems clear that a theory proposed by Wallace alone would have been significantly different to the one the Darwin had developed.

There is also another factor to be taken into account when assessing the claim that in a world without Darwin, Wallace would simply have played an equivalent role. In 1858, Wallace was a relatively unknown naturalist working in the Far East, and he did not return to Britain until 1862. Who would he have sent his paper to, if not Darwin, and would anyone have paid attention? It is hard to imagine either of the two figures, who helped to publish the joint Darwin–Wallace papers, being particularly receptive (Darwin and Wallace 1958). Wallace was certainly a follower of Lyell, but Lyell still did not accept Darwin's theory in 1858 and would have found it hard to grapple with a very short statement of the idea sent to him unexpectedly. The botanist Joseph Hooker, although a convert by 1858, too had resisted Darwin's suggestions for some years, before accepting the theory. In a world without Darwin, he too might have struggled to appreciate the potential significance of what Wallace had written. Bear in mind that even with Darwin's short summary, the joint papers of 1858 attracted very little attention, so if Wallace on his own had been able to get his paper published, its influence would have been minimal. If he had set out to write a more complete statement, it would have taken him years to prepare anything that could have had an impact equivalent to *The Origin of Species*.

If one discounts Wallace, it is hard to think of any other naturalist who might have been in a position to develop the theory of natural selection instead of Darwin. Thomas Henry Huxley wasn't deeply concerned with biogeography, and although he claimed that the idea of natural selection was obvious as soon as it was explained to him, he did not think that it offered a complete explanation of evolution (Bartholomew 1975; Di Gregorio 1984). In contrast, there was a growing sense that some form of natural transmutation would be preferable to the old idea of divine creation, so it is plausible to argue that the general idea of evolution was in the air. James Secord's (2000) work on the reception of Robert Chambers's *Vestiges of the Natural History of Creation* is perhaps the best evidence there is that the basic idea was beginning to be widely discussed by the 1850s. Herbert Spencer had written in support of Lamarck in 1851 and his *Principles of Psychology* in 1855 explained the evolution of the human mind in terms of the inheritance of acquired characteristics. He would have written his *Principles of Biology*, published in 1864, even without Darwin, because it too promoted a Lamarckian approach to evolution; although as Mark Francis (2007) points out, his overall project was very different than Darwin's. In Germany, it seems probable that Ernst Haeckel's commitment to a naturalistic philosophy would have led him toward the morphological evidence for evolution, perhaps a little more slowly without an input from Darwin, and it was Haeckel's *Generelle Morphologie* of 1866, not *The Origin of Species*, that persuaded Huxley to begin using evolution in his scientific work (Di Gregorio 1984).

Evolutionism without Darwin

Thus, I think it is plausible to think that in a world without Darwin, there would have been a gradual acceptance of evolutionism by the scientific community in the course of the 1860s and 1870s. *The Origin of Species* played an important role in precipitating the debate, and without it, the process of conversion would have been more gradual and certainly less traumatic. Those naturalists with relatively radical philosophical positions would have been looking for the opportunity to develop a theory that replaced the old idea of design and divine creation. However, the vast majority had conservative opinions, and it is known that natural selection was seen as a materialistic theory, because it reduced nature to an unplanned system governed by trial and error. The accounts of the Darwinian Revolution are full of examples of complaints about the apparently haphazard nature of evolution in Darwin's scheme and expressions of hope that something more orderly and purposeful would be found (Bowler 1988). If the Darwinian version had not been available, conservative naturalists would have looked for versions of evolutionism that did not challenge their preconceptions, without looking over their shoulder at the bogeyman of extreme materialism. *The Origin of Species* generated a sudden crisis, both because it offered a wealth of new evidence for evolution and because it sought to explain the process by a new and very disturbing mechanism. Without the novel evidence, it would have taken longer for naturalists to face up to the job of constructing a new evolutionary worldview. However, they would have been able to address the problem without the worry of a huge public outcry of a complete destruction of the old faith in a divinely ordered world.

What would this non-Darwinian evolutionism have looked like? We can gain a pretty good idea from the alternatives to the selection theory that *were* developed in our world (Bowler 1983). It should also be noted that the research priorities of most naturalists did not match the program that Darwin outlined, which gives us clues about the priorities as well as the theoretical preconceptions of those involved. The major evolutionary research project of the late nineteenth century was, in any case, something that Darwin himself did not engage in: the attempt to reconstruct the history of life on earth from morphological, paleontological, and biogeographical evidence (Bowler 1996). In a world without Darwin, figures such as Haeckel would have played a more central role, and the project to reconstruct the history of life would have been undertaken with even more enthusiasm. There would have been less focus on the detailed mechanism of evolution, including the issues of variation and heredity, because there would be no need to look for arguments to use against Darwin's new ideas about the raw material of natural selection.

The general outlines of how to conduct phylogenetic research would have been very similar to those we are familiar with today. Evidence from comparative anatomy and embryology would have been used when the fossil record was lacking, and there would have been intense interest in the discovery of new fossils to fill in the gaps. The recapitulation theory would have been prominently in evidence; there would have been the same disputes about the relative value of the different forms of morphological evidence and the same disagreements over whether similar structures were homologies or the products of convergence and parallelism. The most interesting question might be what would have been the role of biogeography, a key strand of evidence in Darwin's argument. Without *The Origin of Species*, figures such as Wallace, Hooker, and Asa Gray would have campaigned to get this line of investigation taken seriously, and it is hard to believe that they would not have succeeded in a world where exploration and colonization were proceeding apace. Other products of the explorers' efforts might have been harder to promote, however. Henry Walter Bates's work on mimicry, for instance, fit so well into the Darwinian paradigm that one suspects it would have received less attention if the selection theory were not an option. Wallace's more detailed work on island biogeography and speciation might have been ignored. An interesting possibility, too complex to discuss here, is whether such topics might have led to a slightly later discovery of the selection theory, perhaps during the 1870s.

Assuming that a major project to reconstruct the history of life on earth began in the 1860s, what would its theoretical foundations have been if the theory of natural selection were not available? There are two ways of looking at this question, and each rest on the different levels at which Darwin forced scientists to rethink their positions.

Adaptation and Common Descent

Darwin took it for granted that one of the major questions to be answered by any workable theory of evolution was how to explain the origin of adaptations. His theory of branching, open-ended evolution was premised on the assumption that the tree of life was produced by the constant divergence of forms from a common ancestor under the pressure of adaptation to different, isolated environments (see Figure 2.1). Natural selection was his preferred explanation of how the process of adaptation worked.

The focus on adaptation was a peculiarly Anglophone concern, prompted to a large extent by the utilitarian foundations of William Paley's design argument. Here the adaptation of structure to function (i.e., their usefulness) was taken as the most obvious indication of the Creator's wisdom and benevolence. However, it has long been recognized that there were other traditions that focused more on the underlying unity of natural groups than on the superficial adaptive modifications of individual species (Rehbock 1983; Russell 1916). This approach flourished in France under the influence of Etienne Geoffroy Saint-Hilaire and in Germany under the influence of the transcendental anatomy created originally by the Naturphilosophen. It was imported into Britain in the 1840s and adapted to the prevailing argument from design by anatomists such as Richard Owen (Rupke 1994). Owen was certainly aware of the importance of functionalism, but he was also concerned that homologies indicate the existence of underlying archetypical forms that imposed a unity on natural groups—a unity that could also be taken as an indication of design by an intelligent Creator. But, the transcendental tradition did not focus exclusively on the adaptive divergence of form from the archetype. It was widely held that many aspects of organic structures could not be explained in terms of adaptation and could only be accounted for by internal laws that governed the ways in which forms could be manifested. Whether or not one believed a Creator designed these laws, they restricted or even directed the process by which embryos could develop, in ways that had nothing to do with adaptation.

When transferred into an evolutionary worldview, the functionalist and the formalist approaches could give very different results. Owen himself became an evolutionist in the 1860s despite his rejection of Darwinism, and it seems reasonable to suppose he would have made the same move even more readily, if there had been no threat to the idea of design. His focus was to recognize the divergence of the main adaptive branches within each major group, in effect conceding that the archetype could be treated as a common ancestor. Yet, his disciple St. George Jackson Mivart was far more willing to attack the whole logic of Darwinism by limiting both the applicability of the theory of common descent and the primacy of adaptation in the effort to understand how life developed (Mivart 1871; Desmond 1982).

Mivart objected to Darwin's idea that variation is largely undirected, so that adaptive modification is the only driving force of evolution. He proposed that there were nonadaptive laws of form governing the sequence of variations that appeared in the course of evolution, opening the way for a totally non-Darwinian model of the history of life in which parallelism played a significant role. For instance, he proposed that the monotremes represented an independent line of development unrelated to the rest of the mammals—the resemblances that had led to them being regarded as primitive mammals were the result of parallel evolution (Mivart 1888). Much of his invective was directed against the selection theory, but his central point would apply against any theory that reduced evolution to an open-ended process governed solely by the pressures of adaptation. So, it is important to bear in mind that in our hypothetical world without Darwin, there were pressures that would have led toward a very different model of how to understand organic relationships and the history of life on earth. Indeed, without the barrage of evidence that Darwin provided for the theory of common descent, these pressures would have been able to manifest themselves more effectively. If Wallace had been trying to promote the model of common descent on his own during the 1860s, he would have been fighting an uphill battle against a widespread assumption that evolution should not be understood solely in terms of how geographically isolated populations adapt to their local environments.

Ernst Mayr used to claim that the theory of common descent was one of Darwin's greatest innovations (Mayr 1991). He was right, but I do not think he was willing to concede the extent of the resistance to that innovation among biologists of the late nineteenth century. The theories of orthogenesis and parallelism that were proposed at the time were all intended to subvert the logic of the claim that underlying similarities of structure were remnants of a common ancestry; instead, they could be explained as independent products of the same variation-trend operating in separate lines of development. The extent of this rival conception of evolution, even today, is hard to overestimate—thus, think of how powerful it would have been in a world in which Darwin's arguments for common descent and the undirected nature of variation were not available. Theories of orthogenesis, as developed by German biologists such as Theodore Eimer and the Americans Edward Drinker Cope and Alpheus Hyatt, all presupposed parallel lines of nonadaptive evolution driven by internally programmed variation trends. Cope (1868) argued that the resemblances used to assign species to a genus were the result, not of descent from a recent common ancestor, but rather of independent lines of evolution within a group reaching the same stage of development at the same time. The question of the relationship between orthogenesis and modern theories of developmental constraints is too complex to enter into here (see Kirkpatrick, Chapter 7 and G. Wagner, Chapter 8). In the twentieth century, Henry Fairfield Osborn could promote the idea of adaptive radiation (Figure 2.2), yet thought that each family in the radiation subsequently developed along rigid orthogenetic lines. But, note that for Osborn adaptive radiation was always rapid. Details of all these theories appear in various other literature (Bowler 1983; see Losos and Mahler, Chapter 15).

Many of the theories were linked to the belief that the succession of predetermined characters appeared in a stepwise fashion, by saltation rather than the accumulation of small-scale variations. Even T. H. Huxley expressed a preference for the discontinuity of variation in the production of new species and hinted at the possibility of laws governing variation. Most of the theories proposed by paleoanthropologists to explain human origins were based on a significant level of parallelism in primate evolution (Bowler 1986). If we want to visualize the evolutionism of a world without Darwin, we just have to look at what happened in the eclipse of Darwinism

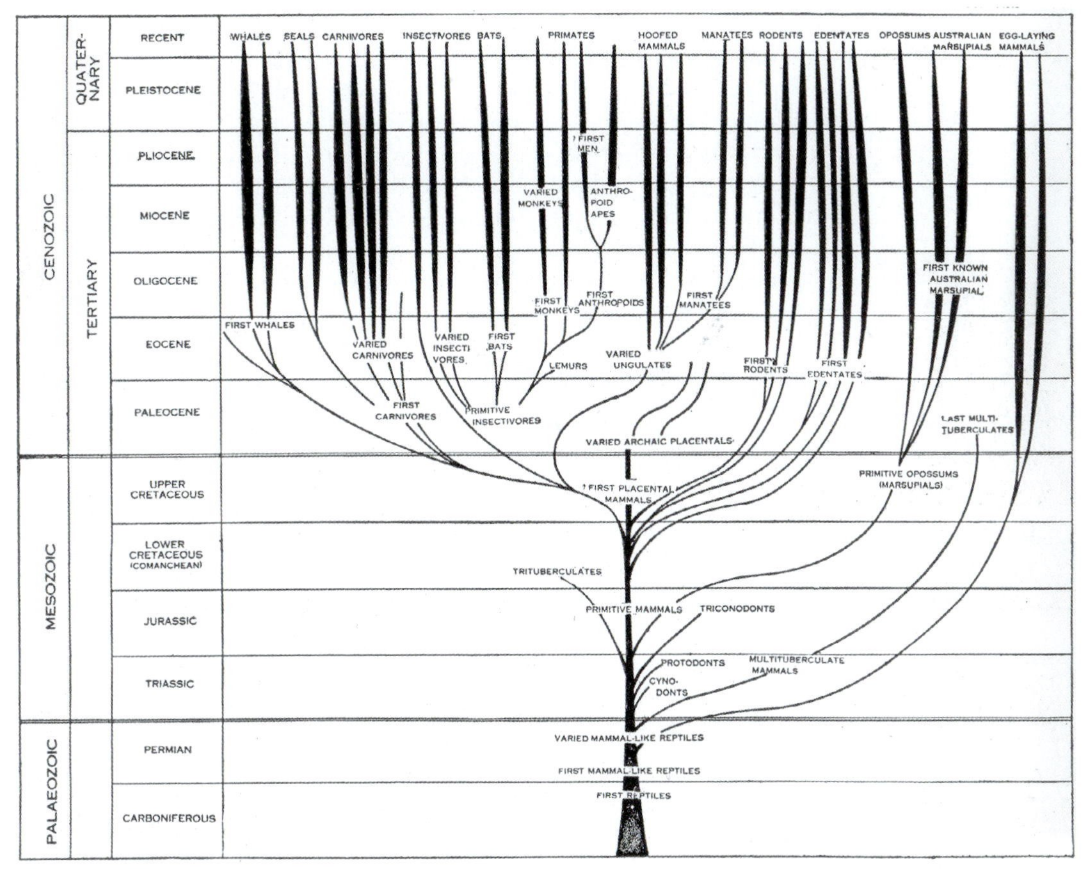

FIG. 114. ANCESTRAL TREE OF THE MAMMALS.

FIGURE 2.2 Phylogenetic Tree of the Mammals Showing Adaptive Radiation Followed by Orthogenetic Evolution within Each Family (From Osborn 1916.)

and imagine those alternative ideas operating unchallenged as the main driving forces of evolutionary thinking.

Mechanisms of Variation without Darwin's Theory

What mechanisms would have been invoked to explain the directed nature of variation? In *Darwin's Watch* Terry Pratchett and his collaborators imagine that Darwin, prevented from travelling on the *Beagle*, becomes a clergyman and writes a book called *The Theology of Species* (Pratchett et al. 2005). This event undermines the progress of science by suggesting that the Creator is directly responsible for shaping the course of evolution, thereby checking efforts to search out a naturalistic mechanism. Whatever you think of the idea that Darwin himself might have written such a book, the fact is that there were many religious scientists who opted to support what is now called theistic evolutionism, in which the course of evolution is, in one way or another, traced back to the direct will of the Creator. This was the argument of Chambers's *Vestiges of Creation*, Owen's theory of derivation, and Mivart's *Genesis of Species*. It was the fallback position adopted by Asa Gray when he finally conceded that his efforts to reconcile natural selection with the argument from design had run into difficulties. It is how Edward Drinker Cope first approached the idea of directed evolution. The suggestion that in a world without Darwin there would have been a greater temptation to explore the idea of evolution by design is thus quite plausible (even without *The Theology of Species*). However, I find it hard to believe that the enthusiasm for theistic evolutionism could have headed off all efforts to search for naturalistic explanations. The radical leaning theorists, who would also have welcomed the idea of evolution, were anxious to eliminate the idea of design and thus would have refused to compromise on this issue. They would not have been prepared to accept evolution without a naturalistic explanation, and in the absence of the selection theory, they would have looked to other possibilities outlined in the following discussion. In the absence of the Darwinian threat, conservative leaning scientists might also have found it easier to envisage naturalistic mechanisms that would generate orderly and purposeful change.

Of course, it was Darwin's selection theory that gave the radical leaning thinkers the opportunity to launch the evolutionary program, even though most of them concluded that it was not an adequate explanation of how the process worked. In a world without Darwin, there was an obvious alternative for those who saw the adaptation of structure to function as a key aspect of the process: the so-called Lamarckian theory of the inheritance of acquired characteristics. In Britain in the 1850s, this theory was already being promoted by Herbert Spencer, and there seems little doubt that Spencer's campaign would have led him to write *Principles of Biology* even without the stimulus of Darwin. He would not have coined the iconic phrase "the survival of the fittest," but his book focuses more on Lamarckism than on selection—and it is important to bear in mind that later in his career Spencer remained a champion of use-inheritance. In Germany, Haeckel too would almost certainly have adopted a Lamarckian approach, as did Eimer for those aspects of his theory that dealt with adaptation. The American neo-Lamarckians, Cope and Hyatt, also moved to adapt use-inheritance as an explanation of how parallel trends could become established, although like Eimer, they retained the belief that much evolution consisted of nonadaptive orthogenesis. Huxley, admittedly, had no time for Lamarckism, but that was because he was more interested in formal relationships—his only interest in the selection theory was as ammunition to use against the churches.

For Huxley and for many others, the preferred naturalistic explanation of evolutionary trends was some form of an internally programmed trend driving variation in one or a few predetermined directions. This idea was commonly associated with the assumption that the processes governing ontogeny could somehow be extended to add on additional stages of growth that would produce individuals constituting a new species. Lamarckism saw characteristics acquired by the organism in response to environmental challenge as the source of the new stages of development. Orthogenesis postulated purely internal and hence nonadaptive trends, which often were supposed to operate in discrete steps. It is because both these mechanisms postulated that variation is an *addition* to individual development that the recapitulation theory became so popular (Gould 1977).

As philosopher of science Ron Amundson (2005) argues, these early alternatives to the Darwinian way of thinking reflect widespread interests in topics that were later marginalized by the rise of modern genetics, but which have now reemerged with the advent of evolutionary developmental biology. I think that in his anxiety to construct a historical pedigree for evo-devo, Amundson underestimates the extent to which the developmental movement of the late nineteenth century veered off in directions that would not be acceptable today. However, his underlying thesis is germane to my present argument, because it suggests that what happened in the late nineteenth century cannot be dismissed as an aberration or scientific blind alley. What I am suggesting for our world without Darwin is, in effect, the emergence in the late nineteenth century of something that would look remarkably like the extreme (i.e., non-selectionist) versions of modern evolutionary developmental biology. Without the challenge of Darwinism, these schools would have been able to retain their hold on the theoretical imagination into the twentieth century. Genetics and the selection theory, when they were eventually developed, would have been incorporated seamlessly into the program as the extreme ideas of parallelism and orthogenesis were toned down in response to further embryological research. The end product would be a synthesis equivalent to that described by Wray (see Wray, Chapter 9) but reached by a different route. It is worth remembering that the very narrow focus on the problems of transmission that characterized genetics in the Anglo-American traditions were not matched in France and Germany, where development remained a key focus throughout the early twentieth century (Burian et al. 1988; Harwood 1993).

Whence the Selection Theory?

This discussion leads to my final topic: in a world without Darwin, when would the theory of natural selection have emerged, and how would the reformation in the thinking on heredity, which we associate with Mendelian genetics, have played out? One point to bear in mind is that it is perfectly possible that the alternative universe would have had the idea that the less successful products of natural evolution (however produced) would eventually be eliminated. Most anti-Darwinians seem to have been perfectly happy with this negative side of selection, and it is hard to believe that it would not have been recognized at some point. As noted, in fact, negative selection may have, more or less, been the model of selection that Wallace was trying to articulate in 1858. The key question is when would the idea of selection, acting on individual variants within the same population, have become a major topic of debate. It is here that I suggest that the absence of Darwin's work might have led to a significant delay. It is conceivable that Wallace and others interested in topics like island biogeography might have been able to promote the idea during the 1870s, yet I suspect that by this time the developmental model of evolution would already have gained enough momentum to ensure its marginalization.

So, when would the selection theory have emerged as a central theme in evolutionism? The answer to this question may be linked to the second question previously mentioned, namely, when would the reformulation of theories of heredity that we associate with Mendelian genetics have occurred? It may well have been developments in the understanding of heredity that paved the way for selection—not, as in our world, the other way around. Without Darwin, there would not have been the intensive study of the nature of variation and heredity of the kind he undertook or that opponents, desperately seeking to undermine the plausibility of his model of natural selection, also engaged in (Gayon 1998). There would have been no focus on the work of the animal breeders—remember, even Wallace did not accept the analogy between natural and artificial selection. There would be no critics of selection (such as Fleeming Jenkin) appealing to the prevailing idea of blending heredity as a way to challenge the logic of Darwin's argument (Jenkin 1867; Hull 1973). August Weismann would not have had a selection theory to encourage him to create the concept of the germ plasm, and Francis Galton would probably not have gone into the topic of heredity at all, without the inspiration provided by his cousin, Darwin. In the world without Darwin the model of soft heredity, which permits the Lamarckian effect, would have had a much easier ride, at least, until carefully designed studies of heredity began. Even then, without employing the analogy to artificial selection, the full power of the selection theory might have been slow to emerge.

Where would the concept of hard heredity have come from? To answer this question, I think we have to invoke wider cultural changes and their ability to channel scientific thought by defining new topics of interest. In the late nineteenth century, the general enthusiasm for possessing an evolutionary worldview certainly seems to resonate with the values of a world in which progress was taken for granted. As mentioned, I believe many elements of what we call social Darwinism required little direct input from the selection theory and would have just as easily drawn their scientific justification from the developmental viewpoint, especially via its Lamarckian components. However, it is known that the 1890s saw a growing suspicion of the ideas that the human race could be improved through environmental control or that character and ability were determined by inheritance (Kevles 1985; see Gould, Chapter 21). The advent of the eugenics movement would certainly have concentrated biologists' attention onto the issue of heredity, and if there were a parallel to the revolt against morphology witnessed in the world that did evolve, it would have encouraged the kind of experimental work that generated models of hard heredity and perhaps also, the concept of the gene. Many of the biologists associated with the rediscovery of Mendel's laws were originally enthusiasts for evolution by saltation, including Hugo De Vries, William Bateson, and Thomas Hunt Morgan (Bowler 1989). Recognizing that evolution might operate in discrete steps paved the way for accepting that heredity was also discontinuous, and saltationism would certainly have been a key feature of the developmentalism that flourished in the non-Darwinian world. As the evidence against the simpler forms of Lamarckism came under suspicion, some form of genetics would have entered science.

The eugenics movement would have had another important consequence. In the absence of the theory of natural selection, the most obvious model upon which to base the argument for a selective breeding policy in humankind would be the work of the animal breeders. In this regard, at last, the concept of selection would have come to the fore, just as it had for Darwin in our world. Highlighting the power of artificial selection in animal breeding at the same time that the evidence for Lamarckism was beginning to look a little shaky would surely have triggered a belated recognition of the possibility that there could be a natural equivalent, driven by the superior reproductive power of those with better-adapted characteristics. Perhaps, those individuals (e.g., R. A. Fisher, J. B. S. Haldane, and Sewall Wright) who put together the synthesis of genetics and Darwinism in our world, would have initiated, in the world without Darwin, a recognition that a natural form of selection was the obvious alternative to the increasingly discredited Lamarckian element of the old developmentalism. However, in our alternative universe, they would have been the *first* to develop a complete theory of natural selection, rather than merely the rejuvenators of an existing model.

The end product would be a school of biology quite similar to what we have today, but reached by a different route. There would have been no rejection of the significance of ontogeny by the new genetics, because the developmental paradigm would have been too powerful, just as it was delineated, including in many European countries in our world. There would have been a smoother transition to a comprehensive evolutionary theory, combining the best elements of developmentalism and selectionism. More importantly, we would have had very different ideas about the wider implications of the various components of the synthesis. Without the specter of materialistic Darwinism haunting the imagination of late nineteenth-century religious thinkers, evolutionism would not have acquired the reputation that encourages many Creationists to regard it as a stepping-stone to atheism. The prospect of an animal origin for humanity would have generated controversy, but we know from the reaction of liberal Christians in our own world that this issue was less divisive when the evolutionary process was thought to be goal-directed. If the selection theory had not been developed until the early twentieth century, it would not have carry the ideological burden of being associated with all the evils of social Darwinism—and some of the alternative theories would have had their true colors revealed, because their role in the promotion of harsh social policies would be all to obvious from their history. There would have been renewed concern about the materialism of the evolutionary process once the selection theory came to the fore—but it would have taken place after the wave of enthusiasm for militarism and imperialism that swept through Western culture, beginning in the last decade of the nineteenth century. The Nazis, who might well exist in the alternative universe, would hardly be inspired by a new theory emphasizing the individualistic nature of the evolutionary mechanism. The whole process by which evolutionism emerged would seem more continuous and less controversial than in our own world, to the extent that we might not be using the term revolution and would have no need for an iconic celebrity to fill Darwin's place as its figurehead.

Literature Cited

Amundson, R. 2005. *The Changing Role of the Embryo in Evolutionary Thought: The Roots of Evo-Devo.* Cambridge University Press, New York.

Bartholomew, M. 1975. Huxley's defence of Darwinism. *Ann. Sci.* 32: 525–535.

Bowler, P. J. 1976. Alfred Russel Wallace's concepts of variation. *J. Hist. Med.* 31: 17–39.

Bowler, P. J. 1983. *The Eclipse of Darwinism: Anti-Darwinian Evolution Theories in the Decades around 1900.* Johns Hopkins University Press, Baltimore.

Bowler, P. J. 1986. *Theories of Human Evolution: A Century of Debate, 1844–1944.* Johns Hopkins University Press, Baltimore.

Bowler, P. J. 1988. *The Non-Darwinian Revolution: Reinterpreting a Historical Myth.* Johns Hopkins University Press, Baltimore.

Bowler, P. J. 1989. *The Mendelian Revolution: The Emergence of Hereditarian Concepts in Modern Science and Society.* Athlone, London.

Bowler, P. J. 1996. *Life's Splendid Drama: Evolutionary Biology and the Reconstruction of Life's Ancestry, 1860–1940.* University of Chicago Press, Chicago.

Bowler, P. J. 2008. What Darwin disturbed: The biology that might have been. *Isis* 99: 560–567.

Bowler, P. J. 2009a. *Evolution: The History of an Idea.* 25th anniversary ed. University of California Press, Berkeley.

Bowler, P. J. 2009b. Darwin's originality. *Science* 323: 223–226.

Brackman, A. C. 1980. *A Delicate Arrangement: The Strange Case of Charles Darwin and Alfred Russel Wallace.* Times Books, New York.

Brooks, J. L. 1983. *Just before the Origin: Alfred Russel Wallace's Theory of Evolution.* Columbia University Press, New York.

Burian, R., J. Gayon, and D. Zallen. 1988. The singular fate of genetics in the history of French biology. *J. Hist. Biol.* 21: 357–402.
Burkhardt, F., D. M. Porter, S. A. Dean, and 5 others (eds.). 2004. *The Correspondence of Charles Darwin*, Volume 14: 1866. Cambridge University Press, Cambridge, UK.
Cope, E. D. 1868. On the origin of genera. *Proc. Acad. Natl. Sci. Philadelphia* 20: 242–300. Reprinted in Cope, E. D. 1887. *The Origin of the Fittest*, pp. 41–123. Macmillan, New York.
Cowley, R. (ed.). 1999. *What If? The World's Foremost Military Historians Imagine What Might Have Been*. Scribner's, New York.
Cowley, R. (ed.). 2001. *More What If? Eminent Historians Imagine What Might Have Been*. Scribner's, New York.
Darwin, C. 1859. *On the Origin of Species by Means of Natural Selection, or the Preservation of Favoured Races in the Struggle for Life*. John Murray, London.
Darwin, C. and A. R. Wallace. 1958. *Evolution by Natural Selection*. Cambridge University Press, Cambridge, UK.
Dempster, W. J. 1996. *Evolutionary Concepts in the Nineteenth Century: Natural Selection and Patrick Matthew*. Pentland Press, Edinburgh.
Desmond, A. 1982. *Archetypes and Ancestors: Palaeontology in Victorian London, 1850–1875*. Blond and Briggs, London.
Desmond, A. and J. R. Moore. 2009. *Darwin's Sacred Cause: Race, Slavery and the Quest for Human Origins*. Allen Lane, London.
Di Gregorio, M. 1984. *T. H. Huxley's Place in Natural Science*. Yale University Press, New Haven.
Ferguson, N. 1997. *Virtual History: Alternatives and Counterfactuals*. Picador, London.
Fogel, R. 1964. *Railroads and American Economic Growth: Essays in Economic History*. Johns Hopkins University Press, Baltimore.
Francis, M. 2007. *Herbert Spencer and the Invention of Modern Life*. Acumen, Stocksfield, UK.
Gayon, J. 1998. *Darwinism's Struggle for Survival: Heredity and the Hypothesis of Natural Selection*. Cambridge University Press, Cambridge, UK.
Gould, S. J. 1977. *Ontogeny and Phylogeny*. Harvard University Press, Cambridge, MA.
Harwood, J. 1993. *Styles of Scientific Thought: The German Genetics Community, 1900–1937*. University of Chicago Press, Chicago.
Hull, D. L. (ed.). 1973. *Darwin and His Critics: The Reception of Darwin's Theory of Evolution by the Scientific Community*. Harvard University Press, Cambridge, MA.
Jenkin, F. 1867. The origin of species. *N. Brit. Rev.* 46: 277–318.
Kevles, D. 1985. *In the Name of Eugenics: Genetics and the Uses of Human Heredity*. Knopf, New York.
Kottler, M. J. 1985. Charles Darwin and Alfred Russel Wallace: Two decades of debate over natural selection. In D. Kohn (ed.), *The Darwinian Heritage: A Centennial Retrospect*, pp. 367–432. Princeton University Press, Princeton.
Mayr, E. 1991. *One Long Argument: Charles Darwin and the Genesis of Evolutionary Thought*. Harvard University Press, Cambridge, MA.
McKinney, H. L. (ed.). 1971. *Lamarck to Darwin: Contributions to Evolutionary Biology*. Coronado Press, Lawrence.
Mivart, St. G. J. 1871. *On the Origin of Genera*. Macmillan, London.
Mivart, St. G. J. 1888. On the possible dual origin of the Mammalia. *Proc. Roy. Soc.* 43: 372–379.

Nicholson, A. J. 1960. The role of population dynamics in natural selection. In S. Tax (ed.), *Evolution after Darwin*, vol. 1, pp. 477–522. University of Chicago Press, Chicago.
Osborn, H. F. 1916. *The Origin and Evolution of Life: On the Theory of the Action, Reaction and Interaction of Energy*. Scribner's, New York.
Pratchett, T., I. Stewart, and J. Cohen. 2005. *The Science of Discworld III: Darwin's Watch*. Ebury Press, London.
Rehbock, P. F. 1983. *The Philosophical Naturalists: Themes in Early Nineteenth-Century British Biology*. University of Wisconsin Press, Madison.
Rupke, N. A. 1994. *Richard Owen: Victorian Naturalist*. Yale University Press, New Haven.
Russell, E. S. 1916. *Form and Function: A Contribution to the History of Animal Morphology*. John Murray, London.
Secord, J. S. 2000. *Victorian Sensation: The Extraordinary Publication, Reception and Secret Authorship of Vestiges of the Natural History of Creation*. University of Chicago Press, Chicago.
Weikart, R. 2004. *From Darwin to Hitler: Evolutionary Ethics, Eugenics and Racism in Germany*. Palgrave Macmillan, New York.

Commentary One

"Where Are We?" Historical Reflections on Evolutionary Biology in the Twentieth Century

Vassiliki Betty Smocovitis

> *It has been claimed again and again in the last thirty years that evolutionary biology is exhausted as a field of research because the synthetic evolutionary theory has supplied all the answers. Let me emphasize, in a variation of Mark Twain's saying, that the news of the death of evolutionary biology is greatly exaggerated (Mayr 1959: 5).*

From a presentist vantage point, it is astonishing to consider Ernst Mayr's thoughts of 1959 (Mayr 1959). Given the honor of delivering the inaugural lecture for the Cold Spring Harbor Symposium on Quantitative Biology on the occasion of the Darwin centenary, he used the opportunity to argue for the continued support of evolutionary biology. Compared to where the field was in 1859, most of the crucial questions relating to evolutionary biology appeared to have been answered. Mayr so feared that this was the case that he evoked Mark Twain's celebrated quip on the report of his own death—it was greatly exaggerated! As Douglas Futuyma shows in Chapter 1, it was an understatement, to be sure. Evolutionary biology continued after 1959 in unimaginably broad directions, incorporating methods, techniques, and insights from so many areas of the biological and other related sciences that parts of it would now be nearly unrecognizable to many workers in 1959, let alone to workers in the late nineteenth century. It is with good reason that some researchers have moved to use the term *evolutionary science* instead of the narrower *evolutionary biology* to describe the area of inquiry into the diversity of life: the field now properly encompasses areas of chemistry, physics, and, provocatively, even some of the social sciences (see Richerson and Boyd, Chapter 20). Its expansion, as Futuyma notes, includes extensions to applied areas of research like computational sciences and the epidemiology of emerging pathogens (see Hillis, Chapter 16).

In this commentary, I touch on some of the major developments in the history of evolutionary biology in order to provide some historical background critical to understanding what was being celebrated in the "year of Darwin": the 150th anniversary of the publication of Darwin's *magnum opus*, *The Origin of Species*, and the 200th year of the anniversary of his birth. I will briefly summarize the major achievements in the late twentieth century, because historians of science have been gaining a deeper understanding of the major features of the history of evolutionary thought from antiquity to the twentieth century. Peter Bowler (2009), for example, has summarized nicely the history of efforts to understand organic change, and Jean Gayon (1998) has tracked the vicissitudes of Darwinian selection theory from its inception through the period of the evolutionary synthesis. What has not been well examined, however, is the history of the discipline itself, which does not properly come into its own until the period of the evolutionary synthesis, approximately in the interval of time between 1930–1950 (Smocovitis 1996). It is frequently forgotten, for example, that Darwin himself was no evolutionary biologist (naturalist would be more accurate) and that his theory was not "evolution by means of natural selection," but instead "descent with modification." Indeed, only the last word of his 1859 book was "evolve," with the term "evolution" gaining currency thanks to the efforts of Darwin's contemporaries, such Herbert Spencer and Thomas Henry Huxley. What Darwin thought and who he was certainly matter, especially to the historian, but modern evolutionary biology has come so far from 1859 that it would be virtually unrecognizable to Darwin and his contemporaries, as would most modern biological science. Modern evolutionary biology, in this reckoning, is a fairly young science, which is only now beginning to reach its potential of integrating a staggering range of disciplines far outside its original domains.

The Darwin Centennial of 1959

That breadth of potential areas of study was already visible in 1959. It could be seen in the Cold Spring Harbor Symposium's invitation list that included nearly two hundred biologists and in the published symposium papers that came out of it. Organized by geneticist Milislav Demerec, the symposium theme was "Genetics and Darwinism in the Twentieth Century." It was heavily represented by geneticists but also included some systematists, paleontologists, and anthropologists, including Ashley Montagu and the more controversial Carlton S. Coon, then engaged in increasingly unpopular racial theories. The emphasis of this celebration was overwhelmingly on the contributions of genetics in understanding evolutionary mechanisms, which was for good reason. Mathematical population genetics and evolutionary genetics had laid the groundwork for the modern synthesis of evolution in the twentieth century—a fact not left without critical notice by the inaugural lecturer, Mayr, who used the opportunity not just to explore the contributions of genetics to Darwinism, but also to point out its limitations. Examining closely the contributions of the mathematical theorists R. A. Fisher, Sewall Wright, and J. B. S. Haldane, Mayr challenged the centrality of what he described famously, indeed notoriously, as mere "beanbag" genetics for its failure to appreciate genetic interactive effects (Mayr 1959). It was a superficial assessment, especially of Wright's contributions (Mayr seemed especially keen to slight Wright, who received the lion's share of attention in the United States), so that it launched a series of heated exchanges with figures like Haldane, who took umbrage at Mayr's mischaracterization of the work (Haldane 1964). Examining the so-called beanbag controversy today, one cannot help but appreciate how far the field has come. Mayr's celebrated critique of mathematical population genetics does highlight (1) the points of friction between areas requiring integration in the middle decades of the twentieth century and (2) the potential impact of too much emphasis in one or another area.

If the Cold Spring Harbor Symposium overwhelmingly emphasized the contributions of genetics and geneticists to the modern understanding of evolution, the even larger event orchestrated by cultural anthropologist Sol Tax in 1959 at the University of Chicago seemed to highlight the contributions of anthropologists and to celebrate unification among the disciplines of knowledge. Clearly, both conferences had included a staggering range of expertise from the physical sciences to the social sciences, but there is little doubt that Tax went out of his way to include social and cultural evolution. As explored elsewhere (Stocking 1968; Smocovitis 1999), anthropology had been so heavily influenced by cultural anthropologists like Franz Boas that attempts to link biology with cultural evolution had been deliberately avoided for the first few decades of the twentieth century. That kind of reduction was viewed as too politically charged, since it introduced discussions of the biology of race—a subject rife with historical controversy, evoking the specter of biological determinism (Stocking 1968). Tax's vision for the celebration stressed the unifying properties of evolution and sought to bring into the fold not just physical but also cultural anthropologists in a way that avoided some of the preceding political controversies (Tax 1960). The special discussion panels, the highlights of the 5-day event, were notable in this regard. From Panel One, titled "The Origin of Life," which brought chemistry, astronomy, and physics into the fold, to Panel Two, entitled "The Evolution of Life," to panels three, four, and five, which were devoted respectively to "Man as an Organism," "The Evolution of Mind," and "Social and Cultural Evolution," the centrality of evolution by natural selection was established as part of what was thought to be a unified evolutionary cosmology or worldview. Some critics, however, viewed the strong consensus projected among the panel members and some of the invited celebrants as part of a growing orthodox doctrine of the synthetic theory (Goudge 1961). Reflecting on the historical consequences of this Darwin centennial, Stephen J. Gould later decried what he viewed as the "hardening" of the synthesis around a selectionist orthodoxy (Gould 1980, 1983; Tattersall 2000).

Clearly, some consensus existed in 1959, but that did not mean that all areas represented in the panels were fully integrated. Fields such as anthropology were incompletely integrated into the synthetic theory of evolution (Smocovitis 1999; Tattersall 2000; Delisle 2007), as was embryology or developmental biology (Waddington 1953; Smocovitis 1996; Amundson 2005). Even parts of paleontology remained unassimilated, as shown by Olson's (1960) vocal dissent at Chicago. Yet, the Chicago conference does reveal the extent to which the unifying powers of evolutionary biology were felt in the middle decades of the twentieth century. That evolution served as a unifying or an organizing principle was a fairly new insight at that time, growing out of "the evolutionary synthesis" and the efforts of a generation of workers in the first few decades of the twentieth century (Mayr and Provine 1980; Provine 1971, 1986; Smocovitis 1996). In contrast, around the turn of the century, the study of evolution was held in disfavor and even thought passé in some communities, relegated to an older natural history tradition because of its methodological lack of rigor. In fact, natural selection, indeed Darwinism itself, was held in disfavor as alternatives to natural selection gained prominence at the turn of the century (Bowler 1983). As a discipline itself, biology was also undergoing internal problems and was rife with debates; it was considered a fragmented, heterogeneous, and immature science that seemed to defy attempts to establish unifying principles (Smocovitis 1996).

The Evolutionary Synthesis and the Convergence of Disciplines 1920–1950

It was for these reasons that Julian Huxley referred to the turn of the century as "the eclipse of Darwin," as he celebrated the "modern synthesis" of evolution in his 1942 book titled *Evolution: The Modern Synthesis*. The synthesis he referred to was between the new science of genetics ("rediscovered" in 1900) and the new understanding of Darwinian selection theory, recovered and fortified through efforts to explain the origin and maintenance of biological diversity. He was building on the work of mathematical population geneticists Fisher, Wright, and Haldane, but also on the efforts of practitioners and field biologists/naturalists, like Theodosius Dobzhansky and E. B. Ford, who provided empirical support for the new theoretical insights that demonstrated the efficacy of natural selection as a mechanism for evolutionary change. Huxley's 1942 book was one of a series of publications that laid the groundwork for the synthetic theory of evolution. The better known of these appeared as monographs or books published by Columbia University Press, because they grew out of the Jesup Lectures at Columbia University. Dobzhansky's (1937) *Genetics and the Origin of Species*, which appeared first, is generally regarded as the single most influential book of twentieth-century evolution, because it created the foundation for evolutionary genetics. It also functioned as a textbook of instruction for an entire generation of evolutionary biologists. Other books in the series included Mayr's (1942) *Systematics and the Origin of Species*, George Gaylord Simpson's (1944b) *Tempo and Mode in Evolution*, and G. Ledyard Stebbins's (1950) *Variation and Evolution in Plants*.

These synthetic books are generally hailed as the defining publications that figured in the modern synthesis of evolution, but it is forgotten that they also were synthetic, drawing on the work of predecessors and contemporaries. Among these important contributors, I would especially include Jens Clausen, who, along with David Keck and William Hiesey, provided crucial experimental insights into distinguishing phenotype from genotype, thereby illuminating the origin and maintenance of variation, elucidating the action of natural selection, and eliminating alternatives like neo-Lamarckism (Smocovitis 2009). Clausen's west coast contemporary, the plant breeder E. B. Babcock, and his many colleagues in genetics at the University of California, Berkeley, also deserve notice for their efforts, which culminated in the first comprehensive phylogenetic treatment of any group of organisms (the genus *Crepis*) and drew on insights and methods from genetics, cytology, systematics, biogeography, and fossil history (Babcock 1947; Smocovitis 2009). Yet another overlooked figure was the German systematist Bernhard Rensch, whose 1947 book *Neuere Probleme der Abstammungslehre* was only translated into English (as *Evolution Above the Species Level*) after the synthesis; his work however, was crucial to figures like Mayr in providing insights into patterns and process of speciation, and his conclusions about macroevolution paralleled Simpson's to a remarkable degree. Finally, Huxley's own efforts to launch a new synthesis and organize his colleagues in Britain are frequently underappreciated. His efforts culminated in 1940 with an edited collection, titled *The New Systematics*, which incorporated genetics and ecology, along with an experimental approach to classical areas like systematics (Huxley 1940). This book and other such efforts emphasized formal discussions of the patterns and processes involved in speciation and helped to establish the biological species concept, as articulated by Dobzhansky, Mayr, and others.

Dobzhansky was inspired, of course, by his celebrated collaboration with Wright (Provine 1986), but he had been predisposed to such collaborative ventures by his training in Russia, where he had been influenced by Russian naturalists–systematists who, following figures like Sergei Chetverikov, sought to assimilate insights from genetics into a populational approach to evolution (Adams 1968, 1980, 1994). When Dobzhansky immigrated to the United States in 1927, he carried what was to remain of the legacy of the ecumenical Russian approaches to genetics and evolution (largely exterminated under Stalin and Lysenko)—an approach that was crucial to his vision of integrating various methods and disciplines in the mid-1930s. It was no accident, in other words, that he was most responsible for the integrative process that resulted in the founding of evolutionary genetics. The title of his book, *Genetics and the Origin of Species*, was designed to directly evoke the synthesis of genetics with Darwinian insights as well as to redress questions into mechanisms of speciation and species definitions left unresolved in Darwin's formulation of 1859 (Dobzhansky 1937). Dobzhansky benefited greatly from his training in Thomas Hunt Morgan's so-called fly group, especially under the tutelage of A. H. Sturtevant, but his intellectual motivation and scientific style clearly reflected the Russian tradition (Provine 1981; Adams 1994).

Both Dobzhansky and Mayr, who immigrated to the United States from Germany in 1931, brought with them the insights of continental workers and gave an international perspective to what Mayr and historian of science William B. Provine later called "the evolutionary synthesis" (Levine et al. 1995; Haffer 2005; Mayr and Provine 1980). Though most of the key participants (later called the "architects" by Mayr in his historical reflections) were active in the United States or Great Britain, contributors came from the international community of evolutionists. Later on, Dobzhansky took his path-breaking insights and methodology to a number of institutions in Latin America, creating centers of evolutionary genetics, along with the help of his many students, post-doctoral associates, visitors, and friends (Glass 1980; Araújo 2004; Araújo 1998; Levine et al. 1995). The legacy of Dobzhansky continues through an international network of collaborators, who continue to study the genetics of natural populations in various species of *Drosophila* (Levine 1995).

The growing consensus on evolution by means of natural selection fueled a number of organizational activities in the mid-1930s to mid-1940s, including an informal group of workers in the San Francisco Bay area in 1935, called "the Biosystematists" (originally named "the Linnaean Club"); the Society for the Study of Speciation, which grew out of the 1939 meetings of the American Association for the Advancement of Science; and the National Research Council-backed Committee on Common Problems of Genetics and Paleontology in 1943, whose name was subsequently amended to Committee on Common Problems of Genetics, Paleontology and Systematics the next year, at the request of Mayr (Smocovitis 1994). During the difficult war years, mimeographed bulletins were disseminated between members as a forum of scientific discussion. Fittingly, it was in the final mimeographed bulletin that Simpson, newly returned from the war, announced that a "field common to the disciplines of genetics, paleontology, and systematics," really did exist and was "beginning to be clearly defined" (Simpson 1944b). With the end of the war, active members of these groups came together in St. Louis in 1946 to found the first formal society for evolutionary biology, The Society for the Study of Evolution (SSE), and the first international journal devoted expressly to study of evolution, which was simply titled *Evolution*. Mayr led efforts to orchestrate and organize the society and to lay down the infrastructure of the new discipline; he served as the first editor of the journal, while Simpson was the first president. Mayr's dominance among evolutionary biologists resulted in part from his energetic management of that journal and his ongoing role in disciplining the study of evolution (Smocovitis 1994). The new discipline was first formally recognized and indeed celebrated, by an impressive number of international workers who attended the famous Princeton meetings of 1947, held on the occasion of the Bicentennial of Princeton University. The proceedings, entitled *Genetics, Paleontology and Evolution* and edited by Glenn L. Jepsen, Simpson, and Mayr (1949), provided explicit reference to the new synthetic discipline. Reflecting the new unified voice of evolution, Hermann J. Muller proclaimed that a "convergence of evolutionary disciplines" had taken place, as he drew an analogy between evolutionary convergence of types and convergence between disciplinary types, such as paleontologists and genetics, to form the "synthetic" type of evolutionist (Muller 1949: 421). A "common ground of theory" served to unite the range of disciplines, because agreement had been reached on the theory's fundamental tenets: (1) the primacy of natural selection as the mechanism for evolutionary change, (2) the gradual rate of change operating on the level of small, individual differences, and (3) the continuum between micro- and macroevolutionary processes. Echoed by figures like Huxley, the newly emerged consensus spoke to a more unified science of evolution, which in turn increasingly served as the unifying principle of the whole of biology (Smocovitis 1996). Small wonder, then, that the year 1959 had special significance for evolutionists and that celebrations in honor of the Darwin centenary were held far and wide. The numbers of evolutionary biologists were growing, in part because people were redefining themselves as evolutionary biologists, instead of ornithologists, botanists, or systematists, and also because of an increase in the number of young people coming into the field. As a consequence of all the attention given to Darwin and evolution, membership in societies like the SSE boomed; Smocovitis (1994) details the membership data and spike after 1959.

Evolutionary Biology in the Post-Sputnik Period

It was not all roses, however. Just as the architects and unifiers began to exert a national and international influence in promoting and building evolutionary biology, involving themselves in teaching initiatives like the Biological Sciences and Curriculum Study (BSCS) and serving in national and international leadership positions, they began to feel an assault from a number of new disciplines that emerged in the wake of efforts to grow science and technology in the midst of the Cold War. (The successful launching of the Soviet space satellite "Sputnik" struck fear into Americans who did not want to fall behind.) First, the new science, known as exobiology, which claimed to understand evolution, drew the scorn of Mayr but especially of Simpson, who wrote a series of celebrated attacks against what he viewed as an illegitimate use of evolutionary principles by those who argued for the "prevalence of humanoids" (Simpson 1964a). Next came a far greater menace, initially posed by the reductionistic and mechanistic new sciences, especially molecular biology, which was fueled by the discovery of the structure of DNA in 1953. It is hard for us to appreciate today how much those new sciences and their supporters were at odds with organismal biologists, like Mayr, Simpson, and even Dobzhansky (Beatty 1990; Smocovitis 1996). The decade of the 1960s in the history of biology was characterized by increasing strife and division in the biological sciences, as molecular biologists began to garner financial support and attention, to the detriment of what became known as "classical biology" (which really meant everything that did not exclusively use molecular or biochemical techniques).

The situation became so dire at places like Harvard University, where Mayr, Simpson, and E. O. Wilson were in close proximity to J. D. Watson and George Wald, that a formal division of biology departments took place beginning in the late 1960s, dividing those who adhered to the study of "organismal biology" and those who fell into the newer areas of "molecular or cell biology and biochemistry." It was with good reason that Wilson devoted an entire chapter of his autobiography (*Naturalist*) to his tumultuous experiences at Harvard in the 1960s and entitled it "Molecular Wars" (Wilson 1994).

The changing divisions and emphases in biology alarmed the architects of the synthesis, including Simpson (1966), who declared a crisis in the biological sciences, and Mayr (1959), who, as early as his 1959 Cold Spring Harbor address, argued for the continued support of evolutionary biology in the wake of the newer area of molecular biology. (In 1963, Mayr made an eloquent call to support classical fields of biology in *Science*.) The changes also worried Dobzhansky (1964, 1966), who addressed the conflict head on in the pages of the *American Naturalist* by asking: "Are naturalists old-fashioned?" and then by giving a thoughtful philosophical analysis of the situation in 1968, envisioning biology in terms of Cartesian and Darwinian halves (Dobzhansky 1968). It fell to him to come up with a workable solution to this crisis by incorporating the DNA molecule as the new unit of evolution. This would serve integrative functions, accounting for both the unity of life (with its reduction to the physical sciences) and the diversity of life (with its connections to organismic biology). His essay (Dobzhansky 1964) exploring both the history and philosophy of the life sciences made for an eloquent plea to preserve the unity of the sciences and to celebrate the biological sciences, located centrally between the physical sciences and the social sciences, and that would answer the "big question" facing each generation: "what is Man?" Dobzhansky, thus, argued forcibly for maintaining a balance between the new fields and for the unifying power of evolution; it was the first time he employed the phrase that gained currency for the next generation of evolutionary biologists, that "nothing makes sense in biology except in the light of evolution" (Dobzhansky 1964: 449; Dobzhansky 1973). In the 1960s, the phrases "unity and diversity of life" and "molecules to man," came to dominate biology textbooks, which often argued for the centrality of evolution (Smocovitis 1996). By the mid-1970s, these book began to employ Dobzhansky's quotation explicitly, often as an epigraph (Dobzhansky et al. 1977).

Conclusions: Reflections on the Future of Evolutionary Biology

Evolutionary biology as a discipline or field survived that period of dissonance, but never officially reached departmental status, though it did form partnerships with ecology and in some places, with behavior, in departments or programs of "ecology and evolutionary biology" or "ecology, evolution, and behavior." Textbooks on evolution, such as *Evolution*, by Dobzhansky et al. (1977) and Futuyma's successful *Evolutionary Biology* (1979) and subsequent *Evolution* (2009) began to be published at this time. Internal debates continued and reached a fevered pitch in the late 1970s and 1980s, as various challenges to the synthetic theory were posed, including: (1) the "paleobiological revolution," launched by a new generation of "young Turks" (Sepkoski and Ruse 2009); (2) the neutral theory of molecular evolution (Kimura 1983), and (3) the work of a growing number of evolutionists keen to finally integrate evolution with genetics and development, as in the new science of evolutionary developmental (evo-devo) biology (Carroll 2005).

Evolutionary biology also survived intact following a staggering number of external assaults, beginning with fundamentalist anti-evolutionary movements that were galvanized by the success of the Darwin celebrations of 1959, especially the one at the University of Chicago. Smocovitis (1999), for example, detailed Huxley's famous speech called the "secular sermon" and described how it fueled anti-evolutionism in America. By the 1970s, as scientific creationism appeared in its pseudo-scientific guise and became more prevalent, anti-evolutionists tried to exclude evolution from secondary school textbooks. Dobzhansky's (1973) assertion that nothing made "sense in biology except in the light of evolution" and similar sentiments (e.g., that evolution was the central organizing or unifying principle of biology) made for a forceful argument for including evolution in science education (Lee 1972).

As we enter the next 50 years of evolutionary biology, we can justly assert that it has survived a number of challenges from within and assaults from outside of science. The field continues to adapt, by incorporating new methods and successfully solving both enduring and new problems across a range of disciplines, which seem to keep increasing with time. What we celebrate, then, is the maturation of a science that persists in fulfilling a promise made by Darwin in 1859 that an understanding of the diversity of life was possible through evolution by means of natural selection.

Literature Cited

Adams, M. B. 1968. The founding of population genetics: Contributions of the Chetverikov school, 1924–1934. *J. Hist. Biol.* 1: 23–39.

Adams, M. B. 1980. Sergei Chetverikov, the Kol'tsov Institute, and the evolutionary synthesis. In E. Mayr and W. B. Provine (eds.), *The Evolutionary Synthesis: Perspectives on the Unification of Biology*, pp. 242–278. Harvard University Press, Cambridge, MA.

Adams, M. B. (ed.). 1994. *The Evolution of Theodosius Dobzhansky*. Princeton University Press, Princeton.

Amundson, R. 2005. *The Changing Role of the Embryo in Evolutionary Thought: Roots of Evo-Devo.* Cambridge University Press, Cambridge, UK.

Araújo, A. M. 1998. A influência de Theodosius Dobzhansky no desenvolvimento da Genética no Brasil. *Episteme* 7: 43–54.

Araújo, A. M. 2004. Spreading the evolutionary synthesis: Theodosius Dobzhansky and genetics in Brazil. *Genet. Mol. Biol.* 27: 467–475.

Babcock, E. B. 1947. *The Genus* Crepis, *I and II.* University of California Publications in Botany, Berkeley.

Beatty, J. B. 1990. Evolutionary anti-reductionism: Historical reflections. *Biol. Phil.* 5: 199–210.

Bowler, P. 1983. *The Eclipse of Darwinism.* Johns Hopkins University Press, Baltimore.

Bowler, P. 2009. *Evolution: The History of an Idea.* 25th anniversary ed. University of California Press, Berkeley.

Carroll, S. B. 2005. *Endless Forms Most Beautiful: The New Science of Evo Devo.* Norton, New York.

Delisle, R. G. 2007. *Debating Humankind's Place in Nature: The Nature of Paleoanthropology.* Prentice Hall, New Jersey.

Dobzhansky, T. 1937. *Genetics and the Origin of Species.* Columbia University Press, New York.

Dobzhansky, T. 1964. Biology, molecular and organismic. *Am. Zool.* 4: 443–452.

Dobzhansky, T. 1966. Are naturalists old-fashioned? *Am. Nat.* 100: 541–50.

Dobzhansky, T. 1968. On Cartesian and Darwinian aspects of biology. *The Graduate Journal* 8: 99–117.

Dobzhansky, T. 1973. Nothing in biology makes sense except in the light of evolution. *Am. Biol. Teacher* 35: 125–129.

Dobzhansky, T., F. J. Ayala, G. L. Stebbins, and 1 other. 1977. *Evolution.* W. H. Freeman, San Francisco.

Futuyma, D. J. 1979. *Evolutionary Biology.* Sinauer Associates, Sunderland, MA.

Futuyma, D. J. 2009. *Evolution*, 2nd ed. Sinauer Associates, Sunderland, MA.

Gayon, J. 1998. *Darwinism's Struggle for Survival: Heredity and the Hypothesis of Natural Selection.* Cambridge University Press, Cambridge, UK.

Glass, B. (ed.). 1980. *The Roving Naturalist: Travel Letters of Theodosius Dobzhansky.* American Philosophical Society, Philadelphia.

Goudge, T. 1961. Darwin's heirs. *U. Toronto Quart.* 30: 246–50.

Gould, S. J. 1980. G. G. Simpson, paleontology and the modern synthesis. In E. Mayr and W. B. Provine (eds.), *The Evolutionary Synthesis: Perspectives on the Unification of Biology*, pp. 153–172. Harvard University Press, Cambridge, MA.

Gould, S. J. 1983. Irrelevance, submission and partnership: The changing role of paleontology in Darwin's three centennials and a modest proposal for macroevolution. In D. S. Bendall (ed.), *Evolution from Molecules to Men*, pp. 347–66. Cambridge University Press, Cambridge, MA.

Haffer, J. 2007. *Ornithology, Evolution and Philosophy: The Life and Science of Ernst Mayr, 1904–2005.* Springer-Verlag, Berlin.

Haldane, J. B. S. 1964. A defense of bean-bag genetics. *Persp. Biol. Med.* 7: 343–359.

Huxley, J. (ed.). 1940. *The New Systematics.* Clarendon Press, Oxford.

Huxley, J. 1942. *Evolution: The Modern Synthesis.* Allen and Unwin, London.

Jepsen, G. L., G. G. Simpson, and E. Mayr (eds.). 1949. *Genetics, Paleontology and Evolution.* Princeton University Press, Princeton.

Kimura, M. 1983. *Neutral Theory of Molecular Evolution.* Cambridge University Press, Cambridge, UK.

Lee, A. 1972. The BSCS position on the teaching of biology. *BSCS Newsletter* 49: 6.

Levine, L. (ed.). 1995. *Genetics of Natural Populations: The Continuing Importance of Theodosius Dobzhansky.* Columbia University Press, New York.

Levine, L., O. Olvera, J. R. Powell, and 5 others. 1995. Studies on Mexican populations of *Drosophila pseudoobscura.* In L. Levine (ed.), *Genetics of Natural Populations: The Continuing Importance of Theodosius Dobzhansky*, pp. 120–139. Columbia University Press, New York.

Lewontin, R. C., J. A. Moore, W. B. Provine, and B. Wallace (eds.). 1981. *Dobzhansky's Genetics of Natural Populations I–XLIII.* Columbia University Press, New York.

Mayr, E. 1942. *Systematics and the Origin of Species.* Columbia University Press, New York.

Mayr, E. 1959. Where are we? *Cold Spr. Harb. Sym.* 24: 1–14.

Mayr, E. 1963. The new versus the classical in science. *Science* 141: 765.

Mayr, E. 1982. *The Growth of Biological Thought.* Belknap Press, Cambridge, MA.

Mayr, E. and W. B. Provine (eds.). 1980. *The Evolutionary Synthesis: Perspectives on the Unification of Biology.* Harvard University Press, Cambridge, MA.

Muller, H. J. 1949. Redintegration of the symposium on genetics, paleontology, and evolution. In G. Jepsen, G. G. Simpson, and E. Mayr (eds.), *Genetics, Paleontology and Evolution*, pp. 421–455. Princeton University Press, Princeton.

Olson, E. C. 1960. Morphology, paleontology and evolution. In S. Tax (ed.), *Evolution after Darwin vol. 1*, pp. 523–45. University of Chicago Press, Chicago.

Provine, W. B. 1971. *The Origins of Theoretical Population Genetics.* University of Chicago Press, Chicago.

Provine, W. B. 1981. Origins of the genetics of natural populations series. In R. C. Lewontin, J. A. Moore, W. B. Provine, and B. Wallace (eds.), *Dobzhansky's Genetics of Natural Populations I–XLIII*, pp. 5–83. Columbia University Press, New York.

Provine, W. B. 1986. *Sewall Wright and Evolutionary Biology*. University of Chicago Press, Chicago.

Rensch, B. 1947. *Neuere Probleme der Abstammungslehre*. Ferdinand Encke Verlag, Stuttgart.

Sepkoski, D. and M. Ruse. 2009. *The Paleobiological Revolution: Essays on the Growth of Modern Paleontology*. The University of Chicago Press, Chicago.

Simpson, G. G. 1944a. Mimeographed bulletins. W. B. Provine Reprint Collection, Marathon, New York.

Simpson, G. G. 1944b. *Tempo and Mode in Evolution*. Columbia University Press, New York.

Simpson, G. G. 1964a. The nonprevalence of humanoids. *Science*: 143: 769–775.

Simpson, G. G. 1964b. Organisms and molecules in evolution. *Science* 146: 1535–1538.

Simpson, G. G. 1966. The crisis in biology. *The Am. Scholar* 36: 363–367.

Smocovitis, V. B. 1994. Organizing evolution: Founding the Society for the Study of Evolution (1939–1950). *J. Hist. Biol.* 24: 241–309.

Smocovitis, V. B. 1996. *Unifying Biology: The Evolutionary Synthesis and Evolutionary Biology*. Princeton University Press, Princeton.

Smocovitis, V. B. 1999. The 1959 Darwin centennial celebration in America. In C. Elliott and P. Abir-Am (eds.), *Osiris Series 2, vol. 14: Commemorative Practices in Science: Historical Perspectives on the Politics of Collective Memory*, pp. 274–323. University of Chicago Press, Chicago.

Smocovitis, V. B. 2009. The "plant *Drosophila*": E. B. Babcock, the genus *Crepis* and the evolution of a genetics research program at Berkeley, 1912–1947. *Hist. Stud. Nat. Sci.* 39: 300–355.

Stebbins, G. L. 1950. *Variation and Evolution in Plants*. Columbia University Press, New York.

Stocking Jr., G. 1968. *Race, Culture, and Evolution: Essays in the History of Anthropology*. The Free Press, New York.

Tattersall, I. 2000. Paleoanthropology: The last half-century. *Evol. Anthro.* 9: 2–16.

Tax, S. (ed.). 1960. *Evolution after Darwin*, 3 vols. University of Chicago Press, Chicago.

Waddington, C. 1953. Epigenetics and evolution. *Symp. Soc. Exp. Bio.* 7: 186–189.

Wilson, E. O. 1994. *Naturalist*. Island Press, New York.

Part II

POPULATIONS, GENES, AND GENOMES

Chapter 3

The Concepts of Population and Metapopulation in Evolutionary Biology and Ecology

Roberta L. Millstein

Contemporary concepts of *population* in ecology and evolutionary biology vary greatly. Biologists rarely defend their choice of population concept and may not even explicitly characterize one. Of the explicit characterizations, some are extremely permissive, whereas others are much less so. Here are some examples to illustrate the diversity of meanings,[1] arranged from most permissive to least permissive:

- "A group of individuals of a single species" (Krebs 1985)
- "A somewhat arbitrary grouping of individuals of a species that is circumscribed according to the criteria of some specific study" (Orians 1973)
- "Group of organisms of the same species living in a particular geographic region" (Lane 1976)
- "All the members of a species that occupy a particular area at the same time" (Arms and Camp 1979)
- "Any group of organisms capable of interbreeding for the most part, and coexisting at the same time and in the same place" (Purves and Orians 1983)
- "A group of conspecific organisms that occupy a more or less well-defined geographic region and exhibit reproductive continuity from generation to generation; it is generally presumed that ecological and reproductive interactions are more frequent among these individuals than between them and members of other populations of the same species" (Futuyma 1986)

Most of the population concepts restrict populations to conspecific organisms; some include space and/or time as criteria, and some incorporate interbreeding or other interactions as criteria. In population genetics,

[1] Most of these examples are taken from Wells and Richmond's (1995) longer list. As will be seen in this chapter, my own concept is closest to that of Futuyma's.

populations are generally characterized (again, assuming they are characterized at all) as a group of interbreeding organisms of the same species; sometimes "in a particular geographic area" (or the equivalent), is added.

Of the concepts listed, the permissive ones are the most problematic; if any grouping[2] of conspecific organisms could constitute a population, then populations could be gerrymandered, that is, their boundaries could be drawn so that the resulting "population" implied a favored conclusion. As I have argued elsewhere (Millstein 2009), if gerrymandered populations were legitimate populations, one could choose a grouping of conspecific individuals so that there was no variation in the trait in question (and thus, no selection or drift) or no fitness differences with respect to the trait in question (and thus, no selection). If gerrymandering were legitimate, then all resulting claims for the presence or absence of selection or drift would be equally correct. Yet, Darwin thought—and most contemporary biologists think—that selection can explain "the mutual relations of all organic beings to each other and to their physical conditions of life" (Darwin 1859, 1964: 80). It is hard to see how a selection process that was so description-relative (as the permissive concepts of population allow) could do any such thing.

More generally, different population and metapopulation delineations yield different answers about which ecological and evolutionary processes are occurring (more on this later in the chapter). Indeed, if populations and metapopulations are real biological entities (and not just human constructions), the wrong concept may cause *mischaracterizations* of ecological and evolutionary processes. As a consequence, inconsistent meanings of the terms population and metapopulation may lead to less than fruitful controversies in which the disputants are not genuinely disagreeing with each other, just using the same terms in different ways. Using the same terms with different meanings also makes it hard to compare results.

Another reason concepts of population and metapopulation are required is that evolutionary processes are often in flux. For example, selection can be operating in one season, and then not in the next, or it can be acting on different traits in different seasons. Thus, Godfrey-Smith's (2009) account of a Darwinian population (i.e., a population that has the conditions necessary for natural selection) needs to be supplemented with the prior notion of a population. Of course, the composition of populations also changes over time, with the addition of new members through birth and immigration and the loss of members through death and emigration. My point is that you cannot track changes in selection without identifying the entity (the population) within which those changes are occurring and that such changes can be too rapid for identification of a population by its selection processes. Indeed, the same population may be many "Darwinian

---

[2] Following Gildenhuys (2009), I use the term grouping rather than group, since group has a technical meaning in evolutionary biology, as in group selection.

populations" over time, because there are different selection processes occurring over the course of a single year.

So, for all of these reasons, it is important to examine the concepts of population and metapopulation in ecology and evolutionary biology in order to find defensible concepts that avoid making ecological and evolutionary processes description-relative. But can a defensible concept be found?

This paper aims to illustrate one of the primary goals of the philosophy of biology—namely, the examination of central concepts in biological theory and practice—through an analysis of the concepts of population and metapopulation in evolutionary biology and ecology. I will first provide a brief background for my analysis, followed by a characterization of my proposed concepts: the causal interactionist concepts of population and metapopulation. I will then illustrate how the concepts apply to six cases that differ in their population structure; this analysis will also serve to flesh out and defend the concepts a bit more. Finally, I will respond to some possible questions that my analysis may have raised and then conclude briefly.

Background

By most accounts, the philosophy of biology is a young discipline that emerged out of the philosophy of science in the 1960s and 1970s.[3] It includes philosophical investigations into biological sciences such as ecology, molecular biology, developmental biology, cognitive ethology, and neuroscience in addition to evolutionary biology. The field has comparatively few textbooks, and articles appear in many different types of journals—philosophy of biology journals, history of biology journals, philosophy of science journals, biology journals, and combinations thereof—making it difficult to characterize. However, I think it is fair to say that much of the work in the philosophy of evolutionary biology, in particular, is captured by the title of a 1996 book by Robert Brandon: *Concepts and Methods in Evolutionary Biology*. In other words, there has been considerable effort devoted to clarification of central terms in evolutionary biology, including the concepts of fitness, species, adaptation, group selection, natural selection, and random drift, exemplified by Keller and Lloyd (1998) as well as considerable effort devoted to analyzing methods of empirical discovery, exemplified by Creath and Maienschein (2000). Both academic philosophers (i.e., scholars in philosophy departments) and biologists practice the philosophy of biology. Indeed, the field has been very much influenced by the philosophical work of biologists such as Michael Ghiselin, Stephen Jay Gould, Richard Lewontin, Ernst Mayr, and Michael J. Wade. Moreover, the best work in the

[3] Byron (2007) provides a review of the standard history and an alternative one.

field has been done by those whose work pays careful attention to historical and contemporary biological practice.

Thus, it is surprising that there has been very little detailed analysis of a concept as central to evolutionary thinking as population, although there are notable exceptions (Goudge 1955; Wells and Richmond 1995; Camus and Lima 2002; Berryman 2002; Gannett 2003; Waples and Gaggiotti 2006; Pfeifer et al. 2007; Gildenhuys 2009). Metapopulation is also a widely used term (albeit one that is less central than population), and yet there has been a similar lack of analysis, again with notable exceptions (Hanski and Gilpin 1991; Harrison 1991; Hanski and Simberloff 1997). The fact that concepts as central as these have received such little philosophical attention contrasts with the massive literature on species concept. In 1969, David Hull remarked that: "the biological literature on the species concept is overwhelmingly large" (Hull 1969: 180), and it has increased significantly since then. Yet, the term population is arguably a far more central concept than species to the study of evolution. After all, populations are at the core of models in population genetics, ecological genetics, population biology, ecology, and evolutionary ecology. (Systematics is perhaps an exception in its use of species over population, but even systematists rely on populations in their analyses.) Populations must evolve before new species evolve; populations are constantly undergoing evolution, whereas the emergence of new species is a far less common event. Populations are also the entities within which abundance and distribution are studied. So, it is strange that the concept of population has received so much less attention. My goal in this paper is not to solve this sociological puzzle, but rather, to go some distance towards rectifying it.

There has also been relatively little written on the history of the concept of population in biology, although there are again notable exceptions (Gerson 1998; Winsor 2000; Gannett 2003; Hey 2009). As Hey describes, the biological use of the term arose out of its statistical use. In statistics, the term originally applied primarily to humans, referring to any set of individuals under investigation (e.g., human females over 50), whereas the biological use of the term refers to a "biological whole" composed of interbreeding individuals (more on this later in the chapter). Hey credits the transition from the statistical use to the biological use to thinkers such as Edward B. Poulton in 1903 (though Poulton does not use the term) and Karl Pearson in 1904. By 1939, biologists were using the term population in different ways, prompting Gilmour and Gregor to coin the term deme, meaning "any specified assemblage of taxonomically related individuals" (Gilmour and Gregor 1939: 333). Their intention was to distinguish between different types of demes. However, deme has come to refer to a collection of interbreeding organisms. Winsor (2000) reveals the irony of this usage as follows. Gilmour and Gregor originally intended the word gamodeme for this purpose, with topodeme and ecodeme referring to "demes occupying specified geographical areas and specified ecological habitats respectively"

(Gilmour and Gregor 1939: 333). However, this suggestion has largely been ignored, and it is simply the term deme that has stuck. Indeed, because the association between the term deme and interbreeding is so strong, and because I will be arguing for a broader term, I will avoid deme in favor of the more open-ended term population.

As for Darwin, one would be hard-pressed to find a contemporary characterization of natural selection that does not refer to changes in populations, and yet the word population does not appear in *The Origin of Species.* Darwin does use the term elsewhere, but primarily to apply to humans. The closest he comes to using the term population in *The Origin of Species* is when he refers to "individuals of the same species inhabiting the same confined locality" (Darwin 1859: 45), and claims that varieties generally arise locally (Darwin 1859: 298).

The history of the concept of metapopulation has been outlined, for example, by Hanski and Gilpin (1991) and Hastings and Harrison (1994). The term metapopulation—a population of local populations—was coined by Levins in 1970. However, Hastings and Harrison trace the general idea (albeit not the term) back to Nicholson and Bailey (1935). More commonly cited precursors for the idea include Wright (1940), as part of his shifting balance model, and Andrewartha and Birch (1954). Levins's metapopulation model is fairly specific; it refers to a population of local populations that go extinct and recolonize, with the local populations being equally spaced and of the same size (it is a deliberately simplified model). However, contemporary usage generally relaxes these strictures in a number of ways: as habitats have become increasingly fragmented as a result of human activities, the development of metapopulation models and their applications have increased in the last several decades. Metapopulation models vary, but at a minimum, most embody metapopulation concepts that allow for some degree of migration or dispersal among local populations. Clearly, however, any metapopulation concept will be parasitic on one's views concerning local populations.

The Causal Interactionist Population Concept

In discussing the concepts of population and metapopulation, my intended foci are evolutionary and ecological contexts, with the understanding that evolutionary factors affect ecological factors and vice versa (as the discipline of population biology recognizes), so that the two cannot be fully disentangled. Other disciplines, such as sociology, biomedicine, and statistics, also utilize population concepts, and perhaps it would be possible to provide a very general notion of population that would accommodate all disciplines. However, I suspect that the specific conceptions are the ones that are most relevant to the practice and understanding of evolution and ecology. In particular, my goal is to describe what are sometimes called local populations of organisms[4] as well as groupings of local populations (i.e., metapopulations). By focusing on populations of *organisms*, I do not mean to deny that other biological entities, such as cells, also can form populations, but rather to suggest that the particular concept I will describe in this chapter would need modification before being used for biological entities other than organisms.

Elsewhere I defend the view that *populations are individuals* (Millstein 2009), drawing inspiration from the Ghiselin-Hull thesis that species are individuals (Ghiselin 1974, 1997; Hull 1976, 1978, 1980). It turns out that once one examines the criteria for individuality (e.g., being a spatiotemporally restricted entity), it is quite straightforward to demonstrate this otherwise oxymoronic-sounding thesis. The thesis that populations are individuals becomes even less controversial if one accepts the view that individuality comes in degrees—a point emphasized by Mishler and Brandon (1987) and others. For example, the key feature that distinguishes an individual from a mere set (which can be any arbitrary collection of individuals, such as the items on my desk) is that individuals are *integrated cohesive entities*.[5] Being an integrated cohesive entity implies that there are causal interactions among the parts of the individual, with the parts having a shared fate (so that what affects one part affects at least some of the other parts). For populations, the parts are organisms; for organisms, the parts are cells. Whatever you take the relevant causal interactions among organisms and cells to be, it will generally be the case that populations are not as integrated and cohesive as organisms are, which may simply mean that populations are not individuals to the same degree that organisms are.

However, for the purposes of this paper, nothing depends on the particular philosophical (or metaphysical) claim that populations are individuals. Indeed, the claim that populations are individuals can be understood equally well as a claim that populations are real entities that act (more or less) as a unit; they are *biological wholes.* Again, the key feature that makes something a biological whole is the presence of causal interactions among the parts. However, since many types of causal interactions exist, it is essential to specify which are the relevant causal interactions for the biological whole in question (Hamilton et al. 2009). Otherwise, the criterion is quite

---

[4] For some species it is difficult to delineate one organism from another. This is an interesting complication that I hope to address in future work. The present analysis addresses only those species for which organisms are reasonably well delineated.

[5] Mishler and Brandon distinguish between integration and cohesion, using the former term "to refer to active interaction among the parts of an entity" and the latter term "to refer to situations where an entity behaves as a whole with respect to some process" such that "all the parts of the entity respond uniformly to some specific process" even if the parts are not interacting (Mishler and Brandon 1987: 400). However, I do not mean to imply this distinction here, and I think that *collective* would be a more appropriate term for what Mishler and Brandon call *cohesion*.

empty, especially if, for instance, infinitesimal gravitational forces count as causal interactions. In this chapter, I will argue that populations are integrated via the *survival and reproductive interactions of organisms*. In an earlier work, I show why other plausible candidates for causal interactions are not suitable (Millstein 2009). Here, I propose to defend my population concept differently, namely by showing its success in illuminating a number of cases and by demonstrating how it straightforwardly yields a metapopulation concept as well.

My proposed *causal interactionist population concept*[6] (with some additional clarifications that follow) is:

- *Populations* (in ecological and evolutionary contexts) consist of at least two conspecific organisms that, over the course of a generation, are *actually* engaged in survival or reproductive interactions, or both.
- The boundaries of the population are the largest grouping for which the rates of interaction are much higher within the grouping than outside (Simon 2002).

From these core ideas, it follows that organisms located in the same spatial area (including recent migrants) are part of the population *if and only if* they are interacting with other conspecifics. Furthermore, if a later grouping is causally connected by survival or reproductive interactions to an earlier grouping, then it is the same population; in this way, populations can be continuous through time.

Both survival and reproductive interactions cover a broad range of interactions.[7] *Reproductive interactions* include not just interbreeding (successful matings) but also unsuccessful matings. After all, unsuccessful matings can have important evolutionary and ecological consequences for the organisms that engage in them, especially if the organism never succeeds in mating. Offspring rearing activities (i.e., interactions between parents and interactions between parents and offspring) can also be included under reproductive interactions, since for many species the offspring produced will not survive without them.[8] As mentioned previously, some concepts of population include only interbreeding (if they include interactions at all); the suggestion here is that such concepts omit important factors relevant to ecology and evolution.

---

[6] This presentation is slightly modified from the one in Millstein (2009).

[7] Social interactions might be a third category; however, my sense is that the social interactions that are relevant for ecology and evolution are those that involve either survival or reproduction (or both). Thus, social interactions will be included under those two categories.

[8] An anonymous reviewer suggests that offspring rearing activities ought to be viewed as survival interactions. While from the perspective of the offspring, these activities are survival interactions, from the perspective of the parents, they are reproductive interactions. Since both types of interactions are included as part of the concept, nothing turns on which way such activities are classified.

Ignoring survival interactions is equally mistaken. Survival interactions played a crucial role in Darwin's thinking; Chapter III of *The Origin of Species*, "Struggle for Existence," is focused almost entirely on them, with reproductive interactions receiving only the briefest mention.[9] Darwin states: "I should premise that I use the term Struggle for Existence in a large and metaphorical sense, including dependence of one being on another, and including (which is more important) not only the life of the individual, but success in leaving progeny"[10] (Darwin 1859: 62). In invoking this "large and metaphorical sense," Darwin provides a wide-ranging list of examples from which we can get a sense for the kind of survival interactions that organisms participate in. Eggs and seeds generally "struggle" the most; Darwin notes especially the seedlings that struggle to germinate in "ground already thickly stocked with other plants." More broadly, organisms can compete for "the same place or food" (i.e., "limited resources"), such that what is taken by one organism becomes unavailable for another. Organisms can also compete directly: "Two canine animals in a time of dearth, may be truly said to struggle with each other [for] which shall get food and live" (Darwin 1859: 62). But survival interactions in the struggle for existence need not involve direct or indirect competition; Darwin also discusses the possibility that "a plant could exist only where the conditions of its life were so favourable that many could exist together, and thus save each other from utter destruction" (Darwin 1859: 70). In other words, organisms of the same species can also "struggle together," or cooperate rather than compete. Again, the point is that survival interactions, such as those invoked by Darwin, are relevant to the cohesiveness of organism groupings and affect their ecological and evolutionary trajectories. Note also that even though asexual organisms do not engage in reproductive interactions per se, they do engage in survival interactions as described here. Thus, according to the

---

[9] On a personal note, this is not a gratuitous reference to Darwin; on the contrary, teaching Chapter III for a philosophy of biology class was the source of my thinking that it was important to include survival interactions in the concept of population.

[10] The phrase "success in leaving progeny" is often quoted, but the frequency with which this sentence is quoted stands in stark contrast to the minuscule amount of space that Darwin used in discussing it. Perhaps, given the emphasis on survival interactions in the rest of the chapter, this passage might be better understood as referring to "success in producing offspring who survive." Indeed, Darwin subsequently remarks: "If an animal can in any way protect its own eggs or young, a small number may be produced, and yet the average stock be fully kept up; but if many eggs or young are destroyed, many must be produced, or the species will become extinct" (Darwin 1859: 66). In any case, "struggle for existence" is probably broader than "survival interactions," since some of the examples that Darwin discusses (e.g., a plant on the edge of the desert struggling for life against the drought) do not seem to involve interactions between organisms.

causal interactionist population concept, asexual organisms can be organized into populations, even multiple populations of the same species. (I will describe one such apparent case later in the paper.)

The causal interactionist population concept leaves out some criteria that other population concepts include. For example, it does not include space or time as a boundary; interactions influence the subsequent fate of the population (and thus, are what matter for ecology and evolution), regardless of whether they occur over large or small stretches of time or space. Indeed, if one were to delimit a grouping in space (a purported population) while excluding some organisms that were frequently interacting with members of the grouping, the predicted future trajectory of that grouping would be very misleading. Including organisms that are not interacting (drawing too large a spatial boundary) would be similarly misleading. For example, the presence of a new adaptive trait among the Torrey pines of San Diego is not likely to affect the Torrey pines of Santa Rosa Island, approximately 190 miles away; thus, one would be mistaken to predict the spread of the trait in Santa Rosa Island (which is not to deny that the trait could occur there, either through an unusual dispersal or a new mutation).[11] Time is excluded from my population concept for similar reasons: if over time, some descendents of earlier population members are no longer interacting with other descendents, it would be misleading to consider them together. Again, the point is that when organisms are interacting, their fates are (to some extent) linked, so that they form a biological whole.

The causal interactionist population concept also does not consider the amount of gene flow relative to selection; the concept of population needs to be independent of selection to avoid circularity, since selection takes place in populations and produces changes in populations. Migration is not included, either, primarily because it does not need to be; if an organism migrates out of the population, then it is no longer interacting with the other organisms. In contrast, if an organism migrates into the population and interacts, then it is part of the population; if it just "passes through," then it is not.[12] So again, interaction is the key criterion. Finally, the concept

---

[11] The Wahlund effect demonstrates a similar point using a more theoretical approach. If two populations are completely isolated, the predicted genotype frequencies of the individual populations differ from that of the two populations pooled together. See, for example, Hartl and Clark's (1989) textbook for discussion.

[12] Slatkin (1987) notes that gene flow (as a result of migration with interbreeding) is difficult to measure and describes two types of measurement methods: direct and indirect. (The following discussion relies heavily on Slatkin's account, leaving out a great deal of the complexity; the reader is referred to Slatkin's excellent discussion for further details.) Direct methods use observations of the frequency and distance of dispersals, together with information about breeding success after dispersal, to infer the amount of gene flow between populations. Indirect methods use allele frequencies or DNA sequence differences with

does not include genetic relatedness, which merely tracks the outcomes of reproductive interactions rather than the interactions themselves.

Even though none of these candidate criteria (space, time, gene flow, migration, genetic relatedness) are included in the causal interactionist population concept, they all may be *indirect indicators* of (or proxies for) populations. This will be seen in the cases that follow.

The causal interactionist population concept also restricts members of the population to *organisms of the same species*. It is my impression that most concepts of local population do likewise. But more importantly, the term *community* can be used for multi-species groupings. Having different terms for each concept can better allow for the possibility of different dynamics in single-species groupings and multi-species groupings.

However, the causal interactionist population concept does not assume any *particular* species concept, excluding only those that define species in terms of populations (which would again introduce circularity). This position is partly to avoid entanglement in the seemingly endless debates over the concept of species, but also to allow for the possibility that there may be more than one legitimate species concept, as Mishler and Brandon (1987) and Ereshefsky (1992), among others, have argued. Thus, the causal interactionist population concept is neutral with respect to species concept; it has some similarities to Templeton's (1989) cohesion species concept, but is neither identical to nor reliant on it. Both Templeton's and my concepts focus on causal processes (or, in my population concept, interactions) rather

---

population genetic models to estimate the level of gene flow that must have occurred to produce those patterns. Slatkin depicts two types of indirect methods. One method uses Wright's F_{ST} statistic to estimate the standardized variance in allele frequencies among local populations. The other method is Slatkin's own; it relies on the frequencies of rare alleles for its calculations. Both methods, according to Slatkin, can be used to estimate Nm, where N is the local effective population size (i.e., the number of breeding organisms in the population) and m is the average rate of immigration in an island model of population structure (where every local population is equally accessible from every other). To estimate m alone, an estimate of N from census data can be performed; N can be used as a measure of how strong drift is likely to be (smaller N = stronger drift). But the key thing to note is that all of these observations, measures, and inferences—dispersal, F_{ST}, N, m, drift—presuppose that the population structure is known. An assertion that dispersal is occurring presupposes that it is known that an organism is leaving one population and migrating to another. When allele frequencies or DNA sequences are sampled, a decision has to be made as to which organisms to sample from; a census of a population again presupposes that it is known which organisms belong and which do not. Thus, although these techniques may help determine population structure by serving as indirect indicators, population is the more basic concept. Population genetic methods do not replace the need to have a concept of population. (Thanks to Douglas Futuyma for encouraging me to clarify the relationship between population genetics approaches and the concept of population, although I am sure there is more to be said here.)

than outcomes, both incorporate a variety of causal processes not limited to reproductive (though not the same set), and both weigh the causal processes differently in different cases (see discussion later in the chapter). However, whereas the cohesion species concept incorporates *potential* processes (potential genetic exchangeability and potential demographic exchangeability), the causal interactionist population concept incorporates only *actual* processes (interactions). Moreover, whereas the cohesion species concept incorporates evolutionary processes, the causal interactionist population concept incorporates processes that give rise to evolutionary processes (e.g., struggle for existence), but not the evolutionary processes themselves.

The causal interactionist population straightforwardly implies a corresponding metapopulation concept.

Causal Interactionist Metapopulation Concept

The causal interactionist metapopulation concept consists of the following elements:

- *Metapopulations* consist of at least two local populations[13] of the same species, linked by migration or dispersal, such that organisms occasionally change which population they are a part of;[14] rates of interaction within local populations are much higher than the rates of interactions among local populations.
- If the rates of interaction within local groupings are *not* significantly higher than the rates of interaction among local groupings, it is a *patchy population*, a term coined by Harrison (1991).

As I mentioned briefly earlier, contemporary meanings of the term metapopulation are much less specific than Levins's meaning; the concept I propose is closer to the contemporary meaning. Migrations/dispersals form the basis of the interactions among local populations. Typically, these interactions are rare, but if they are nonexistent, there is no metapopulation—just a set of unconnected local populations. As these interactions are weak, a metapopulation is much less cohesive—much less a biological whole—than a local population.

As Hanski and Gilpin (1991) point out, the movement of organisms is different at different scales. The movements within a local population are routine feeding and breeding activities, whereas movements from one local population to another are "typically across habitat types which are not suitable for their feeding and breeding activities, and often with substantial risk of failing to locate another suitable habitat patch in which to settle" (Hanski and Gilpin 1991: 7). Thus, the interactions that bind populations together

---

[13] As characterized in the previous section.

[14] As described in the previous section.

are different than the interactions that bind metapopulations. However, the following discussion will show that the types of interactions among the local populations of a metapopulation differ (some focused more on survival and others focused more on reproduction), which will require that the concept be elaborated a bit.

Six Case Studies

We will now turn to six case studies to illustrate the application of these concepts and to show their fruitfulness. Although other types of cases are possible, part of my intent is to demonstrate how the concept handles some seemingly problematic cases. All of these cases will draw upon the published results of particular biologists and will make inferences about population structure[15] based on those results. Thus, it follows that if their results are inaccurate, then my claims concerning the population structure are likely to be inaccurate as well—which is as it should be. Any claim that a particular grouping of organisms forms a particular population structure is an empirical claim, subject to being overturned by better data. The point is not to claim that any grouping of organisms definitely has a particular population structure, but rather to show what sort of population structure would be present if the results of the cited studies are accurate. Moreover, in each case I am not making a claim for the species as a whole (with the possible exception of *Eubalaena australis*), but rather for the species in particular places and at particular times—namely, the places and times of the referenced study. I take this approach not because I think that space and time are part of the population concept, but instead to acknowledge that a species may have one population structure in one place and time and a different population structure in a different place or time. In other words, a given population structure is not a permanent feature of a species, though particular species may tend to form certain types of population structures given their mating and feeding habits with respect to the characteristics of the habitats in which they live.

Case 1: One Continuous Population

Linanthus parryae (desert snow) is a well-studied flowering plant. Perhaps the most famous studies of *L. parryae* occurred in the Mojave Desert in the early 1940s by Epling, Dobzhansky, and Wright (Epling and Dobzhansky 1942; Wright 1943b). In years of heavy rainfall (which there were during this period), the swath of blue and white flowers is "almost like a carpet" (Provine 1986), spread over a large territory. In order to account for the evolution of species such as these, Wright (1943a) developed a "genetic

---

[15] Here I am using the term *population structure* as a general term for the various ways that organisms might be organized: in one population, in several populations, in a metapopulation, or otherwise.

isolation by distance" model. According to Wright's model, when a species has limited mobility (as does *L. parryae*), the parents of any given individual come from a small surrounding region. Using his isolation by distance model, Wright (1943b) estimated that breeding group sizes were approximately one or two dozen productive individuals, with long-range dispersal being a rare event. The genetics of *L. parryae* were not known at that time; the numbers of blue and white flowers were sampled and counted at various locations under the assumption that the differences were genetic in origin.

Is this one population or many? The papers in question are not consistent on that point, sometimes referring to one population with groups, subgroups, or colonies, and sometimes referring to multiple populations. In a case such as this, one might be tempted to say that since there are groupings of interactions within the carpet (as a result of isolation by distance), it follows that there are multiple populations within the carpet. It certainly is not the case that every organism interacts with every other organism, and it would rarely be the case that every organism in *any* population interacted with every other organism. However, the carpet was an area about 80 miles long and on average 10.5 miles wide, with an estimated 10^{10}–10^{11} individual plants (Wright 1943b). It seems pretty clear that many of the plants were not interacting directly.

However, on the causal interactionist population concept, the boundaries of the population are the largest grouping where the rates of interaction are much higher within the grouping than outside it. Despite the likely pockets of density, it does not seem as though there would have been groupings for which interactions were *significantly* higher. Rather, they would have been only somewhat higher, with the densities fairly variable from generation to generation. Thus, if one deploys the causal interactionist population concept, Epling, Dobzhansky, and Wright were studying one continuous population rather than multiple populations (Figure 3.1).

Two considerations support this conclusion. The first is to suggest that while it is clear that many plants were not interacting directly, many would

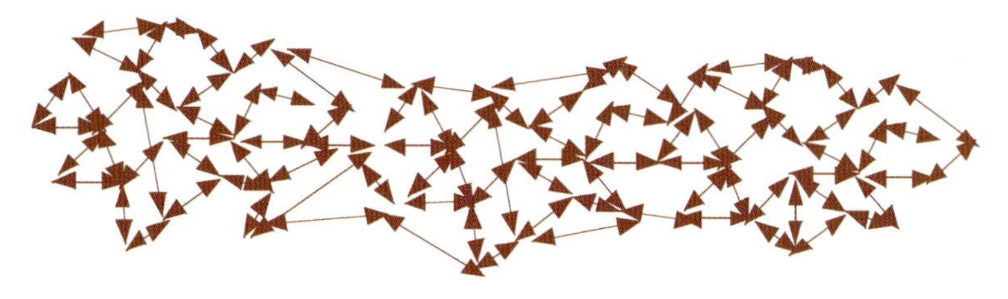

FIGURE 3.1 One Continuous Population Brown arrows represent survival and reproductive interactions. There are places where the interactions are "denser" but no places where the rates of interaction within a grouping are much higher than those outside a grouping.

have been interacting *indirectly*.[16] As an illustration of indirect interaction, suppose plant A takes resources that are no longer available to plant B, which might mean that plant B is not available to interbreed with plant C. Thus, plant A has interacted indirectly (or transitively) with plant C. Without any internal or external isolating mechanisms, such indirect interactions would suggest that the entire carpet was acting as a biological whole. Second, any pockets of density likely would be short-lived, which is not to say that they would not have effects or that the effects could not be ecologically or evolutionarily important.[17] But *sustained* ecological or evolutionary processes would not be expected within those groupings themselves.

Finally, I will make a brief note about evolutionary processes. The authors concluded that drift played a substantial role in the distribution of blue and white flowers. Recent studies, such as one by Schemske and Bierzychudek (2007), have challenged that conclusion. My suggestion is that claims about population structure should be distinct from claims about evolutionary process; that is, assuming the accuracy of Wright's results, in the 1940s in the Mojave Desert *L. parryae* formed a continuous population regardless of whether the distribution of blue and white flowers was due primarily to drift or primarily to selection. (Whether it forms continuous populations in other places and times would be an empirical question.) Indeed, in most of the case studies that follow, biologists made claims concerning the relative importance of selection and drift, but those claims are independent of claims concerning population structure. More precisely, a given population structure does not dictate whether selection and/or drift is operating, though it does dictate the organisms over which these evolutionary processes range.

Case 2: Populations with Only Survival Interactions

One of the reasons to have a population concept that includes survival interactions without requiring reproductive interactions is to enable it to account for the possibility that asexual reproducers form multiple populations. Of course, it may be that some (perhaps even many) asexual reproducers do not form separate populations and instead consist of only one population. The common wisdom for bacteria has been that microorganisms are ubiquitous and the global richness of microbial species is moderate (Finlay and Clarke 1999). However, this view has been challenged recently, suggesting that some bacteria do consist of multiple populations.

---

[16] Note that "interacting indirectly" and factors that serve as "indirect indicators" for other factors are not the same thing, although in this case, there are indirect indicators (the distribution of plants in time and in space, plus what is known about the life cycle of *L. parryae*) that the plants would have been interacting indirectly.

[17] Indeed, these groupings might be groups that could engage in group selection; Shavit (2005) discusses the group concept.

One such challenge comes from a recent study of several *Pseudomonas* (*sensu stricto*) species (Cho and Tiedje 2000). *Pseudomonas* are free-living soil bacteria. Cho and Tiedje examined 85 different *Pseudomonas* genotypes from 38 transect samples of "undisturbed pristine soil" from 10 sites on 4 continents. Their results show that for each of the 85 genotypes, a particular genotype found in one 200-meter transect of a particular site would not be found at the other study sites or other continental regions. Moreover, the majority (91.8%) of genotypes were only found in one transect sample of a site and just 7 of the 85 genotypes were repeatedly isolated from different transect samples of the same site.

Cho and Tiedje state, "This indicates some mixing and dispersal of the genotypes within a site but not between sites and regions" (Cho and Tiedje 2000: 5455). If this statement is correct, their results suggest a remarkable amount of geographic isolation among groupings of particular *Pseudomonas* genotypes.[18] Although there are no reproductive interactions within these groupings, it is reasonable to assume there are survival interactions, such as competition for the same resources. If this is correct, then they form separate populations by the criteria of the causal interactionist population concept (Figure 3.2). Note that there are no direct observations of the sur-

[18] Again, as with all of the cases described, I take no stand on whether the conclusions drawn are accurate; my claims for population structure would be different if the conclusions were different. For this case in particular, Douglas Futuyma (personal communication) suggests that an alternative reason for single-site distributions could be that mutations happen so fast that a genotype gets mutated out of existence soon after it has dispersed to a new location.

FIGURE 3.2 Multiple Populations Green arrows represent survival interactions; there are no reproductive interactions since the bacteria reproduce asexually. Groupings are geographically isolated from one another.

vival interactions, but that the genotype distributions, together with what is known about bacteria, form indirect evidence for the cohesiveness of the groupings.

Case 3: A Simple Metapopulation

Lamotte's 1950s investigation of *Cepaea nemoralis*, an often studied and widely distributed land snail, illustrates the causal interactionist metapopulation concept. In the Aquitaine region in southwestern France, Lamotte found groupings (what he called "colonies") that were 1 to 2 kilometers apart and well isolated chiefly through the requirement of shade, especially in the summer. The numbers of individuals in a grouping ranged from one to several hundred, with some larger. There was some migration between colonies, but in most cases, there was only a very limited amount:

> *In the gaps between the colonies practically no Cepaea are found, or else only a few individuals or little groups of them, here and there, along the hedges running along the field boundaries, or upon shaded banks. The isolated individuals constitute what one may call 'migration trails' between populations and their very scattering shows the low frequency and discontinuity of these exchanges (Lamotte 1959: 66).*

Lamotte describes a paradigmatic example of a metapopulation. The smaller groupings are local populations, with the rates of interaction (presumably both survival and reproductive) within the groupings being very much greater than among the groupings. There are also interactions among the local populations—the infrequent migrations from one to another—and their very infrequency relative to the frequency of interactions within the groupings implies that this is one metapopulation rather than one patchy population. However, the fact that there are *some* interactions among the local populations means that there is some cohesiveness to the whole; that is, if there were no such interactions, there would be no metapopulation, but instead only separate local populations (Figure 3.3).

In the three case studies examined so far, locations of organisms in space have played crucial roles as indirect indicators of causal interactions among organisms, which might lead one to conclude that geographical location ought to be part of the population and metapopulation concepts. However, as will be seen in the subsequent case studies, even though consideration of spatial location is always important, it is not always definitive.

Case 4: A Patchy Population

If the rates of interactions between groupings of organisms are lower, but not *significantly* lower, than the interactions within the groupings, the organisms do not form a metapopulation. Rather, they form one patchy population. A species that is organized into multiple patchy populations is the montane willow leaf beetle, *Chrysomela aeneicollis*, studied in the Sierra Nevada by Rank (1992) and later by Dahlhoff and Rank (2000) and

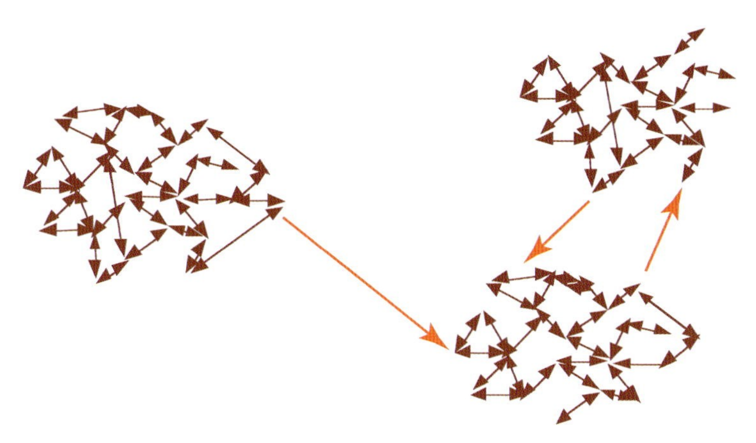

FIGURE 3.3 Multiple Local Populations Forming One Metapopulation The brown arrows that are within the groupings (i.e., the local populations) represent both survival and reproductive interactions, whereas the longer orange arrows between groupings represent migrations from one grouping to another. What makes this a metapopulation is that the latter exist but are much less frequent than the former.

Rank and Dahlhoff (2002). *C. aeneicollis* requires a moist habitat; the beetles can fly but do so only rarely. In the Sierra Nevada, *C. aeneicollis* can be found on willow shrubs; the shrubs are located in numerous physically separated bogs. The bogs themselves are located in one of three drainages, separated by high-elevation ridges that present a challenging (though not insurmountable) barrier for the beetles. Thus, there are groupings (on the shrubs) within groupings (in the bogs) within groupings (in the drainages). Genetic evidence (expressed using a modification of Wright's F_{ST}; see Footnote 12) shows that the differentiation at the drainage level is very much greater than at the bog level, which in turn is greater (but not significantly greater) than that at the shrub level.

With the genetic evidence serving as an indirect indicator of reproductive interactions and with survival interactions being a reasonable inference, these results suggest that each drainage contains a patchy population, with patches at the bog and shrub levels.[19] That is, it appears as though the rates of interaction *within the bushes* are greater (but not very much greater) than the rates of interaction *among the bushes*, and that the interactions *within the bogs* are greater (but not very much greater) than the interactions *among the bogs* (Figure 3.4).

Note that spatial location alone could not have dictated this result. If, for example, it were discovered that, contrary to Rank's (1992) findings, there

[19] Also, given the evidence for minimal migration among the drainages, the three patchy populations in the drainages collectively form a metapopulation.

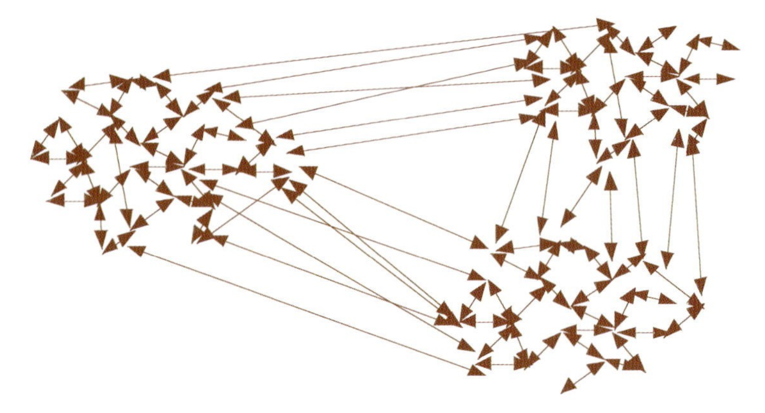

FIGURE 3.4 A Patchy Population The brown arrows represent survival and reproductive interactions as well as regular movements among groupings (which would be followed by further survival and reproductive interactions). The rates of interactions within the three patches shown are only somewhat greater than the rates of interaction among the patches. There is some further patchiness within each of the three patches, but again, it is not significantly greater than the interactions among them.

were significant genetic differentiation among the bogs, which would imply that the rates of interaction within the bogs were much higher than those among the bogs, then that would be evidence that populations formed at the bog level rather than at the drainage level. In other words, in principle, there could be the same distribution of organisms in space, but with different patterns of interaction and thus different entities operating as biological wholes.

Case 5: A Metapopulation of Reproductive Populations

I use the term *metapopulation of reproductive populations* to refer to cases in which organisms mate locally, but struggle (in a Darwinian sense) globally. The study by Kaliszewska et al. (2005) of the Southern Ocean right whale, *Eubalaena australis*, seems to be one such case. *E. australis* has three wintering locations for breeding (i.e., the coastal waters of Argentina, South Africa, and Australia), but in the summer, there are common feeding grounds in the Antarctic. The genetic evidence suggests that whales return to the same breeding grounds with some, but very little, migration among breeding groupings—one or fewer females per decade (Kaliszewska et al. 2005).

This case forms a bit of a puzzle. If one considers survival interactions alone, as occurs in the population concept defended by Gildenhuys (2009), there appears to be only one population because the whales are surely competing intensely at their feeding grounds. However, if one looks at reproductive interactions alone, there appears to be three populations, because there is very little reproductive interaction among the three breeding

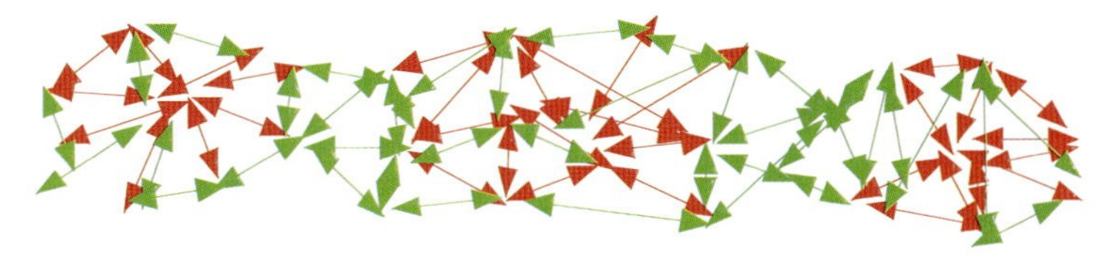

FIGURE 3.5 Multiple Reproductive Populations Forming One Metapopulation The red arrows are reproductive interactions; the green arrows are survival interactions. The organisms engaging in reproductive interactions are also engaging in survival interactions, but there are only survival interactions between these groupings. Thus, the rates of interaction of the former are much greater than those of the latter.

groupings. But considering the two types of interactions together resolves the puzzle, as there are *both* survival and reproductive interactions *within* breeding groupings but *only* survival interactions *between* breeding groupings. Thus, the causal interactions within the three breeding groupings are clearly significantly greater than those among the breeding groupings. In other words, the breeding groupings exhibit more cohesion, operating as a biological whole to a far greater extent than the feeding grouping. It follows that the breeding groupings represent populations (what I call reproductive populations), but the feeding grouping is also a biological whole—it is a metapopulation of reproductive local populations (Figure 3.5).

Note again that while geographical location is relevant, it is not entirely definitive. The whales range over much of the earth's oceans, but what is more significant for future evolutionary outcomes (i.e., for shared fate) is the fact that there are three groupings that hardly ever breed with one another.

Case 6: A Metapopulation of Survival Populations

A *metapopulation of survival populations* is the reverse of a metapopulation of reproductive populations; rather than mate locally and struggle globally, they struggle locally and mate globally. *Gasterosteus aculeatus*, the threespine stickleback fish that inhabits an Alaskan drainage, is an example of this type of metapopulation (Aguirre 2007, 2009). Aguirre found that phenotypes are associated with habitat type (consistent with known adaptations) but not with geographic distances. In contrast, quasi-neutral (microsatellite) genetic differences were associated with geographic distances but not with habitat type.

These results suggest that with respect to reproductive interactions alone, there is one continuous population with isolation by distance, which is similar to the previously discussed case of *Linanthus parryae*. However, with respect to survival interactions, organisms are struggling (again, in a Darwinian sense) most intensely within particular habitats; there are

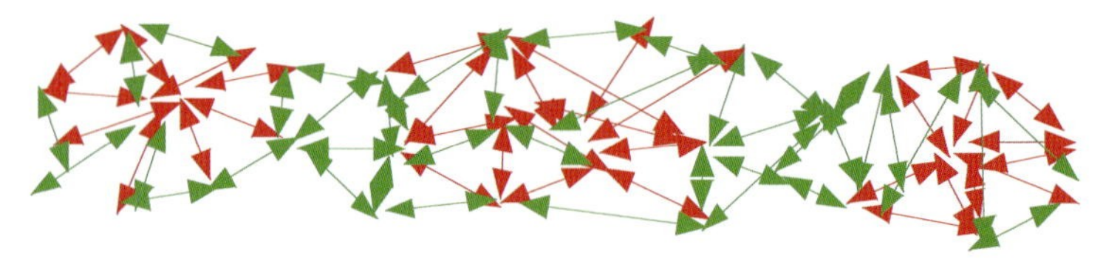

FIGURE 3.6 Multiple Survival Populations Forming One Metapopulation The green arrows are survival interactions; the red arrows are reproductive interactions. The organisms engaging in survival interactions are also engaging in reproductive interactions, but there are only reproductive interactions between these groupings. Thus, the rates of interaction of the former are much greater than those of the latter.

both survival and reproductive interactions within habitat groupings, but primarily reproductive interactions between habitat groupings. Thus, the causal interactions within the habitat groupings are clearly significantly greater than those among the habitat groupings. In other words, the habitat groupings exhibit more cohesion, operating as a biological whole to a far greater extent than the reproductive grouping. It follows that the habitat groupings represent populations (what I call survival populations), but the reproductive grouping is also a biological whole—it is a metapopulation of survival local populations (Figure 3.6).

Some Responses to Possible Questions

As concerns about my proposed causal interaction population concept may have arisen by this point, this section addresses some issues that may be troubling. To begin, one might well ask whether or not there is a continuum between a metapopulation and a patchy population, and if there will be some cases for which it is difficult to determine if a given population structure manifests the former or the latter. The answer is yes, there is a continuum, and the boundaries of the population are the *largest* grouping for which rates of interaction are *much higher* within the grouping than outside. A numerical value could be put on the relative strengths of those interactions (future work might consider how best to do this) to specify where to draw the line, but it is hard to see how any particular value could be defended. Any chosen numerical value would be fairly arbitrary, with little difference between slightly higher and slightly lower values, which implies that in *some* cases (how many remains to be seen) there will be no principled answer as to whether or not there is a metapopulation or a patchy population (alternatively, there could be many cases that are fairly clear-cut). However, this view does *not* imply that populations are arbitrary or that their boundaries can be drawn *anywhere.* It just means that populations are blurry entities, with edges that are not always well defined.

One might also ask if there is a difference between describing a particular population structure as one patchy local population or as a metapopulation of local populations. I have two responses. First, if it were equally correct to call a particular population structure a patchy population or a metapopulation—if one of these characterizations were not a *better* representation of the real world—then some claims concerning evolutionary processes (e.g., selection, drift) could not be judged as superior to other claims. For example, if the Alaskan threespine sticklebacks formed a patchy population (and not a metapopulation as I argued previously), we would calculate the expected average change over the whole population and understand selection in terms of that one overall trend (whatever it would be). But if the sticklebacks formed a metapopulation, then there would be numerous selection processes, with selection differing by local population. The outcomes that would be expected from those two scenarios (e.g., 10, 50, or 100 generations from now) would likely be very different from one another. This shows that different selection processes are implied by different population structures; presumably, one is a better representation of the actual evolutionary phenomena. Indeed, the different population structures yield different predictions, and one set of predictions is likely to be more accurate. However, these judgments can be made only if the difference between a patchy population and a metapopulation is recognized.

Second, regarding the issue of whether it makes a difference if we describe a particular population structure as one patchy local population or as a metapopulation of local populations, there is reason to think that metapopulation models may be distinctive in certain respects, as has been discussed by various authors.[20] In particular, researchers have argued that metapopulation models: (1) describe the ideal conditions for evolution (Wright 1931, 1932), (2) facilitate demonstrations of drift in nature (Lamotte 1959; Millstein 2008), (3) imply simultaneous use of predators in order to control pests (Levins 1969), and (4) predict regional persistence of locally unstable species (Harrison 1991). Collapsing the distinction between metapopulations and patchy populations assumes *a priori* that patchy populations would have these same implications, which they may or may not. By retaining both concepts, the possibility that there are different consequences for patchy population models and metapopulation models is preserved.

Of course, one might acknowledge the need to distinguish between patchy populations and metapopulations without accepting the causal

[20] Hanski and Simberloff (1997) suggest that there may be some cases for which the consequences are "about the same" (e.g., a low rate of long-distance migration as compared to a high rate of short-distance migration). I do not deny this possibility, but would caution that an approach that focuses solely on the consequences of the model regardless of whether or not the assumptions of the model are met could raise problems later—others may misunderstand the simplification or there may be other consequences that are *not* "about the same."

interactionist population and metapopulation concepts. Reasons to accept the concepts argued for in this paper include the following. Obviously, a central feature of these concepts is their focus on *interactions.* Interactions among organisms are more significant than geographical location. Both humans and right whales live widely across the planet (albeit in very different habitats!), but their patterns of interaction differ in evolutionarily important ways. (This chapter has not included a discussion of population structure among humans, a complicated and interesting topic; however, the characterization of relatively distinct breeding groups that have survival interactions among all members would not characterize the present human configuration). The presence of interactions among a grouping of organisms also means that the grouping is acting as an individual (i.e., a biological whole); organisms have a shared fate to a degree. The population concept picks out that individual, and the interactions pick out groupings that are likely to differentiate in the future. For example, barring changes in population structure, we would expect further differentiation among the right whale breeding groupings and the stickleback habitat groupings. In addition to its focus on interactions, another benefit of the causal interactionist population concept is that it does not require reproductive interactions among organisms, so that the concept works for asexual as well as sexual populations. Finally, it does not assume any particular evolutionary process (e.g., selection, drift). Once the population is picked out, then the processes that are acting can be determined. Indeed, by forcing identification of populations (and defense of those identifications) prior to determining their evolutionary and ecological processes, the concept prevents predetermining the outcome of investigations though a convenient choice of population structure.

Conclusion

I hope to have gone a fair way toward establishing the causal interactionist concepts of population and metapopulation, yet there is admittedly much more work that needs to be done. There are more types of cases to be considered, some of which may necessitate modification or elaboration of the concepts, as occurred with the right whale and three-spine stickleback case studies. Another issue that needs to be addressed is the role that time plays in our understanding of populations; it may be that consideration of very short periods of time (taking into account ephemeral interactions) or very long periods of time will require further amendments to the population concept. Finally, there are numerous related concepts, such as organism, group, species, community, and the more general concept of biological individuality, which would enhance the understanding of population concepts. Other philosophers and biologists have worked on these concepts and continue to work on them; progress has been made but more needs to be done.

The length and vehemence of the debate over species concepts in particular (which is understandable, given the complexities of the topic) makes me humble about the prospects for progress on population and metapopulation concepts. However, it is my impression that work in the philosophy of biology continues to improve and that there are many productive interactions between philosophers of biology and biologists (as well as historians of biology). Philosophers of evolutionary biology are more attuned to biological practice than ever before. For those who seek conceptual clarity, it can be a difficult line to walk between the practical demands of theoretical biology and the logical demands of philosophy of biology, and some stray more to one side than the other, striving to find the right balance and perhaps not always achieving it. The fact that researchers struggle with the balance between theoretical biology and philosophy of biology should not be a surprise; each field involves difficult, yet important issues, and thus the intersection does as well. However, if we are to continue to work in a Darwinian vein (always with modifications and enhancements), we must continue to try to resolve them.

Acknowledgments

This chapter has grown out of a paper that preceded it (Millstein 2009), and so in many ways, I also need to thank everyone who assisted me with the earlier paper. Specifically, in regard to the ideas presented here, I would like to thank attendees of the Current Issues in Darwinian Theory workshop, held in Halifax, Nova Scotia (2009), for their exceedingly helpful comments and questions. A last-minute illness prevented me from attending the Darwin 2009 workshop at Stony Brook University, and there is no doubt in my mind that this paper suffers from that absence. Thanks to Michael Bell for good discussion over email as well as a pointer to the stickleback research, and to Nathan Rank for ongoing discussion concerning his montane leaf beetle research. Douglas Futuyma and an anonymous reviewer also provided invaluable comments on the penultimate draft, for which I am very grateful.

Literature Cited

Aguirre, W. E. 2007. *The Pattern and Process of Evolutionary Diversification: Lessons from a Threespine Stickleback Adaptive Radiation*. Ph.D. thesis, Ecology and Evolution, Stony Brook University, Stony Brook.

Aguirre, W. E. 2009. Microgeographical diversification of threespine stickleback: Body shape-habitat correlations in a small, ecologically diverse Alaskan drainage. *Biol. J. Linn. Soc.* 98: 139–151.

Andrewartha, H. G. and L. C. Birch. 1954. *The Distribution and Abundance of Animals*. University of Chicago Press, Chicago.

Arms, K. and P. S. Camp. 1979. *Biology*. Holt, Rinehart, and Winston, New York.

Berryman, A. A. 2002. Population: A central concept for ecology? *Oikos* 97: 439–442.
Brandon, R. N. 1996. *Concepts and Methods in Evolutionary Biology*. Cambridge University Press, Cambridge, MA.
Byron, J. M. 2007. Whence philosophy of biology? *Brit. J. Philos. Sci.* 58: 409–422.
Camus, P. A. and M. Lima. 2002. Populations, metapopulations, and the open-closed dilemma: The conflict between operational and natural population concepts. *Oikos* 97: 433–438.
Cho, J.-C. and J. M. Tiedje. 2000. Biogeography and degree of endemicity of fluorescent *Pseudomonas* strains in soil. *Appl. Environ. Microb.* 66: 5448–5456.
Creath, R. and J. Maienschein (eds.). 2000. *Biology and Epistemology*. Cambridge University Press, New York.
Dahlhoff, E. P. and N. E. Rank. 2000. Functional and physiological consequences of genetic variation at phosphoglucose isomerase: Heat shock protein expression is related to enzyme genotype in a montane beetle. *Proc. Natl. Acad. Sci. USA* 97: 10056–10061.
Darwin, C. [1859] 1964. *On the Origin of Species: A Facsimile of the First Edition*. Harvard University Press, Cambridge, MA.
Epling, C. and T. Dobzhansky. 1942. Genetics of natural populations. VI. Microgeographic races in *Linanthus parryae*. *Genetics* 27: 317–332.
Ereshefsky, M. 1992. Eliminative pluralism. *Philos. Sci.* 59: 671–690.
Finlay, B. J. and K. J. Clarke. 1999. Ubiquitous dispersal of microbial species. *Nature* 400: 828.
Futuyma, D. J. 1986. *Evolutionary Biology*, 2nd ed. Sinauer Associates, Sunderland, MA.
Gannett, L. 2003. Making populations: Bounding genes in space and in time. *Philos. Sci.* 70: 989–1001.
Gerson, E. M. 1998. *The American System of Research: Evolutionary Biology, 1890–1950*. Ph.D thesis, Department of Sociology, University of Chicago, Chicago.
Ghiselin, M. T. 1974. A radical solution to the species problem. *Syst. Zool.* 23: 536–544.
Ghiselin, M. T. 1997. *Metaphysics and the Origin of Species*. SUNY Press, Albany.
Gildenhuys, P. 2009. *A Causal Interpretation of Selection Theory*. Ph.D. thesis, Department of History and Philosophy of Science, University of Pittsburgh, Pittsburgh.
Gilmour, J. S. L. and J. W. Gregor. 1939. Demes: A suggested new terminology. *Nature* 3642: 333.
Godfrey-Smith, P. 2009. *Darwinian Populations and Natural Selection*. Oxford University Press, Oxford.
Goudge, T. A. 1955. What is a population? *Philos. Sci.* 22: 272–279.
Hamilton, A., N. R. Smith, and M. H. Haber. 2009. Social insects and the individuality thesis: Cohesion and the colony as a selectable individual. In J. Gadau and J. Fewell (eds.), *Organization of Insect Societies: From Genome to Sociocomplexity*, pp. 572–589. Harvard University Press, Cambridge, MA.
Hanski, I. and M. Gilpin. 1991. Metapopulation dynamics: Brief history and conceptual domain. *Biol. J. Linn. Soc.* 42: 3–16.
Hanski, I. and D. Simberloff. 1997. The metapopulation approach, its history, conceptual domain, and application to conservation. In I. Hanski and M. Gilpin (eds.), *Metapopulation Biology: Ecology, Genetics, and Evolution*, pp. 5–26. Academic Press, San Diego.

Harrison, S. 1991. Local extinction in a metapopulation context: An empirical evaluation. *Biol. J. Linn. Soc.* 42: 73–88.
Hartl, D. L. and A. G. Clark. 1989. *Principles of Population Genetics*, 2nd ed. Sinauer Associates, Sunderland, MA.
Hastings, A. and S. Harrison. 1994. Metapopulation dynamics and genetics. *Ann. Rev. Ecol. Syst.* 25: 167–188.
Hey, J. 2009. *From Essences to Populations: Unwinding Darwin's Impact on Metaphysics*. Paper read at Current Issues in Darwinian Theory, October 2009, Halifax, NS.
Hull, D. L. 1969. What philosophy of biology is not. *Synthese* 20: 157–184.
Hull, D. L. 1976. Are species really individuals? *Syst. Zool.* 25: 174–191.
Hull, D. L. 1978. A matter of individuality. *Philos. Sci.* 45: 335–360.
Hull, D. L. 1980. Individuality and selection. *Ann. Rev. Ecol. Syst.* 11: 311–332.
Kaliszewska, Z., J. Seger, V. Rowntree, and 14 others. 2005. Population histories of right whales (Cetacea: *Eubalaena*) inferred from mitochondrial sequence diversities and divergences of their whale lice (Amphipoda: *Cyamus*). *Mol. Ecol.* 14: 3439–3456.
Keller, E. F. and E. A. Lloyd (eds.). 1998. *Keywords in Evolutionary Biology*. Harvard University Press, Cambridge, MA.
Krebs, C. J. 1985. *Ecology: the Experimental Analysis of Distribution and Abundance*, 3rd ed. Harper and Roe, New York.
Lamotte, M. 1959. Polymorphism of natural populations of *Cepaea nemoralis*. *Cold Spring Harb. Symp. Q. B.* 24: 65–84.
Lane, T. R. 1976. *Life, the Individual, the Species*. C. V. Mosby Company, St. Louis.
Levins, R. 1969. Some demographic and genetic consequences of environmental heterogeneity for biological control. *Bull. Entomol. Soc. Am.* 15: 237–240.
Levins, R. 1970. Extinction. In M. Gerstenhaber (ed.), *Some Mathematical Problems in Biology*, pp. 77–107. American Mathematical Society, Providence.
Millstein, R. L. 2008. Distinguishing drift and selection empirically: 'The Great Snail Debate' of the 1950s. *J. Hist. Biol.* 41: 339–367.
Millstein, R. L. 2009. Populations as individuals. *Biological Theory: Integrating Development, Evolution and Cognition*. Special Issue: *The Edges and Boundaries of Biological Objects*, M. Haber and J. Odenbaugh (guest eds.). 4: 267–273.
Mishler, B. D. and R. N. Brandon. 1987. Individuality, pluralism, and the phylogenetic species concept. *Biol. Philos.* 2: 397–414.
Nicholson, A. J. and V. A. Bailey. 1935. The balance of animal populations. *Proc. Zool. Soc. Lond.* 3: 551–598.
Orians, G. H. 1973. *The Study of Life*, 2nd ed. Allyn and Bacon, Inc, Boston.
Pearson, K. 1904. Mathematical contribution to the theory of evolution —XII. On a generalised theory of alternative inheritance, with special reference to Mendel's laws. *Philos. Trans. Roy. Soc. Lond. A* 203: 53–86.
Pfeifer, M. A., K. Henle, and J. Settele. 2007. Populations with explicit borders in space and time: Concept, terminology, and estimation of characteristic parameters. *Acta Biotheor.* 55: 305–316.
Poulton, E. B. 1903. What is a species? *Proc. Entomol. Soc. Lond.* 1903: lxxvii–cxvi.
Provine, W. B. 1986. *Sewall Wright and Evolutionary Biology*. University of Chicago Press, Chicago.
Purves, W. K. and G. H. Orians. 1983. *Life: The Science of Biology*. Sinauer Associates, Sunderland, MA.

Rank, N. E. 1992. A hierarchical analysis of genetic differentiation in a montane leaf beetle *Chrysomela aeneicollis* (Coleoptera: Chrysomelidae). *Evolution* 46: 1097–1111.

Rank, N. E. and E. P. Dahlhoff. 2002. Allele frequency shifts in response to climate change and physiological consequences of allozyme variation in a montane insect. *Evolution* 56: 2278–2289.

Schemske, D. W. and P. Bierzychudek. 2007. Spatial differentiation for flower color in the desert annual *Linanthus parryae*: Was Wright right? *Evolution* 61: 2528–2543.

Shavit, A. 2005. The notion of "group" and tests of group selection. *Philos. Sci.* 72: 1052–1063.

Simon, H. A. 2002. Near decomposability and the speed of evolution. *Ind. Corp. Change* 11: 587–599.

Slatkin, M. 1987. Gene flow and the geographic structure of natural populations. *Science* 236: 787–792.

Templeton, A. 1989. The meaning of species and speciation: A genetic perspective. In D. Otte and J. A. Endler (eds.), *Speciation and Its Consequences,* pp. 3–27. Sinauer Associates, Sunderland, MA.

Waples, R. S., and O. Gaggiotti. 2006. What is a population? An empirical evaluation of some genetic methods for identifying the number of gene pools and their degree of connectivity. *Mol. Ecol.* 15: 1419–1439.

Wells, J. V. and M. E. Richmond. 1995. Populations, metapopulations, and species populations: What are they and who should care? *Wildlife Soc. B.* 23: 458–462.

Winsor, M. P. 2000. Species, demes, and the omega taxonomy: Gilmour and the new systematics. *Biol. Philos.* 15: 349–388.

Wright, S. 1931a. Evolution in Mendelian populations. *Genetics* 16: 97–159.

Wright, S. 1931b. The roles of mutation, inbreeding, crossbreeding and selection in evolution. *Proc. Sixth Intern. Congr. Genet.,* pp. 356–366.

Wright, S. 1940. Breeding structure of populations in relation to speciation. *Am. Nat.* 74: 232–248.

Wright, S. 1943a. Isolation by distance. *Genetics* 28: 114–138.

Wright, S. 1943b. An analysis of local variability of flower color in *Linanthus parryae. Genetics* 28: 139–156.

Chapter 4

Evolutionary Genetics: Progress and Challenges

Jianzhi Zhang

Genetics plays a central role in evolutionary biology, because only heritable traits can evolve. Evolution by natural selection can be succinctly described as a consequence of heritable fitness variation among individuals of a population. Charles Darwin clearly understood that a mechanism of inheritance and a source of heritable variation were necessary components of a complete theory of evolution, but both were unknown to him. Although he thought that much heritable variation arises without reference to need, he also thought that traits acquired during the lifetime of an organism could be passed on to the next generation and used this idea of Larmarckian inheritance to formulate his theory. It is of significant historical interest that the revolutionary experiments that eventually led to the revelation of the true mechanism of inheritance had already been continuing for 3 years when Darwin published *The Origin of Species*. Gregor Mendel performed his famous series of pea cross experiments between 1856 and 1863. In 1865, Mendel read his paper entitled, "Experiments On Plant Hybridization," at two meetings of the Natural History Society of Brno. When Mendel published his work in the *Proceedings of the Brno Natural History Society* (1866), he described particulate inheritance in the form of three basic laws: segregation, independent assortment, and dominance. Unfortunately, Mendel's work was unrecognized by the scientific community for the next 34 years, until it was rediscovered in 1900 by Hugo de Vries, Carl Correns, and Erich von Tschermak. The three Mendelian laws of inheritance were later integrated with the chromosomal theory of inheritance, which was developed by Thomas Morgan and colleagues in the 1910s, and formed the core of classical genetics.

It is interesting to ask why Darwin did not know about Mendel's work. First, there is no evidence that Darwin subscribed to the *Proceedings of the Brno Natural History Society*. Second, there were apparently only 11 published references to Mendel's name before 1900, but at least 3 of these references were accessible to Darwin. One of them was in the *Royal Society's*

Catalogue of Scientific Papers (1864–1873), published in 1879, 3 years before Darwin's death. As a member of the Royal Society, Darwin likely had access to this catalogue, but the catalogue gave no indication of the content of Mendel's paper. The second reference was in W. O. Focke's *Die Pflanzen-Mischlinge* (1881), of which Darwin had a copy. Focke apparently did not understand the importance of Mendel's discoveries and placed Mendel among many other plant hybridizers. The pages in which reference was briefly made to Mendel's experiments remain uncut in Darwin's copy of Focke's book. The third reference was in H. Hoffmann's *Untersuchungen zur Bestimmung des Werthes von Species und Varietät* (1869). Like Focke, Hoffmann did not recognize anything exceptional in Mendel's results. Although Darwin annotated this book and cited it in *The Effects of Cross and Self Fertilization* in 1876, he neither referred to Mendel in that work nor annotated the references to Mendel. Finally, given that Mendel's seminal work was completely ignored by the entire scientific community for so long, it is quite likely that Darwin would not have recognized its importance even if he had read it. Interested readers may find this information and more in Kritsky (1973) and Sclater (2006).

Did Mendel know about Darwin's theory and the importance of his own work to that theory? If so, why did he not contact Darwin? There is clear evidence that Mendel read *The Origin of Species* and knew about Darwin's theory. However, Mendel appeared to believe evolution by hybridization rather than evolution by natural selection (Blower 1989). Of course, he would not think that his findings provided a basis for an evolutionary theory that he regarded as incorrect.

In the first half of the twentieth century, Darwinian evolution by natural selection was synthesized with Mendelian genetics by the joint efforts of a wide array of talented evolutionary biologists, such as R. A. Fisher, J. B. S. Haldane, Sewall Wright, Theodosius Dobzhansky, Ernst Mayr, George Simpson, and G. Ledyard Stebbins (see Futuyma, Chapter 1). By the end of the 1950s, the result of this modern synthesis (often called neo-Darwinism) was widely accepted among evolutionary biologists. The basic tenets of neo-Darwinism are that:

1. evolution occurs gradually through mutation, selection, and drift, and it is explained by population genetics theories;
2. discontinuities between species are explained as originating gradually through geographical separation, divergence, and extinction, rather than saltation;
3. natural selection is the primary force driving evolutionary change;
4. genetic variation within populations is abundant and is a key contributor to evolution;
5. microevolution can be extrapolated to explain macroevolution.

In the past half century, genetics has seen extraordinary development that has led to three revolutions, two technological and one conceptual, in evolutionary genetics. Although the basic tenets of neo-Darwinism have not been significantly altered (with one exception discussed later), the understanding of evolution has been drastically widened and deepened. In this article, I review these three revolutions, summarize eight rules of evolutionary genetics that emerged from these revolutions, and discuss five major questions that I believe will be largely answered in the next few decades. I note that evolutionary genetics is an enormous field and my review describes only those subjects that I am familiar with and believe to be important. It is thus, my personal view of the progress and challenges of evolutionary genetics.

Three Revolutions in the Last 50 Years

The Molecular Revolution

The molecular basis of inheritance began to be unraveled in 1944 when Avery and colleagues first demonstrated that DNA is the material of which genes are made. In 1953, Watson and Crick revealed the structure of DNA. In 1958, Crick proposed the central dogma that describes the fundamental information flow from DNA to RNA to protein. Nirenberg and others cracked the genetic code in the early 1960s. By then, the fundamental principles of molecular biology had been established.

The great influence of molecular biology on evolutionary biology started in the second half of the 1960s on three fronts almost simultaneously. First, gel electrophoresis was introduced to detect protein polymorphisms within and between populations (Harris 1966; Hubby and Lewontin 1966). The technique was soon adopted by many population and evolutionary geneticists to survey protein polymorphisms in a great many species, and surprisingly large amounts of molecular polymorphisms were discovered. This empirical research, coupled with theoretical developments, led to the formation of a new field, known as molecular population genetics (Lewontin 1974; Nei 1975). Molecular population genetics gradually moved into the DNA era, as Sanger DNA sequencing and polymerase chain reaction became routine in the late 1980s and early 1990s. Molecular population genetics prompted the study of microevolutionary processes at the molecular genetic level.

Second, amino acid sequences of cytochrome c proteins were used to infer the evolutionary relationships of various available species and the deduced phylogeny was found to be largely consistent with traditional morphology-based phylogenies (Fitch and Margoliash 1967). This success offered the promise that evolutionary relationships among organisms could be inferred from the genetic material itself, which most faithfully records their history. In the subsequent decades, molecular phylogenetic methodologies were developed and thousands of molecular phylogenies were constructed (Hillis et al. 1996; Nei and Kumar 2000; Felsenstein 2004). Today, systematists are mainly using DNA sequences to assemble the Tree of Life that may ultimately include all species on Earth (Cracraft and Donoghue 2004; see Hillis, Chapter 16).

Third, the molecular revolution directly led to the development of the neutral theory of molecular evolution (Kimura 1983), first proposed by Motoo Kimura in 1968 and by King and Jukes in 1969. I will discuss this theory in more detail in a later section.

The molecular revolution also provided tools that allow the identification of the molecular genetic basis of phenotypic evolution, such as the evolution of the body plan, skin color, and visual and olfactory sensitivities, although this fourth influence was unnoticed by most evolutionists until the early 1990s. A good testimony of this influence is the advent of evolutionary developmental (evo-devo) biology, which focuses on the molecular basis of developmental differences within and between species (Raff 1996; Carroll et al. 2001) and is now one of the most rapidly growing fields in evolutionary biology (see Wray, Chapter 9).

The Genomic Revolution

The first complete genome sequence of any free-living organism was determined in 1995 from the pathogenic bacterium *Haemophilus influenzae* (Fleischmann et al. 1995). As of March 2010, 1124 complete genome sequences have been published and at least 5000 genomes are in the process of being sequenced (www.genomesonline.org). With the development of the so-called next-generation sequencing technologies (Mardis 2008; Shendure and Ji 2008), many more genomes are expected to be sequenced (Genome 10K Community of Scientists 2009). In addition to genome sequencing, genomics has also provided an unprecedented amount of functional data at various levels of biological organizations (Gibson and Muse 2004; Pevsner 2009). These include genome-scale data on the concentrations of messenger RNAs, concentrations of proteins, protein half-lives, the stochastic noise of gene expression, genetic interactions, protein–protein interactions, stable protein complexes, protein subcellular localizations, phenotypes and fitness effects of gene deletions, chromatin structures, and various epigenetic modifications.

Genomic data provide substantively more genetic markers for molecular phylogenetics. While multigene phylogenies were still uncommon in systematic studies in the mid-1990s, there are virtually no single-gene phylogenies in publications today unless the purpose is to examine the evolution of the gene itself. Use of the entire genome or all genes in a genome for molecular phylogenetics is no longer rare, and the phylogenomic era has undoubtedly arrived (Delsuc et al. 2005). Genomic data also offer new types of genetic markers that may outperform the single-nucleotide substitutions that have been the primary source of phylogenetic signals in molecular systematics (Rokas and Holland 2000). These new markers include insertions/deletions, transpositions, gene duplications, and other rare genomic events. These events can be sufficiently unique that one is unlikely to encounter homoplasy.

Genomic data allow the identification of the relative importance of alternative evolutionary mechanisms that are known from individual case studies. For example, these data have been used to address whether most stably retained duplicate genes have undergone neofunctionalization, subfunctionalization, or both (He and Zhang 2005). Another example is the relative frequency of intron gains and losses in evolution (Roy and Gilbert 2006). Such questions cannot be tackled without genome-wide data.

Genome data also revealed new mechanisms of evolution and drastically changed the view of the prevalence of certain mechanisms, such as the high incidence of duplication and deletion of genomic segments, commonly resulting in intraspecific variation in copy number (Zhang 2007). Contrary to what was once believed, genomic data also showed that transposons often contribute to the origin of new genes (Nekrutenko and Li 2001).

An Intellectual Revolution: The Neutral Theory of Molecular Evolution

One of the main tenets of the evolutionary synthesis was that natural selection is the primary force of evolution. This assertion was seriously challenged by the neutral theory of molecular evolution, which, in my view, is the only conceptual revolution in the last half-century in evolutionary genetics (and possibly all evolutionary biology). The neutral theory (Kimura 1983) claims that (1) most nucleotide differences between species result from random fixations of neutral mutations, and (2) most intraspecific polymorphisms are also neutral. Based on the evolutionary analysis of three proteins (hemoglobin, cytochrome c, and triosephosphate isomerase), Kimura reported in his classic 1968 *Nature* paper that the rate of nucleotide substitution per generation per mammalian genome is approximately two, which is 600 times what Haldane (1957) considered to be the upper-limit for the rate of adaptive evolution imposed by the cost of natural selection (Kimura 1968). To Kimura, the only solution to this contradiction is that most nucleotide substitutions are not adaptive but neutral.

Fifteen months after Kimura's paper appeared, King and Jukes (1969) published a *Science* article entitled "Non-Darwinian evolution." In this provocative paper, the authors used the available knowledge of molecular biology to argue for the prevalence of neutral genetic changes in evolution. For example, they inferred that nucleotide substitutions are faster at third codon positions than at first and second codon positions. Because a much larger fraction of nucleotide changes at third codon positions than at first and second positions are silent, their result strongly supported the neutral theory. The neutral theory was heatedly debated for most of the following two decades. If it is true, as is widely believed, that most nucleotides in the genomes of complex eukaryotes do not code for useful product and are apparently non-functional, then many or most nucleotide substitutions and polymorphisms in genomes must be unambiguously neutral. Most of the current uncertainty about whether or not the neutral theory is correct concerns the functional part of the genome, including protein-coding, RNA-coding, and regulatory sequences.

It turned out that Kimura made a mistake in his calculation; he assumed that 100% of the mammalian genome codes for proteins, while the best contemporary estimate is ~1.5%. If he had known this, he would have calculated a per genome amino acid substitution rate only nine times Haldane's upper limit, but still consistent with the view that most amino acids substitutions must be neutral. Given the uncertainties associated with the assumed upper limit and the extrapolation of the substitution rate per genome from only three proteins, this difference probably would not have been considered to be a big surprise. This estimate can be further refined with new genome data. A comparison of the mouse and rat orthologous proteins shows that the median sequence identity is 95% (Gibbs et al. 2004). Assuming the two species diverged 18 million years ago (Gibbs et al. 2004), the mean generation time is 1 year, the mean protein size is 450 amino acids, and there are 20,000 protein-coding genes in the mammalian genome, I calculated that the amino acid substitution rate per generation, per genome is 0.0125—about four times Haldane's upper limit. Thus, the majority of amino acid substitutions are still predicted to be neutral, and the observed rate of amino acid substitution would not be incompatible with Haldane's estimate, if a quarter of amino acid substitutions are adaptive.

While the exact fraction of adaptive nucleotide changes in the functional part of the genome is still being debated, there is no doubt that many genetic differences among organisms in the functional part of the genome are neutral or nearly neutral. Today, the neutral theory usually serves as a null hypothesis and adaptation is proposed only when the expectations of neutrality are rejected by significant evidence. This practice strongly contrasts with the situation 50 years ago, when adaptation was the default explanation of almost all biological observations. Although Kimura (1983) expressly limited his neutral theory to molecular sequences and not to phenotypic traits, some authors (Gould and Lewontin 1979) have cited neutrality as a possible alternative to adaptive interpretation of some phenotypic change as well.

Emerging Rules in Evolutionary Genetics

Except for the hypothesis that natural selection is the primary force of evolutionary change, all basic tenets of neo-Darwinism are still intact. Within a framework expanded to include the neutral theory, substantial progress has been made in evolutionary genetics in the last half century. In particular, the molecular and genomic revolutions have provided such an unprecedented amount of data that there is now a much more solid and detailed understanding of evolutionary processes than before (see Kolaczkowski and Kern, Chapter 6). Here, I describe eight emerging rules revealed from the last 50 years of evolutionary genetic studies. Although substantial progress has been made in evolutionary developmental genetics and speciation genetics, I will not discuss these topics, because they are covered in other chapters of the book (see Wray, Chapter 9; Harrison, Chapter 13).

Rule 1: Life Is Fundamentally Conserved

One of the most important observations from molecular genetic and genomic studies is that life is fundamentally conserved at many levels. First, all cellular organisms use DNA as their genetic material. Second, the genetic code is largely conserved across all species, although rare variants do exist (Osawa 1995). Third, the most basic molecular cellular processes, such as DNA replication, transcription, and translation, are largely the same across all species. In addition, a large part of central metabolism is conserved in most organisms, the same signaling pathways exist in many divergent species, and many genes are shared across the three domains of life (see Lane, Commentary 4). Given that the last common ancestor of all extant species lived more than 3 billion years ago (see Lazcano, Chapter 14), the observed level of conservation is astonishing. This extreme conservation suggests that there may be only one or a very small number of ways to construct life and/or that historical contingency plays such a dominant role in evolution that descendants cannot deviate too far from the common ancestor, even with billions of years of modifications.

The high conservation leads to the prediction that purifying selection is the dominant form of natural selection. This prediction is strongly supported by comparative genomic data, which revealed significantly lower rates of nucleotide substitutions that alter the encoded amino acids (nonsynonymous changes) compared to the rates of nucleotide substitutions that do not alter the encoded amino acids (synonymous changes) for the vast majority of genes in a genome (Waterston et al. 2002).

The high conservation across all life forms, especially at molecular and cellular levels, has several significant implications. First, despite the huge diversity of life at the organismal and phenotypic levels, there are universal rules of evolution at the genetic and genomic levels; furthermore, these rules may be discovered by studying a relatively small number of species. Second, the overall conservation in evolution makes phenotypic variations among species particularly interesting, and evolutionary geneticists have been studying both the proximate causes (i.e., molecular genetic and developmental basis) and ultimate causes (i.e., selection or drift) of such variations in the last few decades. Third, many aspects of human biology may be studied using model organisms, such as the mouse *Mus musculus*, fruit fly *Drosophila melanogaster*, nematode *Caenorhabditis elegans*, flowering plant

Arabidopsis thaliana, budding yeast *Saccharomyces cerevisiae*, and bacterium *Escherichia coli*. Thus, funding agencies such as the U.S. National Institutes of Health, whose mission is to improve people's health and save lives, are willing to invest in the study of model organisms.

Rule 2: Chance Plays an Important Role in Evolution

Stochasticity plays roles at many levels of biological organization. At the level of individual organisms, survival, reproduction, and death often have large random (nonselective) components. Molecular cellular processes, such as transcription initiation, protein degradation, protein–protein interaction, and metabolism, depend on biochemical reactions that occur upon random encounters of sometimes small numbers of molecules. For example, there is a high level of intrinsic noise in gene expression revealed by variation in protein expression among isogenic cells under the same condition. This noise arises from stochastic events in processes such as transcription initiation, mRNA degradation, translation initiation, and protein degradation (Raser and O'Shea 2005; Raj and van Oudenaarden 2008). Mutation and recombination, the ultimate sources of genetic variation, are also random events. Finally, random segregation and independent assortment, two of the three Mendelian laws of inheritance, involve stochasticity.

The most fundamental impact of chance on evolution is through random genetic drift, the random sampling of alleles during the reproduction of a finite population. Genetic drift leads to random loss and fixation of alleles that are independent of the relative fitness of those alleles. The neutral theory of molecular evolution (Kimura 1983) asserts that genetic drift accounts for the majority of nucleotide substitutions in evolution. Ample evidence from molecular population genetics and molecular evolution studies supports this view. Because the neutral theory does not deny the occurrence of rare positive Darwinian selection for advantageous alleles in evolution, the occasional identification of positive selection at the molecular genetic level does not reject the theory. In the last decade, a number of authors have tried to quantify the fraction of amino acid substitutions that are adaptive, using large amounts of population genetic data. The results, however, are ambiguous and hard to interpret, in my view. Many, if not all, studies of *Drosophila* indicated that a large fraction (from 30% to 95%) of amino acid substitutions are adaptive (Fay et al. 2002; Smith and Eyre-Walker 2002; Eyre-Walker 2006; Sawyer et al. 2007; Shapiro et al. 2007; Sella et al. 2009). The fraction, however, is found to be very small in humans (Zhang and Li 2005; Eyre-Walker 2006) and effectively zero in the budding yeast (Doniger et al. 2008; Liti et al. 2009). Because genetic drift has a greater impact on evolution in small populations than in large populations (Ohta 1992), one would predict a higher contribution of positive selection in species with larger populations. However, the population size of *Drosophila* is somewhere between that of humans and yeast. So, the results thus far do not make sense. It is unclear whether this inconsistency is due to any peculiarity of the three species, such that they do not respectively represent average species of comparable sizes.

Relative to the strength of selection, genetic drift is expected to play a more important role in small populations associated with large and complex organisms (Ohta 1992), which has led to a nearly neutral explanation of the origin of genome architecture (Lynch 2007). This explanation can be viewed as an extension of the neutral theory (Kimura 1983) and the nearly neutral theory (Ohta 1992) from molecular evolution to genomic evolution. Lynch (2007) argues that complex genomic features, such as the existence of mobile elements, gene families, split genes, and alternative splicing, may have passively originated through non-adaptive processes in small populations. When they first appeared, the features might have been very slightly deleterious, but persisted in organisms with small populations because the purifying selection against them was sufficiently weak. Through long-term evolution, these features became established and may have been modified to perform useful functions, such as the role of alternative splicing in generating multiple proteins of different functions from one gene. This novel hypothesis is certain to stimulate debate and empirical study.

Rule 3: Genomes of Complex Organisms Contain a Lot of Junk DNA

It must have been a big surprise to those who believed that organisms are perfectly or nearly perfectly adapted to their environments to learn that the genomes of many complex organisms, such as vertebrates and flowering plants, contain a large fraction of so-called junk DNA that apparently has no function. For example, only approximately 1.5% of the human genome codes proteins and only about 5% appears to be constrained to various degrees (Waterston et al. 2002). This junk DNA is largely composed of repetitive sequences that originated from transposable elements. It is likely that junk DNA has a very small fitness cost to the host and therefore, is not effectively removed by natural selection, especially in more complex organisms that often have small populations (Lynch 2007).

Is there treasure in the junk DNA? The U. S. National Human Genome Research Institute funded a large project, named The Encyclopedia of DNA Elements (ENCODE), to build a comprehensive parts list of the functional elements of the human genome. The initial analysis of 1% of the human genome provided some surprises (The ENCODE Project Consortium 2007). For example, it was found that the majority of the bases in the human genome are transcribed, although it remains unclear whether the transcription reflects biological function or simply leaky expression. It was also found that many functional elements defined by DNA-protein binding are still seemingly unconstrained across mammalian evolution, suggesting that those regions may not have physiological functions or their loss imposes no fitness reduction. Alternatively, the finding could indicate the existence of a large number of species-specific functional elements in a genome. It is interesting to note that some purportedly functionless elements of the genome may be recruited into protein sequences. For example, it was discovered that about 4% of human proteins contain sequences that originated from transposable elements, which initially resided in introns but were later recruited to become new exons (Nekrutenko and Li 2001). There is also evidence of rare *de novo* gene origination from non-coding sequences (Levine et al. 2006; Chen et al. 2007).

Rule 4: Gene Number Does Not Predict Organismal Complexity

Analogous to the C-value paradox (Gregory 2001), which describes the lack of relationship between organismal complexity and genome size, there is no simple relation between organismal complexity and gene number. There are surprisingly few genes in the human genome (~20,000), compared to *E. coli* (4400), yeast (6000), fruit fly (13,600), nematode (19,000), sea urchin (23,500), *Arabidopsis* (27,500), and rice (41,000). While the C-value paradox has been attributed to the variation in the amount of junk DNA in different genomes, the cause of the lack of correlation between organismal complexity and gene number is yet to be determined. This being said, I emphasize the difficulty in measuring organismal complexity. If we use the number of recognizably different types of cells in an organism as a proxy for organismal complexity, vertebrates are more complex than triploblastic invertebrates, which in turn are more complex than vascular plants (Futuyma 1998). But, gene number is higher in vascular plants than in vertebrates.

Several potential mechanisms could compensate for the low gene number in highly complex organisms, such as vertebrates. First, alternative RNA splicing, prevalent in multicellular organisms, substantially increases the number of different proteins in an organism. It is estimated that over 80% of human genes are alternatively spliced (Matlin et al. 2005). However, a comparison among human, mouse, rat, cow, *D. melanogaster*, *C. elegans*, and *A. thaliana* found comparable frequencies of alternatively spliced genes across species (Brett et al. 2002). It was subsequently suggested that these results were an artifact of the different coverage of expressed sequence tags for the various organisms; when this confounding factor was removed, vertebrates show higher frequencies of alternative splicing than invertebrates (Kim et al. 2007). Nevertheless, because not all spliced forms are functional or functionally distinct, the above finding does not unambiguously demonstrate that vertebrates have substantially more functionally distinct proteins than invertebrates. The key issue in the study of alternative splicing is to estimate the proportion of alternatively spliced forms that are functionally distinct and physiologically useful.

Second, because the potential number of interactions between genes or proteins is much greater than the actual number of genes or proteins, one could hypothesize that organismal complexity depends more on the number of molecular interactions than the number of genes or proteins. Genome-wide protein–protein interactions have been surveyed in a number of model organisms, including yeast, fruit flies, and humans (Beyer et al. 2007). Because of the incompleteness of the data and different biases in different datasets, it is hard to tell at this stage whether the number of protein interactions in a species correlates with the organismal complexity. Genetic interactions, mainly in the form of synthetic lethality or illness, have been examined by simultaneous knock-out or knock-down of pairs of genes in a few model organisms, such as budding yeast, fission yeast, and nematodes (Tong et al. 2004; Boone et al. 2007; Roguev et al. 2008; Tischler et al. 2008). The current data are still far from complete for across-species comparisons.

Finally, gene expression regulation and post-translational modification can potentially increase organismal complexity (see Wray, Chapter 9). However, there is still no good empirical data to test this hypothesis rigorously. For example, no evidence supports the proposal that gene regulation is more complex in vertebrates than in invertebrates and plants. Thus, the molecular genetic basis of organismal complexity remains largely unexplained.

Rule 5: Horizontal Gene Transfers Are Prevalent (at Least in Prokaryotes)

The Origin of Species contained only one figure, which depicted a hypothetical phylogenetic tree of 15 extant taxa. In this tree, genetic information was transmitted vertically from parents to offspring and there was no horizontal transmission of genetic information among different evolutionary lineages. We now know that horizontal gene transfers (HGTs) occur frequently among prokaryotes (Koonin et al. 2001; Gogarten et al. 2009; see Lane, Commentary 4). There is also ample evidence that they occur among eukaryotes and between prokaryotes and eukaryotes (Keeling and Palmer 2008; Gogarten et al. 2009), although the rate of occurrence appears much lower in eukaryotes. HGTs occur through three main mechanisms: transformation, conjugation, and transduction. Transformation refers to the phenomenon that cells from certain species can take up free DNA from their environments. Conjugation is the process by which a living cell transfers genetic material to another cell through the formation of a tube-like structure (i.e., pilus) between cells. Transduction refers to DNA movement from one cell to another by a virus.

HGT was first reported 50 years ago when antibiotic-resistant genes were found to be transferred across bacterial species (Ochiai et al. 1959; Akiba et al. 1960). However, it was not until the late 1990s, when multiple prokaryotic genomes were sequenced and compared, that the prevalence of HGTs in evolution became appreciated. For example, it was reported in one study that 24% of the protein-coding genes in *Thermotoga maritima*, a thermophilic eubacterium, are most similar to archaeal genes (Nelson et al. 1999). Because criteria for identifying probable horizontal gene transfer rely on unusual feature(s) of subsets of genes that distinguish them from the bulk of genes in the genome (Koonin et al. 2001), indications of HGTs remain probabilistic and thus can sometimes be controversial. The current debate centers on the quantitative assessment of the pervasiveness and rate of HGTs (Doolittle 1999; Gogarten et al. 2002; Daubin et al. 2003). Some researchers believe that HGTs are so pervasive and frequent that the Tree of Life (at least in the prokaryotic part) becomes a network of life from which it is neither meaningful nor feasible to reconstruct species phylogenies (Doolittle 1999; Gogarten et al. 2002). More fundamentally, if genes were freely transferred across species, the species concept would collapse. Other researchers believe that a sizeable fraction of genes in the genome are incapable of HGTs, and these genes would allow the reconstruction of a species phylogeny (Daubin et al. 2003; Ciccarelli et al. 2006). For instance, it was proposed in the so-called complexity hypothesis that informational genes, which function in transcription, translation, and related processes, are horizontally transferred with a much lower rate than housekeeping operational genes, because the translational and transcriptional apparatuses are large and complex systems. In this case, a foreign gene is unlikely to be compatible in a system made of native parts (Jain et al. 1999). This hypothesis has received empirical support. For example, in an analysis of 191 species with complete genome sequences, 31 genes that are relatively immune to HGTs were found and all of them are involved in translation (Ciccarelli et al. 2006). A recent study analyzed attempted experimental movement of 246,045 genes from 79 prokaryotic genomes into *E. coli* and identified genes that consistently fail to transfer (Sorek et al. 2007). Interestingly, ribosomal proteins dominate the list of untransferable genes, and toxicity to the host is the primary cause of transfer inhibition (Sorek et al. 2007). Although different genes have different rates of HGT, the question remains whether there is a sufficiently large set of HGT-resistant genes such that a species phylogeny of prokaryotes is both meaningful and reconstructable.

Rule 6: Gene Duplication Is the Primary Source of New Genes

Although many genes involved in the most fundamental molecular cellular processes, such as protein synthesis and DNA replication, are shared among all species (Mushegian and Koonin 1996), there are probably no two species that have exactly the same set of genes. Variation in gene content is a major source of biodiversity. How new genes with novel functions originate has been a fascinating subject to many researchers, and several molecular mechanisms have been proposed. First, exon shuffling combines existing exons between different genes and generates hybrid genes with multiple exons (Gilbert 1978, 1987; Patthy 1995, 1999). The resulting protein thus exhibits additional functions conferred by the newly acquired exons, and the interactions between the amino acids encoded by different exons may also lead to entirely new protein functions. The prevalence of multidomain proteins in high eukaryotes suggests the important contribution of exon shuffling (Patthy 1999). Second, introns (and other noncoding sequences) may, under certain circumstances, be converted to protein-coding sequences (Nekrutenko and Li 2001). Similarly, alternative reading frames or antisense strands of functional genes may sometimes be used as the genetic material for a new gene (Yomo et al. 1992; Golding et al. 1994). However, such events are rare, because of the low probability of the occurrence of long open reading frames from random DNA sequences. The third mechanism is gene sharing. Best known in lens crystallin genes, gene sharing allows one gene to adopt an entirely different function without losing its primary function (Piatigorsky 2007). For instance, in birds and crocodiles, lactate dehydrogenase appears as an enzyme as well as a structural protein in the lens. In theory, a new gene may also arise through a *de novo* process. Although uncommon, several such examples have been reported (Chen et al. 1997; Levine et al. 2006; Chen et al. 2007). Horizontal gene transfer brings new genes from other species to a species, but this process does not generate novel genes. Except for exon shuffling, the other mechanisms seem to have minimal contributions to the origin of new genes with novel functions. However, even exon shuffling cannot account for the high rate of gene origination in evolution. In fact, most new genes were generated through gene duplication.

In 1936, Bridges reported one of the earliest observations of gene duplication in the doubling of a chromosomal band in a *D. melanogaster* mutant that exhibited extreme reduction in eye size (Bridges 1936). Evolutionary biologists quickly realized the potential of gene duplication as a mechanism of evolution of new genes (Stephens 1951). Ohno's seminal book *Evolution by Gene Duplication* (Ohno 1970) further popularized this idea among biologists. However, it was not until the late 1990s, when numerous genomes were sequenced and analyzed, that the widespread prevalence of gene duplication became clear. Virtually every genome sequenced thus far contains a high fraction of duplicate genes, and because ancient duplicate genes are difficult to recognize through sequence comparison, the true percentage in a genome is likely much higher (Zhang 2003). Gene duplication may occur through unequal crossover, retroposition, chromosomal nondisjunction, or polyploidization. These mechanisms are responsible for generating segmental duplication, retroduplication, chromosomal duplication, and genome duplication, respectively (Zhang 2003).

Retroposition was initially thought to create only pseudogenes, because in retroposition the message RNA of a gene is reverse-transcribed into complementary DNA, which is then inserted into the genome randomly. As such, the promoter of the gene is not duplicated along with the coding region of the gene; consequently, the retroduplicate is not expressed. However, recent studies showed that a small fraction of retroduplicates are by chance inserted into introns of existing genes or to genomic regions containing promoters. In such cases, a retroduplicate may become part of a functional gene or a new gene (Long et al. 2003).

A duplicated gene may experience several potential fates even when it is functional and is fixed in a population. The most common fate is pseudogenization, which occurs when the duplicate copy is functionally redundant and therefore is not subject to any selective constraint. Several mechanisms may allow the long-term retention of a duplicate gene. First, increased dosage of certain genes (e.g., ribosomal RNA genes and histone genes) can be beneficial and lead to the retention of duplicate genes even without any change in function or expression (Zhang 2003). Second, the ancestral gene may have multiple functions that are subdivided in the daughter genes, and each of them fixes mutations (perhaps, by genetic drift) that disable some of the ancestral functions (Force et al. 1999; Lynch and Force 2000b). The joint levels of expression and patterns of activity of the two daughter genes are equivalent to those of the single ancestral gene. Consequently, both daughter genes may be stably retained. Third, a duplicate gene may neofunctionalize by acquiring a new function or a new expression pattern, such that the fitness of the organism is enhanced (Ohno 1970). Fourth, it is also possible that a duplicate gene pair experiences subfunctionalization quickly after the duplication, which permits its long-term retention and allows gradual acquisitions of new functions in evolution (He and Zhang 2005). There is no consensus on the relative contributions of these mechanisms that underlie stable retention and evolution of duplicate genes, in spite of the fact that the subject has been studied extensively. The relative roles of natural selection and genetic drift in the fixation and retention of duplicate genes is also contentious (Zhang et al. 1998). While drift is likely important in subfunctionalization, positive selection must be involved in the dosage benefit mechanism and is probable in neofunctionalization. While evolutionary geneticists are interested in the mechanism of duplicate gene retention, molecular biologists tend to focus on the functional similarities and differences among duplicate genes. These two issues are intimately related, as is clear from the above explanation of the different functional alterations invoked in the evolutionary mechanisms.

The most important contribution of gene duplication to evolution is the provision of new genetic material, upon which mutation, drift, and selection act to create either specialized or new gene function. Some of the most exquisite biological responses, such as the adaptive immune system and the olfactory and taste chemosensory systems in vertebrates, rely extensively on duplicate genes that perform similar but distinct functions (Nei et al. 1997; Shi and Zhang 2009). Recent genomic analysis in mutation-accumulation lines of yeasts showed that the spontaneous mutation rate of gene duplication is high (Lynch et al. 2008). Comparative genomics also reveals a high rate of fixed duplications (Lynch 2007). Thus, gene duplication must have contributed greatly to the genetic and phenotypic differences between different evolutionary lineages. Gene and genome duplication may have also directly contributed to speciation through the divergent resolution process, in which the random loss of a redundant gene copy in two populations could result in the missing of both copies in the gametes of their hybrids and thus reproductive isolation by hybrid sterility (Werth and Windham 1991; Lynch and Force 2000a).

Rule 7: Changes in Protein Function and Gene Expression Are Both Important in Phenotypic Evolution

For historical reasons, molecular evolutionary studies have focused more on evolutionary changes in protein sequence and function than changes in gene expression and its regulation. In 1975, King and Wilson reported that humans and chimpanzees have virtually identical protein sequences despite their large phenotypic differences. This observation prompted the authors to propose that changes in gene expression play a more important role in phenotypic evolution than changes in protein function (King and Wilson 1975). This hypothesis has been enormously influential to evolutionary biologists; many were convinced that gene expression changes are more important and have looked for both theoretical and empirical evidence for this hypothesis. We now know that between human and chimpanzee, there are on average about two amino acid differences per protein and more than 70% of proteins are non-identical (Chimpanzee Sequencing and Analysis Consortium 2005; Glazko et al. 2005). So, protein sequence differences between human and chimpanzee are numerous, which can potentially account for many of the phenotypic differences between the two species. Nonetheless, the role of gene expression changes in phenotypic evolution has been documented in many case studies (Wray 2007; Carroll 2008; Stern and Orgogozo 2009; see Wray, Chapter 9).

To answer the question of whether gene expression change is generally more important than protein function change, two research groups recently compiled cases of phenotypic evolution with known genetic mechanisms (Hoekstra and Coyne 2007; Stern and Orgogozo 2008). Although such meta-analyses are valuable in summarizing case studies and providing information about the overall empirical evidence at the present time, caution is needed because case studies are subject to ascertainment biases associated with preferences for certain methods, phenotypes, genes, and types of mutations. The empirical evidence shows that both gene expression change and protein function change are important genetic mechanisms of evolution. It is probably more productive to study whether these two types of genetic mechanisms are disproportionately used for different types of phenotypic evolution than to argue which mechanism is more important. For example, based on case studies and theoretical considerations that a distinction exists in the genetic basis of morphological and physiological evolution, it has been proposed that morphological evolution occurs mainly through gene expression changes, while physiological evolution occurs mainly through protein function changes (Carroll 2005). Because morphology and physiology are intimately connected, it may be difficult to clearly separate them. Nevertheless, one can imagine cases in which physiological changes do not require accompanying morphological changes and vice versa. For example, response to low oxygen levels by modification of the hemoglobin sequence does not involve morphological changes, and wing pigmentation differences in some insects probably do not involve physiological changes. Thus, if morphological and physiological traits are distinguishable, Carroll's hypothesis may be tested best by using genomic data rather than case studies. That is, one could study the genes in which mutations only affect morphological traits and genes in which mutations only affect physiological traits. A recent comparison between the two types of genes in their molecular function and evolutionary pattern lends support to Carroll's hypothesis (Liao et al. 2010).

An alternative approach to studying the genetic mechanisms of phenotypic evolution is experimental evolution (Garland and Rose 2009). Genomic technologies, including high-throughput next-generation sequencing, allow cheap, quick, and accurate determinations of the genome sequences and transcriptomes, including sequencing the starting and end strains from laboratory evolutionary experiments and identifying the mutations that are responsible for the phenotypic (e.g., fitness) changes. Interestingly, two studies, one in *Escherichia coli* (Herring et al. 2006) and the other in yeast (Gresham et al. 2008), showed the prevalence of protein function and copy-number changes in physiological adaptation yet virtually no gene expression changes by *cis*-regulatory sequence alteration. One caveat is that both *E. coli* and yeast contain only short *cis*-regulatory sequences and may not represent complex organisms. Another caveat in microbial experimental evolution is that morphological changes are much more difficult to study than physiological changes.

Rule 8: Intraspecific Genetic Polymorphisms Are Abundant and Largely Neutral

The molecular revolution in evolutionary genetics resulted in a large body of literature on the genetic polymorphisms in hundreds of species. Compared to what neo-Darwinists thought, intraspecific polymorphisms at the DNA and protein sequence levels are astonishingly high (Lewontin 1974; Lynch 2007). For example, even in humans, who have a rather small effective population size estimated at approximately 10^4, the mean nucleotide diversity is on the order of 0.1%, meaning that two randomly chosen alleles of the same gene differ by 1 out of 1000 bases. This level of allelic difference is about 10% of the genetic difference between humans and chimpanzees, although probably no layperson would believe that the human–chimp difference is only 10 times that between two humans.

The neutral theory predicts that, at the mutation-drift equilibrium, neutral nucleotide diversity is given as $\pi = 4N\mu$, in which N is the effective population size and μ is the neutral mutation rate per generation. The best support for the general statement that most nucleotide polymorphisms across the genome are neutral is the clear trend for associations of polymorphism with population size or its surrogates (e.g., organisms with small body sizes have larger populations) (Nei and Graur 1984). This view is so universally accepted that measures of genetic diversity are routinely used as indirect estimates of historical population size (Roman and Palumbi 2003). Furthermore, because the neutral mutation rate μ is higher at synonymous sites than at nonsynonymous sites, the neutral theory also predicts that π is higher at synonymous sites than at nonsynonymous sites, which has been repeatedly shown (Cargill et al. 1999; Moriyama and Powell 1996). These two observations certainly indicate that most intraspecific polymorphisms are best explained by the joint forces of mutation and drift acting on neutral mutations. This finding is expected because: (1) neither deleterious nor advantageous mutations contribute much to the level of intraspecific variations and (2) these alleles are either kept at sufficiently low frequencies or become fixed quickly. Nevertheless, while this global statement is true, there are dozens of case studies in which individual polymorphisms, when examined in detail, appear to be adaptive responses to current conditions (Vasemagi and Primmer 2005; Voight et al. 2006; Linnen et al. 2009; Rebeiz et al. 2009). Detection of selection-maintained polymorphisms has been enhanced by the development of many statistical approaches, such as those associated with coalescence theory (see Kolaczkowski and Kern, Chapter 6; Wakeley, Chapter 5). These individual cases aside, balancing selection advocated by the balance school led by Dobzhansky as a major cause of most genetic polymorphism (Lewontin 1974), does not appear to account for most molecular polymorphisms. Certainly, the form of long-term balancing selection that would lead to even trans-specific polymorphism has been documented in only relatively few genes (Hughes and Nei 1988; Clark and Kao 1991; Cho et al. 2006).

Major Unsolved Questions in Evolutionary Genetics

Although evolutionary genetics has seen rapid progress in the last 50 years, a number of major unsolved questions hamper a complete and accurate understanding of evolutionary processes. Below I describe some of these questions that I think can be largely solved or will at least see significant progress in the next few decades.

Question 1: How Can the Genetic Basis of Macroevolutionary Changes Be Found?

At least in principle, it is no longer challenging to identify the nucleotide substitutions that are responsible for the phenotypic differences between individuals of the same species or closely related species, if only one locus or a small number of loci are involved. In addition to the candidate gene approach, which relies on prior knowledge of the potential roles of candidate genes in controlling a trait, positional cloning and association studies are now routinely used in model organisms and humans. Positional cloning starts from mapping the loci responsible for a trait, which is achieved through linkage analysis in existing pedigrees or designed genetic crosses. The genomic revolution has allowed the identification in model organisms of a wealth of polymorphic genetic markers that can be inexpensively assayed in linkage analysis. Association studies look for genetic markers that are statistically correlated with a phenotypic difference. After the identification of potentially causal mutations, several molecular techniques, such as gene replacement, allow a definitive experimental verification. In the last decade, a number of genetic alterations responsible for microevolutionary phenotypic changes have been identified through the candidate gene approach, positional cloning, and association studies (Johanson et al. 2000; Sucena and Stern 2000; Takahashi et al. 2001; Shapiro et al. 2004; Yoshiura et al. 2006; Zhang 2006; Linnen et al. 2009; Rebeiz et al. 2009; Tung et al. 2009; Wittkopp et al. 2009; Chan et al. 2010).

However, these three approaches are generally difficult to apply to the study of macroevolutionary phenotypic changes, which are often most interesting and amazing, and they include some major subjects of evolutionary developmental biology. Positional cloning is simply not usable because individuals from divergent lineages cannot be crossed. Association studies fail because, statistically, all fixed genetic differences between two species are equally correlated with any fixed phenotypic difference between the species. Only the candidate gene approach may be applied, but its success depends on prior knowledge of gene function and of the extent of conservation of the molecular function and physiological role of a gene during evolution. Another obstacle is that rigorous experimental tests are difficult, even when candidate genes are available, because of large differences in genetic background between divergent species.

A better understanding of developmental pathways and gene–gene interactions in model organisms will help improve knowledge of gene function; much of the success of evolutionary developmental biology is attributable to such critical knowledge. Further, genome-wide systematic comparisons of gene function among divergent model organisms can provide some basic ideas on the conservation of gene function in evolution (Liao and Zhang 2008). These studies will likely offer candidate genes for experimental tests. Development of efficient molecular genetic techniques that allow simultaneous alterations of multiple genes in non-model organisms will also be important.

Question 2: What Is the Molecular Genetic Architecture of Multifactorial Traits?

Most traits are controlled by the developmental expression of many genes (Falconer and Mackay 1996). The molecular genetic basis of phenotypic variation in multifactorial or quantitative traits is usually difficult to discern, although considerable progress in developing the linkage-analysis–based approach to studying quantitative trait variation has been made in model organisms, such as *Drosophila* and yeast (Mackay 2001; Brem et al.

2002; Steinmetz et al. 2002; De Luca et al. 2003; Deutschbauer and Davis 2005; Mackay and Lyman 2005). In addition, genome-wide association studies of many complex human traits (mostly common diseases) have also shown the power of this method in identifying small-effect alleles, but the fact that the identified loci together explain only a small fraction of the heritability of the traits concerned (Manolio et al. 2009) is disappointing. This finding is at least in part explainable by the fact that large-effect mutations that cause diseases are kept at very low frequencies by purifying selection and thus require a sizable sample to be detected statistically. It is not clear whether association studies will be more useful in identifying genes underlying adaptive traits.

In model organisms, such as yeast, nematode, fly, and mouse, the tools are available to experimentally delete many genes individually. It is possible to comprehensively phenotype these gene-deletion strains to gain systematic knowledge about which traits are affected by which genes. For example, a recent yeast study measured 501 morphological traits in each of 4718 nonessential gene deletion haploid strains as well as the wild-type haploid strain by fluorescent imaging (Ohya et al. 2005). Because a gene deletion usually has a larger phenotypic effect than the average natural mutation, the comprehensive phenotyping of deletion strains is likely to reveal gene-trait relationships that are hard to detect in natural populations. Using the gene-trait map obtained from the systematic phenotyping as a guide, one can identify and examine candidate genes in natural populations, which can substantially increase the power of association studies. Thus, even when the purpose is to define the genetic architecture of a multifactorial trait in nature, it is useful to first have a comprehensive gene-trait map from gene deletion strains as a guide.

It is possible that gene-trait relationships in natural populations will be undetectable in gene-deletion strains, such as when a natural mutation results in the gain of a function that is different from a null mutation or deletion. The same mutation may also have different phenotypic effects in different genetic backgrounds as a result of epistasis. Furthermore, the deletion of a gene may not affect any phenotype in an artificial lab environment, but it may impact traits in the natural habitat of an organism. Nevertheless, combining the systematic examination of deletion strains and of natural variants will likely provide a more comprehensive picture of the genes responsible for a multifactorial trait.

Identifying the underlying genes of a multifactorial trait and measuring the size of their phenotypic effect are usually not the final goals of evolutionary genetics. One would also like to know the mechanisms (e.g., signaling pathways) through which certain genes control the trait being studied and explain why the phenotypic effect sizes of these genes are different. It is my contention that in addition to molecular biology, which tells us the properties of the gene products for different alleles, systems biology, which studies interactions among different parts of a system and properties of the system brought about by these interactions, will shed light on these important yet difficult questions.

Question 3: What Is the Genomic Pattern of Epistasis and How Does this Pattern Affect Evolution?

When Bateson coined the term "epistasis" 101 years ago, he meant that the effect of a gene on a trait may be enhanced or masked by one or more other genes (Bateson and Mendel 1909; Phillips 2008). Fisher and other people extended the concept to mean non-independent (non-additive or non-multiplicative) effects of genes (Fisher 1918; Phillips 2008). The direction, magnitude, and prevalence of epistasis is important for understanding many phenomena in gene function and interaction (Hartman et al. 2001; Boone et al. 2007; Phillips 2008), speciation (Coyne 1992), the evolution of sex and recombination (Kondrashov 1988; Barton and Charlesworth 1998), evolution of ploidy (Kondrashov and Crow 1991), mutation load (Crow and Kimura 1979), genetic buffering (Jasnos and Korona 2007), human disease (Cordell 2002; Moore and Williams 2005), and drug–drug interaction (Yeh et al. 2006). Yet, epistasis is arguably the most important but least well understood phenomenon in genetics. Using Fisher's definition of statistical epistasis, recent functional studies have started generating genome-wide epistasis maps in model organisms (Tong et al. 2004; Boone et al. 2007; Roguev et al. 2008; Tischler et al. 2008). These data will provide evidence for general patterns of epistasis, which will in turn allow the testing of many important evolutionary hypotheses that depend on various assumptions about epistasis.

It should be emphasized that statistical epistasis, or non-multiplicative gene effects, may be different from true functional interactions between genes and should be distinguished (He et al. 2010). A recent study by my group revealed a surprisingly high abundance of positive statistical epistasis between deleterious mutations—in other words, two mutations together are not as bad as expected from their individual deleterious effects (He et al. 2010). This finding suggests the need for reevaluation of evolutionary theories that depend critically on overall negative epistasis, such as the mutational deterministic hypothesis of the evolution of sexual reproduction (Kondrashov 1988) and the hypothesis of reduction in mutational load by truncation selection against deleterious mutations (Crow and Kimura 1979).

Question 4: What Is the Genomic Pattern of Pleiotropy and How Does Pleiotropy Affect Evolution?

Pleiotropy refers to the common observation that one gene (or mutation) affects multiple traits. Despite its broad implications in genetics (Wright 1968; Tyler et al. 2009), development (Hodgkin 1998; Carroll 2008), senescence (Williams 1957), disease (Albin 1993; Brunner and van Driel 2004), adaptation (Fisher 1930; Wright 1968; Waxman and Peck 1998; Orr 2000), the maintenance of sex (Hill and Otto 2007), and stabilization of cooperation

(Foster et al. 2004), the genome-wide patterns of pleiotropy are unknown. Due to its central importance in many areas, the implications of pleiotropy have been extensively modeled (Fisher 1930; Turelli 1985; Wagner 1988; Waxman and Peck 1998), but because these theoretical models have virtually no empirical basis, it is unclear whether they are realistic.

Empirical data on pleiotropy are urgently needed to test some of the most fundamental hypotheses in evolutionary genetics. For example, based on Fisher's (1930) geometric model that assumes that a mutation affects all traits of an organism, Orr (2000) derived the formula for the rate of fitness increase during an adaptive walk to the optimum in an organism with n traits. He found that the adaptation rate decreases with n. In other words, complex organisms have lower adaptation rates than simple organisms—a cost of complexity. But Orr's results depend on the assumptions that (1) a mutation affects all traits of an organism and (2) the total phenotypic effect of a mutation is the same in organisms of different levels of complexity. Using a Quantative Trait Locus (QTL) study of mouse skeletal characters, Wagner et al. (2008) recently reported that neither of these assumptions is valid. They found that half of the QTL affects less than 10% of the traits examined and that the mean per-trait effect is larger for those genes influencing more traits (see Wagner, Chapter 9). These results mean that the cost of complexity, while not absent, is significantly lower than Orr's model would suppose. However, the Wagner et al. data set is relatively small (70 traits and 102 QTL), and each of their QTL may include multiple genes. A reanalysis of their data (Hermisson and McGregor 2008) did not find clear evidence that more pleiotropic genes have larger per-trait effects. This example illustrates the importance of collecting pleiotropy data for genes (not QTLs). As previously mentioned, such data can be generated by comprehensive phenotyping of gene deletion strains of model organisms. I expect that many questions related to pleiotropy will have clearer answers in the near future.

Question 5: What Are the Relative Roles of Positive Selection and Genetic Drift in Evolution?

While this question has been investigated and debated since the late 1960s, there is still no agreement among evolutionary biologists. After seeing the genomic data, some researchers are now convinced that most nucleotide substitutions are neutral (Nei 2005), while others believe that most are adaptive and even think that adaptation should now be used as the null hypothesis in explaining evolutionary observations (Hahn 2008). The answer to this question certainly depends on the definition of neutrality. According to Kimura (1983), an allele with an absolute fitness effect $|s| < 0.5/N$, in which N is the effective population size, is effectively neutral. However, Nei (2005) commented that this definition of neutrality is too stringent and suggested that neutrality should be defined by $|s| < 0.5/\sqrt{N}$. If we assume that N is, on average, approximately

10^5 in mammals, the neutrality definition is $|s| < 0.001$ (Nei 2005). The use of Nei's definition would certainly result in an increase of the fraction of substitutions that are regarded as neutral.

In the last two decades, many biologists have tried to identify the action of positive selection on individual genes, especially by inferring selective sweeps and by comparing nonsynonymous and synonymous substitution rates from sequence data (Kreitman and Akashi 1995; Nielsen 2005; Zhang 2010). Experimentalists also tried to identify nucleotide changes that significantly affect gene function or expression and thus potentially affect organismal fitness. While the occurrence of positive selection can be corroborated when the two approaches are consistent (e.g., Sawyer et al. 2005; Zhang 2006), it is important to note that they do not need to be consistent, because each method has its errors and shortcomings and because the two approaches may be measuring quite different things. For example, experimental studies are usually focused on some but not all aspects of the many functions of a gene and can easily miss important functional changes that improve fitness. It has also been reported that some sites that show function-altering substitutions are not found to be under positive selection by statistical analysis. Such inconsistency might be attributable to insufficient statistical power, but it is also possible that the functional changes detected in experiments have no impact on fitness. Experimental tests can sometimes demonstrate selection, but fitness differences that outweigh random drift are often too small to be detected statistically in the lab; moreover, it is frequently difficult to know the environment in which an evolutionary change transpired or to measure all possible components of total fitness.

Because of its central importance in evolutionary biology, the neutralist–selectionist debate is unlikely to be outgrown soon. Fortunately, the debate is becoming more quantitative than qualitative (i.e., about the percentage of fixations that are adaptive and the fitness effects of the fixed changes), and I believe that new findings will continue to emerge from genomic studies.

Outlook

Evolutionary genetics is arguably one of the fastest moving fields in biology. This phenomenon is in part because a complete understanding of any issue in biology requires an explanation of its evolutionary origin that ultimately must reach the level of molecular genetics. In turn, evolutionary genetics has benefited from having close ties with other fields of biology, especially molecular biology and genomics. Not only did evolutionary geneticists quickly adopt new technologies developed in other fields (e.g., protein electrophoresis, DNA sequencing) to address long-standing evolutionary problems (e.g., population genetics, phylogenetics), but they also have quickly identified intriguing evolutionary questions arising from new discoveries in other fields (e.g., origin of introns). Thus, in addition to the core set of long-standing questions described in the previous sections, evolutionary geneticists constantly discover new questions. For example, codon usage bias, first reported 30 years ago (Grantham et al. 1980) and now well documented in many species, is still not fully understood. Codon usage bias is generally thought to be a result of a selection-mutation–drift balance (Bulmer 1991). Here, the word "mutation" refers to mutational bias that causes unequal equilibrium nucleotide frequencies, while the word "selection" refers to translational optimization. However, exactly what is optimized in translation that results in codon usage bias is not so clear. There is unambiguous evidence that codon usage bias is at least partly caused by selection for translational accuracy (Akashi 1994; Drummond and Wilke 2008), but whether it is also caused by selection for translational speed/efficiency remains elusive (Hershberg and Petrov 2008). Another example is the problem of the rate determinants of protein-sequence evolution. Quite surprisingly, about 10 years ago, it was found that the rate of protein-sequence evolution is determined mainly by the gene expression level rather than by the relative importance of the gene, measured by the fitness effect of gene deletion (Hurst and Smith 1999; Pal et al. 2001; Wang and Zhang 2009). Drummond and Wilke (2008) proposed a translational robustness hypothesis to explain why highly expressed proteins evolve more slowly than weakly expressed proteins. They suggest that translational errors often result in protein misfolding, which could be toxic to the cell. Thus, highly expressed proteins are expected to evolve DNA sequences that are more robust to mistranslation-induced misfolding and are thus more conserved than are lowly expressed proteins. There is still no direct evidence for the key assumptions of this provocative hypothesis, but it exemplifies how new evolutionary hypotheses arise from new molecular and genomic information.

Although many of the basic tenets of the neo-Darwinism have remained intact for the last 50 years, evolutionary genetics has made dramatic progress, thanks to the two technological revolutions and one conceptual breakthrough in the last half century. While I expect most of the basic tenets of neo-Darwinism to remain largely unchanged in the next 50 years or more, we can have full confidence that understanding in evolutionary genetics will increase greatly as a result of large amounts of genomic data and the experimental power of molecular biology, functional genomics, and systems biology. If the last 50 years of evolutionary genetics was characterized by deepening the understanding of evolution to the molecular level, the next 50 years will surely see the broadening of our understanding to the genomic scale and the systems level.

Acknowledgments

I thank Meg Bakewell, Walter Eanes, and Doug Futuyma for valuable comments. Research in my lab has been supported by The University of Michigan and U.S. National Institutes of Health.

Literature Cited

Akashi, H. 1994. Synonymous codon usage in *Drosophila melanogaster*: Natural selection and translational accuracy. *Genetics* 136: 927–935.

Akiba, T., K. Koyama, Y. Ishiki, and 2 others. 1960. On the mechanism of the development of multiple-drug-resistant clones of *Shigella*. *Jpn. J. Microbiol.* 4: 219–227.

Albin, R. L. 1993. Antagonistic pleiotropy, mutation accumulation, and human genetic disease. *Genetica* 91: 279–286.

Barton, N. H. and B. Charlesworth. 1998. Why sex and recombination? *Science* 281: 1986–1990.

Bateson, W. and G. Mendel. 1909. *Mendel's Principles of Heredity: A Defence, with a Translation of Mendel's Original Papers on Hybridisation.* Cambridge University Press, Cambridge, UK.

Beyer, A., S. Bandyopadhyay, and T. Ideker. 2007. Integrating physical and genetic maps: From genomes to interaction networks. *Nat. Rev. Genet.* 8: 699–710.

Blower, P. J. 1989. *The Mendelian Revolution.* The Athlone Press, London.

Boone, C., H. Bussey, and B. J. Andrews. 2007. Exploring genetic interactions and networks with yeast. *Nat. Rev. Genet.* 8: 437–449.

Brem, R. B., G. Yvert, R. Clinton, and 1 other. 2002. Genetic dissection of transcriptional regulation in budding yeast. *Science* 296: 752–755.

Brett, D., H. Pospisil, J. Valcarcel, and 2 others. 2002. Alternative splicing and genome complexity. *Nat. Genet.* 30: 29–30.

Bridges, C. B. 1936. The Bar "gene" a duplication. *Science* 83: 210–211.

Brunner, H. G. and M. A. van Driel. 2004. From syndrome families to functional genomics. *Nat. Rev. Genet.* 5: 545–551.

Bulmer, M. 1991. The selection-mutation-drift theory of synonymous codon usage. *Genetics* 129: 897–907.

Cargill, M., D. Altshuler, J. Ireland, and 14 others. 1999. Characterization of single-nucleotide polymorphisms in coding regions of human genes. *Nat. Genet.* 22: 231–238.

Carroll, S. B. 2005. Evolution at two levels: On genes and form. *PLoS Biol.* 3: e245.

Carroll, S. B. 2008. Evo-devo and an expanding evolutionary synthesis: A genetic theory of morphological evolution. *Cell* 134: 25–36.

Carroll, S. B., J. K. Grenier, and S. D. Weatherbee. 2001. *From DNA to Diversity: Molecular Genetics and the Evolution of Animal Design.* Blackwell Scientific, Malden, MA.

Chan, Y. F., M. E. Marks, F. C. Jones, and 13 others. 2010. Adaptive evolution of pelvic reduction in sticklebacks by recurrent deletion of a *Pitx1* enhancer. *Science* 327: 302–305.

Chen, L., A. L. DeVries, and C. H. Cheng. 1997. Evolution of antifreeze glycoprotein gene from a trypsinogen gene in Antarctic notothenioid fish. *Proc. Natl. Acad. Sci. USA* 94: 3811–3816.

Chen, S. T., H. C. Cheng, D. A. Barbash, and 1 other. 2007. Evolution of hydra, a recently evolved testis-expressed gene with nine alternative first exons in *Drosophila melanogaster*. *PLoS Genet.* 3: e107.

Chimpanzee Sequencing and Analysis Consortium. 2005. Initial sequence of the chimpanzee genome and comparison with the human genome. *Nature* 437: 69–87.

Cho, S., Z. Y. Huang, D. R. Green, and 2 others. 2006. Evolution of the complementary sex-determination gene of honey bees: Balancing selection and trans-species polymorphisms. *Genome Res.* 16: 1366–1375.

Ciccarelli, F. D., T. Doerks, C. von Mering, and 2 others. 2006. Toward automatic reconstruction of a highly resolved tree of life. *Science* 311: 1283–1287.

Clark, A. G. and T. H. Kao. 1991. Excess nonsynonymous substitution of shared polymorphic sites among self-incompatibility alleles of Solanaceae. *Proc. Natl. Acad. Sci. USA* 88: 9823–9827.

Cordell, H. J. 2002. Epistasis: What it means, what it doesn't mean, and statistical methods to detect it in humans. *Hum. Mol. Genet.* 11: 2463–2468.

Coyne, J. A. 1992. Genetics and speciation. *Nature* 355: 511–515.

Cracraft, J. and M. J. Donoghue (eds.). 2004. *Assembling the Tree of Life.* Oxford University Press, New York.

Crow, J. F. and M. Kimura. 1979. Efficiency of truncation selection. *Proc. Natl. Acad. Sci. USA* 76: 396–399.

Daubin, V., N. A. Moran, and H. Ochman. 2003. Phylogenetics and the cohesion of bacterial genomes. *Science* 301: 829–832.

Delsuc, F., H. Brinkmann, and H. Philippe. 2005. Phylogenomics and the reconstruction of the tree of life. *Nat. Rev. Genet.* 6: 361–375.

De Luca, M., N. V. Roshina, G. L. Geiger-Thornsberry, and 3 others. 2003. Dopa decarboxylase (*Ddc*) affects variation in *Drosophila* longevity. *Nat. Genet.* 34: 429–433.

Deutschbauer, A. M. and R. W. Davis. 2005. Quantitative trait loci mapped to single-nucleotide resolution in yeast. *Nat. Genet.* 37: 1333–1340.

Doniger, S. W., H. S. Kim, D. Swain, and 4 others. 2008. A catalog of neutral and deleterious polymorphism in yeast. *PLoS Genet.* 4: e1000183.

Doolittle, W. F. 1999. Phylogenetic classification and the universal tree. *Science* 284: 2124–2129.

Drummond, D. A. and C. O. Wilke. 2008. Mistranslation-induced protein misfolding as a dominant constraint on coding-sequence evolution. *Cell* 134: 341–352.

ENCODE Project Consortium. 2007. Identification and analysis of functional elements in 1% of the human genome by the ENCODE pilot project. *Nature* 447: 799–816.

Eyre-Walker, A. 2006. The genomic rate of adaptive evolution. *Trends Ecol. Evol.* 21: 569–575.

Falconer, D. S. and T. F. C. Mackay. 1996. *Introduction to Quantitative Genetics.* Pearson, London.

Fay, J. C., G. J. Wyckoff, and C.-I. Wu. 2002. Testing the neutral theory of molecular evolution with genomic data from *Drosophila. Nature* 415: 1024–1026.

Felsenstein, J. 2004. *Inferring Phylogenies.* Sinauer Associates, Sunderland, MA.

Fisher, R. A. 1918. The correlations between relatives on the supposition of Mendelian inheritance. *Trans. R. Soc. Edinburgh* 52: 399–433.

Fisher, R. A. 1930. *The Genetic Theory of Natural Selection*, 2nd ed. Clarendon, Oxford.

Fitch, W. M. and E. Margoliash. 1967. Construction of phylogenetic trees. *Science* 155: 279–284.

Fleischmann, R. D., M. D. Adams, O. White, and 37 others. 1995. Whole-genome random sequencing and assembly of *Haemophilus influenzae* Rd. *Science* 269: 496–512.

Force, A., M. Lynch, F. B. Pickett, and 3 others. 1999. Preservation of duplicate genes by complementary, degenerative mutations. *Genetics* 151: 1531–1545.

Foster, K. R., G. Shaulsky, J. E. Strassmann, and 2 others. 2004. Pleiotropy as a mechanism to stabilize cooperation. *Nature* 431: 693–696.

Futuyma, D. J. 1998. *Evolutionary Biology*, 3rd ed. Sinauer Associates, Sunderland, MA.
Garland Jr., T. and M. Rose (eds). 2009. *Experimental Evolution: Concepts, Methods, and Applications of Selection Experiments*. University of California Press, Berkeley.
Genome 10K Community of Scientists. 2009. Genome 10K: A proposal to obtain whole-genome sequence for 10,000 vertebrate species. *J. Hered.* 100: 659–674.
Gibbs, R. A., G. M. Weinstock, M. L. Metzker, and 230 others. 2004. Genome sequence of the Brown Norway rat yields insights into mammalian evolution. *Nature* 428: 493–521.
Gibson, G. and S. V. Muse. 2004. *A Primer of Genome Science*, 2nd ed. Sinauer Associates, Sunderland, MA.
Gilbert, W. 1978. Why genes in pieces? *Nature* 271: 501.
Gilbert, W. 1987. The exon theory of genes. *Cold Spring Harb. Symp. Quant. Biol.* 52: 901–905.
Glazko, G., V. Veeramachaneni, M. Nei, and 1 other. 2005. Eighty percent of proteins are different between humans and chimpanzees. *Gene* 346: 215–219.
Gogarten, J. P., W. F. Doolittle, and J. G. Lawrence. 2002. Prokaryotic evolution in light of gene transfer. *Mol. Biol. Evol.* 19: 2226–2238.
Gogarten, M. B., J. P. Gogarten, and L. Olendzenski (eds.). 2009. *Horizontal Gene Transfer: Genomes in Flux*. Humana Press, New York.
Golding, G. B., N. Tsao, and R. E. Pearlman. 1994. Evidence for intron capture: An unusual path for the evolution of proteins. *Proc. Natl. Acad. Sci. USA* 91: 7506–7509.
Gould, S. J. and R. C. Lewontin. 1979. The spandrels of San Marco and the panglossian paradigm: A critique of the adaptationist programme. *Proc. Roy. Soc. Lond. B* 205: 581–598.
Grantham, R., C. Gautier, M. Gouy, and 2 others. 1980. Codon catalog usage and the genome hypothesis. *Nucleic Acids Res.* 8: r49–r62.
Gregory, T. R. 2001. Coincidence, coevolution, or causation? DNA content, cell size, and the C-value enigma. *Biol. Rev. Camb. Philos. Soc.* 76: 65–101.
Gresham, D., M. M. Desai, C. M. Tucker, and 6 others. 2008. The repertoire and dynamics of evolutionary adaptations to controlled nutrient-limited environments in yeast. *PLoS Genet.* 4: e1000303.
Hahn, M. W. 2008. Toward a selection theory of molecular evolution. *Evolution* 62: 255–265.
Haldane, J. B. S. 1957. The cost of natural selection. *J. Genet.* 55: 511–524.
Harris, H. 1966. Enzyme polymorphisms in man. *Proc. Roy. Soc. Lond. B* 164: 298–310.
Hartman, J. L., B. Garvik, and L. Hartwell. 2001. Principles for the buffering of genetic variation. *Science* 291: 1001–1004.
He, X. and J. Zhang. 2005. Rapid subfunctionalization accompanied by prolonged and substantial neofunctionalization in duplicate gene evolution. *Genetics* 169: 1157–1164.
He, X., W. Qian, Z. Wang, and 2 others. 2010. Prevalent positive epistasis in *Escherichia coli* and *Saccharomyces cerevisiae* metabolic networks. *Nat. Genet.* 42: 272–276.
Hermisson, J. and A. P. McGregor. 2008. Pleiotropic scaling and QTL data. *Nature* 456: E3–E4.

Herring, C. D., A. Raghunathan, C. Honisch, and 8 others. 2006. Comparative genome sequencing of *Escherichia coli* allows observation of bacterial evolution on a laboratory timescale. *Nat. Genet.* 38: 1406–1412.
Hershberg, R. and D. A. Petrov. 2008. Selection on codon bias. *Ann. Rev. Genet.* 42: 287–299.
Hill, J. A. and S. P. Otto. 2007. The role of pleiotropy in the maintenance of sex in yeast. *Genetics* 175: 1419–1427.
Hillis, D. M., C. Moritz, and B. K. Mable (eds.). 1996. *Molecular Systematics.* Sinauer Associates, Sunderland, MA.
Hodgkin, J. 1998. Seven types of pleiotropy. *Int. J. Dev. Biol.* 42: 501–505.
Hoekstra, H. E. and J. A. Coyne. 2007. The locus of evolution: Evo devo and the genetics of adaptation. *Evolution* 61: 995–1016.
Hubby, J. L. and R. C. Lewontin. 1966. A molecular approach to the study of genic heterozygosity in natural populations. I. The number of alleles at different loci in *Drosophila pseudoobscura. Genetics* 54: 577–594.
Hughes, A. L. and M. Nei. 1988. Pattern of nucleotide substitution at major histocompatibility complex class I loci reveals overdominant selection. *Nature* 335: 167–170.
Hurst, L. D. and N. G. Smith. 1999. Do essential genes evolve slowly? *Curr. Biol.* 9: 747–750.
Jain, R., M. C. Rivera, and J. A. Lake. 1999. Horizontal gene transfer among genomes: The complexity hypothesis. *Proc. Natl. Acad. Sci. USA* 96: 3801–3806.
Jasnos, L. and R. Korona. 2007. Epistatic buffering of fitness loss in yeast double deletion strains. *Nat. Genet.* 39: 550–554.
Johanson, U., J. West, C. Lister, and 3 others. 2000. Molecular analysis of FRIGIDA, a major determinant of natural variation in *Arabidopsis* flowering time. *Science* 290: 344–347.
Keeling, P. J. and J. D. Palmer. 2008. Horizontal gene transfer in eukaryotic evolution. *Nat. Rev. Genet.* 9: 605–618.
Kim, E., A. Magen, and G. Ast. 2007. Different levels of alternative splicing among eukaryotes. *Nucleic Acids Res.* 35: 125–131.
Kimura, M. 1968. Evolutionary rate at the molecular level. *Nature* 217: 624–626.
Kimura, M. 1983. *The Neutral Theory of Molecular Evolution.* Cambridge University Press, Cambridge, UK.
King, J. L. and T. H. Jukes. 1969. Non-Darwinian evolution. *Science* 164: 788–798.
King, M. C. and A. C. Wilson. 1975. Evolution at two levels in humans and chimpanzees. *Science* 188: 107–116.
Kondrashov, A. S. 1988. Deleterious mutations and the evolution of sexual reproduction. *Nature* 336: 435–440.
Kondrashov, A. S. and J. F. Crow. 1991. Haploidy or diploidy: Which is better? *Nature* 351: 314–315.
Koonin, E. V., K. S. Makarova, and L. Aravind. 2001. Horizontal gene transfer in prokaryotes: Quantification and classification. *Ann. Rev. Microbiol.* 55: 709–742.
Kreitman, M. and H. Akashi. 1995. Molecular evidence for natural selection. *Ann. Rev. Ecol. Syst.* 26: 403–422.
Kritsky, G. 1973. Mendel, Darwin, and evolution. *Am. Biol.Teach.* 35: 477–479.
Levine, M. T., C. D. Jones, A. D. Kern, and 2 others. 2006. Novel genes derived from noncoding DNA in *Drosophila melanogaster* are frequently X-linked and exhibit testis-biased expression. *Proc. Natl. Acad. Sci. USA* 103: 9935–9939.

Lewontin, R. C. 1974. *The Genetic Basis of Evolutionary Change.* Columbia University Press, New York.

Liao, B., M. Weng, and J. Zhang. 2010. Contrasting genetic paths to morphological and physiological evolution. *Proc. Natl. Acad. Sci. USA* 107: 7353–7358.

Liao, B. Y. and J. Zhang. 2008. Null mutations in human and mouse orthologs frequently result in different phenotypes. *Proc. Natl. Acad. Sci. USA* 105: 6987–6992.

Linnen, C. R., E. P. Kingsley, J. D. Jensen, and 1 other. 2009. On the origin and spread of an adaptive allele in deer mice. *Science* 325: 1095–1098.

Liti, G., D. M. Carter, A. M. Moses, and 23 others. 2009. Population genomics of domestic and wild yeasts. *Nature* 458: 337–341.

Long, M., E. Betran, K. Thornton, and 1 other. 2003. The origin of new genes: Glimpses from the young and old. *Nat. Rev. Genet.* 4: 865–875.

Lynch, M. 2007. *The Origins of Genome Architecture.* Sinauer Associates, Sunderland, MA.

Lynch, M. and A. Force. 2000a. Gene duplication and the origin of interspecific genomic incompatibility. *Am. Nat.* 156: 590–605.

Lynch, M. and A. Force. 2000b. The probability of duplicate gene preservation by subfunctionalization. *Genetics* 154: 459–473.

Lynch, M., W. Sung, K. Morris, and 8 others. 2008. A genome-wide view of the spectrum of spontaneous mutations in yeast. *Proc. Natl. Acad. Sci. USA* 105: 9272–9277.

Mackay, T. F. 2001. The genetic architecture of quantitative traits. *Ann. Rev. Genet.* 35: 303–339.

Mackay, T. F. and R. F. Lyman. 2005. *Drosophila* bristles and the nature of quantitative genetic variation. *Philos. Trans. Roy. Soc. Lond. B* 360: 1513–1527.

Manolio, T. A., F. S. Collins, N. J. Cox, and 24 others. 2009. Finding the missing heritability of complex diseases. *Nature* 461: 747–753.

Mardis, E. R. 2008. Next-generation DNA sequencing methods. *Ann. Rev. Genomics Hum. Genet.* 9: 387–402.

Matlin, A. J., F. Clark, and C. W. Smith. 2005. Understanding alternative splicing: Towards a cellular code. *Nat. Rev. Mol. Cell. Biol.* 6: 386–398.

Moore, J. H. and S. M. Williams. 2005. Traversing the conceptual divide between biological and statistical epistasis: Systems biology and a more modern synthesis. *BioEssays* 27: 637–646.

Moriyama, E. N. and J. R. Powell. 1996. Intraspecific nuclear DNA variation in *Drosophila. Mol. Biol. Evol.* 13: 261–277.

Mushegian, A. R. and E. V. Koonin. 1996. A minimal gene set for cellular life derived by comparison of complete bacterial genomes. *Proc. Natl. Acad. Sci. USA* 93: 10268–10273.

Nei, M. 1975. *Molecular Population Genetics and Evolution.* North-Holland, Amsterdam.

Nei, M. 2005. Selectionism and neutralism in molecular evolution. *Mol. Biol. Evol.* 22: 2318–2342.

Nei, M. and D. Graur. 1984. Extent of protein polymorphism and the neutral mutation theory. *Evol. Biol.* 17: 73–118.

Nei, M. and S. Kumar. 2000. *Molecular Evolution and Phylogenetics.* Oxford University Press, New York.

Nei, M., X. Gu, and T. Sitnikova. 1997. Evolution by the birth-and-death process in multigene families of the vertebrate immune system. *Proc. Natl. Acad. Sci. USA* 94: 7799–7806.

Nekrutenko, A. and W.-H. Li. 2001. Transposable elements are found in a large number of human protein-coding genes. *Trends Genet.* 17: 619–621.
Nelson, K. E., R. A. Clayton, S. R. Gill, and 26 others. 1999. Evidence for lateral gene transfer between Archaea and bacteria from genome sequence of *Thermotoga maritima*. *Nature* 399: 323–329.
Nielsen, R. 2005. Molecular signatures of natural selection. *Ann. Rev. Genet.* 39: 197–218.
Ochiai, K., T. Yamanaka, K. Kimura, and 1 other. 1959. Inheritance of drug resistance (and its tranfer) between *Shigella* strains and between *Shigella* and *E. coli* strains (in Japanese). *Hihon Iji Shimpor* 1861: 34.
Ohno, S. 1970. *Evolution by Gene Duplication.* Springer-Verlag, Berlin.
Ohta, T. 1992. The nearly neutral theory of molecular evolution. *Ann. Rev. Ecol. Syst.* 23: 263–286.
Ohya, Y., J. Sese, M. Yukawa, and 22 others. 2005. High-dimensional and large-scale phenotyping of yeast mutants. *Proc. Natl. Acad. Sci. USA* 102: 19015–19020.
Orr, H. A. 2000. Adaptation and the cost of complexity. *Evolution* 54: 13–20.
Osawa, S. 1995. *Evolution of the Genetic Code.* Oxford University Press, Oxford.
Pal, C., B. Papp, and L. D. Hurst. 2001. Highly expressed genes in yeast evolve slowly. *Genetics* 158: 927–931.
Patthy, L. 1995. *Protein Evolution by Exon-Shuffling.* Molecular Biology Intelligence Unit, R. G. Landes Company/Springer, New York.
Patthy, L. 1999. Genome evolution and the evolution of exon-shuffling—a review. *Gene* 238: 103–114.
Pevsner, J. 2009. *Bioinformatics and Functional Genomics*, 2nd ed. Wiley-Blackwell, Hoboken, NJ.
Phillips, P. C. 2008. Epistasis—the essential role of gene interactions in the structure and evolution of genetic systems. *Nat. Rev. Genet.* 9: 855–867.
Piatigorsky, J. 2007. *Gene Sharing and Evolution.* Harvard University Press, London.
Raff, R. A. 1996. *The Shape of Life: Genes, Development, and the Evolution of Animal Form.* University of Chicago Press, Chicago.
Raj, A. and A. van Oudenaarden. 2008. Nature, nurture, or chance: Stochastic gene expression and its consequences. *Cell* 135: 216–226.
Raser, J. M. and E. K. O'Shea. 2005. Noise in gene expression: Origins, consequences, and control. *Science* 309: 2010–2013.
Rebeiz, M., J. E. Pool, V. A. Kassner, and 2 others. 2009. Stepwise modification of a modular enhancer underlies adaptation in a *Drosophila* population. *Science* 326: 1663–1667.
Roguev, A., S. Bandyopadhyay, M. Zofall, and 13 others. 2008. Conservation and rewiring of functional modules revealed by an epistasis map in fission yeast. *Science* 322: 405–410.
Rokas, A. and P. W. Holland. 2000. Rare genomic changes as a tool for phylogenetics. *Trends Ecol. Evol.* 15: 454–459.
Roman, J. and S. R. Palumbi. 2003. Whales before whaling in the North Atlantic. *Science* 301: 508–510.
Roy, S. W. and W. Gilbert. 2006. The evolution of spliceosomal introns: Patterns, puzzles and progress. *Nat. Rev. Genet.* 7: 211–221.
Sawyer, S. A., J. Parsch, Z. Zhang, and 1 other. 2007. Prevalence of positive selection among nearly neutral amino acid replacements in *Drosophila*. *Proc. Natl. Acad. Sci. USA* 104: 6504–6510.

Sawyer, S. L., L. I. Wu, M. Emerman, and 1 other. 2005. Positive selection of primate TRIM5 identifies a critical species-specific retroviral restriction domain. *Proc. Natl. Acad. Sci. USA* 102: 2832–2837.

Sclater, A. 2006. The extent of Charles Darwin's knowledge of Mendel. *J. Biosci.* 31: 191–193.

Sella, G., D. A. Petrov, M. Przeworski, and 1 other. 2009. Pervasive natural selection in the *Drosophila* genome? *PLoS Genet.* 5: e1000495.

Shapiro, J. A., W. Huang, C. Zhang, and 9 others. 2007. Adaptive genic evolution in the *Drosophila* genomes. *Proc. Natl. Acad. Sci. USA* 104: 2271–2276.

Shapiro, M. D., M. E. Marks, C. L. Peichel, and 5 others. 2004. Genetic and developmental basis of evolutionary pelvic reduction in threespine sticklebacks. *Nature* 428: 717–723.

Shendure, J. and H. Ji. 2008. Next-generation DNA sequencing. *Nat. Biotechnol.* 26: 1135–1145.

Shi, P. and J. Zhang. 2009. Extraordinary diversity of chemosensory receptor gene repertoires among vertebrates. In W. Meyerhof and S. Korsching (eds.), *Chemosensory Systems in Mammals, Fishes, and Insects*, pp. 1–23. Springer, Berlin.

Smith, N. G. and A. Eyre-Walker. 2002. Adaptive protein evolution in *Drosophila*. *Nature* 415: 1022–1024.

Sorek, R., Y. Zhu, C. J. Creevey, and 3 others. 2007. Genome-wide experimental determination of barriers to horizontal gene transfer. *Science* 318: 1449–1452.

Steinmetz, L. M., H. Sinha, D. R. Richards, and 5 others. 2002. Dissecting the architecture of a quantitative trait locus in yeast. *Nature* 416: 326–330.

Stephens, S. G. 1951. Possible significances of duplication in evolution. *Adv. Genet.* 4: 247–265.

Stern, D. L. and V. Orgogozo. 2008. The loci of evolution: How predictable is genetic evolution? *Evolution* 62: 2155–2177.

Stern, D. L. and V. Orgogozo. 2009. Is genetic evolution predictable? *Science* 323: 746–751.

Sucena, E. and D. L. Stern. 2000. Divergence of larval morphology between *Drosophila sechellia* and its sibling species caused by cis-regulatory evolution of ovo/shaven-baby. *Proc. Natl. Acad. Sci. USA* 97: 4530–4534.

Takahashi, A., S. C. Tsaur, J. A. Coyne, and 1 other. 2001. The nucleotide changes governing cuticular hydrocarbon variation and their evolution in *Drosophila melanogaster*. *Proc. Natl. Acad. Sci. USA* 98: 3920–3925.

Tischler, J., B. Lehner, and A. G. Fraser. 2008. Evolutionary plasticity of genetic interaction networks. *Nat. Genet.* 40: 390–391.

Tong, A. H., G. Lesage, G. D. Bader, and 47 others. 2004. Global mapping of the yeast genetic interaction network. *Science* 303: 808–813.

Tung, J., A. Primus, A. J. Bouley, and 3 others. 2009. Evolution of a malaria resistance gene in wild primates. *Nature* 460: 388–391.

Turelli, M. 1985. Effects of pleiotropy on predictions concerning mutation-selection balance for polygenic traits. *Genetics* 111: 165–195.

Tyler, A. L., F. W. Asselbergs, S. M. Williams, and 1 other. 2009. Shadows of complexity: What biological networks reveal about epistasis and pleiotropy. *BioEssays* 31: 220–227.

Vasemägi, A. and C. R. Primmer. 2005. Challenges for identifying functionally important genetic variation: The promise of combining complementary research strategies. *Mol. Ecol.* 14: 3623–3642.

Voight, B. F., S. Kudaravalli, X. Wen, and 1 other. 2006. A map of recent positive selection in the human genome. *PLoS Biol.* 4: e72.

Wagner, G. P. 1988. The influence of variation and developmental constraints on the rate of multivariate phenotypic evolution. *J. Evol. Biol.* 1: 44–66.
Wagner, G. P., J. P. Kenney-Hunt, M. Pavlicev, and 3 others. 2008. Pleiotropic scaling of gene effects and the 'cost of complexity'. *Nature* 452: 470–472.
Wang, Z. and J. Zhang. 2009. Why is the correlation between gene importance and gene evolutionary rate so weak? *PLoS Genet.* 5: e1000329.
Waterston, R. H., K. Lindblad-Toh, E. Birney, and 220 others. 2002. Initial sequencing and comparative analysis of the mouse genome. *Nature* 420: 520–562.
Waxman, D. and J. R. Peck. 1998. Pleiotropy and the preservation of perfection. *Science* 279: 1210–1213.
Werth, C. and M. Windham. 1991. A model for divergent, allopatric speciation of polyploid pteridophytes resulting from silencing of duplicate-gene expression. *Am. Nat.* 137: 515–526.
Williams, G. C. 1957. Pleiotropy, natural selection, and the evolution of senescene. *Evolution* 11: 398–411.
Wittkopp, P. J., E. E. Stewart, L. L. Arnold, and 6 others. 2009. Intraspecific polymorphism to interspecific divergence: Genetics of pigmentation in *Drosophila*. *Science* 326: 540–544.
Wray, G. A. 2007. The evolutionary significance of cis-regulatory mutations. *Nat. Rev. Genet.* 8: 206–216.
Wright, S. 1968. *Evolution and the Genetics of Populations.* University of Chicago Press, Chicago.
Yeh, P., A. I. Tschumi, and R. Kishony. 2006. Functional classification of drugs by properties of their pairwise interactions. *Nat. Genet.* 38: 489–494.
Yomo, T., I. Urabe, and H. Okada. 1992. No stop codons in the antisense strands of the genes for nylon oligomer degradation. *Proc. Natl. Acad. Sci. USA* 89: 3780–3784.
Yoshiura, K., A. Kinoshita, T. Ishida, and 36 others. 2006. A SNP in the ABCC11 gene is the determinant of human earwax type. *Nat. Genet.* 38: 324–330.
Zhang, J. 2003. Evolution by gene duplication—an update. *Trends Ecol. Evol.* 18: 292–298.
Zhang, J. 2006. Parallel adaptive origins of digestive RNases in Asian and African leaf monkeys. *Nat. Genet.* 38: 819–823.
Zhang, J. 2007. The drifting human genome. *Proc. Natl. Acad. Sci. USA* 104: 20147–20148.
Zhang, J. 2010. Positive Darwinian selection in gene evolution. In M. Long and H. Gu (eds.), *Darwin's Heritage Today—Proceedings of Darwin 200 International Beijing Conference*, High Education Press, Beijing, in press.
Zhang, J., H. F. Rosenberg, and M. Nei. 1998. Positive Darwinian selection after gene duplication in primate ribonuclease genes. *Proc. Natl. Acad. Sci. USA* 95: 3708–3713.
Zhang, L. and W.-H. Li. 2005. Human SNPs reveal no evidence of frequent positive selection. *Mol. Biol. Evol.* 22: 2504–2507.

Chapter 5

Natural Selection and Coalescent Theory

John Wakeley

The story of population genetics begins with the publication of Darwin's *The Origin of Species* and the tension that followed concerning the nature of inheritance (Darwin 1859). Today, workers in this field aim to understand the forces that produce and maintain genetic variation within and between species. For this purpose, the most direct kind of genetic data are used: DNA sequences, even of entire genomes. The "great obsession" of population genetics (Gillespie 2004a), namely to explain genetic variation, can be traced back to Darwin's recognition that natural selection may occur only if individuals of a species vary, and this variation is heritable. Darwin might have been surprised that the importance of natural selection in shaping variation at the molecular level would be de-emphasized, beginning in the late 1960s, by scientists who readily accepted the fact and importance of his theory (Kimura 1983). The underlying motivation in this chapter is to consider the possible demise of the neutral theory of molecular evolution, which a growing number of population geneticists feel must follow recent observations of genetic variation within and between species (see also Zhang, Chapter 4).

One hundred fifty years after the publication of *The Origin of Species*, researchers are still struggling to fully incorporate natural selection into the modern genealogical models of population genetics. The main goal of this chapter is to present the mathematical models that have been used to describe the effects of positive selective sweeps on genetic variation, as mediated by gene genealogies or coalescent trees. Background material, comprised of population genetic theory and simulation results, is provided in order to facilitate an understanding of these models. A strong thread running throughout is the use of population genetic data to draw conclusions broadly about the process of evolution and its role in shifting ideas about the causes of evolution that have characterized the field at various times, as the ability to sample genetic data has improved.

Theoretical Population Genetics

Provine (1971) described the birth of theoretical population genetics, which originated with Darwin and culminated in the great works of Fisher, Wright, and Haldane. The latter three luminaries established the fundamental dynamics of genetic evolution, that is, of changes in allele frequencies through the interaction of mutation, selection, and random genetic drift. The term "random genetic drift" requires explanation: it is the stochastic side of evolution, which results from the random transmission of genetic material, from one generation to the next in a population, because of Mendelian segregation and assortment as well as the partially unpredictable processes of survival and reproduction. The founding works of this field (Fisher 1918, 1930; Wright 1931; Haldane 1932) remain a crucial part of any advanced education in evolutionary biology (see Futuyma, Chapter 1).

Relevant aspects of the genetic evolution are reviewed below, but one early result deserves to be mentioned here. Consider the probability of fixation of a new mutant allele under the influence of positive natural selection. Initially, every individual has genotype A_1A_1. A mutation produces a new allele, A_2, which gives its carriers an advantage. If A_1A_2 individuals have an average of $1 + s$ offspring and A_2A_2 individuals an average of $1 + 2s$ offspring, relative to A_1A_1 individuals, then the probability that the new mutant allele A_2 goes extinct is approximately $1 - 2s$. This result holds when s is small relative to 1 and the population size, N, is very large (Ns>>1). The formula $1 - 2s$ can be derived using a branching process model, in which each A_2 allele has a Poisson number of descendants with mean $1 + s$ each generation (Haldane 1927; Fisher 1922, 1930), and it can also be obtained using diffusion theory. The probability of fixation is the probability that eventually the entire population will have genotype A_2A_2. In a finite population, the probability of fixation is equal to one minus the probability of extinction, in this case $2s$, which is small. One cannot help but marvel at the possible implications of this result: that the many important adaptations observed in nature might first have gone extinct several times before they became successful and that many, possibly even better, adaptations have not been observed at all, because they were lost despite their selective advantage.

It is remarkable that so much of what Fisher, Wright, and Haldane did in the 1920s and 1930s is still relevant today, given that almost nothing was known at that time about the material bases of heredity, development, and ecology. Although current knowledge of development and ecology is still not sufficient to permit a full evolutionary theory—one that would include the richness between genotype and phenotype and would extend to interactions between individuals and their environment—modern understanding of genetics is quite detailed. As a result, the models of population genetics, have moved away from the simple A_1, A_2, allelic models previously described, to models that include the structure of DNA, the various kinds of mutations, and perhaps most importantly, recombination within as well as between genetic loci. It can be said with some confidence that the fundamental components of genetic evolution are known. As Lynch states: "Many embellishments have been added to the theory, and views have changed on the relative power of alternative evolutionary forces, but no keystone principle of population genetics has been overturned by an observation in molecular, cellular, or developmental biology" (Lynch 2007: 366).

The modern synthesis of the mid-twentieth century was initiated in no small part by the early work of Fisher, Wright, and Haldane. It later involved the wide application of ideas from population genetics to explain the patterns of evolution (Dobzhanshy 1937; Huxley 1942; Mayr 1963), though sometimes without the aid of the vital mathematical models of that field. This period also saw the great development of mathematical theory—largely in the absence of data about the genetic variation the theory purported to explain (Lewontin 1974). Two additional seminal figures of mathematical population genetics from the mid-twentieth century need to be recognized: Malécot and Kimura. Among many important contributions (Nagylaki 1989; Slatkin and Veuille 2002), Malécot introduced the notion of following a pair of alleles backward in time to their common ancestor (Malécot 1941, 1948). This is the basic idea behind the genealogical models of population genetics, which are discussed in detail in the section "Neutral Coalescent Theory." Kimura is best known for the neutral theory of molecular evolution (Kimura 1983), but his place in mathematical population genetics derives from his work on the diffusion theory of allele frequencies.

Diffusion Theory

As the results of diffusion theory are used later in the chapter and the assumptions of coalescent theory and diffusion theory are the same, a brief review of the basic concepts is given here. See Ewens (1979, 2004) for an excellent and thorough treatment.

Diffusion models approximate the dynamics of allele frequencies over time in large populations. The discrete or exact models of population genetics typically imagine a diploid population of constant size N, in which time is measured in discrete units of generations. The number of copies of an allele (e.g., the mutant A_2 in the previous example) must, at any one time, be one of $2N + 1$ possible values: $0, 1, 2,\ldots, 2N - 1, 2N$. If there are k copies of an allele, then the frequency of that allele is $p = k/(2N)$. In a diffusion model, both time and allele frequency are measured continuously: $p \in [0,1]$ and $t \in [0,\infty)$. This is achieved by taking a limit of the dynamics, as N tends to infinity, with time rescaled so that one unit of time in the diffusion model corresponds to $2N$ generations in the discrete model. Intuitively, when N is large, p may assume very many possible values, so there will be little error in measuring allele frequencies continuously. Similarly, a single generation comprises a very small step when time is viewed on the scale of $2N$ generations. Diffusion models allow the computation of many quantities of interest (in order to make predictions, test hypotheses, and estimate parameters), while most discrete models are mathematically intractable.

Discrete models differ in their assumptions about population demography and reproduction, and thus about the dynamics of genetic transmission from one generation to the next. The most commonly used model is the Wright–Fisher model (Fisher 1930; Wright 1931), followed somewhat distantly by the Moran model (Moran 1962). Besides tractability, another advantage of the diffusion approximation is that many different discrete models have the same diffusion limit. Here, "the same" includes the possibility of a constant multiplier of the time scale, so that time is measured in units of $2Nc$ generations. In the Wright–Fisher model, $c = 1$, and in the Moran model, $c = 1/2$, but the mathematical form of the diffusion equations is identical, and it is said that the *effective population size* is $N_e = cN$ diploid individuals (Ewens 1982; Sjödin et al. 2005). Thus, the diffusion approximation of the Wright–Fisher model may be used to illustrate general features of the evolution of populations, knowing that if N is replaced with N_e the results will be valid for other populations that do not conform to the overly simple Wright–Fisher model.

A single effective population size may not exist, as in the case of two populations with little or no gene flow or when the size of the population changes dramatically over time so that the time scale of the diffusion model would also have to change over time. In the interest of brevity and simplicity, populations that deviate so dramatically from the assumptions of the Wright–Fisher model will not be considered here.

On a per-generation basis, the rate of genetic drift, which is the rate at which the frequency of an allele will change at random due to the vagaries of genetic transmission in a population, is equal to $1/(2N)$ in the Wright–Fisher model. The per-generation effects of selection, mutation, and recombination are captured in additional parameters, here denoted as s, u, and r. The definition of s was already provided, and now u and r are defined as the probability of a mutation at a single nucleotide site and the probability of a recombination event between two adjacent nucleotide sites, respectively, between a parent and its offspring. This simple statement of a model leaves out many potentially important things, such as possible variation in these parameters across a genome, among alleles, or through time, and such details should be added to the model later, as needed. In the diffusion limit, in which time is rescaled by $2N$, random genetic drift has rate 1 and the strengths of selection, mutation, and recombination are given by $2Ns$, $2Nu$, and $2Nr$, respectively.

By tradition, the population mutation parameter is defined as $\theta = 4Nu$ or *twice* the population rate of mutation on the diffusion time scale. For consistency, in what follows, the population parameters for selection and recombination will be defined as $\sigma = 4Ns$ and $\rho = 4Nr$. Note, however, that both α and γ are frequently used in place of σ and are often defined as $2Ns$ rather than $4Ns$.

Kimura's (1955a,b) groundbreaking achievement was to obtain the probability density function of the frequency of an allele at any future time, based on its current frequency at a single locus under the influence of natural selection and random genetic drift. Again, it is impossible to make predictions of this sort under most discrete models, in particular the Wright–Fisher model. Kimura's result spurred much further work on diffusion theory, which is reviewed in Ewens (1979, 2004).

Neutral Coalescent Theory

Kimura's use of diffusion theory in the 1950s flowed out of his desire to explore the dynamics of genetic drift, which Wright had promoted as having a dramatic role in evolution. In the 1960s, the focus of population genetics shifted to explaining the new observations of protein-sequence divergence between species and allozyme variation within species (Zuckerkandl and Pauling 1965; Harris 1966; Lewontin and Hubby 1966). These and subsequent data caused a dramatic shift in thinking about the role of natural selection, with Kimura (1968) and others (King and Jukes 1969) suggesting a predominant role for neutral mutations in evolution at the molecular level. Later, this concept was greatly expanded by Ohta (1973, 1992) to include weakly selected, or "nearly neutral" mutations. By emphasizing random genetic drift, the new theories did seem to provide a simple explanation for the observations of the day: that molecular differences between species accumulate surprisingly linearly with time and that natural populations harbor tremendous amounts of genetic variation.

Previously, with little data available, population geneticists had formed two opposing selectionist camps: the classical and balance schools (Dobzhansky 1955). Lewontin (1974) provides a clear analysis of how these gave way to the neutral theory when faced with explaining the observed high levels of polymorphism and divergence, and Crow (2008) recounts the arguments from the key perspective of someone whose career spanned this and other controversies. Lewontin (1974) argued against an unbridled focus on neutrality. He suggested the term neoclassical theory because, although a shockingly large fraction of the functional differences at the molecular level might be invisible to selection, most mutations still are disadvantageous and some or all adaptations must be driven by natural selection. Kimura recognized these points in his concept of the neutral theory: he was ready to accept that approximately 10% of amino acid substitutions between species could be driven by positive selection (Ohta and Kimura 1973) and that 85–95% of nonsynonymous, or amino-acid–changing mutations are substantially deleterious (Kimura 1983).

If one wishes to infer the action of natural selection, then neutrality is the appropriate null hypothesis. Mathematical models soon began to include the assumption that *all* genetic variation was neutral. Importantly, as new types of genetic data became available, models also included increasingly refined assumptions about the mutation structure of variation (Ewens 1972; Ohta and Kimura 1973; Moran 1975; Watterson 1975). It seems inevitable in hindsight that this process would lead to the consideration of the mathematical structure of ancestral relationships among sampled alleles, or gene genealogies. For example, the well-known Ewens sampling formula (Ewens 1972; Karlin and McGregor 1972) clearly has the fundamental structure of gene genealogies under neutrality embedded in it (see Hobolth et al. 2007 and Section 3 of Kingman 1982a). However, a major shift in orientation was required: from the prospective view of classical population genetics to the retrospective view of coalescent theory (Ewens 1990).

The paper by Watterson (1975) is the earliest in which gene genealogies and their relationship to genetic data are easily recognizable. Remarkably, if all variation is selectively neutral, it is possible to model just the ancestors of the sample and ignore the other members of the population. Figure 5.1 shows a hypothetical gene genealogy (or coalescent tree) for a sample of size $n = 6$. It is a binary tree that traces the ancestral lines of the sample back to their most recent common ancestor. Time is measured by vertical distance. The nodes in the tree represent coalescent events, at which a pair of ancestral lines reaches a common ancestor. Each branch in the tree depicts all of the genetic ancestors of particular members of the sample. Therefore, any polymorphisms in the data must be due to mutations along the branches. Watterson used this idea to derive the expectation and variance of the number of polymorphic nucleotide sites in a sample at a locus that does not undergo recombination.

Coalescence and diffusion are inextricably related as *dual* processes (for mathematical details, see Möhle 1999). Under identical assumptions to those made in diffusion theory, but for the moment without selection or recombination, each pair of ancestral lines coalesces independently, with rate equal to 1. In the Wright–Fisher model, this rate follows from the fact that two alleles descend from a common parental allele with probability $1/(2N)$, in each generation looking back. In considering the limit $N \to \infty$, the sample size n is treated as a (finite) constant, which is the reason that all coalescent events occur between pairs of alleles rather than larger numbers.

Because every pair of ancestral lines coalesces with rate equal to one, neutral gene genealogies are random-joining trees. Also, the time, T_i, during which there exist exactly

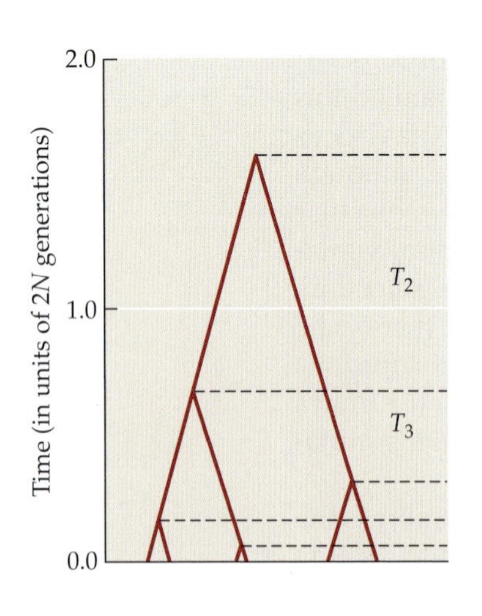

FIGURE 5.1 Example Gene Genealogy of a Sample of Size $n = 6$, with Coalescence Times (T_i on the Right) Drawn to Match Expectations from the Standard Neutral Coalescent

i lines ancestral to the sample is exponentially distributed with mean $2/(i(i-1))$, which is the inverse of the number of possible pairs of i lines. As a result, T_2 tends to be the longest coalescent interval, comprising about half of the time to the most recent common ancestor ($T_{MRCA} = T_n + T_{n-1} + \ldots + T_2$). The gene genealogy in Figure 5.1 is drawn with the times, T_i, equal to their expected values. On the coalescent or diffusion time scale, the expected value of T_{MRCA} is equal to:

$$E(T_{MRCA}) = 2(1 - 1/n) \quad \textbf{(5.1)}$$

This equation can be translated into generations in the Wright–Fisher model by multiplying by $2N$. For large samples, it converges to $4N$ generations, which is not unexpected because this is also the expected fixation time for an allele from forward-time diffusion theory without selection (Kimura and Ohta 1969). With a choice of mutation model, this standard coalescent is an efficient way of predicting patterns of neutral variation. In modeling or simulation, we simply generate the tree and the times, then place mutations randomly along each branch with rate $\theta/2$ per site.

Kingman (1982a,b) gave the mathematical proof of the coalescent process. Independently, biologists were introduced to the theory of gene genealogies and their biological relevance by Hudson (1983a, 1990) and Tajima (1983). Tavaré (1984) helped bridge the gap between the biological and the mathematical models as well as between diffusion theory and coalescent theory. Hudson also studied the effects of recombination, and described how to simulate gene genealogies both with and without recombination (Hudson 1983a, 1983b, 2002). Recombination complicates the coalescent process substantially, but is difficult to brush aside because the per-site rates of mutation and recombination (θ and ρ) appear to be of the same order of magnitude in many species (Table 4.1 in Lynch 2007 gives examples). Thus, a growing number of neutral coalescent approaches to inference take recombination into account (e.g., Becquet and Przeworksi 2009). When natural selection is added to the coalescent, it becomes absolutely critical to include recombination.

Coalescent theory is best known today for having produced a repertoire of tools for statistical inference under the assumption that genetic markers (i.e., polymorphisms) are neutral. In the 1980s, this theory was fueled by the remarkable utility of uniparentally inherited, non-recombining animal mitochondrial DNA for uncovering plausible histories of population expansions and contractions as well as complex patterns of geographic subdivision in many different species (Avise et al. 1987). Appropriate statistical machinery was developed, and work flourished after the introduction of Markov chain Monte Carlo, importance sampling, and Bayesian approaches in computational methods of coalescent-based inference. Stephens and Donnelly (2000), Marjoram and Tavavré (2006), and Felsenstein (2007) together give a comprehensive review of methods. Estimates made using these tools are sensible enough that they have contributed to broad debates about ancient processes and events (e.g., Hey 2005).

Genomic Data and the Modeling Response

The continued development after the 1960s of technologies for measuring genetic variation—from restriction-enzyme digests of mitochondrial DNA (Avise et al. 1979; Brown 1980) to early DNA sequence data (Aquadro and Greenberg 1983; Kreitman 1983)—has led to the massive contemporary genome-sequencing efforts, such as the 1000 Genomes Project (http://www.1000genomes.org) and the Personal Genome Project (http://www.personalgenomes.org). As new data are gathered, paradigms are questioned. At present, genomic polymorphism and divergence data from a growing number of taxa suggest staggering amounts of positive selection. For example, Hahn (2008) cites estimates from *Drosophila melanogaster* and *D. simulans* that 30–94% of amino acid substitutions between species have been driven by positive selection. Halligan et al. (2010) put this figure at 57% for substitutions between the mouse *Mus musculus castaneus* and the rat *Mus famulus*. Large fractions of positively selected substitutions (~50%) have also been reported for noncoding regions in *Drosophila* (Begun et al. 2007). These dramatic observations do not seem to extend to some other well-studied species, in particular *Arabidopsis* (Bustamante et al. 2002) and humans. Sella et al. (2009) summarize a number of studies of humans and conclude that approximately 10% of amino acid substitutions have been driven by positive selection.

The latter figure of approximately 10% is remarkably similar to the initial estimate that was considered broadly consistent with the neutral theory (Ohta and Kimura 1971; Kimura 1983). As Lewontin (1974) pointed out, the discovery that some genetic loci have been the targets of selection does not invalidate the neoclassical view. In addition, Thornton et al. (2007) and Jensen (2009) caution against drawing strong conclusions based on current methodologies and data. Still, Hahn (2008) argues that a new theory is required, one in which selection plays the major role, and Sella et al. (2009) agree that at least some parts of the neutral or neolcassical theory are in dire need of an overhaul (see Kolaczkowski and Kern, Chapter 6).

One major tenet of population genetics, which sits at the base of neutral theory, is clearly not in question: a large fraction of mutations alter function so ruinously that they are extremely unlikely to be observed, either as substitutions between species or as polymorphisms within species. In fact, one of the principal methods of estimating the fraction of positively selected amino acid changes (Smith and Eyre-Walker 2002) is to use the ratio of nonsynonymous to synonymous polymorphisms within species to set a low baseline expectation for the ratio of nonsynonymous to synonymous substitutions between species. Positively selected amino acid changes may then be uncovered, by an excess of nonsynonymous substitutions above the baseline expectation, even if the number of nonsynonymous substitutions per site is much smaller than the number of synonymous substitutions per site. Such considerations lead to sophisticated yet tractable statistical approaches to estimating selection in the case in which sites may be assumed to be independent of one another (Sawyer and Hartl 1992; Sawyer et al. 2003).

Positive natural selection for adaptive traits has been the primary source of excitement among workers studying the genomic effects of selection. In addition to estimating the overall prevalence of positive selection, effort has focused on identifying loci that have been subject to selection. It is possible to identify such loci because the fixation of an advantageous allele at one locus affects loci nearby via a phenomenon known as genetic hitchhiking (Maynard Smith and Haigh 1974; Kaplan et al. 1989). The primary signal of genetic hitchhiking is a reduction in variation around the site of selection, but a number of subtler effects occur as well (Nielsen 2005). The term selective sweep is used loosely to mean the fixation of a positively selected allele or the attendant reduction in variation. Thornton et al. (2007) review genomic scans for recent selective sweeps in *Drosophila* that have identified large numbers of loci. In humans, Williamson et al. (2007) suggest that recent hitchhiking affects 10% of sites in the genome. Although demographic factors can lead to false-positive inferences of selection (Thornton et al. 2007) and divergence and polymorphism data give rather different estimates of the prevalence of selection (Jensen 2009), these recent findings motivate the development of coalescent approaches to modeling selective sweeps (see Kolaczkowski and Kern, Chapter 6).

Selection and Genetic Drift Forward in Time

The models of population genetics are often based on diffusion approximations. But, are the assumptions of diffusion theory reasonable for loci undergoing positive selective sweeps? Diffusion theory assumes that s, u, and r are very small and N is very large. Formally, the limit $N \to \infty$ is taken with $\sigma = 4Ns$, $\theta = 4Nu$, and $\rho = 4Nr$ held constant. However, simulations show that many results of diffusion theory are very accurate for moderate values of the discrete model parameters, such as $N = 100$ and $s = 0.01$. The occurrence of a sweep implies that selection is strong in some sense, so it must be asked specifically whether it is reasonable to use a model in which s is assumed to be much less than one. Estimates from recent selective sweeps suggest that the answer is yes. One example, not from *Drosophila* but from deer mice, was reported recently by Linnen et al. (2009). They estimated $s = 0.0056$ for a recently swept allele affecting pelage color of mice in the Nebraska Sand Hills. Their results are similar to the larger estimates for swept loci in *Drosophila* (Thornton et al. 2007; Sella et al. 2009), so assuming small s appears safe. Still, a sweep certainly indicates that selection has overwhelmed random genetic drift. In the diffusion model, this effect occurs when σ is large. Estimates of σ for swept loci in *Drosophila* range from values in the tens to values in the thousands (Thornton et al. 2007; Sella et al. 2009). In sum, the diffusion with large σ appears to be a good starting point for modeling selective sweeps.

Note that there is an entirely different diffusion model in population genetics (Norman 1975), which may be more appropriate for large σ. Unfortunately, few results are available for this Gaussian diffusion model, so we will not pursue it in what follows. Ewens (1979, 2004) points out that the Gaussian diffusion and the standard one should overlap for certain parameter values (i.e., large values of σ), and this result is illustrated for strong balancing selection and mutation in Wakeley and Sargsyan (2009).

Here, we will define a sweep as the event that a positively selected allele, which starts in frequency $1/(2N)$ as a new mutation, reaches frequency 1 or fixes in the population. For simplicity in what follows, we will assume that all parameters are constant over time. Diffusion theory can tell us about the distribution of trajectories the allele will take on its way to fixation. Knowing this distribution is helpful because many things of interest are functions of the allele-frequency trajectory. For example, the average duration of the sweep is identical to the expected value of the length of the trajectory. Other quantities, such as the probability of coalescence during a sweep or the chance of observing a sweep in a sample of genetic data, also depend on the characteristics of allele-frequency trajectories.

To illustrate the simplest dynamics of genetic evolution and using the emerging estimates from *Drosophila* and humans as a backdrop, let us imagine a locus comprised of a single advantageous site, 1000 deleterious sites, and 1000 neutral sites. For humans, estimates of θ are on the order of 0.001 and estimates of the effective population size are on the order of 10,000. Thus, we will use a Wright–Fisher model with $N = 10^4$ and $u = 2.5 \times 10^{-8}$ as a model for a hypothetical human locus. Then, the total rates of mutation are $\theta_a = 0.001$ for the advantageous site and $\theta_d = \theta_n = 1.0$ (i.e., 1000×0.001) for the deleterious and neutral sites. Let us also assume fairly strong selection, in particular $\sigma_a = 100$ and $\sigma_d = -100$, and with $N = 10^4$, this corresponds to $|s| = 0.0025$ for advantageous and deleterious mutants. Of course, $\sigma_n = 0$. Estimates of θ for *Drosophila* are somewhat more than an order of magnitude greater than those for humans. For computational efficiency, we will obtain the parameters for a hypothetical *Drosophila* locus simply by multiplying the human diffusion-scale parameters by ten, so that $\theta_a = 0.01$, $\theta_d = \theta_n = 10.0$, $\sigma_a = 1000$, and $\sigma_d = -1000$. In the simulations for the present work, this modification was realized by using the same per-generation parameters as for humans but with $N = 10^5$ instead of $N = 10^4$.

This idealized model will serve to generate intuition about selective sweeps and the relative magnitudes of the processes involved. In relation to the estimates of rates of adaptive substitution in humans and *Drosophila*, with these parameters, our model predicts that approximately 9% of substitutions will be driven by positive selection in humans and about 50% of substitutions will be driven by positive selection in *Drosophila*. These percentages are derived, in the usual way, by multiplying the per-generation rates of introduction of each type of mutation ($\theta_a/2$, $\theta_n/2$, $\theta_d/2$) by their probabilities of fixation from diffusion theory,

$$P(\text{fix}) = \begin{cases} \dfrac{1}{2N} & \text{if } \sigma = 0, \\ \dfrac{1-e^{-\sigma/(2N)}}{1-e^{-\sigma}} & \text{if } \sigma \neq 0. \end{cases} \quad (5.2)$$

Note that this is the standard result, which is sufficient for our purposes, and does not include the $s \rightarrow s/(1+s)$ correction suggested by Bürger and Ewens (1995).

Then for humans, we have $P(\text{fix})$ approximately equal to 5×10^{-3}, 5×10^{-5}, and 1.9×10^{-46} for advantageous, neutral, and deleterious mutations, respectively. In our *Drosophila* model, the corresponding values are 5×10^{-3}, 5×10^{-6}, and 2.5×10^{-437}. Note that when $\sigma = 4Ns$ is large and s is small, as is true here, the second case in Equation 5.2 gives $P(\text{fix}) \approx 2s$, which is the classical population genetic result seen previously. The probabilities of fixation of advantageous mutants are thus the same in our hypothetical models for human and *Drosophila*, while the probabilities for neutral mutants differ by a factor of ten due to the difference in population size. In both cases, deleterious mutations are exceedingly unlikely to fix.

Figure 5.2 shows the trajectories of advantageous alleles in simulations for our humans (see Figure 5.2A,B) and *Drosophila* (see Figure 5.2C,D). First, a large number of trajectories was simulated from the introduction of a mutant in a single copy, until the mutant either fixed or went extinct. Then, the

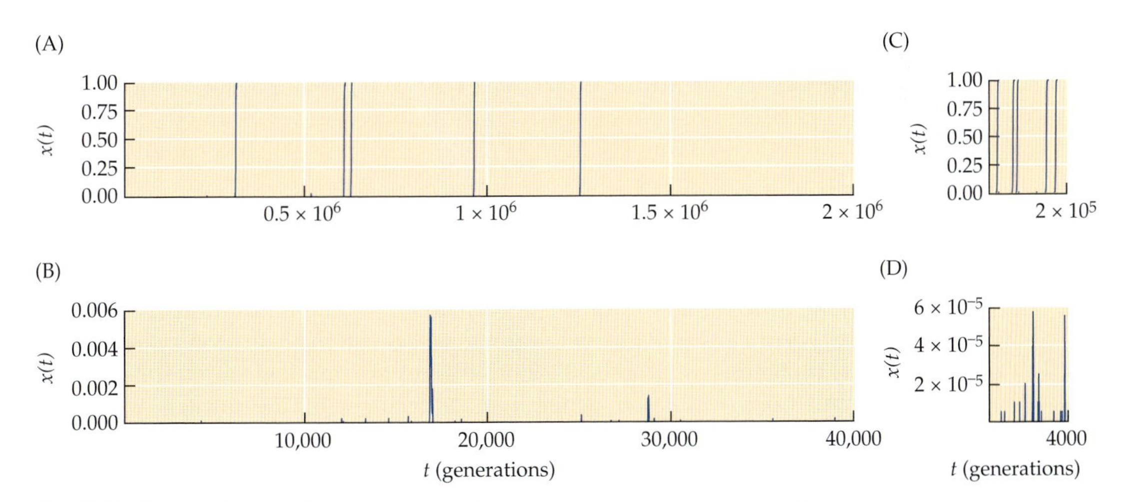

FIGURE 5.2 Simulated Allele-Frequency Trajectories of Advantageous Mutants at the Hypothetical Human and *Drosophila* Loci Panels (A) and (B) show results for humans, and panels (C) and (D) show results for *Drosophila*. (A) and (C) show advantageous mutations that reached high frequencies. (B) and (D) show advantageous mutations that went extinct.

origination times of the mutations were generated using the per-generation population rates of advantageous mutation, $\theta_a/2$. Thus, these simulations are of independent trajectories; they do not take interference between alleles into account. This methodology is reasonable for humans because successful sweeps are fairly well separated in time (see Figure 5.2A) and alleles go extinct quickly when sweeps fail (see Figure 5.2B). It also does not invalidate the qualitative points we will draw from the figure as a whole. Simulations of each trajectory were done according to the discrete Wright–Fisher model, with the parameters above.

Before looking in detail at Figure 5.2, note that our simulations follow the fairly common convention of using one-locus, two-allele dynamics to portray a situation that is probably much more complicated. For example, we have assumed that *every* mutation at the advantageous site is assumed to have the same selection parameter σ_a. Recalling the classical A_1/A_2 model previously described, this assumption would be realized if the average number of offspring of A_2A_2 is mysteriously reset from $1 + 2s$ back to 1 at the conclusion of the sweep, and the next mutation again has selection coefficient s. We have also assumed here that the strength of selection is constant over time, which might not be realistic even within a single sweep. Gillespie (1991, 2004b) has shown repeatedly that key features of the dynamics of fully specified, multi-allele models, in which selection parameters differ among alleles and may change over time, are simply not captured using two-allele approaches, and he has further argued that these shortcomings are fatal to neutral theory explanations of the molecular evolution. For our hypothetical model, they are of somewhat less concern because our focus here is the much shorter time scale of single sweeps.

Figure 5.2 shows the frequency trajectories of advantageous alleles over a period of time during which we expect 1000 advantageous mutations to occur. Sweeps appear as nearly vertical curves, in which the frequency (x) of an allele rises quickly from $1/(2N)$ to 1. Because the probability of fixation is approximately 5×10^{-3} in the models for both humans and *Drosophila*, we expect about five selective sweeps in each. This is exactly what was observed in these particular simulations, but just by chance: in both cases, the number of sweeps is Poisson distributed with a mean of approximately 5. The time it takes to observe 1000 advantageous mutations is 10 times shorter in *Drosophila* than in humans, because the rate of introduction of advantageous mutations ($\theta_a/2$) is 10 times greater. If we ran our *Drosophila* simulations over 2×10^6 generations, as occurred for the humans, we would expect to see 50 sweeps. Time in Figure 5.2 is measured in generations, and accordingly, the panels for *Drosophila* (see Figure 5.2C,D) are one-tenth the length of the panels for humans (see Figure 5.2A,B).

Panels A and C display the entire range of frequencies, and on this scale, only a handful of the trajectories is visible. At our hypothetical locus, with its one positively selected site, we expect 1 advantageous mutation to occur about every 2000 generations in humans and 1 in about every 200 generations in *Drosophila*. Even focusing on much smaller frequencies, as in panels B and D, the trajectories of most alleles are difficult to see. Recall than only about 5×10^{-3} of advantageous mutations will sweep to fixation. The other approximately 99.5% of them go extinct, and they do so very quickly, without ever reaching substantial frequencies. The trajectories of the deleterious alleles at our loci are not shown in Figure 5.2. For the values of σ_d (–100 and –1000), deleterious alleles will essentially never fix in the population. They enter the population and may drift to frequencies of 1% or so, but then are lost.

In our hypothetical model for humans, advantageous and deleterious mutations are not typically expected to have much effect on levels of neutral polymorphism. Either they will never reach appreciable frequencies or they will sweep quickly through the population and only rarely be observed. Sweeps occur on average only every 400,000 generations, while the effective population size, which sets the average time for neutral variation to reach equilibrium levels, is only 10,000. The situation is rather different for our *Drosophila* model, in which sweeps occur at the locus every 40,000 generations and the effective population size is 100,000. In this case, we would expect sweeps to greatly affect levels and patterns of neutral polymorphism.

Methods of inference of selective sweeps often assume that the population has been sampled just at the end of the sweep. In applications, this assumption needs to be justified, because *a priori* the time back to the last sweep is unknown. In our model, it would be roughly exponentially distributed with mean $(P(\text{fix})\theta_a/2)^{-1}$. Sweeps appear to go to completion almost instantaneously on the time scale depicted in Figure 5.2. However, by traveling back to the end of the last sweep that occurred in each case, and changing the timescale, we can see the shape of sweeps. Figure 5.3 shows this for the hypothetical humans (panel A) and *Drosophila* (panel B), with

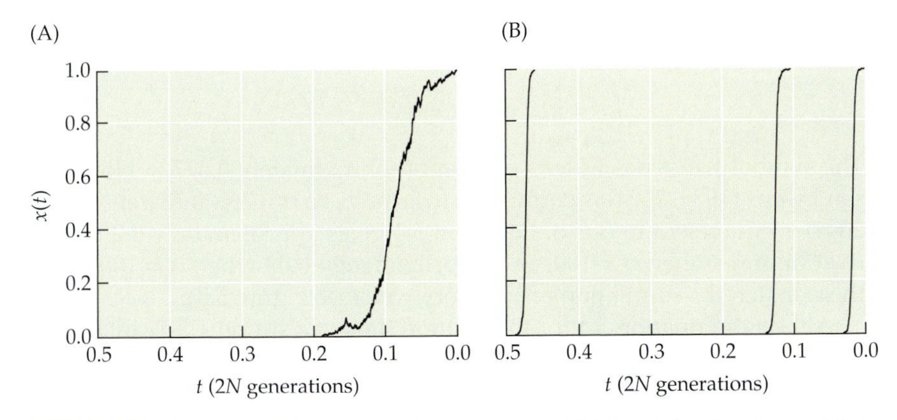

FIGURE 5.3 Example of Population Ancestries, in Which a Selective Sweep Has Just Reached Completion Panel (A) is from the simulations depicted in Figure 5.2A, and panel (B) is from the simulations depicted in Figure 5.2C.

the time scale given in the coalescent or diffusion units of $2N$ generations. In both cases, the total range is 0.5 on the new time scale, which is equivalent to N generations (10,000 for humans and 100,000 for *Drosophila*). Also, time now flows from the moment the population is sampled back into the past, as is the custom in coalescent modeling. In contrast to Figure 5.2, only those trajectories that went to fixation are shown in Figure 5.3.

Figure 5.3 illustrates that sweeps tend to follow sigmoidal trajectories, with allele frequencies changing relatively slowly when $x(t)$ is close to 0 or 1, but moving rapidly through the middle frequencies. With time measured in units of $2N$ generations, more strongly favored alleles will sweep more quickly through the population ($\sigma_a = 100$ in A versus $\sigma_a = 1000$ in B). In addition, in cases like our *Drosophila* model shown in Figure 5.3B, the rate of occurrence of selective sweeps might be such that several will have happened in the recent ancestry of a locus under study. We can recall the results of neutral coalescent theory: (1) that the average time back to the common ancestor for a sample of size $n = 2$ is equal to 1 (i.e., $2N$ generations) and (2) the average time to the most recent common ancestor of all members of a large sample is approximately 2 (i.e., ~$4N$ generations).

The standard diffusion model, with large σ, allows these observations to be quantified. A fundamental result of diffusion theory in population genetics, due to Ewens (1963, 1964), concerns the average time that an allele, which begins in frequency p and sweeps through the population, spends at each frequency x on its way to fixation. The function, called $t^*(x;p)$ and given as equation 5.52 in Ewens (1979) or as equation 5.53 in Ewens (2004) has the interpretation that

$$\int_{x_1}^{x_2} t^*(x;p)dx \tag{5.3}$$

is the average amount of time, on the diffusion time scale that the allele frequency spends in the interval (x_1,x_2) before the allele fixes in the population. Integrating over the entire frequency range gives the expected total sweep time of a new mutant. When σ is large, this may be approximated as

$$t_{\text{fix}} = \int_0^1 t^*(x;p)dx \approx \frac{4(\log(\sigma)+\gamma)}{\sigma} \tag{5.4}$$

The symbol γ above is Euler's constant (approximately 0.5772). Note that s in Ewens (1979, 2004) is equivalent to $2s$ here, so $\alpha = 2Ns$ in Ewens (1979, 2004) is equivalent to our σ. Equation 5.4 gives approximately 0.2 for the fixation time when $\sigma = 100$, and approximately 0.03 when $\sigma = 1000$, and these match the simulation results very well (see Figure 5.3).

Although Equation 5.4 for the fixation time of a strongly advantageous allele starting from a single copy seems to have appeared in the literature only recently (Hermisson and Pennings 2005; Etheridge et al. 2006; Hermisson and Pfaffelhuber 2008), it illustrates something that has been known for several decades. That is, deterministic equations for allele-frequency trajectories can drastically overestimate the amount of time an allele will spend in small frequencies (Ewens 1979: 149). If an advantageous allele is going to sweep to fixation, it must move away from the boundary ($x = 0$) faster than the deterministic equations predict. Still, it is not uncommon to see deterministic results and methods used in this context in the biological literature. The deterministic model (e.g., equation 1.28 in Ewens 1979, but with our s) gives

$$\int_{\frac{1}{2N}}^{1-\frac{1}{2N}} (sx(1-x))^{-1}dx = \frac{2\log(2N-1)}{s} \approx \frac{2\log(2N)}{s} \quad \textbf{(5.5)}$$

for the fixation time, in generations. Equation 5.5 appears in many publications. It may be compared to the diffusion result above by multiplying t_{fix} by $2N$ and rearranging:

$$2Nt_{\text{fix}} \approx \frac{2(\log(2N)+\log(2s)+\gamma)}{s} \quad \textbf{(5.6)}$$

Recall that $\log(a) < 0$ when $0 < a < 1$ and tends to negative infinity as a tends to zero. Even if σ is large, it might not be reasonable to assume that $\log(2N)$ is much greater than both $-\log(2s)$ and γ. For any values of s we are likely to consider, the deterministic result will overestimate the diffusion result. In the human model, the diffusion Equation 5.6 result gives 4146 generations, while the deterministic Equation 5.5 gives 7923 generations. In the *Drosophila* model, the corresponding numbers are 5988 generations and 9765 generations.

The fact that allele-frequency trajectories are sigmoidal is key to understanding coalescent models of selective sweeps because the rates of events in the ancestry of a sample depend on the allele frequencies. As a final point about diffusion models of sweeps before turning to coalescent models, Etheridge et al. (2006) have recently obtained the very interesting result that as σ grows, the fraction of the time that the allele spends in the middle frequencies becomes negligible; see their Lemma 3.1 and note that their α corresponds to our $\sigma/2$. Specifically, the time spent going from frequency ε to $1 - \varepsilon$ becomes negligible, for *any* $0 < \varepsilon < 1$, so that the allele ultimately spends half of t_{fix} in the interval $(0,\varepsilon)$ and the other half in the interval $(1 - \varepsilon,1)$. Because of this, it is possible to make some detailed calculations concerning the approximate behavior of the coalescent process during selective sweeps when σ is large (Etheridge et al. 2006).

The following investigation of coalescent models with selection draws upon the results of diffusion theory just discussed. We will focus on selective sweeps, in particular the effect these have on ancestral processes at nearby neutral loci. It is increasingly clear that the hitchhiking (Maynard Smith and Haigh 1974), which reduces polymorphism levels around the site of a sweep, has affected many loci. For example, Sabeti et al. (2007) listed 22 regions in the human genome where selection appears to have decreased polymorphism over spans of 0.2 to 3.5 Mb, at least in some populations.

Kimura (1983) did not cite Maynard Smith and Haigh (1974), and yet he accepted the estimate that roughly 10% of substitutions might be driven by positive selection (Ohta and Kimura 1971). Again, this is roughly what our hypothetical human model predicts. On the one hand, it is true that polymorphism levels at our human locus should not typically deviate from neutral predictions, because sweeps will occur only about once every $20 \times 2N$ generations. Using the diffusion result for t_{fix}, with $\sigma = 100$, the chance of catching a sweep in progress is only about 1%. On the other hand, if a large number of loci are surveyed, we should not be surprised to find several that have recently been affected by sweep. It is these loci that current genome-wide scans for selection may uncover, and coalescent models are being developed to aid both in their identification and to make estimates of the strength and timing of selection.

Coalescent Models with Selection

When selection operates, for example with two alleles A_1 and A_2, then the population is structured by allelic type such that the average number of offspring of genetic lineages labeled A_1 differs from that of lineages labeled A_2. In order to model the genetic ancestry of a sample, we need to keep track of these labels. With the present concern for selective sweeps as motivation, the following sections outline three different approaches to this problem. A fourth coalescent approach to selection, the ancestral selection graph (Krone and Neuhauser 1997; Neuhauser and Krone 1997), will not be reviewed because it has not been extended to apply to selective sweeps. Hitchhiking will be a key phenomenon in our investigations, because neutral genetic markers contain information about past histories of selection, just as they do about other demographic processes and events. The fact that strongly selected adaptive substitutions have probably occurred at a small minority of sites in the genome means that the bulk of signals of selection will be in patterns of linked variation. Recombination is the process that modulates the effect of linkage, so recombination is fundamental in what follows.

The Structured Coalescent Approach

Hudson and Kaplan (1986) showed how conditioning on the allelic types of a sample alters the coalescent process, in a way that is similar to the effect of geographic structure and migration. Two lineages with the same allelic type may coalesce, but two lineages with different types must wait for mutation to change the type of one or the other. Kaplan et al. (1988) applied this idea to a locus under selection, showing that rates of coalescence and mutation in the ancestral process depend on the frequencies of the two alleles. Hudson and Kaplan (1988) extended the model to describe the coalescent process at a linked neutral locus, conditional on the frequency trajectory at the selected locus. Darden et al. (1989) described the joint process of coalescence at the linked neutral locus and changes in allele frequencies at the selected locus by the standard diffusion. Barton et al. (2004) investigated the model of Darden et al. (1989) more rigorously and found boundary conditions necessary to allow analytical work. Kaplan et al. (1989) considered the specific application of the structured coalescent approach to a strong selective sweep and investigated the effect of sweeps on variation at the linked neutral locus.

The structured coalescent approach has led to a number of useful simulation methods (Slatkin 2001; Kim and Stephan 2002; Przeworski 2003; Coop and Griffiths 2004), in which an allele-frequency trajectory is generated, and then the structured coalescent process is run, conditional on the trajectory. A main goal in developing these simulations is to devise methods of estimating the characteristics of sweeps, such as the selection parameter σ and the time the last sweep began. Kim and Wiehe (2009) review the issues and available software.

A key feature of the structured coalescent approach to selection is that the rate of coalescence within an allelic class depends inversely on the allele frequency. Consider our selectively favored allele A_2, whose frequency is $x(t)$ at time t in the past, measured in units of $2N$ generations. If there are i ancestral lineages of type A_2, then the rate of coalescence between any pair of them is $1/x(t)$, and the total rate is

$$\frac{i(i-1)}{2x(t)} \tag{5.7}$$

If $x(t) = 1$, then the rate is the same as in the standard neutral coalescent. However, if $x(t) < 1$, then the rate is *greater* than in the standard neutral coalescent. The reason is that, when $x(t)$ is smaller, there are fewer possible parents of the i lineages, so the probability of a common ancestor in a single generation is larger. The same notion applies to lineages that possess the A_1 label, but with $1 - x(t)$ instead of $x(t)$.

In considering the effects of linkage, one can imagine a site or locus B that sits at a distance m from the selected locus A. Let m be in units of base pairs and the total scaled rate of recombination be

$$\rho^* = m\rho \tag{5.8}$$

in which ρ is the per-site rate of recombination previously defined before. Recall that $\rho = 4Nr$, and that $\rho/2$ is the rate of recombination between two adjacent base pairs on the coalescent time scale. In defining ρ^* as the product $m\rho$, we have implicitly assumed that m is small enough to ignore interference between cross-over events. Note also that, since (by assumption) variation at locus B is neutral, it is not necessary to specify allelic types at this locus in modeling the genealogy. Rather, as discussed previously, mutations can be placed randomly along on the branches of genealogy after it has been generated.

Each of the members of a sample of size n taken at the B locus will be linked either to an A_1 allele or to an A_2 allele at the selected locus, and the same is true of the ancestral lineages of the sample. It is this linkage that makes the ancestry at the B locus differ from the predictions of the standard neutral coalescent. In modeling the ancestry of the B-locus sample, the appropriate label for each B-locus lineage is the allelic type at the A locus to which it is linked. If i B-locus lineages are linked to A_2 alleles, then the rate of coalescence between each pair is $1/x(t)$ and the total rate is identical to the total rate for the A locus given above. This will be true as long as m is not too large, as it neglects the possibility that both recombination and coalescence occur in a single generation. Crucially, B-locus lineages can switch labels as they are followed back in time. This occurs when a lineage ancestral to the sample was the product of a recombination event in an individual who was heterozygous at the A-locus. If i B-locus lineages are linked to A_2 alleles, then the total rate of this type of event at time t in the ancestral process is

$$i\rho^*(1-x(t))/2 \qquad \textbf{(5.9)}$$

with ρ^* as just defined. If an event of this type occurs, one of the B-locus lineages switches types at the A locus (from A_2 to A_1). To explain the equation above, each of the i lineages hits a recombination event between A and B with rate $\rho^*/2$, but only $1 - x(t)$ of these events occur in heterozygous individuals. There is no additional 2 in the formula, as might be expected given the Hardy–Weinberg proportions implicitly assumed, because we have further assumed that the type of one allele is known. For B-locus lineages that are linked to A_1 alleles, the rate of label switching depends on $x(t)$ instead on $1 - x(t)$.

As suggested previously, B-locus lineages can also escape the sweep due to mutations at the A locus. The probability of this depends on the mutation rate (θ_a) at the A locus but not on distance m between the loci, and over the entire sweep is of order θ_a (Hermisson and Pennings 2005). For simplicity, we will ignore this relatively unlikely possibility.

Figure 5.4A shows a hypothetical gene genealogy of a sample of size $n = 6$ at the B locus, under this structured coalescent model, for a population that has experienced a recent sweep at the A locus. The B-allele–frequency trajectory, shown in purple, is from the simulations already described. Blue squares mark recombination events by which two B-locus lineages were able to escape the sweep by switching labels. As a result, these two members of the sample may carry mutations at the B-locus that occurred in the ancestral population before the sweep. If there is no recombination, then the entire sample will coalesce during the sweep, and any variation in the sample must be due to mutations that occurred since the sweep. The hypothetical time scale in Figure 5.4A can be compared to the standard neutral one in Figure 5.1. The lineages that predate the sweep travel up, out of the figure, because their expected time to common ancestry (see Equation 5.1)

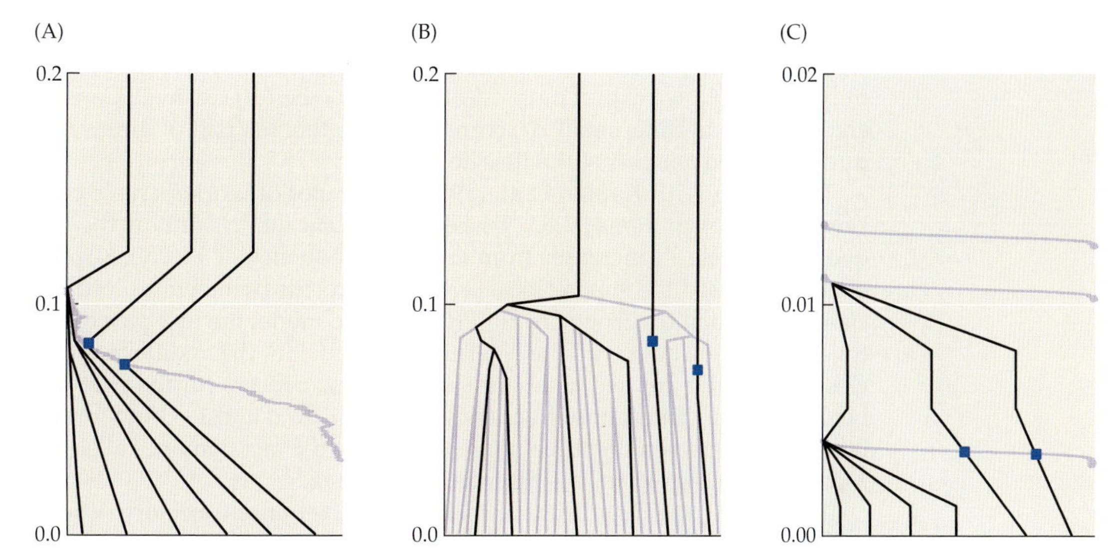

FIGURE 5.4 Hypothetical Gene Genealogies A sample of size $n = 6$ is at a neutral locus, linked to a selected locus, showing the three coalescent approaches to modeling natural selection. Panel (A) depicts the structured coalescent approach, (B) depicts the Yule process approximation, and (C) depicts the multiple-merges coalescent for recurrent selective sweeps. Possible characteristic ranges of time (measured in units of $2N$ generations) for each model are displayed on the vertical axes. Black lines show the ancestry of the sample, while unobserved allele-frequency trajectories and genetic lineages that are not directly ancestral to the sample are shown in purple. Blue squares mark recombination events that allow linked neutral lineages to escape a sweep.

is much greater than the range given in Figure 5.4A. Among these lineages, we would expect to see neutral levels of polymorphism. Thus, polymorphism will be reduced at the B locus only to the extent that extra coalescent events occur during the sweep. Because of the dependence on $\rho^* = m\rho$, larger reductions will occur when locus B is closer to locus A.

In the ancestral process depicted in Figure 5.4A, the frequency of A_2 decreases from 1 down to $1/(2N)$ as we follow it back through the sweep, and then the single A_2 is converted into an A_1 allele by mutation. For the B-locus alleles that are linked to A_2, the rates of coalescence and escape by recombination will *increase* as $x(t)$ decreases. The rate of coalescence increases very dramatically because it depends on $1/x(t)$, while the rate of escape by recombination increases mildly, like $1 - x(t)$. These rates and the changes in $x(t)$ make coalescent analyses of selective sweeps complicated. However, we can see from the large-σ diffusion approximation for t_{fix} (see Equation 5.4) that the duration of a sweep will be small on the coalescent time scale and will become negligible if σ is very large (recall that for $\sigma = 1000$, we have $t_{fix} \approx 0.03$). The results of Etheridge et al. (2006) imply that a strong sweep will be divided fairly neatly into two halves. Because of the way the rates of coalescence and escape by recombination depend on the frequency of A_2, we expect most events to occur when $x(t)$ is small. Considered forward in time, small $x(t)$ corresponds to the first half of the sweep, during the convex part of the trajectory.

Beginning with Kaplan et al. (1989), a number of workers have considered approximations to $x(t)$, based on different models of how the frequency of allele A_2 increases from its initial frequency of $1/(2N)$. Kaplan et al. (1989) used the supercritical branching process (which produced the classical population genetic result $P(\text{fix}) \approx 2s$) to model the first part of the trajectory, followed it with the deterministic model for the middle frequencies, and finally applied a subcritical branching process for the part just before fixation. They chose frequency cutoffs of $10/\sigma$ and $1 - 10/\sigma$ for the boundaries between the three phases, based on the fact that once allele A_2 reaches frequency $10/\sigma$, it is essentially sure to fix.

Later, using the deterministic model over the entire trajectory, Wiehe and Stephan (1993) were able to obtain an analytical result for the decrease in neutral heterozygosity (i.e., for a sample of size $n = 2$). Following considerations of Kaplan et al. (1989), the equation of Wiehe and Stephan (1993) captures the effects of recurrent selective sweeps, which is when a neutral locus is linked to several selected loci that undergo adaptive fixation events at some rate. There has been a great deal of interest in recurrent selective sweeps, and the equation of Wiehe and Stephan (1993) has been used extensively (Jensen 2009; Sella et al. 2009).

Barton (1998) inserted a fourth phase into the trajectory, between the initial branching process and the deterministic model, based on the finding by Otto and Barton (1997) of an acceleration above deterministic predictions over a range of frequencies of A_2. Barton (1998) obtained several new analytical results (also for samples of size $n = 2$), in particular probabilities of identity by descent and, by extension, distributions of pairwise coalescence times.

Eriksson et al. (2008) recently suggested modeling sweeps deterministically, but using the average time that the advantageous allele spends in each frequency class in place of the actual deterministic predictions. The authors employed a Moran model but were, apparently, unaware that some of their results were previously known (Ewens 1963). In the context of the standard diffusion, their method is equivalent to assuming that the fixation event follows the expected trajectory $t^*(x;p)$ exactly. Although Eriksson et al. (2008) offered no mathematical justification for their approach, it has some appeal because $t^*(x;p)$ captures the fact that A_2 moves quickly through the small frequencies.

The Yule Process Approximation

Results for the decrease of heterozygosity are simple and useful, but greater power should be possible in coalescent approaches to samples larger than $n = 2$. For this reason, the computational methods of inference discussed previously for neutral models aim to compute likelihoods for any samples of any size. The likelihood captures all of the information in the data. The goal of simulation-based methods like the one of Coop and Griffiths (2004) is to apply this power of the (structured) coalescent approach to inferences about selection. However, because it is necessary to account for the unknown allele frequency in the population as it changes through time, structured coalescent methods for computing likelihoods are computationally costly. In this section, we consider a promising new model called the Yule process approximation.

In addition to pairwise measures, Barton (1998) investigated the distribution of family sizes descending from a sweep, using simulations. Here, families are the descendants (in the sample) of each lineage that emerges from the sweep, looking backward in time. For example, in Figure 5.4A there are three families, and these have sizes 4, 1, and 1. If we knew the distribution of family sizes, we could derive key quantities, such as the distribution of allele frequencies after a sweep, using coalescent methods rather than forward-time analyses (Kim and Stephan 2002; Kim 2006). The Yule process approximation provides a way to generate the numbers and sizes of families that descend from the sweep, and the times of events in the ancestry of the sample.

Durrett and Schweinsberg (2004, 2005) introduced this approximation through an analysis of selective sweeps in a Moran population model. Etheridge et al. (2006) approached the same problem starting with the standard diffusion, showing that the Yule process approximation applies to a variety of models, in the limit as $N \to \infty$. In a presentation more accessible to biologists, Pfaffelhuber et al. (2006) describe a simulation algorithm for sampling gene genealogies at a neutral locus that is linked to a selected locus, based on a modified version of the Yule process approximation.

Durrett and Schweinsberg (2004, 2005) and Pfaffelhuber et al. (2006) assess the accuracy of the Yule process approximation compared to simulations of the discrete Wright–Fisher model and of the structured coalescent model with a deterministic trajectory. A number of authors, including Braverman et al. (1995), Simonsen et al. (1995), and Przeworski (2002) have used the deterministic model in simulations. In fact, these deterministic simulations are quite accurate for many purposes, especially for small samples. However, Pfaffelhuber et al. (2006) showed that the Yule process approximation gives better predictions for the distribution of family sizes in larger samples.

The Yule process approximation is derived under the assumption that σ is large in the model given, of a neutral locus B sitting near the selected locus A. For very large σ, recall that events in the ancestry of the sample are concentrated in the first half of the sweep (forward in time), when the frequency of A_2 is increasing rapidly. The Yule process approximation is obtained by transforming the diffusion time scale during the sweep by the frequency of allele A_1, $1 - x(t)$. The result is that the second half of the sweep becomes greatly compressed (Pfaffelhuber et al. 2006: Figure 5.1). On this new time scale, the rate of escape by recombination becomes constant along each ancestral lineage. The process of coalescence between *B*-locus lineages that are linked to A_2 alleles follows from the fact that the sample at the *A* locus can be modeled as a random subsample of a larger random tree, called the Yule tree. If we imagine the whole gene genealogy of all the A_2 alleles that do not go extinct, then roughly speaking, the Yule tree is the portion of this genealogy corresponding to the first half of the sweep.

Figure 5.4B depicts the model, with a hypothetical Yule tree drawn in the background (in purple) and the lineages that are ancestral to the sample (in black). In this representation, the time-change $(1 - x(t))$ used in the Yule process approximation has been undone, and the figure is drawn to correspond to the sweep in Figure 5.4A. Blue squares again show recombination events by which two *B*-locus lineages escape the sweep. As the range of time on the vertical axis in 5.4B is the same as in 5.4A, the three lineages that emerge from the sweep again continue up, out of the graph, where they are expected to accrue standard neutral levels of polymorphism. Notice, with respect to Figure 5.4A, the allele-frequency trajectory itself is no longer depicted.

The process that generates the Yule tree is simple enough, but there is no point in describing it here. Note only that it is a binary tree with $\lfloor \sigma \rfloor$ tips, in which $\lfloor \sigma \rfloor$ means the largest integer less than or equal to σ. Importantly, it is not necessary to actually generate the Yule tree, so this approximation relieves us of the detailed, explicit conditioning inherent in the structured coalescent approach, even though we are modeling the same process. However, in generating coalescent times during the sweep using the algorithms in the Appendix of Pfaffelhuber et al. (2006), it is necessary to model events in the Yule tree and consider whether these events occur in the ancestry of the sample. Because of this, simulations of the Yule approximation become slower when σ is larger (Pfaffelhuber et al. 2006), and it might be that other methods are more efficient than the Yule process approximation when σ is very large.

Many mathematical details go into demonstrating the validity of the Yule process approximation and in using it to compute quantities of interest analytically (Etheridge et al. 2006). Note that if none of the *B*-locus lineages escape the sweep, there will be a single family of size n. Further, let a singleton family be a family with just one member, like the two families descending from the blue boxes in Figures 5.4A and B. Etheridge et al. (2006) were able to prove that the probability there will be more than two non-singleton families—one that escapes the sweep (along with some number of singleton families) and one that descends from the original A_2 allele—is of order $1/\log(\sigma)^2$, which tends to zero, albeit slowly, as $N \to \infty$.

Because family sizes are biased toward singletons, approximations have been proposed in which the number of singleton families is a binomial random variable and all remaining lineages descend from the original A_2

allele (Barton 1998; Kim and Nielsen 2004; Pennings and Hermisson 2006; Schweinsberg and Durrett 2005). The heuristic argument for this, which illustrates some important features of strong selective sweeps, is as follows. First, only the first half of the sweep is relevant, and this has a length of approximately $2\log(\sigma)/\sigma$ when σ is very large. Next, the rate of recombination per lineage during this period is effectively $\rho^*/2$ (= $m\rho/2$) per unit of time, because $x(t)$ is very small and $1 - x(t)$ is close to one. Thus, in order for there to be an appreciable effect of recombination during a sweep, the product $\rho^*\log(\sigma)/\sigma$ must also be appreciable. This product, which is equal to $mr\log(\sigma)/s$ is the total rate of escape by recombination for a single B-locus lineage. The probability that a single lineage escapes the sweep is given by

$$1-\exp\left(-\frac{mr\log(\sigma)}{s}\right)=1-\sigma^{-mr/s}=1-(4Ns)^{-mr/s} \tag{5.10}$$

Equation 5.10 is written in terms of the discrete model parameters, so that the effects of each can be seen. One possibly counterintuitive result is that, for a given selection coefficient s and for a locus at a fixed recombination distance from the selected site, as measured by mr, the probability of escape is *greater* when N is greater. Increasing N increases ρ proportionally, but it does not decrease the duration of sweeps proportionally, due to the $\log(\sigma)$ term in the numerator of t_{fix} (see Equation 5.4).

The Coalescent with Multiple Mergers

We can discern three characteristic times in the processes presented in this chapter. The first is the time between new mutations that will fix in the population. The second is the duration of a selective sweep. The third is the neutral coalescence time for a sample to reach its most recent common ancestor. Only the last of these is simple: the neutral coalescence time does not depend on the rates of mutation and recombination or on the selection coefficient. Again, for a large sample it is close to 2 when time is measured in units of $2N$ generations. In this section, we consider the possibility that the time between sweeps is much smaller than this, such that many sweeps may have occurred within 2 units of time. However, we will require that the duration of a sweep is much smaller than the time between sweeps, so that sweeps are non-overlapping. In this case, the ancestry of a sample follows a process that is conceptually similar to the neutral coalescent process, but which is very different in detail.

In other words, in this section, we treat ancestries of samples of size n under the recurrent hitchhiking model mentioned. Coalescence will be driven by the occurrence of selective sweeps rather than by a neutral process of reproduction, by genetic draft rather than by genetic drift (Gillespie 2000). Gillespie (2000, 2001) illustrated this idea with analyses of samples of size $n = 2$, and Nielsen (2005), Hahn (2008), and Sella et al. (2009) promote it as a possible explanation for genomic patterns of variation.

Durrett and Schweinsberg (2005) proved that the gene genealogy at a neutral locus, which is embedded in a genomic region where sweeps occur at some rate and at random locations, will follow a process known as a coalescent with simultaneous multiple mergers. As the name implies, such processes are distinguished from the standard neutral coalescent because more than two lineages may coalesce at the same time. Multiple-mergers coalescent processes were in fact discovered first in neutral population models in which the variance of reproductive success among individuals is very large (Pitman 1999; Sagitov 1999; Schweinsberg 2000; Möhle and Sagitov 2001; Birkner et al. 2005). An introductory look at the wide range of possible behaviors under one particularly simple neutral model of a population, motivated by organisms that reproduce by broadcast spawning, can be found in Eldon and Wakeley (2006).

We can understand the basic idea from Figure 5.4C, without going into the considerable mathematical details of multiple-mergers coalescent models of recurrent selective sweeps. Genetic lineages at the neutral locus travel backward in time, undergoing a possible burst of coalescent events (with associated family sizes) each time a selective sweep happens at a locus in the vicinity of the neutral locus. In Figure 5.4C, four lineages coalesce in the first sweep and two escape by recombination, then the remaining three lineages coalesce during the next sweep. Note the much shorter range of time depicted in Figure 5.4C than in Figures 5.4A and B. In the present case, sweeps hit the population very frequently compared to the rate of coalescence. The rate at which sweeps occur may be difficult to describe. In a simple model, the rate will depend on the rate of advantageous mutations, but in reality it may in turn depend on changing environments and selection pressures (Gillespie 1991, 2000, 2001, 2004b). As a consequence, the rate of sweeps may have little to do with the size of the population, so polymorphism levels will not necessarily be predicted to depend linearly on population size, as they are under standard neutral models. A long-standing observation and conundrum with respect to the neutral theory, is that levels of polymorphism do not in fact increase linearly with estimates of population size (Lewontin 1974; Gillespie 1991; Meiklejohn et al. 2007). Thus, while work on these coalescent models with multiple mergers for selection is still in its infancy, they could help resolve debates about the origin and maintenance of genetic variation.

Hopes for the Future

Darwin could scarcely have imagined the world we live in today. Only the most active imagination could find passages in *The Origin of Species* relevant to coalescent models of natural selection. However, Darwin's fundamental insights are as relevant today as they ever have been. In the field of population genetics, in particular, attention to natural selection as a factor shaping variation at the molecular level has been boosted greatly by recent analyses of genomic data (see Zhang, Chapter 4; Kolaczkowski and Kern, Chapter 6).

This shift comes after a long period, perhaps overly long, in which the neutral models received from Kimura (1983) were the dominant explanatory tools. Given the lack of force of theoretical arguments for the neutral theory (Ewens 1979), the empirical evidence against it, and the fact that the selective models can both provide a better fit to the observations and mimic neutrality itself (Gillespie 1991), the longevity of the neutral theory may be surprising. However, as Crow (2008) points out, the very simplicity of the neutral theory accounts for a lot of its appeal. In large part, this is likely due to the wonderful paper of Kimura and Ohta (1971), which seemingly explained both molecular evolution among species and molecular variation within species as different facets of one relatively simple process.

Neutrality, as a logical assumption of a null model with respect to selection, has also been difficult to reject statistically. Perhaps this has even contributed to the appeal of neutrality. However, Gillespie (1994) has shown that there is low power to detect crucial deviations from neutrality and to distinguish among some selective alternatives to the neutral theory, at least using simple statistical tests. A similar conclusion applies to multiple-mergers coalescent processes (Sargsyan and Wakeley 2008). Another issue, which we touched upon only glancingly in this chapter, is that selective neutrality is but one of several assumptions of any neutral model, and rejecting such a model does not necessarily identify selection as the cause. Although choosing among alternatives to neutrality will be a major challenge—few will "wish to slog through 100 pages of mathematics," as Gillespie puts it, in Chapter 4 of *The Causes of Molecular Evolution* (Gillespie 1991)—some hope is justified that the current rapid pace of research at the interface of theoretical and empirical population genetics will allow more precise inferences to be made.

In closing, it seems appropriate to ask, what is to become of the sophisticated coalescent machinery for making inferences about the demographic history of populations? To what extent is our belief that genetic markers contain information about changes in population size over time or patterns of population structure warranted? Will inferences of migration rates or divergence times have to be reinterpreted in terms of selection? We might hope that our inferential tools can be developed in ways that are robust to the presence of natural selection, even for species in which selection is a dominant force.

Literature Cited

Aquadro, C. F. and B. D. Greenberg. 1983. Human mitochondrial DNA variation and evolution: Analysis of nucleotide sequences from seven individuals. *Genetics* 103: 287–312.

Avise, J. C., J. Arnold, R. M. Ball, E. Bermingham, and 4 others. 1987. Intraspecific phylogeography: The mitochondrial DNA bridge between population genetics and systematics. *Annu. Rev. Ecol. Syst.* 18: 489–522.

Avise, J. C., C. Gilbin-Davidson, J. Laerm, and 2 others. 1979. Mitochondrial DNA clones and matriarchal phylogeny within and among geographic populations of the pocket gopher, *Geomys pinetis*. *Proc. Natl. Acad. Sci. USA* 76: 6694–6698.

Barton, N. H. 1998. The effect of hitch-hiking on neutral genealogies. *Genet. Res. Camb.* 72: 123–133.

Barton, N. H., A. M. Etheridge, and A. K. Sturm. 2004. Coalescence in a random background. *Ann. Appl. Prob.* 14: 754–785.

Becquet, C. and M. Przeworski. 2009. Learning about modes of speciation by computational approaches. *Evolution* 63: 2547–2562.

Begun, D. J., A. K. Holloway, K. Stephens, and 10 others. 2007. Population genomics: Whole-genome analysis of polymorphism and divergence in *Drosophila simulans*. *PLoS Biol.* 5: e310.

Birkner, M., J. Blath, M. Capaldo, and 4 others. 2005. Alpha-stable branching processes and beta-coalescents. *Elec. J. Prob.* 10: 303–325.

Braverman, J. M., R. R. Hudson, N. L. Kaplan, and 2 others. 1995. The hitchhiking effect on the site frequency spectrum of DNA polymorphisms. *Genetics* 140: 783–796.

Brown, W. M. 1980. Polymorphism in mitochondrial DNA of humans as revealed by restriction endonuclease analysis. *Proc. Natl. Acad. Sci. USA* 77: 3605–3609.

Bürger, R. and W. J. Ewens. 1995. Fixation probabilities of additive alleles in diploid populations. *J. Math. Biol.* 33: 557–575.

Bustamante, C. D., R. Nielsen, S. A. Sawyer, and 3 others. 2002. The cost of inbreeding in *Arabidopsis*. *Nature* 416: 531–534.

Coop, G. and R. C. Griffiths. 2004. Ancestral inference on gene trees under selection. *Theor. Pop. Biol.* 66: 219–232.

Crow, J. F. 2008. Mid-century controversies in population genetics. *Ann. Rev. Genet.* 42: 1–16.

Darden, T., N. L. Kaplan, and R. R. Hudson. 1989. A numerical method for calculating moments of coalescent times in finite populations with selection. *J. Math. Biol.* 27: 355–368.

Darwin, C. 1859. *On the Origin of Species by Means of Natural Selection, or the Preservation of Favoured Races in the Struggle for Life.* Murray, London.

Dobzhansky, T. 1937. *Genetics and the Origin of Species*, 1st ed. Columbia University Press, New York.

Dobzhansky, T. 1955. A review of some fundamental problems of and concepts of population genetics. *Cold SH. Q. B.* 20: 1–15.

Durrett, R. and J. Schweinsberg. 2004. Approximating selective sweeps. *Theor. Pop. Biol.* 66: 129–138.

Durrett, R. and J. Schweinsberg. 2005. A coalescent model for the effect of advantageous mutations on the genealogy of a population. *Stochast. Proc. Appl.* 115: 1628–1657.

Eldon, B. and J. Wakeley. 2006. Coalescent processes when the distribution of offspring number among individuals is highly skewed. *Genetics* 172: 2621–2633.

Eriksson, A., P. Fernström, B. Mehlig, and 1 other. 2008. An accurate model for genetic hitchhiking. *Genetics* 178: 439–451.

Etheridge, A., P. Pfaffelhuber, and A. Wakolbinger. 2006. An approximate sampling formula under genetic hitchhiking. *Ann. Appl. Prob.* 16: 685–729.

Ewens, W. J. 1963. The diffusion equation and a pseudo-distribution in genetics. *J. Roy. Stat. Soc. B* 25: 405–412.
Ewens, W. J. 1964. The pseudo-transient distribution and its uses in genetics. *J. Appl. Prob.* 1: 141–156.
Ewens, W. J. 1972. The sampling theory of selectively neutral alleles. *Theor. Pop. Biol.* 3: 87–112.
Ewens, W. J. 1979. *Mathematical Population Genetics*. Springer-Verlag, Berlin. Note: see also the revised and update version, Ewens (2004).
Ewens, W. J. 1982. On the concept of effective size. *Theor. Pop. Biol.* 21: 373–378.
Ewens, W. J. 1990. Population genetics theory—the past and the future. In S. Lessard (ed.), *Mathematical and Statistical Developments of Evolutionary Theory*, pp. 177–227. Kluwer Academic Publishers, Amsterdam.
Ewens, W. J. 2004. *Mathematical Population Genetics, Volume I: Theoretical Foundations*. Springer-Verlag, Berlin.
Felsenstein, J. 2007. Trees of genes in populations. In O. Gascuel and M. Steel (eds.), *Reconstructing Evolution: New Mathematical and Computational Advances*, pp. 3–29. Oxford University Press, Oxford.
Fisher, R. A. 1918. The correlation between relatives on the supposition of Mendelian inheritance. *Trans. Roy. Soc. Edin.* 52: 399–433.
Fisher, R. A. 1922. On the dominance ratio. *Proc. Roy. Soc. Edin.* 42: 321–341.
Fisher, R. A. 1930. *The Genetical Theory of Natural Selection*. Clarendon, Oxford.
Gillespie, J. H. 1991. *The Causes of Molecular Evolution*. Oxford University Press, New York.
Gillespie, J. H. 1994. Alternatives to the neutral theory. In B. Golding (ed.), *Non-Neutral Evolution: Theories and Molecular Data*, pp. 1–17. Chapman & Hall, New York.
Gillespie, J. H. 2000. Genetic drift in an infinite population: The pseudohitchhiking model. *Genetics* 155: 909–919.
Gillespie, J. H. 2001. Is the population size of a species relevant to its evolution? *Evolution* 55: 2161–2169.
Gillespie, J. H. 2004a. *Population Genetics: A Concise Guide*, 2nd ed. Johns Hopkins University Press, Baltimore.
Gillespie, J. H. 2004b. Why $k = 4N_e su$ is silly. In R. Singh and M. Uyenoyama (eds.), *The Evolution of Population Biology—Modern Synthesis*, pp. 181–192. Cambridge University Press, Cambridge, UK.
Hahn, M. W. 2008. Toward a selection theory of molecular evolution. *Evolution* 62: 255–265.
Haldane, J. B. S. 1927. A mathematical theory of natural and artificial selection, part V: Selection and mutation. *Proc. Camb. Phil. Soc.* 23: 838–844.
Haldane, J. B. S. 1932. *The Causes of Natural Selection*. Longmans Green & Co., London.
Halligan, D. L., F. Oliver, A. Eyre-Walker, and 2 others. 2010. Evidence for pervasive adaptive protein evolution in wild mice. *PLoS Genet.* 6: e1000825.
Harris, H. 1966. Enzyme polymorphism in man. *Proc. Roy. Soc. Lond. B* 164: 298–310.
Hermisson, J. and P. S. Pennings. 2005. Soft sweeps: Molecular population genetics of adaptation from standing genetic variation. *Genetics* 169: 2335–2352.
Hermisson, J. and P. Pfaffelhuber. 2008. The pattern of genetic hitchhiking under recurrent mutation. *Elec. J. Prob.* 13: 2069–2106.
Hey, J. 2005. On the number of New World founders: A population genetic portrait of the peopling of the Americas. *PLoS Biol.* 3: e193.

Hobolth, A., M. K. Uyenoyama, and C. Wuif. 2007. Importance sampling for the infinite sites model. *Stat. App. Genet. Mol.* 7, Iss. 1, Art. 32: 1–24.
Hudson, R. R. 1983a. Testing the constant-rate neutral allele model with protein sequence data. *Evolution* 37: 203–217.
Hudson, R. R. 1983b. Properties of a neutral allele model with intragenic recombination. *Theor. Pop. Biol.* 23: 183–201.
Hudson, R. R. 1990. Gene genealogies and the coalescent process. In D. J. Futuyma and J. Antonovics (eds.), *Oxford Surveys in Evolutionary Biology*, vol. 7, pp. 1–44. Oxford University Press, Oxford.
Hudson, R. R. 2002. Generating samples under a Wright-Fisher neutral model of genetic variation. *Bioinformatics* 18: 337–338.
Hudson, R. R. and N. L. Kaplan. 1986. On the divergence of alleles in nested subsamples from finite populations. *Genetics* 113: 1057–1076.
Hudson, R. R. and N. L. Kaplan. 1988. The coalescent process in models with selection and recombination. *Genetics* 120: 831-840.
Huxley, J. S. 1942. *Evolution: The Modern Synthesis*. Allen and Unwin, London.
Jensen, J. D. 2009. On reconciling single and recurrent hitchhiking models. *Genome Biol. Evol.* 1: 320–324.
Kaplan, N. L., T. Darden, and R. R. Hudson. 1988. The coalescent process in models with selection. *Genetics* 120: 819–829.
Kaplan, N. L., R. R. Hudson, and C. H. Langley. 1989. The "hitchhiking effect" revisited. *Genetics* 123: 887–899.
Karlin, S. and J. McGregor. 1972. Addendum to a paper of W. Ewens. *Theor. Pop. Biol.* 3: 113–116.
Kim, Y. 2006. Allele frequency distribution under recurrent selective sweeps. *Genetics* 172: 1967–1978.
Kim, Y. and R. Nielsen. 2004. Linkage disequilibrium as a signature of selective sweeps. *Genetics* 167: 1513–1524.
Kim, Y. and W. Stephan. 2002. Detecting a local signature of genetic hitchhiking along a recombining chromosome. *Genetics* 160: 765–777.
Kim, Y. and T. Wiehe. 2009. Simulation of DNA sequence evolution under models of recent directional selection. *Brief. Bioinform.* 10: 84–96.
Kimura, M. 1955a. Solution of a process of random genetic drift with a continuous model. *Proc. Natl. Acad. Sci., USA* 41: 144–150.
Kimura, M. 1955b. Stochastic processes and the distribution of gene frequencies under natural selection. *Cold SH. Q. B.* 20: 33–53.
Kimura, M. 1968. Evolutionary rate at the molecular level. *Nature* 217: 624–626.
Kimura, M. 1983. *The Neutral Theory of Molecular Evolution*. Cambridge University Press, Cambridge, UK.
Kimura, M. and T. Ohta. 1969. The average number of generations until fixation of a mutant gene in a finite population. *Genetics* 61: 763–771.
Kimura, M. and T. Ohta. 1971. Protein polymorphism as a phase of molecular evolution. *Nature* 229: 467–469.
King, J. L. and T. H. Jukes. 1969. Non-Darwinian evolution. *Science* 164: 788–798.
Kingman, J. F. C. 1982a. On the genealogy of large populations. *J. Appl. Prob.* 19A: 27–43.
Kingman, J. F. C. 1982b. The coalescent. *Stoch. Proc. Appl.* 13: 235–248.
Kreitman, M. 1983. Nucleotide polymorphism at the alcohol dehydrogenase locus of *Drosophila melanogaster*. *Nature* 304: 412–417.
Krone, S. M. and C. Neuhauser. 1997. Ancestral processes with selection. *Theor. Pop. Biol.* 51: 210–237.

Lewontin, R. C. 1974. *The Genetic Basis of Evolutionary Change*. Columbia University Press, New York.
Lewontin, R. C. and J. L. Hubby. 1966. A molecular approach to the study of genic diversity in natural populations II. Amount of variation and degree of heterozygosity in natural populations of *Drosophila pseudoobscura*. *Genetics* 54: 595–609.
Linnen, C. R., E. P. Kingsley, J. D. Jensen, and 1 other. 2009. On the origin and spread of an adaptive allele in deer mice. *Science* 325: 1095–1098.
Lynch, M. 2007. *The Origins of Genome Architecture*. Sinauer Associates, Sunderland, MA.
Malécot, G. 1941. La consaguinité dans une population limitée. *C. R. Acad. Sci., Paris* 222: 841–843.
Malécot, G. 1948. *Les Mathématiques de l'Hérédité*. Masson, Paris. Extended translation: *The Mathematics of Heredity*. W. H. Freeman, San Francisco (1969).
Marjoram, P. and S. Tavaré. 2006. Modern computational approaches for analyzing molecular genetic variation data. *Nat. Rev. Genet.* 7: 759–770.
Maynard Smith, J. M. and J. Haigh. 1974. The hitch-hiking effect of a favourable gene. *Genet. Res. Camb.* 23: 23–35.
Mayr, E. 1963. *Animal Species and Evolution*. Belknap Press, Cambridge, MA.
Meiklejohn, C. D., K. L. Montooth, and D. M. Rand. 2007. Positive and negative selection on the mitochondrial genome. *Trends Genet.* 23: 259–263.
Möhle, M. 1999. The concept of duality and applications to Markov processes arising in neutral population genetics models. *Bernoulli* 5: 761–777.
Möhle, M. and S. Sagitov. 2001. A classification of coalescent processes for haploid exchangeable population models. *Ann. Appl. Prob.* 29: 1547–1562.
Moran, P. A. P. 1962. *Statistical Processes of Evolutionary Theory*. Clarendon Press, Oxford.
Moran, P. A. P. 1975. Wandering distributions and the electrophoretic profile. *Theor. Pop. Biol.* 8: 318–330.
Nagylaki, T. 1989. Gustave Malécot and the transition from classical to modern population genetics. *Genetics* 122: 253–268.
Neuhauser, C. and S. M. Krone. 1997. The genealogy of samples in models with selection. *Genetics* 145: 519–534.
Nielsen, R. 2005. Molecular signatures of natural selection. *Ann. Rev. Genet.* 39: 197–218.
Norman, M. F. 1975. Approximation of stochastic processes by Gaussian diffusions, and applications to Wright–Fisher genetic models. *Siam J. Appl. Math.* 29: 225–242.
Ohta, T. 1973. Slightly deleterious mutant substitutions in evolution. *Nature* 246: 96–98.
Ohta, T. 1992. The nearly neutral theory of molecular evolution. *Ann. Rev. Ecol. Syst.* 23: 263–286.
Ohta, T. and M. Kimura. 1971. On the constancy of evolutionary rate of cistrons. *J. Mol. Evol.* 1: 18–25.
Ohta, T. and M. Kimura. 1973. A model of mutation appropriate to estimate the number of electrophoretically detectable alleles in a finite population. *Genet. Res. Camb.* 22: 201–204.
Otto, S. P. and N. H. Barton. 1997. The evolution of recombination: Removing the limits to natural selection. *Genetics* 147: 879–906.
Pennings, P. S. and J. Hermisson. 2006. Soft sweeps III: The signature of positive selection from recurrent mutation. *PLoS Genet.* 2: e186.

Pfaffelhuber, P., B. Haubold, and A. Wakolbinger. 2006. Approximate genealogies under genetic hitchhiking. *Genetics* 174: 1995–2008.
Pitman, J. 1999. Coalescents with multiple collisions. *Ann. Prob.* 27: 1870–1902.
Provine, W. B. 1971. *The Origins of Theoretical Population Genetics*. University of Chicago Press, Chicago.
Przeworski, M. 2002. The signature of positive selection at randomly chosen loci. *Genetics* 160: 1179–1189.
Przeworski, M. 2003. Estimating the time since the fixation of a beneficial allele. *Genetics* 164: 1667–1676.
Sabeti, P. C., P. Varilly, B. Fry, 9 others, and the International HapMap Consortium. 2007. Genome-wide detection and characterization of positive selection in human populations. *Nature* 449: 913–918.
Sagitov, S. 1999. The general coalescent with asynchronous mergers of ancestral lines. *J. Appl. Prob.* 36: 1116–1125.
Sargsyan, O. and J. Wakeley. 2008. A coalescent process with simultaneous multiple mergers for approximating the gene genealogies of many marine organisms. *Theor. Pop. Biol.* 74: 104–114.
Sawyer, S. A. and D. L. Hartl 1992. Population genetics of polymorphism and divergence. *Genetics* 132: 1161–1176.
Sawyer, S. A., R. J. Kulathinal, C. D. Bustamante, and 1 other. 2003. Bayesian analysis suggests that most amino acid replacements in *Drosophila* are driven by positive selection. *J. Mol. Evol.* 57 Suppl. 1: S154–164.
Schweinsberg, J. 2000. Coalescents with simultaneous multiple collisions. *Elec. J. Prob.* 5: 1–50.
Schweinsberg, J. and R. Durrett. 2005. Random partitions approximating the coalescence of lineages during a selective sweep. *Ann. Appl. Prob.* 15: 1591–1651.
Sella, G., D. A. Petrov, M. Przeworski, and 1 other. 2009. Pervasive natural selection in the *Drosophila* genome? *PLoS Genet.* 5: e1000495.
Simonsen, K. L., G. A. Churchill, and C. F. Aquadro. 1995. Properties of statistical tests of neutrality for DNA polymorphism data. *Genetics* 141: 413–429.
Sjödin, P., I. Kaj, S. Krone, and 2 others. 2005. On the meaning and existence of an effective population size. *Genetics* 169: 1061–1070.
Slatkin, M. 2001. Simulating genealogies of selected alleles in a population of variable size. *Genet. Res. Camb.* 78: 49–57.
Slatkin, M. and M. Veuille. 2002. *Modern Developments in Theoretical Population Genetics: The Legacy of Gustave Malécot*. Oxford University Press, Oxford.
Smith, N. G. and A. Eyre-Walker. 2002. Adaptive protein evolution in *Drosophila*. *Nature* 415: 1022–1024.
Stephens, M. and P. Donnelly. 2000. Inference in molecular population genetics. *J. Roy. Stat. Soc. B* 62: 605–655.
Tajima, F. 1983. Evolutionary relationship of DNA sequences in finite populations. *Genetics* 105: 437–460.
Tavaré, S. 1984. Lines-of-descent and genealogical processes, and their application in population genetic models. *Theor. Pop. Biol.* 26: 119–164.
Thornton, K. R., J. D. Jensen, C. Becquet, and 1 other. 2007. Progress and prospects in mapping recent selection in the genome. *Heredity* 98: 340–348.
Wakeley, J. and O. Sargsyan. 2009. The conditional ancestral selection graph with strong balancing selection. *Theor. Pop. Biol.* 75: 355–364.
Watterson, G. A. 1975. On the number of segregating sites in genetical models without recombination. *Theor. Pop. Biol.* 7: 256–276.

Wiehe, T. H. and W. Stephan. 1993. Analysis of a genetic hitchhiking model, and its application to DNA polymorphism data from *Drosophila melanogaster*. *Mol. Biol. Evol.* 10: 842–854.
Williamson, S. H., M. J. Hubisz, A. G. Clark, and 3 others. 2007. Localizing recent adaptive evolution in the human genome. *PLoS Genet.* 3: e90.
Wright, S. 1931. Evolution in Mendelian populations. *Genetics* 16: 97–159.
Zuckerkandl, E. and L. Pauling. 1965. Evolutionary divergence and convergence in proteins. In V. Bryson and H. J. Vogel (eds.), *Evolving Genes and Proteins.* Academic Press, New York.

Chapter 6

On the Power of Comparative Genomics: Does Conservation Imply Function?

Bryan Kolaczkowski and Andrew D. Kern

Over the past decade, an explosion in complete genome sequencing has been witnessed. This comparative genomics revolution promises to transform our understanding of biology, from the level of DNA and proteins, up the biological hierarchy to populations, and potentially even whole ecosystems. One prospect is that comparative genomics will be able to distill functional information for genomes based on sequence comparison. In principle, this procedure would allow practitioners of evolutionary biology to map phenotypic differentiation among organisms, populations, or species directly down to the level of DNA differences. Thus, the genetic basis for evolutionary change may one day soon be completely understood.

While the notion of being able to trace phenotypic differences onto the genome purely through comparative genomics is an appealing idea, it is—with a few notable exceptions, such as Pollard et al. (2006b)—still a pipe dream. Moreover, the intellectual basis for comparative genomics, which is the simple notion that sequence conservation implies function (Kimura 1983), remains largely untested. This shortfall is a striking omission from the perspective of an evolutionary geneticist. Billions of dollars have been spent, thus far, to sequence the genomes of numerous species (indeed at the time of writing this chapter, a project to sequence more than 10,000 vertebrate genomes has recently been proposed (Genome 10K Community of Scientists 2009). One of the implicit goals of many genome-sequencing projects is finding functional elements via sequence conservation, yet the extent to which purifying selection acts in genomes to conserve sequences of functional elements is still not known. Even if one could reliably identify functional elements via genomic comparison, it is likely that large-scale functional annotation would still prove necessary, as for example occurred in the ENCODE Project (2007).

In this chapter, we review some of the salient advances in evolutionary biology that have been made through comparative genomics. We also provide a unique and strong test of the idea that sequence conservation implies function in genomes. We conclude with a discussion of the prospect of linking evolutionary genomic changes to phenotypic differences between individuals and species using comparative genomics.

What Have We Learned about Evolution from Genome Sequencing?

Thus far, the vast majority of evolutionary information gleaned from comparative genomic sequencing has been about the evolution of genomes themselves, rather than evolution at the functional or phenotypic level (ENCODE Project Consortium 2007; *Drosophila* 12 Genomes Consortium, 2007). These insights have come at two distinct levels: (1) evolution of the whole genome and (2) evolution of individual elements, such as genes, within genomes. Recently, notable progress has been made towards understanding how the regulatory content of genomes evolves between species (Prabhakar et al. 2008; Wittkopp et al. 2009), and an understanding of the ways in which biomolecular interaction networks may evolve is beginning to be formed (Qin et al. 2003; Yeh et al. 2004). In this section, we will concentrate on what comparative genomics has revealed about the evolution of protein-coding genes, arguably the single best understood portion of the genome.

Comparative genomics has provided numerous insights into the ways in which genes and gene content evolve. For some time, it has been known that different genes evolve at different evolutionary rates, most likely resulting from dissimilar levels of selective constraint (Kimura 1983; Gillespie 1986). It is also clear that a gene's characteristic evolutionary rate can regularly vary across lineages (Huelsenbeck and Rannala 1997), possibly due to lineage-specific shifts in protein function. These findings have undermined the validity of the molecular clock (Zuckerkandl and Pauling 1962; Margoliash 1963) and have spurred the development of local- and relaxed-clock methods for dating ancient speciation events (Kumar 2005). It is also apparent that some lineages generally evolve faster than others (Britten 1986; Takano 1998; Andreasen and Baldwin 2001; Nabholz et al. 2008).

Perhaps, one of the most important findings of early comparative genomics was the identification of suites of related genes conserved across large taxonomic ranges—sometimes including all living organisms (Rokas 2008; Erwin 2009). While it may not be surprising that many of the genes involved in basic cellular processes would be highly conserved, the identification of conserved clusters of genes involved in similar processes across organisms has highlighted the high degree of modularity in organismal development and biological function (Gavin et al. 2006; Singh et al. 2008; Kuratani 2009). Possibly, the most well-studied example is the *HOX* cluster—a highly conserved group of genes, both in terms of sequence conservation and linear ordering along the chromosome, that is involved in anterior–posterior axis patterning in early development (McGinnis et al. 1984; Scott and Weiner 1984; Thali et al. 1988; Duboule 1994; see Wray,

Chapter 9). The *HOX* cluster seems to have been modified only slightly across a wide array of animal body plans (Ogishima and Tanaka, 2007; Lanfear and Bromham 2008; Kuraku and Meyer 2009). Indeed, a major emerging theme in genome evolution is the reuse of existing suites of genes in new ways to produce novel phenotypes (Schlosser and Wagner 2004; Snel and Huynen 2004). Modularity in metabolic pathways has also been observed (Peregrín-Alvarez et al. 2009), with evolution reusing sets of pathway components in different ways to alter metabolic output. Although intriguing from a molecular evolutionary perspective, the effects of molecular modularity on functional or phenotypic evolution are largely unknown. A long-standing, and still unanswered, question is whether modularity at the molecular level somehow constrains phenotypic evolution.

Of course, phenotypic novelty cannot arise without molecular change of some sort. One of the aims of comparative genomics has been to determine when novel functional elements arise during genome evolution and where they come from. Whole-genome comparisons have revealed gene and genome duplication as a major source of genetic material that can be acted upon by selection to produce novel molecular functions (Crow and Wagner 2006; Hanada et al. 2009). Historically, this mode of gene-number evolution has been predicted to have great phenotypic effects, perhaps among the most important evolutionary changes (Ohno 1970). From functional studies of gene family evolution, it is clear that evolution can reuse existing genes to generate new molecular functions. Theoretical predictions have been made concerning the ways in which gene duplication can give rise to new functions (Ohno 1970; Kimura and Ohta 1974; Force et al. 1999; Clark 1994; Bridgham et al. 2008), although the precise mechanisms by which this happens are not fully understood. Even less well understood are the functional consequences of whole-genome duplication, although evidence for multiple genome duplications in both animals (Taylor et al. 2003; Dehal and Boore 2005) and plants (Cui et al. 2006) is overwhelming.

The sequencing of each new genome comes with a list of novel lineage-specific genes; genes with no obvious sequence homology to any other known gene from any other organism (Fukuchi and Nishikawa 2004; Begun and Lindfors 2005; Levine et al. 2006; Merkeev and Mironov 2008; Knowles and McLysaght 2009). The origin of such genes remains a mystery. Since there is no homology to other genes in the genome (sometimes even local homology to part of any other gene), these genes are unlikely to have arisen by full- or partial-gene duplication. Similarly, lack of homology to sequences in other organisms tends to rule out horizontal gene transfer as a major source of lineage-specific genes (see Lane, Commentary 4). Are these actual *de novo* genes generated from previously noncoding DNA (Siepel 2009)? Although the mechanisms of novel gene creation by gene duplication and modification are fairly well understood, it is safe to say that we know relatively little about how new genes may be created from nongenic DNA. If *de novo* gene creation is a common feature of genome evolution (Levine et al. 2006; Begun et al. 2007b) or of particular functional importance, though rare, it could be that important functional elements in genomes are actually species-specific, precluding their discovery or examination from a comparative perspective.

A recent comprehensive study of the origins of lineage-specific genes in the *Drosophila melanogaster* subgroup has attempted to quantify the contributions of gene duplication, *de novo* origination from noncoding sequence, retrotransposition, and chimeric gene fusion as generators of new genes (Zhou et al. 2008). While this study found that gene duplication was responsible for the bulk of lineage-specific genes, all other mechanisms examined contributed significantly to gene content.

Similar to the discovery of lineage-specific genes, comparative genomics has revealed groups of genes that have been lost in specific lineages (Krylov et al. 2003; Wang et al. 2006). Most striking is the rampant gene loss and genome compaction typically associated with transitions to parasitic or symbiotic lifestyles (Keeling and Slamovits 2004; Keeling 2004; Delmotte et al. 2006; Hershberg et al. 2007; see Lane, Commentary 4). However, large-scale genome compaction has been observed in free-living, multi-cellular eukaryotes as well (Angrist 1998; Venkatesh et al. 2000). What remains unclear is whether large-scale gene loss is driven by adaptive forces or is a result of shifting evolutionary constraints that no longer require specific molecular functions. It is known, however, that individuals within populations are clearly polymorphic in gene number (Kern and Begun 2008); thus, it may be possible to directly study the mechanisms driving gene loss.

Thus far, comparative genomics has mainly concerned itself with *what* is happening to genes and gene content as the genome evolves. Much less is known about *how* these changes occur through molecular evolutionary processes and very little about *why*. Connecting comparative genomics to population genetic processes, such as natural selection, demographic history, and migration will help to discover the evolutionary forces driving observed changes in genomic sequences.

Inferring Functional Changes from Sequence Evolution

Comparative genomics methods for inferring the functional importance of sequence changes rely heavily on phylogenetic approaches, which provide a natural framework for cross-species comparisons. Although a few studies have considered base composition or other evolutionary parameters (Musio et al. 2002; Ko et al. 2006), most comparative methods focus on changes in site-specific evolutionary rates as an indicator of changes in functional evolution. This focus is justified; evolutionary rates—typically modeled as branch lengths along a phylogenetic tree—give a quantitative estimate of the *amount* of evolution occurring along a given lineage. Quantitatively more evolutionary substitutions indicate a greater potential for function-altering molecular changes, whereas fewer substitutions should indicate conservation of function, which presumably is due to purifying selection.

Among proteins, comparisons of evolutionary rates between nonsynonymous and synonymous changes within a gene sequence, the so-called d_N/d_S ratio, have been extensively used to identify genes evolving under positive selection (Yang 1998; Anisimova and Yang 2007). Relative d_N/d_S rates greater than 1 indicate adaptive evolution, whereas rates less than 1 are taken as evidence for the presence of purifying selection. Modern methods are capable of inferring the presence of positive selection acting on amino-acid replacements along specific lineages and at specific sites in the protein sequence (Zhang et al. 2005). These methods rely on the controversial assumption that synonymous changes are largely neutral. Indeed, the action of selection at synonymous sites has been documented (Akashi 1994; Akashi and Schaeffer 1997; Cutter and Charlesworth 2006) and could potentially undermine this approach if it were found to be widespread.

More recently, the framework of nonsynonymous/synonymous replacement-rate comparisons has been extended to examine selection in noncoding DNA (Wong and Nielsen, 2004). If suitable, nearby neutral sites can be identified, any class of DNA positions can be examined for potential positive selection by comparing the rate of evolution at these specified positions to the rate at nearby neutral positions. Of course, the reliable interpretation of inferred ratios depends absolutely on the reliable identification of suitable neutral positions for comparison. Not only must neutral positions actually be evolving neutrally, they must also have similar mutation rates to the putative non-neutral positions in order to fairly compare the rates of substitution between the two classes.

In addition to evolutionary-rate comparisons between target and neutral positions in the genome, examination of the relative evolutionary rates of different regions across the genome has been used to identify functional regions (Siepel et al. 2005; Kosiol et al. 2008). Sidestepping the requirement for *a priori* identification of neutrally evolving positions, these methods focus on outlier regions that are evolving at rates across the phylogeny that are highly different from the rest of the genome. Currently, methods have been developed using a Hidden Markov Model (HMM) framework to search cross-species genomic sequence alignments for highly conserved regions across the phylogeny, regions that are conserved only within specific lineages, and regions experiencing lineage-specific acceleration of evolutionary rate (Siepel and Haussler 2004; Pollard et al. 2010). Recently, mixed models of rate variation, which effectively relax some of the assumptions underlying HMMs, have also been proposed for examining site-specific evolutionary rate variation (Kolaczkowski and Thornton 2008).

Recent advances in genome sequencing technology, the so-called next-generation genome sequencing, have given rise to the possibility of fully sequencing multiple genomes from different individuals of the same species. A number of such population genomic studies are approaching the first stages of completion. The availability of multiple genomes from the same species promises to bring the rich statistical toolbox of population genetics to bear on the problem of detecting functional elements in genomes. Specifically, the ability of population genetics to make direct inferences regarding the strength and mode of natural selection acting on specific genomic regions in natural populations has the potential to greatly expand the ability to more directly infer functional genomic elements from whole-genome sequence data.

Population genetics has developed a number of simple approaches to finding genomic regions evolving under natural selection based on summary statistics of polymorphism data. Loosely, these can be divided into methods that use summaries of the site frequency spectrum (SFS) (Tajima 1989) and methods that use summaries of patterns of linkage disequilibrium (LD) (Sabeti et al. 2007). These two methods are complementary, as numerous types of selection are predicted to primarily affect different statistical summaries. For example, simple adaptive evolution from a new mutation, a so-called hard sweep, has been shown to directly affect the SFS (Braverman et al. 1995; Przeworski 2002). In contrast, recurrent positive selection and sweeps in progress are more likely to affect patterns of LD around the selected site, which may lead to greater sensitivity of LD-based methods for detecting these types of selection (Kim and Nielsen 2004; Sabeti et al. 2007).

More recently, model-based population genetic methods have been developed to search for signatures of natural selection in genomic data. These techniques are largely based on either a composite likelihood approach or a coalescent framework. Composite likelihood methods (Kim and Stephan 2002) are typically concerned with the effects of selection on the SFS. They calculate the likelihood of the observed SFS around some region of interest under different models of neutral and selective evolution and use these likelihoods to compare the amount of support in the data for each model of the evolutionary process (Kim and Stephan 2002; Nielsen et al. 2005b). One of the major assumptions of composite likelihood—and one known to be violated in real genomic sequence data—is the assumption that the data are independent. It has been shown that violations of this assumption can lead to very high rates of false-positive inferences (Zhu and Bustamante 2005; Jensen et al. 2005).

An alternative to the composite likelihood approach is to use coalescent simulations to examine the fit of the data to different evolutionary scenarios, including various demographic histories, migration, and various types of natural selection. The coalescent framework is a general approach for understanding the evolutionary histories of genes within a population (Hudson 1983); it has been extended to include a wide variety of scenarios, including recombination, demography, migration, and a variety of forms of linked selection (Kaplan et al. 1988; Hudson and Kaplan 1988; Kaplan et al. 1989; Hudson and Kaplan 1995; Hudson 2002). Unfortunately, the calculation of likelihoods under a coalescent model is currently computationally infeasible, so simulations are used to test the fit of empirical data to different coalescent models and test hypotheses about the types of evolutionary forces likely to affect a genomic region of interest.

One approach for inferring the presence of natural selection using coalescent methods is approximate Bayesian computation (ABC) (Pritchard et al. 1999; Beaumont et al. 2002; Przeworski 2003; Jensen et al. 2008). Under this approach, replicate datasets are simulated under a variety of coalescent models with specified parameter values. If the simulated data are similar to the observed data, the simulation model, along with its parameter values, is accepted as an adequate explanation of the data. Typically, similarity between simulated and observed data is calculated using a defined distance function based on summary statistics of the data (Pritchard et al. 1999; Beaumont et al. 2002; Leuenberger and Wegmann 2010). This simulation acceptance/rejection procedure is repeated many times, and the proportion of times that a specified model is accepted is taken as an estimate of the posterior probability support for that model, given the observed data. As an example, if two coalescent models are simulated, one neutral and one with a selection parameter, the proportion of accepted simulations from the selection model represents the statistical support in favor of selection acting on the genomic region being analyzed. Although generally much faster than full likelihood approaches, ABC methods are still computationally demanding, particularly for highly parameterized coalescent models. As a result, most ABC analyses to date have utilized only highly simplified models, with some of the model parameters being estimated directly from the observed data rather than incorporated into the ABC framework. Thus, it remains unclear whether a fully parameterized ABC-coalescent approach represents a viable method for inferring evolutionary parameters from population genomic data.

One of the major challenges for detecting the action of natural selection using population genomic data is that demographic changes and migration from other populations can skew the observed patterns of polymorphism away from those predicted by the standard neutral model, making neutral changes appear as if they were driven by selection (Simonsen et al. 1995). These problems have been dealt with in two ways. First, outlier approaches have been used to find regions of the genome that are extreme compared to the overall genomic background (Akey et al. 2002). The logic behind this approach is that demographic history and/or migration are likely to affect the genome as a whole. Regions that are outliers compared to the genomic background are more likely to have been affected by selection in addition to neutral processes (Lewontin and Krakauer 1973). Although appealing, the power of outlier methods to turn up reliable candidates for adaptive changes has been shown to be poor under reasonable demographic histories (Teshima et al. 2006).

The second approach to dealing with the effects of demographic history has been to build an explicit model of the genome-wide effects of these processes and simultaneously infer the demographic history of a population along with the action of selection (Williamson et al. 2005; Li and Stephan 2006; Boyko et al. 2008; Jensen et al. 2008). By constructing a complete evolutionary model that includes both neutral and selective processes, these methods promise to specifically address how selective versus neutral evolutionary forces shape genome evolution at the population level. Although the initial development of these combined models appears hopeful (Williamson et al. 2005; Boyko et al. 2008), the development of a complete model capable of incorporating a wide range of both neutral and selective effects has not yet been realized.

One of the major problems with any comprehensive population genetics model is that the various evolutionary forces acting on genomes can regularly mimic one another in terms of their signatures. For example, population bottlenecks can generate evolutionary patterns appearing similar to those of positive selection (Simonsen et al. 1995). Any comprehensive model that attempts to simultaneously infer parameters relating to demographic history and selection is likely to encounter a serious parameter identifiability problem. For example, both rampant positive selection and a strong population bottleneck could provide suitable explanations for observed data, so deciding between these alternatives could be difficult in practice. Another major hurdle is dealing with the potentially large number of model parameters in a comprehensive population genetics model. Reliable optimization of such models could require extremely large amounts of data and extensive computational resources.

Large-scale, comparative and population genomic screens have mostly focused on three problems: (1) identifying regions of the genome that have been strongly conserved over large evolutionary time spans, (2) identifying regions that show lineage-specific accelerated evolution, and (3) finding targets of natural selection in genomes. Ultraconserved regions have been identified in humans (Bejerano et al. 2004) and insects (Glazov et al. 2005) using comparative genomic methods. Importantly, it appears as if these regions have been conserved by long-term purifying selection and not by neutral processes (Katzman et al. 2007; Casillas et al. 2007), indicating that they are probably functional. Similarly, human and *D. melanogaster* accelerated regions have been identified using computational methods (Pollard et al. 2006a; Prabhakar et al. 2006; Holloway et al. 2008). Although the population genetic dynamics of *melanogaster*-accelerated genomic regions are largely unknown, functional characterization of human accelerated regions (HARs) has recently begun in earnest (Pollard et al. 2006b; Prabhakar et al. 2008). Initial findings strongly suggested important functional roles for individual HARs (Pollard et al. 2006b; Prabhakar et al. 2008), although recent population genetic analyses of HARs as a whole suggest neutral processes may play a dominant role in HAR evolution (Katzman et al. 2010).

Population genomic screens for the signatures of selection in genomic sequence data have revealed numerous putative targets of adaptive evolution in humans and flies (Nielsen et al. 2005a; Begun et al. 2007a), and a recent study has attempted to identify targets of long-term balancing selection in humans (Andrés et al. 2009). From a population genetics perspective, the identification of the action of natural selection on a given genomic region strongly implies that the region is of functional importance to the organism. However, the specific functions of these identified regions have remained largely unexamined, and the phenotypes associated with them are almost completely unknown. Clearly the marriage of comparative and population genomic methods will lead the way in understanding interesting genomic regions through an evolutionary lens.

Does Conservation Imply Function?

One of the central assumptions of comparative genomics is that conservation of sequence across a phylogeny implies functional constraint (Kimura, 1983). This assumption is certainly intuitively appealing. If a region of the genome were not under some form of long-term evolutionary constraint (i.e., purifying selection), mutation pressure should rapidly drive sequence divergence among species. Although the alternative assumption of long-term mutational cold-spots seems less likely, the assumption that sequence conservation across species implies functional constraint has not been tested.

Here, we use a combination of phylogenetic and population genetic techniques to test the assumption that cross-species conservation implies purifying selection at the genome scale. We first estimated the evolutionary rate for each position in the *D. melanogaster* genome across the 12 *Drosophila* genomes phylogeny (*Drosophila* 12 Genomes Consortium 2007; *Tribolium* Genome Sequencing Consortium et al. 2008) using the method of Yang (1994). A whole-genome MULTIZ (Blanchette et al. 2004) alignment of the 12 *Drosophila* genomes plus three outgroup taxa was obtained from the UCSC genome browser (Hinrichs et al. 2006). We used this alignment to obtain maximum-likelihood estimates of the parameters of the HKY (Hasegawa et al. 1985) model of nucleotide evolution as well as the extent of gamma distributed, among-site rate variation (Yang 1996) across the reference phylogeny using using PAML (i.e., phylogenetic analysis using maximum likelihood) v4.3 (Yang 2007). We employed shape parameter of a 4-category discrete gamma approximation of among-site rate variation to infer site-specific evolutionary rates for every position in the *D. melanogaster* genome. Briefly, the site-specific evolutionary rate for position x_i is given by:

$$rate(x_i) = \sum_{j=1}^{4} r_j P(r_j \mid x_i) \qquad \textbf{(6.1)}$$

In this equation, r_j is the inferred evolutionary rate for gamma category j, and $P(r_j \mid x_i)$ is the posterior probability that site x_i evolves according to rate category j:

$$P(r_j \mid x_i) = \frac{P(x_i \mid r_j)}{\sum_{k=1}^{4} P(x_i \mid r_k)} \quad (6.2)$$

Because of the nature of the gamma rate variation model, the mean rate across all sites is 1.0. We binned sites by evolutionary rate with a bin size of 0.1. This analysis provided a quantitative assessment of site-specific evolutionary conservation across the 12-genomes phylogeny: sites with lower evolutionary rates being considered more conserved, and sites within each bin having relatively equivalent across-species conservation.

Next, we used whole-genome *D. melanogaster* resequencing data from the *Drosophila* Population Genomics Project (DPGP) R1.0 (http://www.DPGP.org) to infer the average strength of selection for sites within each bin. The DPGP R1.0 data consists of 37 lines sampled from Raleigh, NC, USA (Jordan et al. 2007) and balancer-inbred chromosomes (7 chrX, 6 chr2 and 5 chr3) sampled from Malawi, Africa (Begun and Lindfors 2005). Genomes were sequenced using Solexa/Illumina technology (Bentley et al. 2008), assembled against the *D. melanogaster* reference genome R5.20 (Adams et al. 2000) using MAQ 0.6.8 (Li et al. 2008) and filtered to remove repeated sequence. For each position in the DPGP alignment, we inferred the ancestral base using maximum likelihood (ML) ancestral reconstruction (Yang et al. 1995) (again, provided by PAML v4.3), assuming the reference phylogeny. Positions with ML reconstruction posterior probability less than 0.9 were considered potentially unreliable and excluded from the analysis. The ancestral genome sequence was used to polarize single nucleotide polymorphisms (SNPs) into ancestral and derived classes.

We inferred the average scaled strength of selection for all polarized SNPs in each conservation bin using an ML approach derived from the site-frequency spectrum that is conditioned on the observed number of segregating sites in our sample (Bustamante et al. 2001; Williamson et al. 2005). As regions of interest were defined on the basis of their divergence between species, divergence-based ascertainment bias was accounted for, using the method of Kern (2009); in addition, we excluded any SNPs with fewer than 10 samples from North America or fewer than 4 samples from Africa.

If interspecific sequence conservation implies functional constraint, average evolutionary rates across the 12-genomes phylogeny should be strongly correlated with the average strength of selection observed in *D. melanogaster*. Indeed, as Figure 6.1 shows, conservation across the phylogeny is highly predictive of selection strength, although the relationship is not linear. These results strongly suggest that the central assumption of comparative genomics—cross-species conservation implies functional constraint—is indeed valid, at least for these taxa.

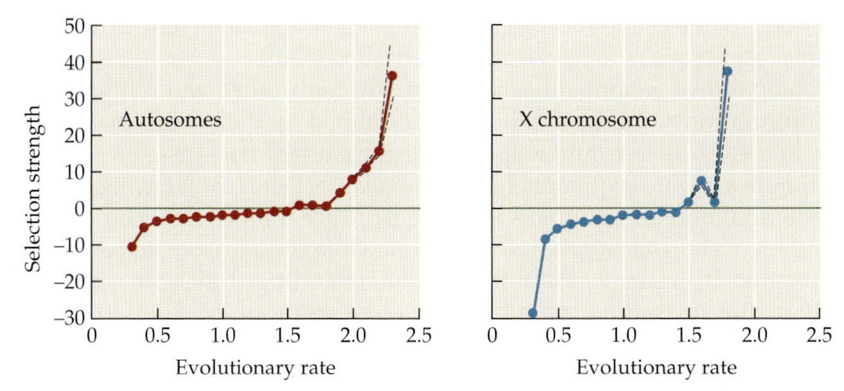

FIGURE 6.1 Sequence Conservation across Phylogenetic Distances Predicts Functional Constraint within a Single Lineage We used maximum-likelihood phylogenetic methods to estimate site-specific evolutionary rates for every position in the *D. melanogaster* genome across the 12 genomes phylogeny. We plotted the site-specific evolutionary rate against the average strength of selection (N_eS) inferred from population genomic data for sites within each rate class. Results are shown separately for autosomal and X-chromosome data. Dashed lines indicate standard error estimates. The horizontal green line indicates inferred neutral evolution.

This result is reassuring, as it confirms our intuition that long-term sequence conservation across species is likely to imply long-term functional constraint at the population level. However, the critical question of whether or not functional constraint at the sequence level implies conservation of biological function remains unanswered. Does the fact that a gene is evolving slowly under long-term purifying selection imply that its biological role has remained unchanged among taxa? It is certainly possible that small changes at the sequence level or changes elsewhere in the genome could have large effects on how a protein functions in an organism. Although systems-based and other types of analyses may be able to address this question at some point in the future, it is important to avoid confounding evolutionary sequence conservation and conservation of biological function when examining genome evolution.

Prospects for Finding Function in Genomes

Although imperfect, positional cloning, genome-wide association, and candidate-gene approaches have been remarkably adept at determining the genetic bases for defined phenotypes of interest (Altshuler et al. 2008). But, what are the phenotypes of interest to evolution? Although selection acts on the phenotype in nature, the signature of historical selective forces is encoded in the heritable genetic material—the genome. Understanding what the genomic signature of selection looks like and developing statistical methods to detect and localize it has been a major goal of comparative genomics and appears to have been largely successful.

Comparative genomics has undoubtedly contributed an exciting array of information about how genomes are shaped by the evolutionary process and is likely to continue to provide a wealth of new insights as more and more genome sequences become available. Dense sequencing within selected taxonomic groups is likely to produce a more nuanced understanding of short-term genome evolution, while comparisons across larger taxonomic distances should refine an understanding of the core genome.

Particularly exciting is the prospect of having multiple individual genomes from single populations. Now that this massive undertaking has begun, the power of population genetics to detect the action of natural selection can be brought to bear on whole-genome data. Indeed, population genetic methods are already being developed and applied to genome-scale datasets, providing a new glimpse into the evolutionary process over short time scales.

It is clear that methodological advances have made solid progress toward detecting the action of selection in comparative and population genomic data. Although technical difficulties remain, such as dealing with the variable data quality and large amounts of missing data in next-generation genomic sequence datasets, the major challenge to any reverse genetic approach to functional genomics is going to be uncovering the phenotypes generated by changes in the genome that have been observed to be driven by natural selection.

The prospects for inferring functional information from genomic sequence data are not clear. The fact is that many of the genomic regions identified as interesting in comparative and population genomic screens have no functional annotation, even in well-studied model organisms. The prospect of extending comparative genomic approaches to non-model organisms and gleaning something of value about their potentially interesting phenotypic evolution seems far off at present. Further, it is unlikely that the true power of comparative genomics to infer functional and phenotypic information will be realized until tight collaborations between computational genomicists and functional molecular biologists are formed to characterize the genomic regions that appear important in cross-species and population level comparisons.

Literature Cited

Adams, M. D., S. E. Celniker, R. A. Holt, and 192 others. 2000. The genome sequence of *Drosophila melanogaster*. *Science* 287: 2185–2195.

Akashi, H. 1994. Synonymous codon usage in *Drosophila melanogaster*: Natural selection and translational accuracy. *Genetics* 136: 927–935.

Akashi, H. and S. W. Schaeffer. 1997. Natural selection and the frequency distributions of "silent" DNA polymorphism in *Drosophila*. *Genetics* 146: 295–307.

Akey, J., G. Zhang, K. Zhang, and 2 others. 2002. Interrogating a high-density SNP map for signatures of natural selection. *Genome Res.* 12: 1805–1814.

Altshuler, D., M. J. Daly, and E. S. Lander. 2008. Genetic mapping in human disease. *Science* 322: 881–888.

Andreasen, K. and B. Baldwin. 2001. Unequal evolutionary rates between annual and perennial lineages of checker mallows (*Sidalcea*, Malvaceae): Evidence from 18s–26s rDNA internal and external transcribed spacers. *Mol. Biol. Evol.* 18: 936–944.

Andrés, A. M., M. J. Hubisz, A. Indap, and 9 others. 2009. Targets of balancing selection in the human genome. *Mol. Biol. Evol.* 26: 2755–2764.

Angrist, M. 1998. Less is more: Compact genomes pay dividends. *Genome Res.* 8: 683–685.

Anisimova, M. and Z. Yang. 2007. Multiple hypothesis testing to detect lineages under positive selection that affects only a few sites. *Mol. Biol. Evol.* 24: 1219–1228.

Beaumont, M. A., W. Zhang, and D. J. Balding. 2002. Approximate Bayesian computation in population genetics. *Genetics* 162: 2025–2035.

Begun, D. J. and H. A. Lindfors. 2005. Rapid evolution of genomic *Acp* complement in the *melanogaster* subgroup of *Drosophila*. *Mol. Biol. Evol.* 22: 2010–2021.

Begun, D., A. Holloway, K. Stevens, and 10 others. 2007a. Population genomics: Whole-genome analysis of polymorphism and divergence in *Drosophila simulans*. *PLoS Biol.* 5: e310.

Begun, D. J., H. A. Lindfors, A. D. Kern, and 1 other. 2007b. Evidence for de novo evolution of testis-expressed genes in the *Drosophila yakuba*/*Drosophila erecta* clade. *Genetics* 176: 1131–1137.

Bejerano, G., M. Pheasant, I. Makunin, and 4 others. 2004. Ultraconserved elements in the human genome. *Science* 304: 1321–1325.

Bentley, D. R., S. Balasubramanian, H. P. Swerdlow, and 192 others. 2008. Accurate whole human genome sequencing using reversible terminator chemistry. *Nature* 456: 53–59.

Blanchette, M., W. J. Kent, C. Riemer, and 9 others. 2004. Aligning multiple genomic sequences with the threaded blockset aligner. *Genome Res.* 14: 708–715.

Boyko, A. R., S. H. Williamson, A. R. Indap, and 11 others. 2008. Assessing the evolutionary impact of amino acid mutations in the human genome. *PLoS Genet.* 4: e1000083.

Braverman, J. M., R. R. Hudson, N. L. Kaplan, and 2 others. 1995. The hitchhiking effect on the site frequency spectrum of DNA polymorphisms. *Genetics* 140: 783–796.

Bridgham, J. T., J. E. Brown, A. Rodríguez-Marí, and 2 others. 2008. Evolution of a new function by degenerative mutation in Cephalochordate steroid receptors. *PLoS Genet.* 4: e1000191.

Britten, R. 1986. Rates of DNA-sequence evolution differ between taxonomic groups. *Science* 231: 1393–1398.

Bustamante, C. D., J. Wakeley, S. Sawyer, and 1 other. 2001. Directional selection and the site-frequency spectrum. *Genetics* 159: 1779–1788.

Casillas, S., A. Barbadilla, and C. M. Bergman. 2007. Purifying selection maintains highly conserved noncoding sequences in *Drosophila*. *Mol. Biol. Evol.* 24: 2222–2234.

Clark, A. G. 1994. Invasion and maintenance of a gene duplication. *Proc. Natl. Acad. Sci. USA* 91: 2950–2954.

Crow, K. and G. Wagner. 2006. What is the role of genome duplication in the evolution of complexity and diversity? *Mol. Biol. Evol.* 23: 887–892.
Cui, L., P. K. Wall, J. H. Leebens-Mack, and 10 others. 2006. Widespread genome duplications throughout the history of flowering plants. *Genome Res.* 16: 738–749.
Cutter, A. D. and B. Charlesworth. 2006. Selection intensity on preferred codons correlates with overall codon usage bias in *Caenorhabditis remanei*. *Curr. Biol.* 16: 2053–2057.
Dehal, P. and J. Boore. 2005. Two rounds of whole genome duplication in the ancestral vertebrate. *PLoS Biol.* 3: 1700–1708.
Delmotte, F., C. Rispe, J. Schaber, and 2 others. 2006. Tempo and mode of early gene loss in endosymbiotic bacteria from insects. *BMC Evol. Biol.* 6.
Drosophila 12 Genomes Consortium. 2007. Evolution of genes and genomes on the *Drosophila* phylogeny. *Nature* 450: 203–218.
Duboule, D. 1994. *Guidebook to the Homeobox Genes*. Guidebook series, Oxford University Press, Oxford.
ENCODE Project Consortium. 2007. Identification and analysis of functional elements in 1% of the human genome by the ENCODE pilot project. *Nature* 447: 799–816.
Erwin, D. H. 2009. Early origin of the bilaterian developmental toolkit. *Phil. Trans. Roy. Soc. Lond. B* 364: 2253–2261.
Force, A., M. Lynch, F. B. Pickett, and 3 others. 1999. Preservation of duplicate genes by complementary, degenerative mutations. *Genetics* 151: 1531–1545.
Fukuchi, S. and K. Nishikawa. 2004. Estimation of the number of authentic orphan genes in bacterial genomes. *DNA Res.* 11: 219–231.
Gavin, A.-C., P. Aloy, P. Grandi, and 29 others. 2006. Proteome survey reveals modularity of the yeast cell machinery. *Nature* 440: 631–636.
Genome 10K Community of Scientists. 2009. Genome 10k: A proposal to obtain whole-genome sequence for 10,000 vertebrate species. *J. Hered.* 100: 659–674.
Gillespie, J. H. 1986. Variability of evolutionary rates of DNA. *Genetics* 113: 1077–1091.
Glazov, E. A., M. Pheasant, E. A. McGraw, and 2 others. 2005. Ultraconserved elements in insect genomes: A highly conserved intronic sequence implicated in the control of *homothorax* mRNA splicing. *Genome Res.* 15: 800–808.
Hanada, K., T. Kuromori, F. Myouga, and 2 others. 2009. Increased expression and protein divergence in duplicate genes is associated with morphological diversification. *PLoS Genet.* 5: e1000781.
Hasegawa, M., H. Kishino, and T. Yano. 1985. Dating of the human-ape splitting by a molecular clock of mitochondrial DNA. *J. Mol. Evol.* 22: 160–174.
Hershberg, R., H. Tang, and D. A. Petrov. 2007. Reduced selection leads to accelerated gene loss in *Shigella*. *Genome Biol.* 8.
Hinrichs, A. S., D. Karolchik, R. Baertsch, and 24 others. 2006. The UCSC genome browser database: Update 2006. *Nucleic Acids Res.* 34: D590–598.
Holloway, A. K., D. J. Begun, A. Siepel, and 1 other. 2008. Accelerated sequence divergence of conserved genomic elements in *Drosophila melanogaster*. *Genome Res.* 18: 1592–1601.
Hudson, R. R. 1983. Properties of a neutral allele model with intragenic recombination. *Theor. Pop. Biol.* 23: 183–201.
Hudson, R. R. 2002. Generating samples under a Wright-Fisher neutral model of genetic variation. *Bioinformatics* 18: 337–338.

Hudson, R. R. and N. L. Kaplan. 1988. The coalescent process in models with selection and recombination. *Genetics* 120: 831–840.
Hudson, R. R. and N. L. Kaplan. 1995. Deleterious background selection with recombination. *Genetics* 141: 1605–1617.
Huelsenbeck, J. and B. Rannala. 1997. Phylogenetic methods come of age: Testing hypotheses in an evolutionary context. *Science* 276: 227–232.
Jensen, J. D., Y. Kim, V. B. DuMont, and 2 others. 2005. Distinguishing between selective sweeps and demography using DNA polymorphism data. *Genetics* 170: 1401–1410.
Jensen, J. D., K. R. Thornton, and P. Andolfatto. 2008. An approximate Bayesian estimator suggests strong, recurrent selective sweeps in *Drosophila*. *PLoS Genet.* 4: e1000198.
Jordan, K. W., M. A. Carbone, A. Yamamoto, and 2 others. 2007. Quantitative genomics of locomotor behavior in *Drosophila melanogaster*. *Genome Biol.* 8: R172.
Kaplan, N. L., T. Darden, and R. R. Hudson. 1988. The coalescent process in models with selection. *Genetics* 120: 819–829.
Kaplan, N. L., R. R. Hudson, and C. H. Langley. 1989. The hitchhiking effect revisited. *Genetics* 123: 887–899.
Katzman, S., A. D. Kern, G. Bejerano, and 5 others. 2007. Human ultraconserved elements are ultraselected. *Science* 317: 915.
Katzman, S., A. D. Kern, K. S. Pollard, and 2 others. 2010. GC-biased evolution near human accelerated regions. *PLoS Genet.* In review.
Keeling, P. J. 2004. Reduction and compaction in the genome of the apicomplexan parasite *Cryptosporidium parvum*. *Dev. Cell.* 6: 614–616.
Keeling, P. J. and C. H. Slamovits. 2004. Simplicity and complexity of *Microsporidian* genomes. *Eukaryot. Cell.* 3: 1363–1369.
Kern, A. D. 2009. Correcting the site frequency spectrum for divergence-based ascertainment. *PLoS ONE* 4: e5152.
Kern, A. D. and D. J. Begun. 2008. Recurrent deletion and gene presence/absence polymorphism: Telomere dynamics dominate evolution at the tip of 3L in *Drosophila melanogaster* and *D. simulans*. *Genetics* 179: 1021–1027.
Kim, Y. and R. Nielsen. 2004. Linkage disequilibrium as a signature of selective sweeps. *Genetics* 167: 1513–1524.
Kim, Y. and W. Stephan. 2002. Detecting a local signature of genetic hitchhiking along a recombining chromosome. *Genetics* 160: 765–777.
Kimura, M. 1983. *The Neutral Theory of Molecular Evolution*. Cambridge University Press, New York.
Kimura, M. and T. Ohta. 1974. On some principles governing molecular evolution. *Proc. Natl. Acad. Sci. USA* 71: 2848–2852.
Knowles, D. G. and A. McLysaght. 2009. Recent de novo origin of human protein-coding genes. *Genome Res.* 19: 1752–1759.
Ko, W.-Y., S. Piao, and H. Akashi. 2006. Strong regional heterogeneity in base composition evolution on the *Drosophila* X chromosome. *Genetics* 174: 349–362.
Kolaczkowski, B. and J. W. Thornton. 2008. A mixed branch length model of heterotachy improves phylogenetic accuracy. *Mol. Biol. Evol.* 25: 1054–1066.
Kosiol, C., T. Vinar, R. R. da Fonseca, and 4 others. 2008. Patterns of positive selection in six mammalian genomes. *PLoS Genet.* 4: e1000144.

Krylov, D., Y. Wolf, I. Rogozin, and 1 other. 2003. Gene loss, protein sequence divergence, gene dispensability, expression level, and interactivity are correlated in eukaryotic evolution. *Genome Res.* 13: 2229–2235.

Kumar, S. 2005. Molecular clocks: Four decades of evolution. *Nat. Rev. Genet.* 6: 654–662.

Kuraku, S. and A. Meyer. 2009. The evolution and maintenance of *HOX* gene clusters in vertebrates and the teleost-specific genome duplication. *Int. J. Dev. Biol.* 53: 765–773.

Kuratani, S. 2009. Modularity, comparative embryology and evo-devo: Developmental dissection of evolving body plans. *Dev. Biol.* 332: 61–69.

Lanfear, R. and L. Bromham. 2008. Statistical tests between competing hypotheses of *HOX* cluster evolution. *Syst. Biol.* 57: 708–718.

Leuenberger, C. and D. Wegmann. 2010. Bayesian computation and model selection without likelihoods. *Genetics* 184: 243–252.

Levine, M. T., C. D. Jones, A. D. Kern, and 2 others. 2006. Novel genes derived from noncoding DNA in *Drosophila melanogaster* are frequently X-linked and exhibit testis-biased expression. *Proc. Natl. Acad. Sci. USA* 103: 9935–9939.

Lewontin, R. C. and J. Krakauer. 1973. Distribution of gene frequency as a test of the theory of the selective neutrality of polymorphisms. *Genetics* 74: 175–195.

Li, H. and W. Stephan. 2006. Inferring the demographic history and rate of adaptive substitution in *Drosophila*. *PLoS Genet.* 2: e166.

Li, H., J. Ruan, and R. Durbin. 2008. Mapping short DNA sequencing reads and calling variants using mapping quality scores. *Genome Res.* 18: 1851–1858.

Margoliash, E. 1963. Primary structure and evolution of Cytochrome C. *Proc. Natl. Acad. Sci. USA* 50: 672–679.

McGinnis, W., M. S. Levine, E. Hafen, and 2 others. 1984. A conserved DNA sequence in homoeotic genes of the *Drosophila Antennapedia* and *Bithorax* complexes. *Nature* 308: 428–433.

Merkeev, I. V. and A. A. Mironov. 2008. Orphan genes: Function, evolution, and composition. *Mol. Biol.* 42: 127–132.

Musio, A., T. Mariani, P. Vezzoni, and 1 other. 2002. Heterogeneous gene distribution reflects human genome complexity as detected at the cytogenetic level. *Cancer Genet, Cytogenet.* 134: 168–171.

Nabholz, B., S. Glemin, and N. Galtier. 2008. Strong variations of mitochondrial mutation rate across mammals—the longevity hypothesis. *Mol. Biol. Evol.* 25: 120–130.

Nielsen, R., C. Bustamante, A. G. Clark, and 10 others. 2005a. A scan for positively selected genes in the genomes of humans and chimpanzees. *PLoS Biol.* 3: e170.

Nielsen, R., S. Williamson, Y. Kim, and 3 others. 2005b. Genomic scans for selective sweeps using SNP data. *Genome Res.* 15: 1566–1575.

Ogishima, S. and H. Tanaka. 2007. Missing link in the evolution of *HOX* clusters. *Gene* 387: 21–30.

Ohno, S. 1970. *Evolution by Gene Duplication.* Springer-Verlag, Berlin.

Peregrín-Alvarez, J. M., C. Sanford, and J. Parkinson. 2009. The conservation and evolutionary modularity of metabolism. *Genome Biol.* 10: R63.

Pollard, K. S., M. J. Hubisz, K. R. Rosenbloom, and 1 other. 2010. Detection of nonneutral substitution rates on mammalian phylogenies. *Genome Res.* 20: 110–121.

Pollard, K. S., S. R. Salama, B. King, and 10 others. 2006a. Forces shaping the fastest evolving regions in the human genome. *PLoS Genet.* 2.

Pollard, K. S., S. R. Salama, N. Lambert, and 13 others. 2006b. An RNA gene expressed during cortical development evolved rapidly in humans. *Nature* 443: 167–172.
Prabhakar, S., J. P. Noonan, S. Paabo, and 1 other. 2006. Accelerated evolution of conserved noncoding sequences in humans. *Science* 314: 786.
Prabhakar, S., A. Visel, J. A. Akiyama, and 10 others. 2008. Human-specific gain of function in a developmental enhancer. *Science* 321: 1346–1350.
Pritchard, J. K., M. T. Seielstad, A. Perez-Lezaun, and 1 other. 1999. Population growth of human Y chromosomes: A study of Y chromosome microsatellites. *Mol. Biol. Evol.* 16: 1791–1798.
Przeworski, M. 2002. The signature of positive selection at randomly chosen loci. *Genetics* 160: 1179–1189.
Przeworski, M. 2003. Estimating the time since the fixation of a beneficial allele. *Genetics* 164: 1667–1676.
Qin, H., H. H. S. Lu, W. B. Wu, and 1 other. 2003. Evolution of the yeast protein interaction network. *Proc. Natl. Acad. Sci. USA* 100: 12820–12824.
Rokas, A. 2008. The origins of multicellularity and the early history of the genetic toolkit for animal development. *Ann. Rev. Genet.* 42: 235–251.
Sabeti, P. C., P. Varilly, B. Fry, and 260 others. 2007. Genome-wide detection and characterization of positive selection in human populations. *Nature* 449: 913–918.
Schlosser, G. and G. P. Wagner. 2004. *Modularity in Development and Evolution.* University of Chicago Press, Chicago.
Scott, M. P. and A. J. Weiner. 1984. Structural relationships among genes that control development: Sequence homology between the *Antennapedia*, *Ultrabithorax*, and *Fushi Tarazu* loci of *Drosophila*. *Proc. Natl. Acad. Sci. USA* 81: 4115–4119.
Siepel, A. 2009. Darwinian alchemy: Human genes from noncoding DNA. *Genome Res.* 19: 1693–1695.
Siepel, A. and D. Haussler. 2004. Combining phylogenetic and hidden Markov models in biosequence analysis. *J. Comp. Biol.* 11: 413–428.
Siepel, A., G. Bejerano, J. S. Pedersen, and 13 others. 2005. Evolutionarily conserved elements in vertebrate, insect, worm, and yeast genomes. *Genome Res.* 15: 1034–1050.
Simonsen, K. L., G. A. Churchill, and C. F. Aquadro. 1995. Properties of statistical tests of neutrality for DNA polymorphism data. *Genetics* 141: 413–429.
Singh, A. H., D. M. Wolf, P. Wang, and 1 other. 2008. Modularity of stress response evolution. *Proc. Natl. Acad. Sci. USA* 105: 7500–7505.
Snel, B. and M. A. Huynen. 2004. Quantifying modularity in the evolution of biomolecular systems. *Genome Res.* 14: 391–397.
Tajima, F. 1989. Statistical method for testing the neutral mutation hypothesis by DNA polymorphism. *Genetics* 123: 585–595.
Takano, T. 1998. Rate variation of DNA sequence evolution in the *Drosophila* lineages. *Genetics* 149: 959–970.
Taylor, J., I. Braasch, T. Frickey, and 2 others. 2003. Genome duplication, a trait shared by 22,000 species of ray-finned fish. *Genome Res.* 13: 382–390.
Teshima, K. M., G. Coop, and M. Przeworski. 2006. How reliable are empirical genomic scans for selective sweeps? *Genome Res.* 16: 702–712.
Thali, M., M. M. Müller, M. DeLorenzi, and 2 others. 1988. *Drosophila* homoeotic genes encode transcriptional activators similar to mammalian OTF-2. *Nature* 336: 598–601.

Tribolium Genome Sequencing Consortium. 2008. The genome of the model beetle and pest *Tribolium castaneum*. *Nature* 452: 949–955.

Venkatesh, B., P. Gilligan, and S. Brenner. 2000. *Fugu*: A compact vertebrate reference genome. *FEBS Lett*. 476: 3–7.

Wang, X., W. Grus, and J. Zhang. 2006. Gene losses during human origins. *PLoS Biol*. 4: 366–377.

Williamson, S. H., R. Hernandez, A. Fledel-Alon, and 3 others. 2005. Simultaneous inference of selection and population growth from patterns of variation in the human genome. *Proc. Natl. Acad. Sci. USA* 102: 7882–7887.

Wittkopp, P. J., E. E. Stewart, L. L. Arnold, and 6 others. 2009. Intraspecific polymorphism to interspecific divergence: Genetics of pigmentation in *Drosophila*. *Science* 326: 540–544.

Wong, W. S. W. and R. Nielsen. 2004. Detecting selection in noncoding regions of nucleotide sequences. *Genetics* 167: 949–958.

Yang, Z. 1994. Maximum-likelihood phylogenetic estimation from DNA sequences with variable rates over sites: Approximate methods. *J. Mol. Evol*. 39: 306–314.

Yang, Z. 1996. Among-site rate variation and its impact on phylogenetic analyses. *Trends Ecol. Evol*. 11: 367–372.

Yang, Z. 1998. Likelihood ratio tests for detecting positive selection and application to primate lysozyme evolution. *Mol. Biol. Evol*. 15: 568–573.

Yang, Z. 2007. PALM 4: Phylogenetic analysis by maximum likelihood. *Mol. Biol. Evol*. 24: 1586–1591.

Yang, Z., S. Kumar, and M. Nei. 1995. A new method of inference of ancestral nucleotide and amino acid sequences. *Genetics* 141: 1641–1650.

Yeh, I., T. Hanekamp, S. Tsoka, and 2 others. 2004. Computational analysis of *Plasmodium falciparum* metabolism: Organizing genomic information to facilitate drug discovery. *Genome Res*. 14: 917–924.

Zhang, J., R. Nielsen, and Z. Yang. 2005. Evaluation of an improved branch-site likelihood method for detecting positive selection at the molecular level. *Mol. Biol. Evol*. 22: 2472–2479.

Zhou, Q., G. Zhang, Y. Zhang, and 7 others. 2008. On the origin of new genes in *Drosophila*. *Genome Res*. 18: 1446–1455.

Zhu, L. and C. D. Bustamante. 2005. A composite-likelihood approach for detecting directional selection from DNA sequence data. *Genetics* 170: 1411–1421.

Zuckerkandl, E. and L. B. Pauling. 1962. Molecular disease, evolution, and genetic heterogeneity. In M. Kasha and B. Pullman (eds.), *Horizons in Biochemistry*, pp. 189–225. Academic Press, New York.

Commentary Two

The Potential for Microorganisms and Experimental Studies in Evolutionary Biology

Daniel E. Dykhuizen

When I entered graduate school in 1965, neither bacterial nor experimental evolution was included in the course of study for an apprentice evolutionary biologist. At that time, there were good reasons for this exclusion. There was no scientific field of microbial phylogeny, and laboratory experimental evolution was generally discouraged, because it was often thought to be trivial and unnatural. Yet, Richard Lewontin, my major professor, encouraged my use of bacteria in a laboratory experiment for my thesis. In this commentary, I argue that microbial and experimental evolution offer untapped opportunities for exciting new insights in evolutionary biology.

Microbiology became an area of serious scientific study only after publication of *The Origin of Species* (Darwin 1859). This is the 150th anniversary of its publication as well as the 128th anniversary of the paper by Robert Koch on the methods to study bacteria in pure culture (Koch 1881). Once pure cultures had been established, bacterial species could be determined and were shown to have stable characteristics. Before that, the few bacterial shapes observed with a microscope had been thought of as different morphological forms of a single species.

Microbiologists tried to fit microorganisms into an evolutionary framework (Woese 1994), but there were too few morphological characters. They gave up by the 1940s and settled for a typological taxonomy, retaining binomial names (van Neil 1946). Only with the advent of DNA hybridization in the late 1960s and later with DNA sequencing could microbiologists establish a taxonomy of bacteria that is rooted in phylogeny (Woese 1987).

Microorganisms were not included in the evolutionary synthesis. In the third edition of *Genetics and the Origin of Species*, Dobzhansky (1951) noted that there had been interesting studies on mutation and selection using bacteria, but the absence of sex was a huge drawback, preventing the study of speciation. The study of bacterial genetics started in the 1950s only after discovery of mechanisms of gene transfer (i.e., sex) between bacterial strains, conjugation in 1946 (Lederburg and Tatum 1946) and transduction in 1951 (Lederburg et al. 1951). Even though transformation had been discovered in 1928 (Griffith 1928), it was not used for genetic analysis until 1951 (Ephrussi-Taylor 1951). Now, sex and speciation are major topics in microbiology.

Microbes should play a larger role in evolutionary biology research, so that the field embraces the entire living world, and this would enhance evolutionary explanations. Most of the history of life on earth is a history of microbes. One can study ancient adaptations that took place before the origin of the metazoans (Zhu et al. 2005) and speculate about the universal ancestor of all cellular life (Woese 1998; Kurland et al. 2007; see Lazcano, Chapter 14). But most importantly, incorporating microorganisms more fully into evolutionary biology presents the opportunity to study multiple solutions to the same evolutionary problems, including, for example, the diversification of metabolic pathways and the evolution of sex.

In bacteria, sex or lateral transfer of DNA between lineages is separated from reproduction, presumably allowing the amount of sex to be optimized, from very little, as in *Borrelia,* to very much, as in *Neisseria.* Bacterial sex usually involves transmission of much less than 1% of the genome and only a few genes. Bacteria are haploid, so sex results in immediate gene replacement and the fitness consequences of that replacement, as compared to transfer of an entire haploid genome from male and female in sexual diploids, with recombinational breakdown happening only in the next generation. The frequency of recombinants or chimeras in natural populations of bacteria will be a function of the rate of recombination and natural selection on these chimeras. Thus, in a background of infrequent recombination, the observed chimeras reflect favorable new gene configurations. In contrast, in species with high recombination rates, chimeric individuals will be common and selection against particular allelic moieties will be revealed by their conspicuous absence (Dykhuizen 2005). Thus, successful recombinants occurring at high frequencies in species with rare sex involve genes with alleles that are under balancing selection. The function of the affected genes will indicate the advantage of sex. In the case of selection against particular recombinants in so-called promiscuous species, the genes are under negative pleitrophic selection and may be drivers of speciation.

While it is clear that the limits of technology and knowledge retarded use of microbial models in evolutionary biology, these are not the reasons for limited use of laboratory experimental evolution in evolutionary biology. Darwin did not consider the possibility that experiments would be important in evolution even though he knew the person who performed the first experimental study of laboratory evolution, the Rev. W. H. Dallinger, president of the Royal Microscopic Society. Dallinger devised an ingenious apparatus to grow protozoa at a controlled temperature; gradually increasing the temperature caused evolution of the protozoa to grow at higher temperatures than the non-adapted ancestor (Dallinger 1887). Rose and Garland (2009) suggest that under Charles Lyell's influence, Darwin expected natural selection to act very slowly, producing observable evolution only over long periods of time. Thus, for the same reason that Darwin's gradualist worldview caused him to overlook the importance of sports (i.e., Mendelian variation), he also failed to appreciate the significance of Dallinger's (1887) early results in experimental microbial evolution.

However, laboratory experiments in both ecology and evolution were common by the 1940s. Dobzhansky used population cages to show that the chromosomal inversions in *Drosophila pseudoobscura* and *D. persimilis,* which had been assumed to be selectively neutral, could represent balanced polymorphism (Wright and Dobzhansky 1946). This result led to the paradigm of widespread balancing selection and to the postulate that analysis of variation at the level of protein electrophoresis and then DNA could yield fundamental insights into the dynamics of natural selection (Lewontin 1974). The neutral theory of Motoo Kimura (Kimura 1983) threatened to destroy this postulate until Kreitman and his collaborators showed that neutral variation could be used to identify genetic variation under balancing selection (Hudson et al. 1987; see Wakeley, Chapter 5). Experimental evolution can have profound consequences for evolutionary theory.

Then, why were laboratory experiments frowned upon in the 1960s? First, uninformative experiments are easy to do. It is legitimate to ask whether the fate of white-eye mutants in population cages offers any insight into the process of selection in natural populations. Second, if one wishes to mimic natural environments in the laboratory, how does one do it? The cell and organism are both definable units, each with an explicit boundary, and can be studied in the laboratory as cells or organisms. Natural selection is about the differential fitness of phenotypes resulting from genotypic differences in a particular environment. Where is the definable boundary of the environment that an organism or population inhabits? For example, the *Tribolium* beetle, as studied by Tom Park and Michael Wade, lives in a seemingly definable environment—flour. However, growing the beetles in small vials of flour radically changed the environment because the beetles could no longer migrate away from each other. The unanticipated and inescapable consequences of the laboratory environment lead to skepticism over the utility of laboratory studies. Instead of studies in macrocosms, experiments in nature and other strategies to avoid this boundary problem have been preferred (e.g., see Endler's 1995 experiments on guppies). While these studies can be very useful, they do not have the experimental control and repeatability of laboratory experiments.

The solution to this criticism is to shift the focus of laboratory-based experiments from mimicking nature to testing theory. Laboratory-based experiments should be thought of as simulations using organisms rather than computers. The experimental design is set up to meet the genetic and environmental assumptions of the theory, even if neither the genetic variation nor the environment is natural. Having met these assumptions, the organism's properties mediate the outcome of the simulation. For example, enzymes with more than one substrate tend to be more polymorphic than those with only one substrate (Singh 1992). This extra variation could be maintained by stabilizing frequency-dependent selection in which one allele is fitter on one substrate and another allele is fitter on the other substrate. Using two different lactose operons in isogenic *Escherichia coli* strains growing on two different beta-galactosides, with one operon fitter on one substrate and the other on the other (thus meeting the assumptions of the theory), it was shown that there was stabilizing frequency-dependent selection that closely matched the theoretical expectation (Lunzer et al. 2002). However, the range of the ratios of the two substrates that allowed coexistence was narrow, less than 10% of the possible range. This finding suggests that in a variable environment, the frequency-dependent selection could prolong the polymorphism, allowing increased observed polymorphism in enzymes that have multiple substrates, but eventually one or the other allele would be lost (i.e., the polymorphism would not be stable over phylogenetic time). While this experiment worked out as expected, it is surprising how often organisms do something completely different from the theoretical expectations (Dykhuizen

1978; Dykhuizen and Dean 2004; Stobel et al. 2008).

A major research program in evolutionary biology designed to understand the entire chain of causation connecting genotype to fitness is underway. The goal is a mechanistic understanding of the causes of natural selection (Dykhuizen 1995; see Hoekstra, Chapter 22). Since bacteria lack diploidy, development, and complicated behavior, this research program will be easier with bacteria than with animals and plants and is already well under way (Dykhuizen and Dean 1990; Weinreich et al. 2006; King et al. 2006; Barrick et al. 2009). Bacteria and laboratory-based experiments will contribute novel perspectives to evolutionary biology in the twenty-first century.

Acknowledgments

I thank Walter Eanes and Nicholas Friedenberg for commenting on this manuscript. This is contribution number 1194 in Ecology and Evolution from Stony Brook University.

Literature Cited

Barrick, J. E., D. S. Yu, S. H. Yoon, and 5 others. 2009. Genome evolution and adaptation in a long-term experiment with *Escherichia coli*. *Nature* 461: 1243–1247.

Dallinger, W. H. 1887. Transactions of the society V. The president's address. *J. Roy. Microscop. Soc.* 1887: 185–199.

Darwin, C. 1859. *On the Origin of Species by Means of Natural Selection, or the Preservation of Favoured Races in the Struggle for Life.* John Murray, London.

Dobzansky, T. 1951 *Genetics and the Origin of Species*, 3rd ed. Columbia University Press, New York.

Dykhuizen, D. 1978. Selection for tryptophan auxotrophs of *Escherichia coli* in glucose-limited chemostats as a test of the energy conservation hypothesis of evolution. *Evolution* 32: 125–150.

Dykhuizen, D. E. 1995. Natural selection and the single gene. In S. Baumberg, J. P. W. Young, E. M. H. Wellington, and J. R. Saunders (eds.), *Population Genetics of Bacteria*, pp. 161–173. *Soc. Gen. Microbiol. Symp. 52.* Cambridge University Press, Cambridge, UK.

Dykhuizen, D. E. 2005. Species numbers in bacteria. *Proc. Cal. Acad. Sci.* 56: 62–71.

Dykhuizen, D. E. and A. M. Dean. 1990. Enzyme activity and fitness: Evolution in solution. *Trends Ecol. Evol.* 5: 257–262.

Dykhuizen, D. E. and A. M. Dean. 2004. Evolution of specialists in an experimental microcosm. *Genetics*. 167: 2015–2026.

Endler, J. A. 1995. Multiple-trait coevolution and environmental gradients in guppies. *Trends Ecol. Evol.* 10: 22–29.

Ephrussi-Taylor, H. 1951. Genetic aspects of transformations in pneumococci. *Cold Spr. Harb. Symp. Quant. Biol.* 16: 445–456.

Griffith, F. 1928. Significance of pneumococcal types. *J. Hyg. Camb.* 27: 113–159.

Hudson, R. R., M. Kreitman, and M. Aguade. 1987. A test of neutral molecular evolution based on nucleotide data. *Genetics* 116: 153–159.

Kimura, M. 1983. *The Neutral Theory of Molecular Evolution*. Cambridge University Press, Cambridge, UK.

King, T., S. Seeto, and T. Ferenci. 2006. Genotype-by environment interactions influencing the emergence of *rpoS* mutations in *Escherichia coli* populations. *Genetics* 172: 2071–2079.

Koch, R. 1881. Zur Untersuchung von pathogenen Organismen. *Mittheilungen aus dem Kaiserlichen Gesundheitsamte* 1: 1–48.

Kurland, C. G., B. Canbaeck, and O. Berg. 2007. The origins of modern proteomes. *Biochimie* 89: 1454–1463.

Lederberg, J. and E. L. Tatum. 1946. Gene recombination in *E. coli*. *Nature* 158: 558.

Lederberg, J., E. M. Lederberg, N. D. Zinder, and 1 other. 1951 Recombination analysis of bacterial heredity. *Cold. Spr. Harb. Symp. Quant. Biol.* 16: 413–443.

Lewontin, R. C. 1974. *The Genetic Basis of Evolutionary Change.* Columbia University Press, New York.

Lunzer, M., A. Natatajan, D. E. Dykhuizen, and 1 other. 2002. Enzyme kinetics, substitutable resources and competition: From biochemistry to frequency-dependent selection in *lac*. *Genetics* 162: 485–499.

Rose, M. R. and T. Garland Jr. 2009. Darwin's other mistake. In T. Garland Jr. and M. R. Rose (eds.), *Experimental Evolution: Concepts, Methods and Applications of Selection Experiments*, pp. 3–13. University of California Press, Berkeley.

Singh, R. S. 1992. A comprehensive study of genetic variation in natural populations of *Drosophilia melanogaster*. V. Structural-functional constraints on protein molecules and enzymes and the levels and patterns of variation among genes. *Genome* 35: 109–119.

Stoebel, D. M., A. M. Dean, and D. E. Dykhuizen. 2008. The cost of expression of *Escherichia coli lac* operon proteins is in the process, not in the products. *Genetics* 178: 1653–1660.

van Neil, C. B. 1946. The classification and natural relationships of bacteria. *Cold Spr. Harb. Symp. Quant. Biol.* 11: 285–301.

Weinreich, D. M., N. F. Delaney, M. A. DePristo, and 1 other. 2006. Darwinian evolution can follow only very few mutational paths to fitter proteins. *Science* 312: 111–114.

Woese, C. R. 1987. Bacterial evolution. *Microbiol. Rev.* 51: 221–271.

Woese, C. R. 1994. There must be a prokaryote somewhere: Microbiology's search for itself. *Microbiol. Rev.* 58: 1–9.

Woese, C. R. 1998. The universal ancestor. *Proc. Natl. Acad. Sci. USA* 95: 11043–11046.

Wright, S. and T. Dobzhansky. 1946. Genetics of natural population. XII. Experimental reproduction of some of the changes caused by natural selection in certain populations of *Drosophila pseudoobscura*. *Genetics* 31: 125–156.

Zhu, G. P., G. B. Golding, and A. M. Dean. 2005. Selective cause of an ancient adaptation. *Science* 307: 1279–1282.

Part III

THE EVOLUTION OF FORM

Chapter 7

Rates of Adaptation: Why Is Darwin's Machine So Slow?

Mark Kirkpatrick

The Origin of Species has at its heart a crushingly simple argument. Darwin (1859) pointed out that if there is both selection and inheritance, then evolution must result. In the last 150 years, it has become clear that the Darwinian machine can in fact work far more efficiently than its discoverer ever dared to suppose. Long-term artificial selection, for example on oil content in corn (Dudley 2004) and milk yield in dairy cattle (Powell and Norman 2006), shows that the mean value of a trait can be changed several-fold in just a few dozen generations. It is now also evident that natural populations can evolve at equally astonishing rates. Responding to selection triggered by a drought in 1977, the mean bill depth of *Geospiza fortis* (Darwin's medium ground finch) on Isla Daphne evolved upwards by about 5% (Grant and Grant 1995). If that rate continued for several more generations, the cumulative change would be enough by some taxonomic standards (Mayr et al. 1953) to declare the origin of a new morphospecies.

Evolutionary biologists are in the embarrassing situation of having a mechanism that works too well. A major challenge is to understand why rates of morphological evolution are typically so much slower than what is possible (Eldredge and Gould 1972). Gingerich (1983) found that rates seen in the fossil record over periods of 10^3 to 10^8 years are typically several orders of magnitude smaller than those seen in selection experiments and short-term evolution in the wild. That conclusion has been corroborated by later work based on fossils (Gingerich 2001; Estes and Arnold 2007; Gingerich 2009) and on comparison of living species in a phylogenetic framework (Lynch 1990; Hendry and Kinnison 1999).

The basic pattern of rates of morphological evolution is illustrated in Figure 7.1. There is a conspicuous lack of points in the upper right region of the plot, showing that the rapid evolution seen in natural and laboratory populations over periods of up to 100 generations (see points at the upper left) are not seen in the fossil record over periods of 10^5 years or more. These results confirm the claim made by Eldredge and Gould (1972): there

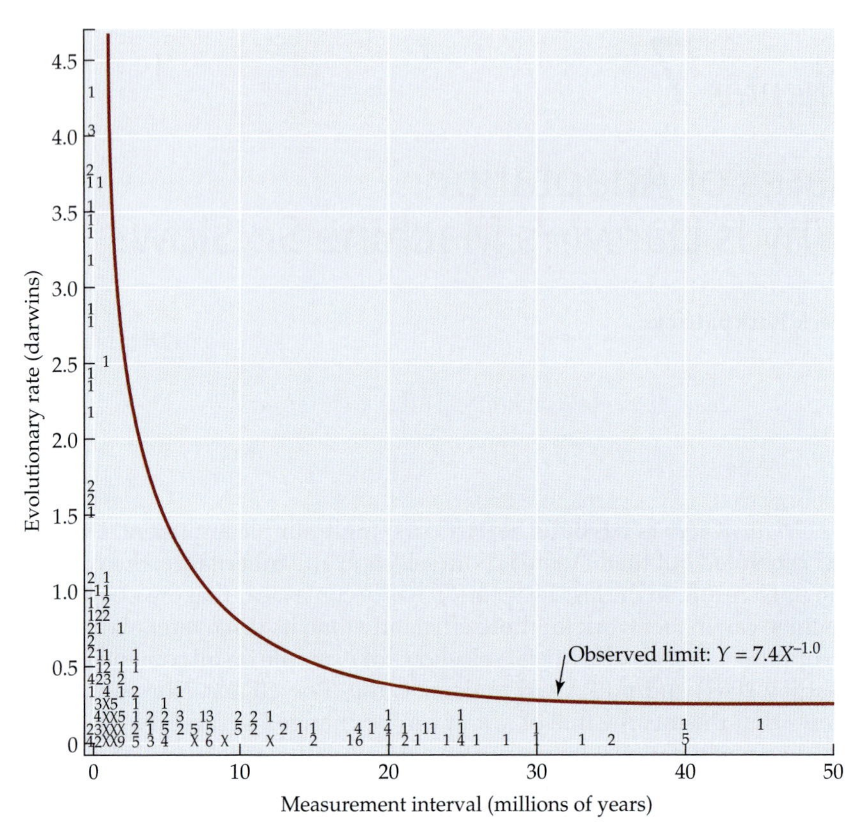

FIGURE 7.1 Evolutionary Rates Plotted against the Interval Over Which the Rate Was Measured Rates are measured in darwins, which are equal to the proportional change in the mean per year, times 10^6 years (Haldane 1949). (Adapted from Gingerich 1983.)

is indeed a pattern of evolutionary stasis on macroevolutionary timescales that must be explained. What is the explanation for that pattern has not yet been fully resolved.

In this paper, I will review four explanations for why observed macroevolutionary rates for morphological traits are often so much smaller than what is possible. The null hypothesis is that directional selection in nature is typically very weak. The three alternative hypotheses, which are not mutually exclusive, are:

H_1: Selection may frequently oscillate in one direction, causing much evolutionary change to be erased and leading to underestimates of rates.

H_2: Limited genetic variation for many traits constrains the response to selection.

H_3: Demographic costs limit the amount of selection that natural populations can sustain.

There are, of course, other schemes for organizing a discussion about evolutionary rates (and other views about what controls them). For alternative perspectives on this topic, the recent papers by Gingerich (2009) and Futuyma (2010) are highly recommended.

H_0: Selection Is Feeble

Evolutionary rates will be small if selection and other forces such as drift are weak. A simple calculation shows just how weak directional selection needs to be in order to produce evolution that is rapid by paleontological standards. Lande (1976) found that evolutionary rates for mammalian teeth in the Tertiary period are compatible with selective mortality rates on the order of one death per million individuals.

That result can be put into context by considering the intensities of selection seen in extant populations. Darwin and generations of biologists that followed him assumed selection was generally so weak that it could not be studied directly. That view changed in the last 30 years, with the development of methods to quantify selection (Lande and Arnold 1983; Endler 1986) and the accumulation of data on many species. Kingsolver et al. (2001b) reviewed over 2500 published estimates of the strength of directional selection in natural populations. They report that the median selection intensity (i.e., the selection gradient standardized by the phenotypic standard deviation) is 0.16, which is several orders of magnitude greater than what is needed to account for observed rates of change over moderate-to-long time periods. The data are likely to be biased upwards, for example, because workers are less likely to choose traits under very weak selection or to publish insignificant results. It seems implausible, however, that the discrepancy between typical selection intensities and evolutionary rates is mainly the result of bias. Thus, it appears we can reject the null hypothesis with some confidence.

H_1: Evolution Is Fickle

Animal and plant breeders apply directional selection towards a single objective. Nature is not so consistent. For instance, in the years before and after the 1977 drought that caused bill size in *Geospiza fortis* to evolve upward, bill size evolved downwards (Grant and Grant 2002). As a result, the net evolutionary change over several generations was close to zero.

Gingerich (1983) suggested that evolutionary reversals are a key to the explanation of the paradox of evolutionary rates. Since then, the case for reversals has been bolstered by statistical analysis of high-resolution fossil sequences. An outstanding example comes from stickleback fishes in a Nevada paleolake, sampled by Bell and colleagues (2006). Their data comprise about 5000 specimens sampled at fine intervals, over a period of some 7000 generations. The data show clear evidence of directional change, over a period of roughly 2000 generations, in the mean values of three skeletal traits after the fish colonized the lake. Subsequently, the traits show fluctuations around a long-term mean, which is interpreted as an adaptive peak (Hunt et al. 2008). In a meta-analysis of over 250 other fossil sequences, Hunt (2007b) found statistical support in about half the sequences for similar patterns of fluctuations around a secular mean. In the majority of the remaining cases, the null model of no change could not be rejected. So, it appears that reversals are a common feature of morphological evolution. They decrease the net amount of directional change that is realized over longer intervals, and so bias downwards estimates of short-term evolutionary rates.

But what causes these reversals? One possibility is random genetic drift. Lande (1976) showed that drift in even moderately large populations (10^4 to 10^5 individuals) can cause the mean value of a quantitative trait to make excursions around a fixed adaptive peak that should be large enough to be detected in fossil sequences. Drift thus provides an appropriate null model for studies seeking positive evidence of selection in the fossil record. Many fossil sequences (Estes and Arnold 2007; Hunt 2007b; Hunt et al. 2008) and patterns of divergence between extant species (Lynch 1990; Hansen 1997) are consistent with models in which fluctuations in mean phenotypes are caused by drift in a constant adaptive landscape.

That agreement, however, is far from proof that drift alone is responsible for the fluctuations. A plausible scenario is that selection drives populations towards an adaptive peak that itself is wandering within a limited phenotypic range—an "adaptive zone" in the parlance of G. G. Simpson (1944). This type of moving selective target can produce the same pattern of evolutionary change as that generated by drift, a constrained random walk that can be modeled by the Ornstein-Uhlenbeck process (Hansen 1997).

Are evolutionary reversals then caused by drift or oscillating selection? Drift is a mathematical certainty, and changing selection would seem to be a biological one. An interesting avenue for future research is to estimate the relative contributions of these two forces. Morphological data alone cannot do that because the amount of drift is determined by the effective population size, which if left as a free parameter can be tuned to fit the data. One might think of using molecular data to estimate effective population sizes from extant populations. Those estimates could not be applied to fossil lineages, of course. They could, however, be used to analyze divergence between living populations and species in a phylogenetic framework. One handicap faced by this approach is that all traces of large evolutionary excursions caused by selection can be erased by reversals and another is that it assumes population sizes remained relatively constant during the period of divergence. Estes and Arnold (2007) recently took a related approach. They fit the data taken from Gingerich's (2001) compendium of fossil sequences to models for phenotypic evolution that include both selection and random genetic drift. Of the simple selection scenarios they considered, the one that fit best involved just a single bout of directional selection towards a new adaptive peak, followed by drift around that peak. This analysis shares the difficulty with the previous approach that each evolutionary rate is estimated from trait means at two points, and so it will not account for intervening evolutionary reversals caused by directional selection.

A possible alternative research strategy might be to integrate high-resolution fossil sequences and data on selection intensities from extant populations. Drift causes a population to wander away from an adaptive peak, which produces directional selection to return. The distribution of resulting selection intensities can be calculated for models in which the peak is constant or moving. One could ask how much movement of the peak (if any) is needed to generate a distribution of directional selection intensities comparable to those typically seen in extant populations.

H_2: Lack of Genetic Fuel

The suggestion by Eldredge and Gould (1972) that morphological stasis might result from constraints engendered a controversy that has engaged paleontologists, morphologists, developmental biologists, and geneticists (Maynard Smith et al. 1985; see Wray, Chapter 9). Constraints can be interpreted from different and complementary perspectives (e.g., developmental and phylogenetic). From a genetic point of view, a constraint results when there is no genetic variation that allows the population to respond to directional selection. Virtually all morphological traits that have been analyzed show substantial levels of additive genetic variation and respond readily to artificial selection (Mousseau and Roff 1987; Houle 1992). Further, long-term selection experiments show that high rates of evolutionary change can be sustained for periods on the order of 10^2 generations (Lynch and Walsh 1998). There is no support for the possibility that genetic variation is exhausted in large outbred populations after a few generations of directional evolution (Hill and Kirkpatrick 2010), and quantitative traits typically have sufficient input of new variation from mutation to rapidly restore what is lost to selection (Keightley and Halligan 2009). These observations have led many evolutionary biologists to reject the idea that constraints play an important role in the pattern of stasis (Charlesworth et al. 1982).

That conclusion, however, may be less secure than many suppose. A first issue is that quantitative genetic variation is only measured on extant phenotypes. The origin of a key morphological innovation, like the pharyngeal jaws in cichlid fishes (Liem 1973; see Losos and Mahler, Chapter 15), is often proposed as the trigger for adaptive radiations. Do other groups of fishes not have pharyngeal jaws because they do not have the requisite mutations to start down that path or because selection does not favor them in other groups of fishes? That question is not accessible to quantitative genetics, because how to identify variation in other taxa that corresponds to a proto-pharyngeal jaw is not known. Some rapid bursts of evolution might well be triggered by innovations that follow the appearance of very rare (even unique) mutations, whether of small or large phenotypic effect.

A second issue is that some kinds of traits may have much less genetic variation than is typical for morphological characters. Hoffmann et al. (2003) studied a physiological trait, desiccation resistance, thought to limit the geographical ranges of species of *Drosophila*. They obtained no response to intense artificial selection to increase the trait mean in the tropical *D. birchii*, despite abundant genetic variation for morphology and at microsatellite loci. Kellerman et al. (2009) expanded that picture by showing that an additional 14 species from the tropics have very low (perhaps zero) genetic variance for cold tolerance as well as desiccation tolerance, while 15 other species with broad geographical distributions show no such signs of constraint. A second rare example of an apparent genetic constraint comes from the winning times of thoroughbred racehorses and greyhound dogs, which have not perceptibly improved in the last 50 years, despite intense artificial selection (Hill and Bunger 2004).

There is yet a third problem with the facile conclusion that there is abundant genetic fuel available for adaptation. It is certainly true that virtually all morphological traits that have been studied have substantial genetic variation. But what is true for single traits may not be true for combinations of traits. When a changed environment suddenly favors a different mean body size, for example, it presumably acts together with directional and stabilizing selection on other parts of the phenotype. Recent research on genetic variation for multivariate phenotypes is challenging the common wisdom that genetic variation is not a limiting factor for rates of adaptation. The next section develops that idea by reviewing recent analyses of multivariate genetic variation. Those analyses lead to a discussion of four case studies that suggest that patterns of genetic variation may in fact influence rates and directions of morphological evolution over macroevolutionary timescales.

The Dimensionality of Genetic Variation

Building on the theory of artificial selection developed by animal breeders, Lande (1979) showed that the evolutionary response in the mean values of a set of quantitative traits to a single generation of selection can be described by a very simple equation. His argument was based on a genetic model that makes several strong assumptions, notably a normal distribution of breeding values, but in fact, the result is also a good approximation under a much wider range of conditions if selection is weak.

For the present discussion, I will use a version of Lande's equation in which the data have been standardized by dividing each measurement by its trait mean, which allows a comparison of traits with very different means (e.g., the mass of a mouse compared with the mass of an elephant). Further, it allows comparison of traits that are measured in incommensurate units (e.g., the mass of a mouse and the length of its femur). These considerations motivated Haldane (1949) to propose the *darwin*, a unit of relative change per million years, in order to measure evolutionary rates on paleontological time scales. While this mean standardization of the data is useful for the analyses that follow, other options are available (Hansen and Houle 2008; Gingerich 2009).

After we have mean-standardized the data, Lande's equation for the response to selection is:

$$\Delta\tilde{\mathbf{z}} = \tilde{\mathbf{G}}\tilde{\beta} = (\mathbf{ERE})\tilde{\beta} \tag{7.1}$$

Here $\Delta\tilde{\mathbf{z}}$ is the vector of evolutionary changes in the trait means over a single generation; $\tilde{\mathbf{G}}$ is the matrix of additive genetic variances and covariances; and $\tilde{\beta}$ is the vector of directional selection gradients. I have added tildes over these quantities as a reminder that we are working with mean-standardized data. In the second line, $\tilde{\mathbf{G}}$ has been decomposed into a product of two matrices (Kirkpatrick 2009). **E** is the diagonal matrix of genetic coefficients of variation, what Houle (1992) called "evolvabilities," and **R** is the matrix of genetic correlations. A convenient benefit of working in mean-standardized units is that rates of evolutionary change calculated from Equation (7.1) can be converted into the units of darwins (see Figure 7.1) simply by multiplying $\Delta\tilde{\mathbf{z}}$ by the factor (generations/10^6/year).

Equation 7.1 shows that the relative selection response depends both on the genetic variation for individual traits (the evolvabilities) and on the genetic correlations between them. As previously suggested, there is a consensus that evolvabilities for single traits are typically large enough that they are not likely to explain a pattern of stasis over even moderate time horizons. But might genetic correlations tell an entirely different story?

Several summary statistics have recently been proposed to quantify the potential constraints imposed by genetic correlations (Hansen and Houle 2008; Kirkpatrick 2009; Pavlicev et al. 2009). Here, I will describe one of the statistics. Consider the effective number of dimensions of the $\tilde{\mathbf{G}}$ matrix, defined as:

$$n_{\mathrm{D}} = \sum_{i=1}^{N} \lambda_i / \lambda_1 \tag{7.2}$$

in which the λ_i are the eigenvalues of $\tilde{\mathbf{G}}$ arranged in order of decreasing size, and N is the number of measured traits (Kirkpatrick 2009). The value of n_{D} measures how evenly the genetic variation is distributed among different combinations of traits. It is appealing for several reasons: (1) it can accommodate data from traits measured in different units, (2) it has a simple intuitive interpretation, and (3) it can be connected with a model for how traits respond to directional selection.

Figure 7.2 visualizes how different distributions of genetic variation affect n_{D}. Values of n_{D} range between 1 and N. It takes on the maximum value when all traits have equal evolvabilities and there are no genetic correlations. In such cases, the population will respond equally rapidly to selection for any combination of traits. At the other extreme, n_{D} takes

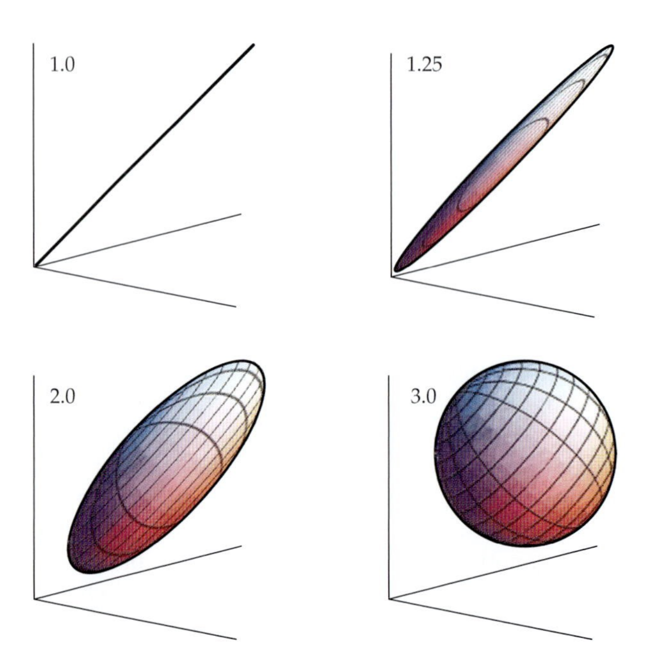

FIGURE 7.2 The Distribution of Genetic Variation in Three Traits for Different Values of n_D, the Effective Numbers of Dimensions In all cases, there is variation for each of the traits, but different degrees of correlation between them. In the examples shown, the ratio of successive eigenvalues is: 1 for $n_D = 3.0$; 0.6 for $n_D = 2.0$; 0.2 for $n_D = 1.25$; and 0 for $n_D = 1.0$.

a value of 1 when all the traits are perfectly correlated. In that case, the population is constrained to evolve in a single dimension within the N-dimensional space. Importantly, this scenario of maximum constraint can occur even when each trait individually has abundant variation (that is, high evolvability).

I found five studies in the literature with the data needed to calculate n_D for a population with five or more traits measured (Kirkpatrick 2009). Three of the data sets are of different linear measures: 10 traits in a fly (McGuigan and Blows 2007), 21 traits in a fish (McGuigan et al. 2005), and 8 traits in beef cattle (Meyer 2005). The other two data sets are on growth trajectories: body mass in mice (Riska et al. 1984) and lactation in dairy cattle (Pander et al. 1993), both as a function of the animal's age. Since the trait value at each age can be regarded as a separate trait, these so-called function-valued characters can be regarded as consisting of an infinite number of dimensions. However, we know that there must be constraints on the form of any growth trajectory, simply because there are limits to how fast any animal can possibly grow or shrink (Kirkpatrick and Lofsvold 1992; Kingsolver et al. 2001a). An attractive feature of the n_D statistic is that it can be used for growth trajectories and other function-valued traits, such as reaction norms.

Results for the effective number of dimensions are striking: they fall in a narrow range between 1.1 (for the dairy cattle) and 1.9 (for the female flies). While the two function-valued data sets give small values for n_D, its value for the three other data sets is not dramatically larger. The implication is that genetic variation is distributed very unevenly, with little scope for evolution to respond to selection for some combinations of traits.

To get a sense of how the effective number of dimensions may limit evolutionary rates, one can do a thought experiment. Begin by assuming that the combinations of traits favored by selection are chosen randomly with respect to the pattern of genetic variation. How rapidly the population will evolve, on average, can then be calculated and compared to the rate it would achieve if selection favored the combination of traits for which there is the most genetic variation (Kirkpatrick 2009). The result is shown in Figure 7.3. The average selection response declines as the number of traits (N) increases and the effective number of dimensions declines. Figure 7.4 shows results for the most extreme case of constraint ($n_D = 1$). Here, the average selection response declines to less than 40% of its maximum, with five traits under selection, and to less than 20% with 25 traits. This finding raises the interesting issue of how many traits actually experience appreciable directional selection. Unfortunately, there currently is no way to estimate that number, even crudely.

One might wonder whether this conclusion results from the way that n_D has been defined. Analyses using other data and other statistical approaches, however, have come to qualitatively similar conclusions: quantitative genetic variation appears to be distributed largely among a relatively

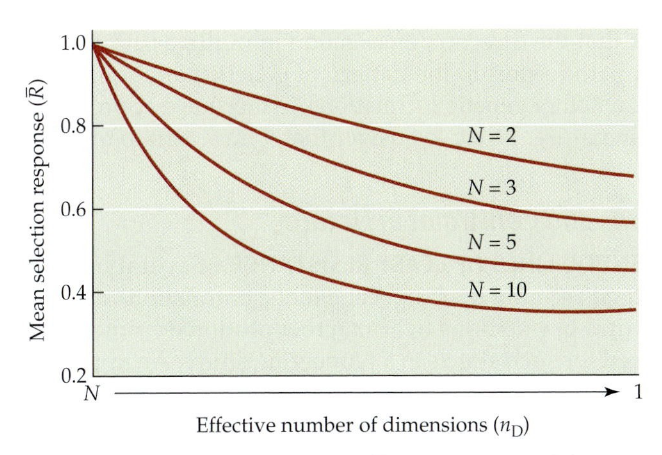

FIGURE 7.3 The Mean Selection Response $\bar{R}$ as a Function of the Degree of Constraint Constraint increases towards the right as the effective number of dimensions (n_D) decreases from its maximum (N) to its minimum (1), and N is the number of traits under selection. $\bar{R}$ is normalized such that its maximum is 1 for a given value of N. (Adapted from Kirkpatrick 2009.)

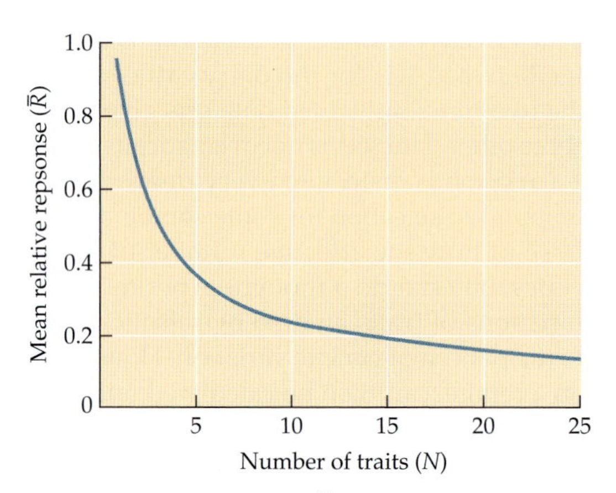

FIGURE 7.4 The Mean Selection Response $\bar{R}$ as a Function of *N* The number of traits under selection, when the effective number of traits $n_D = 1$. (Adapted from Kirkpatrick 2009.)

small number of axes (Kirkpatrick and Lofsvold 1992; Mezey and Houle 2005; Hansen and Houle 2008). A second issue is whether this impression is an artifact of the specific traits and populations that were measured. Perhaps the inclusion of physiological and behavioral traits in addition to morphological ones, for example, would give a different picture. That question will only be answered when additional data sets become available. In any event, the data at hand strongly suggest that a single combination of traits related to overall size dominates patterns of genetic variation for morphological traits. A third possible concern is the assumption used in the calculation that the direction of selection in multivariate space is oriented at random with respect to the pattern of genetic variation. Ultimately, the question is whether genetic correlations, in fact, have appreciable effects on evolution in nature. To try to answer that, I now turn to four case studies.

Correlations and Constraint in Nature

CASE 1: GENETIC LINES OF LEAST RESISTANCE Several lines of evidence have emerged recently that suggest genetic correlations do impact evolution in natural populations over macroevolutionary timescales. Schluter (1996) looked for such effects in a pioneering study. An implication following from Equation (7.1) is that evolutionary trajectories trace paths that are a compromise between the combination of traits favored by selection and the combination that has the most genetic variation. When a population is selected for a new combination of trait values, for example following colonization of a new habitat, evolution in the short term will largely follow the combination of traits for which there is the most variation. This is what Schluter termed the "genetic line of least resistance"; less poetically, it is known as the first principal component (or PC1) of the genetic covariance matrix G. The orientation of PC1 is decided by the genetic correlations between the traits. In later parts of the path towards the new adaptive peak, the trajectory bends away from PC1 and heads more and more directly towards the optimum.

Schluter asked if this pattern could be detected. He found five taxa of vertebrates for which he could estimate PC1 for one of the species, the mean phenotypes of all of them, and the time since the species had last shared a common ancestor. His qualitative prediction was that the phenotypic differences between species that had diverged for the least amount of time should lie more or less along PC1. As the amount of time separating two species increases, the angle between PC1 and the vector of phenotypic divergence between the species should increase. Remarkably, that is what the data show. Schluter used molecular divergence as a proxy for the amount of time separating pairs of species. A rough calibration suggests that genetic correlations are influencing the direction of evolutionary trajectories for a few million years. Several other later studies have likewise found similarity between patterns of genetic and/or phenotypic intraspecific variation and divergence between the interspecific means (Futuyma et al. 1995; Baker and Wilkinson 2003; Begin and Roff 2004; Marroig and Cheverud 2005; Hunt 2007a).

CASE 2: DIVERGENCE IN EVOLVABLE DIRECTIONS The second case study was inspired by Schluter's approach. Hansen and Houle (2008) analyzed the evolution of wing shape among eight species of flies sampled from a clade whose most recent common ancestor lived about 50 million years ago. They measured the differences between the mean phenotypes of a focal species (*Drosophila melanogaster*) and the seven others. They found that these species differences fall out largely along combinations of traits that have large amounts of genetic variation in the focal species. Once again, species appear to have diverged in phenotypic space along axes that are highly "evolvable."

There are open questions about the quantitative conclusions from these two studies (Walker and Bell 2000). Both assume that during evolution of the clade the genetic covariance matrices of the species were similar to that of the extant species from which it was estimated. That assumption has been challenged by a recent analysis of phenotypic variation in the ancestral population for one of the species that Schluter studied (Berner et al. 2010). Secondly, they assume that living populations diverged along evolutionary trajectories that correspond to the differences between them. There remains, however, support for the qualitative view that genetic correlations can affect the rate and direction of phenotypic evolution over substantial evolutionary periods, perhaps 10^6 years or more.

CASE 3: A CONSTRAINT CAUSED BY CORRELATION The third case study gives a microevolutionary perspective. Fundamentally, the discussion is about the relation between what selection favors and what constraints allow. A direct analysis of the issue has not been possible, because a simultaneous picture of genetic variation and selection acting on multivariate phenotypes in the same population is lacking. Blows et al. (2004) have made a unique contribution to the field by doing just that for sexual selection acting on eight cuticular hydrocarbons in male Drosophila serrata. They estimated the genetic covariance matrix for these traits and assayed how they affect male mating success in the lab. All eight hydrocarbons show substantial genetic variance, but importantly there are strong genetic correlations between them. That is, the effective number of dimensions is small, with PC1 accounting for 52% of all genetic variation.

This situation would not represent an evolutionary constraint if selection favored a combination of traits for which there is a lot of genetic variance, but that is not the case. The selection gradient vector points in a direction that is almost orthogonal (at an angle of 75°) to the subspace that contains 99% of the genetic variation (as defined by the first 4-principal components of **G**). In simpler language, selection favors a combination of traits for which there is little or no genetic variation. That lack is because of genetic correlations between the traits, not because the individual traits are not heritable. Once again, the message is that genetic correlations can substantially change the evolutionary response from what might be expected by considering single traits in isolation.

CASE 4: AN ANTAGONISTIC GENETIC CORRELATION FOR FITNESS ITSELF The fourth example deals with genetic correlations for fitness itself. Despite a popular misconception about Fisher's Fundamental Theorem, natural populations seem to harbor respectable amounts of genetic variation for fitness (Burt 1995). Using statistical methods borrowed from animal breeding, over the last decade evolutionary biologists have begun to compile estimates for the genetic variance of fitness in natural populations (Gustafsson 1986; Kruuk et al. 2000; Merila and Sheldon 2000; McCleery et al. 2004; Coltman et al. 2005; Teplitsky et al. 2009).

Two studies have gone further by estimating the genetic correlation between fitness in males and females. In the collared flycatcher, Brommer et al. (2007) estimated the correlation to be –0.85, and in red deer, Foerster et al. (2007) estimated the correlation to be –0.95. Those values should be treated very cautiously since both have large standard errors and neither is significantly different from zero. In contrast, both values are significantly less than 1. These results are consonant with earlier work on a laboratory population of *Drosophila melanogaster* by Chippindale et al. (2001). They found that a sexually antagonistic selection in adult flies essentially zeroes out the lifetime genetic correlation between fitness in males and females.

If these findings were corroborated by further studies, the implications would be profound. A genetic correlation of zero would mean that adaptation in females has no benefit for males, and vice versa. A negative genetic correlation would mean that adaptation in one sex is actually maladaptive for the other. There are further implications, for example, limits to the potential of genetic benefits through mate choice (Qvarnstrom et al. 2006; Brommer et al. 2007). The overall picture is that genetic correlations involving fitness itself may curb rates of adaptation.

Conclusions on Correlations and Constraints in Nature

What can we conclude from these case studies about the role that genetic correlations play in adaptation? Morphological traits individually show abundant genetic variation, but genetic correlations between them have important consequences. They create a small number of axes along which populations respond rapidly to selection and a (potentially much larger) number of axes that contain little or no genetic variation. The uneven distribution of genetic variation among these axes may limit rates of adaptation, over both microevolutionary and macroevolutionary time scales.

Can constraints caused by genetic correlations explain the pattern of stasis illustrated in Figure 7.1? It is difficult to give a quantitative answer for several reasons: we do not know patterns of selection over long time scales, for example, and genetic correlations themselves can evolve. The qualitative impression from this brief review, however, is that genetic correlations can greatly decrease evolutionary rates compared to their potential. In some cases (e.g., growth trajectories), there are obvious absolute constraints in what selection can achieve. For many other multivariate phenotypes, it seems that correlations are likely to decrease selection response, perhaps by as much as one or two orders of magnitude. That is not enough, however, to explain the pattern seen in Figure 7.1. I therefore suspect that correlations contribute to the pattern but are not the full story.

Before leaving the topic of genetic variation, there is one last point worth consideration. An implicit but fundamental assumption of quantitative genetic analysis is that all of the additive genetic variation in a population is available to fuel evolutionary change. In fact, it is likely that many alleles that contribute genetic variation to a quantitative trait have unconditionally deleterious pleiotropic effects (Robertson 1967). Those effects can prevent alleles from spreading to high frequency when the trait is subject to directional selection. Thus the amount of genetic fuel available for adaptation may be even more limited than is generally supposed. How much these pleiotropic effects reduce the genetic potential for adaptation is an important unanswered question.

H_3: Demographic Speed Limits

Darwin (1859) used artificial selection as a simile for natural selection. That simile is still in use, and in fact, plays an important role in the apparent paradox of evolutionary rates: the points in the top left corner of Figure 7.1 come from selection experiments. There are key differences between artificial and natural selection, however, that could draw this simile into question.

Animals and plants under artificial selection typically have unlimited resources and are largely free of parasites, disease, and predators. Those conditions are necessary when very strong selection is applied: a large reproductive excess is needed, because only a small fraction of the population contributes to the following generation. Further, breeders typically use truncation selection (or something similar), which imposes the minimum number of selective deaths for a given intensity of selection.

Life in the wild differs in three important ways. First, it is far less benign. Even in the best of times, the reproductive excess in nature is well below what it is in the lab and barnyard. Second, environmental changes that generate strong directional selection may often be stressful, even to the individuals that do survive and reproduce. Finally, fitness functions in nature never follow a breeder's efficient truncation scheme. Many more selective deaths are therefore needed in nature than in breeding to produce a given amount of evolutionary change.

In sum, strong directional selection in nature may come with a heavy demographic price. Such a heavy price, in fact, that it may only be sustainable for a small number of generations before a population declines to extinction. If so, demographic limits could contribute to the pattern seen in Figure 7.1: populations evolving at high rates do not survive long enough to appear in the upper right part of the graph.

The response of *Geospiza fortis* to the 1977 drought in the Galápagos gives a compelling example. That event caused one of the most rapid evolutionary changes ever documented in a natural population: as mentioned, bill depth increased by about 5% in a single generation (Grant and Grant 1995). During the event, however, the population crashed to 15% of its size before the drought (Boag and Grant 1981). No species can survive long under conditions like those.

What, then, is the maximum evolutionary rate that can be demographically sustained? Theoreticians have struggled with that question since Haldane (1957) first calculated the number of selective deaths required to cause a new advantageous allele to spread from a single copy to fixation. Models for quantitative traits have been developed by recent investigators (Lynch and Lande 1993; Burger and Lynch 1995; Gomulkiewicz and Holt 1995; Gomulkiewicz and Houle 2009). The approach taken here is to assume that a fitness function of a fixed form (typically a Gaussian-shaped adaptive peak) generates both selection and mortality. The population size declines if the mortality caused by selection exceeds a fixed reproductive excess. The results can be used to determine, for example, the critical minimum genetic variance needed for a population to evolve fast enough to avoid extinction under different scenarios for environmental change.

These models share a major limitation: they assume the extra mortality results only from the deviation of phenotypes from some optimal phenotype. Strong selection, however, may often come with ecological conditions that are bad for even the best-adapted individuals. By the end of the 1977 drought, there were very few seeds of any size left for the Galápagos finches to eat (Boag and Grant 1981). Unfortunately, it is not clear if models can provide much useful about the relationship between environmental quality and the intensity of selection. Ultimately, the demographic limits to directional selection may be best determined by empirical approaches.

Conclusions

For more than a century after the publication of *The Origin of Species*, biologists followed Darwin's intuition that evolution was a gradual process too slow to be observed directly. There has been a post-Darwinian revolution with the realization that neither natural selection nor its evolutionary consequences are difficult to see in living populations or in the fossil record. Populations can and often do evolve rapidly over short time scales, sometimes fast enough to change visibly in size and even produce a new species in several generations. It is striking, then, that fossil lineages typically show relatively little change over time scales of 10^3 to 10^7 generations.

Simpson (1944) visualized a clade of closely related species as exploiting an adaptive zone. These zones can come and go, but often remain intact for geologically substantial periods of time. It is difficult to dispute the impression that there really is a niche for planktivorous fish in temperate lakes and another for seed-eating birds in grasslands. The precise locations of the fitness peaks within each zone may change rapidly as the result of shifting constellations of resources, competitors, and predators. In that view, populations chasing local peaks evolve back and forth rapidly, but very rarely stray far from their ecological neighborhood.

That qualitative picture appears consistent with what is known about selection intensities and genetic variation from extant populations and about evolutionary rates on macroevolutionary time scale, at least for morphological traits related to overall size. The picture becomes less clear when trait combinations that can lead to new adaptive zones are considered. It is not clear whether or not sticklebacks could have diversified into the profusion of lifestyles we see in African cichlids had they colonized the Great Rift Lakes. Evolutionary rates in these cases may often be controlled by genetic constraints, for example resulting from genetic correlations, and by a demographic ceiling on the amount of selection that a population can sustain. The relative importance of those factors in limiting the speed of the Darwinian machine has yet to be determined.

Acknowledgments

I thank Mike Bell, Dan Bolnick, Robert Dudley, Anne Duputie, Walt Eanes, Doug Futuyma, Rafael Guerrero, David Houle, and Trevor Price for discussions and constructive comments on the manuscript.

Literature Cited

Baker, R. H. and G. S. Wilkinson. 2003. Phylogenetic analysis of correlation structure in stalk-eyed flies (*Diasemopsis*, Diopsidae). *Evolution* 57: 87–103.

Begin, M. and D. A. Roff. 2004. From micro- to macroevolution through quantitative genetic variation: Positive evidence from field crickets. *Evolution* 58: 2287–2304.

Bell, M. A., M. P. Travis, and D. M. Blouw. 2006. Inferring natural selection in a fossil threespine stickleback. *Paleobiology* 32: 562–577.

Berner, D., W. Stutz, and D. Bolnick. 2010. Foraging trait (co)variances in stickleback evolve deterministically and do not predict trajectories of adaptive diversification. *Evolution*, in press.

Blows, M. W., S. F. Chenoweth, and E. Hine. 2004. Orientation of the genetic variance-covariance matrix and the fitness surface for multiple male sexually selected traits. *Am. Nat.* 163: 329–340.

Boag, P. T. and P. R. Grant. 1981. Intense natural selection in a population of Darwin's finches (Geospizinae) in the Galápagos. *Science* 214: 82–85.

Brommer, J. E., M. Kirkpatrick, A. Qvarnström, and 1 other. 2007. The intersexual genetic correlation for lifetime fitness in the wild and its implications for sexual selection. *PLoS ONE* 2: e744.

Burger, R. and M. Lynch. 1995. Evolution and extinction in a changing environment—a quantitative genetic analysis. *Evolution* 49: 151–163.

Burt, A. 1995. Perspective: The evolution of fitness. *Evolution* 49: 1–8.

Charlesworth, B., R. Lande, and M. Slatkin. 1982. A neo-Darwinian commentary on macroevolution. *Evolution* 36: 474–498.

Chippindale, A. K., J. R. Gibson, and W. R. Rice. 2001. Negative genetic correlation for adult fitness between sexes reveals ontogenetic conflict in *Drosophila*. *Proc. Natl. Acad. Sci. USA* 98: 1671–1675.

Coltman, D. W., P. O'Donoghue, J. T. Hogg, and 1 other. 2005. Selection and genetic (co)variance in bighorn sheep. *Evolution* 59: 1372–1382.

Darwin, C. 1859. *On the Origin of Species by Means of Natural Selection, or the Preservation of Favoured Races in the Struggle for Life.* John Murray, London.

Dudley, J. W. and R. J. Lambert. 2004. 100 generations of selection for oil and protein content in corn. *Plant Breed. Rev.* 24: 79–110.

Eldredge, N. and S. J. Gould. 1972. Punctuated equilibria: An alternative to phyletic gradualism. In T. J. M. Schopf (ed.), *Models in Paleobiology*, pp. 82–115. Freeman, Cooper & Co., San Francisco.

Endler, J. A. 1986. *Natural Selection in the Wild.* Princeton University Press, Princeton.

Estes, S. and S. J. Arnold. 2007. Resolving the paradox of stasis: Models with stabilizing selection explain evolutionary divergence on all timescales. *Am. Nat.* 169: 227–244.

Foerster, K., T. Coulson, B. C. Sheldon, and 3 others. 2007. Sexually antagonistic genetic variation for fitness in red deer. *Nature* 447: 1107–1111.

Futuyma, D. J. 2010. Evolutionary constraint and ecological consequences. *Evolution*, in press.

Futuyma, D. J., M. C. Keese, and D. J. Funk. 1995. Genetic constraints on macroevolution: The evolution of host affiliation in the leaf beetle genus *Ophraella*. *Evolution* 49: 797–809.

Gingerich, P. D. 1983. Rates of evolution—effects of time and temporal scaling. *Science* 222: 159–161.

Gingerich, P. D. 2001. Rates of evolution on the time scale of the evolutionary process. *Genetica* 112: 127–144.

Gingerich, P. D. 2009. Rates of evolution. *Ann. Rev. Ecol. Evol. Syst.* 40: 657–676.

Gomulkiewicz, R. and R. D. Holt. 1995. When does natural selection prevent extinction? *Evolution* 49: 201–207.

Gomulkiewicz, R. and D. Houle. 2009. Demographic and genetic constraints on evolution. *Am. Nat.* 174: E218–E229.

Grant, P. R. and B. R. Grant. 1995. Predicting microevolutionary responses to directional selection on heritable variation. *Evolution* 49: 241–251.

Grant, P. R. and B. R. Grant. 2002. Unpredictable evolution in a 30-year study of Darwin's finches. *Science* 296: 707–711.

Gustafsson, L. 1986. Lifetime reproductive success and heritability: Support for Fisher's Fundamental Theorem. *Am. Nat.* 128: 761–764.

Haldane, J. B. S. 1949. Suggestions as to quantitative measurement of rates of evolution. *Evolution* 3: 51–56.

Haldane, J. B. S. 1957. The cost of natural selection. *J. Genet.* 55: 511–524.

Hansen, T. F. 1997. Stabilizing selection and the comparative analysis of adaptation. *Evolution* 51: 1341–1351.

Hansen, T. F. and D. Houle. 2008. Measuring and comparing evolvability and constraint in multivariate characters. *J. Evol. Biol.* 21: 1201–1219.

Hendry, A. P. and M. T. Kinnison. 1999. Perspective: The pace of modern life: Measuring rates of contemporary microevolution. *Evolution* 53: 1637–1653.

Hill, W. G. and L. Bunger. 2004. Inferences on the genetics of quantitative traits from long-term selection in laboratory and farm animals. *Plant Breed. Rev.* 24: 169–210.

Hill, W. G. and M. Kirkpatrick. 2010. What animal breeding has taught us about evolution. *Ann. Rev. Ecol. Evol. Syst.* 41, in press.

Hoffmann, A. A., R. J. Hallas, J. A. Dean, and 1 other. 2003. Low potential for climatic stress adaptation in a rainforest *Drosophila* species. *Science* 301: 100–102.

Houle, D. 1992. Comparing evolvability and variability of quantitative traits. *Genetics* 130: 194–204.

Hunt, G. 2007a. Evolutionary divergence in directions of high phenotypic variance in the ostracode genus *Poseidonamicus*. *Evolution* 61: 1560–1576.

Hunt, G. 2007b. The relative importance of directional change, random walks, and stasis in the evolution of fossil lineages. *Proc. Natl. Acad. Sci. USA* 104: 18404–18408.

Hunt, G., M. A. Bell, and M. P. Travis. 2008. Evolution toward a new adaptive optimum: Phenotypic evolution in a fossil stickleback lineage. *Evolution* 62: 700–710.

Keightley, P. D. and D. L. Halligan. 2009. Analysis and implications of mutational variation. *Genetica* 136: 359–369.

Kellerman, V., B. van Heerwaarden, C.M. Sgrò, and 1 other. 2009. Fundamental evolutionary limits in ecological traits drive *Drosophila* species distributions. *Science* 325: 1244–1246.

Kingsolver, J. G., R. Gomulkiewicz, and P. A. Carter. 2001a. Variation, selection and evolution of function-valued traits. *Genetica* 112: 87–104.

Kingsolver, J. G., H. E. Hoekstra, J. M. Hoekstra, and 6 others. 2001b. The strength of phenotypic selection in natural populations. *Am. Nat.* 157: 245–261.

Kirkpatrick, M. 2009. Patterns of quantitative genetic variation in multiple dimensions. *Genetica* 136: 271–284.

Kirkpatrick, M. and D. Lofsvold. 1992. Measuring selection and constraint in the evolution of growth. *Evolution* 46: 954–971.
Lande, R. 1976. Natural selection and random genetic drift in phenotypic evolution. *Evolution* 30: 314–334.
Lande, R. 1979. Quantitative genetic analysis of multivariate evolution, applied to brain: Body size allometry. *Evolution* 33: 402–416.
Lande, R. and S. J. Arnold. 1983. The measurement of selection on correlated characters. *Evolution* 37: 1210–1226.
Liem, K. F. 1973. Evolutionary strategies and morphological innovations: Cichlid pharyngeal jaws. *Syst. Zool.* 22: 425–441.
Lynch, M. 1990. The rate of morphological evolution in mammals from the standpoint of the neutral expectation. *Am. Nat.* 136: 727–741.
Lynch, M. and R. Lande. 1993. Evolution and extinction in response to environmental change. In P. M. Kareiva, J. G. Kingsolver, and R. B. Huey (eds.), *Biotic Interactions and Global Climate Change*, pp. 234–250. Sinauer Associates, Sunderland, MA.
Lynch, M. and J. B. Walsh. 1998. *Genetic Analysis of Quantitative Traits.* Sinauer Associates, Sunderland, MA.
Marroig, G. and J. M. Cheverud. 2005. Size as a line of least evolutionary resistance: Diet and adaptive morphological radiation in new world monkeys. *Evolution* 59: 1128–1142.
Maynard Smith, J., R. Burian, S. Kauffman, and 6 others. 1985. Developmental constraints and evolution. *Quart. Rev. Biol.* 60: 265–287.
Mayr, E., E. G. Linsley, and R. L. Usinger. 1953. *Methods and Principles of Systematic Zoology*. McGraw-Hill, New York.
McCleery, R. H., R. A. Pettifor, P. Armbruster, and 3 others. 2004. Components of variance underlying fitness in a natural population of the great tit *Parus major*. *Am. Nat.* 164: E62–E72.
McGuigan, K. and M. W. Blows. 2007. The phenotypic and genetic covariance structure of drosphilid wings. *Evolution* 61: 902–911.
McGuigan, K., S. F. Chenoweth, and M. W. Blows. 2005. Phenotypic divergence along lines of genetic variance. *Am. Nat.* 165: 32–43.
Merilä, J. and B. C. Sheldon. 2000. Lifetime reproductive success and heritability in nature. *Am. Nat.* 155: 301–310.
Meyer, K. 2005. Genetic principal components for live ultrasound scan traits of Angus cattle. *Animal Sci.* 81: 337–345.
Mezey, J. G. and D. Houle. 2005. The dimensionality of genetic variation for wing shape in *Drosophila melanogaster*. *Evolution* 59: 1027–1038.
Mousseau, T. A. and D. A. Roff. 1987. Natural selection and the heritability of fitness components. *Heredity* 59: 181–197.
Pander, B. L., R. Thompson, and W. G. Hill. 1993. Phenotypic correlations among daily records of milk yields. *Indian J. Animal Sci.* 63: 1282–1286.
Pavlicev, M., J. M. Cheverud, and G. P. Wagner. 2009. Measuring morphological integration using eigenvalue variance. *Evol. Biol.* 36: 157–170.
Powell, R. L. and H. D. Norman. 2006. Major advances in genetic evaluation techniques. *J. Dairy Sci.* 89: 1337–1348.
Qvarnstrom, A., J. E. Brommer, and L. Gustafsson. 2006. Testing the genetics underlying the co-evolution of mate choice and ornament in the wild. *Nature* 441: 84–86.
Riska, B., W. R. Atchley, and J. J. Rutledge. 1984. A genetic analysis of targeted growth in mice. *Genetics* 107: 79–101.

Robertson, A. 1967. The nature of quantitative genetic variation. In A. Brink (ed.), *Heritage from Mendel*, pp. 2105–2117. University of Wisconsin Press, Madison.
Schluter, D. 1996. Adaptive radiation along genetic lines of least resistance. *Evolution* 50: 1766–1774.
Simpson, G. G. 1944. *Tempo and Mode in Evolution*. Columbia University Press, New York.
Teplitsky, C., J. A. Mills, J. W. Yarrall, and 1 other. 2009. Heritability of fitness components in a wild bird population. *Evolution* 63: 716–726.
Walker, J. A. and M. A. Bell. 2000. Net evolutionary trajectories of body shape evolution within a microgeographic radioation of threespine sticklebacks (*Gasterosteus aculeatus*). *J. Zool. Lond.* 252: 293–302.

Chapter 8

Evolvability: The Missing Piece in the Neo-Darwinian Synthesis

Günter P. Wagner

The standard model of evolutionary change consists of repeated cycles of random mutation and selection (Figure 8.1). Heritable changes arise by accident, mostly as replication errors, and if these variants convey some reproductive advantage, chances are that such a mutation will be transmitted at a higher rate to the next generation than the original wild type. Thus, the mutant will increase systematically in frequency, which is natural selection. Of course, any actual realization of this process is much more complex and includes the vicissitudes of genetic transmission, demographics, and random sampling (i.e., genetic drift). However, the model of adaptive evolution essentially is one of random mutation and selection. The implicit assumption is that all of evolution, in one way or another, can be traced back to a long sequence of these mutation-selection cycles. That is to say, I and many other evolutionary biologists think that even the most complex and the most perfect organismal structures and functions, like the vertebrate eye, are the result of a sequence of these mutation-selection cycles. The problem—both in terms of an ability to communicate evolution to the general public, as well as in terms of actually delivering on the promise of evolutionary biology to explain all of biology—is that this model is not intuitive. Even though he did not think it was threatening his theory in the end, Darwin felt compelled to write about the plausibility of natural selection as an explanation for the origin of complex organs in his famous quote on the origin of eyes:

> *To suppose that the eye, with all its inimitable contrivances for adjusting the focus to different distances, for admitting different amounts of light, and for the correction of spherical and chromatic aberration, could have been formed by (random mutation and [added by GPW]) natural selection, seems, I freely confess, absurd in the highest possible degree (Darwin 1859: 186).*

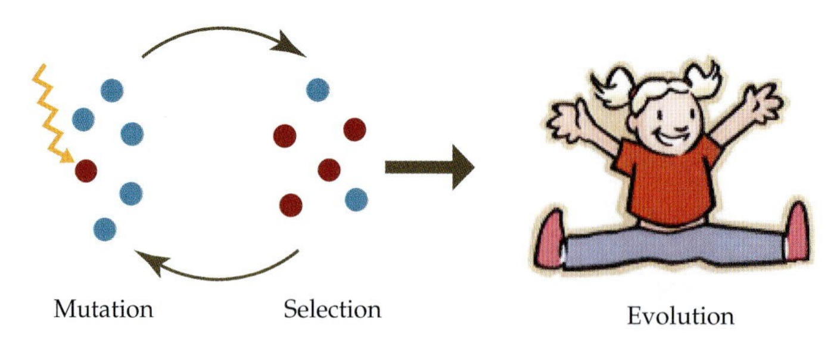

FIGURE 8.1 Mutation-Selection Cycles The standard (cartoon) model of evolution consists of repeated cycles of random mutation and selection, which, after many iterations, lead to the products of evolution, including complex organisms like humans. The problem with this model is that it is not intuitive and ignores issues of evolvability; it is unclear whether and under what boundary conditions this model can explain the origin of complex organisms.

The lack of intuitive appeal of the basic model of adaptive evolution is perhaps best expressed in the metaphor of the tornado in the junkyard that creationist critics of evolutionary biology have contrived to undercut the Darwinan narrative. The story goes like this: *There is raw material, like that available in a junkyard; and, there is a process that causes random change, like a tornado tearing through said junkyard, and out of it comes a 747, soaring into the sky*. The scenario, of course, is impossible, but their point is that equally as implausible as a 747 materializing from a tornado in a junkyard is the prospect that random mutations might be responsible for the origin of complex adaptations, like eyes and brains. Of course, this metaphor is wrong on many levels—not the least of which is that evolution is a stepwise process, as exemplified by the evolution of the eye and the various primitive eye types found in some invertebrate animals. These eye types range from patches of light-sensitive epidermal cells, to light receptors in a groove that allow for a certain amount of direction sensing, to providing vision that uses

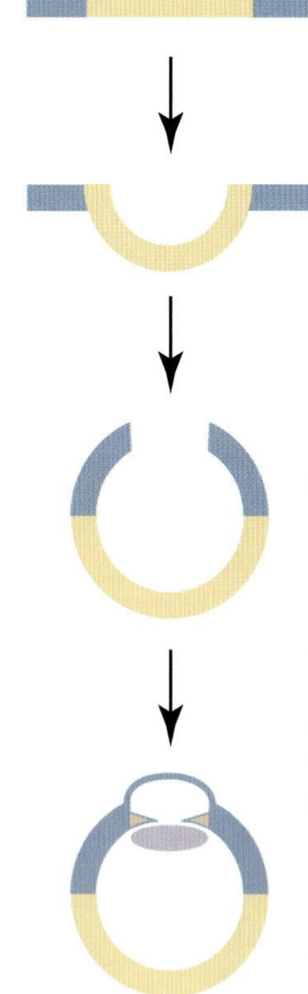

FIGURE 8.2 The Evolution of the Eye The evolution of the eye, which is a paradigm of a complex organ of extreme perfection, most likely was the result of a long process of stepwise changes. The most primitive form of a light-sensory apparatus most likely was a part of the skin that became light sensitive (indicated in yellow). The next likely step was that the light-sensitive patch of cells formed a grove, which allowed a primitive form of directional sensitivity. The first primitive form of an eye that could form an image probably occurred by further invagination of the patch of light-sensitive cells, eventually forming a hollow ball with a small aperture, which could act as a "*camera obscura*." The highest form of a camera eye is found, independently derived, in vertebrates and cephalopods (e.g., squids), which have an adjustable lens, an iris to regulate the amount of light admitted, and a cornea that closes the cavity of the eye to the outside world. All these grades of eye elaboration exist in various invertebrate groups, in particular in mollusks, and are thus also likely the ancestral stages for the origin of the human eye.

a *camera obscura*-type image information, and finally to bestowing a camera-like vision, with lens and iris (Figure 8.2). Hence, adaptive evolution is not one large random event, like a tornado, but instead, a long series of smaller random effects. In addition, this junkyard metaphor does not include the sorting process that natural selection constitutes. However, beyond its obvious shortcomings, there is a deeper problem that ultimately needs to be addressed. It is one that Ronald Fisher addressed in his seminal book, *The Genetical Theory of Natural Selection* (1930): How likely is it that improvements of a complex system could occur by stepwise random mutations? This problem corresponds to one well known in mathematical optimization theory, namely that improvements can become exceedingly difficult (i.e., rare) if the dimensionality of the parameter space to be searched becomes large (see Kirkpatrick, Chapter 7). This phenomenon was named "curse of dimensionality" by Bellman (1957), and in evolutionary biology, Alan Orr called it "the cost of complexity" (Orr 2000). Fisher proceeded to address this question with his so-called "geometric model."

In the geometric model, Fisher assumes that the phenotype of a complex organism can be seen as a point in a high dimensional space, in which the dimensions of that space correspond to the traits of the organism. Complexity, in this context, corresponds to the dimensionality of the phenotype space. The more independent directions of variation that there are for the phenotype, the more complex is the organism and the more difficult it is to realize improvement in fitness that is due to random changes. The intuitive reason for this relationship is that if there are many different ways to change a phenotype, it becomes very unlikely that a random mutation can affect the right combination of traits, in just the right way to improve fitness. As the possible factors that can change a system increase, the chances that any one of the changes will be detrimental, rather than beneficial, also increases. Briefly, Fisher's solution to this problem was to note that the probability of success increases inversely with the magnitude of the mutation effect. The smaller the effect there is, the higher the chance that the mutation will be beneficial. Mutations with infinitesimally small effects have up to a 50% chance of improving fitness (Figure 8.3). This is a simple consequence of multidimensional Euclidian geometry. Hence, mutation effects of small magnitude avoid the curse of dimensionality, at least within the confines of Fisher's geometric model.

There is, however, another problem generated by Fisher's solution. Small mutations lead to higher mutational success rate, but also have a correspondingly smaller effect on the phenotype. This effect was first noted by Ingo Rechenberg (1973), an aeronautic engineer at the Technical University of Berlin, who developed random-based algorithms to solve complex optimization problems in engineering. This branch of engineering is called Evolution Strategy and is associated with a sophisticated mathematical theory about the rate of improvement expected by random change. On the one hand, Rechenberg's work shows that random change is superior to more directed optimization algorithms, and thus he endorses the power of

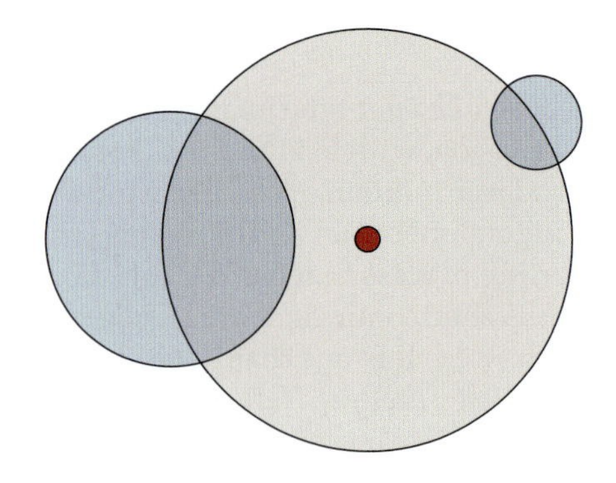

FIGURE 8.3 Fisher's Geometric Model of Adaptation The plane is thought of as representing different phenotypes, such that neighboring points in the plane can easily mutate into each other. The phenotypic optimum is indicated by the red dot, and phenotypes with the same fitness are located on concentric circles around the optimum. The gray circle indicates the set of phenotypes that have higher fitness than the phenotypes on the periphery of the circle. The blue circles represent mutations of phenotypes on the periphery of the circle. The fraction of the blue circle intersecting the gray circle is proportional to the chance that a mutation has higher fitness than the parental phenotype. It is obvious that the fraction of higher fitness phenotypes is smaller for the larger blue circle than that of the smaller blue circle. This difference becomes particularly pronounced if the number of dimensions is not two, but many more. From that geometric observation, Fisher concluded that the probability for success of a mutation increases inversely with the size of the mutation (the diameter of the blue circles). The downside, however, is that mutations with small effect also cause small phenotypic and fitness differences, even though the probability of success is higher.

random mutations and natural selection to solve complex adaptive problems. On the other hand, he notes that there is an intermediate optimal mutational effect size that maximizes progress towards the optimization goal. The problem is that the mutational effect size needs to be closely monitored to allow any progress at all, particularly if the complexity of the problem is high (i.e., if there are many parameters to optimize simultaneously).

From a population genetic perspective, the problem noted by Rechenberg (1973) has another negative consequence. If mutations have a small effect, their effect on fitness must also be small and their chance of becoming fixed in the population will be correspondingly small as well. This problem was first noted by James Crow and Motoo Kimura (1970) and most recently expanded by Allen Orr (2000). Hence, small mutations solve the problem of complex adaptations in terms of probability of success, but do not explain why complex adaptation can happen at a reasonable time scale, because infinitesimally small mutations only lead to infinitesimally small rates of improvement. Hence, the evolution of complex adaptations by random mutations is still an open question, even though there has been more work done in this area than I can review in this chapter. In the remainder of this paper, I want to discuss a number of issues that pertain to the question of how genetic mechanisms can make complex adaptations more likely than may be intuited or as suggested by the geometric models of evolution first introduced by Fisher.

Divide and Conquer: Fisher's Geometric Model Revisited

One of the key assumptions in Fisher's geometric model and one of the reasons for the "cost of complexity" is that each mutation is assumed to potentially affect every trait of the phenotype. Thus, with increasing complexity/dimensionality, the rate of evolution decreases, because the chance of any mutation hitting the right combination of traits becomes small. In the classical population genetic literature, this notion has been called "universal pleiotropy" (Wright 1964) and was supported by anecdotal evidence from large effect mutations. David Stern's (2000) influential paper offers a criticism of universal pleiotropy: he introduces the important distinction between the pleiotropic role of a gene and the pleiotropic effect of a mutation. Many genes, in particular transcription factor genes and signaling molecules, have a large number of pleiotropic roles, meaning that they play a functionally important role in many characters and functions. The number of pleiotropic roles, however, does not determine the number of pleiotropic effects of a mutation. That is the case because the molecular structure of a gene is complex and different parts of the gene are functionally important in different functional contexts. This idea is broadly accepted in the case of many cis-regulatory elements (i.e., noncoding DNA sequences that regulate transcription of a protein-coding gene; Prud'homme et al. 2007). Even transcription factor proteins are modular in this way, which allows changes in the function in one context without affecting the function in other contexts (Lynch and Wagner 2008). Hence, it is important to distinguish between the functional architecture of say, the development process (i.e., what genes are involved in how many developmental processes) and the variational architecture, which is the number of effects a mutation has on the phenotype.

An alternative to the idea of universal pleiotropy is to assume that pleiotropic effects are largely limited to a small subset of traits that correspond to functional subsystems (i.e., modularity *sensu*) (Wagner and Altenberg 1996; Welch and Waxman 2003; Wagner et al. 2007). But until very recently, there was no systematic data about the pleiotropy of the genes underlying quantitative variation, the kind of variation that is potentially adaptive and thus subject to directional natural selection. In recent years, a few studies have been published that reveal a surprisingly consistent answer: the number of pleiotropic effects is small compared to the number of phenotypic traits scored (Brem et al. 2002; West et al. 2007; Albert et al. 2008; Wagner et al. 2008; Zou et al. 2008; Su et al. 2009).

Jane Kenny-Hunt in Jim Cheverud's lab undertook the herculean task of performing quantitative trait loci (QTL) mapping for 70 mostly skeletal characters from all major subsystems of the mouse body: skull, body axis, limbs, and overall body weight (Kenney-Hunt et al. 2008). In an analysis of pleiotropy of these QTL, the loci affect between 1 and 30 traits, with a median of 6 of the 70 traits scored (Wagner et al. 2008). In the same year, a study of stickleback variation mapped QTL for 27 landmarks and found a mean number of 3.5 pleiotropic effects per QTL (Albert et al. 2008). Using the mathematical properties of Fisher's geometric model, Xun Gu, from Iowa State University, developed a method to estimate the number of pleiotropic effects of amino acid substitutions based on the variation

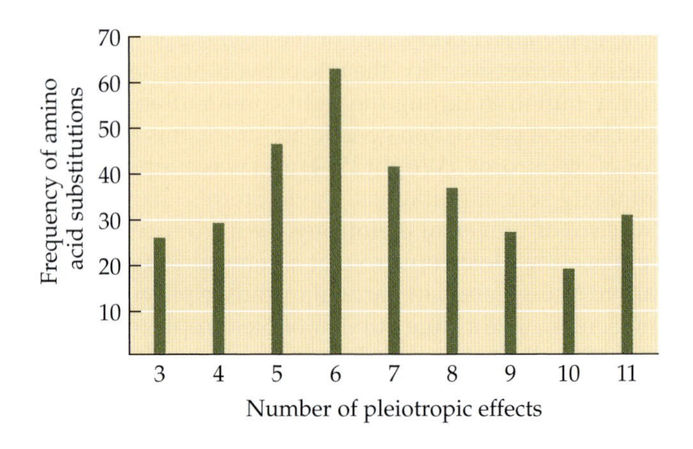

FIGURE 8.4 The Frequency Distribution of the Number of Pleiotropic Effects Caused by Amino Acid Substitution Estimated The results of the study by Su et al. (2009) is typical of many recent studies in which the mean number of pleiotropic effects per fixed mutation is between three and seven for phenotypic characters. This and many other study data suggest that the concept of universal pleiotropy (i.e., each and every mutation affects effectively all phenotypic traits) is probably wrong. The *x*-axis is the number of pleiotropic effects of an amino acid substitution; the *y*-axis is the frequency of amino acid substitutions with the corresponding number of pleiotropic effects.

of substitution rates across lineages. Gu estimated 6 to 7 pleiotropic effects per fixed mutation (Su et al. 2009; Figure 8.4). Perhaps the most highly dimensional phenotypic data available today is mRNA expression data. Brem et al. (2002) have mapped QTL for 570 transcript levels (i.e., expression QTL [eQTL] mapping) and found that the trans-acting eQTL affect 7 to 94 transcripts with a mean of 29. In an eQTL study of the *Arabidopsis* transcriptome, West et al. (2007) found that the vast majority of eQTL (93% of 36,800 eQTL) had small phenotypic effects and only a few are mapped to eQTL hotspots, which affect many genes. Zou et al. (2008) described the phenotypic effects of an RNA interference (RNAi) screen in the early development of *C. elegans* and found a mean number of 7 pleiotropic effects out of 45 traits scored. The message from these data is clear: the average number of pleiotropic effects is a small fraction of any set of traits scored, suggesting that the assumption of universal pleiotropy is wrong. The vast majority of mutations affect a small set of traits.

Low pleiotropy may solve the problem of the "costs of complexity," but only if the pattern of pleiotropy corresponds to co-selected complexes of traits (i.e., traits that serve the same adaptive function; Wagner and Altenberg 1996; Hansen and Wagner 2001; Welch and Waxman 2003). The idea that variational association among traits and functional integration may correspond is an old idea, called morphological integration by Olson and Miller (1958). It is also the basis for Rupert Riedl's (1977, 1978) so-called systems theory of evolution. Whether the pattern of pleiotropy in fact corresponds to functional interdependencies is still unclear (Hallgrimsson et al. 2009). Some preliminary results are consistent with this idea, however. For instance, an analysis of pleiotropic effects of QTL for the mouse mandible shows that they tend to respect the boundary between the two major divisions of the mandible (Mezey et al. 2000), which are the tooth-bearing body and the *ramus ascendus*, at which muscles attach and the mandible articulates with the skull. There are significantly fewer pleiotropic effects that affect both parts of the mandible than expected by chance, given the number of QTL and the number of pleiotropic effects per locus. A reanalysis of these data, however, argues that the localized nature of the pleiotropic effects does not signify discrete modules but rather a continuum of localized effects (Roseman et al. 2009). More studies on a variety of systems are necessary to determine the structural architecture of pleiotropic effects.

In a small-scale QTL study of cichlid mandible morphology, Albertson et al. (2005) showed that there are more QTL shared between the body of the mandible and the closing lever, which are functionally *related*, than between the opening and closing level of the cichlid mandible, which are functionally *unrelated*. While these studies suggest some evidence that pleiotropic effects are patterned to reflect functional interdependencies, as predicted by Riedl more than 30 years ago (1977), more extensive data are required to assess the generality of this pattern.

The questions, of course, remain as to why these patterns of variational and functional integration should correspond to each other and whether they are in fact the result of evolutionary modification. Can natural selection create these patterns? These questions are highly controversial and to date, have not been resolved based on existing data (Love 2003; Sniegowski and Murphy 2006; Lynch 2007; Draghi and Wagner 2008; Parter et al. 2008; Pigliucci 2008).

No comparative QTL data are available to address the question how pleiotropic patterns evolve. Such a study would be very difficult, since it would require linkage maps from a number of species and an independent QTL mapping study for each species. However, the comparison between QTL data and phenotypic correlation patterns shows that correlations and pleiotropy closely match and that phenotypic correlations can be used to indicate the pattern of pleiotropy in comparative studies (Kenney-Hunt et al. 2008).

Accepting phenotypic correlations as reasonable estimates of genetic correlations, comparative morphometic studies can then be used to assess how variational integration changes in evolution. In a pioneering study, Nathan Young and Benedikt Halgrímsson (2005) have compared the correlation patterns within and between the forelimb and hind limb of several mammal species. As expected from the serial homology of the forelimb and hind limb, their corresponding parts are more highly correlated than the non-corresponding parts (e.g., the radius is more highly correlated with the tibia than to the femur). This pattern is found in all quadrupedal mammals that they investigated, but not in bats. The interpretation is that the functional specialization of the forelimb leads to a decrease in the variational integration between the forelimb and hind limb, which would be the expected outcome, if divergent directional selection among characters causes a decrease of pleiotropic effects. This mode of evolution would, in the long run, bring about a correspondence between variational patterns and functional interdependencies. Functionally independent traits, like the forelimbs and hind limbs of bats, will experience independent adaptive episodes and thus, will decrease in correlation, while functionally integrated parts will experience co-selection, supposedly evolving increased correlation and stronger pleiotropic effects. Recently, it was shown that the great apes, including humans, differ from other primates by their higher degree of independence in limb length variation, most likely facilitating the evolution of human-specific limb proportions (Young et al. 2010).

Two questions follow from this scenario:

1. Is there genetic variation in the pleiotropic relationships among traits?
2. Can natural selection act on this variation to shape it in a way that would facilitate evolvability?

To address the first question, James Cheverud developed a QTL mapping method to detect genetic variation in the correlation among quantitative characters (Cheverud et al. 2004). The idea is to test for an association between a hypothetical QTL location and the regression slope between two quantitative characters. These QTL are called *r*QTL, for "relationship QTL," because they influence the variational relationship between two phenotypic traits. Recently, this method was applied to postcranial skeletal characters in the mouse, and it was found that the change in phenotypic regression slope was in part due to differential epistatic effects with other QTL, demonstrating that the change in regression slope was not only phenotypic, but was caused by changes in the pleiotropic effects of interacting loci (Pavlicev et al. 2007; Pavlicev et al. 2008). These two studies showed that, at least in mouse laboratory populations, there is ample genetic variation in the pattern of pleiotropic effects, providing the opportunity for changes in pleiotropy, as suggested by the study of Young and Halgrímsson (2005). The question then is whether and how natural selection acts on this kind of genetic variation. It is easy to show that *r*QTL can be selected if the trait distribution affects fitness, as is the case in truncation selection (Wagner et al. 2008). In a more general model, it can be shown that *r*QTL can be selected directly because of their effect on the speed of adaptation, such that co-selection on two traits favors *r*QTL that increase the genetic correlation between them, and if selection acts on only one trait, then natural selection favors *r*QTL that decrease genetic correlation between them (Pavlicev, Cheverud, and Wagner, unpublished data, 2010). Much more work needs to be done, but so far, it seems that directional selection can modify the variational integration among quantitative traits (i.e., the pattern of genetic correlations), such that functional and variational integration match.

Does Robustness Affect Evolvability?

One factor that may influence evolvability is genetic robustness. Robustness is defined in relative terms, meaning that a genotype is more robust if the average effect of a mutation is smaller than that on another genotype

(Gibson and Wagner 2000) or the fraction of neutral mutations (i.e., the "neutrality") is larger in one genotype compared to another (Nimwegen et al. 1999; Ancel and Fontana 2000). It is now widely accepted that robustness is genetically variable and can evolve, even though it is not clear what selective mechanism determines the robustness of a genotype (Wagner 2005).

There are some theoretical reasons to assume that robustness and evolvability are related. One line of thought is that higher robustness decreases the phenotypic mutation rate and thus has to slow down the response to selection (Ancel and Fontana 2000). The other line of thought is that greater robustness may increase evolvability, since it decreases the incident rate and the size of deleterious pleiotropic effects. Higher robustness thus increases the chance that a mutation that is adaptive with respect to one function will, in fact, have a net positive selective coefficient (Bloom et al. 2006; Wagner 2007). Recent theoretical work shows that both lines of thought are correct, but the relationship between robustness and evolvability depends on the situation. At low robustness, evolvability decreases with increasing robustness, because of the net decrease in phenotypic mutation rate. At high robustness, evolvability increases with further robustness, because of increasing amounts of cryptic variation that can lead to novel phenotypes (Draghi et al. 2010).

There is one empirical study that shows that evolvability can increase with mutational robustness. Robert McBride (from the Paul Turner lab at Yale University) showed that virus strains that were bred to have different levels of mutational robustness also differed with respect to their evolvability to adapt to higher temperature. Within 50 generations, the more robust strains evolved a higher level of fitness under heat shock treatment than the less robust strains (McBride et al. 2008). This study not only shows that increasing mutational robustness can increase evolvability, but also, and perhaps more importantly, it demonstrates that evolvability is a measurable trait, at least under laboratory conditions.

Saving Mutations for Bad Times

Starting in the 1990s, the laboratory of Susan Lindquist published a series of influential papers that introduced the notion of evolutionary capacitance (Nathan et al. 1997; Rutherford and Lindquist 1998; True and Lindquist 2000; Queitsch et al. 2002). The concept is that some molecular mechanisms can regulate phenotypic robustness; thus, these mechanisms either hide and accumulate genetic variation or, under certain conditions, make the phenotype sensitive to this accumulated genetic variation, essentially releasing it when needed. The term capacitance borrows from the phenomenon of electric capacitance, in which a capacitor stores large amounts of charge. This idea was very influential, because it was backed up by two specific molecular mechanisms that could be seen as adaptive capacitors: the heat shock protein 90 (HSP90) and the yeast prion system.

HSP90 is a heat shock protein that stabilizes the structure of other proteins also under physiological conditions in which it stabilizes intrinsically unstable proteins. Both a pharmacological knockdown of the HSP90 protein as well as a knockout of the *Hsp83* gene, leads to the increase of heritable phenotypic variation in such diverse organisms as *Drosophila*, *Arabidopsis*, and yeast (Nathan et al. 1997; Rutherford and Lindquist 1998; Queitsch et al. 2002). There is evidence that a large amount of this variation is not unconditionally deleterious and thus is potentially useful for natural selection (Rutherford 2000). The capacitance idea comes from the fact that HSP90 also effectively gets knocked down under stress conditions, when the amount of misfolded proteins overwhelms the capacity of HSP90 to refold them. What is not clear, however, is what role evolvability plays in the evolution and maintenance of the HSP90 system. The situation is different in the case of the yeast prion system, in which population genetic modeling supports the notion that selection for the evolvability-enhancing effects of this system can play a role in its maintenance. For that reason, I will focus on the yeast prion case.

Prions are proteins that exist in two states, the native and a misfolded prion state. In their native state, prion proteins usually have a physiological function like any other protein. What is special about the misfolded prion state is that its state is infectious, meaning that the presence of prion proteins in a cell induces natively folded protein to also transform to the prion state. In other words, prions act like crystallization nuclei that lead to the accumulation of insoluble protein in the cell, which has serious health consequences, as for instance in the Creutzfeldt-Jakob syndrome. What was remarkable about True and Lindquist's (2000) paper, previously cited, was that the fitness consequences of the prion state for yeast were not uniformly negative, but positive at a high frequency.

True and Lindquist (2000) induced prion formation with salt treatment and compared the growth of these treated yeast strains with native yeast on a variety of environments, including the presence of metals or antibiotics. They found that the outcome of the test strongly depended on the combination of genotypes, presence/absence of the prion, and the environmental stressor (Figure 8.5A). In other words, some genotypes do better in the prion form than as a wild-type phenotype than other genotypes, while other genotypes do better as a wild-type phenotype than in the prion form. Overall, True and Lindquist found that in about 25% of the genotype-by-environment combinations, the prion phenotype had a higher growth rate than the wild-type phenotype. What is the molecular biology of this phenomenon?

The prion True and Lindquist (2000) investigated is a protein from the gene *Sup35*. This protein has two main functional domains (Figure 8.5B). A C-terminal domain is required for the native function of the protein, which is to ensure translational termination at the stop codon. The N-terminal domain is the part of the protein that can misfold into the prion state. The prion activity and the translational termination activity can be separated; the protein can act as translational terminator without the part that can fold into a prion. So, why is the prion activity still there? Following is the model proposed by True and Lindquist (2000).

Because of ongoing gene duplications and a recent genome duplication, the yeast genome is full of pseudogenes. Many pseudogenes lose their function through a premature stop codon rather than because of amino acid substitutions in the encoded protein (Figure 8.5C). Most amino acid substitutions are deleterious, and thus it is less damaging to eliminate a duplicated gene by simply not expressing most of the coding region through premature stop codons. Because in the wild type the sequence after the premature stop codon is not expressed, it accumulates nucleotide substitutions. When Sup35 precipitates as a prion, the likelihood of readthrough translation (i.e., the stop codon can be read as a sense codon) increases and some of the variation that is "hidden" behind the stop codon will be expressed (Figure 8.5D). True and Lindquist (2000) think that these novel proteins endow some genotypes with a selective advantage in some novel environments when the Sup35 is a prion. In other words, the prion phenotype leads to the expression of genetic variation that is not expressed when the Sup35 protein has its wild-type phenotype. Implicit in this model is the assumption that the ability of Sup35 to form prions is a property that is maintained because of its effect on the accumulation and release of cryptic genetic variation—and thus also because of its effects on evolvability.

Masel and Bergman (2003) explored the plausibility of this proposal using a mathematical model of the selective advantage conferred upon a yeast strain by its ability to form prions and parameterized the model using estimates of the critical factors from True and Lindquist's (2000) data. They concluded that it is sufficient that a yeast strain experiences at least one environmental challenge per 10^6 generations. The advantage of being able to express the prion phenotype under stress situations is enough to confer substantial long-term advantage so that the maintenance of the prion-forming segment of the Sup35 protein can be explained by the advantages of evolvability.

The yeast prion system is unique, because there is strong empirical evidence that the prion phenotype can lead to advantages in novel environments, a plausible molecular mechanism that explains its effects, and a population genetic model that supports the notion that this property conveys sufficient long-term advantages to explain its maintenance. These results make it highly likely that selective advantages of evolvability can shape the genome of organisms.

Overview and Outlook

Evolvability—the origin and functional improvement of complex organisms through random genetic change and natural selection—was a matter

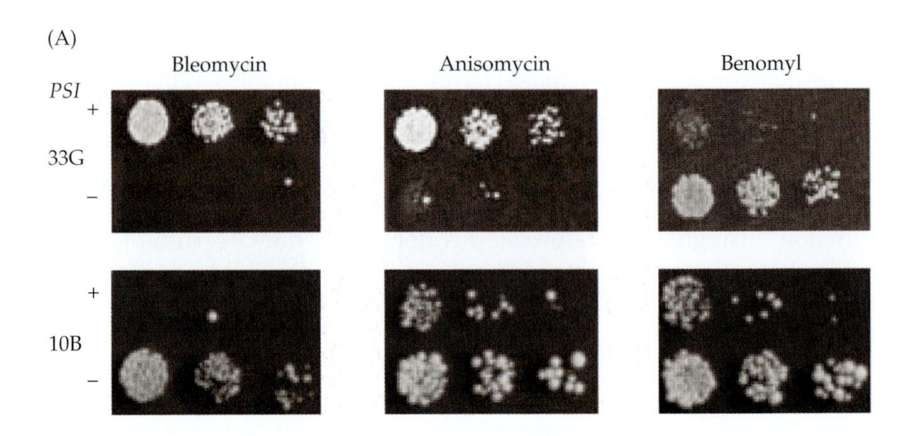
(A)
Bleomycin
Anisomycin
Benomyl
PSI
+
33G
−
+
10B
−

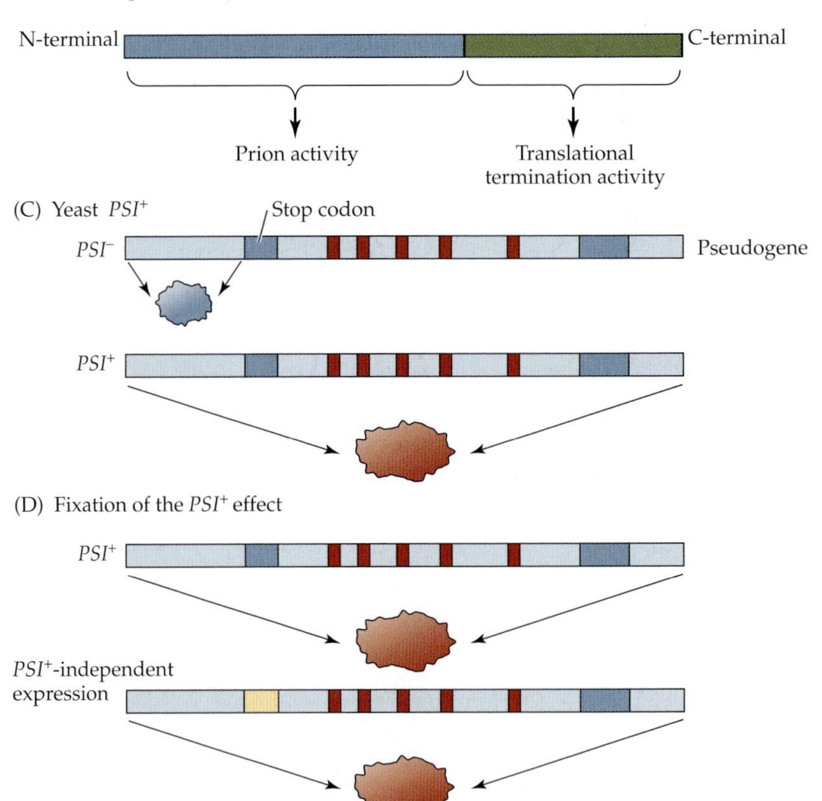
(B) PSI+ is a prion of Sup35
N-terminal
C-terminal
Prion activity
Translational termination activity
(C) Yeast PSI+
Stop codon
PSI−
Pseudogene
PSI+
(D) Fixation of the PSI+ effect
PSI+
PSI+-independent expression

◀ **FIGURE 8.5 The Yeast Prion *PSI*⁺ System and Release of Adaptive Genetic Variation According to True and Lindquist (2000)** (A) Some genotypes can have an advantage in dealing with novel environmental challenges, such as antibiotics, in this case Bleomycin, Anisomycin, or Benomyl. Each panel shows the growth of yeast cells in increasing concentrations of the antibiotic, from left to right. The first row of three colonies illustrates the growth of the genotype in the presence of the prion, and the second row depicts the growth in the wild-type phenotype. The genotype 33G grows better with the prion *PSI*⁺ in the presence of Bleomycin but less well in the presence of Benomyl. In contrast, the genotype 10B has the opposite growth characteristics. This figure shows that the *PSI*⁺ prion reveals different genetic variation in different genotypes and that there is a large chance that this variation is adaptive. (B) The protein that forms the *PSI*⁺ prion is the protein of the *Sup35* gene, which has a wild-type function in translational termination and ensures that an mRNA is translated into a protein only between the start and the stop codons. The Sup35 protein contains two functional domains: a segment that forms the prion at the N-terminal and the part responsible for translational termination activity at the C-terminal. These two activities can be separated. The wild-type function of translational termination can be performed without the (blue) N-terminal part that leads to prion formation. (C) The yeast genome contains numerous pseudogenes that are only partially translated into small peptides (see small blue cloud) because of a premature stop codon (see dark blue rectangle on left). Behind the premature stop codon missense mutations can accumulate without fitness consequences (see red lines), because they are not translated. If *Sup35* forms prions, however, the stop codons are ignored (i.e., a readthrough translation), and the full-length protein, including the formerly hidden mutations, is expressed as protein (see red cloud). It is thought that these novel proteins are the reason why some yeast genotypes do better in the presence of new antibiotics than the wild-type phenotype (see A). (D) If the novel protein conveys a functional advantage, a mutation that removes the premature stop codon (now the yellow rectangle) is favored. A new adaptive trait becomes fixed in this strain and is no longer dependent on the *PSI*⁺ phenotype. (A adapted from True and Lindquist 2000.)

of concern for the founding fathers of evolutionary biology, particularly Darwin and later Fisher. During most of the twentieth century, however, evolvability was largely ignored and only occasionally cropped up in a variety of contexts, including evolutionary engineering (Rechenberg 1973), theories of macroevolution (Riedl 1978), and possibly to explain the maintenance of sex (Maynard-Smith 1978), among others. Towards the end of the twentieth century, however, evolvability started to attract growing attention, and its possible evolution by natural selection started to look increasingly plausible, mostly because of work done in computational modeling studies and reviews of molecular mechanisms (Dawkins 1989; Altenberg 1994; Wagner and Altenberg 1996; Brookfield 2001; Love 2003; Wagner 2005; Gerhart and Kirschner 2007; Crombach and Hogeweg 2008; Draghi and Wagner 2008; Draghi and Wagner 2009; Pigliucci 2008). Most recently, theoretical work on evolvability has started to interact with data showing that it is measurable and that there are plausible mechanisms that affect its expression (see previously cited references). The emerging consensus among researchers investigating mathematical and computational models for complex traits is that directional natural selection leads to a generic increase in evolvability (Masel and Bergman 2003; Kashtan and Alon 2005; Crombach and Hogeweg 2008; Draghi and Wagner 2008; Draghi and Wagner 2009; Parter et al. 2008). We are not yet, however, in a position to describe the necessary and sufficient conditions for the evolution of evolvability, with the same clarity that we can say concisely how natural selection works. However, solving this problem might not be difficult, if more in-depth analyses of computational and experimental examples of the evolution of evolvability are published and commonalities become visible.

All of which is not to say that there are no critics of the notion of evolution and evolvability, even among highly accomplished evolutionary biologists, most prominently Michael Lynch (Sniegowski and Murphy 2006; Lynch 2007). But, critics are good because only with relentless rational criticism will any scientific idea mature and serve the scientific community or society at large. Issues of evolvability are now being discussed in practical contexts and thus, not only deepen the understanding of evolution, but may also enhance the practical usefulness of evolutionary biology (see Gould, Chapter 21). Examples are the evolvability of emerging disease agents, the role of evolvability in determining the invasiveness of species, and the so-called designability of proteins for practical purposes. For instance, understanding the varying evolvabilities of different parts of a virus's coat will allow the designing of vaccines that are not easily evaded by the virus. Hence, at the beginning of the twenty-first century, one can see an increased appreciation of the genetic basis for the evolution of evolvability. The beginnings of an empirical research program on the evolution of evolvability, as well as realize that many problems, both basic and applied, require a deep understanding of evolvability and the factors that determine it.

Literature Cited

Albert, A. Y. K., S. Sawaya, T. H. Vines, and 6 others. 2008. The genetics of adaptive shape shift in stickleback: Pleiotropy and effect size. *Evolution* 62: 76–85.

Albertson, R. C., J. T. Streelman, T. D. Kocher, and 1 other. 2005. Integration and evolution of cichlid mandible: The molecular basis of alternative feeding strategies. *Proc. Natl. Acad. Sci. USA* 102: 16287–16292.

Altenberg, L. 1994. The evolution of evolvability in genetic programming. In J. K. E. Kinnear (ed.), *Advances in Genetic Programming*, pp. 47–74. MIT Press, Cambridge, MA.

Ancel, L. W. and W. Fontana. 2000. Plasticity, evolvability and modularity in RNA. *J. Exp. Zool. (Mol. Dev. Evol.)* 288B: 242–283.

Bellman, R. E. 1957. *Dynamic Programming*. Princeton University Press, Princeton.

Bloom, J. D., S. T. Labthavikul, C. R. Otey, and 1 other. 2006. Protein stability promotes evolvability. *Proc. Natl. Acad. Sci. USA* 103: 5869–5874.

Brem, R. B., G. Yvert, R. Clinton, and 1 other. 2002. Genetic dissection of transcriptional regulation in budding yeast. *Science* 296: 752–755.
Brookfield, J. F. Y. 2001. The evolvability enigma. *Curr. Biol.* 11: R106–R108.
Cheverud, J. M., T. H. Ehrich, T. Y. T. Vaughn, and 3 others. 2004. Pleiotropic effects on mandibular morphology II: Differential epistasis and genetic variation in morphological integration. *J. Exp. Zool. (Mol. Dev. Evol.)* 302B: 424–435.
Crombach, A. and P. Hogeweg. 2008. Evolution of evolvability in gene regulatory networks. *PLoS Comput. Biol.* 4: e1000112.
Crow, J. F. and M. Kimura. 1970. *An Introduction to Population Genetics Theory.* Harper & Row, New York.
Darwin, C. R. 1859. *On the Origin of Species by Means of Natural Selection, or the Preservation of Favoured Races in the Struggle for Life.* John Murray, London.
Dawkins, R. 1989. The evolution of evolvability. In C. Langton (ed.), *Artificial Life: The Proceedings of an Interdisciplinary Workshop on the Synthesis and Simulation of Living Systems,* pp. 202–220. Addison Wesley, Santa Fe.
Draghi, J. and G. P. Wagner. 2008. Evolution of evolvability in a developmental model. *Evolution* 62: 301–315.
Draghi, J. and G. P. Wagner. 2009. The evolutionary dynamics of evolvability in a gene network model. *J. Evol. Biol.* 22: 599–611.
Draghi, J., T. Parsons, G. P. Wagner, and 1 other. 2010. Mutational robustness can increase or decrease evolvability. *Nature* 463: 353–355.
Fisher, R. A. 1930. *The Genetical Theory of Natural Selection.* Clarendon Press, Oxford.
Gerhart, J. and M. Kirschner. 2007. The theory of facilitated variation. *Proc. Natl. Acad. Sci. USA* 104: 8582–8589.
Gibson, G. and G. P. Wagner. 2000. Canalization in evolutionary genetics: A stabilizing theory? *BioEssays* 22: 372–380.
Hallgrímsson, B., H. Jamniczky, N. M. Young, and 4 others. 2009. Deciphering the palimpsest: Studying the relationship between morphological integration and phenotypic covariation. *Evol. Biol.* 36: 355–376.
Hansen, T. F. and G. P. Wagner. 2001. Epistasis and the mutation load: A measurement theoretical approach. *Genetics* 158: 477–485.
Kashtan, N. and U. Alon. 2005. Spontaneous evolution of modularity and network motifs. *Proc. Natl. Acad. Sci. USA* 102: 13773–13778.
Kenney-Hunt, J. P., B. Wang, E. A. Norgard, and 7 others. 2008. Pleiotropic patterns of quantitative trait loci for 70 skeletal traits. *Genetics* 178: 2275–2288.
Love, A. C. 2003. Evolvability, dispositions, and intrinsicality. *Phil. Sci.* 70: 1015–1027.
Lynch, M. 2007. The frailty of adaptive hypotheses for the origins of organismal complexity. *Proc. Natl. Acad. Sci. USA* 104: 8597–8604.
Lynch, V. J. and G. P. Wagner. 2008. Resurrecting the role of transcription factor change in developmental evolution. *Evolution* 62: 2131–2154.
Masel, J. and A. Bergman. 2003. The evolution of the evolvability properties of the yeast prion [PSI^+]. *Evolution* 57: 1498–1512.
Maynard-Smith, J. 1978. *The Evolution of Sex.* Cambridge University Press, Cambridge, UK.
McBride, R. C., C. B. Ogbunugafor, and P. E. Turner. 2008. Robustness promotes evolvability of thermotolerance in an RNA virus. *BMC Evol. Biol.* 8.
Mezey, J. G., J. M. Cheverud, and G. P. Wagner. 2000. Is the genotype-phenotype map modular? A statistical approach using mouse quantitative trait loci data. *Genetics* 156: 305–311.

Nathan, D. F., M. H. Vos, and S. Lindquist. 1997. *In vivo* functions of the *Saccharomyces cerevisiae* Hsp90 chaperone. *Proc. Natl. Acad. Sci. USA* 94: 12949–12956.

Nimwegen, E. V., J. P. Crutchfield, and M. Huynen. 1999. Neutral evolution of mutational robustness. *Proc. Natl. Acad. Sci. USA* 96: 9716–9720.

Olson, E. C. and R. L. Miller. 1958. *Morphological Integration*. University of Chicago Press, Chicago.

Orr, H. A. 2000. Adaptation and the cost of complexity. *Evolution* 54: 13–20.

Parter, M., N. Kashtan, and U. Alon. 2008. Facilitated variation: How evolution learns from past environments to generalize to new environments. *PLoS Comp. Biol.* 4: e1000206.

Pavlicev, M., J. P. Kenney-Hunt, E. A. Norgard, and J. M. Cheverud. 2007. The contribution of differential epistasis to the variation in pleiotropy. *J. Morphol.* 268: 1115–1115.

Pavlicev, M., G. P. Wagner, and J. M. Cheverud. 2009. Measuring evolutionary constraints through the dimensionality of the phenotype: Adjusted bootstrap method to estimate rank of phenotypic covariance matrices. *J. Evol. Biol.* 36: 339–359.

Pigliucci, M. 2008. Is evolvability evolvable? *Nat. Rev. Genet.* 9: 75–82.

Prud'homme, B., N. Gompel, and S. B. Carroll. 2007. Emerging principles of regulatory evolution. *Proc. Natl. Acad. Sci. USA* 104: 8605–8612.

Queitsch, C., T. A. Sangster, and S. Lindquist. 2002. Hsp90 as a capacitor of phenotypic variation. *Nature* 417: 618–624.

Rechenberg, I. 1973. *Evolutionsstrategie*. Friedrich Frommann Verlag, Stuttgart.

Riedl, R. 1977. A systems-analytical approach to macroevolutionary phenomena. *Quart. Rev. Biol.* 52: 351–370.

Riedl, R. 1978. *Order in Living Organisms: A Systems Analysis of Evolution*. Wiley, New York.

Roseman, C. C., J. P. Kenney-Hunt, and J. M. Cheverud. 2009. Phenotypic integration without modularity: Testing hypotheses about the distribution of pleiotropic quantitative trait loci in a continuous space. *Evol. Biol.* 36: 282–291.

Rutherford, S. L. 2000. From genotype to phenotype: Buffering mechanisms and the storage of information. *BioEssays* 22: 1095–1105.

Rutherford, S. L. and S. Lindquist. 1998. Hsp90 as a capacitor for morphological evolution. *Nature* 396: 336–342.

Sniegowski, P. D. and H. A. Murphy. 2006. Evolvability. *Curr. Biol.* 16: R831–R834.

Stern, D. L. 2000. Evolutionary developmental biology and the problem of variation. *Evolution* 54: 1079–1091.

Su, Z., Y. Zeng, and X. Gu. 2009. A preliminary analysis of gene pleiotropy estimated from protein sequences. *J. Exp. Zool. (Mol. Dev. Evol.)* 314B: 115–122.

True, H. L. and S. L. Lindquist. 2000. A yeast prion provides a mechanism for genetic variation and phenotypic diversity. *Nature* 407: 477–483.

Wagner, A. 2005. *Robustness and Evolvability in Living Systems*. Princeton University Press, Princeton.

Wagner, A. 2007. Robustness and evolvability: A paradox solved. *Proc. Roy. Soc. B* 275: 91–100.

Wagner, G. P. and L. Altenberg. 1996. Complex adaptations and the evolution of evolvability. *Evolution* 50: 967–976.

Wagner, G. P., J. P. Kenney-Hunt, M. Pavlicev, and 3 others. 2008. Pleiotropic scaling of gene effects and the "cost of complexity." *Nature* 452: 470–472.

Wagner, G. P., M. Pavlicev, and J. M. Cheverud. 2007. The road to modularity. *Nat. Rev. Genet.* 8: 921–931.
Welch, J. J. and D. Waxman. 2003. Modularity and the cost of complexity. *Evolution* 57: 1723–1734.
West, M. A. L., K. Kim, D. J. Kliebenstein, and 4 others. 2007. Global eQTL mapping reveals the complex genetic architecture of transcription-level variation in *Arabidopsis*. *Genetics* 175: 1441–1450.
Wright, S. 1964. Pleiotropy in the evolution of structural reduction and of dominance. *Am. Nat.* 98: 65–69.
Young, N. M. and B. Hallgrímsson. 2005. Serial homology and the evolution of mammalian limb co-variation structure. *Evolution* 59: 2691–2704.
Young, N. M., G. P. Wagner, and B. Hallgrímsson. 2010. Development and the evolvability of human limbs. *Proc. Natl. Acad. Sci USA* 107: 3400–3405.
Zou, L., S. Sriswasdi, B. Ross, and 3 others. 2008. Systematic analysis of pleiotropy in *C. elegans* early embryogenesis. *PLoS Comp. Biol.* 4: e1000003.

Chapter 9

Embryos and Evolution: 150 Years of Reciprocal Illumination

Gregory A. Wray

Of all the fields of evolutionary biology covered in this book, perhaps none suffers so acutely from an identity crisis as evolutionary developmental biology. The long history of studies of evolutionary developmental biology is remarkable for its heterogeneity: dominant intellectual threads have come and gone several times and the most enduring among them have not proven to be the most illuminating. Indeed, many evolutionary biologists still don't consider evolutionary developmental biology, or evo-devo as it is colloquially known, to be a distinct or perhaps even a legitimate discipline. Even among practitioners, views are inconclusive: one can read reviews and books on the subject of evo-devo biology and still not perceive a central intellectual theme.

What has evo-devo contributed to evolutionary biology in the 150 years since the publication of *The Origin of Species*, and what is it likely to contribute going forward? I will argue in this chapter that for much of its history, evo-devo held only peripheral importance to the notable advances in evolutionary biology, but that it has recently become indispensable for moving forward. The fulcrum of this dramatic shift in relevance has been the impact of molecular biology and developmental genetics on evo-devo, which provided the necessary information and experimental tools to begin identifying and understanding evolutionary changes in developmental mechanisms. Developmental changes have produced much of the variety of form that comprises the grand sweep of biological diversity. The value of evo-devo to evolutionary biology as a whole lies in elucidating the material basis for this relationship: changes in genotype modify development, which in turn alters phenotype. In other words, evo-devo provides the essential and long-neglected link between the evolution of the genome and the evolution of traits.

A Brief History of Intellectual Themes in Evo-Devo

Most serious histories of evo-devo begin with, and focus primarily on, the long arc (some would say shadow) of a single intellectual theme: the temporal parallels between evolutionary history and embryonic development (Gould 1977; Richards 1992; Gilbert 1994; Hall 1999). There is certainly a lot of truth to this perspective. Woven into the long and complex history of ideas about parallels between ontogeny and phylogeny, however, are several other important intellectual themes (Figure 9.1). What follows is an anecdotal rather than serious historical treatment of these themes: the goal here is to highlight the often-distinct intellectual traditions within the history of evo-devo.

Parallels between the Scala Naturae *and Embryonic Development*

As far back as Aristotle, scholars have been struck by parallels between the unfolding of embryogenesis and perceptions of progression in organismal complexity; Gould (1977) and Richards (1992) provide perceptive reviews on the history of these ideas. This parallel was originally framed

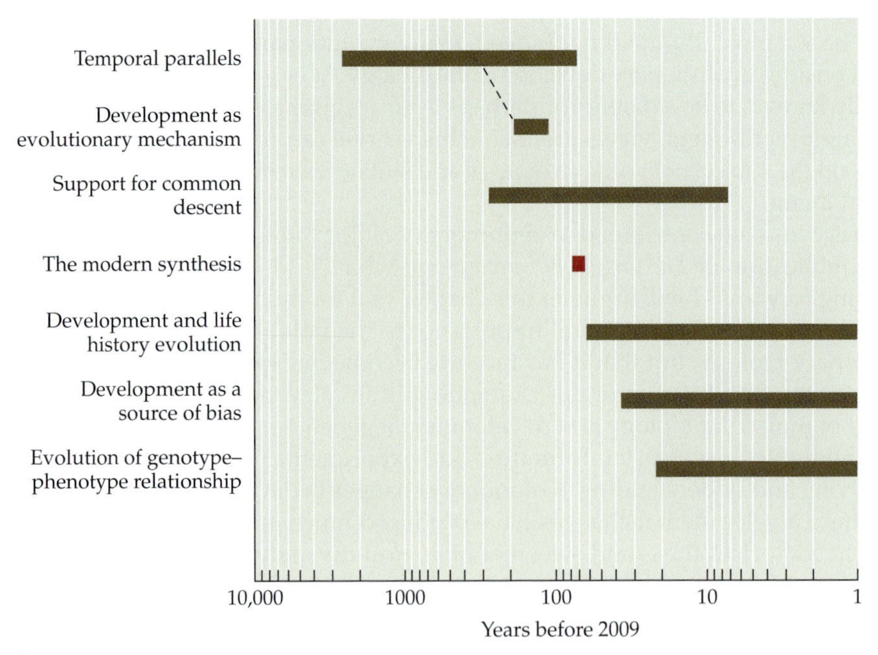

FIGURE 9.1 The Major Intellectual Themes of Evolutionary Developmental (Evo-Devo) Biology Attempts to understand the relationship between embryos and evolution dates back to Aristotle and has inspired a variety of intellectual approaches. Shown here in rather subjective and imprecise terms are some of the major themes. Note that the time scale represents time before the present rather than absolute dates and is logarithmic, so as to provide increasing detail for more recent events.

with respect to the *scala naturae*, or great chain of being, but Johann Meckel and Etienne Serres made the parallel explicitly evolutionary during the early nineteenth century. Karl Ernst von Baer later formulated his eponymous law in non-evolutionary terms: embryos share general similarities at early stages and become progressively distinct as development proceeds. von Baer accepted neither transmutation between species nor Darwinian natural selection, but Ernst Haeckel, another famous nineteenth century embryologist, embraced both. Haeckel's famous phrase "ontogeny recapitulates phylogeny" encapsulated and immortalized the fascination between temporal parallels (Haeckel 1866: 300).

Although Haeckel's assertion was not so much a research agenda as an observation about the natural world, it proved to be a remarkably persistent obsession. Despite the fascination that these parallels exerted over many biologists, they provided little in the way of real insights until the 1970s (Gould 1977). Haeckel wanted to understand why developmental and evolutionary parallels exist and eventually proposed an evolutionary mechanism linking the two processes (as will be discussed), however his particular theory was never widely accepted. Another research agenda, initiated during the 1970s, proposed an ontogenetic criterion for assessing character polarity in phylogenetic reconstruction (Nelson 1978). This criterion posited that character states appearing earlier in development are ancestral to those appearing later. So many exceptions were known that this criterion failed to prove useful (Mabee 1989), and even Haeckel could have provided a compelling list of developmental features that appear out of phylogenetic sequence (Gould 1977; Richards 1992). The most productive insights from temporal parallels came from considering heterochrony in the context of life history evolution, as described later (Gould 1977).

Embryos as Evidence for Common Descent

While still a medical student in Edinburgh, Charles Darwin spent several months studying the reproductive biology of marine invertebrates. This experience is notable for leading to Darwin's first publishable original research (Browne 1995), but it failed to inspire a long-term interest in embryology. When he wrote *The Origin of Species* many years later, Darwin's chief interest in development lay in the evidence that embryos can provide for common descent, citing the presence of gill slits in mammalian embryos as a prominent example. Embryos had no impact on Darwin's thoughts regarding the other major theme in *The Origin of Species*: the mechanism of natural selection—and there is no reason why they should have at the time. Developmental processes are, however, quite closely related to another phenomenon that was important to Darwin, namely the origins of trait variation; however, this conceptual link was not made for another century.

The realization that embryos are often more similar to one another than are the adults to which they give rise (i.e., von Baer's Law) also provided a way to recognize homology among anatomical structures that are otherwise obscure. For instance, Darwin was among the first to discern that barnacles are crustaceans, based on extensive similarities in their larvae that are far less obvious in adults. Another famous phylogenetic affinity revealed by embryos was the close relationship between ascidians (sea squirts) and chordates, in which the two animal groups had very different adult appearance. These were not isolated cases—it was through examining embryos or larvae that many other phylogenetic relationships were first recognized. This phenomenon was particularly true for higher-order relationships among major groups of plants and animals as well as for parasites, whose anatomy is often highly modified. Several clades were originally based on embryological similarities and named accordingly: monocots and dicots (for the organization of nutritive supplies during embryogenesis in flowering plants); Protostomia, Deuterostomia, and Spiralia (for organizational properties of animal embryos); Ecdysozoa (for molting as an essential developmental process); and Hemimetabola and Holometabola (for the absence or presence of radical metamorphosis during insect development).

Recapitulation as an Evolutionary Mechanism

Ernst Haeckel was much taken with Darwin's *The Origin of Species* and fully embraced common descent as an organizing principle for the diversity of life (see Hillis, Chapter 16). He coined the term phylogeny and was among the first biologists to use branching diagrams to propose specific phylogenetic relationships. Darwin famously published only one such figure (see Bowler, Chapter 2, Figure 2.1), and his famous phylogenetic sketch from Notebook B is notable for its singularity among his extensive notes (see Hillis, Chapter 16, Figure 16.1). Like Darwin, Haeckel was interested in understanding evolutionary mechanisms. He viewed recapitulation not simply as a pattern but as a cause of evolutionary change; that is, an evolutionary mechanism to stand beside natural selection in explaining the diversity of life (Richards 1992). Haeckel's Biogenetic Law proposed that traits evolve through terminal addition, the process of appending events on to the end of development, whose outcome is the pattern of recapitulation.

The Biogenetic Law holds that developmental stages of animals alive today are the same stages experienced by the adults of their ancestors. Haeckel extended this logic to the very earliest stages of development, which he proposed were originally free-living adults in the distant past, including Blastea (a blastula-like ancestral organism) and Gastrea (a gastrula-like animal). He proposed that countless instances of terminal addition had extended the development of these simple organisms to produce the much more complex adult organisms that are seen today. This idea never really caught on, in part because terminal addition is really just a pattern, not a mechanism with explanatory power. Perhaps more importantly, however, exceptions to strictly recapitulatory sequences in development are exceedingly common. Haeckel, as mentioned earlier, was well aware of these exceptions, which he called caenogenetic. He recognized both temporal and spatial exceptions, for which he coined the terms heterochrony and heterotopy, respectively.

Integrating Development with Life History Evolution

The Biogenetic Law eventually fell into disrepute because so many exceptions to recapitulation were identified, but it was founded on a very real, general trend. Why is development sometimes strikingly recapitulatory and other times decidedly not? The discovery of the often-bizarre larvae of marine invertebrates during the nineteenth century made this question particularly acute. These were cases in which development seemed to head off on a tangent for a while, producing a distinct, free-living organism, before tacking back towards the more familiar adult morphology. In some invertebrate groups, such as echinoderms, barnacles, bryozoans, ascidians, and flatworms, the difference in appearance is so extreme that their larvae were originally thought to constitute entirely distinct groups of organisms. For a long time, it was not clear why such odd developmental trajectories had evolved. During the 1920s, Walter Garstang was among the first biologists to propose that natural selection acts directly on the way development proceeds. He argued that many seemingly outlandish features of larvae are actually adaptations to ecological demands that are unique to early phases of the life cycle, most importantly nutrition, defense, and settlement (Garstang 1922, 1928). It became clear that development could be altered, sometimes extensively, without causing much change in adult morphology.

From this point forward, development could no longer be viewed simply as a way to produce an adult, but rather as a process shaped extensively by natural selection, such that the developing individual could survive until it reached reproductive age. As a result, development became viewed as a trait, just like any other, subject to modification by natural selection. This recognition opened up an entirely new research program connecting the evolution of life history traits to modifications in development. By the 1970s, many cases could be cited illustrating how simple changes in the timing of developmental processes could alter life histories. Steven Jay Gould argued that particular classes of heterochronies are associated with classic *r*- and *K*-selected life history patterns (Gould 1977). Around the same time, larval biologists became interested in understanding the ecological circumstances that drive developmental changes that can result in life history ramifications (McEdward 1995). During the 1980s, developmental biologists joined this broad research theme, analyzing the cellular and molecular processes that differ among closely related species with divergent life histories (del Piño and Elinson 1983; Cavener 1985; Raff 1987).

The Non-Event: Development and Classical Evolutionary Theory

Developmental biology was not part of the modern synthesis, and there was no reason at the time why it should have been. The framers of the modern synthesis were primarily concerned with uniting genetics and evolutionary biology. They recognized that genes affect traits but did not much concern themselves with how this was accomplished. In an era when the material basis for heredity was still not known, it was reasonable to focus on more immediately tractable questions. The major conceptual and empirical advances in evolutionary biology during the subsequent few decades also occurred without integrating concepts from developmental biology. These advances included the shift to a gene-centric view during the 1950s and the integration of molecular biology during the 1970s and 80s. It was not until the rise of developmental genetics during the 1980s that a meaningful and permanent integration began. The nature and impact of this integration is the focus of the latter part of this chapter.

Development as a Source of Bias in Phenotypic Evolution

Darwin was aware that natural selection has a limited ability to change traits and devoted an enormous amount of time to considering the scope and nature of trait variation and change. The rise of so-called population thinking in evolutionary biology, during the 1940s, identified the finite amount of genetically based variation that exists within a population as one source of constraint on evolutionary change. During the 1970s and 80s, several biologists approached the issue of constraint from a rather different, macroevolutionary perspective. They pointed out, for instance, apparently unoccupied morphospaces among gastropod shell shapes (Raup 1966) and tetrapod digit patterns (Alberch and Gale 1985). These gaps in observed phenotypes occur within large groups with rich fossil records, in the midst of diversification into many other morphologies and despite ample evolutionary time. For these reasons, a lack of genetic variation was not considered a plausible explanation and developmental constraints were instead proposed as the cause. The idea behind developmental constraint is that the organization of developmental processes makes certain evolutionary changes in phenotype more likely than others, rendering some perfectly functional morphologies unobtainable or at least highly unlikely (Maynard Smith et al. 1985). Experimental manipulations demonstrated that development indeed has inherent biases in the phenotypes it produces, when modified, and that these experimentally evoked biases parallel those in phylogenetic diversity and the fossil record (Alberch and Gale 1983).

By the early 1990s, the notion of developmental constraint had lost some of its appeal, in part, because it lacked predictive power: at least in practice, the concept of developmental constraint could only be applied retrospectively to very well studied clades. It lacked explanatory power, delineating limits on the diversity of form without providing any clear understanding of their basis. It became clear that patterns of phenotypic diversity originally attributed to developmental constraint could be the result of other factors, such as negative selection for reduced fitness or geometric and material constraints. There was no obvious way to disentangle these influences (Antonovics and van Tienderen 1991). The related concept of evolvability has largely superseded developmental constraint as a more tractable research program (see Wagner, Chapter 8). John Gerhart and Marc Kirschner proposed several molecular and cellular features of developmental systems that might promote evolvability, including fuzzy interactions and exploratory behavior (Gerhart and Kirschner 1997). Evolvability can be defined much more precisely than could the concept of developmental constraint during the 1980s. A mechanistic understanding of evolvability remains elusive, however, and constitutes a challenge for evo-devo in coming years.

Understanding the Genotype-to-Phenotype Relationship

Developmental biology underwent a revolution with the introduction of saturation mutagenesis screens that collectively identified many genes with critical roles in regulation of pattern-formation and cell-fate specification (Wilkins 1993). As positional cloning revealed the molecular identity of these genes, it became clear that most are involved in regulatory processes. To developmental biologists, this discovery was a satisfying and logical outcome. Development is, fundamentally, a process of regulation that progressively specifies distinct identities among cells over ever finer spatial scales and then directs those cells to adopt distinct functions (Davidson 1986; Wilkins 2002). The marriage of developmental genetics with molecular biology during the 1980s was an intellectual triumph that revealed that the material basis for this complex process involves a distinct set of regulatory proteins, most of which are involved in cell-to-cell signaling (ligands, their receptors, and intracellular transduction systems) or in the regulation of gene expression (transcription factors and co-factors, microRNAs, histone modifiers, and so forth).

It did not take long for comparisons across taxa to result in the addition of an evolutionary dimension. These comparisons began in the late 1980s and rapidly expanded during the 1990s, as described in the following sections. Although the results were surprising and exciting, their evolutionary implications were not immediately obvious. What was clear was that by identifying specific genes that were linked to particular morphological traits, developmental geneticists had paved the way for understanding the material basis for the connection between evolutionary change in the genome and in organismal traits (Raff and Kaufman 1983; Wilkins 2002). It might seem odd that evolutionary biologists had largely ignored this connection, yet this was the case. Even the field of evolutionary genetics, which was ostensibly concerned with understanding this relationship, focused largely on counting and mapping genomic regions that contribute to trait variation; the material basis for the relationship was largely ignored. As described next, a series of events positioned evo-devo to begin uncovering the relationship between genetic and trait evolution.

The Roots of the Evo-Devo Renaissance

Despite the rich and varied history of intellectual themes in evo-devo (as briefly summarized), to all intents and purposes, there was no coherent field of study in evolutionary developmental biology until relatively recently. Few scholars would have identified themselves as working in the area of evolutionary developmental biology prior to the 1990s, and neither conceptual nor empirical advances in developmental biology had contributed much of substance to evolutionary biology before this time. However, a conjunction of prior conditions and contemporary circumstances propelled evolutionary developmental biology into prominence as a distinct and active field of inquiry in the 1990s. In the space of just a few years, it acquired a nickname and witnessed multiple symposia and meeting sessions, a rash of papers in high-profile journals, the launching and rebranding of journals, the posting of numerous targeted job searches, the publication of several books, and a general buzz of excitement.

What lead to this rapid coalescence of identity? An enabling factor was the much earlier publication of *Ontogeny and Phylogeny* by Gould in 1977. Although this book preceded the rise of evo-devo by a decade, it sparked a renewed interest in evolution and development after decades of general neglect, providing intellectual legitimacy and excitement. But, there are three other factors, which began in the 1970s and came together during the 1980s, that also laid the necessary foundations for the evo-devo renaissance.

Gene Regulation and Evolutionary Biology

The first factor was the recognition that regulatory processes form an important substrate for evolutionary change. This recognition began with three very different papers that posited a qualitative distinction in the kinds of traits that regulatory processes and structural changes in proteins can influence. The first was an article by François Jacob and Jacques Monod published shortly after they won the Nobel Prize in Physiology and Medicine. They proposed that: "the genome contains not only a series of blueprints, but a coordinated program of protein synthesis and the means of controlling its execution" and speculated on the evolutionary implications of regulatory sequences (Jacob and Monod 1961: 354). A few years later, Roy Britten and Eric Davidson proposed a model for the organization and evolution of gene regulatory networks, a remarkably far-sighted hypothesis for the time (Britten and Davidson 1969). They also stressed the evolutionary significance of gene regulation (Britten and Davidson 1971). Third, and most famously, Mary-Claire King and Alan Wilson published a paper in 1975, in which they posited that changes in regulatory processes have contributed more to the phenotypic distinctiveness of humans than changes in protein sequence (King and Wilson 1975). All three articles are notable for their visionary, first-principles arguments that were based on very little direct empirical evidence; in each case, these predictions have largely been borne out, although it took decades for empirical studies to catch up.

A fourth publication also played a distinct role in focusing attention on the potential importance of regulatory processes in trait evolution; it was Rudolf Raff and Thomas Kaufman's 1983 book *Embryos, Genes, and Evolution*. Raff and Kaufman drew on discoveries in developmental genetics and molecular biology to argue that mutations affecting regulatory processes can account for many evolutionarily significant trait changes. Raff and Kaufman came from a background in developmental biology and developmental genetics, respectively, and focused attention on understanding how mutations achieve their trait consequences in terms of specific molecular mechanisms. Their book laid out the first concrete approaches for testing this idea, inspiring many biologists to begin searching for regulatory mutations of evolutionary significance.

Developmental Genetics and Evolutionary Biology

The second factor behind the evo-devo renaissance was a greatly expanded understanding of the proteins and molecular mechanisms that regulate development. Forward genetic screens for developmental defects—of the sort pioneered in *Drosophila melanogaster* by Ed Lewis in the 1950s and scaled up enormously by Christiane Nüsslein-Volhard and Eric Wieschaus—provided a way to identify many genes involved in producing interesting phenotypes (Lewis 1978; Nüsslein-Volhard and Wieschaus 1980). Similar, large-scale screens for developmental genes soon followed in other genetically tractable model organisms. This approach was spectacularly successful, and by the mid-1980s, large numbers of genes with critical roles in development had been genetically characterized in detail (Wilkins 1993). Many of these genes had mutant phenotypes that strongly suggested a regulatory function. Particularly dramatic were homeotic genes, for which loss-of-function or gain-of-function mutations produced nearly perfect switches in the developmental fate of individual cells or in entire regions of the body. The most famous examples are mutations in the homeotic genes of *Drosophila*, which transform the identified segments and their associated appendages. Another interesting class of mutations produced heterochronic phenotypes, which shift the timing of developmental events.

During the 1980s, it became possible to isolate and sequence these genes identified in forward genetic screens for developmental roles. It soon became evident that many encoded proteins had regulatory functions, an intellectually satisfying result. The homeotic genes of flies, for instance, were found to encode transcription factors and their DNA-binding domain became known as the homeobox, because it could be boxed as a region of unusually highly similar sequence in alignments of these proteins. Other genes were found to encode cell-surface receptors or the ligands that bind to them. These discoveries quickly lead to detailed characterization of entire signaling systems, including the relay systems that transduce ligand binding on the cell surface into an appropriate response within the cell. The outcome was an enormously expanded and detailed understanding of the molecular mechanisms that regulate development. This knowledge opened the door, for the first time, to the direct study of how regulatory function changes during the course of evolution.

Molecular Function and Evolutionary Biology

The third contributing factor to evo-devo resurgence was the development of molecular methods that allowed investigators to begin analyzing evolutionary changes in gene regulation. At the time (and to a large degree, still today), both population genetics and evolutionary genetics treated genes as rather abstract entities (Figure 9.2). Even molecular evolution, which deals with DNA and protein sequences, often treated these sequences in the abstract; for instance, the field considered all nonsynonymous substitutions equivalently and all synonymous substitutions equivalently. None of these core disciplines within evolutionary biology paid much attention to the molecular mechanisms of regulation or to the gene interactions that lie at the heart of development. The result was some peculiar conceptual biases regarding gene evolution and some large gaps in information. For example, evolutionary biologists primarily studied enzymes and structural proteins rather than the transcription factors, kinases, signaling ligands, and receptors that mediate regulation in cells and embryos. Evolutionary biologists devoted even less attention to noncoding sequences, which were widely considered to be "junk DNA." This approach seems peculiar, in retrospect, because molecular biologists could point to essential functions embedded

FIGURE 9.2 Changing Models of the Gene in Evolutionary Biology (A) Population genetics model, origin circa 1940. At the time of the modern synthesis, the material basis for heredity was not known and genes were represented with letters (see large "A") and conceptualized entirely in the abstract. (B) Evolutionary genetics model, origin circa 1950. The goal of evolutionary genetics was to map traits to physical locations (literally loci, not genes) within the genome and measure their relative contributions to trait variation. Genes were given names based on these traits (e.g., wrinkled seeds). (C) Molecular evolution model, origin circa late 1970s. Once the material basis for a gene and the genetic code were known, it became possible to treat different parts of a gene independently. Considerable emphasis was placed on synonymous and nonsynonymous substitutions, and mutations lying outside codons were largely ignored when evaluating gene function. (D) Evo-devo model, origin circa 1985. For decades, it has been clear that genes are much more complicated than the three other models presented here. Most evolutionary biologists have not incorporated the complexity of genes into their models. The field of evo-devo has focused attention on the importance of incorporating regulatory processes into an understanding of how genes and traits evolve. Shown here are a few of the diverse mechanisms regulating gene expression, most of which are still poorly understood from an evolutionary perspective.

in these sequences and knew a lot about the molecular mechanisms that regulate cellular and developmental processes. As a result, evolutionary biologists were largely ignorant of the portions of the genome that are most likely to harbor mutations influencing many kinds of trait evolution.

To some degree, this inattention to the evolution of gene regulation can be excused based on technological limitations. Several important advances during the 1980s made it easier to assay regulatory function at the molecular level, facilitating evolutionary analyses of regulation. One of the most critical technological advances came with the invention of polymerase chain reaction (PCR), which made molecular cloning far more powerful, fast, and inexpensive. For the first time, it became possible to isolate and sequence orthologous segments of genomic DNA from just about any organism, and

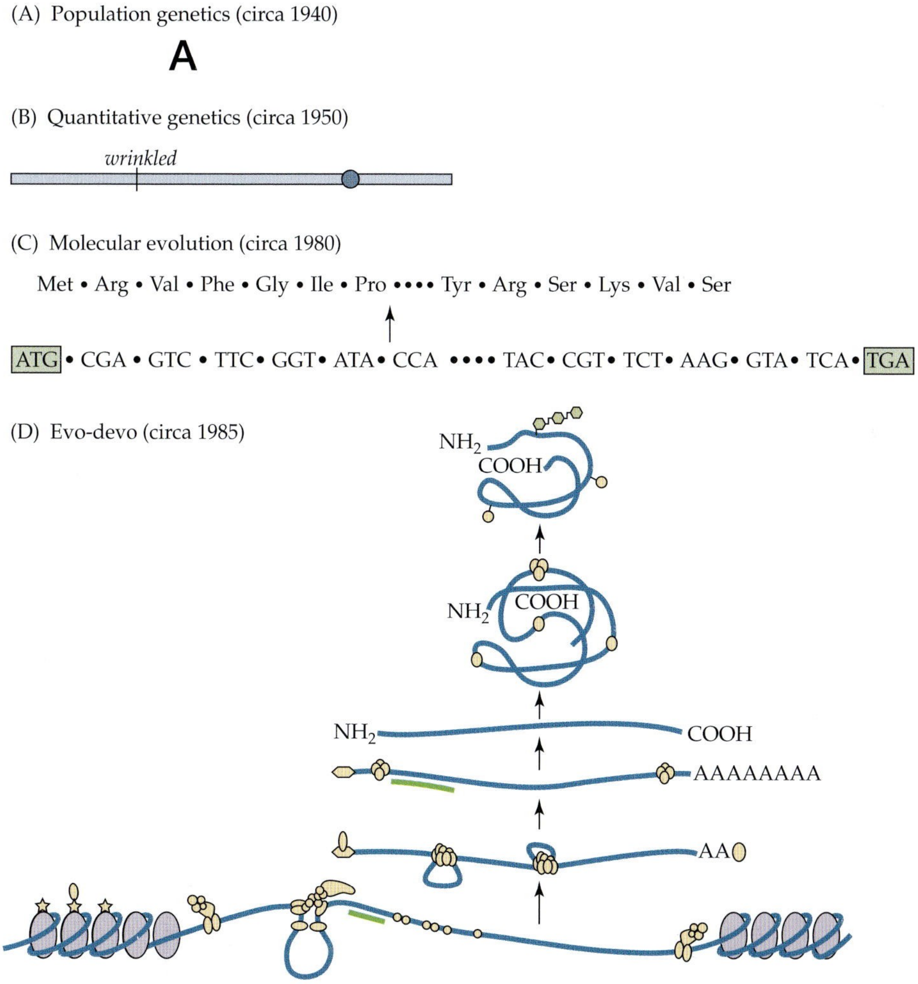

(A) Population genetics (circa 1940)

A

(B) Quantitative genetics (circa 1950)

wrinkled

(C) Molecular evolution (circa 1980)

Met • Arg • Val • Phe • Gly • Ile • Pro •••• Tyr • Arg • Ser • Lys • Val • Ser

ATG • CGA • GTC • TTC • GGT • ATA • CCA •••• TAC • CGT • TCT • AAG • GTA • TCA • TGA

(D) Evo-devo (circa 1985)

NH_2 COOH

NH_2 COOH

NH_2 COOH

AAAAAAAA

AA

within a few years, methods for isolating sequences derived from mRNA also became widely applicable. Another critically important factor was the development of methods for analyzing the spatial distribution of gene expression in embryos: immunolocalization for proteins and *in situ* hybridization for mRNA. Without these technological advances, the results that made the evo-devo renaissance so exciting could not have been gathered.

The Isolated Rise of Evo-Devo

The enabling factors already outlined in this chapter were in place by the mid-1980s, but it was a series of empirical observations, which the technological advances made possible, that truly catalyzed the evo-devo renaissance. Some of those findings were provocative and exciting, but they did not fit comfortably within the conceptual frameworks of either traditional population genetics or molecular evolution. Indeed, it could fairly be said that the conceptual advances that emerged from the evo-devo renaissance during the 1990s were not widely appreciated by many mainstream evolutionary biologists at the time. Evo-devo eventually morphed into something more obviously valuable to evolutionary biologists of all stripes by integrating approaches from population genetics and evolutionary genetics.

Hox *Mania and the Rise of Evo-Devo*

The rise of evo-devo was driven by a series of discoveries that revealed a stunning, wholly unexpected commonality in the genetic basis for virtually every aspect of anatomy throughout the animal kingdom. It is difficult to appreciate in retrospect how revolutionary this discovery was. From just about any perspective, the embryos of the major animal model systems (fly, nematode, and mouse) are about as different as could be imagined: they use different molecular mechanisms to specify primary body axes, the cell biology of early development is utterly different, cell fate specification occurs at very different times, and they give rise to adults with different body plans (Davidson 1991). Until the mid-1980s, the embryos seemed so dissimilar that no one expected them to share so many developmental regulatory genes.

It came as a big surprise to learn that a highly conserved set of genes encodes the basic regulatory machinery of animal development. The first hint was that *Notch*, which directs wing and neural development in *Drosophila*, is the homologue of *lin-12*, which is essential for vulva development in *C. elegans* (Yochem et al. 1988). We now know that these genes encode a cell-surface ligand involved in numerous cell-fate specification events throughout the animal kingdom. More famously, and just a few years later, it became clear that genes of the *Hox* complex—already well known for specifying segment identity along the anteroposterior axis of *Drosophila*—also pattern anteroposterior identity among somites and within the nervous system of vertebrates (Akam 1989). Soon *Hox* genes, which encode transcription factors that regulate gene expression throughout development, had been identified from many other phyla. By the early 1990s, it was clear that many of the developmental regulatory genes identified from genetic screens in model organisms are present in numerous phyla throughout the animal kingdom. This shared set of regulatory genes became known as the "toolkit" of metazoan development.

Hox genes, in particular, became a kind of icon of evo-devo in the 1990s. They are an extreme example of the toolkit concept: not only are the genes themselves present, but their nested pattern of expression along the primary body axis is superficially similar in very different kinds of animals, and gene knock-out models (when available) generate spectacular homeotic phenotypes that correlate with these expression domains. Beyond these features, however, the *Hox* genes exerted an almost mystical attraction, for the simple reason that neither the striking evolutionary conservation of gene order along the chromosome, nor their colinear expression along the body was understood. There was the sense that somewhere in the numerology of *Hox* cluster organization the secrets of an evolutionary animal body plan would emerge—if only one could just figure out what to look for. Many dozens of manuscripts were published listing the membership of *Hox* genes in formerly obscure animal groups (e.g., how many developmental geneticists knew what a priapulid was before the 1990s?). Rainbow-colored diagrams comparing *Hox* cluster organization became a staple of the evo-devo literature for the better part of a decade. Analyses of *Hox* gene expression soon followed, documenting the spatial and temporal distribution during embryogenesis in diverse metazoan groups. The rainbow colors were duly used to match *Hox* orthologs with domains of expression in phylogenetically distant and phenotypically divergent animals.

Parallel studies documented the phylogenetic distribution of other important developmental regulatory genes throughout the animal kingdom. Meanwhile, it became apparent that a parallel situation exists in land plants, with a different family of transcription factors, the MADS-box genes, being essential for floral patterning throughout the angiosperms and present in non-flowering plants as well.

The hope was that information about the expression of developmental regulatory genes would begin to reveal the genetic basis for the evolution of major morphological transitions (Slack et al. 1993; DeRobertis and Sasai 1996; Carroll et al. 2001). The expression domains of these genes, for instance, were thought to provide a reliable guide to identifying homologous anatomical structures among distantly related animals. Knowing what structures are homologous would provide a secure framework for understanding the evolution of major morphological transitions, such as body plans and evolutionary innovations. At a general level, the hope was that evo-devo could address a problem that evolutionary biologists had long ignored: the connection between evolutionary change in the genome and evolutionary change in ecologically relevant traits.

Integration with Paleontology

The exciting discoveries of evo-devo during the 1990s initially had relatively modest impact on the core research paradigms of evolutionary biology, namely molecular evolution, population genetics, and evolutionary genetics. The evo-devo research community in the 1990s drew heavily on phylogenetic methods and, to a much more limited extent, concepts from molecular evolution but paid scant attention to concepts like variation, fitness, selection, or drift. The mainstream evolutionary biology community, for its part, was not thinking much about the evolutionary implications of regulatory change. Evolutionary geneticists, for instance, were primarily concerned with mapping traits, not understanding how genes produced trait variation. As a result, the truly exciting results of the evo-devo renaissance had only a modest intellectual impact on the majority of evolutionary biologists for the better part of a decade.

Paleontologists, however, comprised a notable and prominent exception. The unexpected molecular and genetic commonality of animals with highly divergent morphologies was exciting to anyone interested in big evolutionary transitions. The evolution of animal body plans had long attracted attention as a thorny problem, since phyla are so distinct in overall organization. Renewed attention to the Burgess Shale fossils and the Cambrian Explosion in the 1980s, in particular, sparked interest in the problem among paleontologists (Gould 1989). *Hox* genes and the toolkit of metazoan development provided a fresh and compelling perspective on this puzzle. Early applications included identifying homologous structures and body regions among very distantly related animals (Slack et al. 1993) and reconstructing ancestral organisms based on shared genetic complements (Erwin and Davidson 2002). However, the real excitement lay in trying to identify what kinds of changes in gene sequence or expression might have given rise to changes in body plans (Valentine et al. 1996; Shubin et al. 1997; Carroll et al. 2001). The confluence of interests between evo-devo and paleontology had a positive impact on both fields, which continues today (Levinton 2001).

*The "*Hox *Paradox"*

It is not hard to find evolutionary biologists who feel that evo-devo overreached during the 1990s. Despite all the exciting discoveries, there was a sense of frustration by the end of the decade, as a couple of problems came into focus.

One was that simply cataloging *Hox* gene complements and mapping gene expression domains was not providing any obvious insights into how complex anatomical structures such as body plans evolve. As in so many other areas of biology, the situation proved to be more complex than originally thought. It was soon clear that most regulatory genes are involved in the development of many different, non-homologous structures within a single animal. For instance, *runt*, which encodes a transcription factor, plays several critical roles in at least three very different processes during *Drosophila* development: sex determination, specification of identity along the anteroposterior axis in embryos, and the development of neural cells much later in development. Such realizations made using developmental regulatory genes less useful for identifying homologous organs than initially supposed: if a gene is expressed in several different organs and at different times during development, it can be problematic to understand how these relate to the set of expression domains of different species (Abouheif et al. 1997). This fact did not eliminate the application of gene expression data in the identification of homologous organs, but it did make the exercise more complicated, particularly when comparing distantly related species—which is unfortunately precisely where the most exciting applications lie.

A related problem was that the existence of a shared toolkit meant that simply documenting the presence of a developmental regulatory gene in an animal revealed little about its development or morphology. Almost every metazoan examined had at least one *Notch* gene, and *Hox* genes were found in jellyfish, snails, and sea urchins. What did it mean if the genome of a species contained a few more or less of these genes than other species? No relationship emerged between morphology or body plan organization, on the one hand, and the presence/absence of particular developmental regulatory genes, on the other. In fact, quite the opposite was the case. The shared toolkit raised what has been called the *Hox* Paradox (Wray 2003). How can a shared set of developmental regulatory genes produce such a stunning diversity of animals? In principle, the answer lies in rewiring interactions between developmental regulators and their effectors, or target genes. The multiple developmental roles of individual regulatory genes violates the classical view of genes, which began with Beadle and Tatum's one gene, one enzyme hypothesis and has been retained in mainstream evolutionary biology ever since. When it comes to development, the relationship between genes and molecular function is one-to-many, with most regulatory genes affecting multiple and diverse traits. Further, comparisons among species strongly imply that these gene-to-trait relationships are not nearly as fixed as those between a gene and an enzyme. Instead, developmental genes can be co-opted into novel regulatory functions that bear little resemblance to their prior roles. All of this meant that discovering genes and mapping out their expression domains was not going to be enough—that understanding changes in molecular function and trait association would be essential.

Evo-Devo Starts to Go Mainstream

The evo-devo renaissance encountered these challenges at about the same time. Much of the interest in the evo-devo research community at the time was centered around understanding body plan evolution, which meant comparing data from distantly related organisms, typically species belonging to widely divergent phyla. This approach made the challenges about as difficult to tackle as possible: it was difficult to identify homologous structures, the trait consequences of changes in molecular function were obscured by the many changes that separated the organisms being compared, and making connections to classical molecular evolution and particularly population genetics was difficult.

Two principal solutions to these challenges soon emerged, both quite obvious in retrospect. One was to narrow the phylogenetic scale of comparisons to closely related species and shift attention to more subtle phenotypic differences than body plans. Although these changes had already been happening to a limited extent, even in the 1980s, the enthusiasm of the renaissance was centered on comparisons across phyla. The past decade has seen a pronounced shift towards comparisons within a clade of related species or among multiple populations within a species. At the same time, interest shifted from body plans to traits, like skeletal shape and pigmentation patterns. The second solution was drawing on genetics to establish the link between mutation, its functional impact, and its trait consequences. While evo-devo in the 1990s was notable for the breadth of taxa that were examined, during the first decade of the new millennium the focus shifted back to well-studied model organisms and their close relatives. During this time several powerful new genetic models, such as stickleback fishes, were developed (Kingsley and Peichel 2007). Together, these two operational shifts provided an extremely powerful approach and set the stage for a meaningful integration of evo-devo with the rest of evolutionary biology.

The Central Role of Evo-Devo in Evolutionary Biology

The influence of information and concepts from developmental biology on advances in evolutionary biology have ranged from minor to nonexistent throughout its history—a situation that is changing. Evo-devo is on its way to becoming an essential and well-integrated component of evolutionary biology. This final section outlines some of the ways that evo-devo is contributing to evolutionary biology and will likely continue to do so in the future.

The Central Role of Regulation as the Material Basis for Phenotypic Change

The evo-devo renaissance produced empirical results and conceptual shifts that have altered evolutionary biology in profound ways. Perhaps most fundamentally, the perspective of developmental biology focused attention on the critical and long-ignored issue of how changes in the genome are translated into trait differences. Change in genotype and change in phenotype have been treated as almost entirely distinct processes for most of the history of evolutionary biology. The field of evolutionary genetics has attempted to make this connection in the abstract, principally by asking how many loci underlie a particular trait difference. The critical contribution of evo-devo is the necessity of understanding the material basis of this connection. The premise of evo-devo is that the details matter when considering how a mutation affects organismal traits. What is the precise mutational basis for a trait change? Which molecular and cellular functions are affected by these mutations? How do these changes produce ecologically relevant trait consequences?

One of the significant early results of the evo-devo renaissance was the confirmation of the central role of regulatory changes in trait evolution. Once it was possible to clone genes that emerged from mutant screens for genes affecting development, it became clear that most encode proteins with inherently regulatory functions. Until the evo-devo renaissance, however, the vast majority of empirical information from population genetics and molecular evolution concerned genes with enzymatic or structural function. Very few evolutionary biologists paid much attention to developmental regulatory genes until the evo-devo renaissance. It was not until after 2000 that mutations affecting the expression or proteins these genes encoded became the focus of serious attention. It is still not clear whether the generalizations from studies of molecular evolution, which were built on data from enzymes and structural genes, will hold for regulatory genes.

It has taken several years to grasp the evolutionary implications of the discovery of the shared toolkit for animal development. Yet, the same simple and limited molecular machinery has been repeatedly co-opted into regulating a great variety of traits, which is clear from the fact that a relatively small set of regulatory genes has been recruited repeatedly into new developmental roles during the diversification of plants and animals (Carroll et al. 2001; Wilkins 2002; Wray et al. 2003). This kind of genetic basis for trait evolution is apparently pervasive. It was not, however, anticipated by evolutionary geneticists and instead emerged as a purely empirical result of the evo-devo renaissance. There are interesting implications for thinking about the kinds of mutations that alter traits. No longer are changes in gene function viewed largely through the lens of coding sequences or gene duplication: increasingly attention is being directed towards mutations affecting a variety of regulatory processes, including transcription, splicing, mRNA stability, and the role of noncoding RNAs (Chen and Rajewsky 2007; Wray 2007).

An interesting observation that has been largely ignored by the evolutionary biology community is that most developmental regulatory genes influence a variety of disparate traits (Carroll et al. 2001; Wilkins 2002). For instance, Notch protein is essential for proper development of wings and sensory bristles in *Drosophila* and for proper development of the brain and T-cells within the immune system in mammals. This is another observation that was not predicted by evolutionary geneticists; instead, it emerged directly from developmental genetics and later from the evo-devo renaissance. It is now abundantly clear that many developmental regulatory genes are agnostic with regard to the traits they can influence, in the sense that the same gene can regulate the development of several completely unrelated traits. Unlike enzymes and most structural proteins, which have an obligatory and direct association with a biochemical function, the function of a regulatory gene is simply to influence the function of another gene—the function of that target of regulation determines the trait consequence. At least in principle, any regulatory gene could modify the function of any other gene in the genome (including itself). The essential but poorly understood evolutionary consequence is that regulatory genes are less inherently tied to particular traits than are many structural genes. In particular, regulatory genes can take on new regulatory functions, a phenomenon known as co-option or recruitment (Lowe and Wray 1997; Wilkins 2002). Function acquisition can also happen with genes encoding enzymes and structural proteins, but far fewer examples are known. It is now clear that most developmental regulatory genes influence multiple, unrelated traits; indeed, it is difficult to find a well-studied regulatory gene whose function is dedicated to a single trait.

The Grammar of Mutations Affecting Phenotypic Change

Evolutionary genetics has historically focused on how many genes affect a trait of interest and their relative magnitude of effects. The identity of these genes, the nature of the products they encode, their expression profiles, and the causal mutations that underlie trait differences have not, until relatively recently, been a focus of concern. The evo-devo perspective is that these are not just incidental details. Rather, they provide important insights into the genetic basis for phenotypic change.

One important consideration is the function of the RNA or protein product that the gene encodes. One of the premises of evo-devo is that some kinds of traits are much more likely to evolve through mutations in particular kinds of genes (Raff and Kaufman 1983; Carroll et al. 2001; Wilkins 2002). An early and profound result of developmental genetics was that genes encoding regulatory proteins readily modified loci within the genome for altering morphology, at least based on induced mutations. Another interesting result was that the resulting trait consequences are often highly circumscribed. For instance, homeotic mutations often produce spectacular phenotypic consequences that are precisely limited to a small part of the body. As evo-devo has increasingly incorporated genetics during the past decade, a number of clear examples have emerged in which ecologically relevant phenotypic differences map to known regulatory genes (Wang et al. 1999; Kopp et al. 2000; Skaer and Simpson 2000; Shapiro et al. 2004; Colosimo et al. 2005; Gompel et al. 2005).

A second consideration is how broadly the gene is expressed. Some genes encode products that are essential for life and thus expressed in every cell throughout the life cycle (e.g., core energy metabolism, the machinery of transcription and translation, ion balance, and so forth). Other genes are highly specific to particular tasks and may only be expressed in a single cell type, during a specific part of the life cycle, or under particular environmental conditions (e.g., fertilization components, hormones, stress response effectors, and much more). One might imagine that mutations in broadly expressed genes are more likely to produce highly pleiotropic consequences and thus generally disfavored, while mutations in narrowly expressed genes are sometimes required to change traits associated with their specialized functions. These generalizations may hold true at the extremes of the continuum of universal versus highly specific expression (Carroll et al. 2001; Haygood et al., unpublished). Interestingly, however, most regulatory genes lie somewhere between the two extremes: they are often expressed in several different body regions and cell types across the life cycle and typically contribute to several unrelated organismal traits (Wray et al. 2003). This outcome is not what one would predict, based on first principles from population genetics, because of the attendant pleiotropic exposure. Molecular biology provides the solution, as discussed next.

A third important consideration involves the kind of mutation in the gene, which produces the trait consequence. The most simplistic classification is whether the mutation affects the gene product or not—a classification that is usually expressed as coding versus noncoding or regulatory versus structural. As discussed earlier, this distinction was considered critical on theoretical grounds more than a third of a century ago (Jacob and Monod 1961; Britten and Davidson 1969; King and Wilson 1975). Three important distinctions have emerged as more data have become available: (1) Many functionally important nucleotides in genomes are positioned outside the coding regions of traditional genes. Thus, simply from a quantitative perspective it is important to consider the contribution of these noncoding mutations to trait evolution. Rough estimates suggest that approximately equal numbers of functional nucleotides lie in coding and noncoding regions of the genome (Shabalina and Kondrashov 1999; Andolfatto 2005). (2) The modular nature of many transcriptional regulatory elements provides a molecular mechanism for delimiting pleiotropy (Stern 2000; Davidson 2001). During the past several years, a number of detailed case studies have demonstrated that natural mutations in regulatory elements affect just one aspect of a regulatory gene's function, circumscribing its phenotypic impact (Wray 2007). (3) Some trait changes may only be possible through mutations in regulatory sequences. For instance, regulatory mutations provide a way to change the function of a gene in one tissue, cell type, time in the life cycle, or ecological circumstance, without changing another. Several examples are now known (Shapiro et al. 2004; Gompel et al. 2005; Jeong et al. 2008; Rebeiz et al. 2009; Chan et al. 2010). A broad class of additional cases includes conditional responses to environmental conditions (e.g., the lac operon, heat shock proteins, and immune responses). Phenotypic plasticity, stress responses, inducible defenses, dietary responses, physiological acclimation, and immune function are just a few examples of the many ecologically relevant traits that are inherently regulatory (West-Eberhard 2003). While it is theoretically possible to achieve some of these responses through changes in coding sequences, most known cases involve mutations in noncoding sequences.

Integrating Approaches in Order to Understand Evolutionary Processes

The evo-devo renaissance occurred somewhat in isolation from the rest of evolutionary biology. Many studies published during the 1990s drew on phylogenetic methods to assign orthology among members of multigene families (usually *Hox* complexes), and it was not uncommon to see very simple comparisons of evolutionary rates of sequence change within and outside of DNA-binding domains in orthologous transcription factors from different phyla. By and large, however, the standard analytical frameworks for molecular evolution and population genetics were not widely utilized in evo-devo. The other substantial connection evo-devo made to evolutionary biology during the 1990s was in the field of paleontology, as mentioned earlier. Notably lacking at the time, however, was a connection between evo-devo and the core disciplines of evolutionary biology: population genetics, evolutionary genetics, and molecular evolution. This situation has changed dramatically during the past decade, and evo-devo is now on its way to becoming a well integrated part of contemporary evolutionary biology.

With the increased scrutiny of populations and closely related species by the evo-devo community during the 2000s, the use of analytical methods and questions from traditional population genetics and molecular evolution has become increasingly common. Testing for evidence of natural selection provides insights into how alleles become established or fixed within populations, while sampling functionally relevant variation across multiple populations reveals whether adaptation is based on standing variation or new mutations (Colosimo et al. 2005; Jeong et al. 2008; Chan et al. 2010). Scans addressing selection, at the scale of entire genomes, have identified regulatory genes and noncoding regulatory sequences as frequent targets of positive selection (Gilad et al. 2005; Haygood et al. 2007).

A second area of integration has been in the area of evolutionary genetics of trait variation. Indeed, the field of evolutionary genetics has generally become much more focused on identifying the molecular basis for the traits: both gene function and the nature of causal mutations have become goals of evolutionary genetic analyses. The distinction between evo-devo, evolutionary genetics, and population genetics is blurring considerably (Shapiro et al. 2004; Rebeiz et al. 2009). This trend is likely to continue as evolutionary genetics moves beyond mapping to tackle the interesting problem of what kinds of genomic regions and mutations contribute to trait differences. In a growing number of cases, mutations that produce functionally similar molecular consequences underlie parallel trait evolution

(Gompel et al. 2005; Shapiro et al. 2006; Tishkoff et al. 2007; Tung et al. 2009; Chan et al. 2010). These cases strongly suggest that the genetic and molecular foundations for some trait changes are likely to be highly biased: the causal mutations may typically lie in coding or noncoding sequences and may exert their consequences through certain molecular mechanisms rather than others.

Another area of meaningful integration is with evolutionary ecology. This interest began with the *lac operon* and the birth of a molecular and genetic understanding of gene regulation (Jacob and Monod 1961). Ever since, the ability to modulate transcription in response to changing environmental conditions has been understood as a distinct class of adaptive change in gene function. This aspect of integration between evo-devo and more mainstream fields of evolutionary biology has progressed less than integration with population genetics and evolutionary genetics, but the potential for interesting discoveries is enormous (West-Eberhard 2003; Gilbert and Epel 2008). Evolutionary biologists have largely treated environmental influences on gene function in the abstract, through modeling and assumptions. This situation is rapidly changing now that it is possible to assay gene function across environmental conditions and in the context of different genetic backgrounds. Particular areas of interest include: (1) measuring the relative scope of genetic versus environmental contributions to molecular trait variation and (2) the magnitude of reaction norms at the level of molecular function. Studies focusing on the molecular basis for the evolution of gene-by-environment interactions on traits have begun to appear (Abouheif and Wray 2002; Fay et al. 2004), but not nearly enough attention has been devoted to understanding the influence of the environment on the evolution of development.

Conclusion: Looking to the Future

With the perspective of 150 years that have elapsed since the publication of *The Origin of Species*, it is tempting to look to the future and speculate on how evolutionary biology will change in the coming years. One of the important challenges that evolutionary biology undoubtedly faces is to develop an understanding of the relationship between the evolution of genotype and the evolution of phenotype. These processes have been treated as effectively unrelated, even by evolutionary geneticists who focus on mapping loci rather than understanding the material basis for trait evolution. Evo-devo has begun to fill this void and will continue to do so in the future. The empirical results of the evo-devo renaissance have already demonstrated that patterns of change in genotype cannot be understood without taking into consideration how genes and other elements in the genome function at a molecular level, how they interact with each other, and how they are mechanistically tied to organismal traits through development. Similarly, patterns of change in phenotype cannot be understood without identifying their precise mutational basis, how the mutations alter existing molecular functions, and how those changes modify development to generate phenotypic consequences. Molecular details matter. The days when it was possible to treat genes and mutations in the abstract are over. Real progress in understanding the relationship between genomic and trait evolution will come from an increasingly sophisticated and comprehensive understanding of how specific mutations affect molecular function and how these changes in turn alter organismal traits. After a long and varied history of intellectual themes, evo-devo is once again reinventing itself, this time in a role that has become central to evolutionary biology as a whole.

Literature Cited

Abouheif, E. H., M. Akam, W. J. Dickinson, and 6 others. 1997. Homology and developmental genes. *Trends Genet.* 13: 37–40.

Abouheif, E. H. and G. A. Wray. 2002. The developmental genetic basis for the evolution of wing polyphenism in ants. *Science* 297: 249–252.

Akam, M. 1989. Hox and HOM: Homologous gene clusters in insects and vertebrates. *Cell* 57: 347–349.

Alberch, P. and E. A. Gale. 1983. Size dependence during development of the amphibian foot. Colchicine-induced digital loss and reduction. *J. Exp. Embryol. Morph.* 76: 177–197.

Alberch, P. and E. A. Gale. 1985. A developmental analysis of an evolutionary trend: Digital reduction in amphibians. *Evolution* 39: 8–23.

Andolfatto, P. 2005. Adaptive evolution of non-coding DNA in *Drosophila. Nature* 437: 1149–1152.

Antonovics, J. and P. H. van Tienderen. 1991. Ontoecogenophyloconstraints? The chaos of constraint terminology. *Trends Ecol. Evol.* 6: 166–168.

Britten, R. J. and E. H. Davidson. 1969. Gene regulation for higher cells: A theory. *Science* 165: 349–357.

Britten, R. J. and E. H. Davidson. 1971. Repetitive and non-repetitive DNA sequences and a speculation on the origins of evolutionary novelty. *Quart. Rev. Biol.* 46: 111–138.

Browne, J. 1995. *Charles Darwin: A Biography, Vol. 1—Voyaging*. Knopf, New York.

Carroll, S. B., J. K. Grenier, and S. D. Weatherbee. 2001. *From DNA to Diversity: Molecular Genetics and the Evolution of Animal Design*. Blackwell, Malden.

Cavener, D. R. 1985. Coevolution of the *glucose dehydrogenase* gene and the ejaculatory duct in the genus *Drosophila. Mol. Biol. Evol.* 2: 141–149.

Chan, Y. F., M. E. Marks, F. C. Jones, and 13 others. 2010. Adaptive evolution of pelvic reduction in sticklebacks by recurrent deletion of a *Pitx1* enhancer. *Science* 327: 302–305.

Chen, K. and N. Rajewsky. 2007. The evolution of gene regulation by transcription factors and microRNAs. *Nat. Rev. Genet.* 8: 93–103.

Colosimo, P. F., K. E. Hosemann, S. Balabhadra, and 7 others. 2005. Widespread parallel evolution in sticklebacks by repeated fixation of Ectodysplasin alleles. *Science* 307: 1928–1933.

Davidson, E. H. 1986 *Gene Activity in Early Development,* 3rd ed. Academic Press, New York.

Davidson, E. H. 1991. Spatial mechanisms of gene regulation in metazoan embryos. *Development* 113: 1–26.
Davidson, E. H. 2001. *Genomic Regulatory Systems: Development and Evolution*. Academic Press, San Diego.
del Piño, E. M. and R. P. Elinson. 1983. A novel developmental pattern for frogs: Gastrulation produces an embryonic disk. *Nature* 306: 589–591.
DeRobertis, E. M. and Y. Sasai. 1996. A common plan for dorsoventral patterning in Bilateria. *Nature* 380: 37–40.
Erwin, D. H. and E. H. Davidson. 2002. The last common bilaterian ancestor. *Development* 129: 3021–3032.
Fay, J. C., H. L. McCullough, P. D. Sniegowski, and 1 other. 2004. Population genetic variation in gene expression is associated with phenotypic variation in *Saccharomyces cerevisiae*. *Genome Biol.* 5: R26.
Garstang, W. J. 1922. The theory of recapitulation: A critical restatement of the biogenetic law. *Zool. J. Linn. Soc. Lond.* 35: 81–101.
Garstang, W. J. 1928. The morphology of the Tunicata, and its bearing of the phylogeny of the Chordata. *Quart. J. Microsc. Sci.* 87: 103–193.
Gerhart, J. and M. Kirschner. 1997. *Cells, Embryos, and Evolution: Toward a Cellular and Developmental Understanding of Phenotypic Variation and Evolutionary Adaptability*. Blackwell, Malden.
Gilad, Y., A. Oshlack, G. K. Smyth, and 2 others. 2005. Expression profiling in primates reveals a rapid evolution of human transcription factors. *Nature* 440: 242–245.
Gilbert, S. F. (ed.). 1994. *A Conceptual History of Modern Embryology*. Johns Hopkins Press, Baltimore.
Gilbert, S. F. and D. Epel. 2008. *Ecological Developmental Biology*. Sinauer Associates, Sunderland, MA.
Gompel, N., B. Prud'homme, P. J. Wittkopp, and 2 others. 2005. Chance caught on the wing: *cis*-regulatory evolution and the origin of pigment patterns in *Drosophila*. *Nature* 433: 481–487.
Gould, S. J. 1977. *Ontogeny and Phylogeny*. Belknap Press, Cambridge, MA.
Gould, S. J. 1989. *Wonderful Life: The Burgess Shale and the Nature of History*. Norton, New York.
Hall, B. K. 1999. *Evolutionary Developmental Biology*, 2nd ed. Springer, Berlin.
Haygood, R., C. C. Babbitt, O. Fédrigo, and 1 other. 2010. Strong contrasts between adaptive coding and noncoding changes during human evolution. *Proc. Natl. Acad. Sci. USA*. In review.
Haygood, R., O. Fédrigo, B. Hanson, and 2 others. 2007. Promoter regions of many neural- and nutrition-related genes have experienced positive selection during human evolution. *Nat. Genet.* 39: 1140–1144.
Jacob, F. and J. Monod. 1961. Genetic regulatory mechanisms in the synthesis of proteins. *J. Mol. Biol.* 3: 318–356.
Jeong, S., M. Rebeiz, P. Andolfatto, and 3 others. 2008. The evolution of gene regulation underlies a morphological difference between two *Drosophila* sister species. *Cell* 132: 783–793.
King, M. C. and A. C. Wilson. 1975. Evolution at two levels in humans and chimpanzees. *Science* 188: 107–116.
Kingsley, D. M. and C. L. Peichel 2007. The molecular genetics of evolutionary change in sticklebacks. In S. Östlund-Nilsson, I. Mayer, and F. A. Huntingford (eds.), *Biology of the Three-Spined Stickleback*, pp. 41–81. CRC Press, Boca Raton.

Kopp, A., I. Duncan, D. Godt, and 1 other. 2000. Genetic control and evolution of sexually dimorphic characters in *Drosophila*. *Nature* 408: 553–559.

Levinton, J. 2001. *Genetics, Paleontology, and Macroevolution*. Cambridge University Press, New York.

Lewis, E. B. 1978. Gene complex controlling segmentation in *Drosophila*. *Nature* 276: 565–570.

Lowe, C. J. and G. A. Wray. 1997. Radical alterations in the roles of homeobox genes during echinoderm evolution. *Nature* 389: 718–721.

Mabee, P. M. 1989. An empirical rejection of the ontogenetic polarity criterion. *Cladistics* 5: 409–416.

Maynard Smith, J., R. Burian, S. Kauffman, and 6 others. 1985. Developmental constraints and evolution. *Quart. Rev. Biol.* 60: 265–287.

McEdward, L. R. (ed.). 1995. *Ecology of Marine Invertebrate Larvae*. CRC Press, Boca Raton.

Nelson, G. J. 1978. Ontogeny, phylogeny, and the biogenetic law. *Syst. Zool.* 22: 87–91.

Nüsslein-Volhard, C. and E. Wieschaus. 1980. Mutations affecting segment number and polarity in *Drosophila*. *Nature* 287: 795–801.

Raff, R. A. 1987. Constraint, flexibility, and phylogenetic history in the evolution of direct development in sea urchins. *Dev. Biol.* 119: 6–19.

Raff, R. A. and T. C. Kauffman. 1983. *Embryos, Genes, and Evolution*. MacMillan, New York.

Raup, D. M. 1966. Geometric analysis of shell coiling: General problems. *J. Paleontol.* 40: 1178–1190.

Rebeiz, M., J. E. Pool, V. A. Kassner, and 2 others. 2009. Stepwise modification of a modular enhancer underlies adaptation in a *Drosophila* population. *Science* 326: 1663–1667.

Richards, R. J. 1992. *The Meaning of Evolution: The Morphological and Ideological Construction of Darwin's Theory*. University of Chicago Press, Chicago.

Shabalina, S. A. and A. S. Kondrashov. 1999. Patterns of selective constraint in *C. elegans* and *C. briggsae* genomes. *Genet. Res.* 74: 23–30.

Shapiro, M. D., M. A. Bell, and D. S. Kingsley. 2006. Parallel genetic origins of pelvic reduction in sticklebacks. *Proc. Natl. Acad. of Sci. USA* 103: 13753–13758.

Shapiro, M. D., M. E. Marks, C. L. Peichel, and 5 others. 2004. Genetic and developmental basis of evolutionary pelvic reduction in threespine sticklebacks. *Nature* 428: 717–723.

Shubin, N., C. Tabin, and S. B. Carroll. 1997. Fossils, genes, and the evolution of animal limbs. *Nature* 388: 639–648.

Skaer, N. and P. Simpson. 2000. Genetic analysis of bristle loss in hybrids between *Drosophila melanogaster* and *D. simulans* provides evidence for divergence of *cis*-regulatory sequences in the *achaete-scute* gene complex. *Dev. Biol.* 221: 148–167.

Slack, J. M., P. W. Holland, and C. F. Graham. 1993. The zootype and the phylotypic stage. *Nature* 361: 490–492.

Stern, D. L. 2000. Perspective: Evolutionary developmental biology and the problem of variation. *Evolution* 54: 1079–1091.

Tishkoff, S. A., F. A. Reed, A. Ranciaro, and 16 others. 2007. Convergent adaptation of human lactase persistence in Africa and Europe. *Nat. Genet.* 39: 31–40.

Tung, J., A. E. Primus, A. Bouley, and 3 others. 2009. Evolution of a malaria resistance gene in wild primates. *Nature* 460: 388–391.

Valentine, J. W., D. H. Erwin, and D. Jablonski. 1996. Developmental evolution of metazoan body plans: The fossil evidence. *Dev. Biol.* 173: 373–381.
Wang, R. L., A. Stec, J. Hey, and 2 others. 1999. The limits of selection during maize domestication. *Nature* 398: 236–239.
West-Eberhard, M. J. 2003. *Developmental Plasticity and Evolution*. Oxford University Press, Oxford.
Wilkins, A. S. 1993. *Genetic Analysis of Animal Development*. Wiley-Liss, Inc., New York.
Wilkins, A. S. 2002. *The Evolution of Developmental Pathways*. Sinauer Associates, Sunderland, MA.
Wray, G. A. 2003. Transcriptional regulation and the evolution of development. *Int. J. Dev. Biol.* 47: 675–684.
Wray, G. A. 2007. The evolutionary significance of *cis*-regulatory mutations. *Nat. Rev. Gen.* 8: 206–216.
Wray, G. A., M. W. Hahn, E. Abouheif, and 4 others. 2003. The evolution of transcriptional regulation in eukaryotes. *Mol. Biol. Evol.* 20: 1377–1419.
Yochem, J., K. Weston, and I. Greenwald. 1988. The *Caenorhabditis elegans lin-12* gene encodes a transmembrane protein with overall similarity to *Drosophila* Notch. *Nature* 335: 547–550.

Part IV

ADAPTATION AND SPECIATION

Chapter 10

Tradeoffs and Negative Correlations in Evolutionary Ecology

Anurag A. Agrawal, Jeffrey K. Conner, and Sergio Rasmann

Hairless dogs have imperfect teeth; long-haired and coarse-haired animals are apt to have, as is asserted, long or many horns; pigeons with feathered feet have skin between their outer toes; pigeons with short beaks have small feet, and those with long beaks large feet. Hence, if man goes on selecting, and thus augmenting, any peculiarity, he will almost certainly unconsciously modify other parts of the structure, owing to the mysterious laws of the correlation of growth (Darwin 1859: 11–12).

...as Goethe expressed it, 'in order to spend on one side, nature is forced to economise on the other side.' I think this holds true to a certain extent with our domestic productions: if nourishment flows to one part or organ in excess, it rarely flows, at least in excess, to another part; thus it is difficult to get a cow to give much milk and to fatten readily. The same varieties of the cabbage do not yield abundant and nutritious foliage and a copious supply of oil-bearing seeds. When the seeds in our fruits become atrophied, the fruit itself gains largely in size and quality. In our poultry, a large tuft of feathers on the head is generally accompanied by a diminished comb, and a large beard by diminished wattles. With species in a state of nature it can hardly be maintained that the law is of universal application; but many good observers, more especially botanists, believe in its truth. I will not, however, here give any instances, for I see hardly any way of distinguishing between the effects, on the one hand, of a part being largely developed through natural selection and another and adjoining part being reduced by this same process or by disuse, and, on the other hand, the actual withdrawal of nutriment from one part owing to the excess of growth in another and adjoining part (Darwin 1859: 147).

Why Are We Interested in Tradeoffs?

Tradeoffs have played a prominent role in evolutionary thinking for many reasons, most of which are directly tied to the factors that limit the adaptive potential of organisms. Why is it that few plants are free from herbivory and that most animals cannot tolerate polar and equatorial climates? The answer accepted by most biologists is that tradeoffs present limits to adaptation (Futuyma and Moreno 1988). As the above quotations from *The Origin of Species* show, Darwin anticipated this argument as well as two concepts that are still prominent today. He clearly understood that: (1) genetic correlations (a term he did not use) are common and can cause evolutionary responses in traits not under direct selection and (2) correlations can be caused by resource allocation tradeoffs.

One profound biological consequence of tradeoffs involves their role in biodiversity. First, there is heritable genetic variation for most traits of organisms, ranging from morphology to life history. What maintains this genetic diversity within species? A leading hypothesis is that tradeoffs in fitness-enhancing traits, including tradeoffs across environments, maintain genetic diversity. Second, species diversity is likely maintained by tradeoffs in species' traits (Clark et al. 2007; see Losos and Mahler, Chapter 15). Analogous to the process operating within species, tradeoffs across environments limit the niche breadth and geographical range of a species, thus generating and maintaining species diversity.

Two kinds of tradeoffs are distinguished, based on whether they involve one or multiple traits (Box 10.1). A one-trait tradeoff occurs when there is opposing selection on a single trait by different selective agents (including different environments) or through different components of fitness (Figure 10.1A). Most examples of stabilizing selection are due to one-trait tradeoffs. In humans, higher birth weight increases postpartum infant survival, while babies that are too large are more likely to die during childbirth (Karn and Penrose 1951). The ovipositors of parasitoid wasps cannot reach flies in larger galls produced by larvae of the goldenrod gall fly, but these larger galls are found and attacked more frequently by birds (Weis and Gorman 1990). Mauricio and Rausher (1997) demonstrated stabilizing selection on glucosinolate concentration in *Arabidopsis thaliana*, resulting from a balance between selection for increased defense in the presence of herbivores and the cost of glucosinolate production (see Berenbaum and Schuler, Chapter 11). Additionally, polymorphisms are often maintained by opposing selection in two

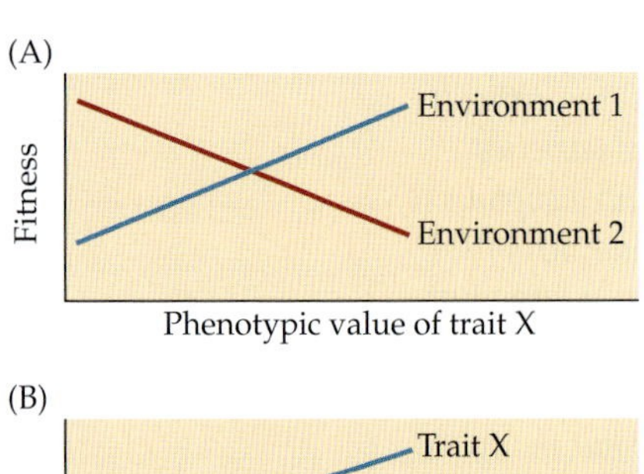

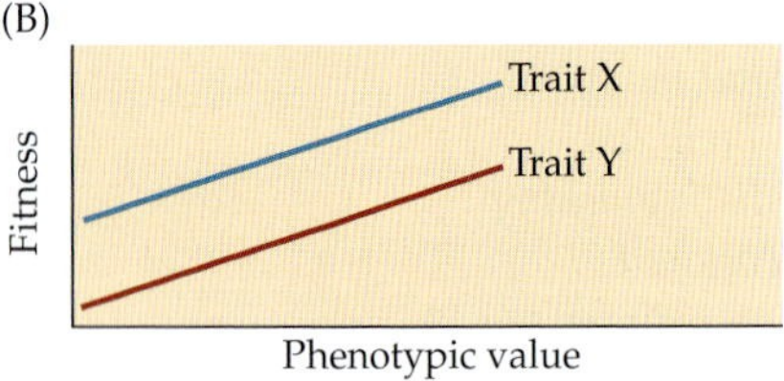

(C)
Trait X
Trait Y
Limiting resource

FIGURE 10.1 Depictions of Natural Selection Scenarios that Result in an Evolutionary Tradeoff (A) A one-trait tradeoff caused by opposing selection resulting from different environments, selective agents, and/or fitness components. (B,C) A two-trait tradeoff caused by consistent directional selection on two traits that share a limiting resource.

BOX 10.1

DEFINITIONS OF TERMS AND CONCEPTS RELATING TO TRADEOFFS AND ADAPTIVE CORRELATIONS

Definitions of key terms used in this chapter are provided in this box. Because some of these terms have a wide range of usage in the literature, definitions supplied are as specific as possible to reduce confusion.

Adaptive negative correlation: A negative correlation (within or among species) that is generated by fitness benefits of not expressing two traits simultaneously. For example, traits that are each costly but functionally redundant may show adaptive negative correlations. The signature of an adaptive negative correlation is negative correlational selection on the two traits that interact to determine fitness.

Correlated evolution: Repeated (parallel or convergent) evolution of an association between two traits across species; it is typically tested by a phylogenetically independent contrast or the generalized least squares equivalent method (Pagel 1999).

Genetic correlation: A measure of the degree to which two traits are affected by the same locus or loci as a result of pleiotropy or linkage disequilibrium. Selection on one trait produces an evolutionary change in all traits that have an additive genetic correlation with the selected trait (Conner and Hartl 2004).

Tradeoff: Any case in which fitness cannot be maximized because of competing demands on the organism, which can take the form of opposing selection on one trait (i.e., one-trait tradeoff) or of selection to increase two or more traits that share a limiting resource (i.e., multiple-trait tradeoff). A one-trait tradeoff may or may not arise due to the allocation of a limiting resource.

Trait hierarchy: Phenotypic traits can be defined at a number of hierarchical levels—each level dependent on a number of traits at lower levels. For example, the form of an enzyme encoded by a gene is a phenotype, as is a physiological function like metabolic rate that depends on a number of enzymes. A number of different physiological functions affect morphological traits like height, and physiology and morphology together can affect behavioral phenotypes such as courtship. Finally, all these lower level traits can affect high-level life history traits like survival and reproduction, which determine the ultimate trait of individual fitness (Conner and Hartl 2004).

environments, indicative of one-trait tradeoffs (Futuyma 1997). Although studies employing reciprocal transplants that find local adaptation imply a tradeoff, the traits under selection are often not identified and so the nature of the tradeoff is unknown.

A multiple-trait tradeoff occurs when two or more traits (including fitness components), which are both under directional selection to increase, share a limiting resource (Figure 10.1B,C). In other words, a tradeoff occurs when multiple traits that compete for resources are favored by natural selection. Examples of multiple-trait tradeoffs include flower size versus number (Worley and Barrett 2000), offspring size versus number (Messina and Fox 2001), and levels of different defensive chemicals that share a common precursor (Berenbaum et al. 1986). One-trait and multiple-trait tradeoffs are fundamentally different, because in a one-trait tradeoff there is opposing selection, while in a multiple-trait tradeoff selection acts to increase allocation to all traits. Both types of tradeoffs are also sometimes referred to as costs (Futuyma 1997).

There are many uses of the term tradeoff in the literature. In this chapter, attention is focused on tradeoffs that take one of three forms: (1) costs of functional traits, such as defense against enemies; (2) life-history tradeoffs within environments; and (3) adaptation to alternative environments (including the evolution of specialization, niche breadth, and range limits.

Life history tradeoffs typically involve multiple traits, while costs of functional traits and adaptation to alternative environments can occur through either one or multiple traits. Much of the work on adaptation to alternative environments does not examine specific traits, but rather it measures fitness or a high-level trait (e.g., damage by herbivores) in each environment (see "trait hierarchy" in Box 10.1). For example, Fry (1996) has defined a tradeoff as any case in which traits that increase fitness on one host are detrimental on others (i.e., phenotypic focus) or any case for which there is no (homozygous) genotype with maximal fitness in both environments (i.e., genotypic focus) (Fry 2003). These fitness tradeoffs could be caused by opposing selection on a single trait in the two environments or by two traits that share a limiting resource and are both selected in the same direction in the two environments. An example of a one-trait tradeoff across environments is coat color in mice: darker colors are favored in woodlands and lighter colors in open beach habitats (Hoekstra et al. 2004).

An example of a multiple-trait tradeoff in adaptation to alternative environments comes from work by Charles Fox and colleagues on seed beetle (*Stator limbatus*) life histories. Increased egg size and number are presumably both favored in all environments but are negatively genetically correlated (Fox et al. 1997). Larger egg size is under stronger positive selection on more resistant host plants (*Cercidium floridum*) compared to less resistant plants (*C. microphyllum* or *Acacia greggii*) (Fox and Mousseau 1996). Thus, adaptation to alternative host plants is mediated by a genetic tradeoff between egg size and number (Fox et al. 1997). Ultimately, beetles collected from *C. floridum* populations lay fewer and larger eggs than beetles collected from susceptible plant populations (Fox and Mousseau 1998). Remarkably, in addition to the genetic differentiation between populations of beetles, an adaptive maternal effect has evolved such that individual beetles alter egg size (and number) appropriately on hosts of varying resistance (Fox et al. 1997).

Many hypotheses in evolutionary ecology assume that natural selection influences organisms' allocation of limiting resources to fitness components and fitness-enhancing traits (Figure 10.2). It may very well be that life history traits, which typically use a large fraction of the total resources available, must trade off. Nonetheless, much work on tradeoffs examines the relationship between low-level traits, requiring few resources, and high-level traits, contributing to fitness (see Figure 10.2). Demonstrating that tradeoffs are a source of constraint is exceedingly challenging, for a number of reasons that are detailed in the following sections.

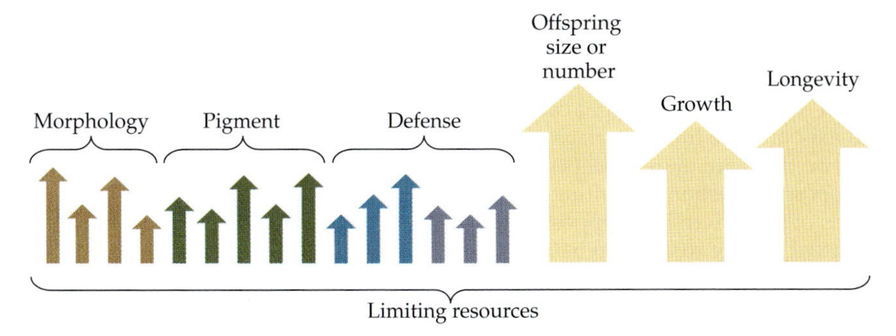

FIGURE 10.2 A Model of Resource Allocation Arrows represent different traits, with their width signifying the amount of the organism's resource budget used and height indicating relative link to fitness. Larger arrows are typically higher-level traits comprised of many lower-level traits (see Box 10.1). Colors represent classes or types of traits (e.g., yellow indicates life history traits, while blues represent defensive traits).

Interpreting Correlations

Although it is intuitive that tradeoffs must exist (allocation of limiting resources, fitness cannot be infinite), it has been surprisingly difficult to obtain strong evidence for tradeoffs within species, where they are actually operating (van Noordwijk and de Jong 1986; Fry 2003). We have compelling evidence for life history tradeoffs (Rose and Charlesworth 1981; Schluter et al. 1991; Stearns 1992; Messina and Fox 2001; Roff 2002) but perhaps less for adaptations to different environments or for specific functional traits such as morphology. We emphasize that strong evidence for a negative correlation between traits, even two traits that are both under positive directional selection, does not necessarily indicate a tradeoff, because negative correlations could also be caused by developmental or physiological linkages that are not due to adaptation or shared limiting resources or by adaptation, specifically natural section favoring a negative correlation (i.e., adaptive negative correlation, see following).

A very popular evolutionary ecological shorthand has been to study the correlations (phenotypic, genetic, or species correlations) between two traits presumed both to be positively associated with fitness. Such correlations are interpreted in three main ways: adaptive, constraining, or as evidence of a tradeoff. In many cases, these interpretations have little or no support beyond the presence of the correlation itself. This finding is particularly true for positive correlations with size-related traits, because genes that increase size through any number of actions (e.g., resource accrual, rates of cell growth and division, physiological processes) will pleiotropically affect the dimensions and numbers of myriad traits in an organism. Given recent examples of rapid, independent evolution of pairs of traits that are positively genetically correlated, such as Beldade et al. (2002) and others to be discussed in this chapter, one should not assume that a correlation

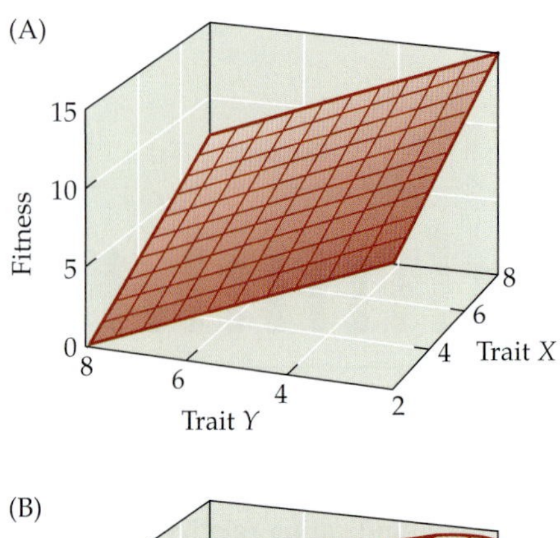

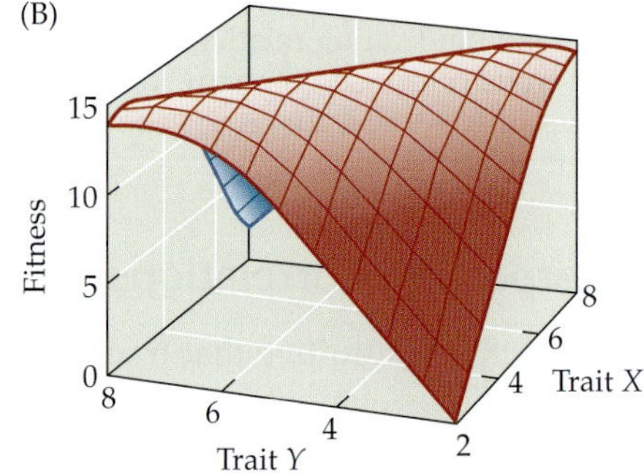

FIGURE 10.3 Hypothetical Depictions of Directional Selection and Negative Correlational Selection on a Pair of Traits In both cases there are two traits *X* and *Y*, with the smallest values of each in the lower right corner. Fitness is on the vertical axis, and the fitness surface is depicted in red. (A) The two traits are under independent directional selection in opposite directions, resulting in the evolution of the means of the two traits in opposite directions (i.e., not a negative genetic correlation between the traits). The lines on the fitness surface parallel to each trait axis are all parallel to each other, showing that there is no interaction between the traits in determining fitness and, thus, no correlational selection. The elevation and intercept are altered by changes in the other trait, but the selection gradient (slope) that measures the strength of selection does not. (B) Negative correlational selection, in which the two traits interact to determine fitness. The fitness surface is a ridge, with equally high fitness stemming from larger values of one trait associated with smaller values of the other. Valleys of low fitness occur when both traits have low (front, right) or high (back, left) values. Note that the lines on the surface parallel to each trait axis change as the value of the other trait changes, depicting a true interaction between the traits in determining fitness.

causes a meaningful constraint (although this has not been adequately tested with negative correlations). The presence of a negative genetic correlation between two traits that both are reasonably assumed to be under positive directional selection (e.g., fitness components) can be reasonably interpreted as a tradeoff, but this interpretation is strengthened by actual evidence for selection and evidence of a shared limiting resource.

The type of selection that will lead to adaptive correlations is called correlational selection, which is fundamentally distinct from directional selection on the individual traits (Figure 10.3). In correlational selection, there is a ridge in the fitness surface, so that several different combinations of the two traits lead to high fitness (along the ridge), while at least two combinations lead to low fitness (see Figure 10.3B). In directional selection, only one combination leads to the highest fitness and one combination leads to the lowest fitness (e.g., low values of one trait and high values of the other and vice versa; see Figure 10.3A). The key point is that with correlational selection there is no consistent pattern of selection on each trait individually; selection on the one trait depends on the value of the other trait and vice versa. Correlational selection is estimated by a significant cross-product (interaction) term between two traits in a Lande–Arnold (1983) selection gradient analysis (i.e., multiple regression) (see Figures 10.3 and 10.4).

As a potential example of an adaptive negative correlation, van der Meijden et al. (1988) argued that resistance to and tolerance of herbivory were alternative strategies. Plant species that are able to resist herbivory (i.e., are not attacked) do not experience strong selection to tolerate herbivory (i.e., have little fitness impact from damage); and vice versa, species that are tolerant are not expected to experience strong selection for resistance. This pattern has been borne out for genotypes within plant species (Fineblum and Rausher 1995; Stowe 1998; Pilson 2000; Fornoni et al. 2003) as well as animal genotypes (Raberg et al. 2007), although it is certainly not universal (Núñez-Farfán et al. 2007). Resistance and tolerance to other forms of stress (e.g., frost, herbicide application) have similarly shown evidence of selection for a negative correlation between the two as a result of alternative fitness peaks favoring high resistance and low tolerance or low resistance and high tolerance (Agrawal et al. 2004; Baucom and Mauricio 2008) (Figure 10.4). This effect is probably because the traits are both redundant and costly, resulting in selection for a negative correlation between the two.

It is currently unclear whether or not adaptive correlations and correlational selection are common. Alternative organismal strategies may result

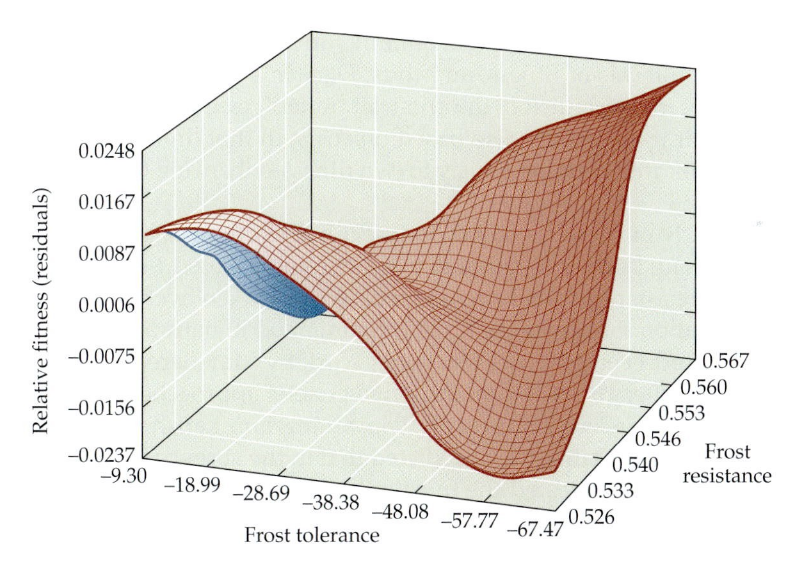

FIGURE 10.4 Fitness Surface Depicting Negative Correlational Selection on Resistance and Tolerance to Frost Damage in Wild Radish (*Raphanus raphanistrum*) Data were taken from a field experiment on 75 paternal half-sibling families subjected to a natural, catastrophic hard frost late in the spring. The two fitness peaks occur at high frost tolerance and low resistance (near left corner) and high resistance and low tolerance (far right); note that lowest tolerance is on the right. Intermediate values of both traits led to intermediate levels of fitness (the "saddle" in the center of the figure). (Adapted from Agrawal et al. 2004.)

from adaptive correlations (i.e., costly redundant traits) or from fundamental tradeoffs (see previous discussion of Fox's beetles). Distinguishing adaptive negative genetic correlations from tradeoffs is straightforward if data on selection are available, but this is not the case for most examples in the literature.

Detecting Tradeoffs: Great Successes and Hurdles

Perhaps the modern version of tradeoffs in life history theory stems from Lack's (1947) study of clutch size in altricial birds, in which he hypothesized that stabilizing selection results from decreased survival of offspring in larger clutches under resource-limiting conditions (a one-trait tradeoff; see Figure 10.1). This work laid the foundation for many studies, for example by Ricklefs (1977) that employed a comparative analysis to examine costs of reproduction associated with clutch size. Major reviews on the topic include Stearns (1992) and Roff (2002). Over a century before Lack, in *The Origin of Species,* Darwin (1859: 148) summarized the view that most adaptations are costly: "If under changed conditions of life a structure before useful becomes less useful, any diminution, however slight, in its development, will be seized on by natural selection, for it will profit the individual not to have its nutriment wasted in building up an useless structure." Costs are defined in terms of fitness, and thus, Darwin's conceptualization can be thought of in the context of the one-trait tradeoff (see Figure 10.1A). The multiple-trait tradeoffs are also based on costs, in that limiting resources cannot be allocated simultaneously to two fitness-enhancing traits (see Figure 10.1C).

In the study of plant resistance to insect herbivores, costs of resistance traits have long been invoked to explain two interrelated issues: (1) why completely resistant plants have not evolved and taken over, and (2) why genetic variation for resistance traits is maintained in natural populations (Whittaker and Feeny 1971). Pioneering work by May Berenbaum, Ellen Simms, and colleagues used quantitative genetic approaches to estimate costs of resistance (Berenbaum et al. 1986; Simms and Rausher 1987, 1989). Using monocarpic plants, they predicted that in the absence of herbivores, the genotypes that invested the most in resistance traits would have the lowest lifetime reproduction. In at least a few cases, opposing natural selection on the same trait was demonstrated by growing plants in environments with herbivores versus without herbivores (Berenbaum et al. 1986; Mauricio and Rausher 1997; Shonle and Bergelson 2000). However, many other studies failed to find costs (Simms 1992; Bergelson and Purrington 1996; Strauss et al. 2002). The emerging paradigm is that costs are often dependent on the environment in two ways, which have been referred to as ecological costs (Strauss et al. 2002). First, costs are likely to be most detectable under competitive or otherwise stressful conditions in which resources become especially limiting. Many experiments designed to detect costs are conducted under benign conditions and thus, may underestimate costs. Second, even if a trait is not a significant energetic drain on the organism, it may reduce fitness-enhancing interactions. For example, highly defended plants may have reduced visitation by pollinators (Strauss et al. 1999).

Another success story in the detection of costs has been the tradeoff between investment in flight muscles and fecundity in insects (Zera and Harshman 2001). Many insect species are naturally polymorphic, with macropterous (normal winged) and micropterous (reduced wings and/or flight muscles) forms—a polymorphism that is thought to have evolved because flight muscles are costly and the benefit of flight is low in some environments (e.g., high, local resource availability and low predation). Studies have consistently shown phenotypic and genetic negative correlations between flight structures and fecundity, indicative of a tradeoff, because these traits are not redundant.

Evolution of Insect Host Range

One of the great challenges in the study of plant–animal interactions has been the question of host specialization in herbivorous insects, an example of the broader problem of ecological specialization (Futuyma and Moreno 1988; Fry 2003). Well over 50% of herbivorous insects feed on plants in a single genus, a highly restricted subset of the available host species (Schoonhoven et al. 2005). Tradeoffs have been the long-held explanation for host restriction, on the supposition that adaptations to the defenses of one host species detract from the ability to cope with defenses of other hosts. A jack-of-all-trades is master of none.

Somewhat counterintuitively, researchers have often used generalist herbivores to test for tradeoffs in host use—possibly because it is difficult to force specialists to feed on non-hosts. Mackenzie (1996) reported that among 77 clones of the black bean aphid (*Aphis fabae*), a tradeoff in fecundity was found on one out of three pairwise combinations of the three hosts. Three independent selection experiments with two-spotted spider mites (*Tetranychus urticae*) found that lines adapted to novel and somewhat toxic hosts showed a tradeoff in fitness on the original host (Gould 1979; Fry 1996; Agrawal 2000), although this was detectable only by the loss of adaptation to the novel host when the adapted mites were reverted onto the original host for several generations. In a fourth study of these mites, Yano et al. (2001), like previous authors, found no reduction of fecundity on the original host but reported that adaptation to the novel host was associated with reduced male ability to compete for mates.

Specialization and tradeoffs have also been studied in oligophages (e.g., herbivores restricted to one plant family). Evidence comes from the oligophagous pea aphid (*Acyrthosiphon pisum*), for which Via and Hawthorne

(2002) have identified the genetic basis of a tradeoff in performance on two legume hosts. Nonetheless, many other systems have failed to find evidence for tradeoffs (James et al. 1988; Forister et al. 2007; Futuyma 2008). For example, Ueno et al. (1995) used specialist *Epilachna spp.* beetles to test hypotheses about the evolution of host shifts and specialization. They employed the approach of assessing genetic variation for performance on the typical host plant and the host plants of close beetle relatives. Where heritable variation was found, there was essentially no evidence for tradeoffs in the use of alternative hosts.

Difficulties in Detecting Multiple-Trait Tradeoffs within Species

Because a multiple-trait tradeoff is likely to be reflected in a negative correlation between traits at some level in the trait hierarchy, much effort, mainly in the 1980s, was directed towards discovering negative genetic correlations, especially in the study of host use by herbivorous insects (Futuyma and Moreno 1988). Many of these studies failed to find negative correlations (Rausher 1984; Futuyma and Philippi 1987; James et al. 1988; Karowe 1990; Fox 1993; Forister et al. 2007; Ferrari et al. 2008; Agosta and Klemens 2009). However, a negative correlation between two traits might not be found even when there is some underlying, fundamental tradeoff between these traits.

The basic idea is that stronger positive relationships can mask underlying tradeoffs. This point was first made for phenotypic relationships by van Noordwijk and de Jong (1986), who showed graphically how variation in resource acquisition could create a positive correlation between two traits, even when there is a tradeoff in the allocation of that resource between the traits (Figure 10.5). Houle (1991) extended this general concept to genetic correlations, showing theoretically that if there were more loci involved in resource acquisition than in allocation of that resource, the genetic correlation could be positive in spite of a fundamental allocation tradeoff.

If negative phenotypic or genetic correlations among traits that are both positively related to fitness are found, it is excellent evidence for a multiple-trait tradeoff. In many cases in the literature, the traits are fitness components such as survival or fecundity, so the relationship to fitness is clear, but in other cases the positive relationship of the traits to fitness should be demonstrated empirically, ideally using Lande–Arnold selection gradients (Lande and Arnold 1983) and/or experimental manipulation (Conner and Hartl 2004). What about the cases in which positive phenotypic correlations are found between traits that are expected to trade off, based on knowledge of their biology? The first step is to estimate the genetic correlations, as these are not affected by the environmental correlations (van Noordwijk and de Jong 1986). This approach could be implemented using sibling analysis, but a more reliable method of testing for genetic correlation is

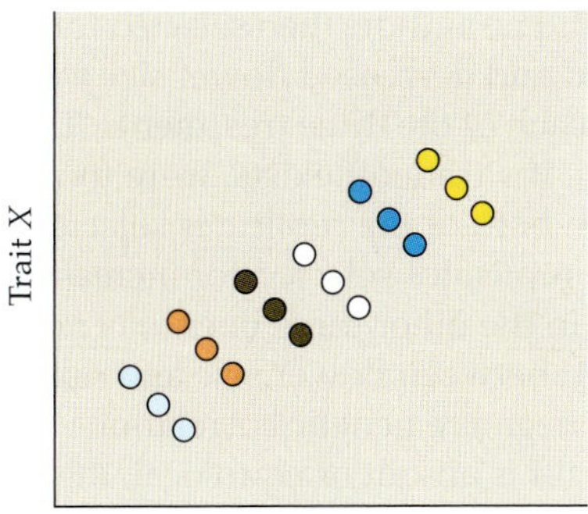

FIGURE 10.5 How a Tradeoff in the Allocation of Resources to Two Traits Can Be Masked across Scales Within each color, a tradeoff is represented, which could be within a genotype or across genotypes within a species. Different colors represent alternative scales (e.g., resource environments, genotypes with altered resource acquisition, or different species). Three scenarios discussed in this chapter could fit the pattern. First, van Nordwijk and de Jong (1986) proposed that a strong positive environmental correlation caused by variation in resource acquisition could create a positive phenotypic correlation between two traits, even when there is a tradeoff between the two traits. For example, in resource rich environments, egg size and number may increase, even though the two traits show a negative correlation within a resource environment. Houle (1991) extended this concept to genetic correlations, showing that if there is more genetic variation in resource acquisition than allocation, the genetic correlation could also be positive in spite of a tradeoff. We add to these scales by suggesting that positive species correlation may occur if species vary in their total acquisition despite genetic tradeoffs that occur within species (see Figure 10.8).

artificial selection (Bell and Koufopanou 1986; Conner 2003; Fry 2003). We suggest imposing artificial selection on one of the traits and testing for an evolutionary response in the opposite direction in the other trait. Ideally, this evaluation would be conducted in all four treatment combinations (i.e., selection for increased and decreased values of each trait individually) with replication. However, if there were more genetic variation for resource acquisition than for allocation, then the correlated responses would still be expected to be positive. In this case, one could attempt to control resource acquisition, if this can be measured.

A study of the expected tradeoff between flower size and number in water hyacinth (*Eichhornia paniculata*) serves as a good illustration for both this approach and the often equivocal case of tradeoffs (Worley and Barrett 2000). Using a maximum-likelihood pedigree analysis on the base population, the genetic correlation between flower size and number was estimated as weakly but significantly *positive* (0.18) and was nonsignificantly negative after leaf area and flowering time were included in the model, in an attempt to reduce variance in resource acquisition; thus, there was no initial evidence for a tradeoff. The authors then selected for increased flower number and both increased and decreased flower size for two generations, with two replicate lines of each of the three treatments. The predicted correlated responses to selection, if a tradeoff exists, were found in half of the six selection lines; that is, in both of the replicates selected for decreased flower size and in one of the replicates selected for increased flower number. The response to selection in the other three lines was not significant. Thus, the evidence for a tradeoff between flower size and number is equivocal.

If the evidence for negative genetic correlations between two traits that are expected to trade off is absent or equivocal, the next step in testing for tradeoffs is to dissect the traits more finely, both phenotypically and genetically. One possibility is to examine traits at a lower level in the trait hierarchy (see Box 10.1), because high-level traits, such as fitness components, plant damage by herbivores, or insect performance on a given host plant, are affected by more gene loci, thus increasing the probability that resource acquisition loci mask the postulated allocation tradeoff. For example, the genetic correlations among the physiological and/or morphological traits that determine fitness components, herbivore resistance, or host use could be examined with artificial selection. Selection experiments on lower-level traits that provide herbivore resistance have often successfully demonstrated tradeoffs (Ågren and Schemske 1993; Zangerl and Berenbaum 1997; Siemens and Mitchell-Olds 1998; Stowe 1998; Marak et al. 2003). For a more complete understanding of any tradeoff, it is necessary to uncover the gene loci underlying the complex traits. From there, it is possible to determine whether there are negatively pleiotropic resource allocation loci, and if so, how they function and how large their effects are on higher-level phenotypic traits, relative to acquisition loci. Although these kinds of studies are still very difficult even in genetic model organisms, they are becoming more feasible with rapid advances in sequencing and other molecular genetic technologies. One example of this general approach is a quantitative trait loci (QTL) analysis of fitness components in a cross between two ecotypes of wild barley (Verhoeven et al. 2004); the investigators found QTL for fitness traits in each environment, but no case in which different alleles at these QTL were favored in the two environments. In other words, there was a lack of evidence for tradeoffs at the level of small sections of chromosomes. Even here, it may be that tradeoffs might be found at the level of individual genes, or for lower-level traits, as all the traits examined in this study were high-level components of fitness.

Evidence that Genetic Correlations Cause Evolutionary Constraint

Even if a negative genetic correlation between two fitness-enhancing traits exists, it is unclear how strong an evolutionary constraint it would cause. For a pair of traits, only a genetic correlation coefficient of –1 would prevent

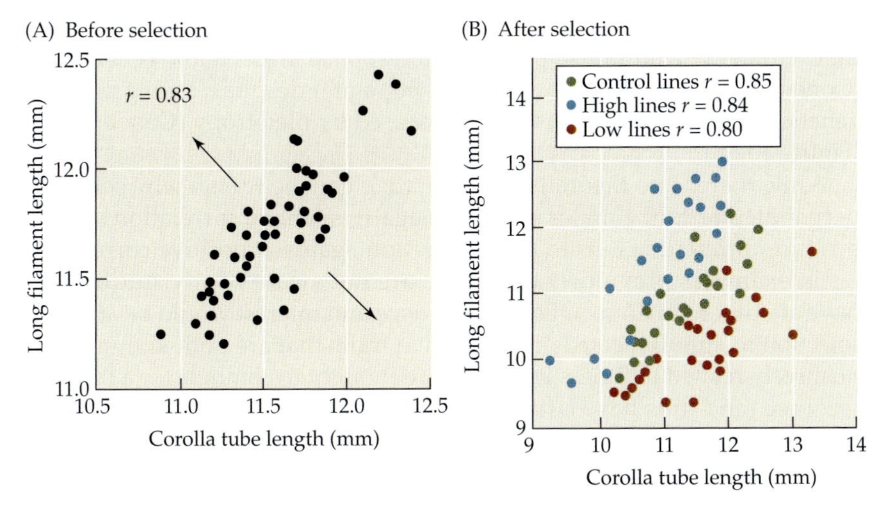

FIGURE 10.6 Artificial Selection Perpendicular to the Major Axis of the Correlation Between Filament and Corolla Tube Lengths in Wild Radish (*Raphanus raphanistrum*) (A) The genetic correlation in the original population. The arrows show the direction of selection in the high and low anther exsertion lines; note that these are in the direction of least variation in bivariate space. Each point is the mean of all offspring of one sire from a nested half-sibling design (Conner and Via 1993). The correlation of these sire family means is an estimate of the additive genetic correlation. (B) Results after five or six generations of selection (Conner et al., unpublished). Selection has moved the elliptical cloud of points in the directions of the arrows in (A) without changing the shape of the ellipse, that is, the correlation within each group. Each point is the mean for a full-sibling family; the resulting correlations are broad-sense genetic correlations that include covariance due to dominance and maternal effects. The difference in estimation methods is responsible for the greater range of values in panel (B) than in panel (A). (Adapted from Conner 2003.)

evolution of larger values of both traits (Via and Lande 1985; Houle 1991). Genetic correlations greater than –1 would slow down, but not prevent, a response to selection, because there is at least some independent genetic variation for both traits (Figure 10.6A). This is a crucial point—if there is any genetic variation in the direction of selection, evolution *can* occur (although it may or may not occur over a timescale concordant with environmental change). There is also convincing evidence that positive genetic correlations may not even slow down response to strong selection, because when artificial selection is applied perpendicular to the major axis of the correlation, rapid evolution occurs. For example, artificial selection in *Bicyclus anynana* butterflies has produced independent evolution of a variety of positively correlated traits, including forewing versus hindwing-spot size (Beldade et al. 2002) and forewing area and body size (Frankino et al. 2005), although selection on wing-spot color composition failed to produce independent evolution (Allen et al. 2008). Artificial selection on filament and corolla tube lengths in wild radish produced a response in just a few generations (Conner 2003; Conner et al., unpublished) again despite a strong positive genetic correlation between the traits caused by pleiotropy (Conner 2002) (Figure 10.6B).

While there is no theoretical quantitative genetic reason why selection perpendicular to the major axis of a negative genetic correlation should produce different results than the selection against a positive correlation (as in examples previously cited), we are not aware of any studies that have attempted to do so. Of most relevance and interest would be artificial selection to simultaneously increase two traits that are both known to be positively related to fitness in nature, are thought to compete for a limiting resource (and thus have a fundamental tradeoff between them), and are known to be negatively genetically correlated; to our knowledge, a study of this description has not been attempted. Life history traits are prime candidates for such experiments, and their close ties to fitness may make correlations between life history traits likely constraints.

The lack of constraint seen in artificial selection studies on pairs of traits that are positively correlated might also be explained if constraints are not pairwise, but fundamentally multivariate (Blows and Hoffmann 2005; Walsh and Blows 2009). The argument here is that while there is some genetic variation perpendicular to the major axis in most or all pairs of traits (i.e., the genetic correlation is not 1 or –1), there may be dimensions in multivariate space where genetic variation is completely lacking. Indeed, this may be the dimension in which there is directional or stabilizing selection that has depleted the available variation. While there is some laboratory evidence for selection on multivariate axes (Brooks et al. 2005; Van Homrigh et al. 2007), the examples to date are for complex traits (e.g., cuticular hydrocarbons in *Drosophila* spp., components of a cricket call) for which the direction of selection would be difficult to predict. In the cuticular hydrocarbon example, genetic variance is lacking along the multivariate axis upon which sexual selection is exerted by females (Van Homrigh et al. 2007), and artificial selection on male mating success for 10 generations failed to produce a response (McGuigan et al. 2008), as would be predicted if the hydrocarbons are the major determinant of male mating success. To our knowledge, artificial selection has not been applied directly to a multivariate axis.

In an analogous fashion, it is possible that tradeoffs are often not pairwise, but involve multiple traits simultaneously. Many key limiting resources (carbon, nitrogen, phosphorus, amino acids, water) are allocated among many traits simultaneously or sequentially in an organism. Thus, two traits that are relatively minor sinks for a resource might not tradeoff with each other, but might together tradeoff with a more major resource sink (see Figure 10.2). This is an area for future study.

Macroevolutionary Approaches to Studying Tradeoffs and Correlations

As defined here, the terms tradeoffs and adaptive correlations are microevolutionary phenomena (see Box 10.1). Nonetheless, tradeoffs and correlations within ancestral species may influence the patterns of divergence across closely related descendant species. Whereas an intraspecific pattern may reflect a tradeoff or adaptive correlation, interspecific patterns may represent different adaptive solutions to a tradeoff or be biased by genetic correlation. A pattern of tradeoffs may be more evident across species than within species, because there has been substantial time for selection to create larger relative differences in trait means—an extension of the fact that artificial selection may be a more powerful way to detect negative correlations than sibling analysis. An example may be *r*- versus *K*-life–history strategies in closely related species, for which there are many examples among plants and insects (Gadgil and Solbrig 1972; Stearns 1992; Roff 2002; Mooney et al. 2008).

One way that microevolutionary processes might determine macroevolutionary patterns across species is what Schluter called evolution along genetic lines of least resistance (Schluter 1996), whereby genetic correlations within ancestral species bias the phenotypic divergence of correlated traits among descendent species, so that species occupy a restricted area of bivariate space. A number of studies have now tested whether plant and animal species tend to diverge mainly along the trajectory predicted by genetic or phenotypic correlations within one of these species, and most studies find that this is the case, but that a few species do diverge substantially from this predicted trajectory (Schluter 1996; Hansen and Houle 2008; Marroig and Cheverud 2005; Hunt 2007; Conner 2006). For example, Schluter's original study (1996) showed that macroevolutionary patterns of morphological diversification in stickleback fish, birds, and mice conformed to patterns of within-species genetic correlations. Therefore, genetic correlations may bias the direction of macroevolutionary divergence, but this bias can be broken, presumably when selection or drift is strong enough.

A hypothesis in plant defense evolution has been that shared precursors limit the production of diverse types of beneficial defenses (i.e., an allocation tradeoff) (Berenbaum et al. 1986; Gershenzon and Croteau 1992; Agrawal et al. 2002). Alternatively, changes in the level of a common precursor may cause similar effects in the expression of multiple products by simply changing the overall flux through the pathway, causing the levels of the products to be positively correlated (Martens and Mithofer 2005), which is another example of variance in acquisition being greater than variance in allocation. Most research on this topic has been on model species in which the flow of specific compounds can be followed, or competition for a particular enzymatic precursor can be identified (Keinanen et al. 1999; Kao et al. 2002; Laskar et al. 2006; Scalliet et al. 2006). However, an approach that considers patterns across species addresses a different question about the long-term persistence and convergent evolution of particular compounds or the associations between branches of biosynthetic pathways (Liscombe et al. 2005; Pelser et al. 2005; Agrawal 2007). In other words, are physiological tradeoffs more or less evident as species diversify? Do associations between biosynthetic pathways persist over evolutionary time, or are they eroded by natural selection or nonadaptive processes that reduce their interdependence?

In a study of phenolics and cardenolides (two classes of plant defenses) across 35 species of milkweed, strong evidence was found for integration among phenolic classes and among flavonoids (a class of phenolics) and cardenolides (Agrawal et al. 2009b). Within the phenolics, caffeic acid derivatives and flavonoids share *p*-coumaric acid as a precursor, and there appears to be evolutionary competition for this precursor (i.e., species have evolved to different points along the tradeoff). In contrast, cardenolides and flavonoids, which are both constructed with products from the acetate–malonate pathway (Andersen et al. 2006), show positively correlated interspecific expression. This latter result suggests that milkweed species have evolved changes in the flux through the acetate–malonate pathway, resulting in concordant shifts in flavonoids and cardenolides. We do not interpret these results to mean there is no physiological competition for precursors, but rather that species are evolving to different levels of acquisition of the precursor (Houle 1991).

Another long-standing hypothesis in plant–herbivore interactions is that plants that possess one type of highly effective defense should lack others (i.e., an adaptive negative correlation based on costs and redundancy). Within a species, this outcome may result in an adaptive negative correlation as previously discussed. Nonetheless, intraspecific analyses rarely find significant genetic correlations between defensive traits (Koricheva et al. 2004). A comparative approach has been taken to examine the macroevolutionary correlations between putatively defensive traits, although very few have employed phylogenetically informed methods. Most of these studies do not support a negative correlation model (Steward and Keeler 1988; Twigg and Socha 1996; Heil et al. 2002; Rudgers et al. 2004; Agrawal and Fishbein 2006). In an example from our own research, two defensive traits of milkweeds show no genetic correlation within one species but are positively correlated across species (Figure 10.7). Thus, one can only conclude that the redundancy model of negative correlations between resistance traits is too simple, especially in the context of variable environments and multiple herbivores.

(A)

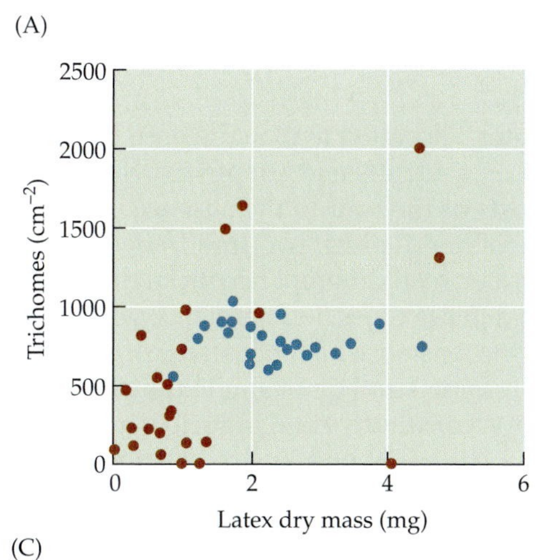

(B)

(C)

FIGURE 10.7 Micro- to Macroevolution of Plant Defense Traits (A) Latex and trichomes of milkweed show no genetic correlation within species (*Asclepias syriaca*, 23 full-sibling families in a common garden, blue circles: $P = .75$), but show a positive correlation across 24 species of *Asclepias* (red circles: $P < .01$ in both phylogenetically corrected and uncorrected analyses. Both data sets are from field experiments at the same site, during the same year, and with at least five replicates per genetic family or species. Armbruster et al. (2004) have argued that such correlations across but not within species are suggestive of adaptation. Despite the lack of a correlation across genetic families, variation in latex and trichomes was negatively correlated with herbivore damage for *A. syriaca* (Agrawal 2005). These two traits might have a synergistic impact on reducing herbivory. For example, when eggs of the monarch butterfly *Danaus plexippus* hatch, young caterpillars often graze a bed of trichomes before puncturing the leaf surface (B). Once the caterpillars puncture the leaves, they encounter pressurized latex (C). A substantial fraction of monarch caterpillars die in latex (see arrow). (A, data from Agrawal 2005 and Agrawal and Fishbein 2006; adapted from Agrawal 2007; photos by Anurag Agrawal.)

Case Study of Adaptive Allocation to Plant Defense across Scales

As discussed throughout this chapter, allocation to plant defense has been one of the major areas of research on tradeoffs and adaptive correlations. One-trait tradeoffs (i.e., fitness costs of the trait in the absence of herbivores, but benefits in the presence of herbivores), two-trait tradeoffs (i.e., competition for a shared precursor between different beneficial defenses), and adaptive correlations between defense types (i.e., resistance versus tolerance) have all been proposed. One axis on which negative correlations have been predicted is in the expression strategy of single traits. For example, based in plant defense theory, constitutive and inducible resistance have long been predicted to show a pattern of negative correlation within and across species (Rhoades 1979; Brody and Karban 1992; Koricheva et al. 2004). The basis for this hypothesis is that plants investing in high levels of constitutive defense (i.e., traits that are always expressed) will experience minimal attack and need not be inducible following herbivory (Zangerl and Bazzaz 1992). Conversely, cases for which the probability of attack is unpredictable, plants may be expected to invest relatively little in constitutive defense but to show high levels of inducibility following attack.

The predicted negative correlation between constitutive and induced resistance could be adaptive if the traits are each beneficial, redundant, and costly. Despite some support for this negative genetic correlation (Koricheva et al. 2004), several statistical issues have plagued the accurate assessment of the relationship, which is typically estimated as a species- or family-mean correlation between investment in constitutive traits and inducibility and defined as the absolute increase in the same traits after herbivore damage (Morris et al. 2006).

We have been studying the relationship between constitutive and induced cardenolide production in milkweeds (*Asclepias* spp.) (Rasmann et al. 2009; Rasmann and Agrawal, unpublished data) (Figure 10.8). Cardenolides disrupt the sodium and potassium flux in animal cells, making them remarkably potent toxins, with no known primary function in plants (Malcolm 1991). In response to foliar herbivory by specialist monarch butterfly caterpillars, most species of *Asclepias* induce increases in cardenolide expression; similarly, root herbivory by larvae of specialist cerambycid beetles (*Tetraopes* spp.) typically increases cardenolide concentrations in roots. In both above- and below-ground tissues, we have now repeatedly found the striking pattern of a negative genetic correlation between constitutive and induced cardenolides within species and a positive association between constitutive cardenolides and induction across species (see Figure 10.8).

The intraspecific data could be interpreted in two ways. The negative genetic correlation between constitutive and induced cardenolides in *A. syriaca* could be an adaptive negative correlation. That is, constitutive and induced cardenolides may be redundant (because inducing higher levels of

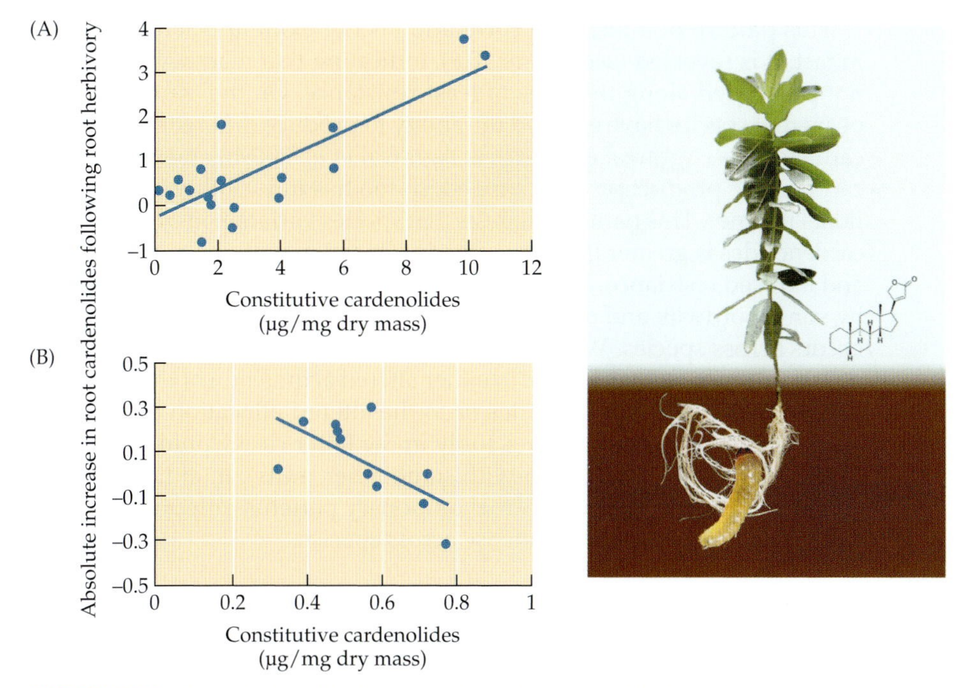

FIGURE 10.8 The Relationship between Constitutive and Induced Cardenolides in the Roots of Milkweeds The relationship is presented across (A) 18 species of *Asclepias* and (B) 11 full-sibling families of *Asclepias syriaca*. Slopes are significant after being corrected for various statistical biases (Morris et al. 2006), including phylogenetic nonindependence. Similarly, divergent relationships have been found for above-ground leaf cardenolide concentrations in response to monarch caterpillar herbivory. (From Rasmann et al. 2009; Rasmann and Agrawal, unpublished data; photo by Sergio Rasmann.)

cardenolides in high constitutive genotypes may provide little additional defense) and costly. We currently do not have data to address either of these suppositions. Alternatively, constitutive and induced cardenolides may exhibit a two-trait tradeoff. That is, the traits may not be redundant (i.e., both may be under positive selection), and limiting resources (i.e., the absolute total amount of cardenolides) prevent both high constitutive cardenolides and inducibility from being simultaneously attained. Indeed, in both the root data presented here (see Figure 10.8B) and in work on cardenolides in leaves (Bingham and Agrawal, submitted 2010), it does appear that *A. syriaca* genotypes produce a maximum level of cardenolides that is modulated by increasingly low levels following damage, maintaining intermediate levels following damage, or decreasing cardenolides after damage. The fact that the highest constitutive cardenolide genotypes decrease cardenolides following damage is surprising but has been consistently observed in our experiments.

This pattern of a negative correlation is not exhibited across species; in fact, it is reversed (see Figure 10.8), indicating that macroevolution has not proceeded along the lines of least resistance. On the contrary, it appears that species have evolved the ability to produce different amounts of cardenolides (Agrawal et al. 2009a; Agrawal et al. 2009b) and that as total cardenolide production has changed, so too has inducibility in a proportional manner. This pattern suggests that species variation in production of cardenolides is greater than variation in allocation patterns to constitutive and induced resistance. The logic here is exactly the same as that described by van Noordwijk and de Jong (1986), except that the positive association occurs across species. We note that the species differences in total cardenolide production may be due to either altered acquisition of the resources needed to produce cardenolides or to the altered allocation of the same set of resources to cardenolides (only in the latter case would one expect to find a negative correlation with some other traits). In either case, the countervailing micro- and macroevolutionary patterns reflect major shifts in total plant investment in defense across species.

Conclusion

We have emphasized several overarching themes in the study of tradeoffs. First, tradeoffs can either act through a single trait selected in opposite directions by different selective agents, fitness components, or environments, or they can act through multiple traits that compete for a shared limiting resource. Second, variation in acquisition of a limiting resource can be greater than variation in allocation of that resource, resulting in a positive, rather than negative, correlation between traits that actually trade off. This outcome can occur (1) at the phenotypic level within populations, if positive environmental correlations are greater than negative genetic correlations; (2) at the genetic level within populations, if genetic variation in acquisition is greater than genetic variation in allocation; or (3) at the macroevolutionary level, if interspecific variation in acquisition is stronger than negative genetic correlations within species. Third, negative correlations by themselves can be difficult to interpret, as they could either represent a tradeoff or an adaptation, with the latter being more likely in traits that are both redundant and costly. Finally, future work on tradeoffs should both move down lower and up higher in the trait hierarchy, by identifying individual gene loci and physiological processes directly involved in resource allocation and by determining the effects of these loci and processes on as comprehensive a fitness measure as possible.

Acknowledgments

We thank Doug Futuyma for stimulating us to think about tradeoffs and for detailed comments on the manuscript. We also thank Susan Cook, Alexis

Erwin, Monica Geber, Amy Hastings, Raffica La Rosa, Brian Lazzaro, Anne Royer, Mike Stastny, and Marjorie Weber for comments and discussion. During the writing of this chapter, we were supported by NSF DEB 0447550 and 0950231 to AAA, NSF DBI 0638591 and DEB-0919452 to JKC, and a postdoctoral fellowship from the Swiss National Science Foundation PA0033-121483 to SR.

Literature Cited

Agosta, S. J. and J. A. Klemens. 2009. Resource specialization in a phytophagous insect: No evidence for genetically based performance trade-offs across hosts in the field or laboratory. *J. Evol. Biol.* 22: 907–912.

Agrawal, A. A. 2000. Host range evolution: Adaptation of mites and trade-offs in fitness on alternate hosts. *Ecology* 81: 500–508.

Agrawal, A. A. 2005. Natural selection on common milkweed (*Asclepias syriaca*) by a community of specialized insect herbivores. *Evol. Ecol. Res.* 7: 651–667.

Agrawal, A. A. 2007. Macroevolution of plant defense strategies. *Trends Ecol. Evol.* 22: 103–109.

Agrawal, A. A. and M. Fishbein. 2006. Plant defense syndromes. *Ecology* 87: S132–S149.

Agrawal, A. A., J. K. Conner, and J. R. Stinchcombe. 2004. Evolution of plant resistance and tolerance to frost damage. *Ecol. Lett.* 7: 1199–1208.

Agrawal, A. A., M. Fishbein, R. Halitschke, and 3 others. 2009a. Evidence for adaptive radiation from a phylogenetic study of plant defenses. *Proc. Natl. Acad. Sci. USA* 106: 18067–18072.

Agrawal, A. A., A. Janssen, J. Bruin, and 2 others. 2002. An ecological cost of plant defense: Attractiveness of bitter cucumber plants to natural enemies of herbivores. *Ecol. Lett.* 5: 377–385.

Agrawal, A. A., J.-P. Salminen, and M. Fishbein. 2009b. Phylogenetic trends in phenolic metabolism of milkweeds (*Asclepias*): Evidence for escalation. *Evolution* 63: 663–673.

Ågren, J., and D. W. Schemske. 1993. The cost of defense against herbivores: An experimental study of trichome production in *Brassica rapa*. *Am. Nat.* 141: 338–350.

Allen, C. E., P. Beldade, B. J. Zwaan, and 1 other. 2008. Differences in the selection response of serially repeated color pattern characters: Standing variation, development, and evolution. *BMC Evol. Biol.* 8: 94.

Andersen, Ø. M. and K. R. Markham (eds.). 2006. *Flavonoids: Chemistry, Biochemistry and Applications*. CRC Press/Taylor & Francis, Boca Raton.

Armbruster, W. S., C. Pelabon, T. F. Hansen, and 1 other. 2004. Floral integration, modularity and accuracy: Distinguishing complex adaptations from genetic constraints. In M. Pigliucci and K. Preston (eds.), *Phenotypic Integration: Studying the Ecology and Evolution of Complex Phenotypes*, pp. 23–49. Oxford University Press, New York.

Baldwin, I. T. and W. Hamilton. 2000. Jasmonate-induced responses of *Nicotiana sylvestris* results in fitness costs due to impaired competitive ability for nitrogen. *J. Chem. Ecol.* 26: 915–952.

Baucom, R. S. and R. Mauricio. 2008. Constraints on the evolution of tolerance to herbicide in the common morning glory: Resistance and tolerance are mutually exclusive. *Evolution* 62: 2842–2854.

Beldade, P., K. Koops, and P. M. Brakefield. 2002. Developmental constraints versus flexibility in morphological evolution. *Nature* 416: 844–847.

Bell, G. and V. Koufopanou. 1986. The cost of reproduction. *Oxford Surveys Evol. Biol.* 3: 83–131.

Berenbaum, M. R., A. R. Zangerl, and J. K. Nitao. 1986. Constraints on chemical coevolution: Wild parsnips and the parsnip webworm. *Evolution* 40: 1215–1228.

Bergelson, J. and C. B. Purrington. 1996. Surveying patterns in the cost of resistance in plants. *Am. Nat.* 148: 536–558.

Blows, M. W. and A. A. Hoffmann. 2005. A reassessment of genetic limits to evolutionary change. *Ecology* 86: 1371–1384.

Brody, A. K. and R. Karban. 1992. Lack of a tradeoff between constitutive and induced defenses among varieties of cotton. *Oikos* 65: 301–306.

Brooks, R., J. Hunt, M. W. Blows, and 3 others. 2005. Experimental evidence for multivariate stabilizing sexual selection. *Evolution* 59: 871–880.

Clark, J. S., M. Dietze, S. Chakraborty, and 4 others. 2007. Resolving the biodiversity paradox. *Ecol. Lett.* 10: 647–659.

Conner, J. and S. Via. 1993. Patterns of phenotypic and genetic correlations among morphological and life history traits in wild radish, *Raphanus raphanistrum*. *Evolution* 47: 704–711.

Conner, J. K. 2002. Genetic mechanisms of floral trait correlations in a natural population. *Nature* 420: 407–410.

Conner, J. K. 2003. Artificial selection: A powerful tool for ecologists. *Ecology* 84: 1650–1660.

Conner, J. K. 2006. Ecological genetics of floral evolution. In L. D. Harder and S. C. H. Barrett (eds.), *The Ecology and Evolution of Flowers*, pp. 260–277. Oxford University Press, New York.

Conner, J. K. and D. L. Hartl. 2004. *A Primer of Ecological Genetics*. Sinauer Associates, Sunderland, MA.

Darwin, C. 1859. *On The Origin of Species By Means of Natural Selection, or The Preservation of Favoured Races in The Struggle for Life*. John Murray, London.

Falconer, D. S. and T. F. C. Mackay. 1996. *Introduction to Quantitative Genetics*. Longman, Essex, England.

Ferrari, J., S. Via, and H. C. J. Godfray. 2008. Population differentiation and genetic variation in performance on eight hosts in the pea aphid complex. *Evolution* 62: 2508–2524.

Fineblum, W. L. and M. D. Rausher. 1995. Tradeoff between resistance and tolerance to herbivore damage in a morning glory. *Nature* 377: 517–520.

Forister, M. L., A. G. Ehmer, and D. J. Futuyma. 2007. The genetic architecture of a niche: Variation and covariation in host use traits in the Colorado potato beetle. *J. Evol. Biol.* 20: 985–996.

Fornoni, J., P. L. Valverde, and J. Nú ez-Farfán. 2003. Quantitative genetics of plant tolerance and resistance against natural enemies of two natural populations of *Datura stramonium*. *Evol. Ecol. Res.* 5: 1049–1065.

Fox, C. W. 1993. A quantitative genetic analysis of oviposition preference and larval performance on two hosts in the bruchid beetle, *Callosobruchus maculatus*. *Evolution* 47: 166–175.

Fox, C. W. and T. A. Mousseau. 1996. Larval host plant affects fitness consequences of egg size variation in the seed beetle *Stator limbatus*. *Oecologia* 107: 541–548.

Fox, C. W. and T. A. Mousseau. 1998. Maternal effects as adaptations for transgenerational phenotypic plasticity in insects. In T. A. Mousseau and C. W. Fox (eds.), *Maternal Effects as Adaptations*, pp. 159–177. Oxford University Press, New York.

Fox, C. W., M. S. Thakar, and T. A. Mousseau. 1997. Egg size plasticity in a seed beetle: An adaptive maternal effect. *Am. Nat.* 149: 149–163.
Frankino, W. A., B. J. Zwaan, D. L. Stern, and 1 other. 2005. Natural selection and developmental constraints in the evolution of allometries. *Science* 307: 718–720.
Fry, J. D. 1996. The evolution of host specialization: Are trade-offs overrated? *Am. Nat.* 148: S84–S107.
Fry, J. D. 2003. Detecting ecological trade-offs using selection experiments. *Ecology* 84: 1672–1678.
Futuyma, D. J. 1997. *Evolutionary Biology*, 3rd ed. Sinauer Associates, Sunderland MA.
Futuyma, D. J. 2008. Specialization, speciation, and radiation. In K. Tilmon (ed.), *Specialization, Speciation, and Radiation: The Evolutionary Biology of Herbivorous Insects*, pp. 36–150. University of California Press, Berkeley.
Futuyma, D. J. and G. Moreno. 1988. The evolution of ecological specialization. *Ann. Rev. Ecol. Syst.* 19: 207–234.
Futuyma, D. J. and T. E. Philippi. 1987. Genetic variation and covariation in response to host plants by *Alsophila pometaria* (Lepidoptera: Geometridae). *Evolution* 41: 269–279.
Gadgil, M. and O. T. Solbrig. 1972. Concept of *r*-selection and *k*-selection—evidence from wild flowers and some theoretical considerations. *Am. Nat.* 106: 14–31.
Gershenzon, J. and R. Croteau. 1992. Terpenoids. In G. A. Rosenthal and M. R. Berenbaum (eds.), *Herbivores: Their Interactions With Secondary Plant Metabolites, 2nd ed., Vol. I: The Chemical Participants*, pp. 165–219. Academic Press, San Diego.
Gould, F. 1979. Rapid host range evolution in a population of the phytophagous mite *Tetranychus urticae* Koch. *Evolution* 33: 791–802.
Hansen, T. F. and D. Houle. 2008. Measuring and comparing evolvability and constraint in multivariate characters. *J. Evol. Biol.* 21: 1201–1219.
Heil, M., T. Delsinne, A. Hilpert, and 5 others. 2002. Reduced chemical defence in ant-plants? A critical re-evaluation of a widely accepted hypothesis. *Oikos* 99: 457–468.
Hoekstra, H. E., K. E. Drumm, and M. W. Nachman. 2004. Ecological genetics of adaptive color polymorphism in pocket mice: Geographic variation in selected and neutral genes. *Evolution* 58: 1329–1341.
Houle, D. 1991. Genetic covariance of fitness correlates—what genetic correlations are made of and why it matters. *Evolution* 45: 630–648.
Hunt, G. 2007. Evolutionary divergence in directions of high phenotypic variance in the ostracode genus *Poseidonamicus*. *Evolution* 61: 1560–1576.
James, A. C., J. Jakubczak, M. P. Riley, and 1 other. 1988. On the causes of monophagy in *Drosophila quinaria*. *Evolution* 42: 626–630.
Kao, Y. Y., S. A. Harding, and C. J. Tsai. 2002. Differential expression of two distinct phenylalanine ammonia-lyase genes in condensed tannin-accumulating and lignifying cells of quaking aspen. *Plant. Physiol.* 130: 796–807.
Karn, M. N. and L. S. Penrose. 1951. Birth weight and gestation time in relation to maternal age, parity and infant survival. *Ann. Eugenic.* 16: 147–164.
Karowe, D. N. 1990. Predicting host range evolution—colonization of *Coronilla varia* by *Colias philodice* (Lepidoptera, Pieridae). *Evolution* 44: 1637–1647.
Keinanen, M., R. Julkunen-Tiitto, P. Mutikainen, and 3 others. 1999. Trade-offs in phenolic metabolism of silver birch: Effects of fertilization, defoliation, and genotype. *Ecology* 80: 1970–1986.
Koricheva, J., H. Nykanen, and E. Gianoli. 2004. Meta-analysis of trade-offs among plant antiherbivore defenses: Are plants jacks-of-all-trades, masters of all? *Am. Nat.* 163: E64–E75.

Lack, D. 1947. The significance of clutch size. *Ibis* 89: 302–352.
Lande, R. and S. J. Arnold. 1983. The measurement of selection on correlated characters. *Evolution* 37: 1210–1226.
Laskar, D. D., M. Jourdes, A. M. Patten, and 3 others. 2006. The *Arabidopsis* cinnamoyl CoA reductase irx4 mutant has a delayed but coherent (normal) program of lignification. *Plant J.* 48: 674–686.
Liscombe, D. K., B. P. MacLeod, N. Loukanina, and 2 others. 2005. Evidence for the monophyletic evolution of benzylisoquinoline alkaloid biosynthesis in angiosperms. *Phytochemistry* 66: 1374–1393.
MacKenzie, A. 1996. A trade-off for host plant utilization in the black bean aphid, *Aphis fabae*. *Evolution* 50: 155–162.
Malcolm, S. B. 1991. Cardenolide-mediated interactions between plants and herbivores. In G. A. Rosenthal and M. R. Berenbaum (eds.), *Herbivores: Their Interactions With Secondary Plant Metabolites, 2nd ed., Vol. I: The Chemical Participants,* pp. 251–296. Academic Press, San Diego.
Marak, H. B., A. Biere, and J. M. M. Van Damme. 2003. Fitness costs of chemical defense in *Plantago lanceolata* L.: Effects of nutrient and competition stress. *Evolution* 57: 2519–2530.
Marroig, G. and J. M. Cheverud. 2005. Size as a line of least evolutionary resistance: Diet and adaptive morphological radiation in new world monkeys. *Evolution* 59: 1128–1142.
Martens, S. and A. Mithofer. 2005. Flavones and flavone synthases. *Phytochemistry* 66: 2399–2407.
Mauricio, R. and M. D. Rausher. 1997. Experimental manipulation of putative selective agents provides evidence for the role of natural enemies in the evolution of plant defense. *Evolution* 51: 1435–1444.
McGuigan, K., A. V. Homrigh, and M. W. Blows. 2008. An evolutionary limit to male mating success. *Evolution* 62: 1528–1537.
Messina, F. J. and C. W. Fox. 2001. Offspring size and number. In C. W. Fox, D. A. Roff, and D. J. Fairbairn (eds.), *Evolutionary Ecology: Concepts and Case Studies,* pp. 113–127. Oxford University Press, New York.
Mole, S. 1994. Trade-offs and constraints in plant-herbivore defense theory: A life-history perspective. *Oikos* 71: 3–12.
Mole, S. and A. J. Zera. 1993. Differential allocation of resources underlies the dispersal-reproduction trade-off in the wing-dimorphic cricket, *Gryllus rubens*. *Oecologia* 93: 121–127.
Mole, S. and A. J. Zera. 1994. Differential resource consumption obviates a potential flight fecundity trade-off in the sand cricket (*Gryllus firmus*). *Func. Ecol.* 8: 573–580.
Mooney, K. A., P. Jones, and A. A. Agrawal. 2008. Coexisting congeners: Demography, competition, and interactions with cardenolides for two milkweed-feeding aphids. *Oikos* 117: 450–458.
Morris, W. F., M. B. Traw, and J. Bergelson. 2006. On testing for a tradeoff between constitutive and induced resistance. *Oikos* 112: 102–110.
Núñez-Farfán, J., J. Fornoni, and P. L. Valverde. 2007. The evolution of resistance and tolerance to herbivores. *Ann. Rev. Ecol. Evol. Syst.* 38: 541–566.
Pagel, M. 1999. Inferring the historical patterns of biological evolution. *Nature* 401: 877–884.
Pelser, P. B., H. de Vos, C. Theuring, and 3 others. 2005. Frequent gain and loss of pyrrolizidine alkaloids in the evolution of *Senecio* section Jacobaea (Asteraceae). *Phytochemistry* 66: 1285–1295.

Pilson, D. 2000. The evolution of plant response to herbivory: Simultaneously considering resistance and tolerance in *Brassica rapa*. *Evol. Ecol.* 14: 457–489.
Raberg, L., D. Sim, and A. F. Read. 2007. Disentangling genetic variation for resistance and tolerance to infectious diseases in animals. *Science* 318: 812–814.
Rasmann, S., A. A. Agrawal, A. C. Erwin, and 1 other. 2009. Cardenolides, induced responses, and interactions between above and belowground herbivores in the milkweeds (*Asclepias* spp). *Ecology* 90: 2393–2404.
Rausher, M. D. 1984. Tradeoffs in performance on different hosts: Evidence from within- and between-site variation in the beetle *Deloyala guttata*. *Evolution* 38: 582–595.
Rausher, M. D. 1992. Natural selection and the evolution of plant-insect interactions. In B. D. Roitberg and M. B. Isman (eds.), *Insect Chemical Ecology: An Evolutionary Approach*, pp. 20–88. Chapman & Hall, New York.
Reznick, D. 1985. Costs of reproduction: An evaluation of the empirical evidence. *Oikos* 44: 257–267.
Rhoades, D. F. 1979. Evolution of plant chemical defense against herbivores. In G. A. Rosenthal and D. H. Janzen (eds.), *Herbivores: Their Interaction With Secondary Plant Metabolites*, pp. 3–54. Academic Press, New York.
Ricklefs, R. E. 1977. Evolution of reproductive strategies in birds—reproductive effort. *Am. Nat.* 111: 453–478.
Roff, D. A. 2002. *Life History Evolution*. Sinauer Associates, Sunderland, MA.
Rose, M. R. and B. Charlesworth. 1981. Genetics of life-history in *Drosophila melanogaster*. II. Exploratory selection experiments. *Genetics* 97: 187–196.
Rudgers, J. A., S. Y. Strauss, and J. F. Wendel. 2004. Trade-offs among anti-herbivore resistance traits: Insights from Gossypieae (Malvaceae). *Am. J. Bot.* 91: 871–880.
Scalliet, G., C. Lionnet, M. Le Bechec, and 10 others. 2006. Role of petal-specific orcinol O-methyltransferases in the evolution of rose scent. *Plant Physiol.* 140: 18–29.
Schluter, D. 1996. Adaptive radiation along genetic lines of least resistance. *Evolution* 50: 1766–1774.
Schluter, D., T. D. Price, and L. Rowe. 1991. Conflicting selection pressures and life-history trade-offs. *Proc. Roy. Soc. Lond. B* 246: 11–17.
Schoonhoven, L., J. van Loon, and M. Dicke. 2005. *Insect-Plant Biology*, 2nd ed. Oxford University Press, New York.
Shonle, I. and J. Bergelson. 2000. Evolutionary ecology of the tropane alkaloids of *Datura stramonium* L. (Solanaceae). *Evolution* 54: 778–788.
Siemens, D. H. and T. Mitchell-Olds. 1998. Evolution of pest-induced defenses in *Brassica* plants: Tests of theory. *Ecology* 79: 632–646.
Simms, E. L. 1992. Costs of plant resistance to herbivores. In R. S. Fritz and E. L. Simms (eds.), *Plant Resistance to Herbivores and Pathogens: Ecology, Evolution, and Genetics*, pp. 392–425. University of Chicago Press, Chicago.
Simms, E. L. and M. D. Rausher. 1987. Costs and benefits of plant resistance to herbivory. *Am. Nat.* 130: 570–581.
Simms, E. L. and M. D. Rausher. 1989. The evolution of resistance to herbivory in *Ipomoea purpurea*. II. Natural selection by insects and costs of resistance. *Evolution* 43: 573–585.
Stearns, S. C. 1992. *The Evolution of Life Histories*. Oxford University Press, Oxford.
Steward, J. L. and K. H. Keeler. 1988. Are there trade-offs among antiherbivore defenses in *Ipomoea* (Convolvulaceae)? *Oikos* 53: 79–86.

Stowe, K. A. 1998. Experimental evolution of resistance in *Brassica rapa*: Correlated response of tolerance in lines selected for glucosinolate content. *Evolution* 52: 703–712.
Strauss, S. Y., J. A. Rudgers, J. A. Lau, and 1 other. 2002. Direct and ecological costs of resistance to herbivory. *Trends Ecol. Evol.* 17: 278–285.
Strauss, S. Y., D. H. Siemens, M. B. Decher, and 1 other. 1999. Ecological costs of plant resistance to herbivores in the currency of pollination. *Evolution* 53: 1105–1113.
Twigg, L. E. and L. V. Socha. 1996. Physical versus chemical defense mechanisms in toxic *Gastrolobium*. *Oecologia* 108: 21–28.
Ueno, H., N. Fujiyama, I. Yao, and 2 others. 2003. Genetic architecture for normal and novel host-plant use in two local populations of the herbivorous ladybird beetle, *Epilachna pustulosa*. *J. Evol. Biol.* 16: 883–895.
van der Meijden, E., M. Wijn, and H. J. Verkaar. 1988. Defense and regrowth, alternative plant strategies in the struggle against herbivores. *Oikos* 51: 355–363.
Van Homrigh, A., M. Higgie, K. McGuigan, and 1 other. 2007. The depletion of genetic variance by sexual selection. *Curr. Biol.* 17: 528–532.
van Noordwijk, A. J. and G. de Jong. 1986. Acquisition and allocation of resources—their influence on variation in life-history tactics. *Am. Nat.* 128: 137–142.
Verhoeven, K. J. F., T. K. Vanhala, A. Biere, and 2 others. 2004. The genetic basis of adaptive population differentiation: A quantitative trait locus analysis of fitness traits in two wild barley populations from contrasting habitats. *Evolution* 58: 270–283.
Via, S. and D. J. Hawthorne. 2002. The genetic architecture of ecological specialization: Correlated gene effects on host use and habitat choice in pea aphids. *Am. Nat.* 159: S76–S88.
Via, S. and R. Lande. 1985. Genotype-environment interaction and the evolution of phenotypic plasticity. *Evolution* 39: 505–522.
Walsh, B. and M. W. Blows. 2009. Abundant genetic variation + strong selection = multivariate genetic constraints: A geometric view of adaptation. *Ann. Rev. Ecol. Evol. Sys.* 40: 41–59.
Weis, A. E. and W. L. Gorman. 1990. Measuring selection on reaction norms—an exploration of the *Eurosta-Solidago* system. *Evolution* 44: 820–831.
Whittaker, R. H. and P. P. Feeny. 1971. Allelochemics: Chemical interactions between species. *Science* 171: 757–770.
Worley, A. C. and S. C. H. Barrett. 2000. Evolution of floral display in *Eichhornia paniculata* (Pontederiaceae): Direct and correlated responses to selection on flower size and number. *Evolution* 54: 1533–1545.
Yano, S., J. Takabayashi, and A. Takafuji. 2001. Trade-offs in performance on different plants may not restrict the host plant range of the phytophagous mite, *Tetranychus urticae*. *Exp. Appl. Acarol.* 25: 371–381.
Zangerl, A. R. and F. A. Bazzaz. 1992. Theory and pattern in plant defense allocation. In R. S. Fritz and E. L. Simms (eds.), *Plant Resistance to Herbivores and Pathogens: Ecology, Evolution, and Genetics*, pp. 363–391. University of Chicago Press, Chicago.
Zangerl, A. R. and M. R. Berenbaum. 1997. Cost of chemically defending seeds: Furanocoumarins and *Pastinaca sativa*. *Am. Nat.* 150: 491–504.
Zera, A. J. and L. G. Harshman. 2001. The physiology of life history trade-offs in animals. *Ann. Rev. Ecol. Syst.* 32: 95–126.

Chapter 11

Elucidating Evolutionary Mechanisms in Plant–Insect Interactions: Key Residues as Key Innovations

May R. Berenbaum and Mary A. Schuler

Although Charles Darwin wrote prolifically on a spectacular range of biological subjects, he is best remembered for his tremendously influential book *The Origin of Species by Means of Natural Selection, or The Preservation of Favoured Races in the Struggle for Life* (Darwin 1859). The impact of the book has been so profound that Darwin's other transformative contributions are often overlooked. For example, in writing about "Darwin's enduring legacy" for the journal *Nature*, Padian (2008) describes Darwin's 1862 book *On the Various Contrivances by which British and Foreign Orchids are Fertilised by Insects and on the Good Effects of Intercrossing*, as "one of Darwin's lesser-known books." In fact, Padian suggests that the single most important contribution of this book was its dramatic example of the predictive power of evolutionary theory—Darwin's prediction that an African orchid with a corolla "eleven and a half inches long" must be pollinated by "moths with probosces capable of extension to a length of between ten and eleven inches!" (Darwin 1862: 198). Indeed, within 40 years of the publication of *On the Various Contrivances*, a sphingid moth from Madagascar with a proboscis of the appropriate length was described and named *Xanthopan morganii praedicta* by Rothschild and Jordan (1903) (although the subspecies was later synonymized out of existence).

Actually, Darwin wrote over 20 papers and 2 books on pollination and in the process, did more to bring the study of pollination into mainstream science than anyone else. Recognition of plant sexuality dates back to Camerarius (1694), and was the basis for the system of plant classification used by Linnaeus in his *Systema Naturae* (1758), but the role of animal mutualists in pollen transfer was first explicated by Christian Konrad Sprengel. In his 1793 book *Das entdeckte Geheimnis der Natur im Bau und in der Befruchtung der Blumen*, Sprengel marshaled evidence that floral color patterns and nectar production function primarily to attract insects. Although the book was initially favorably received (Zepernick and Meretz 2001), Sprengel died in obscurity and pollination remained a marginalized subject.

Darwin, however, rehabilitated Sprengel's reputation in the process of formulating an evolutionary perspective on plant pollination. Darwin's friend and colleague Thomas H. Huxley, in *Natural History as Knowledge Discipline and Power* (1856), wondered: "Who has ever dreamed of finding an utilitarian purpose in the forms and colors of flowers, in the sculpture of pollen-grains..." (Huxley 1856: 311). But in Darwin's worldview, colored by the nascent theory of natural selection, the forms and colors of flowers must have some utilitarian purpose. Starting in 1857 and for 20 years thereafter, he devoted much of his scholarly attention to ascertaining those utilitarian purposes. Within a few years, his pollination studies so influenced his thoughts on the theory of natural selection that pollinator–plant interactions were mentioned in the introduction to *The Origin of Species*. As an example of:

> *how the innumerable species, inhabiting this world have been modified, so as to acquire that perfection of structure and coadaptation which justly excites our admiration,* he notes that the misseltoe [sic mistletoe] has *flowers with separate sexes absolutely requiring the agency of certain insects to bring pollen from one flower to the other (Darwin 1902: 3).*

Darwin continued to pursue his studies of pollination and in his 1862 book, *On the Various Contrivances*, he discounted natural theology and confirmed the sexual life of flowers and the contributions of animal pollinators to plant reproduction. Together, *The Origin of Species* and *On the Various Contrivances* provided the foundation for the field of pollination biology.

The rise of pollination biology, in turn, influenced the study of other forms of plant–insect interactions. Darwin's *The Origin of Species*, translated into German by paleontologist Heinrich Georg Bronn in 1860, inspired a new focus on external environmental influences on plant evolution and initiated a shift from laboratory-based to field-based research among young German botanists (Cittadino 1990). In 1876, one such botanist, Anton Kerner von Marilaun, published *Die Schutzmittel der Blüthen gegen unberufene Gäste,* reinterpreting much of floral morphology that had been carefully detailed by German plant anatomists over three-quarters of the nineteenth century in the context of evolutionary theory. Darwin provided an enthusiastic preface for the English translation of Kerner's book, remarking that: "the conclusion that flowers are not only delightful from their beauty and fragrance, but display most wonderful adaptations for various purposes" (Kerner 1876: 1).

Among the "wonderful adaptations" described by Kerner was the use of noxious substances by plants to defend against herbivory. Apparently, across all of his writings, Darwin's sole discussion of plant–insect coadaptation in the context of chemical defense was his citation of Kerner's book in his 1877 book, *The Different Forms of Flowers on Plants of the Same Species*:

> *...the ray-florets of the Compositæ ...often contain matter which is excessively poisonous to insects, as may be seen in the use of flea-powder, and in the case of Pyrethrum, M. Belhomme has shown that the ray-florets are more poisonous than the disc-florets in the ratio of about three to two. We may therefore believe that the ray-florets are useful in protecting the flowers from being gnawed by insects.... Kerner in his interesting essay ('Die Schutzmittel der Blüthen gegen unberufene Gäste,' 1875: 19) insists that the petals of most plants contain matter which is offensive to insects, so that they are seldom gnawed, and thus the organs of fructification are protected (Darwin 1877: 6).*

Natural Selection Meets Plant Chemistry

Darwin's suggestion that natural selection accounts for the diversity of plant morphological and chemical attributes, that had been meticulously documented by a generation of German botanists, profoundly influenced fellow German botanist Ernst Stahl, in Jena. In his remarkably perceptive *Pflanzen und Schnecken Eine biologische Studie über die Schutzmittel der Pflanzen gegen Schneckenfrass* (1888) (translation: Plants and Snails, A Biological Study Concerning Means of Protection in Plants against Damage by Snails), Stahl noted that, in contrast to plant coadaptation with pollinators, little attention had been paid to "protective devices against attack by animals," through which a plant "if it cannot resist the attack of the surrounding animal world, it can at least resist extermination" (Stahl 1888; Cittadino 1990, translation). Stahl cited Chapter 3 of Darwin's *The Origin of Species* to bolster his argument that animals can act as selective agents on plants (e.g., in this chapter, Darwin recounts that of 357 seedlings sprouting in a small area, 295 were "destroyed by snail and insects"). Stahl also argued that even if the means of defense are not externally obvious, "all the wild plants studied, even the seemingly defenseless, possess means of protection against attack by animals" (Stahl 1888; Cittadino 1990, translation: 89). In his later work, Stahl proposed that (1) reciprocal adaptation characterizes the relationship between a specialist herbivore and its host plant, and (2) plant compounds that are unpalatable to generalists may stimulate feeding in a specialist.

> *Here the objection has to be countered that substances like tannins, bitter compounds, essential oils, alkaloids, etc.—whose role in plant chemistry is almost completely obscure—would exist as essential components of metabolism in a complete absence of animals. That these compounds are essential components of metabolism should not be denied as well as their presence in plants before they were subjected to natural selection by herbivores. However, their current quantitative design, their distribution within plant organs, their often preferred peripheral accumulation, and,*

> *particularly, their early appearance can exclusively be understood by the impact of the animal kingdom surrounding plants. Moreover, even the idea should not be denied that the quality of all these compounds in respect to smell, taste, toxicity and thus chemical composition must be affected by the selection pressure of the animal kingdom. The variability of plants not only concerns morphology but also metabolic processes. Humans obtained by breeding of inconspicuous, tasteless wild fruit species—I just recall the pears—a rich variety of differently smelling fruits which at least in respect to their aroma have different chemical composition. It can be assumed that in the same way under the selection pressure of herbivores, plant constituents with improved deterrent or detrimental properties against herbivores are created (Translated by Hartmann 2008: 5).*

Stahl's remarkable insights did not stimulate a great deal of work, in part because successors felt that his selectionist interpretations were teleological but also because sophisticated understanding of the chemical mediation of interactions between plants and insects was hampered by operational constraints and methodological limitations. After seven decades of languishing, however, Stahl's work played an important role in reactivating interest in plant–herbivore interactions.

Gottfried Fraenkel, a German-born entomologist, moved to London in 1933 to avoid religious persecution. During World War II, as part of the war effort, he studied the nutritional requirements of stored-product pest insects (at the time a major concern for the military). These studies convinced him that "the basic food requirements of many insects…were essentially identical, and similar to those of higher organisms" (Fraenkel 1984). After emigrating to the United States in 1947, he broadened his studies to examine plant–herbivore interactions and the nutritional requirements of insect herbivores. Commenting on a striking dichotomy in the dietary constituents ingested by insects, he stated in his 1953 contribution to the Proceedings of the 14th International Congress of Zoology that:

> *In this report, I am going to suggest that leaves from different plants differ relatively little in chemical composition as far as the nutritional needs of insects for food substances proper are concerned, i.e., protein, carbohydrates, fat, minerals, sterols, and vitamins, and that the specificity of food for various insects is based on the presence of chemical compounds which serve as chemical stimuli for the sensory organs of smell and taste… (Fraenkel 1953: 90).*

In 1959, he published a paper in the journal *Science* that resurrected Stahl in much the same way that Darwin resurrected Sprengel a century earlier (Fraenkel 1959). He summarized eight decades of phytochemical experimentation to support Stahl's contention that secondary substances play a critical role in mediating interactions with herbivores, thereby providing a *raison d'être* for secondary plant substances. In the process, he provided an English translation of a key paragraph from *Pflanzen und Schnecken* and thus introduced Stahl to the English-speaking world:

> *We have long been accustomed to comprehend many manifestations of the morphology [of plants]...as being due to the relations between plants and animals, and nobody, in our special case here, will doubt that the external mechanical means of protection of plants were acquired in their struggle...with the animal world...In the same sense, the great differences in the nature of chemical products [Excrete] and consequently of metabolic processes, are brought nearer to our understanding, if we regard these compounds as means of protection, acquired in the struggle with the animal world. Thus, the animal world which surrounds the plants deeply influenced not only their morphology, but also their chemistry (Stahl, as translated by Fraenkel 1959: 1470).*

Fraenkel proposed that "reciprocal adaptive evolution in the feeding habits of insects and in the biochemical characteristics of plants" is responsible for the process by which erstwhile repellents evolve to become attractants and feeding stimulants (Fraenkel 1959: 1466).

Key Innovations and Adaptive Radiations

According to Fraenkel (1984), the novel concept presented in his *Science* paper was greeted with "icy silence" by the scientific community. It did, however, resonate with Ehrlich and Raven (1964), who built on Fraenkel's ideas and proposed a stepwise scenario to account for reciprocal adaptive selection. According to this scenario, plants randomly acquire novel chemical compounds via mutation and recombination, which serve as key innovations (see Losos and Mahler, Chapter 15) and confer protection against herbivores. The plants then enter a new adaptive zone (Simpson 1944) until such time that random mutation or recombination lead to a physiological or biochemical key innovation in an insect lineage that confers resistance against a toxin, allowing this lineage to utilize a novel resource, enter an adaptive zone, diversify, and ultimately use the erstwhile defense compound as an attractant or recognition factor. Whittaker and Feeny (1971) subsequently likened this stepwise process to a continual "coevolutionary arms race" between insects and their host plants.

The nature of those key innovations defied easy characterization, and for years, they were described primarily as complex suites of phenotypic traits related to the ability to utilize particular host plants (Berenbaum et al. 1996), such as acceptance or rejection behaviors and growth inhibition or concentration-dependent mortality. Putative key plant innovations (i.e., chemical defenses) had been identified chemically for a long time, and even if the evidence was circumstantial, there at least was ample evidence of their effects on insect behavior and performance. Janzen's (1980) call for evidence of reciprocal selection in recognizing coevolutionary interactions led to the application of quantitative genetics methods to measure heritability of defense and resistance traits as well as the intensity and direction of selection and evolutionary responses (Rausher 1996). Evidence accumulated that documented genetic contributions to variation in plant defense and insect resistance, revealing the raw material on which selection can act (Berenbaum and Zangerl 1992). Experimental manipulations also convincingly demonstrated the selective impact of insect herbivory on plant chemistry. Mauricio and Rausher (1997), for example, documented the selective impact of herbivores on two defense traits (i.e, total glucosinolate concentration and trichome density) in *Arabidopsis thaliana* (mouse-ear cress or thale cress) under field conditions by experimentally manipulating herbivore loads.

As genetic studies progressed, characterization of the defensive traits proceeded apace. A substantial body of experimental work culled out the defensive function of a wide variety of phytochemicals, and for many of these, the biochemical pathways were schematized and the enzymes identified (Rosenthal and Berenbaum 1992). Similarly, detoxification enzymes were documented in a variety of insect herbivores (Brattsten 1992). Thus, specific traits mediating coevolutionary interactions could be identified, just as morphological innovations associated with adaptive radiations had been (Schluter and McPhail 1992; Schluter et al. 1997).

What remained was characterizing the genes themselves—a major advance that could provide insights into how reciprocal responses come about in an appropriate evolutionary timeframe. Advances in studies of morphological adaptation illustrated the value of gene characterization. Genetic inroads into developmental evolutionary biology have made possible a mechanistic understanding of morphological adaptations regarding environmental selection pressures (see Wray, Chapter 9). A relatively small number of regulatory mutations in major developmental control genes can produce rapid anatomical changes in structures under selection in natural populations, without loss of function of these genes in other processes (Bell et al. 1993; Shapiro et al. 2004; Chan et al. 2010).

In the past decade, genomic sequencing has provided useful tools for mechanistically defining the innovations that underlie adaptive radiations (see Kolaczkowski and Kern, Chapter 6). In 2000, the first sequenced plant genome, *Arabidopsis thaliana*, quickly provided evidence of the process by which novel phytochemicals can arise. With Fraenkel (1959) having remarked on the centrality of mustard oil glycosides (glucosinolates) in mediating interactions with the specialist fauna of the Brassicaceae, Kliebenstein et al. (2001) identified two oxoglutarate-dependent dioxygenase genes, *AOP2* and *AOP*, involved in biosynthesis of glucosinolates. Coded for by closely related genes, the protein AOP2 mediates the transformation of methylsulfinylalkyl glucosinolates to alkenyl glucosinolates and AOP3 catalyzes the transformation of methylsulfinylalkyl glucosinolates to hydroxyalkyl products. Comparisons between these genes have indicated that they arose from a recent gene duplication that led to a key biochemical innovation that enhanced resistance to herbivores by extending the range of glucosinolates produced. For their part, herbivores associated with *A. thaliana* and other brassicaceous plants also yielded evidence that key biochemical innovations lead to enhanced resource utilization efficiency and adaptive radiation (Wheat et al. 2007). The glucosinolate defense system of crucifers depends on the action of the enzyme myrosinase on glucosinolates, with myrosinase-mediated hydrolysis yielding toxic isothiocyanates. An examination of the lepidopteran family Pieridae revealed that only those members of the family associated with glucosinolate-producing plants possess the capacity to convert mustard oils to nontoxic nitriles, rather than toxic isothiocyanates, via a unique set of so-called nitrile-specifier proteins. Further molecular analyses provided evidence that multiple-domain nitrile-specifier proteins arose in this insect lineage (the subfamily Pierinae) approximately 10 million years ago, shortly after the origin of the Brassicales host plants, by a process of gene and domain duplication (Wheat et al. 2007; Fischer et al. 2008). This key innovation of the evolution of nitrile-specifier proteins is thought to have led to an increase in pierine speciation rates relative to sister lineages. The presumed mechanism is that gene duplicates acquired novel functions during a period of relaxed selection and neofunctionalization. According to this analysis, these genes are still evolving rapidly, as would be expected for genes under variable selection by herbivores.

Key Innovations and Cytochrome P450s

Although genome sequencing has provided insights into the process by which biochemical innovations may have arisen, molecular modeling of the resulting protein structures can shed light on the precise nature and extent of genetic changes that can lead to enhanced resistance to plant chemical defenses. One particular gene superfamily has proved to be uniquely well-suited to providing insights to the reciprocal adaptive evolutionary process by examining protein and genome changes: the cytochrome P450 monooxygenases (P450s). P450s are members of a superfamily of membrane-bound heme proteins with over 11,500 sequenced members, with little absolute protein sequence identity except for the conserved heme-binding signature motif F—G—C—G (Nelson 2009). Despite limited conservation in their primary sequences, this highly diverse group of proteins catalyzes nicotinamide adenine dinucleotide phosphate (NADPH)-associated reductive cleavages of oxygen to generate hydrophilic (and hence more excretable) metabolites and water. The proliferation of these genes and proteins reflects their functional versatility in mediating a wide range of oxidative transformations (de Montellano 2005). Thus, P450s contribute to biosynthetic and catabolic reactions in both angiosperm plants and herbivorous insects (as well as most other organisms).

Genome projects revealed substantial variation in the inventory of P450 genes, with each genome containing a range of P450 gene families varying in sequence, regulation, genome structure, and function (Nelson 2009). In the available sequenced insect genomes, P450 gene numbers within individual genomes range from 46 in *Apis mellifera* (honey bee; Honey Bee Genome Sequencing Consortium 2006) to 134 in *Tribolium castaneum* (red flour beetle; *Tribolium* Genome Sequencing Consortium 2008). The number of P450 genes in angiosperm plant genomes is higher, from 245 in *Arabidopsis thaliana* to 332 in *Oryza sativa* (rice) (Nelson et al. 2008). Gonzalez and Nebert (1990) suggested that the P450 superfamily began to diversify over 400 million years ago, concomitant with the colonization of terrestrial habitats by plants and herbivorous animals. Rapid diversification may have been a consequence of reciprocal selection pressures whereby plants evolved biosynthetically novel defense compounds (which in terrestrial environments are more likely to be lipophilic than in aquatic environments) and insects (and other herbivores) overcame the toxins with novel oxidative detoxification pathways, a process these authors described as "plant-animal warfare" (Gonzalez and Nebert 1990).

Multiple gene duplication events, in both plant biosynthetic pathways and insect detoxification pathways, have allowed P450s to acquire new functions in the presence of novel environmental selective agents while retaining ancestral metabolic capabilities (Roth et al. 2007). Thomas (2007) refined this view by differentiating between phylogenetically stable and unstable P450 genes. Whereas stable genes are associated with endogenous functions involving highly conserved substrates, phylogenetically unstable genes are associated with highly variable environmental selection pressures, including those involving xenobiotic toxicity. Genes associated with such exogenous selection pressures tend to undergo frequent duplication events and birth–death evolution, thereby contributing disproportionately to P450 genome inventories. Thus, whereas the CYP2 clade, which encodes enzymes involved in the biosynthesis of hormones, comprises 6 genes in the *Drosophila melanogaster* (fruit fly) genome, 9 genes in the *Apis mellifera* genome, and 8 genes in the *Tribolium castaneum* genome, the CYP3 clade, which encodes xenobiotic metabolizing enzymes, comprises 36 genes in the *D. melanogaster* genome, 28 genes in the *A. mellifera* genome, and 70 genes in the *T. castaneum* genome. Although orthologs are readily identifiable in the stable genes with highly conserved substrates, orthology is essentially nonexistent among the highly diverse, unstable genes. Identifying their functions requires knowledge of the exogenous environmental chemicals that act as selective agents.

After duplication events, acquisition of novel functions in P450 evolution has been suggested to result from bouts of positive selection for divergence interspersed with bouts of purifying selection for maintenance of critical functions. Across the length of the P450 coding sequence, particular regions are key to the acquisition of novel functions. Based on sequence alignments and comparisons with known crystal structures, Gotoh (1992) recognized six substrate recognition sites (SRS) in human P450s of the CYP2 family. Since that time, site-directed mutagenesis in these substrate recognition sites has provided ample evidence that they contribute significantly to function (Domanski and Halpert 2001)—even a single amino acid substitution can change the reactivity and substrate profile for a P450 (Amichot et al. 2004).

Such crucial residues in these SRS regions thus contribute to functional diversification and are less likely to be conserved during the process of subfunctionalization or neofunctionalization (Lee 2008). In analyzing P450s in ten vertebrate genomes, Thomas (2007) evaluated 12 *CYP2D* genes from rodents that arose by recent gene duplication and found that the specific amino acid sites most likely to show evidence of strong positive selection are in substrate-binding regions or substrate channel regions. Three-dimensional models based on P450s with known crystal structure revealed that positive selection tends to reduce the distance from the heme to the bound ligand (warfarin, in the case of CYP2D), thereby increasing metabolic efficiency. Examining P450-mediated metabolism of plant chemicals within a group of closely related insect species with well-characterized host plant utilization patterns can provide insights into how P450s evolve and diversify in response to environmental selective agents. The furanocoumarins—phytochemicals restricted in distribution to Apiaceae, Rutaceae, and a handful of other plant families—are toxic to a wide variety of organisms due to their ability, in the presence of ultraviolet light, to form an excited state and bind covalently to pyrimidine bases in DNA (Berenbaum 1991). These compounds are formed by two parallel biosynthetic pathways, leading in one case to compounds with the furan ring attached at the 6,7 position (linear furanocoumarins) and in the other case to compounds with the furan ring attached at the 7,8 positions (angular furanocoumarins) (Figure 11.1). Numerous P450 enzymes are involved in the biosynthesis of these compounds. In parsnip (*Pastinaca sativa*), the sequence of psoralen synthase (CYP71AJ3), a key enzyme in the biosynthesis of linear furanocoumarins, and angelicin synthase (CYP71AJ4), a key enzyme in the biosynthesis of angular furanocoumarins, share 70% identity across the coding region but only 40% identity in the substrate recognition sites. Thus, these two biosynthetic genes may have evolved via a recent gene duplication event and subsequent differentiation, leading to the evolution and diversification of two structurally and toxicologically distinct groups of phytochemical defenses (Larbat et al. 2009).

Although furanocoumarins are toxic to a wide variety of insects, several groups feed exclusively on such plants. The genus *Papilio* (swallowtail butterflies) is typical of herbivorous insects in that most species in the genus are oligophagous; over 75% of the 200+ species in the genus feed on host plants in the furanocoumarin-containing families Rutaceae and Apiaceae (Berenbaum 1983). One widespread characteristic of P450-

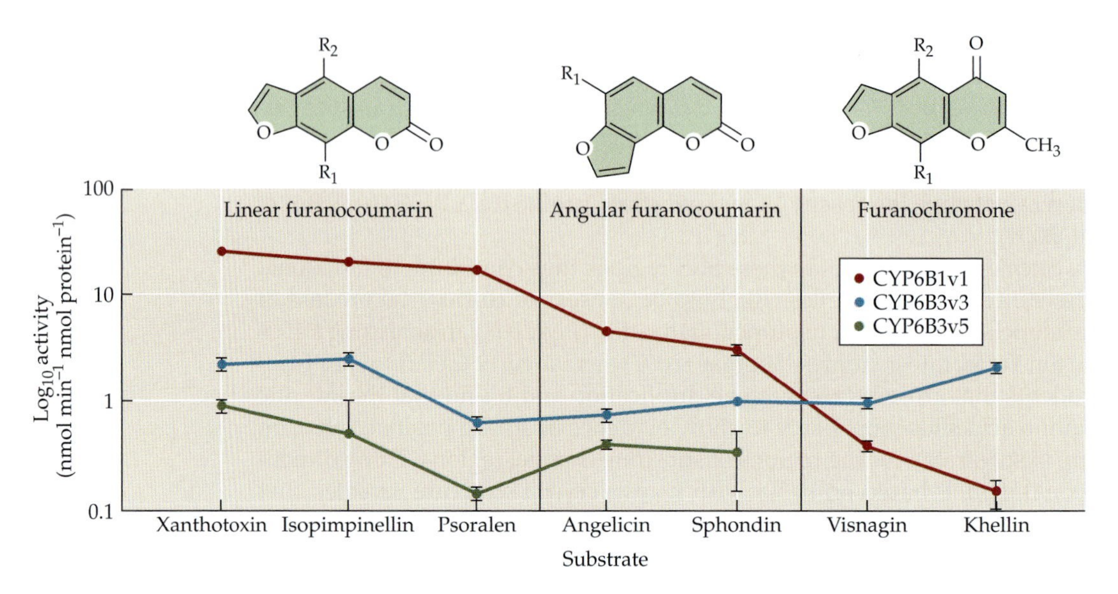

FIGURE 11.1 Comparison of Catalytic Activity of Allelic Variants of CYP6B1 and CYP6B3 toward Host Phytochemicals (Data from Wen et al. 2006.)

mediated metabolism is that catalytically active P450 enzymes are often substrate-inducible; P450s in *Papilio* display differences both in constitutive activity against furanocoumarins and for inducibility in response to furanocoumarin ingestion (Berenbaum 1995). In general, these differences are associated with the evolutionary frequency of ecological exposure to furanocoumarins. Constitutive P450-mediated activity against xanthotoxin, a linear furanocoumarin, is high in *Papilio polyxenes*, a specialist on Apiaceae containing furanocoumarins. In contrast, P450-mediated metabolism of furanocoumarins is undetectable in *P. troilus*, a specialist on Lauraceae, which lack furanocoumarins. It is intermediate in *Papilio glaucus*, which, as the most polyphagous species in the genus, is exposed to furanocoumarins in just one of its many host plants (Cohen et al. 1992). Similarly, both constitutive and induced activity against angular furanocoumarins in *P. polyxenes*, which rarely encounters these compounds in its host plants, is lower than constitutive and induced activity against the frequently encountered linear furanocoumarins (Cohen et al. 1992).

Key Innovations in Diversification, Subfunctionalization, and Neofunctionalization

The black swallowtail *P. polyxenes* feeds exclusively on furanocoumarin-containing herbaceous host plants in the families Apiaceae and Rutaceae (including *Pastinaca sativa*, the species from which CYP71AJ3 and CYP71AJ4 were isolated). In 1992, the furanocoumarin-inducible enzyme CYP6B1 was identified as the principal detoxification mechanism for furanocoumarins in this species. This enzyme detoxifies most linear furanocoumarins with high levels of activity and angular furanocoumarins with lower levels of activity (Cohen et al. 1992; Ma et al. 1994; Wen et al. 2003, 2006). Multiple allelic variants at the *CYP6B1* locus are highly inducible by the linear furanocoumarin xanthotoxin and less inducible by the angular furanocoumarin angelicin (rarely encountered in host plants) (Hung et al. 1995a,b; Petersen et al. 2001). Molecular models of the CYP6B1 protein reveal that the catalytic site contains four key aromatic amino acids (Phe116, His117, Phe371, Phe484) that contribute to the establishment of a resonant network; this network stabilizes the catalytic site and plays an essential role in maintaining the orientation of furanocoumarin substrates (Chen et al. 2002; Baudry et al. 2003) (Figure 11.2). Other key residues are Ala113 and Ile115; they contribute to both the degree of substrate specificity and the rate of product exit from the catalytic site (Pan et al. 2004; Wen et al. 2005).

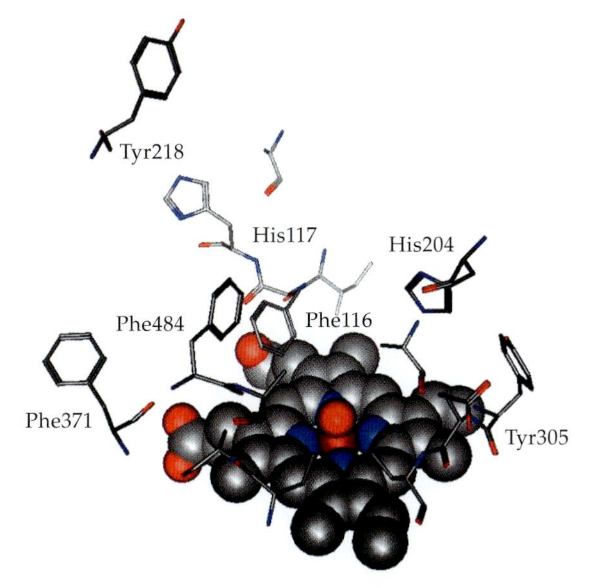

FIGURE 11.2 Molecular Model of CYP6B1v1 from the Furanocoumarin Specialist *Papilio polyxenes* It reveals that the catalytic site contains key aromatic amino acids (Phe116, His117, Phe371, Phe484) that contribute to establishing a resonant network that stabilizes the catalytic site and plays an essential role in maintaining the orientation of furanocoumarin substrates. In CYP6B4 from *Papilio glaucus*, a more functionally generalized enzyme from a more polyphagous species than *Papilio polyxenes*, replacement of Phe371 and Phe484 by aliphatic amino acids opens up the catalytic site to a greater diversity of substrates. (Adapted from Baudry et al. 2003.)

CYP6B3, a second P450 locus encoding an enzyme sharing 88% sequence identity with CYP6B1, including the aromatic amino acid resonant network, is inducible by a wider range of furanocoumarins in a greater variety of tissues than is the CYP6B1 locus (Peterson et al. 2001). The CYP6B1 and CYP6B3 proteins metabolize linear and angular furanocoumarins, but their activities toward these differ in magnitude and rank order (Wen et al. 2006). Whereas CYP6B3 is less efficient at metabolizing both linear and angular furanocoumarins, it displays greater activity against another group of biosynthetically distinct secondary metabolites, the furanochromones (Wen et al. 2006; see Figure 11.1), which are found in several of the apiaceous host species of *P. polyxenes* (Borges et al. 2008). Given the extent of their sequence identity (88%), these two genes likely arose from a relatively recent gene duplication event and have undergone independent but complementary changes such that together the two duplicates expanded upon the metabolic capabilities of the ancestral gene (but without loss of the ancestral function).

A comparison of seven CYP6B3 allelic variants with three CYP6B1 variants in *P. polyxenes* reveals different patterns of substitution in SRS regions in comparison to the overall coding sequence (CDS). The three CYP6B1 variants display 29 nucleotide substitutions, with two-thirds (19 of 29) of these being synonymous; no nonsynonymous substitutions occur in SRS regions (which disproportionately affect catalytic activity). This analysis suggests that the catalytic activity (against furanocoumarins, present in all known hosts) of CYP6B1 is so essential that there are multiple constraints on its variability and that it was the subject of intense purifying selection after the duplication event that led to the evolution of CYP6B3. In contrast, nucleotide heterogeneity is considerably greater among CYP6B3 variants, with a total number of 88 nucleotide substitutions. Almost half (40 of 88 or 45%) of the nucleotide substitutions in CYP6B3 are nonsynonymous; of the 40 nonsynonymous substitutions, 3 occur in SRS regions. Thus, although also under purifying selection after the duplication event, CYP6B3 appears to have evolved with fewer constraints, particularly in its SRS regions, consistent with neofunctionalization after duplication (and in the process facilitating colonization of apiaceous host plants that produce the comparatively rarer furanochromones).

Key Innovations in the Evolution of Polyphagy

In contrast to the vast majority of *Papilio* species, two species, *P. glaucus* and *P. canadensis* (the eastern and Canadian tiger swallowtails), are polyphagous, utilize multiple host plant families, and rarely, in the case of *P. glaucus*, or never, in the case of *P. canadensis*, feed on plants containing furanocoumarins (Li et al. 2003). The furanocoumarin-inducible CYP6B4 protein in *P. glaucus* metabolizes a broader range of both linear and angular furanocoumarins than the CYP6B1 and CYP6B3 proteins in the specialist

P. polyxenes (Li et al. 2003), although the CYP6B4 protein generally has lower activity against these substrates. The putative closest relative of the polyphagous *glaucus* group, *P. multicaudatus* (Caterino and Sperling 1999), represents a presumptive intermediate oligophagous ancestral condition, feeding on three host plant families—only one of which produces furanocoumarins, with the vast majority of *Papilio* species representing the almost certain ancestral condition of feeding exclusively on furanocoumarin-containing host plants. The furanocoumarin-inducible CYP6B33 protein in *P. multicaudatus* metabolizes at least two angular furanocoumarins and four linear furanocoumarins with activities intermediate between those of the specialist CYP6B1 and the generalist CYP6B4 proteins (Mao et al. 2006a, 2007b).

Sequence comparisons of *Papilio* CYP6B proteins suggest that only a few mutations in the catalytic site effect fundamental changes in their metabolic activities, leading to acquisition of a greater range of substrates without complete loss of ancestral furanocoumarin-metabolizing activities (Li et al. 2003; Wen et al. 2006; Mao et al. 2007b). Molecular models for the CYP6B1, CYP6B4, and CYP6B33 proteins were developed and docked with the linear furanocoumarin xanthotoxin (Baudry et al. 2003; Li et al. 2003). In CYP6B1 from *P. polyxenes*, Phe116 and His117 in SRS1, Phe371 in SRS5, and Phe484 in SRS6, by forming the (aforementioned) resonant network, they stabilize the catalytic site and accommodate xanthotoxin and other linear furanocoumarins. In CYP6B33 and CYP6B4 from *P. multicaudatus* and *P. glaucus*, two of these residues (Phe116 and His117 in SRS1), are conserved, whereas two others (Phe371 in SRS5 and Phe484 in SRS6) are replaced by alternate hydrophobic and hydrophilic residues (Mao et al. 2007b). Variations at position 481 can have effects on substrate specificity in enzymes lacking an aromatic amino acid at position 484, which in CYP6B1 helps to stabilize the catalytic site (Chen et al. 2002). At this position in the catalytic site, the nonconservative switch of Ser-to-Lys is predicted to increase the distance from the heme-bound oxygen to the attack site on the furan ring of xanthotoxin without otherwise profoundly altering the catalytic site. At this distance, both CYP6B4 and CYP6B33 metabolize xanthotoxin at slower rates than that of CYP6B1 (Wen et al. 2006), where the distance from the heme-bound oxygen to the furan ring is predicted to be significantly shorter (Baudry et al. 2003). Just one or two amino acid substitutions in critical positions reduce the efficiency of these CYP6B enzymes for furanocoumarin metabolism. However, by enlarging the active pocket, these changes expand the diversity of substrates that can be accommodated; for instance, with CYP6B17, the polyphagous, *P. glaucus,* can metabolize ethoxycoumarin, a hydroxycoumarin that cannot be metabolized by CYP6B1 of the oligophagous *P. polyxenes* (Li et al. 2003).

Thus, within the *Papilio* lineage evolving from specialization on furanocoumarin-containing plants to polyphagy, the loss of host plant specialization that characterizes the *glaucus* group may have been facilitated at least in part by a limited series of mutational changes in P450 enzymes that reduce specificity and enlarge the active pocket, to possibly allow the acquisition of novel substrates without loss of ancestral furanocoumarin metabolic activities. Although total enumeration of xenobiotic-metabolizing CYP6B proteins awaits sequencing, current accumulated work suggests that the most polyphagous species in the *Papilio* lineage have the largest inventory of furanocoumarin-inducible P450s with the broadest substrate capacity (but, with respect to particular substrates, generally lower activity).

Key Innovations in Population Differentiation and Phenotype Matching

In contrast to *Papilio* species, which specialize on foliage of species across the family Apiaceae that contain primarily linear furanocoumarins, the parsnip webworm *Depressaria pastinacella* (Lepidoptera: Oecophoridae) feeds only on reproductive structures of two apiaceous genera, *Pastinaca* and *Heracleum*, both of which produce both linear and angular furanocoumarins in abundance. This oligophagous caterpillar metabolizes furanocoumarins at rates nearly ten-fold higher than do *Papilio* species (Berenbaum 1990). Certain furanocoumarins in wild parsnip (*P. sativa*) function as resistance factors against webworms (Berenbaum 1990; Zangerl and Berenbaum 1993). Webworm damage selects for increased concentrations of three furanocoumarins, including xanthotoxin, bergapten, and sphondin (Berenbaum et al. 1986; Zangerl and Berenbaum 1993). Genotypes with high levels of these furanocoumarins experience lower fitness in the absence of herbivores than genotypes with lower furanocoumarin content (Zangerl and Berenbaum 1997), indicative of the cost of producing these compounds and of tradeoffs between defense and fitness (see Agrawal et al., Chapter 10). These three furanocoumarins act as resistance factors, at least in part, because webworms are less well equipped to metabolize these compounds than they are other furanocoumarins in their host plants. The extraordinary reciprocity of this intimate association in North America (where the insect essentially has only one other very closely related host plant, and the plant has very few other enemies) is manifested in the phenomenon of chemical phenotype matching, in which the furanocoumarin profile of the parsnip plants in a given population corresponds closely to the furanocoumarin-metabolizing capacity of the webworms in the population infesting those plants (Berenbaum and Zangerl 1998; Zangerl and Berenbaum 2003).

The furanocoumarin-inducible CYP6AB3 protein from *D. pastinacella*, for example, is highly specialized for metabolism of imperatorin, which is the most abundant furanocoumarin in hosts of webworms across their range (Mao et al. 2006b). CYP6AB3 brings about this metabolic transformation via epoxidation but evidently is incapable of metabolizing any other linear or angular furanocoumarin so far tested, making it among the most specialized xenobiotic-metabolizing P450s yet characterized. The only other known substrate for CYP6AB3 is myristicin (Mao et al. 2008), an essential oil phenylpropanoid, which synergizes the toxicity of co-occurring furanocoumarins in less specialized herbivores by competitively inhibiting the P450s that metabolize them (Nitao 1989). The fact that myristicin can be metabolized by this otherwise specialized P450 highlights the extent to which the oligophagous *D. pastinacella* has evolved mechanisms that can circumvent synergistic inhibition. Cloning and heterologous expression of the allelic variants CYP6AB3*v*2 and CYP6AB3*v*1 reveal that they differ significantly in the rate of metabolism of imperatorin (Mao et al. 2007a). Comparisons of the NADPH consumption rates for these variants indicate that the CYP6BA3*v*2 protein utilizes this electron source and the P450 redox partner more rapidly. Molecular modeling of the five amino acid differences between these variants and their potential interactions with P450 reductase suggests that replacement of Val92 on the proximal face of CYP6AB3*v*1 with Ala92 in CYP6AB3*v*2 affects interactions with P450 reductase so as to enhance its catalytic activity. Thus, replacement of a single amino acid can increase activity so substantially that herbivore populations may be able to respond rapidly to selection-mediated increases in host plant defense profiles. Chemical phenotype matching of the sort observed in the interaction between parsnip webworms and its principal North American host plant, *P. sativa*, may thus be a matter of a relatively small number of strategic amino acid replacements (Zangerl and Berenbaum 2003; Zangerl et al. 2008).

Making the Most of Genomic Approaches

Colonization of biochemically novel host plants clearly involves more than a few amino acid substitutions in xenobiotic-metabolizing enzymes: multiple enzyme systems are involved in detoxification alone (Claudianos et al. 2006) and host-finding and host evaluation involve other multigene families (Berenbaum 1990), including olfactory receptor genes, gustatory receptor genes, and odorant-binding protein genes, which have undergone birth-and-death evolution akin to that displayed by the P450 superfamily (Egsontia et al. 2007). Despite the complexity, sequenced genomes and bioinformatic techniques make the molecular characterization of host-shift gene suites a distinct possibility, although inferring the phenotypic consequences of nucleotide changes remains challenging. At present, sequenced genomes are available for only two true herbivorous insects: (1) the monophagous leaf-feeder *Bombyx mori* (domesticated silkworm; Xia et al. 2004), which is restricted to *Morus alba* (common mulberry), and (2) the astonishingly polyphagous *Tribolium castaneum* (red flour beetle), which is essentially omnivorous in its ability to consume almost any kind of dried plant material (*Tribolium* Sequencing Consortium, 2008). Comparisons between taxa separated by many millions of years are difficult to interpret—such are the shortcomings of model organisms. Next-generation sequencing technologies, however, will soon make whole-genome sequencing of lineages constituting adaptive radiations affordable, and new analytic methods for identifying molecular networks should be capable of identifying mechanisms by which genomes are remodeled in the course of adaptive evolution (Chen et al. 2008).

For over 150 years, natural selection and evolution have provided a robust framework for understanding the selective pressures influencing the distribution and abundance of the chemical compounds of plants as well as the reciprocal adaptive process by which these chemical compounds influence the behavior and physiology of plant consumers. Understanding the genetic and genomic underpinnings of the key innovations leading to host shifts in phytophagous insects can provide insights into the ecological circumstances promoting host shifts and fostering evolutionary constraints on host utilization patterns (a subject too involved to discuss in this chapter). Just as the characterization of phytochemical structures provided insights and explanations as to why certain insect herbivores are associated with particular plant groups, characterization of the structure and function of the proteins that allow herbivores to cope with these phytochemicals can provide insights and explanations as to how these associations came to be formed. At the moment, relatively few ecologists and evolutionary biologists are inclined to delve deep enough into molecular modeling to find answers to coevolutionary puzzles. It is worth pointing out, though, that after Fraenkel (1959) suggested that phytochemical distribution and abundance could provide insights into patterns of host plant utilization, a cadre of ecologists, plant biologists, and evolutionary biologists acquired the necessary training to become functional phytochemists. Proteins may be the next group of chemicals to play a central role in understanding chemically mediated coevolution.

Lepidopterous herbivores are distinctive among plant-feeding insects in that different life stages interact with the chemistry of their host plants in fundamentally dissimilar ways. With few exceptions, host plant finding and assessment among lepidopterans is largely the province of adult females, while consumption and digestion is the province of larval stages. This dichotomy between adult behavioral preference and larval physiological performance as well as the timing and nature of adaptation to plant chemistry throughout the life cycle in Lepidoptera is key to longstanding debates over the evolution of specialization in general. Over 60 years ago, Dethier (1948) suggested that: "the first barrier to be overcome in the insect–plant relationship is a behavioral one. The insect must sense and discriminate before nutritional and toxic factors become operative" (98). Thus, Dethier argued for the primacy of adult preference, or detection and response to kairomonal cues in host plant shifts. In contrast, Ehrlich and Raven (1964) reasoned that "after the restriction of certain groups of insects to a narrow range of food plants, the formerly repellent substances of these plants might...become chemical attractants," arguing for the primacy of shared allomonal phytochemistry and larval detoxification in initiating host shifts (with the evolution of kairomonal responses following) (602). Oviposition mistakes, in which females lay eggs on host plants that do not support larval growth (Berenbaum 1981), are cited as evidence for the importance of behavioral cues in initiating host shifts; colonization of novel host plants that share the same range of defense compounds is claimed as evidence of the primacy of larval performance in the process (review by Berenbaum 1990).

In view of the fact that the physiological and behavioral changes necessary to effect a shift to a chemically different host plant arise by random mutation, primacy is in a sense irrelevant. Host shifts are completed when the necessary behavioral and physiological traits are in place, irrespective of the order in which they occur (Berenbaum 1990). Moreover, each of these traits is, itself, complex and likely to involve multiple genetic changes. Multigene families, such as the cytochrome P450s, play critical roles in mediating behavioral preference (e.g., by degrading chemical signals bound to receptors on sensory structures) and physiological performance (e.g., by detoxification of phytochemical defenses) in herbivorous insects in general and Lepidoptera in particular. Host plant finding and assessment by ovipositing females are mediated by volatile long-distance attractants and contact oviposition stimulants, the detection of which depends on the large gene families of olfactory receptors and odorant-binding proteins. By contrast, ingestion and processing of host plant tissues by caterpillars likely involve detoxification of toxins by the principal detoxification enzymes, including cytochrome P450 monooxygenases, glutathione-S-transferases, and carboxylesterases, as well as by olfactory receptors and odorant-binding proteins for detection of phytochemical feeding stimulants. Genomic comparisons of closely related congeners with divergent hosts, using patterns and mechanistic studies of protein structure and function, can provide a blueprint for the number and extent of genetic changes that allow colonization of chemically novel host plants. Such a blueprint is of use in a practical context, such as designing ways to incorporate host plant resistance into integrated pest management programs (see Gould, Chapter 21), but also in understanding opportunities and constraints in the structure and sustainability of natural communities.

Perhaps more important than what genomics can bring to the study of plant–insect interactions is what ecological and evolutionary studies can bring to the study of insect (and plant) genomics (see Rest, Commentary 8). Particularly for environmental response genes (i.e., the phylogenetically unstable lineages that diversify in response to environmental selective forces), gene functions simply cannot be documented in a laboratory. Identifying the full range of substrates of a particular enzyme demands knowledge of the chemical context in which the enzyme evolved. The past century of studies of ecology, genetics, and evolution of plant–insect interactions thus will be invaluable in tackling the daunting work of understanding genomes.

Acknowledgments

We thank the editors for their patience and encouragement and NSFDEB-0816616 for support for some of the work reported in this chapter.

Literature Cited

Amichot, M., S. Tarès, A. Brun-Barale, and 3 others. 2004. Point mutations associated with insecticide resistance in the *Drosophila* cytochrome P450 *Cyp6a2* enable DDT metabolism. *Eur. J. Biochem.* 271: 1250–1257.

Baudry, J., W. Li, L. Pan, and 2 others. 2003. Molecular docking of substrates and inhibitors in the catalytic site of *CYP6B1*, an insect cytochrome P450 monooxygenase. *Prot. Eng.* 16: 577–587.

Bell, M. A., G. Ortí, J. A. Walker, and 1 other. 1993. Evolution of pelvic reduction in threespine stickleback fish: A test of competing hypotheses. *Evolution* 47: 906–914.

Berenbaum, M. R. 1981. An oviposition "mistake" by *Papilio glaucus* (Papilionidae). *J. Lepid. Soc.* 35: 75.

Berenbaum, M. R. 1983. Coumarins and caterpillars: A case for coevolution. *Evolution* 37: 163–179.

Berenbaum, M. R. 1990. Evolution of specialization in insect-umbellifer associations. *Ann. Rev. Entomol.* 35: 319–343.

Berenbaum M. R. 1991. Coumarins. In G. Rosenthal and M. Berenbaum (eds.), *Herbivores: Their Interactions with Secondary Plant Metabolites*, pp. 221–249. Academic Press, New York.

Berenbaum, M. R. 1995. Chemistry and oligophagy in the Papilionidae. In J. M. Scriber (ed.), *Swallowtail Butterflies: Ecology and Evolutionary Biology*, pp. 27–38. Scientific Publishers, Gainseville.

Berenbaum, M. R. and A. R. Zangerl. 1992. Genetics of physiological and behavioral resistance to host furanocoumarins in the parsnip webworm. *Evolution* 46: 1373–1384.

Berenbaum, M. R. and A. R. Zangerl. 1998. Chemical phenotype matching between a plant and its insect herbivore. *Proc. Natl. Acad. Sci. USA* 95: 13743–13748.

Berenbaum, M. R., C. Favret, and M. A. Schuler. 1996. On defining "key innovations" in an adaptive radiation: Cytochrome P450s and Papilionidae. *Am. Nat.* 148: S139–S155.

Berenbaum, M. R., A. R. Zangerl, and J. K. Nitao. 1986. Constraints on chemical coevolution: Wild parsnips and the parsnip webworm. *Evolution* 40: 1215–1228.

Borges, M. L., L. Latterini, F. Elisei, and 4 others. 2008. Photophysical properties and photobiological activity of the furanochromones visnagin and khellin. *Photochem. Photobiol.* 67: 184–191.

Brattsten, L. B. 1979a. Biochemical defense mechanisms in herbivores against plant allele chemicals. In G. A. Rosenthal and D. H. Janzen (eds.), *Herbivores: Their Interaction with Secondary Plant Metabolites*, pp. 199–271. Academic Press, New York.

Camerarius, R. J. 1694. *De Sexu Plantarum Epistola* [Letter on the Sex of Plants]. Translated into German by M. Mobius, in Ostwald's *Klassiker der ExaktenWissenschaften* [Classics of the Exact Sciences]. Engelmann, Leipzig.

Caterino, M. S. and F. A. H. Sperling. 1999. *Papilio* phylogeny based on mitochondrial cytochrome oxidase I and II genes. *Mol. Phy. Evol.* 11: 122–137.
Chan, Y. F., M. E. Marks, F. C. Jones, and 13 others. 2010. Adaptive evolution of pelvic reduction in sticklebacks by recurrent deletion of a *Pitx1* enhancer. *Science* 327: 302–305.
Chen, J.-S., M. R. Berenbaum, and M. A. Schuler. 2002. Amino acids in SRS1 and SRS6 are critical for furanocoumarin metabolism by *CYP6B1v1*, a cytochrome P450 monooxygenase. *Insect Mol. Biol.* 11: 175–186.
Chen, Y., J. Zhu, P. Y. Lum, and 19 others. 2008. Variations in DNA elucidate molecular networks that cause disease. *Nature* 452: 429–435.
Cittadino, E. 1990. *Nature As the Laboratory: Darwinian Plant Ecology in the German Empire, 1880–1900*. Cambridge University Press, Cambridge, UK.
Claudianos, C., H. Ranson, R. Johnson, and 5 others. 2006. A deficit of detoxification enzymes: Pesticide sensitivity and environmental response in the honey bee. *Insect Mol. Biol.* 15: 615–636.
Cohen, M. B., M. A. Schuler, and M. R. Berenbaum. 1992. A host-inducible cytochrome P450 from a host-specific caterpillar: Molecular cloning and evolution. *Proc. Natl. Acad. Sci. USA* 89: 10920–10924.
Consortium HGS. 2006. Insights into social insects from the genome of the honeybee *Apis mellifera*. *Nature* 443: 931–949.
Darwin, C. R. 1859. *On the Origin of Species by Means of Natural Selection, or the Preservation of Favoured Races in the Struggle for Life*. John Murray, London.
Darwin, C. R. 1862. *On the Various Contrivances by which British and Foreign Orchids are Fertilised by Insects, and on the Good Effects of Intercrossing*. John Murray, London.
Darwin, C. R. 1877. *The Different Forms of Flowers on Plants of the Same Species*. John Murray, London.
Darwin, C. R. 1902. *On the Origin of Species: A Facsimile of the First Edition*. Grant Richards, London.
De Montellano, P. R. 2004. *Cytochrome P450: Structure, Mechanism and Biochemistry*. Springer, New York.
Dethier, V. G. 1948. *Chemical Insect Attractants and Repellents*. H. K. Lewis, London.
Domanski, T. L. and J. R. Halpert. 2001. Analysis of mammalian cytochrome P450 structure and function by site-directed mutagenesis. *Curr. Drug Metab.* 2: 117–137.
Egsontia, P., A. P. Sanderson, M. Cobb, and 3 others. 2007. The red flour beetle's large nose: An expanded odorant receptor gene family in *Tribolium castaneum*. *Insect Biochem. Mol. Biol.* 38: 387–397.
Ehrlich, P. R. and P. H. Raven. 1964. Butterflies and plants: A study in coevolution. *Evolution* 18: 586–608.
Feyereisen, R. 1999. Insect P450 enzymes. *Ann. Rev. Entomol.* 44: 507–533.
Fischer, H. M., C. W. Wheat, D. G. Heckel, and 1 other. 2008. Evolutionary origins of a novel host plant detoxification gene in butterflies. *Mol. Biol. Evol.* 25: 809–820.
Fraenkel, G. S. 1953. The nutritional value of green plants for insects. In *Transactions of the IXth International Congress of Entomology*, pp. 90–100. W. Junk, The Hague.
Fraenkel, G. S. 1959. The *raison d'être* of secondary plant substances. *Science* 129: 1466–1470.
Fraenkel, G. 1984. Citation classic: The *raison d'etre* of secondary plant substances. *Current Contents/Life Sci.* 11: 18.

Gonzalez, F. J. and D. W. Nebert. 1990. Evolution of the P450 gene superfamily: Animal-plant "warfare," molecular drive and human genetic differences in drug oxidation. *Trends Genet.* 6: 182–186.
Gotoh, O. 1992. Substrate recognition sites in cytochrome P450 family 2 (CYPB) proteins inferred from comparative analyses of amino acid and coding nucleotide sequences. *J. Biol. Chem.* 267: 83–90.
Hartmann, T. 2008. The lost origin of chemical ecology in the late 19th century. *Proc. Natl. Acad. Sci. USA* 105: 4541–4546.
Honey Bee Genome Sequencing Consortium 2006. Insights into social insects from the genome of the honeybee *Apis mellifera. Nature* 443: 931–949.
Hung, C.-F., T. L. Harrison, M. R. Berenbaum, and 1 other. 1995a. *CYP6B3*: A second furanocoumarin-inducible cytochrome P450 expressed in *Papilio polyxenes. Insect Mol. Biol.* 4: 149–160.
Hung, C.-F., H. Prapaipong, M. R. Berenbaum, and 1 other. 1995b. Differential induction of cytochrome P450 transcripts in *Papilio polyxenes* by linear and angular furanocoumarins. *Insect Biochem. Mol. Biol.* 25: 89–99.
Huxley, T. H. 1856. On natural history, as knowledge, discipline, and power. *Proc. Roy. Inst.* 2: 187–195.
Janzen, D. H. 1980. When is it coevolution? *Evolution* 34: 611–612.
Kerner, A. 1876. *Schutzmittel der Blüthen gegen unberufene Gäste* [Protective Means of Flowers against Unbidden Guests]. Wagner'sche Universitäts-Buchhandlung, Innsbruck (English translation 1878).
Kliebenstein, D. J., V. M. Lambrix, M. Reichelt, and 2 others. 2001. Gene duplication in the diversification of secondary metabolism: Tandem 2-oxoglutarate-dependent dioxygenases control glucosinolate biosynthesis in *Arabidopsis. Plant Cell* 13: 681–693.
Larbat, R., A. J. Hehn, S. Hans, and 5 others. 2009. Isolation and functional characterization of CYP71AJ4 encoding for the first P450 monooxygenase of angular furanocoumarin biosynthesis. *J. Biol. Chem.* 284: 4776–4785.
Lee, T.-S. 2008. Reverse conservation analysis reveals the specificity determining residues of cytochrome P450 family 2 (CYP 2). *Evol. Bioinform. Online* 4: 7–16.
Li, W., M. A. Schuler, and M. R. Berenbaum. 2003. Diversification of furanocoumarin metabolizing cytochrome P450s in two papilionids: Specificity and substrate encounter rate. *Proc. Natl. Acad. Sci. USA* 100: 14593–14598.
Linnaeus, C. 1758. *Systema naturae per regna tria naturae: Secundum classes, ordines, genea, species, cum characteribus, differentiis, synoymis, locis* 10th ed. Holmiae (Laurentii Salvi).
Ma, R., M. B. Cohen, M. R. Berenbaum, and 1 other. 1994. Black swallowtail (*Papiliopolyxenes*) alleles encode cytochrome P450s that selectively metabolize linear furanocoumarins. *Arch. Biochem. Biophys.* 310: 332–340.
Mao, W., M. A. Berhow, A. R. Zangerl, and 2 others. 2006a. Cytochrome P450-mediated metabolism of xanthotoxin by *Papilio multicaudatus. J. Chem. Ecol.* 32: 523–536.
Mao, W., S. Rupasinghe, A. R. Zangerl, and 2 others. 2006b. Remarkable substrate-specificity of *CYP6AB3* in *Depressaria pastinacella*, a highly specialized caterpillar. *Insect Mol. Biol.* 15: 169–179.
Mao, W., S. Rupasinghe, A. Zangerl, and 2 others. 2007b. Allelic variation in the *Depressaria pastinacella* CYP6AB3 protein enhances metabolism of plant allelochemicals by altering a proximal surface residue and potential interactions with cytochrome P450 reductase. *J. Biol. Chem.* 282: 10544–10-552.

Mao, W., M. A. Schuler, and M. R. Berenbaum. 2007a. Cytochrome P450s in *Papilio multicaudatus* and the transition from oligophagy to polyphagy in the Papilionidae. *Insect Mol. Biol.* 16: 481–490.
Mao, W., A. R. Zangerl, M. R. Berenbaum, and 1 other. 2008. Metabolism of myristicin by *Depressaria pastinacella CYP6AB3v2* and inhibition by its metabolite. *Insect Biochem. Mol. Biol.*38: 645–651.
Mauricio, R. and M. D. Rausher. 1997. Experimental manipulation of putative selective agents provides evidence for the role of natural enemies in the evolution of plant defense. *Evolution* 51: 1435–1444.
Nelson, D. R. 2009. The cytochrome P450 homepage. *Hum. Genomics* 4: 59–65.
Nelson, D. R., R. Ming, M. Alam, and 1 other. 2008. Comparison of cytochrome P450 genes from six plant genomes. *Trop. Plant Biol.* 1: 216–235.
Nitao, J. K. 1989. Enzymatic adaptation in a specialist herbivore for feeding on furanocoumarin-containing plants. *Ecology* 70: 629–635.
Padian, K. 2008. Darwin's enduring legacy. *Nature* 451: 632–634.
Pan, L., Z. Wen, J. Baudry, and 2 others. 2004. Identification of variable amino acids in the SRS1 region of *CYP6B1* modulating furanocoumarin metabolism. *Arch. Biochem. Biophys.* 422: 31–41.
Petersen, R. A., A. R. Zangerl, M. R. Berenbaum, and 1 other. 2001. Expression of *CYP6B1* and *CYP6B3* cytochrome P450 monooxygenases and furanocoumarin metabolism in different tissues of *Papilio polyxenes* (Lepidoptera: Papilionidae). *Insect Biochem. Mol. Biol.* 31: 679–690.
Rausher, M. D. 1996. Genetic analysis of coevolution between plants and their natural enemies. *Trends Genet.* 12: 212–217.
Rosenthal, G. A. and M. R. Berenbaum (eds.). 1992. *Herbivores: Their Interaction with Secondary Plant Metabolites*. Academic Press, New York.
Roth, C., S. Rastogi, L. Arvestad, and 4 others. 2007. Evolution after gene duplication: Models, mechanisms, sequences, systems, and organisms. *J. Exp. Zool. (Mol. Dev. Evol.)* 308B: 58–73.
Rothschild, W. and K. Jordan. 1903. A revision of the Lepidopterous family Sphingidae. *Novit. Zool.* 9 (Suppl.): 1–972.
Schluter, D. and J. D. McPhail. 1992. Ecological character displacement and speciation in sticklebacks. *Am. Nat.* 140: 85–108.
Schluter, D., T. D. Price, A. Ø. Mooers, and 1 other. 1997. Likelihood of ancestor states in adaptive radiation. *Evolution* 51: 1699–1711.
Shapiro, M. D., M. E. Marks, C. L. Peichel, and 5 others. 2004. Genetic and developmental basis of evolutionary pelvic reduction in threespine stickleback. *Nature* 428: 717–723.
Simpson, G. G. 1944. *Tempo and Mode in Evolution*. Columbia University Press, New York.
Sprengel, C. K. 1793. *Das entdeckte Geheimnis der Natur im Bau und in der Befruchtung der Blumen*. [*The Revealed Secret of Nature in the Structure and Fruiting of Flowers*]. Friedrich Vieweg, Berlin.
Stahl, E. 1888. Pflanzen und Schnecken: Eine biologische Studie über die Schutzmittel der Pflanzen gegen Schneckenfrass [Plants and snails: A biological study concerning means of protection in plants against damage by snails]. *Jenaer Zeitschr. Medizin Naturwissenschaften* 22: 557–684.
Thomas, J. H. 2007. Rapid birth-death evolution specific to xenobiotic cytochrome P450 genes in vertebrates. *PLoS Genet.* 3: 720–727.
Tribolium Genome Sequencing Consortium. 2008. The genome of the model beetle and pest *Tribolium castaneum*. *Nature* 452: 949–955.

Wen, Z., J. Baudry, M. R. Berenbaum, and 1 other. 2005. Ile115Leu mutation in the SRS1 region of an insect cytochrome P450 (*CYP6B1*) compromises substrate turnover via changes in a predicted product release channel. *Prot. Eng. Design Select.* 18: 191–199.

Wen, Z., L. Pan, M. R. Berenbaum, and 1 other. 2003. Metabolism of linear and angular furanocoumarins by *Papilio polyxenes CYP6B1* coexpressed with NADPH cytochrome P450 reductase. *Insect Biochem. Mol. Biol.* 33: 937–947.

Wen, Z., S. Rupasinghe, G. Niu, and 2 others. 2006. *CYP6B1* and *CYP6B3* of the black swallowtail (*Papilio polyxenes*): Adaptive evolution through subfunctionalization. *Mol. Biol. Evol.* 23: 2434–2443.

Wheat, C. W., H. Vogel, U. Wittstock, and 3 others. 2007. The genetic basis of a plant-insect coevolutionary key innovation. *Proc. Natl. Acad. Sci. USA* 104: 20427–20431.

Whittaker, R. H. and P. P. Feeny. 1971. Allelochemics: Chemical interactions between species. *Science* 171: 757–770.

Xia, Q. Y., Z. Y. Zhou, C. Lu, and 92 others. 2004. A draft sequence for the genome of the domesticated silkworm (*Bombyx mori*). *Science* 306: 1937–1940.

Zangerl, A. R. and M. R. Berenbaum. 1993. Plant chemistry, insect adaptations to plant chemistry, and host plant utilization patterns. *Ecology* 74: 47–54.

Zangerl, A. R. and M. R. Berenbaum. 1997. Cost of chemically defending seeds: Furanocoumarins and *Pastinaca sativa*. *Am. Nat.* 150: 491–504.

Zangerl, A. R. and M. R. Berenbaum. 2003. Phenotype matching in wild parsnips and parsnip webworms: Causes and consequences. *Evolution* 57: 806–815.

Zangerl, A. R., M. C. Stanley, and M. R. Berenbaum. 2008. Selection for chemical trait remixing in an invasive weed after reassociation with a coevolved specialist. *Proc. Natl. Acad. Sci. USA* 105: 4547–4552.

Zepernick, B. and W. Meretz. 2001. Christian Konrad Sprengel's life in relation to his family and his time. On the occasion of his 250th birthday. *Willdenowia* 31: 141–152 *http://www.bgbm.org/willdenowia/w-pdf/w31-1Zepernick+Meretz.pdf.*

Chapter 12

Behavioral Ecology: The Natural History of Evolutionary Theory

Hanna Kokko and Michael D. Jennions

Instincts are as important as corporeal structures for the welfare of each species, under its present conditions of life. Under changed conditions of life, it is at least possible that slight modifications of instinct might be profitable to a species; and if it can be shown that instincts do vary ever so little, then I can see no difficulty in natural selection preserving and continually accumulating variations of instinct to any extent that was profitable. It is thus, as I believe, that all the most complex and wonderful instincts have originated (Darwin 1859: 209).

In Chapter 7 of *The Origin of Species* Darwin confidently stated that "instincts" could evolve under selection in the same way as could the morphological characters of domesticated and wild animals and plants, which he had discussed in earlier chapters. His views on the evolution of behavior agree surprisingly well with those of modern-day behavioral ecologists. In this chapter, we track some of the developments that led to behavioral ecology, discuss why it is often seen as a separate field from evolutionary biology, and assess whether this distinction is warranted. Finally, we comment on the potential consequences for the future of this field.

Darwin used the term instincts for behaviors that appear to require little or no learning and then wrote something that is unlikely to have satisfied ethologists, who, a century later, made their careers distinguishing between innate and learned behaviors:

I will not attempt any definition of instinct. It would be easy to show that several distinct mental actions are commonly embraced by this term; but everyone understands what is meant, when it is said that instinct impels the cuckoo to migrate and to lay her eggs in other bird's nests (Darwin 1859: 207).

Darwin wanted to distinguish between instincts and habits (i.e., learned behaviors). Doing so allowed him to separate the evolution of behavior through Lamarckian inheritance of acquired traits (e.g., learned acts), which he considered rare, from selection acting on variation in the *propensity* to act in a certain way in a given environment (i.e., instincts). In short, Darwin argued that behavior could evolve by exactly the same selection process that he had described for morphological traits. Unfortunately, Darwin's ignorance of the rules of inheritance led him to prefer Lamarckian inheritance in subsequent editions of *The Origin of Species* (Bowler 1989). Nonetheless, his initial writing shows that he recognized that certain behaviors could be treated as traits, even if distinguishing learned and innate behavior is difficult compared to morphological variation. From the outset, Darwin argued that the evolution of behavior by natural selection is worthy of scientific investigation.

The kinship between Darwin and behavioral ecologists is based on at least two shared traits. The first one is an emphasis on the study of plants and animals in nature, supplemented by studies in captivity. The second is a curious tension between attraction to studying the quirky or bizarre (e.g., the various contrivances by which orchids are fertilized, the sex lives of barnacles, the formation of vegetable mould) and the search for generalities. For both Darwin and behavioral ecologists the tension is resolved whenever the theory of natural selection transforms the unintelligible into the explicable. A widely acknowledged strength of Darwin's case was his accumulation of evidence from multiple sources. This approach continues in behavioral ecology to this day. Despite the high-profile status of a few model species (e.g., blue tits, red deer, collared flycatchers, and guppies), every year behavioral ecologists publish studies on species of which little is known so far.

The Fate of Behavioral Studies after Darwin

Darwin was interested in behavior throughout his life, with a highlight being the publication of *The Expression of the Emotions in Man and Animals* in 1872, but culminating in a letter to *Nature* on cuckoo behavior less than a year before he died in 1882. After Darwin's death, however, behavioral studies made only a minor contribution to the advancement of evolutionary biology. Indeed, rediscovery of Mendel's work and the rise of genetics brought a temporary decline in the value placed on Darwin's insights (i.e., "the eclipse of Darwinism," according to Bowler 1989) that arguably lasted until the 1940s. Darwin never understood how inheritance worked, and some of his writings (especially credence of Lamarckism in later editions of *The Origin of Species*) contradicted the findings of early geneticists. Then, in 1918, Fisher reconciled the conflict between the continuous variation of most traits that interest field biologists with the discrete inheritance of Mendelian genetics (Fisher 1918). Over the next 20 years, rapid progress in population and quantitative genetics produced an integrated theoretical framework that illuminated the roles of selection, drift, and mutation in evolution of gene frequencies and phenotypes. The emphasis of population genetics on differential survival of variants reinstated Darwinian theory.

Darwin's view that natural selection is central to evolution was embodied by what Huxley (1942) termed "the modern synthesis." This synthesis incorporated input from luminaries like G. G. Simpson, E. Mayr, and T. Dobzhansky working in paleontology, systematics, and genetics, respectively. However, natural history, which had so delighted Darwin, was much reduced (Gardner 2010). The mathematical theory underlying the modern synthesis was couched in abstract terms. For example, although Fisher's fundamental theorem or Wright's adaptive landscapes could have been illustrated using examples from natural history, they were usually presented as the products of mathematical abstraction from patterns detected by geneticists. There were no data on gene flow among populations or on the genetic basis of variation in wild animals. The natural environment became almost erased from the theory of natural selection.

This is not to say that the evolution of instincts was entirely neglected, but Fisher and his colleagues showed little inclination to work out the mathematics of specific selection scenarios that are everyday events in nature (e.g., when to fight and when to cooperate; when to feed and when to hide). Wider acceptance of the scientific value of behavioral field studies had to wait.

The Rise of Ethology

The more recent origins of behavioral ecology can be traced back to the merger of behaviorism, comparative psychology, and ethology. Behaviorists, led by J. B. Watson (1924), attempted to remove introspection from psychology and restrict analyses to observed behaviors. Despite the limits of their approach when dealing with species for which a theory of the mind seems essential (e.g., work on vervet monkeys by Cheney and Seyfarth 1990), their strict empiricism ensured that behavioral studies became increasingly quantitative and less anthropomorphic. Ironically, however, behavioral ecology is awash with apparent anthropomorphism, yet this practice is usually little more than stating that selection favors animals that act as though they were consciously trying to achieve a goal (e.g., forage optimally). Behaviorists provided the first tools to identify and quantify behavioral traits.

Comparative psychologists embraced the experimental approach that subsequently proved critical for progress in behavioral ecology. Debate among comparative psychologists about the goals of their research ensured the proper use of phylogenetic methods and selection of study organisms. Was the research goal to seek generalities or taxon-specific differences, to reconstruct the history of a behavior or to identify general mechanisms of adaptation? Answering this question offered insight into the appropriate use of model species. Were these species stand-ins that could be used to explore human behavior or simply representatives of a higher taxon? This debate also clarified the need to tailor experimental tests to the species, rather than using a standard methodology to place species along a scale of nature. All these issues are reviewed in a landmark paper by Hodos and Campbell (1969). However, most contemporary behavioral ecologists see their immediate ancestors as ethologists (especially if Americans who conduct field-based animal behavior studies are included).

Ethologists, such as the Nobel Prize winners Konrad Lorenz and Niko Tinbergen, rose to prominence in the post-war period. They shared Darwin's approach to the study of behavior: (1) observe animals (if possible, in the wild), (2) perform simple experiments (with which Darwin's books are peppered), and (3) make inferences about nature. Ethologists differed from Darwin, however, in that most were interested in *how* biological systems generate behavior. They were curiously uninterested in questions about the adaptive value of behavior that had so strongly motivated Darwin's research. Ethologists paid lip service to natural selection but made little effort to link their results to evolutionary theory. For example, Tinbergen's (1963), landmark paper that explicitly deals with the survival value of behaviors does not cite Fisher, Wright, or Haldane—or, for that matter, Darwin. Interestingly, however, he notes that "[i]t is through [Lorenz's] interest in survival value that he appealed so strongly to naturalists, to people who saw the whole animal in action in its surroundings" (Tinbergen 1963: 417).

Interest in survival value is the hallmark of modern behavioral ecology. With hindsight, however, the efforts of early ethologists to understand behavioral adaptations appear unambitious and theoretically naïve. They studied traits with relatively obvious survival value, but behavioral ecologists rely far more on evolutionary theory to generate more sophisticated questions than did the ethologists to produce answers that were not previously obvious. For example, in colonies of eusocial ants, how is the fitness of workers and queens affected by the offspring sex ratio for reproductives? What behavioral strategies arise if there is queen–worker conflict (Trivers and Hare 1976)? Finally, might behavioral strategies explain why queens mate many times on their nuptial flight before founding a colony? Behavioral ecology is more than a showcase to assemble disconnected facts on the natural history of innumerable species. It can resolve challenging questions about the evolution of conflict and cooperation and what behaviors will evolve in response to selection when organisms compete for limited resources.

The Four Questions

The claim that ethologists were uninterested in "why" questions is slightly unfair. Tinbergen (1963) enumerated three problems to determine whether a trait is adaptive. First, identifying how natural selection has shaped a trait is not equivalent to determining its current survival value (which is all that currently can be measured). This debate is ongoing and was recapitulated decades later in a report contrasting definitions of adaptation in regard to the selective history of a trait (Orzack and Sober 2001) versus its current survival value (Reeve and Sherman 1993). Second, it is more difficult for a morphological trait to infer whether the absence of a behavioral trait will affect survival. Third, there are major practical difficulties in manipulating behavior. Experimental manipulation is usually required to identify causation, and it is generally easier to alter morphological than behavioral traits. These concerns led many ethologists to sidestep adaptive questions about behavior.

Arguably, however, Tinbergen's (1963) most enduring contribution was the way in which he codified the famous "four questions" in biology. Giving due credit, he noted that Julian Huxley had identified three major problems that biologists tackle: *survival value*, *evolution*, and *causation*. To these Tinbergen added a fourth that is of special relevance to behavior: *ontogeny*. Tinbergen's first two questions are known as "ultimate" and the latter as "proximate" (Dewsbury 1992). These problems are best illustrated by examples of the questions they generate and the kinds of answers that are provided.

1. *What is the survival value or function of the behavior?* What is the behavior *for*, and why does it increase survival (or, in the case of sexually selected traits, why does it elevate fertilization success)? For example, humans avoid incest because inbreeding compromises offspring viability (see following discussion).
2. *Evolution: When did the behavior evolve into its current form?* As Tinbergen (1963) noted, behavior rarely leaves a fossil record. However, if a behavior is mapped onto a phylogeny, one can infer its origin for the study species and possibly use a molecular clock to date the event. Analysis of the evolution of a behavior is also problematic if it is ubiquitous (e.g., inbreeding avoidance). This characteristic might imply that it evolved once in the distant past, but a real problem emerges if several mechanisms generate identical behavioral outcomes. For example, most birds and mammals avoid inbreeding but probably do so using very different sensory cues. Then we need to understand proximate mechanisms to distinguish homologous and analogous behaviors.
3. *Causation: What factors are required for the behavior to occur?* In migratory birds, shorter day length triggers physiological changes that heighten the likelihood of departure. With inbreeding avoidance, chemical cues may decrease sexual motivation towards kin (Sherborne et al. 2007; Paterson and Hurst 2009).

4. *Ontogeny: How does the behavior develop?* For example, being raised together triggers incest avoidance in humans, regardless of true kinship (Lieberman et al. 2007). In other species, notably birds, inbreeding avoidance is thought to arise through sexual imprinting on opposite-sex parents (but see Schielzeth et al. 2008) or familial imprinting (Penn and Potts 1998) and subsequent avoidance of mates with similar phenotypes.

Ignore Your Instincts: The Birth of Behavioral Ecology

Ethologists emphasized behaviors that were stereotyped, species-specific, did not require practice, and appear fully formed in isolated animals. Lehrman (1953) famously challenged these aspects of Lorenz's theory of instinctive behavior, because they diminished any role for ontogeny, which was unsatisfactory to comparative psychologists investigating learning (i.e., organism–environment interactions). Currently, biologists dismiss any nature–nurture dichotomy as misleading and identify a continuum of behaviors that vary by degree for phenotypic plasticity. Nevertheless, misunderstandings still abound (Pinker 2002; Ridley 2003). Innate and instinctive are terms now rarely used because of their historical baggage, but behaviors can be described as innate if they occur in most of the natural environments in which an organism may develop (i.e., show minimal phenotypic plasticity). In some temperate birds, for example, all individuals migrate in response to shorter day length cues. Migration is therefore considered innate, even though there are conceivable artificial environments in which migration might not occur.

Semantics aside, the existence of instincts is crucial for behavioral ecologists: animals must vary in their behavioral propensities for evolution by selection. However, behavioral ecologists differ from ethologists (as characterized by Lehrman 1953) in two key respects. First, behavioral ecologists focus on behavioral variation among animals and discount the restrictive concept that only stereotyped behaviors are innate. Darwin himself readily attributed significance to the fact that three cats, "according to Mr. St. John," each preferred to hunt for different prey. Identifying variation is a prerequisite to measuring selection on behavior. Second, behavioral ecologists acknowledge that behavioral variation can be generated by environmental differences (as a result of learning *and* other factors), but they are less interested in the ontogeny of variation. It is simply assumed that, at some point, selection on heritable variants has generated adaptive behavior (e.g., regardless of whether this is due to selection among different learning strategies or fixed responses to specific stimuli). Research is directed towards investigating the survival value of different behavioral phenotypes.

An emphasis on stereotyped behavior rather than on variation meant that ethology was relatively uninformed by evolutionary theory, which invokes selection, drift, and gene flow. Of course, behavioral work with important ramifications for evolutionary theory was still conducted. For example, research on *Drosophila* reproductive behavior by Bateman (1948) and Maynard Smith (1956) later proved pivotal to the development of sexual selection and parental investment theory (Trivers 1972; Lande 1981). Pioneering field studies of natural selection on morphological and life history traits showed how one might integrate behavioral studies and evolutionary theory. Indeed, field studies of natural selection by E. B. Ford, Phillip Sheppard, A. J. Cain, Bernard Kettlewell, and David Lack began to transform ethology. Eventually, trained ethologists like Richard Dawkins, Bert Hölldobler, Jack Bradbury, and many others helped establish the new field of behavioral ecology.

Was there a key event that transformed ethology into behavioral ecology? Many suggest that the tipping point was the *levels of selection debate*. The modern synthesis was established by the 1940s, but even in the 1960s, biologists still presented muddled arguments to explain particular adaptations. Poor mathematical training meant that general messages from theoretical population genetics were ignored in analyses of the evolution of specific traits. This problem might not have arisen if Fisher and others had extended their general theory to develop formal models for, say, how male traits evolve when females choose mates. They did not, and many ethologists thereafter published studies that invoked species-level advantage or even stranger *just-so* stories to explain behavior. One example is the claim that male fiddler crabs (Figure 12.1) wave in synchrony to burn off excess energy (Gordon 1958). Recent research by behavioral ecologists shows that synchrony is a byproduct of female preference for males that wave first (Reaney et al. 2008). Lorenz himself repeatedly wrote that instincts curb aggression and limit damage to conspecifics in species with dangerous weaponry, to the benefit of the species. Even Julian Huxley argued that ritualized fighting evolved to limit intraspecific damage (Huxley 1966).

Although both morphological and behavioral traits were misattributed to species and group-level selection, this was a particular problem for behavioral traits. This bias is not surprising because behavior offers clear examples of apparent individual restraint (e.g., helping another adult rear offspring) that evoke group-selectionist thinking. The potential damaging effects of behavior on others are also more conspicuous than those that arise from morphology. We think this distinction helped behaviorists to become sensitive to levels of selection: it ensured that ethologists, albeit belatedly, embraced the insights of modern evolutionary theory sooner than, for example, physiologists or neurobiologists.

Fuzzy thinking about levels of selection came to a head when Vero Wynne-Edwards published *Animal Dispersion in Relation to Social Behavior* in 1962. This book offered adaptive explanations for behaviors that allegedly benefited the group at a cost to the individual. For example, night roosting by corvids was hypothesized to occur so that birds could estimate their numbers and avoid over-breeding and depleting the local food supply.

FIGURE 12.1A–D Appreciating Variation among Individuals Was One of Darwin's Key Insights Ethologists later largely ignored variation within species, while behavioral ecologists returned it to center stage. All of these photographs depict males of the fiddler crab *Uca capricornis*. Interestingly, its seemingly irrelevant carapace pattern variation has functional importance. Males use these distinctive patterns to discriminate between foreign and neighboring females (female carapaces are similarly variable). (From Detto et al. 2006; photographs © Tanya Detto.)

Similar group selection arguments had been expressed previously, such as in W. C. Allee's work, which fared better in ecology (the "Allee effect") than evolutionary biology, because his views in the latter case (e.g., those on animal sociality) were strongly informed by group selection (Allee 1938, 1951). The arguments against group selection were clarified by G. C. Williams (1966) in *Adaptation and Natural Selection: A Critique of Some Current Thought*. Although his book was not written in direct response to Wynne-Edwards, it lucidly explained why natural selection is more efficient at the individual than group level. Meanwhile, W. D. Hamilton (1963, 1964a,b) had quietly worked out the theory to show that selection can promote altruistic behavior to benefit kin.

Intriguingly, Tinbergen was in a meeting in Oxford, alongside A. J. Cain, J. Maynard Smith, and D. Lack, in which the neologism "kin selection" was coined to distinguish Hamilton's explanation from group selection (Kohn 2004). The term was first used in print by Maynard Smith (1964). This confluence of events brought evolution back into the mainstream of behavioral studies. Conflict and cooperation due to differences in relatedness, which are so evident in social interactions, were suddenly legitimate topics of inquiry, because they could be discussed within a robust theoretical framework. The tedious descriptive activities of drawing ethograms and describing fixed action patterns were replaced by active engagement with biology's greatest discovery—natural selection. The burst of creativity that followed is seen in landmark papers by Trivers (collected in Trivers 2002), Maynard Smith and Price (1973), Parker (1970, 1974), and Zahavi (1975), among many others. The excitement of the decade was captured in Richard Dawkins' (1976) *The Selfish Gene,* which viewed selection through the prism of levels of selection and inclusive fitness and emphasized behavioral questions.

Learning to Count and Not Being Neutral

The appeal of behavioral ecology lies in a challenge: can we identify how selection shaped specific behavioral traits? Behavioral ecologists have become increasingly mathematically literate and embraced appropriate theoretical tools (e.g., the comparative method, kin selection, optimality, and game theory) to answer such questions. The field is no longer constrained by theoretical naivete but rather by the empirical challenges common to any study that attempts to measure Darwinian fitness (Kruuk and Hill 2008).

In its early years, behavioral ecology faced strong criticism. Some arose from political concerns that a science based on behavior's being heritable was being misused to justify sexism and racism (Segerstråle 2000). More serious scientific concerns than those arose because behavioral ecology used adaptationism as a theoretical framework to generate predictions. Initially, the main methods used to predict the outcome of selection were optimality or game theory, in which animals interacted so that the optimum depends on the responses of others. Identification of optima requires simplifying assumptions: the environment is constant, selection is strong, the focal trait is subject to selection, and the focal trait is heritable. It is easy to conflate these modeling assumptions with the belief that selection in the real world actually generates optimally designed organisms. Gardner (2010) has noted that: "this cannot be defended as an accurate view of the natural world." We agree, but we reject the more polemic claim of critics of the adaptationist program that it invariably leads to "adaptive story-telling" (Gould and Lewontin 1979). Optimality provides a powerful tool to generate testable predictions. It is noteworthy that a similarly extreme position is taken by those studying molecular evolution (Bromham 2008). The assumption of neutrality, even though selection obviously occurs, has also led to testable predictions and advanced knowledge (see Zhang, Chapter 4). Philosophers of science seek perfect solutions, but empiricists and theoreticians know that simplifying assumptions and "elegant lies" can yield new insights (Kokko 2007).

We acknowledge, however, a distinction between neutrality and optimality as starting points. The former is arguably a better *null model*. If stochastic processes are correctly identified, which is not always easy (e.g., sampling subdivided populations can create spurious patterns; Nachman 2006; see Millstein, Chapter 3), then any deviation from the null expectation tells us that selection is operating. In contrast, an optimality approach combines two assumptions. First, it is assumed that the system is in equilibrium. This assumption is the *null* because it is simpler and better defined than the myriad, possible disequilibrium conditions. The second assumption is that selection is optimizing. Any deviation from an expected trait value could be due to either or both assumptions being unwarranted. A challenging practical problem is to identify the trait is being optimized and factors that constrain it. Fitness in the wild is hard to measure; so, behavioral ecologists assume that organisms face a series of quasi-independent problems that involve optimization of outcomes correlated with fitness (e.g., kilojoules consumed/hour). Unlike molecular ecologists studying neutrality, they cannot easily refute optimality. Instead, they have to ask which particular adaptive explanation provides the best fit to the data. Moreover, it can be a major intellectual task even to predict the optimal solution, and it is surprisingly easy to misidentify it by ignoring feedback processes and multiple routes to increase inclusive fitness (as we will review when we investigate inbreeding avoidance). Behavioral ecology is a strange mixture of Popperian falsification (i.e., explanation *A* does not fit) (Popper 2003) and corroboration (i.e., explanation *B* provides a better fit than explanations *C* or *D*). The complications that ensue are best explored in a case study.

Weavers, Cuckoos, and Cowbirds

Darwin believed that natural selection can fine-tune behavior, a view shared by behavioral ecologists. He speculated that interspecific brood parasitism evolved because of positive selection on cuckoos with an instinct to engage more often in such acts than to rear their own chicks (Darwin 1859). But why do hosts (the birds feeding cuckoo chicks) not always detect and reject cuckoo eggs? Davies (2000) offers two reasons why hosts might accept cuckoo eggs as their own. First, suitable variants that can reject cuckoo eggs might never have arisen and been exposed to natural selection (or confer such a small fitness advantage that they are lost due to genetic drift in small populations). Second, an overly suspicious host could erroneously discard its own eggs (i.e., a "false positive").

The first explanation could be true but it appears inadmissible whenever hosts reject foreign eggs. This explanation also offers no quantitative test, except, perhaps, that host rejection might be less likely in small populations. The second explanation, however, creates many independent, testable predictions, because rejection behavior will depend on costs and benefits that might vary spatially and temporally for several reasons. For example, one can predict that selection will alter egg rejection thresholds so that hosts reject dissimilar (i.e., non-mimetic of host) eggs more often when brood parasites are abundant (i.e., the ratio of false negatives to false positives declines). Davies quipped that to test this theory, one needs to conduct an experiment where brood parasites are removed for hundreds of years to see how host rejection thresholds evolve.

In fact, this experiment was unwittingly conducted by slave traders in the Caribbean. African village weaver birds (*Ploceus cucullatus*) were imported to Hispaniola in the eighteenth century and soon became established (Cruz and Wiley 1989). In their native range, these weavers show great variation in egg color and spotting among females, but highly repeatable egg patterning within females, which allows females to readily detect a foreign egg that differs from their own eggs (Lahti and Lahti 2002). In their home range, weaver nests are parasitized by Diederik's cuckoo (*Chrysococcyx capricus*), but in Hispaniola they were freed from brood parasitism for over 200 years. Then, in the late 1960s, shiny cowbirds arrived in Hispaniola. By the early 1980s, they were established as brood parasites on many birds, including weavers (Post and Wiley 1977). These historic events created two natural experiments: (1) relaxed selection against foreign egg detection, formerly imposed by brood parasitism by Diederik's cuckoo, and (2) renewed selection for foreign egg detection upon the arrival of cowbirds.

Over the last 20 years, several studies of egg recognition by weavers have produced the following key findings. An experiment that placed real eggs in weaver nests in 1982 found no marked differences in rejection rates (10–18%) whether the egg was non-mimetic (dissimilar color or spotting pattern), mimetic, or a cowbird egg (Cruz and Wiley 1989). These rejection levels were, however, far lower than those for non-mimetic real eggs in African weaver populations (58–73%). Assuming that the African population represents the ancestral level of egg discrimination, this study implies that weavers freed from parasitism had evolved a reduced ability to recognize foreign eggs. Cruz and Wiley (1989) predicted that the presence of cowbirds would cause rejection rates to increase again. Indeed, studies in the 1990s found far higher levels of rejection of artificial eggs resembling cowbird or non-mimetic weaver eggs (68–89%), compared to 25% for mimetic eggs (Robert and Sorci 1999), and similarly, in the 2000s higher levels of rejection were found for non-mimetic real weaver eggs (64%; Lahti 2006). Finally, recent studies (1999–2002) using real eggs show a 23% rejection rate for mimetic eggs, 33–62% for non-mimetic eggs and 85% for cowbird eggs (Cruz et al. 2008). In addition, this recent dataset shows that rejection of cowbird eggs was lower in southwestern Hispanola (73%), where cowbirds are rare, than in the central or northern areas, where they are common (90–96%).

There are two temporal trends that need to be explained. First, using extant African populations as a benchmark, why had egg rejection declined so markedly when Cruz and Wiley (1989) conducted their study? Second, why had the rate of rejection increased so much when Robert and Sorci (1999) and Lahti (2006) conducted their studies? It is tempting to argue that the former reflects microevolution due to selection against alleles favoring egg discrimination and rejection because of the unwarranted risk of rejecting one's own eggs (i.e., false positives). The rate of the recent increase in egg rejection after the arrival of cowbirds is, however, impossibly high: a microevolutionary model can only reproduce it in a best-case scenario with perfect heritability and other unlikely assumptions (Robert and Sorci 1999). This suggests that learning (phenotypic plasticity) to reject foreign eggs based on nesting success (cowbird chicks reduce host nestling survival) is partly or entirely responsible for the currently observed variation in egg rejection rates.

Although behavioral ecologists would like to establish the relative roles of learning and microevolution, there is a sense in which it is irrelevant. The more general point is that weavers appear to exhibit a behavior that improves their reproductive success when the environment changes. If avoidance of brood parasitic eggs is learned, then one can simply focus on selection for phenotypic plasticity. Our inability to distinguish between these two possibilities is unsatisfactory, especially when the processes involved in each are poorly understood. For example, the claim that rejection of brood-parasite eggs is learned is based on the premise that females learn the appearance of their own eggs (i.e., imprint). But is imprinting evolutionarily stable if a female's first brood is likely to contain a foreign egg? To date, no one has produced a theoretical model to scrutinize the claim that learning will cause egg rejection ability to covary with brood parasitism rates. Although we have estimates of some important parameters for a microevolutionary model (Robert and Sorci 1999), we lack accurate estimates of the parameters needed to test the learning hypothesis—let alone to make predictions if reality involves both learning and microevolution.

There is clearly ample scope for further empirical work. For instance, one could conduct learning experiments to test whether initial clutch imprinting occurs. It might be possible to perform common garden experiments on inexperienced birds from northern and southern Hispaniola and from Africa to test for difference in egg discrimination ability in the absence of learning opportunities. Of course, the genetic architecture underlying the traits studied or whether dispersal disrupts local adaptation are entirely unknown. Indeed, the Hispaniola weavers illustrate several characteristics of research in behavioral ecology: emphasis on adaptive explanations, use of non-model organisms, and lack of experimental standardization. We consider each of these in turn.

Emphasis on Adaptation

Almost all behavioral ecologists prefer questions concerning function and survival or reproductive success to other questions. Their primary interest is usually in deciding whether a trait increases lifetime reproductive success or inclusive fitness, and if so, why. There is a secondary interest in how a behavior develops (ontogeny). Determining the proximate cues involved is often left to those in other disciplines. Many behavioral ecologists show little interest in identifying the genes that underlie a trait. They usually make the phenotypic gambit that the underlying genetic architecture of a trait will not prevent the most successful strategy identified by optimality or game theory from reaching fixation (Grafen 1991). Nonetheless, there has been a recent growth in interest in the quantitative genetics of behavioral traits in natural populations (Kruuk and Hill 2008).

Unfortunately, an emphasis on adaptive explanations can produce untenable conclusions. For example, it might be shown in weavers that learning egg types does not occur or that the heritability of foreign egg rejection is too low to account for the observed rate of microevolution, given the strength of selection (Robert and Sorci 1999). This occurrence is not a fatal flaw, however, as most scientists eventually rely on consistency with findings in other disciplines to avoid errors (e.g., nothing prevents a comparative psychologist from testing the learning abilities of Hispaniola weavers). It is also important to recall that scientists in other disciplines make mistakes. Similarly, we know little about the proximate mechanisms that permit mothers to bias offspring sex ratios in species with chromosomal sex determination (Pike and Petrie 2003). This makes it possible to dismiss cases in which sex ratios deviate from 1:1 as sampling errors, but numerous studies nevertheless show that such deviations tend to be in the direction that enhances maternal fitness (West 2009). It seems more prudent to assume that reproductive biologists have yet to identify the mechanisms to bias sex than to maintain that the mechanisms do not exist.

Tinbergen's four questions do not form a hierarchy based on relative importance. Many would agree, however, that Darwin's discovery and subsequent exploration of natural selection was an unprecedented intellectual achievement. In our experience, the path from hypothesis to robust answer is more perilous when asking ultimate than proximate questions. This finding is especially true for the evolution of social behaviors, in which the actions and reactions of several players must be simultaneously considered.

Is it really more difficult to study adaptation than other levels of causation? In some respects, the answer is no. Behavioral ecologists often use very simple, elegant experiments to test whether selection has produced fine-scale refinement in how animals adjust their behavior to perceived changes in the costs and benefits of different actions. On the one hand, using little more than string, glue, and a nail, Detto et al. (2010) tethered fiddler crabs (*Uca annulipes*) next to each other to demonstrate that males can evaluate their size, relative to intruders and neighbors, to decide when to cooperate with a neighbor to form a territorial coalition and repel an intruder. On the other hand, behavioral ecologists battle to produce explanations where each component of the argument is established beyond reasonable doubt. The problem can be illustrated by returning to a topic that we used to highlight Tinbergen's questions on inbreeding avoidance. This avoidance behavior is common but not universal. An obvious hypothesis is that it prevents the production of inbred offspring, because they have low viability. However, closer inspection shows that we must account for a potential kin-selected advantage to inbreeding (Kokko and Ots 2006). A sister that mates with her brother increases her inclusive fitness if he would otherwise have been unlikely to mate. The brother increases his direct reproductive success, but there is kin-selected cost, because he also lowers his sister's reproductive success if she could have bred with a less closely related male and produced more viable offspring.

It is insufficient to document lower viability for inbred offspring to show that the observed level of inbreeding avoidance is adaptive. One must at least quantify the likelihood of different types of matings and the associated offspring fitness. Both tasks are a challenge. Inbreeding depression is often underestimated under benign laboratory conditions (Keller and Waller 2002; Joron and Brakefield 2003), and long-term studies consistently show temporal variation in selection in the wild (Robinson et al. 2008). A cost of inbreeding that is undetectable in most years might be massive when a new disease arises or a drought occurs (Ross-Gillespie et al. 2007). Even if inbreeding effects on specific traits are quantified, one still has to predict the net effect on lifetime reproductive success of offspring. This prediction is especially difficult with males if there is sexual selection that is due to competition to attract mates or repel rivals and inbreeding affects secondary sexual traits (Drayton et al. 2007). Likewise, quantifying the likelihood of different types of matings is difficult if inbreeding avoidance itself generates non-random spacing of kin (e.g., sex-biased dispersal). There is then a need to model what happens if inbreeding avoidance were to become less common. Although these are ultimately logistic issues, they should not obscure the fundamental theoretical challenge. It is easy to produce inappropriate models that produce inaccurate fitness sums.

We hope this example shows that even explaining a seemingly clear-cut adaptive behavior can tax the best of minds and require complex models. Fortunately, many behavioral ecologists excel at working out the kinds of decisions animals have to make and what information their decisions might require. When combined with a good grasp of levels of selection and feedback, researchers can generate plausible predictions about which traits should be favored by selection, even when the scenarios being considered make anthropomorphism difficult. For example, once the rules guiding filial cannibalism by male fish are known, it is predictable that females will preferentially spawn with males that are already caring for other females' eggs. The rules differ from those used by humans, but they make intuitive sense.

The case of inbreeding avoidance shows, however, that the use of species-sensitive anthropomorphism can still be misleading. Thus, formal theoretical models are essential to corroborate seemingly plausible verbal arguments (for a good example that exposed flawed thinking see Queller 1997). It took the appearance of a formal model (Kokko and Ots 2006) before many behavioral ecologists realized that their intuitions were wrong and that genetic theory can predict conditions in which at least one sex benefits from inbreeding, even if the resulting offspring survive poorly. Intriguingly, in botanical studies, analogous results have been accepted for far longer (Fisher 1941), while the same message has been presented several times in the animal literature but has gone largely unnoticed (Waser et al. 1986). It is tempting to conclude that a reduced empathy of humans for the problems faced by plants increased reliance on genetic theory to work out which traits selection favors. This outcome is a mixed blessing. The quick route to creative (and often correct) ideas about which traits natural selection might favor is for naturalists to envisage the options open to organisms. The caveat is that behavioral ecologists should then build formal models to confirm that these ideas are valid.

A fundamental issue with using current modeling approaches in evolution is how large a role one assigns to selection. Many chapters in this volume reveal an on-going debate about the relative roles of adaptive and non-adaptive processes in shaping evolutionary trajectories at all levels of biological organization. There are remarkable similarities between the optimality debate and current discussion of the role of neutrality and selection as determinants of evolutionary change at the molecular level (see Zhang, Chapter 4; see Kolaczkowski and Kern, Chapter 6). A starting premise for molecular evolutionary biologists is that selection is absent (neutrality), while behavioral ecologists assume that it is all powerful (optimality). Both views caricature reality. Most behavioral ecologists appear uninterested in resolving the issue (Orzack 2010), although the foundations of an answer were laid by Grafen (2007), who demonstrated the formal link between optimization of fitness by organisms and gene frequency changes under natural selection (Gardner 2010). McNamara et al. (2003) offer equivalent work in the field of sexual selection. The reason why behavioral ecologists accept optimality is that most traits appear well designed to ensure survival and reproduction, which justifies invoking a powerful role for selection. At the same time, it is sensible to recognize that other forces exist and that true optimization is rare. In recent years, there has been greater acknowledgement that local optima are rarely reached due to gene flow between populations under different selective pressures, genetic drift in finite populations, or most notably in the last decade, rapid environmental change (often human-induced).

So what does a pragmatist do? Many behavioral ecologists sidestep the issue of strong adaptationism by testing mutually exclusive hypotheses whenever possible. This approach stems from Karl Popper's view that science requires falsifiable predictions (Popper 2003). Gould and Lewontin's (1979) phrase "just-so stories," borrowed from Kipling's (1902) fanciful accounts for how conspicuous animals acquired their signature features, implies that adaptationism leads to irrefutable statements. If, however, falsifiable hypotheses are proposed one can restore credibility that one is doing valid science. Indeed, we believe that there has been a recent proliferation of tests of sometimes absurdly trivial "hypotheses" by behavioral ecologists. There is nothing wrong in stating predictions that follow from different sets of assumptions, particularly those derived from explicit mathematical models. It is, however, naïve to believe that hypothesis testing alone can resolve the problem.

The vulnerability of adaptationism is that most competing hypotheses, with the exception of the standard null hypothesis that a trait has no effect on fitness, still assume that selection has optimized behavior. In our view, the problem is not insurmountable. Research in other fields with more ready access to genomic data shows that selection shapes traits (see Kolaczkowski and Kern, Chapter 6). Behavioral ecologists must simply acknowledge that they work under the premise that selection has molded the behavior of interest, while simultaneously trying to determine the relevant constraints, which might include the genetic architecture of traits (Blows 2007). In the absence of this combination of assumptions, it is impossible to even begin to identify likely sources of selection (Birkhead and Monaghan 2010).

There Is No Such Thing as a Model Organism

The cuckoo story is typical of behavioral ecology: uncoordinated questions arise haphazardly and are answered idiosyncratically and incompletely. Differences among studies in where, when, and how they are conducted make it impossible to identify a single factor responsible for variation in outcomes (Jennions et al. 2010a). For example, is the difference in egg rejection found by Robert and Sorci (1999) and Cruz and Wiley (1989) due to the use of natural versus artificial eggs, from the studies being conducted in southern versus eastern Hispaniola, or because cowbirds had been established for 15 years longer by the time of the later study? Similarly, Africa and Hispaniola differ in many ways. To interpret this inadvertent experiment as a comparison of equivalent populations with and without brood parasites is an unwarranted simplification, albeit a creative one. Lahti (2005) also noted that variability in egg patterning has evolved in Hispaniola, which affects the interpretation of the same egg rejection tests between the two populations on the island. Despite these caveats, we suspect that most readers are sympathetic to Cruz and Wiley's (1989) original interpretation that reduced egg rejection was a response to counter-selection. One reason is that most evolutionary biologists accept that selection shapes phenotypes, so large changes in traits that affect fitness are unlikely to be solely attributable to chance. In addition, other lines of evidence seem to corroborate the interpretation: greater egg rejection after the arrival of cowbirds, population differences associated with the spread of cowbirds, evolution of egg patterns in both Hispaniola, and in a second introduced weaver population in Mauritius (Lahti 2005). Each line of evidence is quite weak, but together they bolster confidence that changes in egg rejection rates are an adaptive response to prevailing brood parasitism levels.

The Hispaniola weaver studies are a reminder that introduced species provide exceptional opportunities to study evolution (Shine 2010). So why are village weavers not being turned into a model system? The short answer is that there are too many other fascinating species to study. There is a limited set of themes in behavioral ecology, and one might ask why behavioral ecologists have not picked a few species that are each best suited to explore each theme. Other subdisciplines within evolutionary biology have developed model species (e.g., *Drosophila melanogaster* in evolutionary genetics) with great success. The answer is that behavioral ecologists are sensitive to the role of history and contingency in evolution and feel compelled to study the spectrum of life to ensure that their generalizations are universal and important differences among taxa are detected. If only model species are studied, there is a legitimate concern that we would be misled.

Behavioral ecologists follow Darwin's lead by studying diverse taxa to identify higher-level patterns. This practice inevitably leads to a lack of specialization, and individual studies are shallower as a result. Accordingly, behavioral ecologists have adopted the comparative methods initiated by Felsenstein (1985) to study the evolution of behavioral traits (Mank et al. 2005) and their by-products (e.g., sexual dimorphism due to mating preferences; Dunn et al. 2001). There has also been rapid growth recently in the use of meta-analysis. In ecology and evolution, many researchers are prepared to sacrifice greater certainty at the study level in exchange for detecting broader trends (Jennions et al. 2010a,b). Behavioral ecologists acknowledge the correlational nature of much of their evidence for selection, including that from comparative analyses. Indeed, formal selection analyses are inherently correlational, albeit with every effort made to control statistically for co-variation among traits (Lande and Arnold 1983). Interpretation is always based on plausibility, functional arguments, and the congruence of independent evidence. The same questions always arise: what other known factors might covary with the trait I am studying? Could factors other than selection create the observed differences? Do patterns consistent with a common selective pressure occur within and among species and populations?

Intriguingly, limitations to how evidence from a single study should be interpreted are more obvious in some areas of biology than others. For example, many readers would immediately have dismissed the weaver data as nearly worthless, if we had described it as a study using a phylogenetic tree with two branches and a single pairwise comparison. Conversely, we suspect people are more likely to be convinced by the conclusions of a molecular study using Kimura's neutral theory to test for selective sweeps based on genomic comparison of *Drosophila* from a source and an introduced population (A. Kern, personal communication). The context seems to matter, even when the issues are nearly identical. The molecular story may seem more convincing than the weaver one because hundreds of bases have been sequenced, but the true level of replication in both cases is at the population (i.e., a single comparison).

In evolutionary biology, innumerable theoretical models have been developed using only a few empirical facts. The ability to generalize is the core of most science (Dunbar 1995), but it should not cause a discounting of contingency. A truly satisfactory account of evolution will explain the general processes that drive change, and account for the amazing diversity that arises due to local contingencies. The general processes involved *and* the resultant outcomes are equally worthy of study (Futuyma 1998), which is why it is inappropriate to study only a few species.

It is probably fair to acknowledge that the average behavioral ecology study may be less reliable than studies in other areas of biology. It is a constraint imposed by the need to measure numerous variables, to examine multiple species, and limited resources. Some have suggested that behavioral ecology would be better served if studies were more often replicated (Palmer 2000; Kelly 2006). An opposing view is that the outcome of individual studies does not need to be known with great certainty if the real interest lies in knowing how widely individual findings can be extrapolated (Jennions et al. 2010b).

A statistical analogy is appropriate. Like all researchers, behavioral ecologists prefer models with analytical solutions, but it is occasionally necessary to resort to computer simulations. Simulation studies of stochastic events pose difficulties because one cannot state with certainty what will happen given a specific set of parameter values. The exploration of an initially unknown virtual world proceeds most efficiently when the simulation is run once per x-axis value (denoting an adjustable parameter), rather than many times for a single value (Figure 12.2; Kokko 2007). The advantage of replication is that, for the same computing time, one obtains an estimate of the response function shape as well as its variance. This approach provides an intriguing contrast with that used in empirical experiments. Empiricists are typically taught to investigate just a few treatments and replicate within each. The difference in approach is odd, as identical time and resources limitations apply. A strong argument can be made for looking at many treatment values and reducing replication per treatment in experiments and simulations alike. Indeed, it has been pointed out that choice of an overly narrow range of treatments can yield no variation in outcome and that a regression approach is often superior to analysis of variance of a few levels (Cottingham et al. 2005).

The analogy with the approach of behavioral ecologists should be apparent. If many studies are conducted on a few model species, very accurate

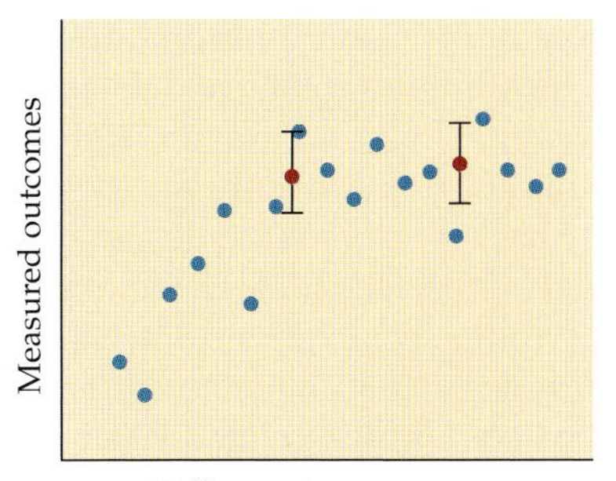

FIGURE 12.2 What if Model Systems Trade-Off with Diversity? The same amount of research effort could be spent on finding and measuring the location of more blue dots or on increasing certainty regarding the precise location of the red dots. We argue that a combination of approaches will prove healthiest in the long term.

information will eventually be available. A few model species are probably also sufficient to study highly conserved traits (see Zhang, Chapter 4; Kolaczkowski and Kern, Chapter 6), though such conservatism might be revealed as illusory when other species are investigated. The more important point is that each species combines traits shared among species with numerous idiosyncratic features. It is surely worth asking what governs which traits will evolve, where, and why? It is impossible to draw such generalities from a single study, no matter how often it is replicated (Futuyma 1998).

In addition, we already know from a purely practical perspective, that different species allow us to address different theoretical questions. For example, sex allocation theory advanced rapidly, in part, because of a focus on haplodiploid species in which there is a simple mechanism for females to bias offspring sex (i.e., whether or not to fertilize eggs) (West 2009). Had only mice and *Drosophila* been studied, the extent to which selection fine-tunes offspring sex ratios would probably have been overlooked and the interaction between kin selection and sex ratios would not have been explored.

There is also a related point to be made about the wider intellectual framework and how fields interact. If it were not for behavioral ecologists asking how selfish genes produce organisms with selfless behavior, microbial interactions that involve amazing elements of altruism would probably not have been interpreted in this exciting intellectual light (Foster et al. 2007).

Maintaining Standards

Limited standardization across studies in behavioral ecology than in the rest of evolutionary biology is an obvious source of variation among outcomes. For example, Cruz et al. (2008) criticized earlier weaver bird studies for using artificial rather than real eggs. Compared with many others within evolutionary biology, the methods of behavioral ecologists are unbelievably diverse, and techniques are often short-lived. However, the enforced use of identical methodologies could be actively misleading when working with many taxa (Hodos and Campbell 1969). For example, it makes no sense to test kin discrimination ability using pheromones in species that are highly visual. In practice, most methodological decisions reflect time, funding, and personnel constraints. Even so, greater uniformity could be achievable at little cost, but this effort is currently hampered by a *laissez faire* attitude. This is unfortunate because a common design can make questions more amenable to meta-analysis and facilitate detection of biologically relevant sources of variation in results. For example, the use of an experimental design by Wedell and Tregenza (1998) to test the genetic benefits of polyandry has been widely replicated. The design made it easier to argue that polyandry increases offspring survival (Simmons 2005). Even here, however, small additions to the basic experimental design and alternate statistical analyses can make it difficult to extract common effect sizes from studies that are ostensibly asking the same general question (Jennions, unpublished data).

The diversity of approaches employed by behavioral ecologists extends to their mathematical methodologies, as well. Unlike much of theoretical population genetics, there is liberal use of theoretical tools from other disciplines, primarily economics, engineering, and politics; these include game theory (Hamilton 1967; Maynard Smith and Price 1973), optimality (reviewed by McNamara et al. 2001), dynamic optimization (reviewed by Clark and Mangel 2000), individual-based simulations (reviewed by DeAngelis and Mooij 2005), and adaptive dynamics (reviewed by Dercole and Rinaldi 2008). Other theoreticians use modeling tools from the synthesis of population and quantitative genetics but have extended them to ask questions about specific behavioral traits (Lande 1981; Kirkpatrick 1982). In combination, this multi-pronged approach has allowed the logic of compelling verbal arguments to be confirmed, refuted, or refined (e.g., from Trivers 1972 to Kokko and Jennions 2008; or from Zahavi 1975 to Grafen 1990). There have also been important contributions by theoreticians exploring how these theoretical tools are related (Marrow et al. 1996; Page and Nowak 2002; Champagnat et al. 2006). They show an underlying unity behind different methods, which confirms that this methodological diversity is healthy. It ensures that behavioral ecologists have access to the most efficient tool for a given task (Dieckmann and Doebeli 2005; Kokko 2007).

Progress in a field can be retarded when researchers feel constrained to take a single theoretical approach. For example, quantitative geneticists have only recently started to model interacting phenotypes, whereby gene expression depends on the social context, leading to indirect genetic effects (Moore and Pizzari 2005; Bijma et al. 2007a,b). There is nothing inherently wrong with this approach—but, we are, perhaps, not alone in suspecting that the reason quantitative geneticists only now have begun to include phenomena that other methods tackled decades ago is related to the relatively tedious account-keeping they require. The emphasis researchers place on particular biological questions is always dictated by how easily a method can be used to answer them. It is worth repeating Medawar's

(1969) maxim that science is the art of the soluble. Proponents of different theoretical tools may disregard the importance of factors that are hard to incorporate using their favorite approach. For example, game theorists rarely consider gene flow, and it is similarly challenging to get a quantitative geneticist interested in complicated social interactions that have a temporal component. Use of diverse modeling approaches, each making slightly different assumptions, has accelerated the speed by which theory has developed and complementary empirical studies have been initiated by behavioral ecologists.

The willingness of behavioral ecologists to borrow methods from other disciplines extends far beyond mathematics. It is remarkable that DNA fingerprinting first resulted in a criminal conviction in the United Kingdom in 1986, and by 1987, behavioral ecology studies of birds, using fingerprinting, were already being published (Wetton et al. 1987; Burke and Bruford 1987). Sometimes methods fail (e.g., using fluctuating asymmetry to identify high-quality individuals). Sometimes the initial use is naïve but eventually becomes sophisticated enough to be useful (e.g., field studies of immunocompetence). There is little doubt of the importance for behavioral ecologists to remain engaged with methodological developments in other fields of biology and to forge collaborative links because the whole organism is central to their analysis. It is arguably harder to create links between those working on specialized aspects of specific organ systems or gene expression. Being a *jack-of-all trades* is the blessing but *master of none* might be the curse for the behavioral ecologist.

Despite its merits, there is a real danger that funding decisions now favor a narrow focus away from behavioral ecology. The elephant in the room is undoubtedly the dramatic rise in molecular genetics, with all the opportunities it has created to investigate important, previously unanswerable problems in evolutionary biology as well as applications in health and agriculture. Darwin made fundamental contributions without any knowledge of genetics, but studies that fail to incorporate genomic data today run the risk of being stigmatized as outdated. While important new evolutionary insights will undoubtedly come from the genomic revolution, it would be a tragedy if we lost our ability to root the results of genomic research in the phenotypic and ecological context in which information recorded in the genome evolved. In the near future it is reasonable to expect that molecular biology will become sufficiently accessible for many species that genotype–phenotype mapping will be more easily achieved than, for example, measuring fitness in the wild (see Hoekstra, Chapter 22). The recently proposed program to sequence 10,000 vertebrate genomes (Genome 10K Community of Scientists 2009) illustrates the speed with which such changes are occurring. The steady reduction in sequencing costs makes it clear that many genetic tools will become akin to modern statistics. The value of such tools, however, will depend on the ability of researchers to connect their molecular findings to phenotypes and the ecological contexts in which traits have evolved. Accomplishments in behavioral ecology have proven that it has a crucial role to play in this enterprise.

Conclusions

In the Australian Snowy Mountains, a male grasshopper, *Kosciuscola tristis* (Figure 12.3), shows a reversible color change, from dark green to bright blue, when its body temperature exceeds 25°C (K. Umbers, personal communication). Does this color change affect male behavior in a way that elevates survival and/or reproductive success? The color change does not bring about a clear benefit in the efficacy of thermoregulation, it is no more likely to induce predator attacks than dark green; brighter males do not disproportionately win fights over duller males, and artificially bluer males are no more attractive to females (K. Umbers, personal communication). This grasshopper is a decidedly non-model study organism. We worry about the fate of students who start a career studying a largely unknown species ending up with an unresolved mystery, such as this grasshopper. We also salute their determination.

FIGURE 12.3 Male Grasshopper *Kosciuscola tristis* At cooler body temperatures the beautiful blue color of this grasshopper disappears, revealing a drab green pigment. Nobody knows why. (© Kate Umbers.)

Instead of the case study above, it was tempting to close this chapter with an example of a behavioral study of a non-model organism that provided profound evolutionary insights, because many such studies exist. However, highlighting such rare cases would create a dishonest portrait of our science. Predicting scientific progress is as difficult as studying natural selection in a novel environment; it is hard to predict in advance who will be the fittest. Similarly, once a minimum quality threshold is exceeded, it is rarely possible to predict which project in either pure or applied science will yield the greatest intellectual or material advance. To ensure that such highly productive studies emerge, there must be sufficient standing variation at the onset. Using this analogy does not imply that we recommend funding projects without estimating their likely value. However, funding agencies increasingly run the risk of artificially restricting variation in an attempt to predict the most profitable path *given what is currently known*. It is thus fitting to end with a salutary quote from a geneticist at the conclusion of a review on the impressive success of his colleagues in detecting selection at the molecular level:

> *Despite the extensive evidence for selection in different regions of the genome and in different species, there are still relatively few examples where the functional significance of allelic variants is well understood in a particular environmental setting. Functional studies that dissect the biochemical consequences of genetic variation, combined with phenotypic studies of the fitness effects of functional differences will help us to understand the evolutionary significance of genetic variation. Such studies will help make the connection between ecological and evolutionary timescales (Nachman 2006: 118).*

Over a decade ago, Douglas J. Futuyma (1998) expressed the wish that we abandon the dichotomy between naturalists and scientists, pointing out that the best biologists have always been both. With more efficient research tools at hand than in 1998, the task of integrating field studies with modern technologies should be easier, not harder (see Rest, Commentary 8).

Acknowledgments

Funding was provided by the Academy of Finland and the Australian Research Council. We thank Kate Umbers for allowing us to cite her unpublished results. Doug Futuyma, Isobel Booksmythe, Andrew Kahn, Kate Umbers, and Marlene Zuk provided helpful comments for which we are grateful.

Literature Cited

Allee, W. C. 1938. *The Social Life of Animals*. William Heinemann, London.

Allee, W. C. 1951. *Cooperation among Animals with Human Implications*. Henry Schuman, New York.

Bateman, A. J. 1948. Intra-sexual selection in *Drosophila*. *Heredity* 2: 349–368.

Bijma, P., W. M. Muir, and J. A. M. Van Arendonk. 2007a. Multilevel selection 1: Quantitative genetics of inheritance and response to selection. *Genetics* 175: 277–288.

Bijma, P., W. M. Muir, E. D. Ellen, and 2 others. 2007b. Multilevel selection 2: Estimating the genetic parameters determining inheritance and response to selection. *Genetics* 175: 289–299.

Birkhead, T. R. and P. Monaghan. 2010. Ingenious ideas—the history of behavioral ecology. In D. F. Westneat and C. W. Fox (eds.), *Evolutionary Behavioral Ecology*, pp. 3–15. Oxford University Press, Oxford.

Blows, M. W. 2007. A tale of two matrices: Multivariate approaches in evolutionary biology. *J. Evol. Biol.* 20: 1–8.

Bowler, P. J. 1989. *Evolution: The History of an Idea*. University of California Press, Berkeley.

Bromham, L. 2008. *Reading the Story in DNA: A Beginner's Guide to Molecular Evolution*. Oxford University Press, Oxford.

Burke, T. and M. W. Bruford. 1987. DNA fingerprinting in birds. *Nature* 327: 149–152.

Champagnat, N., R. Ferrière, and S. Méléard. 2006. Unifying evolutionary dynamics: From individual stochastic processes to macroscopic models. *Theor. Pop. Biol.* 69: 297–321.

Cheney, D. L. and R. M. Seyfarth. 1990. *How Monkeys See the World*. Princeton University Press, Princeton.

Clark, C. W. and M. Mangel. 2000. *Dynamic State Variable Models in Ecology: Methods and Applications*. Oxford University Press, Oxford.

Cochran, G. and H. Harpending. 2009. *The 10,000 Year Explosion*. Basic Books, New York.

Cottingham, K. L., J. T. Lennon, and B. L. Brown. 2005. Knowing when to draw the line: Designing more informative ecological experiments. *Front. Ecol. Environ.* 3: 145–152.

Cruz, A. and J. W. Wiley. 1989. The decline of an adaptation in the absence of a presumed selection pressure. *Evolution* 43: 55–62.

Cruz, A., J. W. Prather, J. W. Wiley, and 1 other. 2008. Egg rejection behavior in a population exposed to parasitism: Village weavers on Hispaniola. *Behav. Ecol.* 19: 398–403.

Darwin, C. 1859. *On the Origin of Species by Means of Natural Selection, or the Preservation of Favoured Races in the Struggle for Life*, John Murray, London.

Davies, N. B. 2000. *Cuckoos, Cowbirds and Other Cheats*. T. & A. D. Poyser, UK.

Dawkins, R. 1976. *The Selfish Gene*. Oxford University Press, Oxford.

DeAngelis, D. L. and W. M. Mooij. 2005. Individual-based modeling of ecological and evolutionary processes. *Ann. Rev. Ecol. Evol. Syst.* 36: 147–168.

Dercole, F. and S. Rinaldi. 2008. *Analysis of Evolutionary Processes: The Adaptive Dynamics Approach and Its Applications*. Princeton University Press, Princeton.

Detto, T., P. R. Y. Backwell, J. Hemmi. and 1 other. 2006. Visually mediated species and neighbour recognition in fiddler crabs (*Uca mjoebergi* and *Uca capricornis*). *Proc. Roy. Soc. Lond. B* 273: 1661–1666.

Detto, T., M. D. Jennions, and P. R. Y. Backwell. 2010. When and why do territorial coalitions occur? Experimental evidence from a fiddler crab. *Amer. Nat.* 175: E119–125.
Dewsbury, D. A. 1992. On the problems studied in ethology, comparative psychology, and animal behavior. *Ethology* 92: 89–107.
Dieckmann, U. and M. Doebeli. 2005. Pluralism in evolutionary theory. *J. Evol. Biol.* 18: 1209–1213.
Drayton, J., J. Hunt, R. Brooks, and 1 other. 2007. Sounds different: Inbreeding depression in sexually selected traits in the field cricket *Teleogryllus commodus. J. Evol. Biol.* 20: 1138–1147.
Dunbar, R. 1995. *The Trouble with Science.* Harvard University Press, Cambridge, MA.
Dunn, P. O., L. A. Whittingham, and T. E. Pitcher. 2001. Mating systems, sperm competition, and the evolution of sexual dimorphism in birds. *Evolution* 55: 161–175.
Felsenstein, J. 1985. Phylogenies and the comparative method. *Amer. Nat.* 125: 1–15.
Fisher, R. A. 1918. The correlation between relatives on the supposition of Mendelian inheritance. *Trans. Roy. Soc. Edin.* 52: 399–433.
Fisher, R. A. 1941. Average excess and average effect of a gene substitution. *Ann. Eugen.* 11: 53–63.
Foster, K. R., K. Parkinson, and C. R. L. Thompson. 2007. What can microbial genetics teach sociobiology? *Trends Genet.* 23: 74–80.
Futuyma, D. J. 1998. Wherefore and whither the naturalist? *Am. Nat.* 151: 1–6.
Gardner, A. 2010. Adaptation as organism design. *Biol. Lett.* 5: 861–864.
Genome 10K Community of Scientists. 2009. Genome 10k: A proposal to obtain whole-genome sequence for 10,000 vertebrate species. *J. Hered.* 100: 659–674.
Gordon, H. R. S. 1958. Synchronous claw-waving of fiddler crabs. *Anim. Behav.* 6: 238–241.
Gould, S. J. and R. C. Lewontin. 1979. The spandrels of San Marco and the Panglossian paradigm: A critique of the adaptationist programme. *Proc. Roy. Soc. Lond. B* 205: 581–598.
Grafen, A. 1990. Biological signals as handicaps. *J. Theor. Biol.* 144: 517–546.
Grafen, A. 1991. Modelling in behavioral ecology. In J. R. Krebs and N. B. Davies (eds.), *Behavioral Ecology: An Evolutionary Approach,* 3rd ed., pp. 5–31. Blackwell, Oxford.
Grafen, A. 2007. The formal Darwinism project: A mid-term report. *J. Evol. Biol.* 20: 1243–1254.
Hamilton, W. D. 1963. The evolution of altruistic behaviour. *Amer. Nat.* 97: 354–356.
Hamilton, W. D. 1964a. The genetical evolution of social behaviour. I. *J. Theor. Biol.* 7: 1–16.
Hamilton, W. D. 1964b. The genetical evolution of social behaviour. II. *J. Theor. Biol.* 7: 17–52.
Hamilton, W. D. 1967. Extraordinary sex ratios. *Science* 156: 477–488.
Hodos, W. and C. B. G. Campbell. 1969. *Scala naturae*: Why there is no theory in comparative psychology. *Psych. Rev.* 76: 337–350.
Huxley, J. S. 1914. The courtship habits of the great crested grebe (*Podiceps cristatus*); with an addition to the theory of sexual selection. *Proc. Zool. Soc. Lond.* 1914: 491–562.

Huxley, J. S. 1923. Courtship activities in the red-throated diver (*Colymbus stealla-tus* Pontopp); together with a discussion on the evolution of courtship in birds. *J. Linn. Soc.* 35: 253–291.
Huxley, J. S. 1942. *Evolution: The Modern Synthesis.* Allen & Unwin, London.
Huxley, J. S. 1966. Introduction. *Phil. Trans. Roy. Soc. Lond. B* 251: 249–271.
Jennions, M. D., C. J. Lortie, and J. Koricheva. 2010a. Using meta-analysis to test ecological and evolutionary theory. In J. Koricheva, J. Gurevitch, and K. Mengersen (eds.), *Handbook of Meta-analysis in Ecology and Evolution.* Princeton University Press, Princeton, in press.
Jennions, M. D., C. J. Lortie, and J. Koricheva. 2010b. Meta-analysis and interpreting the scientific literature. In J. Koricheva, J. Gurevitch, and K. Mengersen (eds.), *Handbook of Meta-analysis in Ecology and Evolution.* Princeton University Press, Princeton, in press.
Joron, M. and P. M. Brakefield. 2003. Captivity masks inbreeding effects on male mating success in butterflies. *Nature* 424: 191–194.
Keller, L. F. and D. M. Waller. 2002. Inbreeding effects in wild populations. *Trends Ecol. Evol.* 17: 230–241.
Kelly, C. D. 2006. Replicating empirical research in behavioral ecology: How and why it should be done but rarely ever is. *Quart. Rev. Biol.* 81: 221–236.
Kipling, R. 1902. *Just So Stories.* MacMillan & Co., London.
Kirkpatrick, M. 1982. Sexual selection and the evolution of female choice. *Evolution* 36: 1–12.
Kohn, M. 2004. *A Reason for Everything.* Faber and Faber, London.
Kokko, H. 2007. *Modelling for Field Biologists and Other Interesting People.* Cambridge University Press, Cambridge, UK.
Kokko, H. and M. D. Jennions. 2008. Parental investment, sexual selection and sex ratios. *J. Evol. Biol.* 21: 919–948.
Kokko, H. and I. Ots. 2006. When not to avoid inbreeding. *Evolution* 60: 467–475.
Kruuk, L. E. B. and W. G. Hill. 2008. Introduction: Evolutionary dynamics of wild populations: The use of long-term pedigree data. *Proc. Roy. Soc. Lond. B* 275: 593–596.
Lahti, D. C. 2005. Evolution of bird eggs in the absence of cuckoo parasitism. *Proc. Natl. Acad. Sci. USA* 102: 18057–18062.
Lahti, D. C. 2006. Persistence of egg recognition in the absence of cuckoo brood parasitism: Pattern and mechanism. *Evolution* 60: 157–168.
Lahti, D. C. and A. R. Lahti. 2002. How precise is egg discrimination in weaverbirds? *Anim. Behav.* 63: 1135–1142.
Lande, R. 1981. Models of speciation by sexual selection on polygenic traits. *Proc. Natl. Acad. Sci. USA* 78: 3271–3275.
Lande, R. and S. J. Arnold. 1983. The measurement of selection on correlated characters. *Evolution* 37: 1210–1226.
Lehrman, D. S. 1953. A critique of Konrad Lorenz's theory of instinctive behaviour. *Quart. Rev. Biol.* 28: 337–363.
Lieberman, D., J. Tooby, and L. Cosmides. 2007. The architecture of human kin detection. *Nature* 445: 727–731.
Lorenz, K. 1948. *King Solomon's Ring.* Methuen, London.
Mank, J. E., D. E. L. Promislow, and J. C. Avise. 2005. Phylogenetic perspectives in the evolution of parental care in ray-finned fishes. *Evolution* 59: 1570–1578.
Marrow, P., R. A. Johnstone, and L. D. Hurst. 1996. Riding the evolutionary streetcar—where population genetics and game theory meet. *Trends Ecol. Evol.* 11:445–446.

Maynard Smith, J. 1956. Fertility, mating behaviour and sexual selection in *Drosophila suboscura*. *J. Genet.* 54: 261–279.
Maynard Smith, J. 1964. Group selection and kin selection. *Nature* 201: 1145–1147.
Maynard Smith, J. and G. R. Price. 1973. The logic of animal conflict. *Nature* 246: 15–18.
McNamara, J. M., A. I. Houston, and E. J. Collins. 2001. Optimality models in behavioral biology. *Siam Rev.* 43: 413–466.
McNamara, J. M., A. I. Houston, M. Marques dos Santos, and 2 others. 2003. Quantifying male attractiveness. *Proc. Roy. Soc. Lond. B* 270: 1925–1932.
Medawar, P. B. 1969. *The Art of the Soluble: Creativity and Originality in Science.* Penguin Books, London.
Moore, A. J. and T. Pizzari. 2005. Quantitative genetic models of sexual conflict based on interacting phenotypes. *Amer. Nat.* 165: S88–S97.
Nachman, M. W. 2006. Detecting selection at the molecular level. In C. W. Fox and J. B. Wolf (eds.), *Evolutionary Genetics: Concepts and Case Studies*, pp. 103–118. Oxford University Press, Oxford.
Orzack, S. 2010. Box 2.1: Optimality models. In D. F. Westneat and C. W. Fox (eds), *Evolutionary Behavioral Ecology*, pp. 26–27. Oxford University Press, Oxford.
Orzack, S. H. and E. Sober (eds.). 2001. *Adaptationism and Optimality.* Cambridge University Press, Cambridge, UK.
Page, K. M. and M. A. Nowak. 2002. Unifying evolutionary dynamics. *J. Theor. Biol.* 219: 93–98.
Palmer, A. R. 2000. Quasireplication and the contract of error: Lessons from sex ratios, heritabilities and fluctuating asymmetry. *Ann. Rev. Ecol. Syst.* 31: 441–480.
Parker, G. A. 1970. Sperm competition and its evolutionary consequences in insects. *Biol. Rev.* 45: 525–567.
Parker, G. A. 1974. Assessment strategy and evolution of fighting behaviour. *J. Theor. Biol.* 47: 223–243.
Paterson, S. and J. L. Hurst. 2009. How effective is recognition of siblings on the basis of genotype? *J. Evol. Biol.* 22: 1875–1881.
Penn, D. and W. Potts. 1998. MHC-disassortative mating preferences reversed by cross-fostering. *Proc. Roy. Soc. Lond. B* 265: 1299–1306.
Pike, T. W. and M. Petrie. 2003. Potential mechanisms of avian sex manipulation. *Biol. Rev.* 78: 553–574.
Pinker, S. 2002. *The Blank Slate.* Penguin Books, London.
Popper, K. R. 2003. *Conjectures and Refutations: The Growth of Scientific Knowledge*, 2nd ed. Routledge, London.
Post, W. and J. W. Wiley. 1977. Shiny cowbird in West-Indies. *Condor* 79: 119–121.
Queller, D. C. 1997. Why do females care more than males? *Proc. Roy. Soc. Lond. B* 264: 1555–1557.
Reaney, L. T., R. A. Sims, S. W. M. Sims, and 2 others. 2008. Experiments with robots explain synchronized courtship in fiddler crabs. *Curr. Biol.* 18: R62–R63.
Reeve, H. K. and P. W. Sherman. 1993. Adaptation and the goals of evolutionary research. *Quart. Rev. Biol.* 68: 1–32.
Ridley, M. 2003. *Nature via Nurture.* The Fourth Estate, London.
Robert, M. and G. Sorci. 1999. Rapid increase of host defence against brood parasites in a recently parasitized area: The case of village weavers in Hispaniola. *Proc. Roy. Soc. Lond. B* 266: 941–946.

Robinson, M. R., J. G. Pilkington, T. H. Clutton-Brock, and 2 others. 2008. Environmental heterogeneity generates fluctuating selection on a secondary sexual trait. *Curr. Biol.* 18: 751–757.
Ross-Gillespie, A., M. J. O'Riain, and L. F. Keller. 2007. Viral epizootic reveals inbreeding depression in a habitually inbreeding mammal. *Evolution* 61: 2268–2273.
Schielzeth, H., C. Burger, E. Bolund, and 1 other. 2008. Sexual imprinting on continuous variation: Do female zebra finches prefer or avoid unfamiliar sons of their foster parents? *J. Evol. Biol.* 21: 1274–1280.
Segerstråle, U. 2000. *Defenders of the Truth*. Oxford University Press, Oxford.
Sherborne, A. L., M. D. Thom, S. Paterson, and 5 others. 2007. The genetic basis of inbreeding avoidance in house mice. *Curr. Biol.* 17: 2061–2066.
Shine, R. 2010. The ecological impact of invasive cane toads (*Bufo marinus*) in Australia. *Quart. Rev. Biol.*, in press.
Simmons, L. W. 2005. The evolution of polyandry: Sperm competition, sperm selection, and offspring viability. *Ann. Rev. Ecol. Evol. Syst.* 36: 125–146.
Tinbergen, N. 1963. On aims and methods of ethology. *Z. Tierpsychol.* 20: 410–433.
Trivers, R. L. 1972. Parental investment and sexual selection. In B. Campbell (ed.), *Sexual Selection and the Descent of Man*, pp. 136–179. Aldine, Chicago.
Trivers, R. L. 2002. *Natural Selection and Social Theory: Selected Papers of Robert Trivers*. Oxford University Press, Oxford.
Trivers, R. L. and H. Hare. 1976. Haplodiploidy and evolution of social insects. *Science* 191: 249–263.
Waser, P. M., S. N. Austad, and B. Keane. 1986. When should animals tolerate inbreeding? *Am. Nat.* 128: 529–537.
Watson, J. B. 1924. *Behaviorism*. W. W. Norton, New York.
Wedell, N. and T. Tregenza. 1998. Benefits of multiple mating in the cricket *Gryllus bimaculatus*. *Evolution* 52: 1726–1730.
West, S. A. 2009. *Sex Allocation*. Princeton University Press, Princeton.
Wetton, J. H., R. E. Carter, D. T. Parkin, and 1 other. 1987. Demographic study of a wild house sparrow population by DNA fingerprinting. *Nature* 327: 147–149.
Williams, G. C. 1966. *Adaptation and Natural Selection: A Critique of Some Current Evolutionary Thought*. Princeton University Press, Princeton.
Wynne-Edwards, V. C. 1962. *Animal Dispersion in Relation to Social Behaviour*. Oliver & Boyd, Edinburgh.
Zahavi, A. 1975. Mate selection—a selection for a handicap. *J. Theor. Biol.* 53: 205–214.

Chapter 13

Understanding the Origin of Species: Where Have We Been? Where Are We Going?

Richard G. Harrison

Diversity, discontinuity, and adaptation (i.e., good fit) characterize the world's biota and motivate interest in the evolutionary process. A common measure of diversity is simply the number of discrete lineages or species within a defined clade or community. Beetles are diverse, birds much less so. Because diversity is a function of origination and extinction rates, how new species arise (i.e., how single lineages split into two or more independent daughter lineages) has been a central and frequently controversial issue in evolutionary biology.

In spite of the title, Darwin's *The Origin of Species* did not directly confront the question of how new species arise. Darwin (1859) focused on how and why change over time occurs, and the so-called one long argument of *The Origin of Species* is primarily an explanation for the origin of new form and function, rather than the origin of discrete lineages. This interpretation is not new, nor is it universally accepted (Mallet 2008). Much hinges on what is meant by species and speciation and whether speciation is viewed as equivalent to the origin of intrinsic barriers to gene exchange. If the process of speciation is considered a distinct subset of the more inclusive process of evolutionary divergence, then the primary focus for Darwin was on understanding the latter. However, many passages in *The Origin of Species* do make clear that Darwin was thinking about how new species arise, about the geographical context in which this might occur, and about the factors that promote and constrain the origin and persistence of species.

Determining whether distinct forms should be ranked as varieties, subspecies, or species was a difficult problem for Darwin, and amount of difference was an important criterion in determining the ranking. "These differences," he noted, "blend into each other in an insensible series; and a series impresses the mind with the idea of an actual passage" (Darwin 1859: 51). Thus, the origin of new species is simply one manifestation of change over time. But what drives change over time and what allows distinct lineages to form and to remain separate? Darwin's answer to the first question is that observed differences would not be due to chance and instead must often be the result of a deterministic force: natural selection.

But how does natural selection lead to different outcomes in different lineages, and why do we not see many transitional forms, but instead discrete species? Here the answers are less clear. Darwin did invoke geographic isolation: "...and in such islands distinct species might have been separately formed without the possibility of intermediate varieties existing in the intermediate zones" (Darwin 1859: 174); but, in other passages, he seemed less certain that isolation is a prerequisite for divergence: "Although I do not doubt that isolation is of considerable importance in the production of new species, on the whole I am inclined to believe that largeness of area is of more importance" (Darwin 1859: 105). Elsewhere in *The Origin of Species*, he pointed to competition as an explanation for the absence of transitional forms: "...both the parent and all the transitional varieties will generally have been exterminated by the very process of formation and perfection of the new form" (Darwin 1859: 172). Darwin recognized that persistence of discrete forms in sympatry cannot be assumed, because of competition and/or hybridization, and he suggested that ecological and behavioral differences might allow co-existence: "...varieties of the same animal can remain distinct, from haunting different stations, from breeding at slightly different seasons, or from varieties of the same kind preferring to pair together" (Darwin 1859: 103). Finally, he discussed the problem of hybrid sterility, which he recognized could not be produced directly by natural selection. He argued that sterility is a by-product of divergence, and that it "is not a specially acquired or endowed quality, but is incidental on other acquired differences" (Darwin 1859: 245).

Where Have We Been?

Many issues about the origin of species that Darwin confronted in *The Origin of Species* emerged as prominent themes during and after the modern synthesis. Among these were: (1) the nature of species; (2) the geographic context for speciation (is sympatric speciation possible?); (3) the role of chance (genetic drift) in population divergence; (4) the role of natural and/or sexual selection in the origin of reproductive isolation; and (5) the relative importance and rate of accumulation of prezygotic and postzygotic barriers.

Biological Species Concept

Ernst Mayr (1942, 1963) and Theodosius Dobzhansky (1937, 1940) had profound influences on how evolutionary biologists thought about species and speciation in the second half of the twentieth century. Both were strong proponents of the biological species concept, which emphasized interbreeding within species and reproductive isolation between species and, therefore, focused attention on phenotypic traits that contribute to barriers to gene flow. Mayr and Dobzhansky frequently referred to these trait differences as "isolating mechanisms," and at least Dobzhansky saw their importance as preventing hybridization between divergent forms, because "transpecific hybridization would weaken the reproductive potential of the populations engaged in such hybridization and gene exchange" (Dobzhansky 1970: 312). Reproductive isolation is a solution to this problem, allowing sexually reproducing organisms to "avoid disruption of their genetic systems by interspecific hybridization" (Dobzhansky 1970: 312). The emphasis on gene flow, and the potentially negative consequences of hybridization, strongly influenced how most evolutionary biologists viewed the speciation process.

Geography of Speciation: Allopatric and Sympatric

Combined with the emphasis on gene flow and reproductive barriers was a strong assertion (most emphatically from Mayr) that geographic (i.e., allopatric) speciation was the predominant, perhaps almost the exclusive, mode of speciation. Mayr was dismissive of sympatric speciation: "It is rather discouraging to read this perennial controversy because the same old arguments are cited again and again in favor of sympatric speciation, no matter how decisively they have been disproved previously" (Mayr 1963: 450–451). His attitude, and support provided by a later paper by Futuyma and Mayer (1980), had the effect of elevating sympatric speciation to a *cause célèbre*, defended passionately by its proponents.

Theoretical models, beginning with Maynard Smith (1966) and including the often-cited and very influential paper of Felsenstein (1981), demonstrated that sympatric speciation was possible and described the constraints and opportunities. Subsequent models have elaborated and extended the scenarios for sympatric divergence (Dieckmann and Doebeli 1999; Kondrashov and Kondrashov 1999; Gavrilets 2004). These models, together with empirical data from a limited number of systems—*Rhagoletis pomonella* (Bush 1969), Cameroon cichlids (Schliewen et al. 1994), and others—have made believers out of most skeptics, but believers who generally think that sympatric speciation is rare and that certain groups (e.g., phytophagous insects, fish in crater lakes) may be particularly prone to speciation in the absence of extrinsic barriers to gene flow. Gradually, support has grown for the notion that sympatric speciation does occur.

At the same time, there has been increased recognition that the phylogenetic and population genetic signatures of speciation may be complex and that detailed analyses of multi-locus data will be required to infer demographic and geographic histories (Nichols 2001; Hey and Nielsen 2004; Knowles 2009; Nielsen and Beaumont 2009). It is essential to recognize that the speciation process plays out over time and space; both selection coefficients and rates of gene exchange may vary in time, space, and across the genome. It is also possible that the geography of origins will often be obscured by subsequent population histories.

Parapatric Speciation

Parapatric speciation, "semi-geographic speciation" of Mayr (1942), occupies a middle ground between allopatric and sympatric speciation. Narrowly construed, parapatric speciation implies a specific geographic distribution for diverging populations: that they have adjacent or contiguous (but non-overlapping) ranges (Futuyma 2009). Parapatric speciation occurs along environmental gradients, and conditions for its occurrence are much less restrictive than for sympatric speciation (Gavrilets 2004). Endler (1977) brought parapatric divergence into the spotlight by suggesting that many natural clines and hybrid zones may have formed, not as a result of secondary contact, but through differentiation in a continuous series of populations (i.e., primary intergradation) (Slatkin 1973). Endler argued that, because the two histories produce the same pattern of variation at the loci under selection, they cannot easily be distinguished. The debate about hybrid zone origins mirrored the debate about the geography of speciation, and Endler's book challenged the established dogma of allopatric speciation and secondary contact.

Subsequent reviews (Barton and Hewitt 1985; Hewitt 1989, 2000; Harrison 1990) used verbal arguments and empirical data to demonstrate that many hybrid zones in the temperate zones are the result of post-glacial, secondary contact and that clear examples of parapatric speciation are rare. However, these authors were careful to point out that accounting for the current distribution of differentiated taxa does not inform directly how, when, and where differences between these taxa arose. Losos and Glor (2003) provide a somewhat different perspective on inferring the geography of speciation. In theory, sympatric speciation could give rise to daughter lineages that become allopatric and then encounter each other in secondary contact. The history of these lineages subsequent to their divergence might well obscure the circumstances of their origin. One lesson learned is that studies of current hybrid zones will not easily resolve debates about modes of speciation, although they can provide important insights into the nature of barriers to gene exchange and the genetic architecture of such barriers (see the following discussion).

Founder Effects and Genetic Drift

The role of genetic drift in divergence and speciation also generated heated debate among evolutionary biologists in the second half of the twentieth century. Mayr's verbal models of peripatric speciation envisioned genetic revolutions in peripheral populations as a consequence of founder events or population bottlenecks (Mayr 1963); such models involved loss of genetic variation and traversing deep fitness valleys on an adaptive landscape. Motivated primarily by observations of Hawaiian *Drosophila*, where most species are single island endemics and thus most colonization events must result in speciation, Carson (1971, 1975) proposed the founder-flush or flush-crash models (a cluster of somewhat different histories). Using an explicit population genetics approach, Templeton followed with the transilience model, in which speciation involves the instability of the intermediate stage and is characterized by overcoming some selective barrier (Templeton 1980, 1981). Carson and Templeton (1984) provide a comparison of the three distinct models. These models stimulated a number of different fly labs to look at the outcome of single or repeated founder events in *Drosophila* population cages (Powell 1978; Dodd and Powell 1985; Ringo et al. 1985; Galiana et al. 1993; Moya et al. 1995; Rundle et al. 1998). Results were mixed, but in the end, did not provide support for founder effect or founder-flush speciation.

The several founder effect models were different in detail and sophistication, but none ever convinced the broader community that founder effect speciation, in any of its many guises, was an important source of new species (Barton and Charlesworth 1984; Barton 1989; Gavrilets 2004). Templeton (2008), however, published an opposing view. Founder events can have major impacts on genetic diversity, but the probability that such events will lead to reproductive isolation is low (Gavrilets 2004). An exception may be the origin of some types of chromosomal rearrangements, for example, centric fusions in house mice (Searle 1993; Hauffe and Searle 1998; Pialek et al. 2001). However, the more general proposition that chromosomal speciation is a major factor in animal speciation (White 1969, 1978) is not widely accepted, and Coyne and Orr conclude that "we know of no compelling evidence for chromosomal speciation in animals other than mammals" (Coyne and Orr 2004: 265). However, chromosome rearrangements clearly are important for the persistence of divergent forms, but their importance may be due to suppression of local recombination rather than contributions to hybrid unfitness (Rieseberg 2001; Noor et al. 2001; Ortíz-Barrientos et al. 2002; Navarro and Barton 2003; Feder et al. 2003; Kulathinal et al. 2009).

Instead of crossing deep valleys on a rugged adaptive landscape, populations may often evolve along ridges of high-fitness genotypes, thereby avoiding the improbable crossing of a deep fitness valley (Gavrilets 2004). Dobzhansky–Muller incompatibilities are simple (two-locus) examples of evolution along high fitness ridges; in this case, alleles at two different loci arise and spread to fixation in different populations. In the genetic background in which they arise, the two alleles are neutral (perhaps even slightly advantageous), but when the two alleles occur together, incompatibility leads to reduced viability or fertility.

Natural Selection and Sexual Selection

If genetic drift is relegated to a relatively minor role in the origin of species, the important drivers of speciation are most likely natural and sexual selection (Coyne and Orr 2004). Both types of selection almost certainly contribute to the phenotypic differences that underlie reproductive isolation, either indirectly (as a by-product of divergence in allopatry) or directly (in reinforcement of post-zygotic barriers or disruptive selection leading to sympatric divergence). In by-product models, populations may evolve in different directions as a consequence of divergent selective pressures (different environments). However, even when selective pressures are uniform, outcomes may differ depending on the order in which mutations appear. The importance of so-called mutation order speciation remains unclear (Schluter 2009).

In many instances natural and sexual selection act independently (and sometimes in opposition). In other cases, natural selection may drive changes in phenotypes that are important for mate recognition, and so natural and sexual selection work in concert (Schluter 2000). The current enthusiasm for so-called ecological speciation has focused attention on the importance of divergent natural selection (Rundle and Nosil 2005; Schluter and Conte 2009). However, some models suggest that sexual selection is more effective in generating linkage disequilibrium and, as a result, is more powerful than natural selection in causing a gene pool to fission (Kirkpatrick and Ravigné 2002). Evidence can certainly be found in many lineages that both divergent natural selection and divergent sexual selection have driven speciation. However, defining the characteristics of individual lineages or life histories that predispose organisms to divergence through natural versus sexual selection is an issue that needs to be addressed more carefully.

Reinforcement models invoke a direct role for selection in speciation; hybrids are less fit than both parental types, favoring traits that directly result in positive assortative mating or that allow mate discrimination. Reinforcement has remained a persistent leitmotif in the speciation literature—enjoying episodes of great popularity punctuated by periods of doubt (Howard 1993; Liou and Price 1994; Kirkpatrick and Servedio 1999; Servedio and Noor 2003). Currently, theory suggests that reinforcement can occur (and surely with greater likelihood than sympatric speciation [Kirkpatrick and Ravigné 2002; Gavrilets 2004]), although other outcomes of hybridization (e.g., fusion) are more likely in some circumstances. Coyne and Orr's analysis of patterns of speciation in *Drosophila* (Coyne and Orr 1989, 1997) did much to encourage proponents of reinforcement, but specific case histories in which reinforcement has been clearly demonstrated remain remarkably few (Coyne and Orr 2004).

Barriers to Gene Exchange

Most evolutionary biologists might agree that the relative importance of different sorts of barriers and the chronological order in which such barriers arise will be lineage-specific and should be tested for particular clades or groups (Gleason and Ritchie 1998; Mendelson 2003; Ramsey et al. 2003; Martin and Willis 2007). Nonetheless, each of us is strongly influenced by the patterns we see in the taxa we study. Dobzhansky focused much of his attention on systems of complementary genes and (as previously mentioned) on the disruption of gene interactions that could arise as a simple by-product of divergence in allopatry. He adopted Wright's imagery of a rugged, adaptive landscape and visualized species as residing on adaptive peaks, each species on a different peak. Recombination will disrupt the "at least tolerably harmonious system of genes and chromosome structures" and therefore "a majority, and probably a vast majority, of the new patterns are discordant, and fall in the adaptive valleys" (Dobzhansky 1937: 229). Dobzhansky (1940) was also a strong proponent of reinforcement models, and argued that post-zygotic barriers accumulate in allopatry and are subsequently reinforced in secondary contact by pre-zygotic barriers that arise as a result of selection against hybrids. Such models make clear predictions about the order in which barriers arise. Dobzhansky's outlook influenced a large community of *Drosophila* population geneticists, who directed their attention to Dobzhansky–Muller incompatibilities and to the genetics of post-zygotic isolation. Because *Drosophila* provided the most useful model system for investigating the genetic basis of speciation, "Dobzhansky–Muller genetic incompatibilities took center stage in the genetic analysis of speciation" (Via 2009: 9939). This bias persisted until recently (Coyne and Orr 1998), but a major shift in emphasis and focus is now ongoing (see following discussion).

Classifying barriers to gene exchange as either "pre-" or "post-" requires choice of an event or life history stage that is the dividing line. In some classifications, the divide is mating; in other classifications, the divide is zygote formation (fertilization). Of course, mating and zygote formation are often displaced in time, and lots may happen between mating (spawning in many marine invertebrates or shedding of pollen in plants) and fertilization. Interest has grown rapidly in identifying and characterizing the set of barriers that are post-mating, but pre-zygotic (Howard 1999; Swanson et al. 2006). These would include reduced sperm viability (or poor sperm competitive ability) and reduced affinity of sperm and egg surface proteins for their counterparts. Strikingly, it turns out that proteins that are transferred from males to females during copulation or that mediate sperm–egg interactions often exhibit rapid adaptive evolution, suggesting that post-copulatory sexual selection or sexual conflict are potential sources of rapid divergent selection on these proteins (Swanson et al. 2001; Swanson and Vacquier 2002; Wolfner 2002; Andres et al. 2006). Thus, a current classification of barriers to gene exchange would include three categories: pre-mating; post-mating, pre-zygotic; and post-zygotic. Further subdivisions are possible, for instance, pre-mating barriers may be those that prevent potential mates from meeting (temporal and ecological), those that operate to prevent mating even when potential mates meet (behavioral), and those that directly interfere with mating (mechanical). In contrast to the chronological order in which they arose, which is difficult to determine, it is easy to define the order in which barriers currently operate (Ramsey et al. 2003). The problem is that there are very few systems for which data exist that allow investigators not only to define the full range of barriers but also to estimate the extent to which each barrier reduces gene flow.

The history of ideas about speciation in the twentieth century reveals long periods during which dogma prevailed, episodes of heated confrontation between conflicting schools of thought, and the constant reworking of old ideas in the context of new theories and new data. Views on speciation may have been unduly influenced by a few dominant figures and by a few dominant model systems. This influence is now beginning to wane, and studies of speciation in the 21st century will not only take advantage of new sources of data but will obtain these data from a remarkable diversity of fungi, plants, and animals.

Where Are We Going?

The past 10 years have witnessed a spate of books on speciation and diversification (Schluter 2000; Coyne and Orr 2004; Gavrilets 2004; Grant and Grant 2007; Price 2007; Losos 2009), together with a rapidly expanding literature on the genetic and molecular genetic basis of species differences. New theoretical models, new model systems, new molecular approaches, and new terminology appear with great regularity, and there is the sense that the field is poised to make important discoveries. Here I focus on current issues and controversies that have caught the attention of an evolutionary biology community that, in trying to throw off the shackles of the Mayr–Dobzhansky era, remains divided with respect to how to interpret the history of ideas and which pieces of the old dogma are worth salvaging.

Species and Speciation

Debates and discussions about species concepts and definitions at one time appeared frequently in the evolutionary biology and systematic biology literature. Some viewed these debates as counterproductive: "It is clear that the arguments will persist for years to come, but equally clear that, like barnacles on a whale, their main effect is to retard slightly the progress of the field" (Coyne 1992: 290). Those who viewed discussions of species concepts as unproductive are no doubt delighted that the public debate has subsided. However, placing the process of speciation in the context of different species concepts provides a useful framework for a research agenda.

Harrison (1998) suggested that the process of divergence and speciation could be thought of as a passage through a series of stages, each reflecting a distinct view of what is a species. Figure 13.1 summarizes this view of the speciation process. The figure includes three definitions of species that represent a temporal series in the accumulation of genetic differences between populations. The populations are either isolated by extrinsic barriers so that $m = 0$ or they diverge in the face of gene flow. The first stage, *phylogenetic species*, is simply populations that are fixed for different alleles or traits (Nixon and Wheeler 1990); these differences need not have consequences for interbreeding or hybrid viability/sterility. *Biological species* are genetically isolated by one or more barriers to gene exchange (i.e., at least one of the possibly many fixed differences) must contribute to reproductive

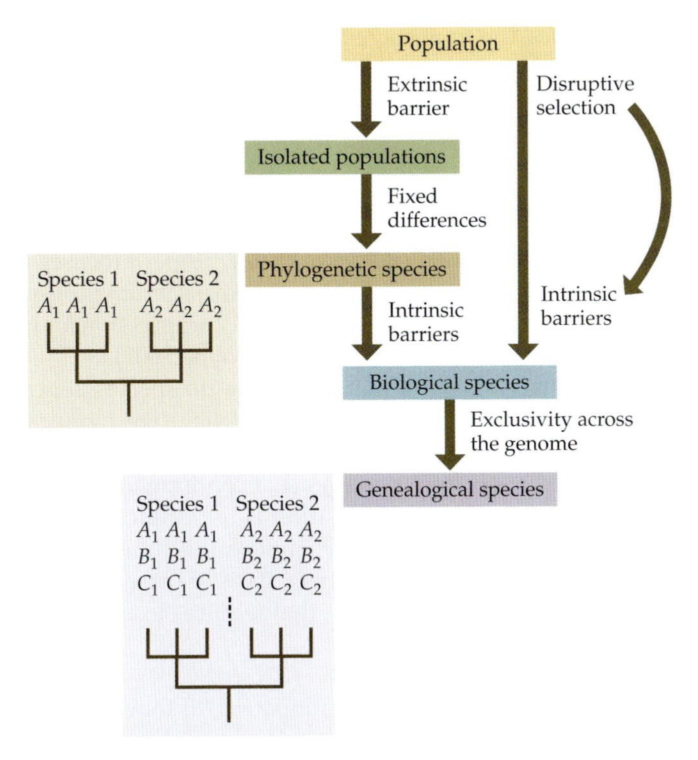

FIGURE 13.1 The Speciation Process as a Passage through Three Stages, Each Represented by a Different Species Concept On the left, a single population is divided into two isolated populations by an extrinsic barrier, the populations diverge and accumulate one or more fixed trait differences, giving rise to phylogenetic species. Among the fixed trait differences will appear some that lead to intrinsic barriers to gene exchange (i.e., pre- or post-zygotic barriers). To the extent that these barriers lead to reproductive isolation, the populations become biological species. Finally, over long time periods, the two populations become differentiated across the genome (i.e., they become exclusive groups for genes *A*, *B*, *C*, etc.) and therefore are genealogical species. On the right, the same sequence is shown for populations that diverge in the face of gene flow. In such cases, among the first fixed differences that appear must be those that result in intrinsic barriers to gene flow; disruptive selection catalyzes the direct transition from a single population into biological species. (Adapted from Harrison 1998.)

isolation (pre-zygotic or post-zygotic) (Dobzhansky 1937, 1970; Mayr 1942). Most studies of the speciation process focus on the origin of intrinsic barriers to gene exchange (i.e., on when diverging lineages become biological species). When there is no gene flow between diverging populations, trait differences that result in intrinsic barriers may arise at any time (relative to other trait differences). However, when gene flow is present, traits that confer reproductive isolation and traits subject to divergent selection will be the first to differentiate (see Figure 13.1).

Genealogical species, the final stage in a speciation process, are exclusive groups of individuals, that is, groups in which all individuals within the group are more closely related to each other than they are to individuals outside the group (Baum and Shaw 1995). Fixed differences imply exclusivity, but only for the surrounding genome region. Genealogical species are (in the extreme) exclusive groups all across the genome. For studies of speciation, the genealogical species concept provides a valuable complement to the biological species concept (Baum and Shaw 1995). It focuses attention on gene genealogies, the mosaic nature of genomes, and the consequences of random lineage sorting, selective sweeps, and introgression for interpreting recent evolutionary history (Hudson and Coyne 2002).

Although there is no need to agree on a single species concept (and surely, little hope that biologists will), it is essential to recognize that speciation is a process that plays out over time and to identify components that distinguish speciation from divergence. I continue to subscribe to the notion that the unique aspect of speciation is the origin of reproductive isolation or barriers to gene exchange. Therefore, it is the ecological, physiological, behavioral, molecular, and genetic basis of these barriers, and the selective forces that have molded them, that require close attention.

Geography and Gene Flow

Mayr defined two forms or species as allopatric: "if they do not occur together, that is that they exclude each other geographically" (Mayr 1942: 149). Allopatric speciation involves divergence of geographically isolated populations. Speciation without geographic isolation is sympatric speciation; it occurs "within a single local population, that is within a single interbreeding unit" (Mayr 1942: 189). A straightforward interpretation of allopatric and sympatric is that they represent the ends of a gene flow continuum, $m = 0$ (allopatric) and $m = 0.5$ (sympatric), in which m is the proportion of individuals exchanged between two populations each generation. Such a definition abandons the explicitly spatial, but somewhat ambiguous, relationships implied by the terms sympatric and allopatric.

Parapatric speciation initially implied divergence along a geographic transect or gradient, among a continuous series of populations. Clines or hybrid zones could be the result of parapatric divergence (Endler 1977). Like sympatric and allopatric, parapatric was a term that specified an explicit spatial arrangement of populations (Smith 1965). More recently, parapatric speciation has sometimes been interpreted as including speciation in all cases in which $0 < m < 0.5$ between diverging populations. This usage is problematic, because many spatial arrangements, other than two entities having ranges that are contiguous (the original meaning of parapatry), can result in $0 < m < 0.5$.

Butlin et al. (2008) suggested that the allopatric, parapatric, sympatric classification is unsatisfactory because it attempts to represent a continuum with three discrete categories. In so doing, they say, it emphasizes the extremes. Furthermore, because speciation is a process, values of m between diverging populations may change over time. Explicit recognition of this temporal component has spawned terms such as "allo-parapatric" or "allo-sympatric" (Coyne and Orr 2004).

Fitzpatrick et al. (2008, 2009) recently critiqued usage of sympatric speciation and concluded that because most populations have some spatial structure and because structure may change during the period of divergence (speciation), sympatric speciation is not a useful term. They advocated an explicit population genetic approach, in which all speciation events are considered in the context of a divergence with gene flow model (recognizing that values of m will be unique for each case and that m will sometimes be zero). They also suggested that an "obsession with identifying true cases of sympatric speciation" (Fitzpatrick et al. 2008: 1457) is not a productive way to move forward—a statement that elicited a strong response from a cadre of authors who have published on systems that are considered possible examples of sympatric speciation (Mallet et al. 2009). Although they agreed that explicit consideration of gene flow is appropriate, Mallet et al. (2009) contended that sympatry should be retained as an important term in the speciation literature. Their arguments reflect a desire to continue the allopatric/sympatric debate, so that "the classical argument about whether sympatric speciation...is common in nature..." can be resolved (Mallet et al. 2009: 2332).

I side with those who view the allopatric/parapatric/sympatric classification of speciation with increasing skepticism. The speciation process could, instead, be described in terms of the amount of gene flow at the time populations begin to diverge (m_i) and how extrinsic barriers to gene flow change over time. Extrinsic barriers reduce encounter rates between members of diverging lineages but are not a product of trait differences between those lineages. (Extrinsic barriers can be, and often are, dependent on organismal traits; for example, riverine barriers may be important for small mammals but not for birds.) Figure 13.2 presents several hypothetical clade-specific frequency distributions for m_i.

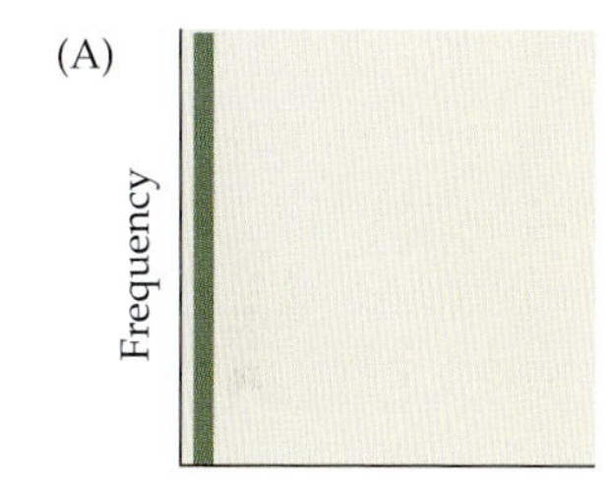

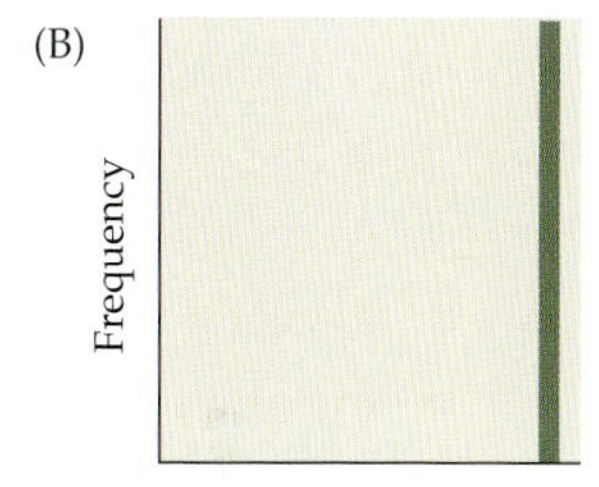

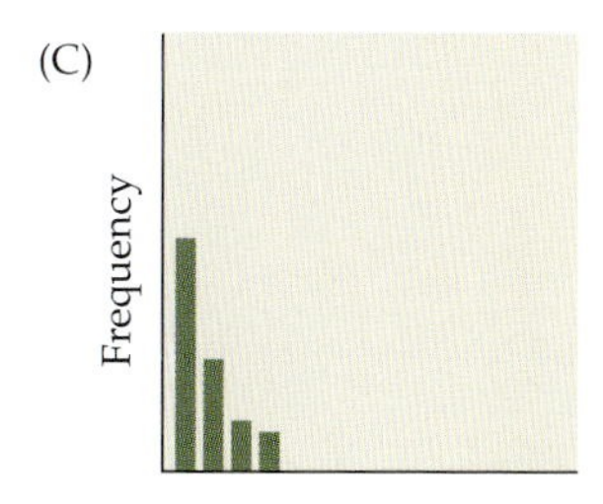

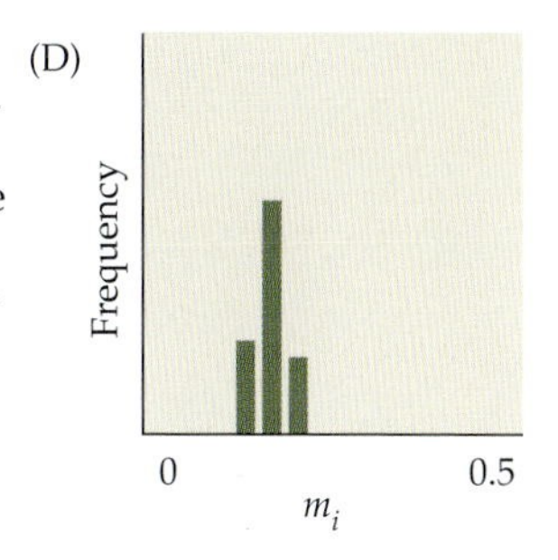

FIGURE 13.2 Hypothetical Clade-Specific Frequency Distributions of the Extent of Gene Exchange, m_i, between Population(s) at the Time They Begin to Diverge to Become Distinct Sister Species m_i is the gene exchange at $t = 0$, before divergence begins. (A) Pure allopatric speciation. ($m_i = 0$ for all cases of population divergence that give rise to distinct species.) (B) Pure sympatric speciation. All new species arise from divergence within single, randomly mating populations. ($m_i = 0.5$ for all population divergence resulting in distinct species.) (C) Populations diverge into species when $m_i < x$, that is, in the absence of gene flow or in the presence of limited amounts of gene flow. (D) Diverging populations are never completely isolated by extrinsic barriers, but speciation occurs in the presence of some gene flow. If gene flow is too high, populations do not diverge.

Mallet et al. argue that: "...no individual is truly in demic sympatry with its neighbour, nor, conversely, likely to be in complete allopatry either" (Mallet et al. 2009: 2335). It is true that initial divergence with $m_i = 0.5$ may be rare, although models of pure sympatric speciation should have this as the initial condition; in contrast, the other end of the continuum ($m_i = 0$) could be common following vicariance events. Although gene flow may show a gradual decline as a barrier forms (not a sudden decline as assumed in many models), the important point is that m does go to zero and presumably remains there for long periods (e.g., marine taxa on either side of the Isthmus of Panama). Such a scenario can be represented as a graph of extrinsic barrier strength (b_e) as a function of time since initial divergence (Figure 13.3). If we accept extrinsic barrier strength and gene flow at initial divergence as our metrics, we can then ask two questions: (1) what are the (lineage-specific) frequency distributions of m_i between diverging populations that ultimately become good species (see Figure 13.2); and (2) how does b_e change over the course of the speciation process (see Figure 13.3). Of course, intrinsic and extrinsic barriers are sometimes difficult to distinguish, the spatial structure of a system of populations can change over time, and temporal variation in the environment can influence dispersal rates between pairs of populations. However, replacing geography with gene flow encourages use of coalescent models of isolation with migration or of divergence with gene flow to generate estimates of gene flow, population sizes, and time since divergence (Nielsen and Wakeley 2001; Hey and Nielsen 2004). It makes good sense to retain allopatric, parapatric, and sympatric as descriptors of geographic patterns. But, the potential ambiguity of these terms (and

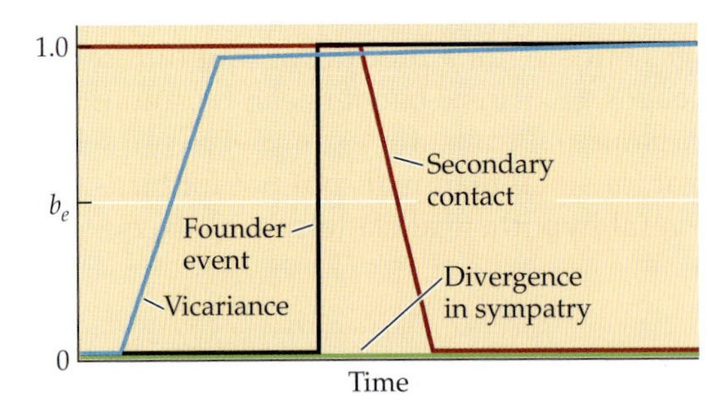

FIGURE 13.3 Change in Extrinsic Barrier Strength (b_e) as a Function of Time Shown are trajectories for populations that diverge following vicariance events (light blue) or founder events (black), populations that diverge when $m = 0.5$ (green), and populations that initially diverge in allopatry and then come together in secondary contact (red). For both vicariance and secondary contact, increases or decreases in extrinsic barrier strength are imagined to occur gradually over time, whereas for founder events, gene flow is reduced to zero at the time of the founder event. Most speciation histories will be far more complex than those shown here.

the apparently shifting landscape of terminology) argues for leaving behind the classic allopatric/sympatric speciation debate and moving on.

The argument to retain a focus on sympatric speciation includes a redefinition of conditions under which speciation is considered sympatric. Mallet et al. (2009) even coin a new term, "mosaic sympatry," to describe populations that occupy different patches within a region and therefore will be sympatric, but have $m < 0.5$. However, no explanation is provided for how populations came to be different in the first place, such that they occupy different patches where they are sympatric. What is the initial condition for sympatric speciation (Kirkpatrick and Ravigné 2002)? Can m_i be < 0.5 and speciation still be sympatric?

The spatial arrangement described as mosaic sympatry can also be produced through divergence in allopatry and subsequent secondary contact. Indeed, Howard (1986) and Harrison (1986), working on two different cricket systems, used the term "mosaic hybrid zone" to describe just such a situation. When examples of mosaic sympatry are encountered, it is important to know the value of m during the period that lineages diverge in patch occupancy, which is, again, a matter of choice of initial condition. Did populations come to be patch specialists in the presence or absence of gene flow? Pea aphids (*Acyrthosiphon pisum*) are often touted as a possible example of sympatric speciation (Via 2001, 2009), and populations of clover and alfalfa appear to exemplify mosaic sympatry (Via 1999). But pea aphids in North America almost certainly represent a mosaic hybrid zone produced by secondary contact, given that the two host races of pea aphids were introduced into North America and that pea aphids on alfalfa and clover in Europe do not appear to be sister lineages in a phylogeny that includes multiple host races (Peccoud et al. 2009).

In the evolutionary biology literature, allopatric speciation (or at least allopatric divergence) has always been viewed as the default option or null hypothesis. Because allopatric speciation is universally acknowledged to be important, models and model systems that challenge the necessity for $m_i = 0$ (divergence with gene flow models) garner the most attention (and regularly get published in *Science* and *Nature*). Indeed, as Coyne and Orr suggest: "Allopatric speciation appears so plausible that it hardly seems worth documenting" (Coyne and Orr 2004: 123). Abandoning the allopatric/sympatric contrast will alter the dynamics of debates about speciation and may result in less attention to the classic arguments about the possibility/probability of pure sympatric speciation. To my mind, this approach will be liberating, although I recognize that colleagues who have been fighting the sympatric speciation battle for many years will probably disagree.

Ecology and Speciation: What Exactly Is Ecological Speciation?

A major role for ecology in the speciation process is not a novel idea in the evolutionary biology literature. Mayr emphasized the importance of ecological factors and divergent natural selection in both sympatric and geographic models of speciation, but suggested that the two models "differ in the sequence in which the steps of the speciation process follow each other" (Mayr 1963: 451). In geographic speciation, "ecological factors... [play] their major role after the populations have become geographically separated. According to the theory of sympatric speciation the splitting of the gene pool itself is caused by ecological factors..." (Mayr 1963: 451). The literature is full of examples of closely related species that differ in ecological characteristics that directly influence the extent of gene flow between them; such differences include traits that are related to habitat and resource association and seasonal phenology—traits that almost certainly reflect responses to divergent natural selection.

Why then do we encounter the argument that: "It has taken evolutionary biologists almost until now to realize that [Darwin]...was probably correct in asserting that new species originate by natural selection" (Schluter and Conte 2009: 9955) or that: "The role of natural selection in speciation, first described by Darwin, has finally been widely accepted" (Via 2009: 9939)? What were we thinking before we came to this realization? Did most evolutionary biologists believe that the dominant modes were speciation by genetic drift, speciation that was due to divergence under uniform selection (or mutation order speciation), and polyploid speciation (Schluter (2001)? Is this a debate between proponents of divergent natural selection and divergent sexual selection?

Ecological speciation is defined as "the evolution of reproductive isolation between populations as a result of ecologically-based divergent natural selection" (Schluter and Conte 2009: 9955; see also Rundle and Nosil 2005; Hendry 2009). Although the term applies to divergence in the absence of gene flow, many or most of the systems that have been touted as examples of ecological speciation are also presumed examples of divergence in the face of gene flow (and some have been cited as possible instances of sympatric speciation). What is new about ecological speciation is that it proposes that ecological divergence that is due to divergent (disruptive) natural selection is the first barrier to arise. For nearly all of the cited examples, barriers maintained by ongoing divergent selection are the only impediments to gene flow. As a consequence, the ecological speciation literature rarely mentions examples of ecologically distinct taxa for which current sympatry is likely a consequence of secondary contact. It is again instructive to examine the hybrid zone literature, in which there are numerous examples of ecological differences between hybridizing entities (species, strains, races, etc.) that play a central role in defining hybrid zone structure and the extent of gene exchange. An early and incomplete summary of such examples is in Harrison (1990: 108, Table 3). Additional examples come from recent discussions of mosaic hybrid zones, which are, by definition, zones in which the interacting taxa exhibit patchy distributions as a consequence of environmental (ecological) heterogeneity, for example, plants in different habitats (Arnold and Bennett 1993; Wang et al. 1997), crickets on different soil types (Rand and Harrison 1989; Ross and Harrison 2002, 2006), or amphibians in ephemeral and permanent ponds (McCallum et al. 1998; Vines et al. 2003). Although most of these examples do not directly demonstrate the linkage between divergent selection and differentiated phenotypes (e.g., through reciprocal transplant experiments), it is hard to avoid the conclusion that ecological differentiation is a major barrier to gene exchange in many hybrid zones.

The ecological speciation literature repeatedly alludes to a relatively few examples of well-studied case histories from nature: fish in post-glacial lakes, pea aphids, other insect–host plant systems (walking sticks, leaf beetles, *Rhagoletis* flies), and *Heliconius* butterflies. These are marvelously interesting systems, with robust and diverse data sets that have been carefully analyzed, but these systems are surely not a representative sample of biodiversity. I do not mean to suggest that ecological speciation (in a broad sense) is rare; in fact, divergent selection resulting in ecological differentiation, which in turn impacts patterns of gene exchange, is probably very common. However, to return to the distinction made by Mayr (previously mentioned), it appears that the term ecological speciation is really being reserved for those cases where divergent natural selection is the initial, and for some period, the only barrier to gene exchange. Lineages that diverge in allopatry often accumulate multiple differences that reduce gene flow (including ecological differences), so it is difficult or impossible to cite these as examples of what we might call "narrow-sense ecological speciation."

A direct comparison between two study systems may be instructive. Previously, I suggested that pea aphids on clover and alfalfa form a mosaic hybrid zone where they are found in agricultural landscapes in North America (Figure 13.4A). The field crickets *Gryllus firmus* and *G. pennsylvanicus* also form a mosaic hybrid zone in eastern North America, in which the mosaic is determined by a strong association of the two species with different soil

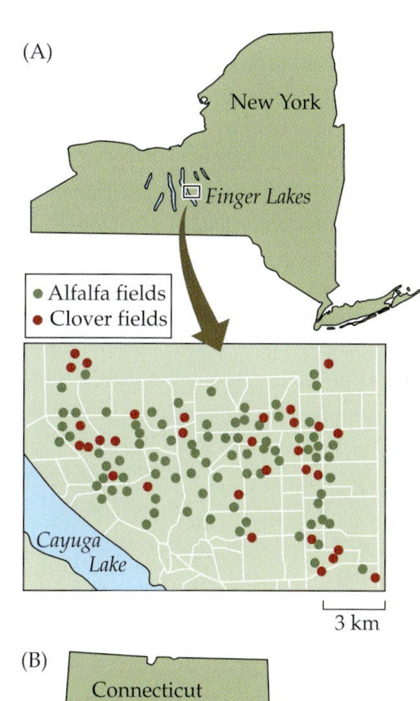

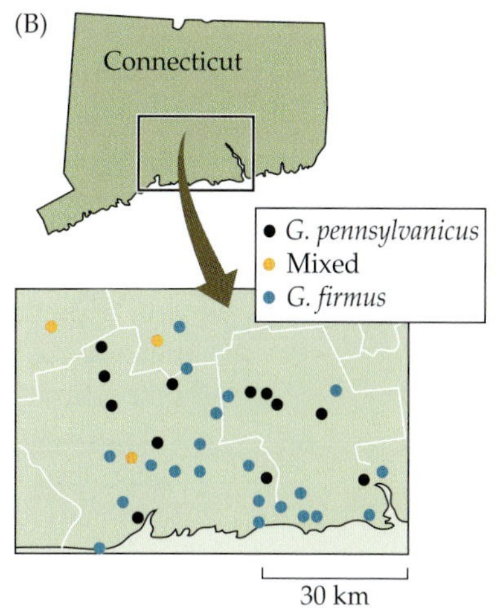

FIGURE 13.4 Comparison of Two Mosaic Hybrid Zones (A) Distribution of fields of alfalfa and clover in upstate New York. These fields are occupied by two different genotypes of pea aphids (*Acyrthosiphon pisum*). (B) Patchy distribution of two hybridizing field crickets (*Gryllus firmus* and *G. pennsylvanicus*) in Connecticut. The patchy distribution is determined by an underlying heterogeneity in soil type (i.e., sand versus loam). (A, adapted from Via 1999; B, adapted from Harrison and Rand 1989.)

types (Figure 13.4B). Pea aphids are regularly cited as an example of ecological speciation, but no one has ever used that term to describe the cricket system, although it is clear that ecological differences contribute to reproductive isolation (Harrison 1985; Rand and Harrison 1989; Harrison and Bogdanowicz 1997; Ross and Harrison 2002). Perhaps the major reason for this difference is that the only documented barrier between the pea aphids is the difference in acceptance of and performance on the two host plants—a barrier that is maintained by ongoing divergent selection. In contrast, the crickets show evidence of multiple barriers. In addition to temporal isolation that is due to life cycle differences along an elevation gradient and habitat isolation that is due to soil association, differences in mate choice or preference (Harrison and Rand 1989; Maroja et al. 2009b) and a one-way barrier to fertilization (Harrison 1983; Maroja et al. 2008) are also important. In spite of these many differences, the cricket species are estimated to have diverged only about 200,000 years ago (Maroja et al. 2009a).

So, ecological speciation is used primarily to describe situations in which single barriers to gene exchange, maintained by divergent natural selection, dominate. Via (2009) refers to two stages of ecological speciation, the first involving ecological differentiation, and the second differentiation in other traits resulting either from drift or selection. As she suggests, the scenario reverses the order in which barriers to gene exchange are assumed to occur in other speciation models (e.g., reinforcement). Most of the taxa for which ecological speciation has been proposed as the appropriate model are still in the first stage. But the second stage may be critically important if species are to persist in a spatially and temporally heterogeneous environment. If strong natural selection maintains current barriers, then changes in selection pressures, as a result of environmental change, may cause the barriers to disappear. Such changes have apparently occurred in one of the lakes that harbored both the limnetic and benthic forms of lake sticklebacks (Gow et al. 2006). Changes in water quality altered the probability of gene flow, and what had been two distinct populations became a hybrid swarm. A similar example of speciation reversal has been documented in African cichlids (Seehausen et al. 1997; 2008), again as a result of changes in water quality. My former colleague at Yale University, Charles Remington, used to talk about speciation in terms of "irreversible barriers to future fusion." Few barriers are truly irreversible, but because of the importance of natural selection in maintaining many ecological barriers and because the strength and direction of selection may change over time, ecological barriers (and behavioral barriers that depend on single communication modalities) may be particularly fragile and potentially transient.

Studies of ecological speciation have focused attention on a number of issues that were not previously much discussed in the speciation literature. First is evidence that post-zygotic isolation may often be ecologically dependent, that hybrids may be less fit because "they fall between ecological niches" (Rundle and Whitlock 2001: 198) rather than because of intrinsic incompatibilities between parental genomes. Studies of ecological speciation also highlight the occurrence of parallel speciation, which provides important evidence of the link between ecological differentiation and reproductive barriers (Rundle et al. 2000). Finally, there is now good evidence that divergent natural selection can lead directly to phenotypic differences that influence mate preference, such as *Heliconius* wing patterns (Jiggins et al. 2001) or stickleback morphology (Rundle et al. 2000).

What role does narrow-sense ecological speciation play in the long-term generation of biodiversity? Many lineages may persist for relatively short periods, remaining distinct as a result of divergent natural selection. Whether such lineages usually pass through stage 2 and become what many of us would recognize as good species has not yet been resolved. What is clear is that ecology plays a major role in speciation—both in the origin and persistence of species. We have known that fact for a very long time, and to claim otherwise is to ignore a substantial literature. The challenge now is to understand the genetic basis of ecologically important traits/phenotypes, to document how these distinct phenotypes perform in relevant natural environments, and to show to what extent gene flow is reduced between these phenotypic classes. With that information, it becomes possible to describe both the history of natural selection and the link between ecological divergence and reproductive isolation.

The Mosaic Genome

It is now widely recognized that the genomes of sexually reproducing organisms are mosaics of different histories—the pattern of the mosaic dictated by the history of recombination and selection. A direct consequence is that the forces that have shaped and are shaping current patterns of variation will differ across the genome. Local selective sweeps will reduce variation in a chromosomal region surrounding the selected locus; the size of the region depends on the strength of selection (which determines the rate of the sweep) and the amount of recombination (which allows linked regions to escape from further effects of the sweep). When diverging populations are in contact, genome regions that include genes that contribute to reproductive barriers or that are under divergent selection will tend to remain distinct, whereas gene flow (introgression) at neutral loci may erase patterns of differentiation.

Barton and Hewitt explicitly made the argument that the extent of gene exchange between hybridizing taxa will depend on the genome region: "Strict application of the biological species concept might lead to different results for different loci; perhaps one can only define groups of actually or potentially interbreeding natural populations... at the gene level" (Barton and Hewitt 1981: 119; Bazykin 1969). The concept of differential introgression was widely accepted and discussed in the hybrid zone literature. Species boundaries were described as semipermeable (Key 1968; Harrison 1986; Harrison and Rand 1989) and isolation was understood to be a property of individual genes or gene regions (Harrison 1990, 1991). Initial evidence for differential introgression came from comparisons of morphological, allozyme, and mitochondrial DNA (mtDNA) data (Harrison 1990, 92, Table 1). Studies of hybridization in *Helianthus* not only clearly identified chromosome regions that introgressed significantly more or less than expected, but demonstrated that patterns of differential introgression were consistent across three presumably independent contact zones (Rieseberg et al. 1999).

The notion of a mosaic genome also emerged from comparisons among multiple gene genealogies for closely related groups of species. Such comparisons often reveal discordant patterns, with some loci showing species to be reciprocally monophyletic or exclusive, whereas haplotypes at other loci do not sort by species (Beltrán et al. 2002; Machado and Hey 2003; Dopman et al. 2005; Putnam et al. 2007; Andres et al. 2008). Simple protocols for estimating allele frequencies for single-nucleotide polymorphisms (SNPs) and/or indels afford the opportunity to scan the genome for so-called F_{ST} outliers (Emelianov et al. 2004; Grahame et al. 2006; Wood et al. 2008; Via and West 2008). However, Noor and Bennett (2009) set out some caveats. As with comparison of genealogies from different gene regions, the goal is to identify regions that have become (presumably through selective lineage sorting) or remain (in the face of introgression) differentiated. This approach is obviously most effective when much or most of the genome shows little or no differentiation between populations or species, which will be true relatively early in the speciation process.

The new genetic and genomic data have led to a proliferation of terminology: we now have the "genic view of species" (Wu 2001; Wu and Ting 2004), "genomic islands of speciation" (Turner et al. 2005), the "genetic mosaic of speciation" (Via and West 2008), and "heterogeneous genomic divergence" (Nosil et al. 2009). All basically refer to the fact that the genome mosaic includes "islands of divergence" (Nosil et al. 2009) that are candidate gene regions to include one or more genes that contribute to reproductive isolation. How many islands of divergence are there, how large are such islands, and how are they distributed across the genome? Are these islands of divergence or are they islands of exclusivity? That is, do we expect these gene regions to exhibit fixed allelic differences between species?

One of the first papers to focus on divergence islands, by Ting et al. (2000), examined patterns of differentiation around the *Odysseus* (*OdsH*) locus, a gene known to contribute to hybrid male sterility between *Drosophila simulans* and *D. mauritiana*. A pattern of exclusivity seen at *OdsH* broke down within a few kilobases of the gene, creating an initial impression that islands of divergence might be very small. Ting et al. commented that "the hitchhiking process...must have been relatively ineffectual over a longer distance" (Ting et al. 2000: 5316). Several recent papers cite Charlesworth et al. (1997), in which it is argued that in subdivided populations with moderately strong local selection (one allele favored in one patch type, a second allele favored in the other patch type), islands of divergence can be large (Via and West 2008; Via 2009; Nosil et al. 2009). Large islands persist because effective recombination is potentially much less than recombination predicted on the basis of map distance and random mating; selection against individuals with the wrong allele allows linkage disequilibrium to persist over larger map distances. Via and West (2008) indeed argue that between two pea aphid host races, they see large islands of differentiation (based on F_{ST}); in contrast, other recent studies suggest that such islands may be much smaller (Yatabe et al. 2007; Wood et al. 2008; Smadja et al. 2008). In *Anopheles* mosquitoes, differences between two reproductively isolated chromosomal forms are restricted to three small regions, which reveal fixed sequence differences between forms (Turner et al. 2005).

Regions of locally restricted recombination may be particularly important in speciation (Butlin 2005). Chromosomal inversion differences that are fixed between species suppress local recombination in hybrids and enforce linkage disequilibrium. Inverted regions are therefore predicted (and found) to harbor genes underlying reproductive isolation (Noor et al. 2001; Rieseberg 2001; Navarro and Barton 2003; Feder et al. 2003). Gene genealogies that show *Drosophila pseudoobscura* and *D. persimilis* to be exclusive groups are consistently associated with inverted regions (Machado and Hey 2003). In general, because gene regions that already harbor barrier genes have reduced levels of gene flow, further divergence is more likely in these regions. Thus, in the face of hybridization and introgression, loci contributing to reproductive isolation, and therefore exhibiting divergence in allele frequencies, will tend to be clustered (Gavrilets 2004).

A search for outlier loci or patterns of genealogical exclusivity (fixed differences), when combined with detailed linkage maps, provides a mechanism for identifying gene regions (ultimately genes) that are potentially important for differential adaptation and speciation. An alternative approach is to identify candidate genes that might account for the phenotypic differences responsible for reproductive isolation. Both approaches have great potential, and the next decade will no doubt witness the discovery of many barrier or speciation genes—genes that currently contribute to reproductive isolation and that may have been involved in initial divergence (Noor and Feder 2006). To date, the list of such genes is relatively short and is heavily populated with genes responsible for hybrid inviability or sterility in the *Drosophila melanogaster* species group. Recent additions to the list include genes involved in Dobzhansky–Muller incompatibilities in *Drosophila* (Brideau et al. 2006; Tang and Presgraves 2009). But, also characterized have been genes that affect phenotypic traits subject to strong natural selection, such as genes that determine beak morphology in Galapagos finches (Abzhanov et al. 2004, 2006), armor plate patterning in threespined sticklebacks (Shapiro et al. 2004; Colosimo et al. 2005), and pelage color in mice (Hoekstra and Nachman 2003; Hoekstra et al. 2006; Steiner et al. 2007,

2009). To the extent that adaptive divergence leads to reproductive isolation, some of these phenotypic traits may also be involved in speciation.

The current preoccupation with finding *the gene* has some potential pitfalls. Although a focus on the genotype and gene expression is appropriate in the age of genomics and transcriptomics, the essential starting point for studies of speciation must be a clear and comprehensive understanding of the phenotypic basis for reproductive isolation and (if possible) inferences about the order in which different barriers arose. Of course, the latter is often not possible, even for pairs of recently diverged species. Trait differences that contribute to current reproductive isolation may: (1) have arisen subsequent to complete interruption of gene flow; (2) be unimportant, because there is now a suite of earlier-acting barriers; (3) operate only in certain contexts or parts of the range; and (4) be reversible if environments change. Before investing in next-generation sequencing and microarrays, evolutionary biologists need to be certain to target the relevant phenotypes.

Studies of the genetic basis of speciation have been biased by the pull of the possible. Certain classes of genes, certain groups of taxa, and the genetic basis of certain phenotypes are easier to characterize, leading to the inevitable circumstance that what can be done takes priority over what should be done. Whole genome/transcriptome sequences are now routinely obtained for organisms that are models in ecology and evolution, and proteomics allows identification of the components of important biological systems. There is no longer a need to rely so heavily on *Drosophila* and *Mus*; instead, the number (and phylogenetic extent) of lineages for which we can integrate genetics and molecular/developmental biology with ecology, physiology, and behavior can be expanded.

Conclusions

The understanding of speciation models and mechanisms has become far more sophisticated in the last half century. We have carefully defined the realm of the possible (from theory) and characterized small outposts in the realm of the actual (from data). Ernst Mayr and Theodosius Dobzhansky are still cited regularly, but their influence has perhaps begun to wane. Their writings serve as a persistent frame of reference, an established orthodoxy that many continue to challenge. In fact, the orthodoxy dissolved long ago. Evolutionary biologists now recognize that the process of speciation can proceed along many different trajectories, in the face of gene flow and in its absence, with divergence driven by natural selection, sexual selection, and/or genetic drift. We differ in how we weight the importance of these contributors (e.g., what does the plot in Figure 13.2 really look like?). Currently, much of the speciation literature focuses on models and mechanisms that challenge the traditional focus on allopatry, post-zygotic barriers, hybrid zones, and reinforcement. Ecological speciation, hybrid speciation, and speciation driven by sexual conflict are important modes of speciation in the new order. There is no doubt that each has a place in generating diversity, but the relative roles of the current stable of challengers remains unclear.

Finally, we should not forget that much of what is now being said has been said before, perhaps in different words, and with reference to different systems. It is particularly appropriate, as part of a celebration of the 150th anniversary of *The Origin of Species,* to acknowledge the wealth of ideas and information that can be found in the older literature. In many cases, the questions we ask are not new; it is the opportunities to answer those questions that have changed dramatically.

Acknowledgments

I thank the past and present members of the Harrison lab, for educating me in the ways of many different organisms and insuring that I maintained a broad perspective. The University of St. Andrews, School of Biological Sciences, provided me with a home on sabbatical leave, during which I brooded about most of the issues that I discuss in this chapter. Also, many thanks to the organizers of the Darwin 2009 meeting at Stony Brook, a wonderfully interactive and stimulating encounter. Mohamed Noor and Doug Futuyma provided insightful comments on an earlier draft of this chapter. Finally, thanks to the National Science Foundation and United States Department of Agriculture for funding my research on crickets and corn borers, which are excellent model systems for studying the origin of species.

Literature Cited

Abzhanov, A., W. P. Kuo, C. Hartmann, and 3 others. 2006. The calmodulin pathway and the evolution of elongated beak morphology in Darwin's finches. *Nature* 442: 563.

Abzhanov, A., M. B. Protas, R. Grant, and 2 others. 2004. Bmp4 and morphological variation of beaks in Darwin's finches. *Science* 305: 1462.

Andres, J. A., L. S. Maroja, S. M. Bogdanowicz, and 2 others. 2006. Molecular evolution of seminal proteins in field crickets. *Mol. Biol. Evol.* 23: 1574–1584.

Andres, J. A., L. S. Maroja, and R. G. Harrison. 2008. Searching for candidate speciation genes using a protemomic approach: Seminal proteins in field crickets. *Proc. Roy. Soc. B* 275: 1975–1983.

Arnold, M. L. and B. D. Bennett. 1993. Natural hybridization in Louisiana irises: Genetic variation and ecological determinants. In R. Harrison (ed.), *Hybrid Zones and the Evolutionary Process*, pp. 115–139. Oxford University Press, New York.

Barton, N. H. 1989. Founder effect speciation. In D. Otte and J. A. Endler (eds.), *Speciation and its Consequences*, pp. 229–256. Sinauer Associates, Sunderland, MA.

Barton, N. H. and B. Charlesworth. 1984. Genetic revolutions, founder effects, and speciation. *Ann. Rev. Ecol. Syst.* 15: 133–164.
Barton, N. H. and G. M. Hewitt. 1981. Hybrid zones and speciation. In W. R. Atchley and D. S. Woodruff (eds.), *Evolution and Speciation*, pp. 109–145. Cambridge University Press, Cambridge, UK.
Barton, N. H. and G. M. Hewitt. 1985. Analysis of hybrid zones. *Ann. Rev. Ecol. Syst.* 16: 113–148.
Baum, D. A. and K. L. Shaw. 1995. Genealogical perspectives on the species problem. In P. C. Hoch and A. G. Stevenson (eds.), *Experimental and Molecular Approaches to Plant Biosystematics*, pp. 289–303. Missouri Botanical Garden, St. Louis.
Bazykin, A. D. 1969. Hypothetical mechanism of speciation. *Evolution* 23: 685–687.
Beltrán, M., C. D. Jiggins, V. Bull, and 4 others. 2002. Phylogenetic discordance at the species boundary: Comparative gene genealogies among rapidly radiating *Heliconius* butterflies. *Mol. Biol. Evol.* 19: 2176–2190.
Brideau, N. J., H. A. Flores, J. Wang, and 3 others. 2006. Two Dobzhansky-Muller genes interact to cause hybrid lethality in *Drosophila*. *Science* 314: 1238–1239.
Bush, G. L. 1969. Sympatric host race formation and speciation in frugivorous flies of the genus *Rhagoletis* (Diptera, Tephritidae). *Evolution* 23: 237–251.
Butlin, R. K. 2005. Recombination and speciation. *Mol. Ecol.* 14: 2621–2635.
Butlin, R. K., J. Galindo, and J. W. Grahame. 2008. Sympatric, parapatric or allopatric: The most important way to classify speciation? *Phil. Trans. Roy. Soc. B.* 363: 2997–3007.
Carson, H. L. 1971. Speciation and the founder principle. *Stadler. Symp.* 3: 51–70.
Carson, H. L. 1975. The genetics of speciation at the diploid level. *Am. Nat.* 109: 83–92.
Carson, H. L. and A. R. Templeton. 1984. Genetic revolutions in relation to speciation phenomena: The founding of new populations. *Ann. Rev. Ecol. Syst.* 15: 97–131.
Charlesworth, B., M. Nordborg, and D. Charlesworth. 1997. The effects of local selection, balanced polymorphism, and background selection on equilibrium patterns of genetic diversity in subdivided populations. *Genet. Res.* 70: 155–174.
Colosimo, P. F., K. E. Hosemann, S. Balabhadra, and 7 others. 2005. Widespread parallel evolution in sticklebacks by repeated fixation of ectodysplasin alleles. *Science* 307: 1928–1933.
Coyne, J. A. 1992. Much ado about species. *Nature* 357: 289–290.
Coyne, J. A. and H. A. Orr. 1989. Patterns of speciation in *Drosophila*. *Evolution* 43: 362–381.
Coyne, J. A. and H. A. Orr. 1997. "Patterns of speciation in *Drosophila*" revisited. *Evolution* 51: 295–303.
Coyne, J. A. and H. A. Orr. 1998. The evolutionary genetics of speciation. *Phil. Trans. Roy. Soc. Lond. B* 353: 287–305.
Coyne, J. A. and H. A. Orr. 2004. *Speciation*. Sinauer Associates, Sunderland, MA.
Darwin, C. 1859. *On the Origin of Species by Means of Natural Selection, or the Preservation of Favoured Races in the Struggle for Life*. John Murray, London.
Dieckmann, U. and M. Doebeli. 1999. On the origin of species by sympatric speciation. *Nature* 400: 354–357.
Dobzhansky, T. 1937. *Genetics and the Origin of Species*. Columbia University Press, New York.

Dobzhansky, T. 1940. Speciation as a stage in evolutionary divergence. *Am. Nat.* 74: 302–321.

Dobzhansky, T. 1970. *Genetics of the Evolutionary Process*. Columbia University Press, New York.

Dodd, D. M. B. and J. R. Powell. 1985. Founder-flush speciation: An update of experimental results with *Drosophila*. *Evolution* 39: 1388–1392.

Dopman, E. B., L. Peréz, S. M. Bogdanowicz, and 1 other. 2005 Consequences of reproductive barriers for genealogical discordance in the European corn borer. *Proc. Natl. Acad. Sci. USA* 102: 14706–14711.

Emelianov, I., F. Marec, and J. Mallet. 2004. Genomic evidence for divergence with gene flow in host races of the larch budmoth. *Proc. Roy. Soc. Lond. B* 271: 97–105.

Endler, J. 1977. *Geographic Variation, Speciation, and Clines*. Princeton University Press, Princeton.

Feder, J. L., J. B. Roethele, K. Filchak, and 2 others. 2003. Evidence for inversion polymorphism related to sympatric host race formation in the apple maggot fly, *Rhagoletis pomonella*. *Genetics* 163: 939–953.

Felsenstein, J. 1981. Skepticism towards Santa Rosalia, or why are there so few kinds of animals? *Evolution* 35: 124–138.

Fitzpatrick, B. M., J. H. A. Fordyce, and S. Gavrilets. 2008. What, if anything, is sympatric speciation? *J. Evol. Biol.* 21: 1452–1459.

Fitzpatrick, B. M., J. H. A. Fordyce, and S. Gavrilets. 2009. Pattern, process, and geographic modes of speciation. *J. Evol. Biol.* 22: 2342–2347.

Futuyma, D. J. 2009. *Evolution*, 2nd ed. Sinauer Associates, Sunderland, MA.

Futuyma, D. J. and G. C. Mayer. 1980. Non-allopatric speciation in animals. *Syst. Zool.* 29: 254–271.

Galiana, A., A. Moya, and F. J. Ayala 1993. Founder-flush speciation in *Drosophila pseudoobscura*: A large-scale experiment. *Evolution* 47: 432–444.

Gavrilets, S. 2004. *Fitness Landscapes and the Origin of Species*. Princeton University Press, Princeton.

Gleason, J. M. and M. G. Ritchie. 1998. Evolution of courtship song and reproductive isolation in the *Drosophila willistoni* complex: Do sexual signals diverge the most quickly? *Evolution* 52: 1493–1500.

Gow, J. L., C. L. Peichel, and E. B. Taylor. 2006. Contrasting hybridization rates between sympatric three-spined sticklebacks highlight the fragility of reproductive barriers between evolutionarily young species. *Mol. Ecol.* 15: 739–752.

Grahame, J. W., C. S. Wilding, and R. K. Butlin. 2006. Adaptation to a steep environmental gradient and an associated barrier to gene exchange in *Littorina saxatilis*. *Evolution* 60: 268–278.

Grant, P. R. and B. R. F. Grant. 2007. *How and Why Species Multiply: The Radiation of Darwin's Finches*. Princeton University Press, Princeton.

Harrison, R. G. 1983. Barriers to gene exchange between closely related cricket species. I. Laboratory hybridization studies. *Evolution* 37: 245–251.

Harrison, R. G. 1985. Barriers to gene exchange between closely related cricket species. II. Life cycle variation and temporal isolation. *Evolution* 39: 244–259.

Harrison, R. G. 1986. Pattern and process in a narrow hybrid zone. *Heredity* 56: 347–359.

Harrison, R. G. 1990. Hybrid zones: Windows on evolutionary process. *Oxford Surv. Evol. Biol.* 7: 69–128.

Harrison, R. G. 1991. Molecular changes at speciation. *Ann. Rev. Ecol. Syst.* 22: 281–308.

Harrison, R. G. 1998. Linking evolutionary pattern and process. In D. J. Howard and S. H. Berlocher (eds.), *Endless Forms*, pp. 19–31. Oxford University Press, New York.

Harrison, R. G. and S. M. Bogdanowicz. 1997. Patterns of variation and linkage disequilibrium in a field cricket hybrid zone. *Evolution* 51: 493–505.

Harrison, R. G. and D. M. Rand. 1989. Mosaic hybrid zones and the nature of species boundaries. In D. Otte and J. A. Endler (eds.), *Speciation and its Consequences*, pp. 111–133. Sinauer Associates, Sunderland, MA.

Hauffe, H. C. and J. B. Searle. 1998. Chromosomal heterozygosity and fertility in house mouse (*Mus musculus domesticus*) from northern Italy. *Genetics* 150: 1143–1154.

Hendry, A. P. 2009. Ecological speciation! Or the lack thereof? *Can. J. Fish. Aquat. Sci.* 66: 1383–1398.

Hewitt, G. M. 1989. The division of species by hybrid zones. In D. Otte and J. A. Endler (eds.), *Speciation and its Consequences*, pp. 85–110. Sinauer Associates, Sunderland, MA.

Hewitt, G. M. 2000. The genetic legacy of the Quartenary ice ages. *Nature* 405: 907–913.

Hey, J. and R. Nielsen. 2004. Multilocus methods for estimating population sizes, migration rates and divergence time, with applications to the divergence of *Drosophila pseudoobscura* and *D. persimilis*. *Genetics* 167: 747–760.

Hoekstra, H. E. and M. W. Nachman. 2003. Different genes underlie adaptive melanism in different populations of pocket mice. *Mol. Ecol.* 12: 1185–94.

Hoekstra, H. E., R. J. Hirschmann, R. J. Bundey, and 2 others. 2006. A single amino acid mutation contributes to adaptive color pattern in beach mice. *Science* 313: 101–104.

Howard, D. J. 1986. A zone of overlap and hybridization between two ground cricket species. *Evolution* 40: 34–43.

Howard, D. J. 1993. Reinforcement: Origin, dynamics, and fate of an evolutionary hypothesis. In R. G. Harrison (ed.), *Hybrid Zones and the Evolutionary Process*, pp. 46–69. Oxford University Press, New York.

Howard, D. J. 1999. Conspecific sperm and pollen precedence and speciation. *Ann. Rev. Ecol. Syst.* 30: 109–132.

Hudson, R. R. and J. A. Coyne. 2002. Mathematical consequences of the genealogical species concept. *Evolution* 56: 1557–1565.

Jiggins, C. D., R. E. Naisbit, R. L. Coe, and 1 other. 2001. Reproductive isolation caused by colour pattern mimicry. *Nature* 411: 302–305.

Key, K. H. L. 1968. The concept of stasipatric speciation. *Syst. Zool.* 17: 14–22.

Kirkpatrick, M. and V. Ravigné. 2002. Speciation by natural and sexual selection: Models and experiments. *Am. Nat.* 159: S22–S35.

Kirkpatrick, M. and M. R. Servedio 1999. The reinforcement of mating preferences on an island. *Genetics* 151: 865–884.

Knowles, L. 2009. Statistical phylogeography. *Ann. Rev. Ecol. Evol. Syst.* 40: 593–612.

Kondrashov, A. S. and F. A. Kondrashov. 1999. Interactions among quantitative traits in the course of sympatric speciation. *Nature* 400: 351–354.

Kulathinal, R. J., L. S. Stevison, and M. A. F. Noor. 2009. The genomics of speciation in *Drosophila*: Diversity, divergence, and introgression estimated using low-coverage genome sequencing. *PLoS Genet.* 5: e1000550.

Liou, L. W. and T. D. Price 1994. Speciation by reinforcement of prezygotic isolation. *Evolution* 48: 1451–1459.

Losos, J. B. 2009. *Lizards in an Evolutionary Tree*. University of California Press, Berkeley.
Losos, J. B. and R. E. Glor. 2003. Phylogenetic comparative methods and the geography of speciation. *Trends Ecol. Evol.* 18: 220–227.
Machado, C. A. and J. Hey. 2003. The causes of phylogenetic conflict in a classic *Drosophila* species group. *Proc. Roy. Soc. Lond. B* 270: 1193–1202.
Mallet, J. 2008. Mayr's view of Darwin: Was Darwin wrong about speciation? *Biol. J. Linn. Soc.* 95: 3–16.
Mallet, J., A. Meyer, P. Nosil, and 1 other. 2009. Space, sympatry, and speciation. *J. Evol. Biol.* 22: 2332–2341.
Maroja, L. S., J. A. Andrés, and R. G. Harrison. 2009a. Genealogical discordance and patterns of introgression and selection across a cricket hybrid zone. *Evolution* 63: 2999–3015.
Maroja, L. S., J. A. Andrés, J. R. Walters, and 1 other. 2009b. Multiple barriers to gene exchange in a field cricket hybrid zone. *Biol. J. Linn. Soc.* 97: 390–402.
Maroja, L. S., M. E. Clark, and R. G. Harrison. 2008. *Wolbachia* plays no role in the one-way reproductive incompatibility between the hybridizing field crickets *Gryllus firmus* and *G. pennsylvanicus*. *Heredity* 101: 435–444.
Martin, N. H. and J. H. Willis 2007. Ecological divergence associated with mating system causes nearly complete reproductive isolation between *Mimulus* species. *Evolution* 61: 68–82.
Maynard Smith, J. 1966. Sympatric speciation. *Am. Nat.* 100: 637–650.
Mayr, E. 1942. *Systematics and the Origin of Species*. Columbia University Press, New York.
Mayr, E. 1963. *Animal Species and Evolution*. Harvard University Press, Cambridge, MA.
McCallum, C. J., B. Nürnberger, N. H. Barton, and 1 other. 1998. Habitat preference in a *Bombina* hybrid zone in Croatia. *Evolution* 52: 227–239.
Mendelson, T. C. 2003. Sexual isolation evolves faster than hybrid inviability in a diverse and sexually dimorphic genus of fish (Percidae: *Etheostoma*). *Evolution* 57: 317–327.
Moya, A., A. Galiana, and F. J. Ayala. 1995. Founder-effect speciation theory: Failure of experimental corroboration. *Proc. Natl. Acad. Sci. USA* 92: 3983–3986.
Navarro, A. and N. H. Barton. 2003. Accumulating post-zygotic isolation genes in paraptry: A new twist on chromosomal speciation. *Evolution* 57: 447–459.
Nichols, R. 2001. Gene trees and species trees are not the same. *Trends Ecol. Evol.* 16: 358–364.
Nielsen, R. and M. A. Beaumont 2009. Statistical inferences in phylogeography. *Mol. Ecol.* 18: 1034–1047.
Nielsen, R. and J. Wakeley 2001. Distinguishing migration from isolation: A Markov chain Monte Carlo approach. *Genetics* 158: 885–896.
Nixon, K. C. and Q. D. Wheeler. 1990. An amplification of the phylogenetic species concept. *Cladistics* 6: 211–223.
Noor, M. A. F. and S. M. Bennett 2009. Islands of speciation or mirages in the desert? Examining the role of restricted recombination in maintaining species. *Heredity* 103: 439–44.
Noor, M. A. F. and J. L. Feder. 2006. Speciation genetics: Evolving approaches. *Nat. Rev. Genet.* 7: 851–861.
Noor, M. A. F., K. L. Grams, L. A. Bertucci, and 1 other. 2001. Chromosomal inversions and the reproductive isolation of species. *Proc. Natl. Acad. Sci. USA* 98: 12084–12088.

Nosil, P., D. J. Funk, and D. Ortiz-Barrientos. 2009. Divergent selection and heterogeneous genomic divergence. *Mol. Ecol.* 18: 375–402.
Ortiz-Barrientos, D., J. Reiland, J. Hey, and 1 other. 2002. Recombination and the divergence of hybridizing species. *Genetica* 116: 167–178.
Peccoud, J., A. Ollivier, M. Plantegenest, and 1 other. 2009. A continuum of genetic divergence from sympatric host races to species in the pea aphid complex. *Proc. Natl. Acad. Sci. USA* 106: 7495–7500.
Pialek, J., H. C. Hauffe, K. M. Rodreiguez-Clark, and 1 other. 2001. Raciation and speciation in house mice from the Alps: The role of chromosomes. *Mol. Ecol.* 10: 613–625.
Powell, J. R. 1978. The founder-flush speciation theory: An experimental approach. *Evolution* 32: 465–474.
Price, T. 2007. *Speciation in Birds*. Roberts and Company, Greenwood Village, CO.
Putman, A. S., J. M. Scriber, and P. Andolfatto. 2007. Discordant divergence times among Z chromosome regions between two ecologically distinct swallowtail butterfly species. *Evolution* 61: 912–927.
Ramsey, J., H. D. Bradshaw, and D. W. Schemske. 2003. Components of reproductive isolation between the monkeyflowers *Mimulus lewisii* and *M. cardinalis* (Scrophulariaceae). *Evolution* 57: 1520–1534.
Rand, D. M. and R. G. Harrison. 1989. Ecological genetics of a mosaic hybrid zone: Mitochondrial, nuclear, and reproductive differentiation of crickets by soil type. *Evolution* 43: 432–449.
Rieseberg, L. H. 2001. Chromosomal rearrangements and speciation. *Trends Ecol. Evol.* 16: 351–358.
Rieseberg, L. H., J. Whitton, and K. Gardner. 1999. Hybrid zones and the genetic architecture of a barrier to gene flow between two sunflower species. *Genetics* 152: 713–727.
Ringo, J., D. Wood, R. Rockwell, and 1 other. 1985. An experiment testing two hypotheses of speciation. *Am. Nat.* 126: 642–661.
Ross, C. L. and R. G. Harrison 2002. A fine-scale spatial analysis of the mosaic hybrid zone between *Gryllus firmus* and *Gryllus pennsylvanicus*. *Evolution* 56: 2296–2312.
Ross, C. L. and R. G. Harrison. 2006. Viability selection on overwintering eggs in a field cricket mosaic hybrid zone. *Oikos* 115: 53–68.
Rundle, H. D. and P. Nosil. 2005. Ecological speciation. *Ecol. Letters* 8: 336–352.
Rundle, H. D. and M. C. Whitlock. 2001. A genetic interpretation of ecologically dependent isolation. *Evolution* 55: 198–201.
Rundle, H. D., A. O. Mooers, and M. C. Whitlock. 1998. Single founder-flush events and the evolution of reproductive isolation. *Evolution* 52: 1850–1855.
Rundle, H. D., L. Nagel, J. W. Boughman, and 1 other. 2000. Natural selection and parallel speciation in sympatric sticklebacks. *Science* 287: 306–308.
Schliewen, U. K., D. Tautz, and S. Pääbo. 1994. Sympatric speciation suggested by monophyly of crater lake cichlids. *Nature* 368: 629–632.
Schluter, D. 2000. *The Ecology of Adaptive Radiation*. Oxford University Press, Oxford.
Schluter, D. 2001. Ecology and the origin of species. *Trends Ecol. Evol.* 16: 372–380.
Schluter, D. 2009. Evidence for ecological speciation and its alternative. *Science* 323: 737–741.
Schluter, D. and G. L. Conte 2009. Genetics and ecological speciation. *Proc. Natl. Acad. Sci. USA* 106: 9955–9962.

Searle, J. B. 1993. Chromosomal hybrid zones in eutherian mammals. In R. G. Harrison (ed.) *Hybrid Zones and the Evolutionary Process*, pp. 309–353. Oxford University Press, New York.

Seehausen, O., G. Takimoto, D. Roy, and 1 other. 2008. Speciation reversal and biodiversity dynamics with hybridization in changing environments. *Mol. Ecol.* 17: 30–44.

Seehausen, O., J. J. M. van Alphen, and F. Witte. 1997. Cichlid fish diversity threatened by eutrophication that curbs sexual selection. *Science* 277: 1808–1811.

Servedio, M. R. and M. A. F. Noor. 2003. The role of reinforcement in speciation: Theory and data. *Ann. Rev. Ecol. Syst.* 34: 339–364.

Shapiro, M. D., M. E. Marks, C. L. Peichel, and 5 others. 2004. Genetic and developmental basis of evolutionary pelvic reduction in threespine sticklebacks. *Nature* 428: 717–723.

Slatkin, M. 1973. Gene flow and selection in a cline. *Genetics* 75: 733–756.

Smadja, C., J. Galindo, and R. K. Butlin. 2008. Hitching a lift on the road to speciation. *Mol. Ecol.* 17: 4177–4180.

Smith, H. M. 1965. More evolutionary terms. *Syst. Zool.* 14: 57–58.

Steiner, C. C., H. Römpler, L. M. Boettger, and 2 others. 2009. The genetic basis of phenotypic convergence in beach mice: Similar pigmentation patterns but different genes. *Mol. Biol. Evol.* 26: 35–45.

Steiner, C. C., J. N. Weber, and H. E. Hoekstra. 2007. Adaptive variation in beach mice caused by two interacting pigmentation genes. *PLoS Biol.* 5: 1880–1889.

Swanson, W. J. and V. D. Vacquier. 2002. The rapid evolution of reproductive proteins. *Nat. Rev. Genet.* 3: 137–144.

Swanson, W. J., A. G. Clark, H. M. Waldrip-Dail, and 2 others. 2001. Evolutionary EST analysis identifies rapidly evolving male reproductive proteins in *Drosophila*. *Proc. Natl. Acad. Sci. USA* 95: 4051–4054.

Swanson, W. J., T. M. Panhuis, and N. L. Clark. 2006. Rapid evolution of reproductive proteins in abalone and *Drosophila*. *Phil. Trans. Roy. Soc. B* 361: 261–268.

Tang, S. and D. C. Presgraves 2009. Evolution of the *Drosophila* nuclear pore complex results in multiple hybrid incompatibilities. *Science* 323: 779–782.

Templeton, A. R. 1980. The theory of speciation via the founder principle. *Genetics* 94: 1011–1034.

Templeton, A. R. 1981. Mechanisms of speciation—a population genetics approach. *Ann. Rev. Ecol. Syst.* 12: 23–48.

Templeton, A. R. 2008. The reality and importance of founder speciation in evolution. *BioEssays* 30: 470–479.

Ting, C.-T., S.-C. Tsaur, and C.-I. Wu 2000. The phylogeny of closely related species as revealed by the genealogy of a speciation gene, *Odysseus*. *Proc. Natl. Acad. Sci. USA* 97: 5313–5316.

Turner, T. L., M. W. Hahn, and S. V. Nuzhdin. 2005. Genomic islands of speciation in *Anopheles gambiae*. *PLoS Biol.* 3: 1572–1578.

Via, S. 1999. Reproductive isolation between sympatric races of pea aphids. I. Gene flow restriction and habitat choice. *Evolution* 53: 1446–1457.

Via, S. 2001. Sympatric speciation in animals: the ugly duckling grows up. *Trends. Ecol. Evol.* 16: 381–390.

Via, S. 2009. Natural selection in action during speciation. *Proc. Natl. Acad. Sci. USA* 106: 9939–9946.

Via, S. and J. West 2008. The genetic mosaic suggests a new role for hitchhiking in ecological speciation. *Mol. Ecol.* 17: 4334–4345.

Vines, T. H., S. C. Kohler, M. Thiel, and 5 others. 2003. The maintenance of reproductive isolation in a mosaic hybrid zone between the fire-bellied toads *Bombina bombina* and *B. variegata*. *Evolution* 57: 1876–1888.
Wang, H., E. D. McArthur, S. C. Sanderson, and 2 others. 1997. Narrow hybrid zone between two subspecies of big sagebrush (*Artemisia tridentata*: Asteraceae). IV. Reciprocal transplant experiments. *Evolution* 51: 95–102.
White, M. J. D. 1969. Chromosomal rearrangements and speciation in animals. *Ann. Rev. Genetics* 3: 75–98.
White, M. J. D. 1978. *Modes of Speciation*. W. H. Freeman and Company, San Francisco.
Wolfner, M. F. 2002. The gifts that keep on giving: Physiological functions and evolutionary dynamics of male seminal proteins in *Drosophila*. *Heredity* 88: 85–93.
Wood, H. M., J. W. Grahame, S. Humphrey, and 2 others. 2008. Sequence differentiation in regions identified by a genome scan for local adaptation. *Mol. Ecol.* 17: 3123–3135.
Wu, C.-I. 2001. The genic view of the process of speciation. *J. Evol. Biol.* 14: 851–865.
Wu, C.-I. and C. T. Ting. 2004. Genes and speciation. *Nat. Rev. Genet.* 5: 114–122.
Yatabe, Y., N. C. Kane, C. Scotti-Saintagne, and 1 other. 2007. Rampant gene exchange across a strong reproductive barrier between the annual sunflowers, *Helianthus annuus* and *H. petiolaris*. *Genetics* 175: 1883–1893.

Commentary Three

Ecology in Evolutionary Biology

Mark A. McPeek

The recent, rapid advances of DNA sequencing technologies have clearly been a boon for the study of molecular evolution, evolutionary genetics, and evolutionary developmental biology (see Kolaczkowski and Kern, Chapter 6; Wray, Chapter 9). The entire genome or transcriptome of any species can be rapidly and inexpensively sequenced, and the time and money required continues to decrease precipitously (see Hoekstra, Chapter 22). While the genetic basis of important traits under natural selection for a few classic examples are being determined, these exemplars are typically single genes of major effect (Geffeney et al. 2005; Hoekstra et al. 2006; Chan et al. 2010). However, most characters under natural selection in the wild have a complicated genetic basis, and unraveling the signatures of these complicated genetic architectures in genomic data will be quite difficult (see Kirkpatrick, Chapter 7; G. Wagner, Chapter 8).

This flood of genetic data will also be of little use to evolutionary biologists unless the ecological context of the organism's place in its environment is well understood. Ecologists still have a tremendous amount of work ahead to develop a comprehensive understanding of the ecological dynamics of evolutionary processes at both the micro and macro scales. Ecological dynamics are the proximate causes of natural selection, genetic drift, and gene flow, and the resulting evolutionary changes in populations and species in turn shape their ecological dynamics. We are only beginning to understand the feedback between ecological and evolutionary dynamics.

In this year that the 150th anniversary of the publication of *The Origin of Species* is celebrated, there is no doubt of the importance of natural selection in shaping the biological world around us. The central mechanism of natural selection is defined by how the phenotype of an organism interacts with its environment to determine its survival, growth, and fecundity; the differential demographic success of individuals with different phenotypes is the basis of natural selection (Darwin 1859). Consequently, the primary information needed to understand evolution by natural selection is the set of phenotypic traits on which selection acts as well as the form and intensity of that selection. A rich catalog of natural selection in the wild is being developed (Endler 1986; Kingsolver et al. 2001; Siepielski et al. 2009), and the basic ecological context in which selection acts is plain to see. The emergence of a more sophisticated conceptual framework for the ecological context of evolutionary interactions is also becoming apparent (Thompson 2005).

However, I would argue that there is still a lack of fundamental understanding of how ecological interactions with the abiotic environment as well as within and among species generate the evolutionary dynamics of natural selection and of how responses to natural selection in turn shape broader ecological patterns of species distributions and abundances. Much of what is known about the operation of natural selection in the wild is largely based on only partial information about systems. For example, most of our measurements of natural selection in real populations are not based on the total fitness of the organism, but rather on selection that is due to one fitness component (e.g., selection based on survival to reproductive age), selection over only part of the life cycle (e.g., selection from the seedling state onward), or an estimate of selection derived from a surrogate measure for fitness (e.g., growth rate). These shortcomings are primarily due to the extreme logistical difficulty in measuring what really needs to be measured—that is, the relationship between fitness measured over the entire life cycle of the organism and all the phenotypic traits that influence fitness. Ecologists need the same scale of technical advances for rapidly quantifying all sorts of morphological, physiological, and behavioral traits (selection does not act only on body size and morphology) for many individuals (Houle 2010) as exists for rapidly quantifying the entire nucleotide sequences of scores of individuals in the past decade.

A far more sophisticated link between ecological dynamics and evolutionary process also needs to be developed. At present, fitness is typically thought of as a fixed property of the environment against which a population or species evolves. However, most selection agents are dynamical ecological entities, be they physical environmental factors (e.g., temperature, which is a continuously varying environmental property), abiotic resources (e.g., minerals, like phosphorus and nitrogen, which are depleted and renewed in a system), or other species (e.g., biotic resources, predators, mutualists, parasites). For example, imagine a prey species adapting to a predator. When the predator is very rare, the fitness differences that make the prey with different phenotypes differentially vulnerable will be very small, because almost no prey are dying by means of predation. As the predator becomes more common, the strength of selection will increase, because the fitness differences among the prey will increase as more prey are killed. Thus, the shape of the fitness surface for the prey changes with: (1) the abundance of the predator, and (2) when the predation rate depends on the prey's abundance and the abundances of alternative prey (as is usually the case). These ecological dynamics all generate density and frequency dependence in fitness that regulates populations within a community. Yet, there is still a relatively poor understanding of the dynamics and outcomes of density- or frequency-dependent natural selection. The conceptual development of these fundamental links between demography, ecological dynamics, and natural selection has been slow, but recent theoretical and empirical progress suggests that biologists may be on the cusp of developing a true integration of population and community ecology with the micro- and macroevolutionary dynamics of natural selection (Pelletier et al. 2007; McPeek 2008; Jones et al. 2009).

Likewise, genetic drift and gene flow are fundamentally driven by ecological dynamics. The rate of genetic drift is governed by current and long-term trends in population size across a species' range (i.e., the reasons that particular species are common or rare locally and the history of population bottlenecks and expansions), and these dynamics are driven by the ecological processes regulating population size as well as extrinsic factors (e.g., climate change). Gene flow is a byproduct of dispersal among populations that may be accomplished by passive movement or active choice based on fitness relationships. The signatures of these ecological processes also are written into the genomes of species through their effects on shaping the coalescent histories of every gene in a genome and the phylogeographic structure of genetic variability across the range of each species.

Technological advances in recent years have provided a flood of genomic data for the theoretical machinery that Lewontin (1974) described. However, genomic data cannot address many of the major questions in evolutionary biology. But, even for those that they can address, deciphering the evolutionary patterns embedded in genomic data still requires a fundamental understanding of the interaction of the whole organism with its environment.

Literature Cited

Chan, Y. F., M. E. Marks, F. C. Jones, and 13 others. 2010. Adaptive evolution of pelvic reduction in sticklebacks by recurrent deletion of a *Pitx1* enhancer. *Science* 327: 302–305.

Darwin, C. 1859. *On the Origin of Species by Means of Natural Selection, or the Preservation of Favoured Races in the Struggle for Life.* Murray, London.

Endler, J. A. 1986. *Natural Selection in the Wild.* Princeton University Press, Princeton.

Geffeney, S. L., E. Fujimoto, E. D. Brodie III, and 2 others. 2005. Evolutionary diversification of TTX-resistant sodium channels in a predator-prey interaction. *Nature* 434: 759–763.

Hoekstra, H. E., R. J. Hirschmann, R. J. Bundey, and 2 others. 2006. A single amino acid mutation contributes to adaptive color pattern in beach mice. *Science* 313: 101–104.

Houle, D. 2010. Numbering the hairs on our heads: the shared challenge and promise of phenomics. *Proc. Natl. Acad. Sci. USA* 107 (suppl. 1): 1793–1799.

Jones, L. E., L. Betz, S. P. Ellner, and 3 others. 2009. Rapid contemporary evolution and clonal food web dynamics. *Philos. Trans. Roy. Soc. London B* 364: 1579–1591.

Kingsolver, J. G., H. E. Hoekstra, J. M. Hoekstra, and 6 others. 2001. The strength of phenotypic selection in natural populations. *Am. Nat.* 157: 245–261.

Lewontin, R. C. 1974. *The Genetic Basis of Evolutionary Change.* Columbia University Press, New York.

McPeek, M. A. 2008. The ecological dynamics of clade diversification and community assembly. *Am. Nat.* 172: E270–E284.

Pelletier, F., T. Clutton-Brock, J. Pemberton, and 2 others. 2007. The evolutionary demography of ecological change: Linking trait variation and population growth. *Science* 315: 1571–1574.

Siepielski, A. M., J. D. Dillattista, and S. M. Carlson. 2009. It's about time: The temporal dynamics of phenotypic selection in the wild. *Ecol. Lett.* 12: 1261–1276.

Thompson, J. N. 2005. *The Geographic Mosaic of Coevolution.* University of Chicago Press, Chicago.

Part V

DIVERSITY AND THE TREE OF LIFE

Chapter 14

The Origin and Early Evolution of Life: Did It All Start in Darwin's Warm Little Pond?

Antonio Lazcano

Like his scientific predecessors Erasmus Darwin and Jean-Baptiste de Lamarck, Charles Darwin was convinced that plants and animals arose naturally from simple non-living inorganic compounds. Nevertheless, he carefully avoided discussing this possibility in his books. In private, however, he was much less restrained, as shown by the letter that he sent in February 1871 to Francis Hooker, in which he famously wrote:

> *...it is often said that all the conditions for the first production of a living organism are now present, which could ever have been present. But if (and oh what a big if) we could conceive in some warm little pond with all sorts of ammonia and phosphoric salts,–Light, heat, electricity &c. present, that a protein compound was chemically formed, ready to undergo still more complex changes, at the present day such matter w^{d} be instantly devoured, or absorbed, which would not have been the case before living creatures were formed (Figure 14.1).*

Although Darwin's reluctance to address publicly the origin of life surprised many of his friends and followers, the idea that living organisms were the historical outcome of gradual transformations of lifeless matter became widespread soon after the publication of *The Origin of Species*. This view soon merged with the emergent fields of biochemistry and cell biology, leading to proposals in which the origin of protoplasm was equated with the origin of life. Some of these hypotheses considered life to be an emergent feature of nature and attempted to understand its origin by introducing principles of historical explanation. But, most of these explanations went unnoticed, in part because they were incomplete, speculative schemes largely devoid of direct evidence and not subject to fruitful experimental testing (Lazcano 2010).

Feb 1 /71

Beckenham Down.
~~Bromley.~~
Kent. S.E.

My dear Hooker

I return the pamphlet, which I have been very glad to read.— It will be a curious discovery if Mr Lowne's observation that boiling does not kill certain moulds is proved true; but then how on earth is the absence of all living things in Pasteur's experiment to be accounted for?— I am always delighted to see a word in favour of Pangenesis, which some day, I believe, will have a resurrection. Mr Dyer's paper strikes me as a very able Spencerian production.— It is often said that all the conditions for the first production of a living organism are now present, which could ever have been present.— But if (& oh what a big if) we could conceive in some warm little pond with all sorts of ammonia & phosphoric salts,— light, heat, electricity &c present, that a protein compound was chemically formed, ready to undergo still more complex changes, at the present day such matter wd be instantly devoured, or absorbed, which would not have been the case before living creatures were formed.—

Henrietta makes hardly any progress, & God knows when she will be well.—

I enjoyed much the visit of you four gentlemen, i.e. after the Saturday night, when I thought I was quite done for.—

Yours affect

C. Darwin

FIGURE 14.1 Darwin's Letter to Hooker On February 1st, 1871 Darwin wrote to Hooker describing his views on the appearance of life in a "warm little pond." Like most of Darwin's handwriting, it is hard to read, but the "big if," which is his cautious reminder of the lack of evidence, can be easily identified in the text. Parts of the letter were included by Francis Darwin as a footnote in a volume in which he published some of his father's letters. In 1969, Melvin Calvin published both the transcription and the facsimile, garnering the attention of the origins-of-life community. (From Darwin Archive, Cambridge University Library.)

This situation changed during the 1920s, when Alexander I. Oparin, a young Russian biochemist, proposed that life had been preceded by a lengthy period of abiotic syntheses and accumulation of organic compounds that had taken place soon after the Earth was formed (Oparin 1924). Trained both as a plant biochemist and as an evolutionary biologist, it was impossible for Oparin to reconcile his Darwinian credence with the concepts of a gradual, slow evolution from the simple to the complex or with the commonly held suggestion that life had emerged already endowed with an autotrophic metabolism that included photosynthetic pigments, enzymes, and the ability to synthesize organic compounds from CO_2.

Based on the simplicity and ubiquity of fermentative reactions and on a detailed analysis of chemical synthesis and astronomical observations, Oparin attempted a theoretical reconstruction of the conditions of the primitive Earth and the evolution of organic molecules into pre-cellular systems, from which anaerobic cells evolved, nourishing themselves from the prebiotic soup. Some time after, the English geneticist John B. S. Haldane independently proposed a somewhat similar scheme, suggesting that a CO_2-rich atmosphere had facilitated the formation of organic compounds, and assuming that viruses represented an intermediate step in the transition from the prebiotic broth to the first cells (Haldane 1928). The discussion of the emergence of life was thus transformed into a workable multidisciplinary research program, although experimental reconstructions of the prebiotic environment did not start until the early 1950s, when Stanley L. Miller demonstrated the possibilities of synthesizing organic compounds under what were then considered primitive Earth conditions (Miller 1953).

On the Antiquity of Life

It is unlikely that the paleontological record will provide direct data on how life first appeared. There is no geological evidence of the environmental conditions on the Earth at the time of the origin of life, nor any fossil register of the evolutionary processes that preceded the appearance of the first cells. Direct information is lacking not only on the composition of the terrestrial atmosphere during this period, but also on the temperature, ocean pH values, and other general and local environmental conditions that may or may not have been important for the emergence of life. The attributes of the first living organisms are also unknown. They were probably simpler than any cell now alive and may have lacked not only protein-based catalysis, but also perhaps even the familiar genetic macromolecules, with their ribose-phosphate backbones.

We cannot approach this issue from a cladistic perspective, since all possible intermediates that may have once existed have long since vanished. Comparative genomics provide important clues on very early stages of biological evolution, but the applicability of this approach cannot be extended beyond a threshold that corresponds to a period of cellular evolution in which ribosome-mediated protein synthesis was already in operation (Becerra et al. 2007). Older stages are not yet amenable to this type of analysis, and the organisms at the base of universal phylogenies are ancient species, not primitive unmodified microbes.

Minor differences in the basic molecular processes of the three main cell domains, (i.e., the Bacteria, Archaea, and Eukarya) can be distinguished, but all known organisms share, in essence, the same genetic code, as well as the same basic features of genome replication, gene expression, basic anabolic reactions, and membrane-associated ATPase-mediated energy production. It is unlikely that the high conservation of these traits among all known living beings reflects intense lateral gene transfer or biased sampling. The most likely explanation is that the molecular details of these universal processes provide direct evidence of the monophyletic origin of all known forms of life and of intense selection, while their variations can be easily explained as the outcome of divergent processes from an ancestral life form, *fons et origo* of all contemporary organisms. When and how did such ancestral form arise?

Unfortunately, it is not possible to assign a precise chronology to the origin and earliest evolution of cells. Indeed, the recognition that life is a very ancient phenomenon runs parallel to the limits imposed by a scarce Archaean geological record that becomes increasingly blurred going back in time, with very few rocks older than 3.5 billion years. Those that remain have been so extensively altered by metamorphic processes that any direct evidence of life predating this limit has apparently been largely obliterated, and most of the sediments that have been preserved have been metamorphosed to a considerable extent (Schopf 2001).

Nevertheless, there is evidence that life emerged on Earth as soon as it was possible to do so. Such rapid development speaks for the relatively short timescale required for the origin and early evolution of life on Earth and suggests that the critical factor may have been the presence of liquid water, which began to accumulate as soon as the planet's surface cooled down (Figure 14.2). Although the biological origin of the microstructures present in the 3.5×10^9 years Apex cherts of the Australian Warrawoona formation (Schopf 1993) has been disputed, the weight of evidence favors the idea that life already existed during those times (Altermann and Kazmierczak 2003). This possibility has been reinforced by the recent report of

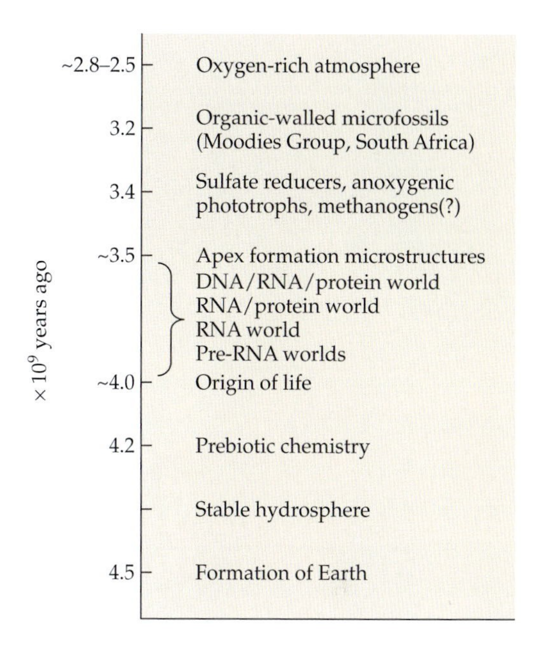

Figure 14.2 A Possible Timeline for the Origin and Early Evolution of Life (Adapted from Becerra et al. 2007.)

a complex array of organic-walled microfossils in the 3.2 billion-year-old Moodies Group deposits in South Africa (Javaux et al. 2010).

Isotopic fractionation data and other biomarkers support the possibility of a metabolically diverse Archaean microbial biosphere. The proposed timing of the onset of microbial methanogenesis based on the low ^{13}C values in methane inclusions found in hydrothermally precipitated quartz in the 3.5 billion-year-old Dresser Formation in Australia (Ueno et al. 2006) has been challenged (Lollar and McCollom 2006). However, sulfur isotope investigations of the same site indicate biological sulfate-reducing activity (Shen et al. 2001), and analyses of 3.4×10^9 million-year-old South African cherts suggest that they were inhabited by anaerobic photosynthetic prokaryotes in a marine environment (Tice and Lowe 2004). These results support the idea that the early Archaean Earth was teeming with prokaryotes, which included anoxygenic phototrophs, sulfate reducers, and perhaps, even methanogenic archaea (Canfield 2006).

Prebiotic Chemistry and Darwin's Warm Pond

The hypothesis of chemical evolution is supported not only by a number of laboratory simulations, but also by a wide range of astronomical observations and the analysis of samples of extraterrestrial material. These extraterrestrial samples include the existence of organic molecules of potential prebiotic significance in interstellar clouds and cometary nuclei as well as of small molecules of considerable biochemical importance that are present in carbonaceous chondrites. The copious array of amino acids, carboxylic acids, purines, pyrimidines, hydrocarbons, and other molecules that have been found in the 4.5×10^9-year-old Murchison meteorite and other carbonaceous chondrites gives considerable credibility to the idea that comparable syntheses took place in the primitive Earth (Ehrenfreund et al. 2002).

The first successful synthesis of organic compounds under plausible primordial conditions was reported in 1953, when Stanley L. Miller, a graduate student of the University of Chicago working with Harold C. Urey, studied the action of electric discharges over a mixture of CH_4, NH_3, H_2, and H_2O, using different experimental settings (Figure 14.3). The outcome

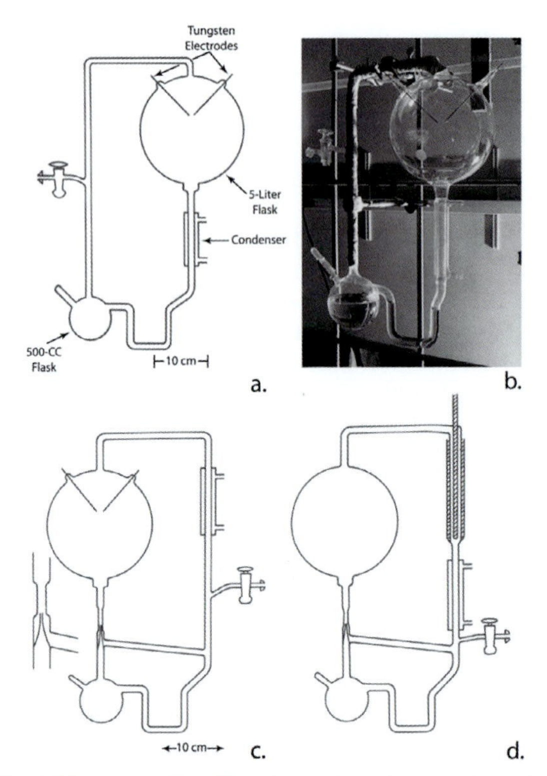

Figure 14.3 The Different 1953 Miller–Urey Experiments In order to simulate the interactions between a primitive, reducing atmosphere and an ocean, Miller (1953) designed three different experimental settings in which he simulated the interaction between an atmosphere and an ocean. In the classical Miller–Urey experiment (upper part), a spark discharge between the tungsten electrodes simulated lightning. The apparatus depicted in lower left part of the figure incorporated an aspirating nozzle attached to the water-containing flask in order to inject a jet of steam and gas into the spark flask of the apparatus, in effect simulating a "volcanic steam eruption." The apparatus shown in the lower right part incorporated the aspirator device but used a silent discharge instead of electrodes. (From Johnson et al. 2008.)

was a complex mixture of monomers, which included racemic mixtures of several proteinic amino acids in addition to hydroxy acids, urea, and other molecules (Miller 1953). Analysis of the original Miller samples using contemporary analytical techniques demonstrated a much wider range of organic compounds (Johnson et al. 2008). Prebiotic synthesis of the amino acids reported by Miller largely followed the Strecker synthesis of the 1850s, which involved aqueous-phase reactions of highly reactive intermediates (Figure 14.4). The rates of these reactions depend on temperature, pH, and hydrogen cyanide (HCN), ammonia (NH_3), and aldehyde concentrations. The reactions are rapid on a geological time scale; the half-lives for the

$$\text{(A)}\quad RCHO + NH_3 + HCN \underset{}{\overset{K}{\rightleftarrows}} RCH(NH_2)CN + H_2O$$

$$RCH(NH_2)CN + 2H_2O \xrightarrow{k} RCH(NH_2)COOH + NH_3$$

$$\text{(B)}\quad RCHO + HCN \overset{H}{\rightleftarrows} RCH(OH)CN$$

$$RCH(OH)CN + 2H_2O \xrightarrow{h} RCH(OH)COOH + NH_3$$

Figure 14.4 The Strecker Synthesis of Amino Acids and Hydroxy Acids Detection of hydrogen cyanide, ammonia, and carbonyl compounds (aldehydes or ketones) indicated that Miller was observing the result of the so-called Strecker synthesis, which forms aminonitriles. (A) These are highly reactive intermediates that undergo irreversible hydrolysis to form amino acids. (B) Depending on the concentration of ammonia in the reaction mixture, varying amounts of hydroxy acids are produced as well. Miller found essentially the same reactions, with larger relative amounts of hydroxy acids being formed in a reaction mixture containing less ammonia. (From Bada and Lazcano 2009.)

hydrolysis of the intermediate products in the reactions, aminonitriles and hydroxy nitriles, are less than a thousand years at 0° C, and there are no known slow steps (Miller and Lazcano 2002). Other mechanisms that could have been involved in the prebiotic synthesis of amino acids include the Bucherer–Berg pathway and the hydrolysis of HCN polymers, which can also be expected to have been quite efficient (Cleaves et al. 2008). These findings suggest that the origin and early evolution of life took place on a geologically short timescale.

The remarkable ease by which adenine (Figure 14.5) can be synthesized by the aqueous polymerization of ammonium cyanide demonstrated the significance of HCN and its derivatives in prebiotic chemistry (Oró 1960). The prebiotic importance of HCN has been further substantiated by the discovery that the hydrolytic products of its polymers include amino acids, purines, and orotic acid, which is a biosynthetic precursor of uracil (Ferris et al. 1978). The reaction of urea with cyanoacetylene or cyanoacetaldehyde, which is a hydrolytic derivative of HCN (Ferris et al. 1968), leads to high yields of cytosine and uracil, especially under simulated evaporating pond conditions, which increase the urea concentration (Robertson and Miller 1995).

The ease of formation in one-pot reactions of amino acids, purines, and pyrimidines supports the contention that these molecules were present in the prebiotic broth. Laboratory simulations suggest that urea, alcohols, and sugars—formed by the non-enzymatic condensation of formaldehyde, a wide variety of aliphatic and aromatic hydrocarbons, urea, carboxylic acids, and branched and straight fatty acids, including some that are membrane-forming compounds—were also components of the primitive soup. Accordingly, it is reasonable to assume that the prebiotic soup must have

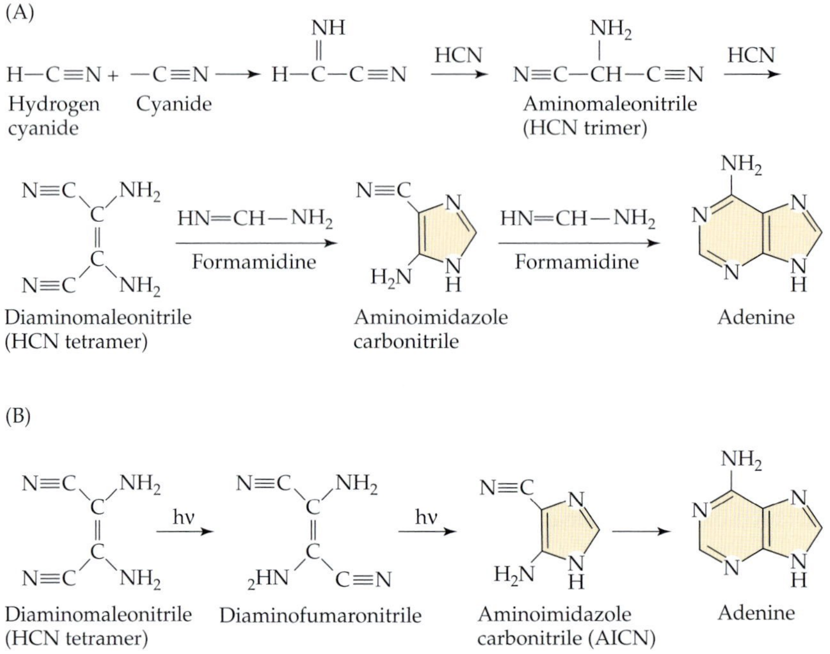

(A)

$$H{-}C{\equiv}N + {}^{-}C{\equiv}N \longrightarrow H{-}C({=}NH){-}C{\equiv}N \xrightarrow{HCN} N{\equiv}C{-}CH(NH_2){-}C{\equiv}N \xrightarrow{HCN}$$

Hydrogen cyanide · Cyanide · Aminomaleonitrile (HCN trimer)

Diaminomaleonitrile (HCN tetramer) $\xrightarrow[\text{Formamidine}]{HN{=}CH{-}NH_2}$ Aminoimidazole carbonitrile $\xrightarrow[\text{Formamidine}]{HN{=}CH{-}NH_2}$ Adenine

(B)

Diaminomaleonitrile (HCN tetramer) $\xrightarrow{h\nu}$ Diaminofumaronitrile $\xrightarrow{h\nu}$ Aminoimidazole carbonitrile (AICN) $\longrightarrow$ Adenine

Figure 14.5 Abiotic Synthesis of Adanine from Hydrogen Cyanide Adenine is easily synthesized as a result of the polymerization of hydrogen cyanide (HCN) (Oró 1960). (A) If concentrated solutions of ammonium cyanide are refluxed for a few days, significant yields of adenine are observed, together with 4-aminoimidazole-5 carboxamide. (B) As demonstrated by James P. Ferris and Leslie E. Orgel, the addition of formamidine can be bypassed with a two-photon photochemical rearrangement of diaminomaleonitrile, which with sunlight, readily converts to amino imidazole carbonitrile with a high yield. (From Bada and Lazcano 2009.)

been a bewildering organic chemical mixture, but it could not have included all the compounds or the molecular structures found today, even in the most ancient extant forms of life, nor could the first organisms have sprung completely assembled from the precursors present in the primitive oceans. The fact that a number of chemical constituents of contemporary forms of life can be synthesized non-enzymatically under laboratory conditions does not necessarily imply that they were essential for the origin of life or were available in the primitive environment.

The basic tenet of the heterotrophic theory of the origin of life is that the maintenance and reproduction of the first living entities depended primarily on prebiotically synthesized organic molecules. There has been no shortage of discussion about how the formation of the primordial soup took place, but it is very unlikely that any single mechanism can account for the wide range of organic compounds that may have accumulated on the primitive Earth. Very likely, the prebiotic soup was formed by contributions from endogenous syntheses involving the anoxic interface of the atmosphere/ocean system, metal sulfide-mediated synthesis in deep-sea vents, and exogenous sources, such as comets, meteorites, and interplanetary dust. This eclectic view does not beg the issue of the relative significance of the different sources of organic compounds, but simply recognizes the wide variety of potential sources of the raw material required for the emergence of life.

The existence of different abiotic mechanisms by which biochemical monomers can be synthesized under plausible prebiotic conditions is well established. Of course, not all prebiotic pathways are equally efficient, but the wide range of experimental conditions under which organic compounds can be synthesized, including the formation of amino acids in a CO_2-rich atmosphere (Cleaves et al. 2008), demonstrates that primitive syntheses of the building blocks of life are robust; that is, the abiotic reactions leading to them do not take place under a narrow range defined by highly selective reaction conditions, but rather under a wide variety of experimental settings (Figure 14.6).

The remarkable coincidence between the molecular constituents of living organisms and those synthesized in prebiotic experiments is too striking to be fortuitous, and the robustness of this type of chemistry is supported by the occurrence of most of these biochemical compounds in the 4.5 billion-year-old Murchison carbonaceous chondrites and in carbon-rich meteorites (Ehrenfreund et al. 2002). How life first evolved is not known, but analysis of carbonaceous chondrites and the laboratory simulations of the primitive Earth suggest that prior to the emergence of the first living systems the prebiotic environment was endowed with: (1) a large suite of organic compounds of biochemical significance; (2) many organic and inorganic

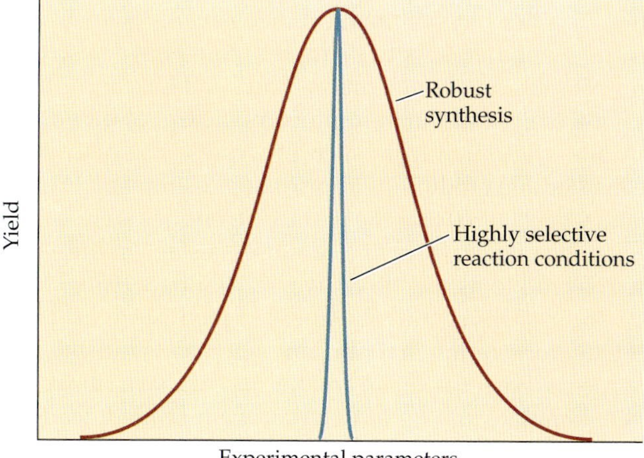

Figure 14.6 The Robustness of Prebiotic Syntheses Although not all prebiotic pathways are equally efficient, the wide range of experimental conditions under which organic compounds can be synthesized demonstrates that prebiotic syntheses of the building blocks of life are robust. Quite surprisingly, the abiotic reactions that form organic molecules do not take place under a narrow range defined by highly selective reaction conditions, but rather under a wide variety of experimental settings. (Adapted from Miller and Lazcano 2002.)

catalysts (such as cyanamide, metallic ions, sulfur-rich minerals and clays); (3) purines and pyrimidines (i.e., the potential for template-dependent polymerization reactions); and (4) membrane-forming compounds.

How Did the RNA World Originate?

The leap from biochemical monomers and small oligomers to living cells is enormous. There is a major gap between the current descriptions of the primitive soup and the appearance of simple chemical systems capable of replication with variation. Solving this issue is essential to our understanding of the origin of the biosphere: regardless of the chemical complexity of the prebiotic environment, life could not have evolved in the absence of a genetic replicating mechanism insuring the maintenance, stability, and diversification of its basic components. The organic material that may have accumulated on the early Earth before life existed very likely consisted of a wide array of different types of compounds, including many of the simple compounds that play a major role in biochemistry today. How these abiotic organic constituents were assembled into polymers and then into the first living entities is currently one the most challenging areas of research in the study of the origin of life.

It is reasonable to assume that condensation reactions leading to polymers also took place in the primitive Earth. Synonymous terms like "primitive soup," "primordial broth," or "Darwin's warm little pond," have led in some cases to major misunderstandings, including the simplistic image of an organic-rich worldwide ocean, abounding in self-replicating molecules. The term "warm little pond," which has long been used for convenience, refers to parts of the hydrosphere where the accumulation and interaction of the products of prebiotic synthesis may have taken place—including oceanic sediments, intertidal zones, shallow ponds, fresh water lakes, lagoons undergoing wet-and-dry cycles, and eutectic environments (e.g., glacial ponds)—and where evaporation or other physicochemical mechanisms (such as the adherence of biochemical monomers to active surfaces) could have raised local concentrations and promoted polymerization (Bada and Lazcano 2009).

It is very unlikely, however, that the ubiquitous nucleic acid-based genetic system of extant life originated from such processes. The discovery of ribozymes gave considerable credibility to prior suggestions by Woese (1967), Crick (1968), and Orgel (1968) that the first living entities were based on RNA, as both the genetic material and as a catalyst—a stage in the early evolution of life now referred to as the "RNA world" (Gilbert 1986; Joyce 2002). There are many indications of the robustness of the RNA world hypothesis, including the demonstration that ribosomal peptide synthesis is a ribozyme-mediated process (Ban et al. 2000) and the report that a ribozyme that catalyzes the RNA-template joining of RNA can be modified, leading to two ribozymes that catalyze each other's synthesis from a total of four oligonucleotide substrates (Lincoln and Joyce 2009). These cross-replicating catalytic RNAs undergo self-sustained, exponential amplification in the absence of proteins or other biological materials.

Although the existence of an early stage during which RNA played a much more conspicuous role in biological processes is widely accepted, the problems associated with the synthesis and accumulation of the RNA components have led to the idea that this molecule was not a direct outcome of prebiotic evolution. For instance, the ribose component of RNA is very unstable which makes its presence in the prebiotic milieu unlikely. The presence of lead hydroxide (Zubay 1998) and borate minerals is known to stabilize ribose (Ricardo et al. 2004), and cyanamide is known to react with ribose to form a stable bicyclic adduct (Springsteen and Joyce 2004). However, in order to be involved in the polymerization reactions leading to RNA, ribose would likely have to be present in solution, where it would be prone to decomposition. How RNA came into being remains an open question. These difficulties have led to proposals of pre-RNA worlds, in which informational macromolecules with backbones (and, perhaps, even nucleobases) that were different from those of extant nucleic acids may have also been endowed with catalytic activity (i.e., with phenotype and genotype also residing in the same molecule). Thus, a simpler genetic self-replicator must have come first and several possible contenders have been suggested. The hypothetical precursor(s) of RNA would have had the capacity to catalyze reactions and to store information, although the component nucleobases and the backbone that held the polymer together were not necessarily the same as those in modern RNA and DNA.

Unfortunately, the chemical nature of these hypothetical genetic polymers and the catalytic agents that may have formed the pre-RNA worlds are completely unknown and can only be surmised. Modified nucleic acid backbones have been synthesized, which either incorporate a different version of ribose or lack it altogether. Experiments on nucleic acid with hexoses instead of pentoses and on pyranoses instead of furanose, suggest that a wide variety of informational polymers is possible, even when restricted to sugar phosphate backbones (Eschenmoser 1999). There are other possible substitutes for ribose, like threose nucleic acids (TNA), in which the backbone is made up of L-threose connected by 3′, 2′ phosphodiester bonds and forms double helical structures through Watson-Crick base pairing with complimentary strands of themselves, or with RNA and DNA. Other nucleic acid analogues that have been extensively studied are the peptide nucleic acids (PNAs), which have a polypeptide-like backbone of achiral 2-amino-ethyl-glycine, to which nucleic acid bases are attached by an acetic acid (Nielsen 1993). Such molecules form very stable complementary duplexes, both with themselves and with nucleic acids, and because they have

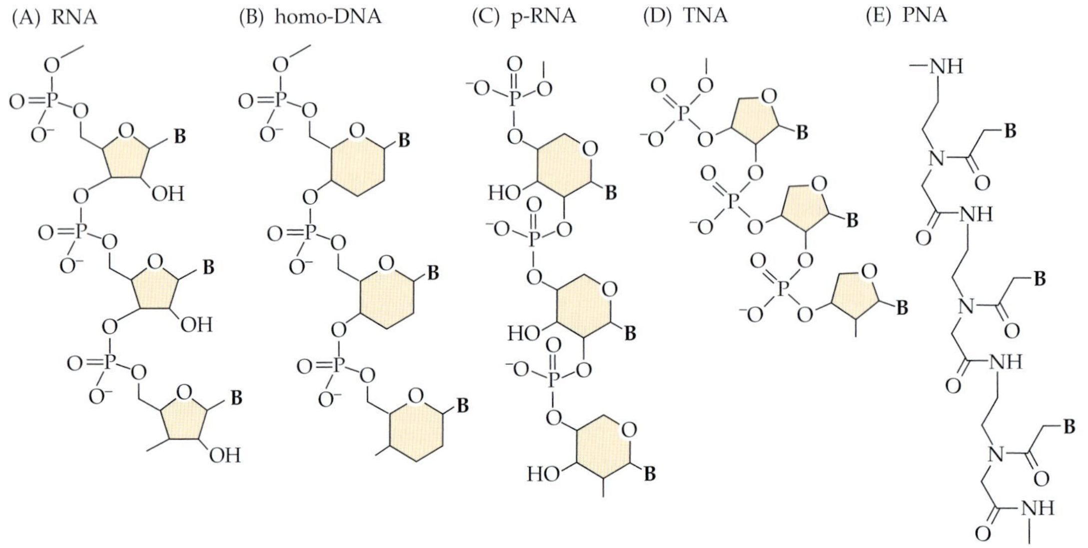

Figure 14.7 Alternative Genetic Polymers Chemical synthesis of nucleic acid analogues with backbones different from those of nucleic acids has provided useful laboratory models of the molecules that may have bridged the gap between the prebiotic soup and the earliest living systems. In spite of the different types of backbones, like RNA (A) and DNA (B), they can form double-stranded polymeric structures held together by Watson-Crick base pairing. (From Orgel 2004.)

basically the same functional groups as in RNA, it is possible that they may also be endowed with catalytic properties (Figure 14.7).

Identification of adenine, guanine, uracil, and other nucleobases in the Murchison meteorite supports the idea that these bases were present in the primitive environment. However, it is likely that other heterocycles capable of forming hydrogen bonding were also available. The Watson–Crick base-pair geometry permits more than the four usual nucleobases, and simpler genetic polymers may not only have lacked the sugar–phosphate backbones, but may also have depended on alternative, nonstandard hydrogen bonding patterns. The search for experimental models of pre-RNA polymers will be rewarding but difficult; it requires the identification of potentially prebiotic components and demonstration of their non-enzymatic, template-dependent polymerization as well as a coherent hypothesis of how they may have catalyzed the transition to an RNA world.

Toward the First Cells

The reports of selection of ribozyme catalysts or of RNA viruses under *in vitro* conditions demonstrates that RNA molecules can evolve, but few would consider them alive. As shown by the endless discussions on the nature of viruses, some systems may not be "half-alive," but they can exhibit some of the properties that are normally associated with living entities. Is it possible to establish a point in time when the difference between complex networks of molecules of abiotic origin and the truly primordial, first living system took place? The recognition that life is the outcome of an evolutionary process, constrained by the laws of physics and chemistry, implies that many properties associated with living systems, such as replication, self-assemblage, or catalysis, are also found in nonliving entities (Lazcano 2008).

There are many definitions of the RNA world, but the discovery of ribozymes does not imply that wriggling, autocatalytic nucleic acid molecules, prepared to be used as primordial genes, were floating in the primitive oceans or that the replicative RNA-based systems emerged completely assembled from simple precursors present in the prebiotic broth. While it is true that gene-first proposals of the origin of life do not require enclosure within compartments, the emergence of life may be best understood in terms of the dynamics and evolution of complex networks of chemical replicating entities. Whether such polymolecular entities were enclosed within membranes is not yet clear, but given the prebiotic availability of amphiphilic compounds, this may well have been the case (Deamer 2002). Indeed, the evidence supporting the presence of lipidic molecules in the prebiotic environment and their natural ability to self-organize into vesicular compartments underlines the significance of theoretical models using simple cells associated with an evolving ribozymic RNA polymerase as well as the importance of increasingly sophisticated laboratory models of precellular systems (Szostak et al. 2001; Deamer and Dworkin 2005; Mansy et al. 2008).

For obvious methodological reasons, experimental simulations of prebiotic events have concentrated on the empirical analysis of single variables. However, the study of specific conditions, including the laboratory simulation of localized environments such as volcanic islands, tidal zones, and microenvironments, including liposomes, mineral surfaces, and volcanic ponds (which could have been prevalent in the primitive environment), are likely to yield promising results. This assertion is not purely speculative. For instance, there is experimental evidence showing that the interactions between liposomes and different water-soluble polypeptides lead to major changes in the morphology and permeability of liposomes of phosphatidyl-L-serine, and to a transition of poly-L-lysine from a random coil into an α-helix that exhibits hydrophobic bonding with the lipidic phase (Hammes and Scullery 1970). It is thus reasonable to assume that the association and interplay of different biochemical monomers and oligomers in more complex experimental settings would lead to physicochemical properties not exhibited by their isolated components (Deamer et al. 2002).

The Early Evolution of Protein Biosynthesis

Until a few years ago, the origin of the genetic code and of protein synthesis was considered synonymous with the appearance of life itself. This is no longer a dominant point of view: four of the central reactions involved in protein biosynthesis are catalyzed by ribozymes, and their complementary nature suggests that they first appeared in an RNA world (Kumar and Yarus 2001), that is, that ribosome-catalyzed, nucleic acid-coded protein synthesis is the outcome of Darwinian selection of protein-free, RNA-based biological systems and not of mere physicochemical interactions that took place in the prebiotic environment.

It is likely that a wide assortment of abiotically formed oligopeptides were present in the prebiotic environment (Bada and Lazcano 2009). Although they may have facilitated a number of prebiotic processes, they are not ancestrally related to extant proteins, whose origin can only be understood within the framework of the catalytic abilities of RNA molecules. It is possible that the ribonucleotide moieties of many coenzymes (i.e., small organic molecules that provide a varied set of reactive groups to catalytic proteins) are in fact molecular fossils of a primordial metabolic state and reflect recruitment processes that diversified the catalytic abilities of ribozymes, providing new venues for an increasingly complex RNA-world metabolism (Orgel and Sulston 1971; White 1976).

It is possible that an equivalent process involving abiotically synthesized amino acids would have led to the acquisition of functional side-chains that diversified the catalytic abilities of ribozymes. The experimental evidence demonstrating that ribozymes can mediate amino acid activation, aminoacyl-RNA synthesis, peptide-bond formation, and RNA-based coding suggests ribosome-mediated protein synthesis first evolved in an RNA world (Kumar and Yarus 2001). In fact, the demonstration that ribosomal peptide synthesis is a ribozyme-catalyzed reaction makes it almost certain that there was once an RNA world and that protein biosynthesis is one of its evolutionary outcomes. The extraordinary structural and functional complexity of extant ribosomes must have been preceded by simpler structures. The observations that have shown that peptide-bond formation occurs in a highly conserved site devoid of proteins (Ban et al. 2000; Nissen et al. 2000)—formed by two 60-ribonucleotide, L-shaped RNA core units that appear to be the outcome of an early duplication (Agmon 2009)—are consistent with the hypothesis that protein biosynthesis first evolved in an RNA world and that the original proto-ribosome lacked proteins.

As shown by reports on the incorporation of unusual amino acids, like pyrrolysine, into contemporary proteins (Heiheman et al. 2009), primitive protein biosynthesis may have involved the polymerization of amino acids available on the primitive broth but no longer seen in proteins. Since it is unlikely that primitive ribosome-mediated protein biosynthesis produced highly specific catalytic oligopeptides from the very beginning, it has been argued that the first oligopeptides synthesized by proto-ribosomes were selected because of their role in stabilizing functional conformations in the RNA world. If this is the case, then the analysis of RNA-binding domains (Delaye and Lazcano 2000) and of highly invariant small amino acids sequences that exhibit a surprising degree of conservation, such as GHVDHGKT, DTPGHVDF, and GAGKSTL (Goto et al. 2002), may be among the oldest recognizable motifs found in extant databases and may provide insights on the nature of ancestral proteins.

It is possible that by the time RNA-based life appeared on Earth, many of the components of the primordial soup to sustain life had become largely exhausted. Thus, the origin of simple metabolic-like pathways must have been in place in order to ensure a supply of the ingredients needed to sustain the existence of the primitive living entities. The ultimate origin of metabolism, however, is an open issue. If self-sustaining reaction chains arose on the early Earth, they could have played an important role in enriching the prebiotic soup in components not readily synthesized by other abiotic reactions or delivered from space, but the lack of simple continuity between the biosynthetic and the (possible) prebiotic pathways makes it unlikely that the extant anabolic pathways arose from prebiotic transformations (Lazcano and Miller 1999).

For instance, abiotic amino acid formation occurs by way of the Strecker synthesis or Bucherer-Berg reaction (Cleaves et al. 2008), which are very different processes than transamination and the reverse Krebs cycle. The prebiotic synthesis of purines is from HCN (Oró 1960) and not from glycine, formate, and NH_3. Only the amino imidazole carboxamide ribotide in the biosynthetic pathway is similar to the amino imidazole carbonitrile synthesized in the prebiotic pathways. There are few additional examples of chemical similarities between putative abiotic synthesis and biosynthetic processes. However, they do not necessarily indicate an evolutionary continuity between prebiotic chemistry and biochemical pathways, but may reflect chemical determinism (Lazcano 2010). These processes are similar because they may be the unique way in which given reactions can take place.

Analysis of the increasingly large database of completely sequenced cellular genomes, from the three major domains that define the set of the most conserved protein-encoding sequences, has shown the preeminence not only of components of the transcription and translation machineries, but also of additional molecules involved in RNA metabolism (i.e., sequences that synthesize, degrade, or interact with RNA and ribonucleotides) (Figure 14.8). In fact, conserved sequences related to biosynthetic pathways include those encoding putative phosphoribosyl pyrophosphate synthetase and thioredoxin, which participate in nucleotide metabolism, suggesting that these vestiges of an RNA/protein world are part of an essential and highly conserved pool of protein domains that preceded the emergence of DNA genomes (Becerra et al. 2007).

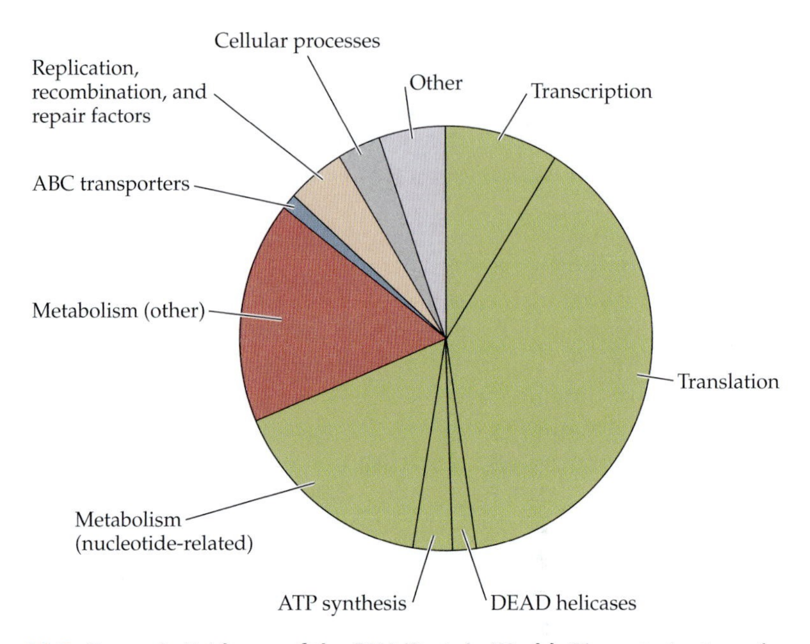

Figure 14.8 Genomic Evidence of the RNA/Protein World Characterization of the gene complement of the last common ancestor of extant life shows that the set of sequences is dominated by different putative ATPases and by molecules involved in gene expression and RNA metabolism. Green area indicates gene sequences that code for proteins involved in RNA metabolism. DEAD-type RNA helicase and enolase genes, which are known to be part of the RNA degradosome, are as conserved as many transcription and translation genes. Conserved sequences related to biosynthetic pathways include those encoding putative phosphoribosyl pyrophosphate synthase and thioredoxin, which participate in nucleotide metabolism. (Adapted from Delaye et al. 2005.)

The Origin of DNA Genomes

Although the nature of the predecessor(s) of the RNA world are completely unknown and can only be surmised, it is reasonable to assume that the extreme chemical instability of catalytic and replicative polyribonucleotides may have been the main limitation of the RNA world. In fact, the chemical lability of RNA implies that primordial ribozymes must have been very efficient in carrying out self-replication reactions in order to maintain an adequate inventory of molecules needed for survival. Survival of nucleic acids is limited by the hydrolysis of phosphodiester bonds (Lindahl 1993), and the stability of Watson–Crick helices (or their pre-RNA equivalents) is strongly diminished by high temperatures. The greatest problem for an RNA-based biosphere is the reduced thermal stability on the geologic timescale of ribose and other sugars (Larralde et al. 1995), but the situation is equally bad for pyrimidines, purines, and some amino acids. It is possible that RNA was stabilized by the interaction with polyamines, some clay minerals, polycyclic aromatic hydrocarbons, and a few other components of the prebiotic environment. However, its instability may have been one of the primary reasons underlying the transition to the extant DNA/RNA/protein world in which, because the increased stability of the genetic molecules, survival would have been less dependent on polymer stability.

The sequence similarities shared by many ancient, large proteins found in all three domains (Becerra et al. 2007) suggest that considerable fidelity existed in the operative genetic system of their common ancestor, but such fidelity is unlikely to be found in RNA-based genetic systems (Lazcano et al. 1992). When did the transition from RNA to DNA cellular genomes take place? As is often the case with metabolic innovations, the dating cannot be documented from geological data. However, since all extant cells are endowed with DNA genomes, the most parsimonious conclusion is that this genetic polymer was already present in the last common ancestor (LCA), or cenancestor, that existed prior to the divergence of the three primary domains: Bacteria, Archaea, and Eukarya, as mentioned. Delaye et al. (2004) discuss the fact that although there have been a number of suggestions that the LCA (or its equivalent) was endowed with genomes formed by small sized RNA molecules or hybrid RNA/DNA genetic system, there are manifold indications that double-stranded DNA genomes of monophyletic origin had become firmly established prior to the divergence of the three primary domains.

In contrast with other energetically favorable biochemical reactions (such as hydrolysis of the phosphodiester backbone or the transfer of amino groups), the direct removal of the oxygen from the 2′-C ribonucleotide pentose ring to form the corresponding deoxy-equivalents is a thermodynamically much less favored reaction. This is a major constraint that strongly reduces the likelihood of multiple, independent origins of biological ribonucleotide reduction. In fact, although the demonstration of the monophyletic origin of ribonucleotide reductases (RNR; which are the enzymes involved in the biosynthesis of deoxyribonucleotides from ribonucleotide precursors) is greatly complicated by their highly divergent primary sequences and the different mechanisms by which they generate the substrate 3′-radical species required for the removal of the 2′-OH group. However, sequence analysis and biochemical characterization of RNR from the three primary biological domains has confirmed their structural similarities, which speaks to their ultimate monophyletic origin (Freeland et al. 1999).

Molecular Cladistics and Early Cell Evolution

The extraordinary similarities at very basic biochemical and genetic levels among all known life forms can be interpreted as propinquity of descent (i.e., all organisms are of monophyletic origin). Although complex, multigenic traits must have evolved through a series of simpler states, no evolutionary intermediate stages or ancient simplified versions of ATP production, DNA replication, or ribosome-mediated protein synthesis have been discovered in extant organisms. The molecular details of these universal processes not only provide direct evidence of the monophyletic origin of all extant forms of life, but also imply that the sets of genes encoding the components of these complex traits were frozen a long time ago; thus, major changes in them are strongly selected against.

Although no evolutionary intermediate stages or ancient simplified versions of the basic biological processes have been discovered in contemporary organisms, the differences in the structure and mechanisms of gene expression and replication among the three main cell lineages provide insights on the stepwise evolution of the replication and translational apparatus, including some late steps in the development of the genetic code. It is thus feasible to distinguish the origin of life problem from a whole series of other issues, often confused, that belong to the domain of the evolution of microbial life.

It is true that no ancient incipient stages or evolutionary intermediates of these molecular structures have been detected, but the existence of graded intermediates can be deduced. Clues to the evolution of protein synthesis are provided by paralogous genes, which are sequences that diverge not through speciation, but rather after a duplication event. For instance, in all known cells, protein synthesis requires the presence of two homologous elongation factors (EF), EF-Tu and EF-G, which are GTP-dependent enzymes. Evidence of their common ancestry indicates that prior to the duplication event that produced them, a more primitive, less-regulated version of protein synthesis was taking place with only a single ancestral elongation factor.

Protein sequence comparisons have confirmed the role that many ancient gene duplications have played in the evolution of genomes (Becerra and Lazcano 1998). Clues to the genetic organization and biochemical complexity of primitive entities from which the cenancestor evolved may be derived from the analysis of paralogous gene families. The number of sequences that have undergone such duplications before the divergence of the three lineages includes genes coding for a variety of enzymes that participate in widely different processes, such as translation, DNA replication, CO_2 fixation, nitrogen metabolism, and biosynthetic pathways. Whole-genome analysis has revealed the impressive expansion of sequences involved in membrane transport phenomena, such as ABC transporters, P-type ATPases, and ion-coupled permeases (Clayton et al. 1997). Structural studies of proteins provide evidence of another group of paralogous duplications. A number of enzymes, including protein disulfide oxidoreductase (Ren et al. 1998), the large subunit of carbomoyl phosphate synthetase (Alcántara et al. 2000), and HisA, a histidine biosynthetic isomerase (Alifano et al. 1996), are formed by two homologous modules, arranged in tandem. The structure and spatial arrangement of the genes indicates that the size and structure of a number of proteins are the evolutionary outcome of paralogous duplications followed by gene fusion events that took place prior to the divergence of the three primary kingdoms.

A third group of paralogous genes can be recognized. All cells are endowed with different sets formed by a relatively small number of paralogous sequences. The list includes, among others, the pair of homologous genes encoding the EF-Tu and EF-G elongation factors (Iwabe et al. 1989). Another is formed by the duplicated sequences encoding the F-type ATPase hydrophilic α and β subunits (Gogarten et al. 1989). No cell is known to be endowed with only one EF factor or only one type of F-type ATPases hydrophilic subunit. However, the extraordinary conservation of these duplicates implies that the LCA was preceded by a simpler cell with a smaller genome, in which only one copy of each of these genes existed (i.e., preceded by cells in which protein synthesis involved only one elongation factor) and with ATPases with limited regulatory abilities.

Paralogous families of metabolic genes provide strong support for the proposal that anabolic pathways were assembled by the recruitment of primitive enzymes that could react with a wide range of chemically related substrates—the so-called patchwork assembly of biosynthetic routes (Jensen 1976). Such relatively slow, nonspecific enzymes may have represented a mechanism by which primitive cells with small genomes could have overcome their limited coding abilities. The high levels of genetic redundancy detected in all sequenced genomes imply not only that duplication has played a major role in the accretion of the complex genomes found in extant cells, but also that prior to the early duplication events revealed by the large protein families, simpler living systems existed and lacked the large sets of enzymes and sophisticated regulatory abilities of contemporary organisms. How early cells overcame the bottlenecks imposed by such limitations is still an open problem that can be addressed experimentally.

Conclusions

If the origin of life is seen as the evolutionary transition between the nonliving and the living, then it is meaningless to attempt to draw a strict line between these two worlds (Lazcano 2008). We remain lamentably ignorant about major portions of the processes that preceded life, but there is strong evidence of an evolutionary continuum that seamlessly joins the prebiotic synthesis and accumulation of organic molecules in the primitive environment, with the emergence of self-sustaining, replicative chemical systems capable of undergoing Darwinian evolution. In other words, the appearance of life on Earth should be seen as the evolutionary outcome of a process and not of a single, fortuitous event.

The origin of life will always be shrouded in mystery. As in other areas of evolutionary biology, answers to questions on the origin and nature of the first life forms can only be regarded as explanatory, rather than definitive and conclusive. This conclusion does not imply, however, that our explanations can be dismissed as purely speculative, but rather that the issue should be addressed conjecturally, in an attempt to construct, not a mere chronology, but rather a coherent historical narrative, woven together by a large number of miscellaneous observational findings and experimental results (Kamminga 1986). We should not feel overwhelmed with the sheer scale of the questions to be solved but encouraged by the ingenuity with which those issues can be confronted.

Acknowledgments

This chapter was completed during a sabbatical leave in which I enjoyed the hospitality of Professor Jeffrey L. Bada and his associates at the Scripps Institution of Oceanography, University of California, San Diego. Support from a UC Mexus-CONACYT Fellowship is gratefully acknowledged.

Literature Cited

Agmon, I. 2009. The dimeric proto-ribosome: Structural details and possible implications on the origin of life. *Int. J. Sci.* 10: 2921–293.

Alcántara, C., J. Cervera, and V. Rubio. 2000. Carbamate kinase can replace in vivo carbamoyl phosphate synthetase. Implications for the evolution of carbamoyl phosphate biosynthesis. *FEBS Lett.* 484: 261–64.

Alifano, P., R. Fani, P. Lio, and 4 others. 1996. Histidine biosynthetic pathway and genes: Structure, regulation, and evolution. *Microbiol. Rev.* 60: 44–69.

Altermann, W. and J. Kazmierczak. 2003. Archean microfossils: A reappraisal of early life on Earth. *Res. Microbiol.* 154: 611–617.

Bada, J. L. and A. Lazcano. 2009. The origin of life. In M. Ruse and J. Travis (eds.), *The Harvard Companion of Evolution*, pp. 49–79. Belknap Press, Cambridge, MA.

Ban, N., P. Nissen, J. Hansen, and 2 others. 2000. The complete atomic structure of the large ribosomal subunit at 2.4 Å resolution. *Science* 289: 905–920.

Becerra, A. and A. Lazcano. 1998. The role of gene duplication in the evolution of purine nucleotide salvage pathways. *Orig. Life Evol. Biosph.* 28: 539–53.

Becerra, A., L. Delaye, A. Islas, and 1 other. 2007. Very early stages of biological evolution related to the nature of the last common ancestor of the three major cell domains. *Ann. Rev. Ecol. Evol. Syst.* 38: 361–379.

Canfield, D. E. 2006. Biochemistry: Gas with an ancient history. *Nature* 440: 426–427.

Clayton, R. A., O. White, K. A. Ketchum, and 1 other. 1997. The genome from the third domain of life. *Nature* 387: 459–462.

Cleave, J. H., J. H. Chalmers, A. Lazcano, and 2 others. 2008. Prebiotic organic synthesis in neutral planetary atmospheres. *Orig. Life Evol. Biosph.* 38: 105–155.

Crick, F. H. C. 1968. The origin of the genetic code. *J. Mol. Biol.* 39: 367–380.

Deamer, D. W. and J. P. Dworkin. 2005. Chemistry and physics of primitive membranes. In P. Walde (ed.), *Prebiotic Chemistry: From Simple Amphiphiles to Protocell Models*, pp. 1–27. Springer, Berlin.

Deamer, D. W., J. P. Dworkin, S. A. Sanford, and 2 others. 2002. The first cell membranes. *Astrobiology* 2: 371–382.

Delaye, L. and A. Lazcano. 2000. RNA-binding peptides as molecular fossils. In J. Chela-Flores, G. Lemerchand, and J. Oró (eds.), *Origins from the Big-Bang to Biology: Proceedings of the First Ibero-American School of Astrobiology*, pp. 285–88, Klüwer, Dordrecht.

Delaye, L., A. Becerra, and A. Lazcano. 2004. The nature of the last common ancestor. In L. R. de Pouplana (ed.) *The Genetic Code and the Origin of Life*, pp. 34–47. Landes Bioscience, Georgetown.

Delaye, L., A. Becerra, and A. Lazcano. 2005. The last common ancestor: What's in a name? *Orig. Life Evol. Biosph.* 35: 537–554.

Ehrenfreund, P., W. Irvine, L. Becker, and 10 others. 2002. Astrophysical and astrochemical insights into the origin of life. *Rep. Prog. Phys.* 65: 1427–1487.

Eschenmoser, A. 1999. Chemical etiology of nucleic acid structure. *Science* 284: 2118–2124.

Ferris, J. P., P. D. Joshi, E. H. Edelson, and 1 other. 1978. HCN, a plausible source of purines, pyrimidines, and amino acids on the primitive Earth. *J. Mol. Evol.* 11: 293–311.

Ferris, J. P., R. P. Sanchez, and L. E. Orgel. 1968. Studies in prebiotic synthesis. III. Synthesis of pyrimidines from cyanoacetylene and cyanate. *J. Mol. Biol.* 33: 693–704.

Freeland, S. J., R. D. Knight, and L. F. Landweber. 1999. Do proteins predate DNA? *Science* 286: 690–692.

Gilbert, W. 1986. The RNA world. *Nature* 319: 618.

Gogarten, J. P., H. Kibak, P. Dittrich, and 7 others.1989. Evolution of the vacuolar H^+-ATPase: Implications for the origin of eukayotes. *Proc. Natl. Acad Sci. USA* 86: 6661–6665.

Goto, N., K. Kurokawa, and T. Yasunaga. 2002. Finding conserved amino acid sequences among prokaryotic proteomes. *Genome Inform.* 13: 443–444.

Hammes, G. G. and S. G. Schullery. 1970. Structure of molecular aggregates. II. Construction of model membranes from phospholipids and polypeptides. *Biochemistry* 9: 2555–2558.

Heiheman, I. U, P. O'Donoghue, C. Madinger, and 4 others. 2009. The appearance of pyrrolysine in $tRNA^{His}$ guanylyltransferase by neutral evolution. *Proc. Natl. Acad. Sci. USA* 106: 21103–21108.

Iwabe, N., K. Kuma, M. Hasegawa, and 2 others. 1989. Evolutionary relationship of archaebacteria, eubacteria, and eukaryotes inferred from phylogenetic trees of duplicated genes. *Proc. Natl. Acad. Sci. USA* 86: 9355–9359.

Javaux, E. J., C. P. Marshall, and A. Bekker. 2010. Organic-walled microfossils in 3.2-billion-year-old shallow marine siliciclastic deposits. *Nature* 463: 934–938.

Jensen, R. A. 1976. Enzyme recruitment in evolution of new function. *Ann. Rev. Microbiol.* 30: 409–425.

Johnson, A. P., H. J. Cleaves, J. P. Dworkin, and 3 others. 2008. The Miller volcanic spark discharge experiment. *Science* 322: 404.

Joyce, G. F. 2002. The antiquity of RNA-based evolution. *Nature* 418: 214–221.

Kamminga, H. 1986. Historical perspective: The problem of the origin of life in the context of developments in biology. *Orig. Life Evol. Biosph.* 18: 1–10.

Kumar, R. K. and M. Yarus. 2001. RNA-catalyzed amino acid activation. *Biochemistry* 40: 6998–7004.

Larralde, R., M. P. Robertson, and S. L. Miller. 1995. Rates of decomposition of ribose and other sugars: Implications for chemical evolution. *Proc. Natl. Acad. Sci. USA* 92: 8158–8160.

Lazcano, A. 2008. What is life? A brief historical overview. *Chem. Biodiv.* 5: 1–15.

Lazcano, A. 2010. Historical development of origins of life. In D. W. Deamer and J. Szostak (eds.), *Perspective on the Origins of Life.* Cold Spring Harbor Press, Cold Spring Harbor, NY., in press.
Lazcano, A. and S. L. Miller. 1999. On the origin of metabolic pathways. *J. Mol. Evol.* 49: 424–431.
Lazcano, A., G. E. Fox, and J. Oró. 1992. Life before DNA: The origin and early evolution of early Archean cells. In R. P. Mortlock (ed.), *The Evolution of Metabolic Function*, pp. 237–295. CRC Press, Boca Raton.
Lincoln, T. A. and G. F. Joyce. 2009. Self-sustained replication of an RNA enzyme. *Science* 232: 1229–1232.
Lindhal, T. 1993. Instability and decay of the primary structure of DNA. *Nature* 362: 709–715.
Lollar, B. S. and T. M. McCollom. 2006. Biosignatures and abiotic constraints on early life. *Nature* 444: E18.
Mansy, S. S., J. P. Schrum, M. Krishnamurthy, and 3 others. 2008. Template-directed synthesis of a genetic polymer in a model protocell. *Nature* 454: 122–125.
Miller, S. L. 1953. Production of amino acids under possible primitive Earth conditions. *Science* 117: 528.
Miller, S. L. and A. Lazcano. 2002. Formation of the building blocks of life. In J. W. Schopf (ed.), *Life's Origin: The Beginnings of Biological Evolution*, pp. 78–112. University of California Press, Berkeley.
Nielsen, P. 1993. Peptide nucleic acid, PNA, a model structure for the primordial genetic material? *Orig. Life Evol. Biosph.* 23: 323–327.
Nissen, P., J. Hansen, N. Ban, and 2 others. 2000. The structural bases of ribosome activity in peptide bond synthesis. *Science* 289: 920–930.
Orgel, L. E. 1968. Evolution of the genetic apparatus. *J. Mol. Biol.* 38: 381–392.
Orgel, L. E. 2004. Prebiotic chemistry and the origin of the RNA world. *Crit. Rev. Biochem. Mol. Biol.* 39: 99–123.
Orgel, L. E. and J. E. Sulston. 1971. Polynucleotide replication and the origin of life. In A. P. Kimball and J. Oró (eds.), *Prebiotic and Biochemical Evolution*, pp. 5–16. North-Holland Publ. Co., Amsterdam.
Oró, J. 1960. Synthesis of adenine from ammonium cyanide. *Biochem. Biophys. Res. Commun.* 2: 407–412.
Peretó, J., J. L. Bada, and A. Lazcano. 2009. Charles Darwin and the origins of life. *Orig. Life Evol. Biosph* 39: 395–406.
Ren, B., G. Tibbelin, D. de Pascale, and 3 others. 1998. A protein disulfide oxidoreductase from the archaeon *Pyrococcus furiosus* contains two thioredoxin fold units. *Nat. Struct. Biol.* 7: 602–611.
Ricardo, A., M. A. Carrigan, A. N. Olcott, and 1 other. 2004. Borate minerals stabilize ribose. *Science* 303: 196.
Robertson, M. P. and S. L. Miller. 1995. An efficient prebiotic synthesis of cytosine and uracil. *Nature* 375: 772–774.
Schopf, J. W. 1993. Microfossils of the early Archaean Apex chert, new evidence for the antiquity of life. *Science* 260: 640–646.
Schopf, J. W. 2001. *The Cradle of Life: The Discovery of Earth's Earliest Fossils.* Princeton University Press, Princeton.
Shen, Y., R. Buick, and D. E. Canfield. 2001. Isotopic evidence for microbial sulphate reduction in the early Archaean era. *Nature* 410: 77–81.
Springsteen, G. and G. F. Joyce. 2004. Selective derivatization and sequestration of ribose from a prebiotic mix. *J. Amer. Chem. Soc.* 126: 9578–9583.

Szostak, J. W., D. P. Bartel, and P. L. Luisi. 2001. Synthesizing life. *Nature* 409: 387–390.
Tice, M. M. and D. R. Lowe. 2004. Photosynthetic microbial mats in the 3,416-Myr-old ocean. *Nature* 431: 549–552.
Ueno, Y., K. Yamada, N. Yoshida, and 2 others. 2006. Evidence from fluid inclusions for microbial methanogenesis in the early Archaean era. *Nature* 440: 516–519.
White, H. B. 1976. Coenzymes as fossils of an earlier metabolic state. *J. Mol. Evol.* 7: 10–104.
Woese, C. R. 1967. *The Genetic Code: The Molecular Basis for Gene Expression.* Harper and Row, New York.
Zubay, G. 1998. Studies on the lead-catalyzed synthesis of aldopentoses. *Orig. Life Evol. Biosph.* 28: 12–26.

Commentary Four

The Genomic Imprint of Endosymbiosis

Christopher E. Lane

The origin of biological matter and living cells is one of the more perplexing issues in evolutionary biology (see Lazcano, Chapter 14). Theories abound to explain how prokaryotic and eukaryotic cells first formed (Embley and Martin 2006), but experimental evidence to support one hypothesis over another is difficult to come by. Based on cellular and geologic evidence, what can be generally agreed upon, however, is that prokaryotes predate eukaryotes and that some degree of cellular fusion between evolutionarily divergent members (i.e., a eubacterium and an archaean) resulted in the origin and radiation of eukaryotic lineages. There is little consensus with regard to the order of events or constituents that gave rise to the first eukaryote, but at the very least, it is known that all extant eukaryotes are descendant from an single endosymbiotic event that gave rise to the mitochondrion (Lang et al. 1999). It is also abundantly clear that multiple endosymbioses have imported and propagated photosynthesis among eukaryotic lineages (Palmer 2003). Like the mitochondrion, a single endosymbiotic event brought photosynthesis into eukaryotes via the uptake and retention of a cyanobacterium, but once established, endosymbiotic relationships between two eukaryotes have moved plastids throughout the Tree of Life a minimum of seven times (Archibald 2009). Despite the numerous cell fusions that have occurred throughout the history of eukaryotic evolution, until recently, little attention has been paid to the genomic consequences of endosymbiosis and the effect it might have on the ability to reconstruct the origin and subsequent evolution of eukaryotes. In this commentary, the mosaic nature of eukaryotic nuclei and the problems they create for the study of ancient evolutionary events will be introduced.

Symbiosis to Endosymbiosis

Symbioses can be found nearly everywhere in nature, allowing at least two organisms to exploit a niche together that they would not be able to utilize alone. In some cases, this relationship continues for so long that it becomes obligate for both parties, and the host begins to control aspects of the symbiont lifecycle, such as cell division. Once the host exerts this level of control over the symbiont, there is typically a cascade of genomic effects for the symbiont, resulting in the loss of many genes and metabolic pathways necessary for free living, and the relationship becomes endosymbiotic (Wilcox et al. 2003).

The most ubiquitous endosymbionts are hardly recognizable as formerly free-living cells, but both plastids (or chloroplasts) and mitochondria are the result of ancient relationships between free-living cells (Bhattacharya et al. 2007). In the case of the mitochondrion, available evidence points to a single event that occurred when an alpha-proteobacterium was incorporated into the ancestor of eukaryotes and became an essential part of the cell (Lang et al. 1999). Despite early reports of mitochondrion-lacking parasites, upon closer scrutiny no eukaryote has been found to lack a mitochondrion or an organelle derived from a mitochondrion, such as a hydrogenosome or mitosome (Roger and Silberman 2002). Plastids, however, have a much more difficult past to reconstruct.

The majority of plastids can be separated into either primary or secondary varieties. Primary plastids are those found in the Archaeplastida lineage (Adl et al. 2005), which includes glaucophytes, red algae, and green algae (including land plants) and are surrounded by two bounding membranes. The plastids in this group are all derived from a cyanobacterial ancestor (Figure 1A)

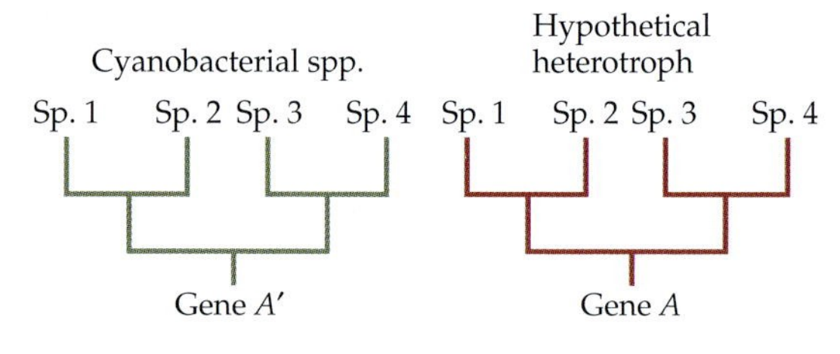

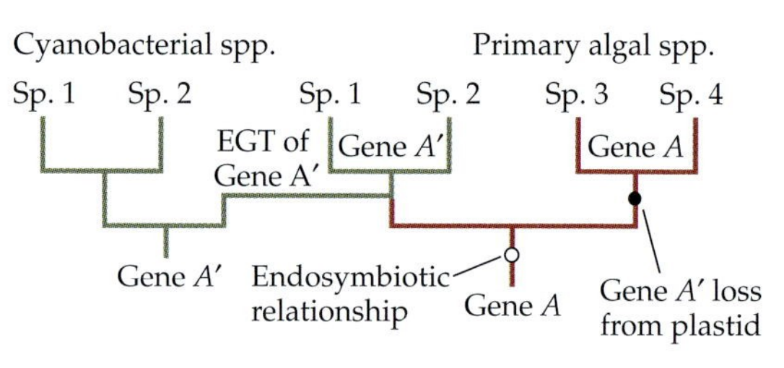

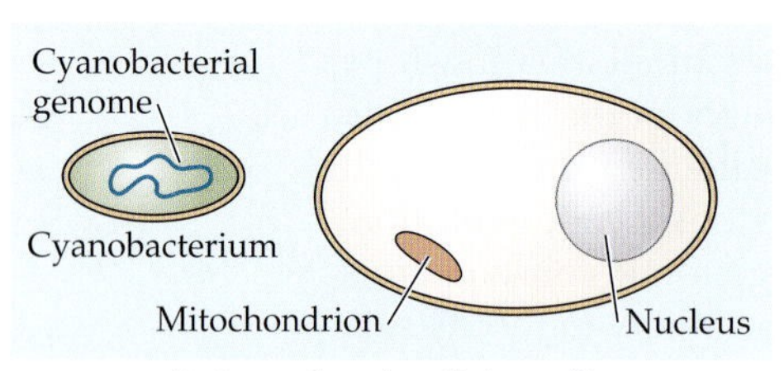

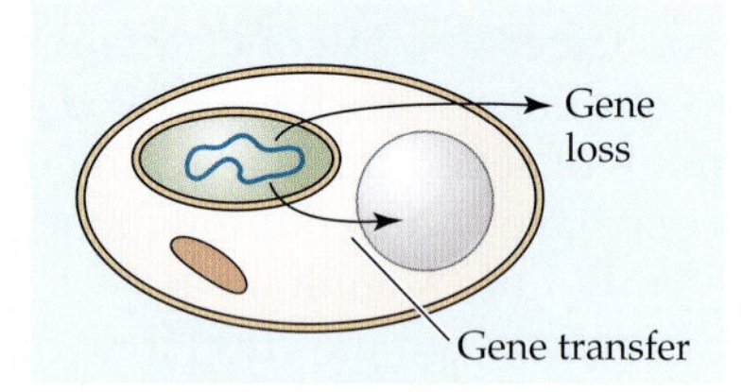

FIGURE 1 The Effect of Endosymbiotic Gene Transfer (EGT) on Phylogenetic Reconstruction (A) Genes are typically inherited in a vertical fashion, from ancestor to descendant, and their history can be used to reconstruct the evolutionary relationships between species. (B) When an endosymbiont is taken in, genes can be transferred to the host nucleus and, in some cases, replace the native copy (see Sp. 1 & 2). In related taxa (see Sp. 3 & 4), the gene may be lost from the plastid without being transferred to the host nucleus. Instances of EGT obscure the evolutionary history of species when the gene history is used as a proxy for the evolution of species.

that became an endosymbiont within a formerly heterotrophic eukaryote (Palmer 2003). After the different primary plastid-containing lineages split from one another, the process of plastid acquisition played itself out again, but this time in a fusion of two eukaryotic cells. A green algal cell has been converted to a secondary plastid in two distinct lineages (euglenophytes in the jakobid supergroup and chlorarachniophytes in the Rhizaria), and the plastid in the hypothetical ancestor of the chromalveolate supergroup is derived from a red algal cell (Archibald 2009). Therefore, on at least three occasions either an entire red or green algal cell has been converted into a photosynthetic organelle by a formerly heterotrophic eukaryote, spawning a novel photosynthetic lineage. These secondary plastids are surrounded by three or four membranes, and in addition to the euglenophytes and chlorarachniophytes previously mentioned, they persist in diverse lineages, such as brown algae, diatoms, haptophytes, cryptophytes, and dinoflagellates, as well as in a reduced and non-photosynthetic form in apicomplexans, such as the causative agent of malaria, *Plasmodium falciparum*. Multiple tertiary endosymbioses have spread plastids even further, but only in a few species of dinoflagellates. Regardless of their prokaryotic or eukaryotic origin, however, all plastids contain genomes that encode only a small fraction of their original gene complement.

The Genomic Consequences of Endosymbioses

The most gene-rich plastid genome known is that of the red alga *Porphyra purpurea*, which encodes a mere 259 single-copy genes (Reith and Munholland 1995). In comparison, even the smallest cyanobacterial (the ancestors of the red algal plastid) genomes encode over 2000 genes (Kettler et al. 2007)—nearly ten times the number found in the plastid with the greatest number of genes. Secondary plastids have similarly small plastid genomes, but they were once free-living eukaryotes that lost their mitochondria, their nuclei (i.e., except for cryptophytes and chlorarachniophytes), and all of the genetic information encoded within those organelles. Where has it all gone?

Much of the endosymbiont's genetic information is redundant with the host's and has been lost as superfluous to its new lifestyle (see Figure 1). However, the proteins required for the plastid to function outnumber those encoded on plastid genomes and are, in fact, nuclear encoded. Large numbers of proteins are made in the cytosol and targeted to both the plastid and mitochondrion in a variety of intricate membrane transport systems (Gould et al. 2008). The majority of organelle-targeted genes in the nucleus have been transferred from the organelles over time (Figure 1B), providing the host greater control over its function, but organelle-derived genes are also a source of genetic novelty for the host.

Based on genome surveys, it appears that as many as one-in-five nuclear genes may have an endosymbiotic origin (see Figure 1B) in organisms like the primary plastid-containing plant, *Arabidopsis thaliana* (Martin et al. 2002), and in the secondary plastid-containing diatom, *Phaeodactylum tricornutum* (Moustafa et al. 2009). Strikingly, over half of these genes are predicted to have cytosolic, and not plastid, functions (Martin et al. 2002). Endosymbiont genes can either provide novel cellular pathways or replace the native host copy, and even core housekeeping genes do not appear to be immune to replacement in large numbers in some organisms (Ahmadinejad et al. 2007). This reality, underappreciated until very recently, complicates our ability to reconstruct deep eukaryotic evolutionary history based on gene sequences (see Hillis, Chapter 16), because the genes commonly used in broad-scale, multi-gene analyses may have conflicting histories *within* certain organisms, let alone between lineages.

New phylogenetic tools are being developed to separate conflicting signals in large data sets and partition the resulting trees so that conflicting phylogenies can be independently evaluated. These methods (Pisani et al. 2007; Moustafa and Bhattacharya 2008) are being used to untangle the various genetic inputs in nuclei in an attempt to reconstruct the earliest events in the history of eukaryotic evolution, but much work remains. However, one of the greatest contributions of the genomics era has been an improved understanding of the makeup of nuclear genomes and their chimeric nature.

Literature Cited

Adl, S. M., A. G. B. Simpson, M. A. Farmer, and 25 others. 2005. The new higher level classification of eukaryotes with emphasis on the taxonomy of protists. *J. Eukaryot. Microbiol.* 52: 399–451.

Ahmadinejad, N., T. Dagan, and W. Martin. 2007. Genome history in the symbiotic hybrid *Euglena gracilis*. *Gene* 402: 35–39.

Archibald, J. M. 2009. The puzzle of plastid evolution. *Curr. Biol.* 19: R81–R88.

Bhattacharya, D., J. M. Archibald, A. P. M. Weber, and 1 other. 2007. How do endosymbionts become organelles? Understanding early events in plastid evolution. *BioEssays* 29: 1239–1246.

Embley, T. M. and W. Martin. 2006. Eukaryotic evolution, changes and challenges. *Nature* 440: 623–630.

Gould, S. B., R. R. Waller, and G. I. McFadden. 2008. Plastid evolution. *Ann. Rev. Plant Biol.* 59: 491–517.

Kettler, G., A. Martiny, K. Huang, and 11 others. 2007. Patterns and implications of gene gain and loss in the evolution of *Prochlorococcus*. *PLoS Genet.* 3: e231.

Lang, B. F., M. W. Gray, and G. Burger. 1999. Mitochondrial genome evolution and the origin of Eukaryotes. *Ann. Rev. Genet.* 33: 351–397.

Martin, W., T. Rujan, E. Richly, and 7 others. 2002. Evolutionary analysis of *Arabidopsis*, cyanobacterial, and chloroplast genomes reveals plastid phylogeny and thousands of cyanobacterial genes in the nucleus. *Proc. Natl. Acad. Sci. USA* 99: 12246–12251.

Moustafa, A., B. Beszteri, U.-G. Maier, and 3 others. 2009. Genomic footprints of a cryptic plastid endosymbiosis in diatoms. *Science* 324: 1724–1726.

Moustafa, A. and D. Bhattacharya. 2008. PhyloSort: A user-friendly phylogenetic sorting tool and its application to estimating the cyanobacterial contribution to the nuclear genome of *Chlamydomonas*. *BMC Evol. Biol.* 8: 6.

Palmer, J. D. 2003. The symbiotic birth and spread of plastids: How many times and whodunit? *J. Phycol.* 39: 4–11.

Pisani, D., J. A. Cotton, and J. O. McInerney. 2007. Supertrees disentangle the chimerical origin of eukaryotic genomes. *Mol. Biol. Evol.* 24: 1752–1760.

Reith, M. and J. Munholland. 1995. Complete nucleotide sequence of the *Porphyra purpurea* chloroplast genome. *Plant Mol. Biol. Rep.* 13: 333–335.

Roger, A. J. and J. D. Silberman 2002. Cell evolution: Mitochondria in hiding. *Nature* 418: 827–829.

Wilcox, J. L., H. E. Dunbar, R. D. Wolfinger, and 1 other. 2003. Consequences of reductive evolution for gene expression in an obligate endosymbiont. *Mol. Microbiol.* 48: 1491–1500.

Chapter 15

Adaptive Radiation: The Interaction of Ecological Opportunity, Adaptation, and Speciation

Jonathan B. Losos and D. Luke Mahler

Darwin may have been the first to describe adaptive radiation when, contemplating the variety of finches that now bear his name, he remarked: "Seeing this gradation and diversity of structure in one small, intimately related group of birds, one might really fancy that from an original paucity of birds in this archipelago, one species has been taken and modified for different ends" (Darwin 1845: 380). Since Darwin's time, naturalists and evolutionary biologists have been fascinated by the extraordinary diversity of ecology, morphology, behavior, and species richness of some clades, but interest in adaptive radiation has surged in recent years, largely as a result of three developments.

First, evolutionary ecologists have focused on the mechanisms that produce adaptive radiation. Careful study of ecological divergence within and among populations has yielded a wealth of information about how natural selection leads to evolutionary diversification (Schluter 2003; Nosil and Crespi 2006; Grant and Grant 2008a) (Figure 15.1). Schluter's (2000) seminal work, *The Ecology of Adaptive Radiation* was a watershed in the field, building upon and extending in important ways Simpson's (1953) *The Major Features of Evolution* published a half-century earlier.

Second, the explosion of molecular phylogenetics in the last two decades (see Hillis, Chapter 16) has revealed the diversification histories of countless clades and has provided the raw material for a renaissance of adaptive radiation studies. Molecular research has offered surprising discoveries about the history and magnitude of many adaptive radiations, such as the vangids of Madagascar (Yamagishi et al. 2001) (Figure 15.2), the corvoids of Australia (Sibley and Ahlquist 1990; Barker et al. 2004), the cichlids of Lake Victoria (Meyer et al. 1990; Seehausen 2006), the lobeliads of Hawaii (Givnish et al. 2009) and a plethora of others. In each of these cases, the great ecological and morphological diversity of a group had been thought to be the result of independent colonization events from multiple, differently adapted ancestral lineages. Instead, new molecular phylogenies revealed

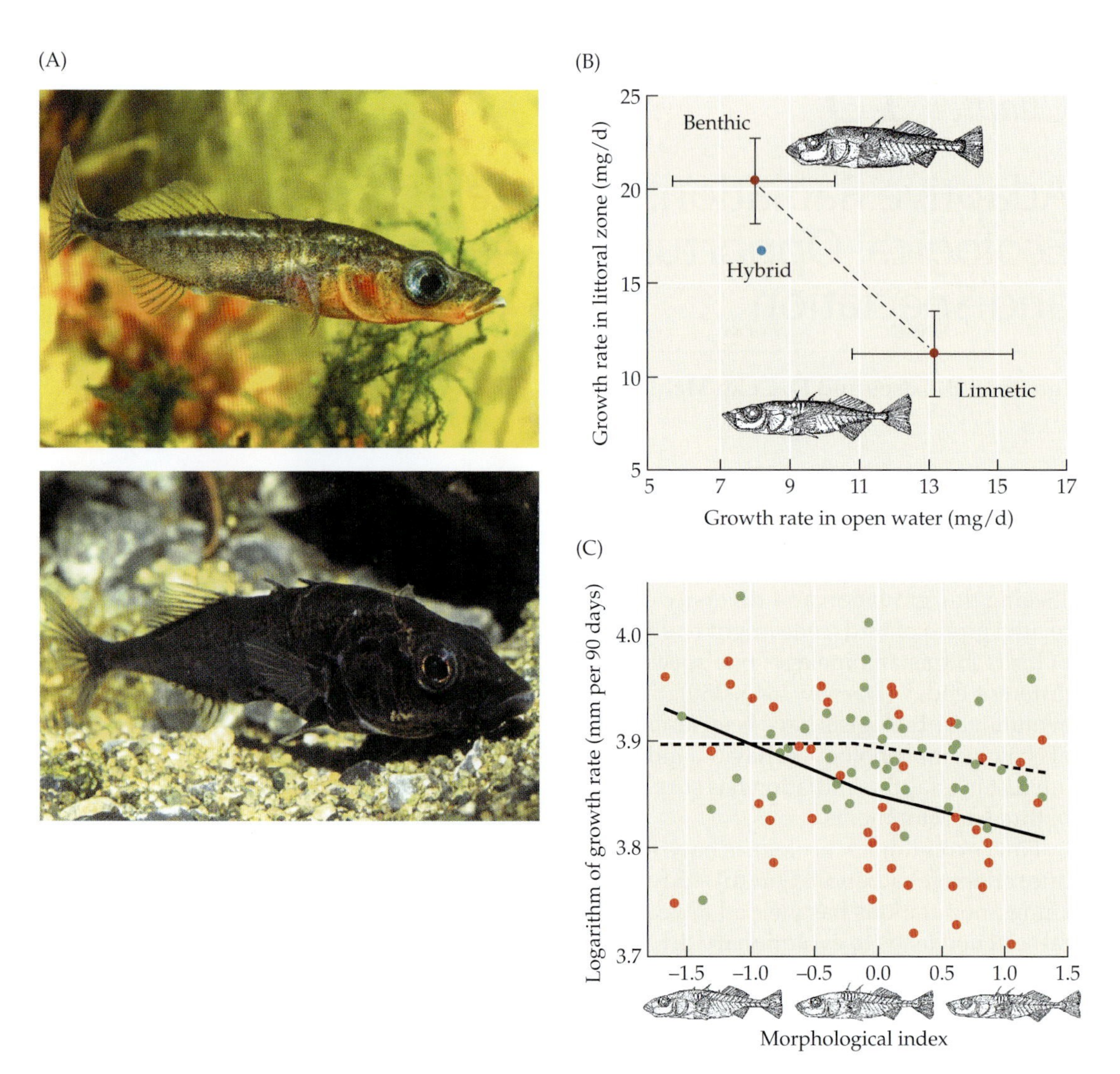

FIGURE 15.1 Natural Selection and Adaptation in British Columbian Threespine Sticklebacks (*Gasterosteus aculeatus*) (A) A number of lakes in British Columbia contain two species of sticklebacks, one that is slender-bodied, with a small mouth, eats zooplankton, and occupies open water (limnetic, above); the other is larger, deeper-bodied, has a large mouth, and eats invertebrates near the lake bottom (benthic, below). (B) Each species has higher growth rates in its own habitat, and hybrids are inferior in both habitats. (C) Natural selection (using growth rate as a proxy) on a highly variable hybrid population favors more benthic-like individuals in the presence of the limnetic species (red dots and solid line), but not in their absence (green dots and dashed line). Such studies have shed light on the role of ecology and natural selection in driving divergence during adaptive radiation. (A, from Rundle and Schluter 2004, photos © Ernie Cooper; B, adapted from Schluter 1995; C, adapted from Schluter 1994.)

FIGURE 15.2 Malagasy Vangids Molecular phylogenetic study indicates that the vangids represent a monophyletic group, rather than being members of several different families with closest relatives elsewhere. (From Yamagishi et al. 2001.)

that the great diversity in these groups is the result of *in situ* evolution, that is, adaptive radiation (see references previously cited).

Not only have many previously unknown adaptive radiations been discovered with molecular data, but also time-calibrated phylogenetic trees of extant taxa allow classic hypotheses of adaptive radiation to be tested in new ways (Glor 2010). Previously, the primary way to estimate the tempo of macroevolution was to measure it directly using high-quality paleontological data (see Wagner, Chapter 17), which are available for relatively few lineages. However, nucleotide sequence data now provide an alternative means of reconstructing phylogenetic relationships and the timing of lineage divergence events. With time-calibrated phylogenies, one may ask questions such as whether the occurrence of adaptive radiation is correlated with historical events (e.g., mass extinctions, changes in climate) or whether the pace of diversification decreases through time, as often is expected of an adaptive radiation (Box 15.1).

Third, experimental studies of microbial evolution have added a new dimension to the study of adaptive radiation (see Dykhuizen, Commentary 2). Such studies bring the benefits of experimental control, large sample sizes and replication, and the ability to not only track lineages through the diversification process, but also to freeze ancestral taxa and subsequently resurrect them to interact with their descendants (Lenski and Travisano 1994). Although thus far primarily based on studies of short-lived asexually reproducing organisms diversifying under simplified ecological conditions (a situation that is changing), microbial evolution studies have permitted experimental tests of many of the basic hypotheses of adaptive radiation, allowing microevolutionary processes to be directly and experimentally connected to macroevolutionary outcomes and often confirming predictions of the adaptive radiation model (Kassen 2009) (Figure 15.3).

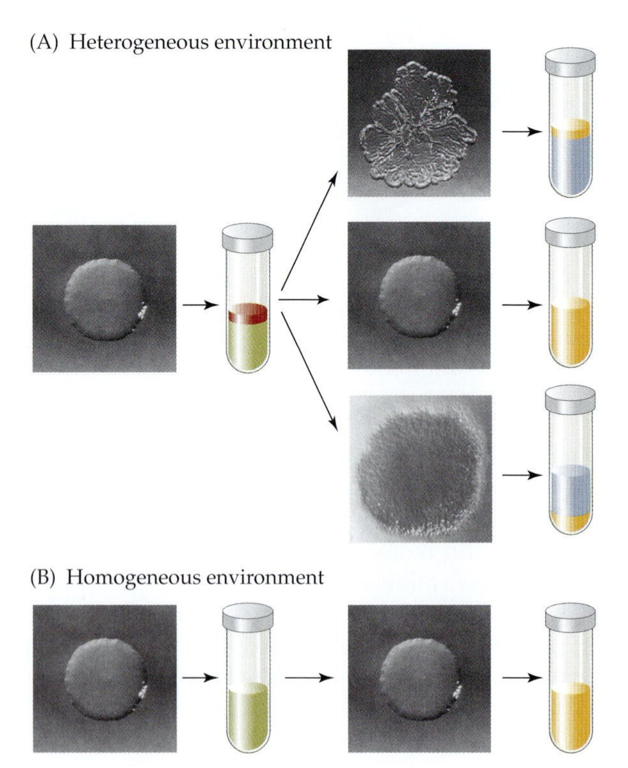

FIGURE 15.3 Experiments on the Evolutionary Diversification of the Bacterium *Pseudomonas fluorescens* Resources were distributed either heterogeneously (solution with red and green strata in A) or homogeneously (uniform green solution in B) distributed. After several days, bacteria in the heterogeneous environments repeatedly diverged into the same three morphotypes, which interact negatively and differ in resource use (the yellow shading on the right depicts typical resource use for each morphotype; the different forms tend to thrive as a surface film, in solution, or along the substrate, respectively). By contrast, only one morphotype occurred in the homogenous treatment. (Adapted from Rainey and Travisano 1998.)

BOX 15.1 TESTING FOR CHARACTERISTIC PATTERNS OF ADAPTIVE RADIATION

With time-calibrated phylogenies, researchers can investigate whether lineage diversification patterns match patterns expected from the ecological process of adaptive radiation. These approaches focus on how the pace of lineage or phenotypic diversification alters over time or with changing ecological conditions, such as ecological opportunity. The most common approach to testing for the signature of adaptive radiation is to construct models in which parameters describe changes in the tempo of diversification as a result of ecological conditions. Alternative ecological models may be compared to each other, to non-ecological models, or to a null model in which diversification proceeds at a constant rate.

Tests for the signature of adaptive radiation have a rich pedigree in quantitative paleontology, in which numerous studies have tracked the rise and fall of diversity and disparity in relation to mass extinction events, the evolution of key innovations, and colonization of new regions (Simpson 1953; Sepkoski 1978; Foote 1997, 1999; reviewed in Erwin 2007; see Foote, Chapter 18) (Figure 1A,B).

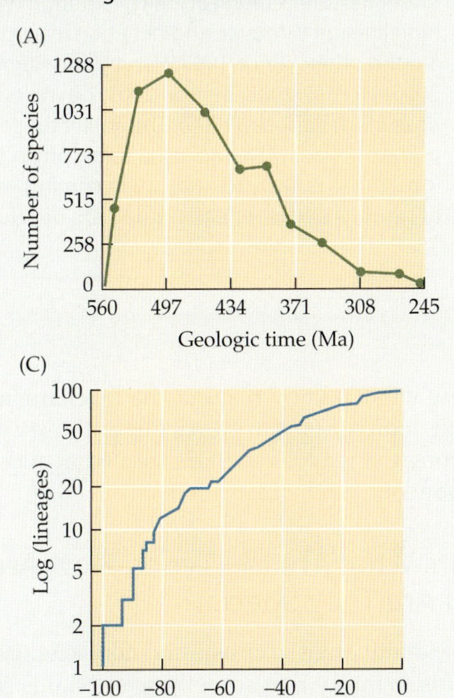

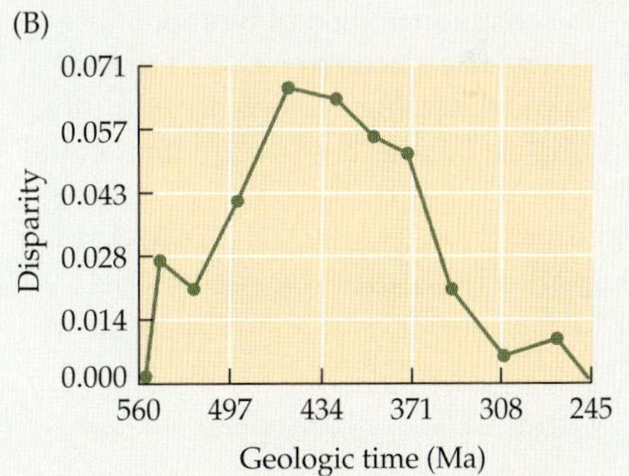

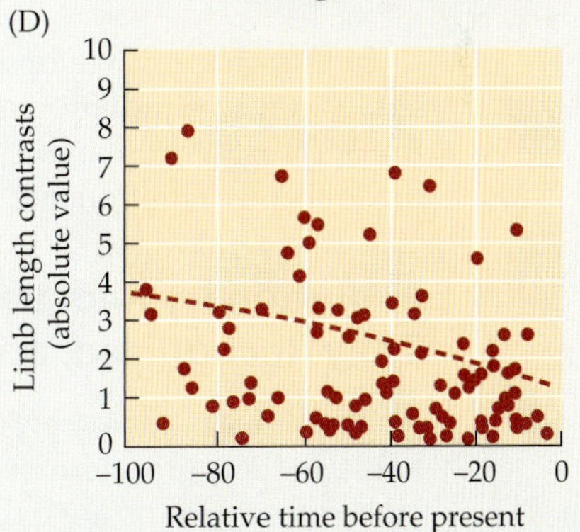

FIGURE 1 The "Early Burst" Pattern of Adaptive Radiation In the fossil record, an early burst may be detected by tracking the number of species and the total disparity of a radiation over geological time. In trilobites, species richness (A) and disparity (B) both peaked early in the history of the clade. Phylogenetic comparative methods may be used to detect an early burst from data on extant taxa. Greater Antillean *Anolis* lizards exhibit both a rapid early accumulation of lineages, depicted as a concave lineages-through-time plot in (C), as well as an early burst in phenotypic evolution, evident as larger independent contrasts early in the radiation (D). (A,B adapted from Foote 1993; C,D adapted from Mahler et al. 2010.)

(continued)

BOX 15.1 *Continued*

The most commonly used phylogenetic test evaluates the early burst model, in which early diversification is explosive as lineages rapidly adapt to new ecological roles but slows as opportunities disappear due to ecological saturation (Nee et al. 1992; Rabosky and Lovette 2008; Phillimore and Price 2009; reviewed in Glor 2010). The majority of tests of the "early burst" model have investigated patterns of lineage diversification, but similar models have recently been developed for phenotypic traits as well (Freckleton and Harvey 2006; Agrawal et al. 2009; Harmon et al. 2010, in press; Mahler et al. 2010, in revision) (Figure 1C,D). Support for this model varies—the common finding of temporally declining lineage diversification has led some to proclaim a strong role for ecological opportunity in regulating cladogenesis, although the sample of clades studied may be biased (McPeek 2008; Phillimore and Price 2008; Ricklefs 2009). Among phenotypic studies of the early burst model (which, to date, are far fewer than studies focusing on species richness), some studies report declining rates of evolution with diminishing ecological opportunity (Freckleton and Harvey 2006; Agrawal et al. 2009; Mahler et al. 2010, in revision), whereas Harmon et al. (2010) find only limited support for the early burst model in a wide survey of animal radiations.

A key feature of the model-fitting approach is the flexibility provided by the variety of models that may be compared when evaluating adaptive radiation hypotheses. In addition to the early burst model, a diversity of alternative models are available for fitting data, including radiation under flexible ecological limits, radiation with high lineage turnover, and radiation in stages, among others (Price 1997; Harvey and Rambaut 2000; McPeek 2008; Benton 2009; Gavrilets and Vose 2009; Rabosky 2009). The ability to identify patterns of adaptive radiation using phylogenetic methods is likely to improve in coming years as new and more refined models are developed.

In this chapter, we review what is known about the mechanisms that drive adaptive radiation and, more importantly, highlight those areas requiring further research. Along the way, we will discuss what constitutes an adaptive radiation and how one can be identified.

Evolutionary Radiation: What Are the Types, How Are They Recognized, and Are They Special?

Adaptive radiations draw the attention of scientists and non-scientists alike because their grandeur seems to imply that something special is responsible: these groups are extraordinary and require explanation, invocation of some special attribute, either intrinsic or external, that can explain why these particular clades have diversified to such an extreme extent (Box 15.2). But, adaptive radiation describes only one part of the spectrum of evolutionary radiations. Although distinguishing among types of evolutionary radiation may involve making arbitrary distinctions (Olson and Arroyo-Santos 2009), doing so provides a useful framework for further study of the key features of radiations.

BOX 15.2
ARE ADAPTIVE RADIATIONS EXCEPTIONAL?

We suggest that the term adaptive radiation should be reserved for those clades exhibiting exceptional ecological and phenotypic disparity (see Figure 15.4). The rationale for this argument is that it is the unusually great degree of disparity in these clades that requires explanation—by identifying such clades, researchers can focus on them to understand what has triggered their extraordinary evolutionary diversification. Implicit in this approach is the need to develop statistical methods to separate those clades that constitute adaptive radiations from those that do not.

This approach can be criticized on two counts. First, it creates an arbitrary dichotomy in what is most likely a continuous distribution (Olson and Arroyo-Santos 2009). That is, the degree of adaptive disparity of clades is surely continuously distributed. How can one draw a line and say that all clades with a greater amount of disparity than the threshold constitute adaptive radiations and that all with even a slightly lesser amount, are not?

An alternative approach to the threshold-based approach would be to quantify disparity for a sample of clades and investigate whether the degree of disparity is statistically related to a factor, such as degree of ecological opportunity (presuming it could be quantified), that has been hypothesized to drive adaptive radiation. Increasingly, tools for investigating the relationship between such factors and patterns of ecological diversification are being developed and employed (Olson and Arroyo-Santos 2009; Glor 2010).

The second criticism of this definition of adaptive radiation is based on the view that adaptive radiation is a process as well as an outcome (just as one might argue that adaptation is both a process and an end-result). This argument suggests that the same processes are involved in adaptive diversification whether the result is an enormous adaptive radiation, such as African Rift Lake cichlids, or a small clade of species slightly morphologically differentiated to adapt to minor differences in habitat use. Because both cases are the result of the same process of cladogenesis plus adaptive divergence driven by natural selection, this view would suggest that all such clades, no matter how disparate, should be considered adaptive radiations.

Such a view would render the term adaptive radiation meaningless. The vast majority of clades are composed of species that exhibit at least a small degree of phenotypic disparity that has arisen as a result of adaptive divergence driven by natural selection. Assuming this is true, almost all clades would be adaptive radiations, and the term would have little utility in identifying clades of special interest. We feel that this not only neuters the term "adaptive radiation," but also departs from its use throughout the history of evolutionary biology.

One way out of this problem might be to restrict the term adaptive radiation to clades in which adaptive diversity arose in a burst of evolution early in a clade's history, with subsequent deceleration in the rate of evolution. Indeed, many definitions of adaptive radiation include the proviso that radiation must occur quickly (Givnish 1997). This view, however, also has problems. Either the definition must include clades that achieve only modest disparity as long as they accumulated it early, or it must rely on an arbitrary disparity threshold.

From our perspective and that of many others (Givnish 1997), the important aspect of adaptive radiation is the disparity produced by the clade (Foote 1997; Erwin 2007), rather than the pace at which it accumulates. In some models of adaptive radiation (Harvey and Rambaut 2000), an adaptive radiation may unfold at a rather steady pace; indeed, ecological opportunity may ap-

(*continued*)

BOX 15.2 *Continued*

pear suddenly in some cases (e.g., colonization of an unoccupied island) but more gradually in others (e.g., co-radiation with another clade). We prefer to reserve the question of timing of diversification as a hypothesis to be tested among adaptive radiations, rather than a criterion for deciding whether a clade constitutes an adaptive radiation or not. At the least, if one takes the more restrictive definition that includes timing, then another term is needed for those clades that produce exceptional disparity but in a non-explosive way, perhaps such as Simpson's (1953) mostly forgotten "progressive occupation of adaptive zones."

Evolutionary radiation results in the production of two components of diversity—species richness and phenotypic diversity (often termed "disparity" to avoid confusion with "species diversity"). Adaptive radiation is a type of evolutionary radiation, emphasizing the extent of phenotypic differentiation among members of a clade as species adapt to use different ecological resources; we henceforth refer to this as "adaptive disparity." Although a wide variety of definitions of adaptive radiation have been proposed (Givnish 1997), Futuyma's (1998) definition seems to capture the sense of most of these: "evolutionary divergence of members of a single phylogenetic lineage into a variety of different adaptive forms." More specifically, we propose that the term adaptive radiation should refer to those clades that exhibit an exceptional extent of adaptive disparity. Definitional issues and methods for identifying adaptive radiations are reviewed in Box 15.3.[1]

Much of the literature on evolutionary radiations equates species richness with adaptive radiation. Indeed, many of the world's most remarkable and celebrated radiations are rich in both species and adaptive form, including the African Rift Lake cichlids, Hawaiian *Drosophila*, Caribbean *Anolis* lizards and, at a higher level, beetles and angiosperms. Such lineages undoubtedly represent adaptive radiations and suggest that species proliferation and ecological radiation occur hand in hand. Although often true, this need not be the case.

Some clades are exceptionally diverse phenotypically and constitute adaptive radiations, despite having unexceptional species diversity (Figure 15.4). For example, the lizard clades Pygopodidae and Cordylidae both contain great ecological and morphological disparity despite being species poor (Webb and Shine 1994; Branch 1998). Groups such as these are commonly neglected in studies of adaptive radiation, but deserve more

[1] Integral to the concept of adaptive radiation is the concept of adaptation itself. Evolutionary divergence can occur for reasons other than adaptive differentiation. Thus, investigation of the adaptive basis of trait differentiation is essential in any study of adaptive radiation. Arnold (1994), Larson and Losos (1996), and McPeek, Commentary 3 (in this volume) provide reviews of adaptation and how it can be studied. For the purposes of discussion here, we will assume that phenotypic differences among species are adaptively based.

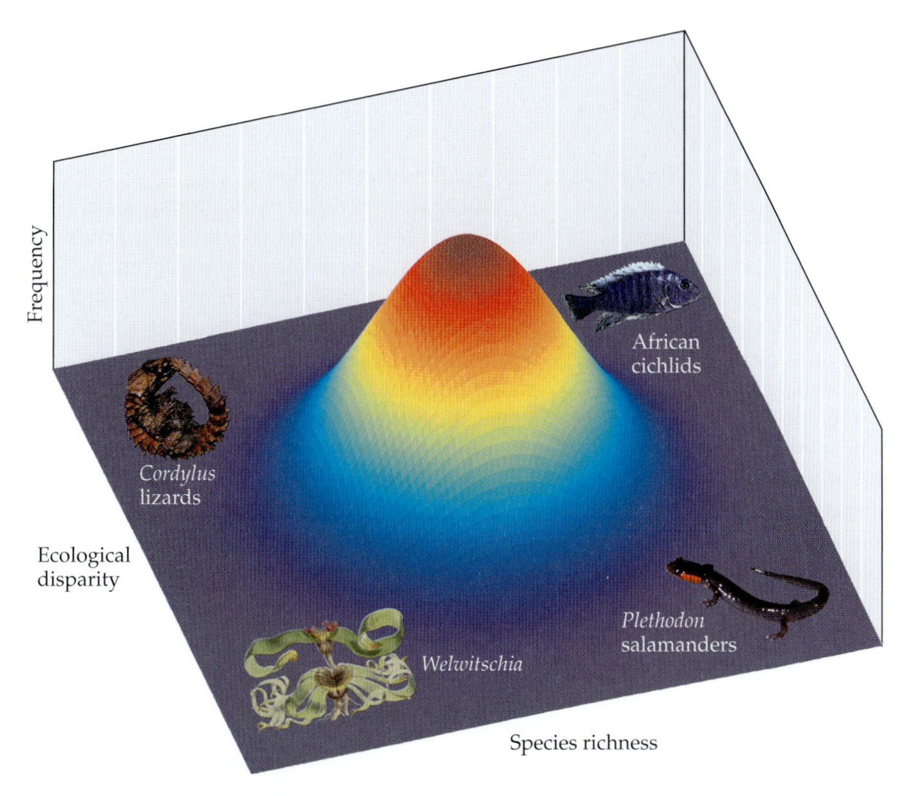

FIGURE 15.4 The Axes of Evolutionary Radiation Clades can be diverse in both species richness and ecological variety, termed disparity. Clades that have exceptional ecological disparity are adaptive radiations, whether they have great (e.g., African lake cichlids) or little species richness (e.g., cordylid lizards). Non-adaptive radiations are those that are exceptional in species richness, but not in ecological disparity, such as plethodontid salamanders. Some clades are exceptionally non-disparate and species-poor, such as *Welwitschia*.

attention, particularly in studies of the role of speciation in adaptive radiation (see subsequent discussion).

Conversely, other clades contain a great number of species, but little ecological or phenotypic disparity. For example, the eastern North American slimy salamanders (*Plethodon*) are a species-rich clade that diversified rapidly early in their evolutionary history. However, these salamander species are distributed almost entirely allopatrically and are ecologically similar (Kozak et al. 2006). Such clades have been termed non-adaptive radiations (Gittenberger 1991; Kozak et al. 2006; Rundell and Price 2009). Great species richness without substantial phenotypic disparity could arise if a clade was predisposed to speciate in ways that do not involve adaptive divergence, as might result from allopatric isolation in similar environments or from the operation of sexual selection occurring in divergent ways in different populations.

BOX 15.3
METHODS FOR IDENTIFYING EXCEPTIONAL CLADES

If exceptional diversity is a key feature of adaptive radiation, statistical methods are needed that can test whether a radiation is exceptional. Such a test must demonstrate that a clade contains more disparity than would be expected under a neutral model of diversification, or that it is exceptionally disparate compared to other radiations.

In principle, such questions could be addressed either by examining the fossil record or by using phylogenetic methods. Both of these methods have advantages and disadvantages. Paleontological methods directly track changes in diversity over time. By contrast, with phylogenetic methods using extant taxa, macroevolutionary changes must be inferred. As such, results arising from these methodologies are only as sound as their assumptions, and inadequate models for reconstructing evolutionary history may lead to both Type I and Type II errors in testing adaptive radiation hypotheses (Revell et al. 2005; see Foote, Chapter 18). Paleontological studies incorporate extinct taxa, and although phylogenetic methods exist for estimating the influence of extinction (Nee et al. 1994; Kubo and Iwasa 1995), they often suffer from low statistical power (Bokma 2009; Quental and Marshall 2009; Rabosky 2009). In contrast, the phylogenetic approach has several distinct advantages. In particular, phylogenetic studies permit investigation of long-term evolutionary patterns in taxa that are also well-studied ecologically—inferences about the ecology of fossil taxa can be unreliable, particularly when the taxa do not have comparable extant counterparts, and morphological homoplasy may also undermine accurate estimation of the relationships of fossil taxa in paleontological studies. Moreover, the quality of the fossil record varies greatly among taxa; phylogenetic methods can be used even for taxa with little or no fossil record. Of course, the most reliable insights into the large-scale pattern of adaptive radiation will be those that are well supported by both paleontological and phylogenetic investigations.

Statistical approaches to the identification of exceptional clades have concentrated almost entirely on patterns of lineage diversification rather than phenotypic diversification. The first lineage diversification models, in which speciation occurred as a uniform stochastic process, produced surprising results (Raup et al. 1973; Slowinski and Guyer 1989). The resulting phylogenies tended to be topologically unbalanced, suggesting that large differences in clade species richness could occur by chance alone (Guyer and Slowinski 1993; Barraclough and Nee 2001; Nee 2001, 2006). Nonetheless, many clades in the Tree of Life are exceptionally species-rich (or species-poor), even compared to the highly variable neutral expectation. For instance, at a very broad scale, Alfaro et al. (2009) identified nine such clades among all vertebrates.

Of course, while such lineage diversification models are useful for testing whether species radiations are exceptional, they do not necessarily identify adaptive radiations, which are distinguished by ecological and morphological disparity. Paleontologists have long used measures of disparity to trace patterns of phenotypic evolution through time, and changes in disparity in fossil lineages have been pivotal in testing and refining adaptive radiation theory in recent decades (Foote 1997; Roy and Foote 1997; Ciampaglio et al. 2001; Erwin 2001, 2007).

To date, few studies of extant taxa have attempted to identify those clades that are exceptionally disparate using phylogenetic methods (Losos and Miles 2002). Direct comparisons of disparity among taxa are problematic, because disparity is a function of the rate of evolution, the age of a clade, and the phylogenetic topology of the clade (O'Meara et al. 2006) (Figure 1). One approach is to estimate the rate of phenotypic evolution in a phylogenetic context and to compare rates among clades (Collar et al. 2005; Harmon et al. 2008; Pinto et al. 2008).

BOX 15.3 *Continued*

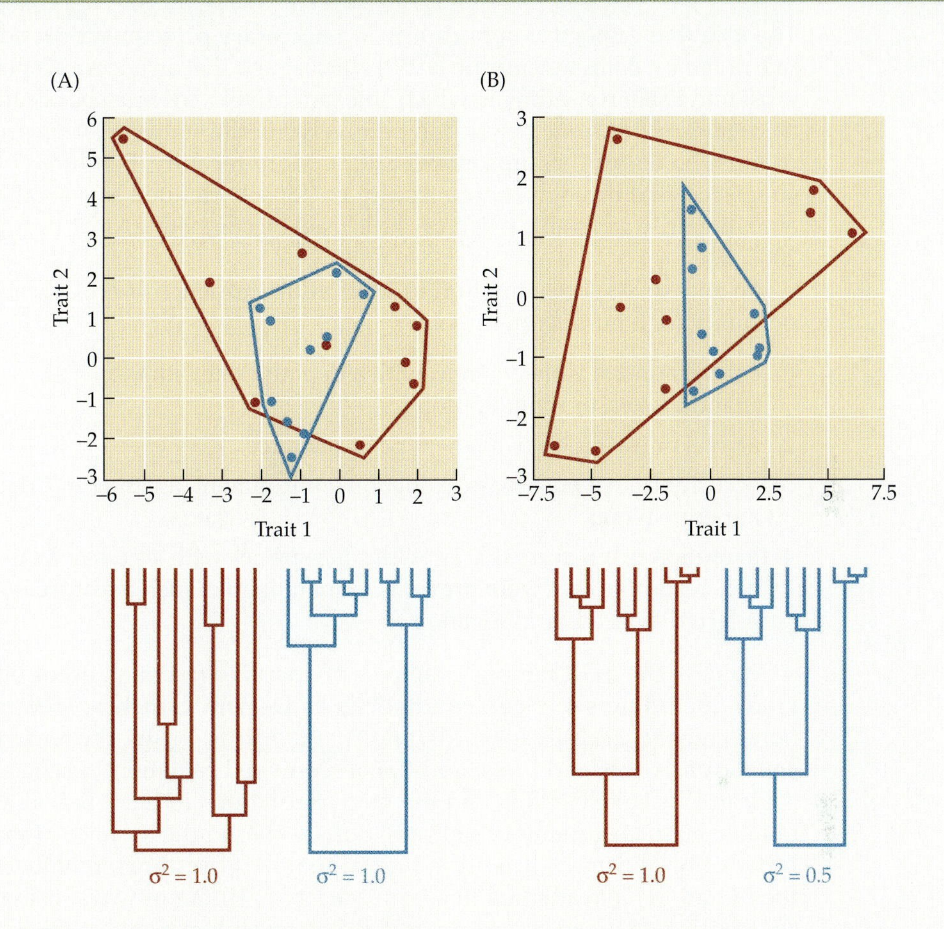

FIGURE 1 The Effect of Phylogenetic Structure and Rate of Evolution on Expected Disparity Under a Brownian motion model of evolution, disparity among extant taxa is a function of the rate of phenotypic evolution as well as of the structure of the phylogenetic tree relating those taxa. In (A), a phenotypic trait evolved at a constant rate on phylogenetic trees that were identical in branching order, but different in the relative timing of divergence. In the first phylogeny, lineage diversification was concentrated early, while in the second phylogeny, lineage divergences were concentrated late. Although the rate of trait evolution was the same for each phylogeny, greater disparity resulted when divergences were concentrated earlier, because traits evolved independently for longer periods of time. In (B), a trait evolved on identical phylogenies, but the rate of evolution (σ^2) differed by a factor of two between the two phylogenies. In this case, when the rate of evolution was higher, trait disparity was greater. In both (A) and (B), disparity differences are depicted graphically, using minimum convex polygons for the evolved trait values, and data in the upper panels are colored to match the phylogenies and rates under which they were generated.

Ecological Opportunity and Adaptive Radiation

The idea that ecological opportunity is a necessary prerequisite for adaptive radiation dates to Simpson (1953) who argued that an ancestral species must find itself in a setting in which "the [adaptive] zone must be occupied by organisms for some reason competitively inferior to the entering group or must be empty" (Simpson 1953: 207). More recently, Schluter (2000) suggested that ecological opportunity is "loosely defined as a wealth of evolutionarily accessible resources little used by competing taxa" (Schluter 2000: 69).

An ancestral species might find itself in the presence of ecological opportunity for a number of reasons:

1. colonization of isolated areas with a depauperate biota, such as islands, lakes, or mountaintops;
2. arrival or evolution of a new type of resource;
3. occurrence in a post-mass extinction environment, again with a depauperate biota;
4. evolution of a feature that provides the species with access to available resources that were previously unattainable; such a feature is referred to as a key innovation[2].

Many of the most famous examples of adaptive radiation occur on islands, including such iconic radiations as Darwin's finches, Hawaiian honeycreepers and silverswords, and *Anolis* lizards. Lakes, the terrestrial counterparts of islands, host additional renowned radiations, such as the African Rift Lake cichlids and the many radiations in Lake Baikal. One notable feature of many island radiations is that in the absence of many types of organisms normally found in mainland settings, members of the radiation have adapted in a wide variety of different ways, using resources that on the mainland are utilized by taxa not present on the island (Carlquist 1974; Leigh et al. 2007; Losos and Ricklefs 2009). The absence of many types of predators on islands may also play a role, by allowing organisms to use resources and habitats previously unavailable because of predation threat (Carlquist 1974; Schluter 1988; Benkman 1991). As a result, the ecological and phenotypic disparity of island radiations is often much greater than their mainland relatives (Carlquist 1974; Schluter

[2] A key innovation is defined as a trait that allows a species to interact with the environment in a fundamentally different way (Miller 1949; Liem 1974). For a critical discussion of the concept of key innovations, see Cracraft (1990) and Donoghue (2005). In recent years, the term has also been used to refer to a trait that increases the rate of species diversification (Heard and Hauser 1995; Hodges and Arnold 1995; Sanderson and Donoghue 1996; Ree 2005). This definition, however, introduces a different concept; conflating the two under the same name is confusing (Hunter 1998). A new term is needed for the latter phenomenon.

2000; Lovette et al. 2002; Figure 15.5). This oft-repeated phenomenon clearly indicates the role of ecological opportunity in spurring adaptive radiation.

Species may also experience ecological opportunity without moving to a new area if new resources appear, by immigration or evolution, where a

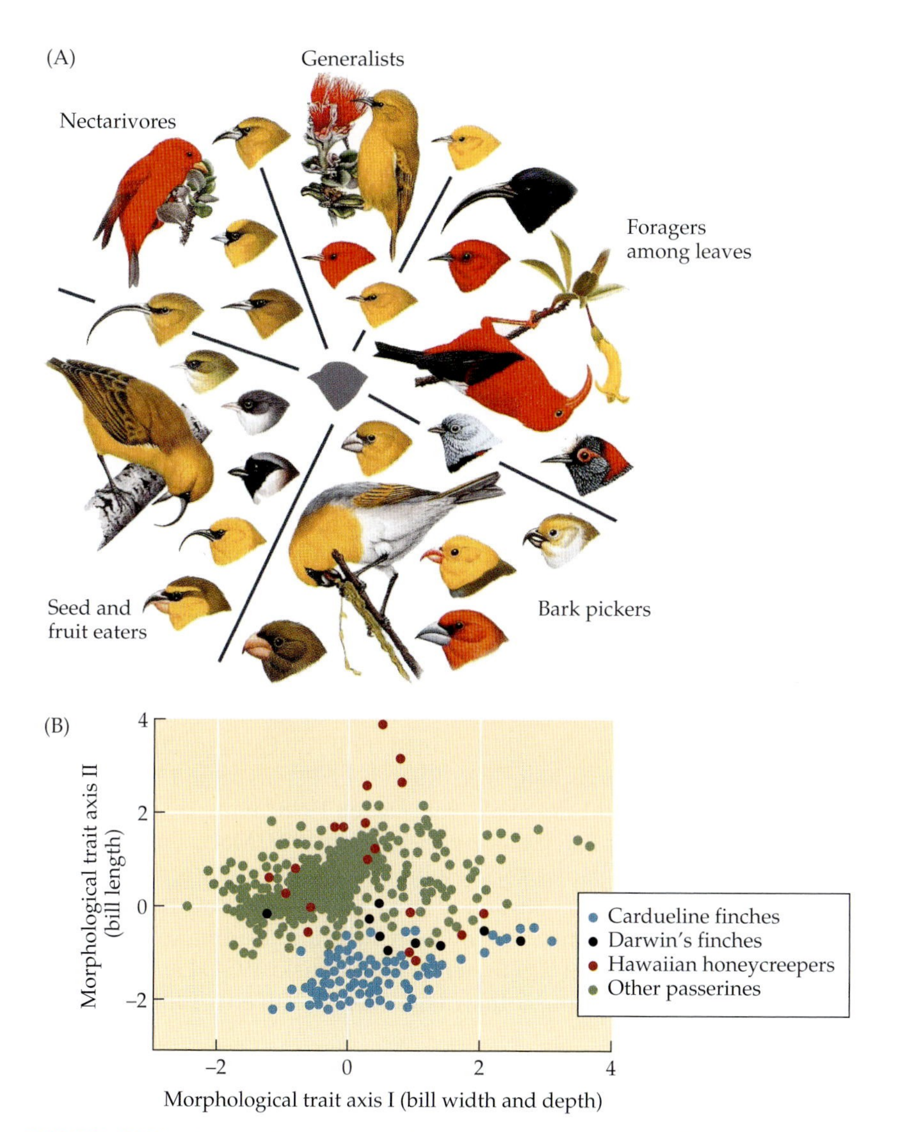

FIGURE 15.5 Adaptive Radiation of Island Birds (A) Hawaiian honeycreepers exhibit stunning morphological disparity in bill traits which corresponds to diversity in resource use. (B) The disparity of honeycreepers surpasses that of their mainland relatives, the cardueline finches and rivals that of all passerine birds. Another island radiation, Darwin's finches, also exhibits substantial disparity. In both island radiations, species occupy a diverse range of ecological niches utilized by species in many different families on the mainland. (A, reproduced with permission from Pratt 2005; B, adapted from Lovette et al. 2002 and Losos and Ricklefs 2009.)

clade already occurs. For example, the radiation of horses is often attributed to the spread of grasslands in the Miocene (MacFadden 1992). Similarly, the plant genus *Espeletia* has radiated extensively in paramo habitats at high elevations in the northern Andes (Monasterio and Sarmiento 1991; Rauscher 2002; Hooghiemstra et al. 2006). The Andes are young, and presumably, the ancestral *Espeletia* rode the rising mountain chain, adapting to new high-elevation habitats as they appeared. Hughes and Eastwood (2006) infer a similar history for Andean *Lupinus*.

Ecological opportunity also appears in the aftermath of mass extinctions (Erwin 2001, 2007). Surviving taxa often radiate rapidly and relatively quickly attain similar disparity to that seen before the extinction event. In some cases, clades radiate into parts of morphological space that previously were unoccupied by members of that clade (Foote 1999), whereas in other cases a surviving subclade expands its disparity to encompass parts of morphological space previously occupied by other subclades that perished in the mass extinction (Foote 1996; see Foote, Chapter 18; Ciampaglio 2002; McGowan 2004; Friedman 2010). The high rates of phenotypic evolution exhibited after mass extinctions by these clades strongly support the role of ecological opportunity in spurring radiation.

The evolution of a key innovation can allow access to previously unattainable resources, providing the stimulus for adaptive radiation. For example, the evolution of wings in bats allowed them to prey upon a wide range of flying insects probably unavailable to their earthbound ancestors. Many other examples of key innovations have been proposed, although such hypotheses are usually difficult to test because they represent unique historical events. One way around this difficulty is to look for putative key innovations that have evolved multiple times to see if they have repeatedly led to adaptive radiation (Mitter et al. 1988; de Queiroz 2002); several examples, such as toe pads in lizards (Russell 1979; Larson and Losos 1996), phytophagy in insects (Mitter et al. 1988; Farrell 1998), wings in vertebrates, and pharyngeal jaws in fish (Stiassny and Jensen 1987; Mabuchi et al. 2007), seem to pass this test.[3] However, key innovations do not necessarily lead to adaptive radiation; the evolution of such a trait may allow the species to interact with the environment in new ways without leading to substantial evolutionary diversification (Fürsich and Jablonski 1984; Levinton 1988; de Queiroz 2002).[4] For example, aardvarks have evolved a suite of skeletal

---

[3] All of these examples are cases in which taxa with the putative key innovation exhibit substantially greater disparity than their sister taxa, although this observation has not been statistically tested. Instead, quantitative tests in these papers have compared species richness, rather than disparity.

[4] Similarly, it is possible that the evolution of the same key innovation could lead to radiation in some clades and not others, depending on the context in which it evolves. Consequently, failure to find a statistical relationship between evolution of a putative key innovation and adaptation radiation does not indicate the

modifications permitting a termitophagous existence, but this specialization has not led to evolutionary diversification (Hunter 1998; see also Baum and Larson 1991 on *Aneides* salamanders).

Although not usually discussed in these terms, the evolution of mutualistic interactions may also function in a manner analogous to that of a key innovation by providing access to resources that were previously inaccessible (see Lane, Commentary 4). For example, the great diversity of herbivorous insects and vertebrates would not be possible if it were not for their mutualistic microbial gut inhabitants that allow them to digest cellulose (Janson et al. 2008). Similarly, the great variety among modern corals is likely in part the result of a mutualism between scleractinian corals and their endosymbiotic zooxanthellae, which provides photosynthetic energy in return for protection and nutrition (Stanley 1981).

The concept of ecological opportunity is straightforward and intuitive. Given the number of examples in which ecological opportunity (achieved in the four ways just enumerated) has led to adaptive radiation, there can be little doubt that it is an important trigger to adaptive radiation. Nonetheless, we may wonder whether ecological opportunity is a prerequisite for adaptive radiation.

Can Adaptive Radiation Occur in the Absence of Pre-Existing Ecological Opportunity?

Adaptive radiation could occur in the absence of pre-existing ecological opportunity either by members of a clade wresting resources away from other taxa that had been using them or by creating their own opportunity.

COMPETITIVE REPLACEMENT OF ONE CLADE BY ANOTHER Much of the older paleontological literature is peppered with proposals that one group has radiated by outcompeting another, usurping its resources and forcing it into evolutionary decline (e.g., the decline of mammal-like reptiles putatively as a result of the purported superiority of dinosaurs in locomotion or thermoregulation; Bakker 1968). However, very few of these cases hold up to close scrutiny. Usually, when one group replaces another ecologically similar group, the explanation is that the first group went extinct, followed by radiation of the second group (Rosenzweig and McCord 1991; Benton 1996; Brusatte et al. 2008). In a few cases, the fossil record does support the argument that one group has radiated at the expense of a clade that previously utilized the same resources. Perhaps, the best example is the interaction between cheilostome and cyclostome bryozoans, in which cheilostomes have outcompeted cyclostomes in local interactions over the course of many millions of years, all the while diversifying adaptively, while cyclostomes decreased in diversity (Sepkoski et al. 2000).

---

evolution of the trait did not lead to adaptive radiation in any of the clades in which it occurs (de Queiroz 2002).

This phenomenon of one group preventing evolutionary radiation of another has been termed niche incumbency and is seen in experimental microbial systems as well as in the fossil record (Brockhurst et al. 2007). It parallels the ecological phenomenon of the priority effect, which occurs when one species can prevent another from becoming established in a community by dint of prior occupancy (MacArthur 1972; Chase 2007). Moreover, the observation that the introduced invasive species almost never causes the global extinction of native species by outcompeting them[5] lends support to the idea that a radiating clade is unlikely to drive another clade to extinction through competitive interactions (Simberloff 1981; Davis 2003; Sax et al. 2007).

CAN ADAPTIVE RADIATIONS CREATE THEIR OWN OPPORTUNITY? It would seem, then, that ecological opportunity is usually necessary for adaptive radiation and that in most well-studied cases, such opportunity exists prior to adaptive radiation. However, a rarely considered alternative possibility is that lineages create their own opportunity as they radiate. Such self-propagation of an evolving radiation could occur in two ways.

First, as a clade radiates, the increased number of co-occurring clade members may create opportunities for exploitation by other species. The standard view among evolutionary biologists is that ecological opportunity decreases through the course of a radiation as niches are filled. However, an alternative view is that the more species that occur in a community, the more opportunity there is for other species to take advantage of them through predation, parasitism, mutualism, or other processes (Whittaker 1977; Tokeshi 1999; Erwin 2008). Most communities are composed of species from many different clades, but in cases (usually on islands) in which members of a single radiation are extremely diverse ecologically, the possibility exists for an evolutionary component to this view: as the clade radiates, it may create additional opportunities, spurring further radiation, thus creating further opportunities, and so on.

Implicit in this hypothesis is the view that interactions driving adaptive radiation occur not only as competition for resources, but also between species on different trophic levels. Indeed, adaptive radiations are known in which some members of the radiation prey on other members. Among African Rift Lake cichlids, for example, some species prey on others by eating their young, plucking their eyeballs, or rasping scales off their body. Social parasites in hymenopterans are often closely related and are sometimes the sister taxa of the species they parasitize, which is referred to as "Emery's rule" (Bourke and Franks 1991; Savolainen and Vepsalainen 2003). These observations certainly indicate the possibility that adaptive radiations create additional opportunity as they unfold. Although Schluter

---

[5] The same is not true of introduced predators, which are responsible for scores of species extinctions (Davis 2003; Sax et al. 2007).

(2000) suggested this idea a decade ago,[6] no empirical work has addressed this question; however, theoretical literature on the evolution of food webs is beginning to develop (Ingram et al. 2009).

Radiations may also be self-propagating when two clades co-radiate. Just as a clade may continually create its own ecological opportunity as it radiates, two clades may reciprocally generate opportunity for each other as they coevolve. Such coevolutionary radiations could take many forms. For example, the radiation of one group, perhaps driven by interspecific competition or predation, may create new ecological opportunities for a second clade that utilizes members of the first as a resource. In turn, radiation of the first group may continue if a species evolves some adaptation that frees it from attack by the second group, allowing it to utilize resources that were previously inaccessible. This last scenario is the basis of Ehrlich and Raven's (1964) famous escape and radiation theory of plant–herbivore coevolution, and plays an important part in Vermeij's (1987) theory of "evolution and escalation" between predators and prey. In other cases, co-radiations may simply result if species in one clade are each specialized to use a single member of a second clade; in this case, adaptive radiation in the host clade may be mirrored by radiation in their specialists. The recent divergence of the parasitic wasp, *Diachasma alloeum*, in response to divergence of its host, the apple maggot fly, *Rhagoletis pomonella*, illustrates how such matched divergence may unfold (Forbes et al. 2009). A similar process may have contributed to the extraordinary parallel radiations of figs and fig wasps as well as of *Glochidion* trees and *Epicephala* moths (Herre et al. 1996; Weiblen and Bush 2002; Kato et al. 2003).

As with the role of mutualism in adaptive radiation, the processes driving coevolutionary radiations are the same as in other radiations, but the synergistic interactions among co-radiating clades are distinctive. Such coevolutionary radiations may be particularly important in plant–herbivore systems (Farrell and Mitter 1994, 1998; Roderick and Percy 2008; Winkler and Mitter 2008; see Berenbaum and Schuler, Chapter 11).

Adaptive radiations may create their own opportunity in a second way: members of a radiating clade may alter their environment, creating ecological opportunities that did not previously exist. The concept of "ecosystem engineering" refers to the role that organisms play in altering their physical environment (Jones et al. 1994, 1997); ecosystem engineers, such as corals or rainforest trees, change the physical environment in ways that allow the existence of new species, potentially even leading to the evolution of ecological types that otherwise would not exist (Erwin 2008). Although several studies have documented that ecosystem engineering in one clade may trigger adaptive radiation in another clade (e.g., corals and tetraodontiform

[6] And Wilson in 1992 proposed: "what I like to call the test of a complete adaptive radiation: the existence of a species specialized to feed on other members of its own group, other products of the same adaptive radiation" (Wilson 1992: 118).

fishes; Alfaro et al. 2007), few examples from nature document members of an adaptive radiation altering the environment in such a way to create ecological opportunities that have then been utilized by other members of their own radiation. One possible example is the evolution of lobeliad plants in Hawaii. Among the first plant clades to arrive and radiate in this archipelago, lobeliads diversified to fill ecological roles ranging from canopy trees to shrubs, epiphytes, and vines—one-eighth of the Hawaiian flora in total (Givnish et al. 2009). Given their early arrival and the extent of their ecological diversification, it is plausible that some lobeliads (e.g., shade-dependent shrubs) radiated in microhabitats created by other members of the radiation (e.g., woody trees), a hypothesis that could be tested with further phylogenetic analyses.

Laboratory studies of microbial adaptive radiation also support a role for ecosystem engineering driving diversification. Several studies have shown that the waste product of an ancestral microbial species created a food source subsequently used by a second type that evolved from its ancestor (Kassen et al. 2009).

Does "Ecological Opportunity" Have More than Heuristic Value?

The conclusion of the preceding discussion is that ecological opportunity is usually necessary for adaptive radiation, but whether we can predict *a priori* if a clade will radiate is not clear for two reasons. First, ecological opportunity, though usually necessary, may not be sufficient for radiation. Second, ecological opportunity may be extremely difficult to define objectively *a priori* and, as a result, the concept may have limited utility for predicting when clades would be expected to radiate adaptively, as opposed to explaining retrospectively why they did so.

Many clades fail to radiate despite apparently abundant ecological opportunity. For example, the Galápagos and Hawaii are famous for their bird radiations (finches and honeycreepers, respectively), but many other bird lineages on these islands have failed to produce adaptive radiations, including mockingbirds on the Galápagos (Arbogast et al. 2006) and thrushes on Hawaii (Lovette et al. 2002). The same is true of many other island taxa. Radiation may not occur in such circumstances for several reasons:

1. The perception of ecological opportunity may be mistaken.
2. Inability to access or utilize resources (e.g., from the lack of genetic variation or phenotypic plasticity that could produce phenotypes capable of taking advantage of novel available resources).
3. Lack of speciation: if, for some reason, speciation cannot occur (as discussed subsequently), then adaptive radiation cannot result.
4. Inability to diversify ecologically: even when speciation can occur, if the resulting descendant species are unable to diverge phenotypically to specialize on different resources, then a clade may not be able to radiate adaptively. Lack of such evolvability (Liem 1974; Vermeij

1974; Cheverud 1996; Wagner and Altenberg 1996; Gerhart and Kirschner 1998; Rutherford and Lindquist 1998; see G. Wagner, Chapter 8) may occur for a variety of reasons, often referred to as "evolutionary constraints" (see Wray, Chapter 9).

The first hypothesis that one might pose about a clade that has failed to radiate could invoke lack of ecological opportunity, but it is not clear how one would test this idea.[7] A way of doing so might involve estimating selection on an adaptive landscape (Fear and Price 1998; Schluter 2000; Arnold et al. 2001). The existence of unutilized adaptive peaks might suggest that a species had the opportunity to diversify and occupy those peaks. Of course, the existence of multiple adaptive peaks on a landscape does not guarantee that selection would push a clade to diversify to produce species occupying all of these peaks: speciation must occur, and the landscape itself will change when other species are present. Research of this type has rarely been conducted, the most thorough being studies on Darwin's finches (Schluter and Grant 1984; Schluter 2000; see also Case 1979). Development of these sorts of ideas is needed to make ecological opportunity a fully operational and predictive concept.

As a result, currently ecological opportunity is only recognizable after the fact, as a plausible and often surely correct explanation for why a clade radiated.[8] As such, the concept may have great heuristic value in understanding what causes adaptive radiation, but it may have little operational value in the absence of a radiation to predict whether radiation could occur.

Adaptive Divergence, Speciation, and Adaptive Radiation

Adaptive radiation has two components: proliferation of species (speciation) and divergence of species into different ecological niches (Losos 2009; see Harrison, Chapter 13). Two important questions concern: (1) whether any process other than natural selection could produce adaptive divergence and (2) whether speciation and adaptation are causally connected.

Divergence of species to utilize different aspects of the environment could occur in two ways, either with genetic drift playing a large role or by divergent natural selection. Although drift by itself would not be expected to lead

[7] In this regard, one might suggest that the concept of ecological opportunity suffers from the same problems as the empty niche concept (Chase and Leibold 2003). Both empty niches and ecological opportunity are difficult to identify in the absence of species that fill or take advantage of them. Moreover, one might question whether resources are ever truly unutilized. How often, for example, is some type of food resource not eaten by any organism? If nothing else, they are a resource for decomposers, which is the reason for the wording of the definitions of ecological opportunity presented earlier.

[8] Ecological opportunity is also an explanation for why some species exhibit exceptionally broad ecological and phenotypic diversity, which is arguably the first step in adaptive radiation (Parent and Crespi 2009).

to adaptive change, in an adaptive landscape, it is possible that it could move a population off one adaptive peak and into the domain of another (the basis of Wright's 1932 famous shifting balance theory). Alternatively, drift could move a population along an adaptive ridge, from one high point in an adaptive landscape to another equally high point, assuming that the two points are connected by a ridge of equally high fitness (Schluter 2000). While these scenarios could ultimately result in the evolution of a suite of adaptively differentiated species, they have received relatively little empirical support. Rather, the standard and widely accepted view is that adaptive radiation is driven by divergent natural selection, in which species diverge as they adapt to use different parts of the environment. Most theories of adaptive radiation assume that a trade-off exists, such that enhanced adaptation to use one part of the environment comes with a concomitant cost of decreased adaptation to another part of the environment. Evidence for such trade-offs is strongly implied by work on polymorphisms and local adaptation (Schluter 2000), although the specific traits involved are often unknown (see Agrawal et al., Chapter 10).

With regard to the second question, speciation and adaptive divergence could be related in a number of ways:

1. Allopatric speciation could occur with adaptive divergence.
2. Allopatric speciation could occur without adaptive divergence, followed by adaptive divergence when species secondarily establish sympatry.
3. Some degree of adaptive divergence and evolution of reproductive isolation could occur in allopatry, followed by enhanced adaptive divergence and completion of the speciation process in sympatry (if not completed in allopatry).
4. Speciation could occur in sympatry accompanied by adaptive divergence. With respect to adaptive radiation, adaptive divergence is usually invoked as an integral part of the speciation process, though in theory, sympatric speciation could occur in non-adaptive ways (e.g., by polyploidy), followed by adaptive divergence, which would occur as in the first scenario in this list.

Few workers have suggested that most or all of the adaptive divergence during adaptive radiation evolves in allopatry (option 1). Rather, sympatric divergence, either during or after speciation (options 2, 3, and 4), is the primary, though not exclusive, focus of theoretical and empirical discussion.

Sympatric speciation driven by disruptive selection is in a way the most biogeographically parsimonious explanation for the occurrence of a clade of co-occurring, ecologically differentiated species, because it does not require the invocation of one or more rounds of range contraction and expansion to permit allopatric speciation followed by current-day sympatry. Nonetheless, the possibility of sympatric speciation of this sort is highly controversial, and probably the majority of workers consider the prerequisites for it to occur to be very stringent and likely to be met by few organisms in most settings (Coyne and Orr 2004; Gavrilets 2004; Bolnick and Fitzpatrick 2007; see Harrison, Chapter 13). In a few examples, the case for adaptive radiation by sympatric speciation is strong. The most convincing is the occurrence of two clades of ecologically differentiated cichlid fishes, each one in different volcanic crater lakes in Cameroon (Schliewen et al. 1994). The monophyly of the clades (comprised of 9 and 11 species) makes *in situ* speciation far more plausible than the alternative of many colonization events followed by extinction of related forms outside the lakes (which would make the lake species monophyletic with respect to extant species). The lakes are small and homogeneous, so it is hard to imagine an allopatric phase in the speciation process, leading to the conclusion that sympatric speciation likely occurred.

Many clades of insects, some extremely species-rich, are composed of species adapted to specialize on different host plants (see Berenbaum and Schuler, Chapter 11). In many cases, multiple—sometimes many—clade members occur sympatrically. Some workers in this area consider the evolution of sympatric host-specialist species to be most readily explicable by sympatric speciation (Berlocher and Feder 2002; Drès and Mallet 2002; Abrahamson and Blair 2008). However, controversy over the theoretical likelihood of sympatric speciation also pertains to host race speciation (Coyne and Orr 2004; Gavrilets 2004; Bolnick and Fitzpatrick 2007). Futuyma (2008) argued that the empirical evidence in support of this view is not generally strong and suggests that host shifts in allopatry are a likely alternative possibility. Although the literature on these insects usually is not couched in terms of adaptive radiation (exceptions include Després and Cherif 2004; Price 2008; Roderick and Percy 2008), many host-specialist complexes surely are adaptive radiations, and if sympatric speciation is a common mode of speciation in these groups, then they may represent examples of adaptive radiation by sympatric speciation.

The extreme alternative to adaptive divergence during sympatric speciation is speciation in allopatry without adaptive ecological divergence. Subsequently, the new species become sympatric and as a result of divergent natural selection, adapt to different ecological niches (i.e., ecological character displacement). Speciation without adaptive divergence could occur if allopatric populations diverge and become reproductively isolated as a result of genetic drift or if sexual selection pressures in the two populations lead to the evolution of different mating preferences (Gittenberger 1991; Price 2008). One difficulty with these scenarios is that, if resources are limiting, ecologically undifferentiated species may compete when they become sympatric and thus, may not be able to coexist long enough for evolutionary divergence to occur (MacArthur and Levins 1967; Slatkin 1980; Gomulkiewicz and Holt 1995).

An intermediate possibility emphasizes the role that different selective environments may play in causing allopatric populations to diverge. Studies in both the laboratory and in nature clearly indicate that isolated populations experiencing different selective pressures are more likely to evolve reproductive isolation as an incidental result of adaptive differentiation (Rice and Hostert 1993; Funk et al. 2006; Funk and Nosil 2008) (Figure 15.6). Consequently, speciation is more likely to occur, or at least be initiated, by populations experiencing different selective pressures. Moreover, if such populations become sympatric, the evolved differences in ecology increase the possibility that the populations can coexist long enough for character displacement to occur, leading to greatly enhanced ecological differentiation that permits coexistence. This scenario corresponds to the archipelago model of adaptive radiation (Grant and Grant 2008a; Price 2008).

Because species proliferation is required for adaptive radiation to occur, taxa that are more likely to speciate may also be more likely to adaptively radiate. Thus, factors that predispose taxa to speciate, such as a mating system emphasizing female choice, may be linked to adaptive diversification (Schluter 2000; Rundell and Price 2009). In the same vein, the evolution of a trait that leads to enhanced speciation rates may be the trigger that promotes adaptive radiation in a group, if ecological opportunity is already present.

Overall, the role of speciation in adaptive radiation is poorly understood. In recent years, "ecological speciation"—the idea that divergent adaptation to different environments leads to the speciation—has attracted considerable attention (for insightful reviews, see Hendry 2009 and see Harrison, Chapter 13). There is no doubt that adaptation to different environments by allopatric populations enhances the likelihood that those populations will become reproductively isolated, and this model is a key stage in the standard model for adaptive radiation on islands. Whether or

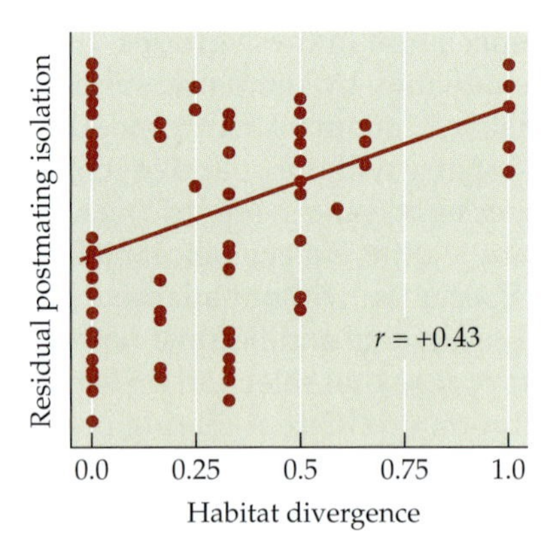

FIGURE 15.6 The Relationship between Divergent Adaptation and Reproductive Isolation The more dissimilar two species are in habitat use, the greater the degree of reproductive isolation that has evolved between them. Each point represents the degree of difference between two closely related species or populations. (Adapted from Funk et al. 2006.)

not ecological speciation in sympatry (i.e., sympatric speciation) commonly occurs and leads to adaptive radiation is an open question. Similarly, the role of nonadaptive modes of speciation, such as founder effect speciation, in adaptive radiation is also controversial (Coyne and Orr 2004; Gavrilets 2004; Futuyma 2005; Price 2008). More generally, the extent to which adaptive radiation is limited by the production of new species is unclear and thus the degree to which factors that promote speciation may be important indirect promoters of adaptation radiation remains uncertain.

Adaptation to Different Aspects of the Environment in the Absence of Interspecific Interactions

Regardless of how speciation occurs, what ecological mechanisms drive adaptive divergence? Divergent natural selection is the underlying cause, but the question is whether the driving force behind such selection is simply adaptation to different aspects of the environment or whether interspecific interactions change the selective landscape, producing divergent selection that would not occur in the absence of the interacting species (Schluter 2000).[9]

It is easy to envision how two allopatric populations experiencing different environments would evolve different adaptations. Moreover, because the environment and hence, the adaptive landscape is rarely identical in different places, it is possible that two adaptive peaks that are separated by an adaptive valley in one location may be connected by an adaptive ridge in another. Suppose a population on one island occupied the lower of two adaptive peaks, which were separated by an adaptive valley, so that the higher peak was not attainable. If members of that population colonized a second island on which the peaks were connected by an adaptive ridge, then the population would adapt to the higher peak. Subsequently, if members of the second population colonized the first island, they would occur on the higher peak, producing sympatry of the species on different peaks (Figure 15.7).

In theory, this sort of scenario, in which adaptive divergence occurs entirely in allopatry, could lead to adaptive radiation, producing species adapted to many different aspects of the environment. Most views of adaptive radiation do not take such an extreme view, though proponents of the role of interspecific interactions envision allopatric differentiation as the initial step in divergence, setting the stage for subsequent, much more substantial divergence in sympatry that is driven by sympatric interactions (i.e., character displacement).

An alternative view is that divergent selection in sympatry can split an initially homogeneous population into distinct, reproductively isolated and

---

[9] In this discussion, we consider the food an organism eats to be part of the environment. By interspecific interactions, we refer to interactions among species on the same trophic level or between the focal species and other species on higher trophic levels.

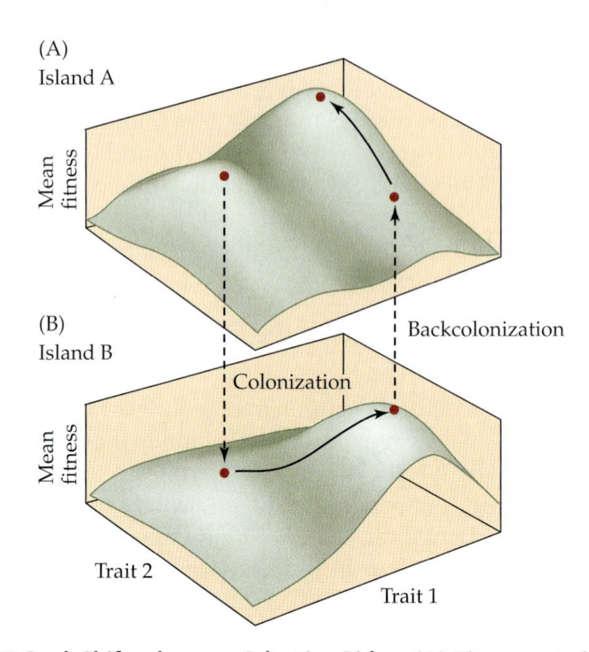

FIGURE 15.7 Peak Shifts along an Adaptive Ridge (A) The ancestral species occurs on the lower adaptive peak on Island A. Because an adaptive valley lies between the two peaks, natural selection cannot drive the population to the second peak. (B) However, on Island B, an adaptive ridge connects the ancestral position to a higher peak, and when individuals immigrate to Island B from Island A, the new population on Island B evolves a new morphology. Subsequently, the population on Island B evolves reproductive isolation. When individuals from Island B then recolonize Island A, they may evolve to the second peak on that island, leading to the coexistence of two species—one on each adaptive peak. (Adapted from Schluter 2000.)

adaptively differentiated populations. Some variants of this model envision intraspecific resource competition or interspecific interactions as the driving force, but many prominent proposals, particularly those concerning host–plant specialists (see following discussion), simply invoke adaptation to different aspects of the environment, requiring a tradeoff such that disruptive selection produces a bimodal distribution of phenotypes. As just discussed, the prerequisites for reproductive isolation to evolve in such a scenario are very strict, and controversy still exists on how likely it is to occur, even in host–plant specialized insects.

Adaptation to Different Aspects of the Environment as a Result of Interspecific Interactions

Interspecific Competition

Dating back to the seminal work by Simpson (1953), which focused primarily on vertebrates, a commonly held view is that adaptive radiation primarily results from interspecific competition for resources, leading to character displacement and adaptation to different resources (Schluter 2000; Grant and Grant 2008a). Repeated numerous times, this process leads to sympatric coexistence of species adapted to a variety of different ecological niches (i.e., an adaptive radiation).

Although character displacement was controversial in the 1970s and 1980s, a growing consensus exists that it is an important evolutionary phenomenon (Schluter 2000; Dayan and Simberloff 2005; Pfennig and Pfennig 2009). Experimental studies have confirmed that interspecific competition can lead to strong divergent selection (Schluter 1994; Schluter 2003), and the process of character displacement has been directly observed in the field in Darwin's finches (Grant and Grant 2006) (Figure 15.8) as well as in laboratory studies (e.g., Barrett and Bell 2006; Tyerman et al. 2008; Kassen 2009). It seems safe to conclude that interspecific competition-driven character displacement is a common means by which adaptive radiation occurs.

The question is, then, whether interactions other than competition can lead to adaptive radiation, and if so, whether they have been important in driving adaptive radiations throughout the history of life.

Predation, Parasitism, and Herbivory

In ecological terms, predation, parasitism, and herbivory are similar processes in that they refer to individuals of one species directly consuming members of another species (hence, we refer to all as "predators" in the following discussion). These processes could produce divergent natural selection pressures when species adapt to different predators or when species adapt in divergent ways to the same predator. Moreover, species initially preyed upon by the same predator may diverge so as not to share the same predator. Although predation may spur divergence and diversification, it also can hinder it, and the extent to which predation plays an important role in driving adaptive radiation remains poorly understood (Vamosi 2005; Langerhans 2006).

One can easily envision how species occurring in different places, with different predation regimes, would face selection to evolve different phenotypes. Several recent studies have demonstrated the importance of predation in driving such divergence. For example, both mosquitofish and damselflies exhibit differences in their behavior, habitat use, and morphology depending on which predators are present in the lakes in which they occur (McPeek et al. 1996; Stoks et al. 2003; Langerhans et al. 2007).

In sympatry, predation-driven selection may be divergent for several reasons. First, multiple ways may exist to avoid predation. For example, *Timema* walking sticks that use different host plants have diverged to enhance crypsis on the differently colored plants; manipulative experimental studies demonstrate that selection is divergent when predators are present (Nosil and Crespi 2006) (Figure 15.9). Among garter snakes, disruptive selection favors striped individuals that rapidly flee from predators and

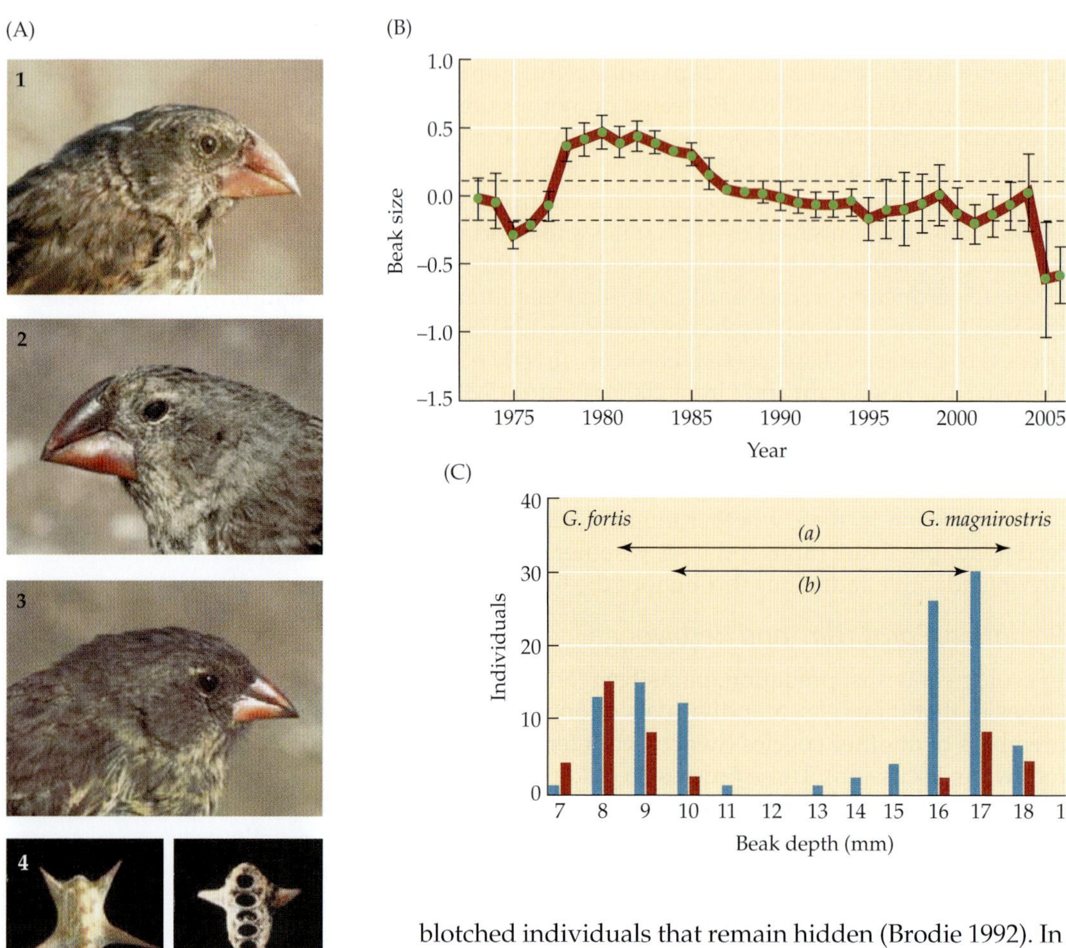

blotched individuals that remain hidden (Brodie 1992). In theory, species evolving different anti-predator adaptations might not differ in any other way with regard to resource or habitat use, in which case ecological opportunity might not be involved in adaptive divergence. However, most documented examples include correlated shifts in other ecological and behavioral aspects; the evolution of body armor, for example, has consequences for locomotion, which in turn may affect where and how an animal can forage (Bergstrom 2002; Losos et al. 2002). Consequently, in this predator-driven scenario for adaptive radiation, ecological opportunity would still be required; multiple distinct habitats or resources to which different prey species could adapt would be necessary, and these different niches could not already be preempted by other species.

Selection may also favor sympatric species to diverge in habitat use so as to avoid being preyed upon by the same predator. If prey species are preyed upon by the same predator, then under some circumstances, increased population size of one of the prey species would lead to increased

FIGURE 15.8 Character Displacement in Darwin's Finches (A) Daphne Major, in the Galápagos Islands, harbors both the large ground finch, *Geospiza magnirostris* (A2), and the medium ground finch, *Geospiza fortis* (A1, A3), the latter of which exhibits substantial variability in beak shape. Only large-beaked birds can eat large seeds, such as those from *Tribulus cistoides* (A4). (B) During the drought of 1977, when only the medium ground finch occurred on the island, small seeds were rapidly consumed and only large seeds remained. Selection strongly favored large-beaked medium ground finches (as indicated by the calculation of selection gradients, which are not shown here), and the population evolved larger beak size. Another drought occurred in 2003 and 2004; however, in the intervening years, the large ground finch had colonized Daphne Major. During this drought, the large ground finches monopolized the larger seeds. Mortality was very high in both species, and in the medium ground finch, smaller-beaked birds that could eat the few remaining small seeds were favored, and the population evolved smaller beak size, the opposite of what occurred in the absence of the large-beaked ground finch. Beak size units represent scores on the first axis of a principal components analysis of six bill measurements. (C) Thus, in the 2003–2004 drought, selection favored smaller-beaked birds in the medium ground finch and larger-beaked birds in the large ground finch, and the phenotypic distributions before the drought (indicated by blue bars) and afterwards (indicated by red bars) can be compared; as a result, differences in beak size between the two species were greater after the drought (arrow a) than before (arrow b), making this a classic example of character displacement. (A, photos from Grant and Grant 2006; B, adapted from Grant and Grant 2006; C, adapted from Grant and Grant 2008b.)

population size of the predator; thus, the population size of the other prey species would decrease, as they are preyed upon by the greater number of predators. The result is that a negative relationship would exist between the population sizes of the prey species, just as would occur through interspecific competition (Holt 1977). Assuming the predator species is not able to function equally successfully in all parts of the environment, prey species may diverge to use different resources or habitats, if they are available, and thus no longer share predators. Subsequently, prey species would adapt to the different habitats or resources they were utilizing, producing the same outcome as competition-driven character displacement: an adaptive radiation driven by "competition for enemy free space" (Jeffries and Lawton 1984) or "apparent competition" (Holt 1977).

In theory, adaptive radiation also could result from predation-driven divergent selection by the means just outlined. However, we are aware of few purported cases. The diversity of some tropical butterfly clades, involving multiple mimicry complexes and a suite of other ecological and behavioral differences, might be one example (Elias et al. 2008).

Processes Driving Radiation: Conclusions

The role of interspecific competition in driving evolutionary radiation is well established and likely to be of paramount importance. Other ecological processes may be important, either directly (e.g., predation, herbivory,

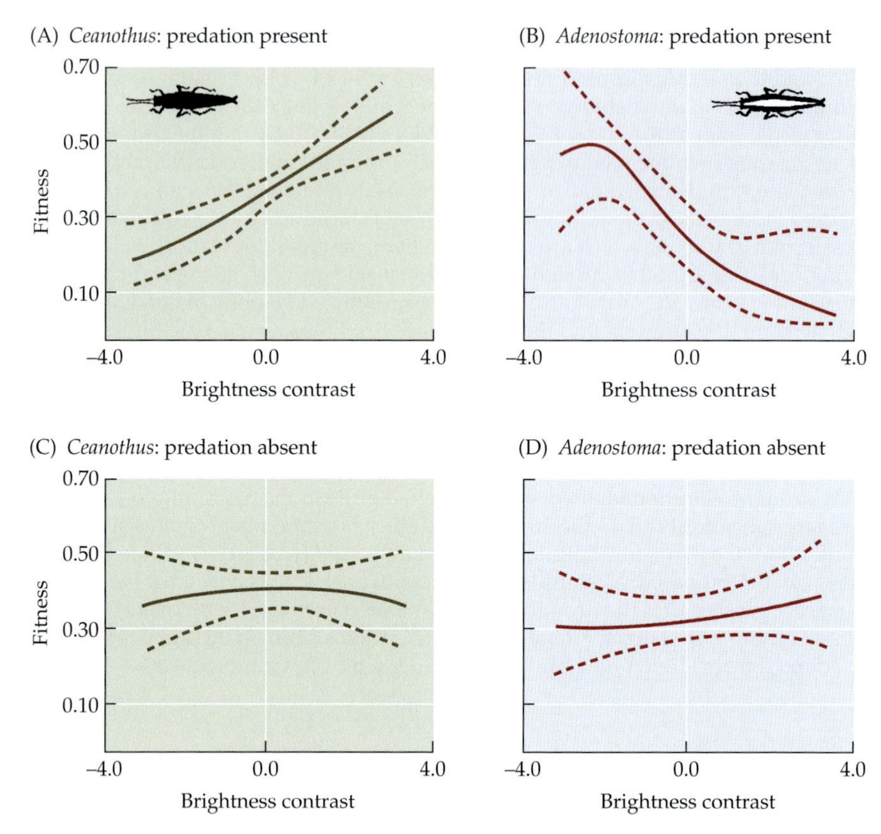

FIGURE 15.9 Selection for Background Matching in the Walking Stick, *Timema cristinae* Experimental studies of selection reveal that in the presence of avian predators, dull stripes, and bright bodies ("brightness contrast" is body brightness minus stripe brightness) are favored in the *Ceanothus* ecotype (A) and bright stripes and dull bodies in the *Adenostoma* ecotype (B). In the absence of predators, selection does not differ between insects on the two bushes (C and D). (Adapted from Nosil and Crespi 2006.)

and parasitism) or indirectly (e.g., mutualism and coevolution), by creating the opportunity for other processes to operate. However, empirical evidence demonstrating the role of other ecological processes is sparse. To some extent, this paucity of case studies, may be a result of the historical focus on competition as a driver of character displacement, but attention among both theorists and ecologists to other mechanisms has occurred for long enough now that one might expect more examples to have been documented. The lack of such examples would seem to be indicative of their limited importance relative to that of direct competition. Certainly, the evolution of mutualisms and coevolution are events that can trigger adaptive radiation. How frequently they occur is unclear.

Laboratory experiments on microbial evolution support these conclusions (see Dykhuizen, Commentary 2). Kassen (2009) reviewed the rapidly growing literature in this field and found overwhelming support for the role of interspecific competition as the primary driver of adaptive radiation. The addition of predators as an experimental treatment occasionally enhanced adaptive radiation but only when the variety of resources was limited. In situations with a wide variety of resources, the presence of predators slowed, rather than enhanced, the rate of adaptive diversification (Meyer and Kassen 2007; see also Benmayor et al. 2008). Other possible triggers for adaptive radiation, such as ecosystem engineering or the evolution of mutualisms, also rarely are important (Kassen 2009). The correspondence between the results from experimental laboratory studies and the empirical literature from nature suggests that interspecific competition is the primary driver of adaptive radiation.

Conclusions and Future Directions

Many of what we consider the classic ideas about adaptive radiation are well supported. In particular, ecological opportunity usually is the key to adaptive radiation, and interspecific competition often is the driving force behind it. Nonetheless, these are generalizations, and further work is needed to understand the relative frequency and significance of alternatives. In many cases, other possibilities, such as the role of predation in driving adaptive radiation or the extent to which radiations create their own ecological opportunity, have been little explored. Whether further work will alter the bigger picture of our understanding of adaptive radiation remains to be seen.

We also conclude that while ecological opportunity is typically necessary, it is not sufficient for adaptive radiation, as many clades seemingly in the presence of opportunity fail to radiate. Several factors likely contribute, but at present, these are too poorly understood to predict with any certainty whether adaptive radiation will occur in the presence of ecological opportunity. Why, for example, did Darwin's finches, but not mockingbirds, radiate in the Galápagos? Experimental diversification studies have come to the forefront in filling this gap; the combination of replication, strict environmental control, and fine-scale genetic characterization of lineages makes for a promising approach to dissect the nature of historical contingency in adaptive radiation (Lenski and Travisano 1994; Fukami et al. 2007; Blount et al. 2008).

While we have offered a review of the current knowledge pertaining to adaptive radiation, numerous basic questions about adaptive radiation remain unanswered, and recent conceptual and technological advances promise to lead adaptive radiation research into exciting new directions. Toward this end, we suggest that new approaches that take advantage of both conceptual and technological advances may be instrumental in moving forward understanding of adaptive radiation in years to come.

Our review has focused, for the most part, on the role of external environmental factors in shaping adaptive radiation. However, over the last several decades, evolutionary biologists have debated the role that internal constraints—manifested in limitations and directionality in the availability of phenotypic variation upon which selection can operate—play in limiting and directing evolutionary change (Gould 2002).

The role of such constraints on adaptive radiation is now being addressed in two distinctive ways. On one hand, a number of researchers have suggested that interspecific hybridization can provide enhanced variation that may be critical in allowing extensive adaptive diversification. On the other hand, the burgeoning field of genomics is now at last permitting evolutionary biologists to truly understand the relationship between genetic change and phenotypic response.

Hybridization

Several authors have recently suggested that hybridization among closely related species may play an important role in generating adaptive diversity during the early stages of adaptive radiation (Rieseberg et al. 1999; Seehausen 2004; Grant and Grant 2008a,b; Mallet 2009). Increasingly sophisticated methods are being developed to distinguish past gene flow from incomplete lineage sorting, permitting detection of historical hybridization in a phylogenetic framework (Hey and Nielsen 2004; Hey et al. 2004; Kubatko 2009), which may facilitate investigation of the contribution of hybridization to adaptive diversification. Nonetheless, directly measuring the effect of hybridization on divergence in the wild is often difficult because of the low frequency of hybridization events and the infeasibility of tracking hybrid lineages over many generations, although Grant and Grant (2008b, 2009) provide remarkable examples in Darwin's finches. However, future studies might assess the role of hybridization in adaptive radiation by testing whether radiations that exhibit historical signatures of hybridization exhibit greater adaptive diversity than radiations in which there is little evidence of hybridization.

Genomics

Thanks to the ever-decreasing cost of genome sequencing, the ability to examine the genomes of multiple members of an adaptive radiation will soon be readily available. With such information, researchers will be able to examine the extent to which genetic architecture constrains and directs the pathways by which adaptive diversification has occurred. Combined with studies of natural selection in the field, we soon will have the ability to fully integrate genetics and the study of natural selection to understand how and why adaptive radiation has occurred. Experimental studies of this sort have already been conducted (e.g., microbial laboratory experiments: Bantinaki et al. 2007 and Spencer et al. 2007; sticklebacks: Barrett et al. 2008) and no doubt will soon become commonplace. The next 10–20 years should prove extremely enlightening as the integration of genomic and selection studies, in a phylogenetic context, ushers in a golden age for the study of adaptive radiation.

The Impact of Adaptive Radiation on Communities and Ecosystems

Finally, virtually all research on adaptive radiation has investigated the influence of ecological factors on adaptive diversification, but until recently, few have asked how the process of adaptive radiation may alter the structure of communities and ecosystems. Adaptive radiation typically involves a dramatic change in the ecological diversity of a lineage, usually within a local setting. Diversification may influence community interactions, alter food webs, and ultimately affect nutrient and energy flow in ecosystems, and radiations have been implicated in major ecological changes in Earth's history (e.g., the rise and proliferation of autotrophs changing the atmospheric oxygen concentration roughly 2 billion years ago). However, only recently has research specifically sought to quantify the influence of evolutionary diversification on community structure and ecosystem function (Loueille and Loreau 2005, 2006; Harmon et al. 2009; Ingram et al. 2009). Continued research in this direction may result in a richer understanding of the complex and interactive relationship between ecological and evolutionary diversity.

Acknowledgments

This chapter benefited from discussions with and critiques by: Adam Algar, Mike Bell, Joel Cracraft, Doug Futuyma, Luke Harmon, Mark McPeek, Trevor Price, Bob Ricklefs, John Thompson, and Peter Wainwright.

Literature Cited

Abrahamson, W. G. and C. P. Blair. 2008. Sequential radiation through host-race formation: Herbivore diversity leads to diversity in natural enemies. In K. J. Tilmon (ed.), *Specialization, Speciation, and Radiation: The Evolutionary Biology of Herbivorous Insects*, pp. 188–202. University of California Press, Berkeley.

Adams, D. C., C. M. Berns, K. H. Kozak, and 1 other. 2009. Are rates of species diversification correlated with rates of morphological evolution? *Proc. Roy. Soc. Lond. B* 276: 2729–2738.

Agrawal, A. A., M. Fishbein, R. Halitschke, and 3 others. 2009. Evidence for adaptive radiation from a phylogenetic study of plant defenses. *Proc. Natl. Acad. Sci. USA* 106: 18067–18072.

Alfaro, M. E., F. Santini, and C. D. Brock. 2007. Do reefs drive diversification in marine teleosts? Evidence from the pufferfish and their allies (Order Tetraodontiformes). *Evolution* 61: 2104–2126.

Alfaro, M. E., F. Santini, C. Brock, and 5 others. 2009. Nine exceptional radiations plus high turnover explain species diversity in jawed vertebrates. *Proc. Natl. Acad. Sci. USA* 106: 13410–13414.

Arbogast, B. S., S. V. Drovetski, R. L. Curry, and 6 others. 2006. The origin and diversification of Galápagos mockingbirds. *Evolution* 60: 370–382.

Arnold, S. J., M. E. Pfrender, and A. G. Jones. 2001. The adaptive landscape as a conceptual bridge between micro- and macroevolution. *Genetica* 112/113: 9–32.

Bakker, R. T. 1968. The superiority of dinosaurs. *Discovery* 3: 11–22.

Bantinaki, E., R. Kassen, C. G. Knight, and 3 others. 2007. Adaptive divergence in experimental populations of *Pseudomonas fluorescens*. III. Mutational origins of wrinkly spreader diversity. *Genetics* 176: 441–453.

Barker, K. F., A. Cibois, P. Schikler, and 2 others. 2004. Phylogeny and diversification of the largest avian radiation. *Proc. Natl. Acad. Sci. USA* 101: 1040–1045.

Barraclough, T. G. and S. Nee. 2001. Phylogenetics and speciation. *Trends Ecol. Evol.* 16: 391–399.

Barrett, R. D. H. and G. Bell. 2006. The dynamics of diversification in evolving *Pseudomonas* populations. *Evolution* 60: 484–490.

Barrett, R. D. H., S. M. Rogers, and D. Schluter. 2008. Natural selection on a major armor gene in threespine stickleback. *Science* 322: 255–257.

Baum, D. A. and A. Larson. 1991. Adaptation reviewed: A phylogenetic methodology for studying character macroevolution. *Syst. Zool.* 40: 1–18.

Benkman, C. W. 1991. Predation, seed size partitioning and the evolution of body size in seed-eating finches. *Evol. Ecol.* 5: 118–127.

Benmayor, R., A. Buckling, M. B. Bonsall, and 2 others. 2008. The interactive effects of parasites, disturbance, and productivity on experimental adaptive radiations. *Evolution* 62: 467–477.

Benton, M. J. 1996. On the nonprevalence of competitive replacement in the evolution of tetrapods. In D. Jablonski, D. H. Erwin, and J. Lipps (eds.), *Evolutionary Paleobiology*, pp. 185–210. University of Chicago Press, Chicago.

Benton, M. J. 2009. The red queen and the court jester: Species diversity and the role of biotic and abiotic factors through time. *Science* 323: 728–732.

Bergstrom, C. A. 2002. Fast-start swimming performance and reduction in lateral plate number in threespine stickleback. *Can. J. Zool.* 80: 207–213.

Berlocher, S. H. and J. L. Feder. 2002. Sympatric speciation in phytophagous insects: Moving beyond controversy? *Ann. Rev. Entom.* 47: 773–815.

Blount, Z. D., C. Z. Borland, and R. E. Lenski. 2008. Historical contingency and the evolution of a key innovation in an experimental population of *Escherichia coli*. *Proc. Natl. Acad. Sci. USA* 105: 7899–7906.

Bokma, F. 2009. Problems detecting density-dependent diversification on phylogenies. *Proc. Roy. Soc. Lond. B* 276: 993–994.

Bolnick, D. I. and B. M. Fitzpatrick. 2007. Sympatric speciation: Models and empirical evidence. *Ann. Rev. Ecol. Evol. Syst.* 38: 459–487.

Bourke, A. F. G. and N. R. Franks. 1991. Alternative adaptations, sympatric speciation and the evolution of parasitic, inquiline ants. *Biol. J. Linn. Soc.* 43: 157–178.

Branch, W. R. 1998. *Field Guide to Snakes and Other Reptiles of Southern Africa*. Ralph Curtis Books, Sanibel Island, FL.

Brockhurst, M. A., N. Colegrave, D. J. Hodgson, and 1 other. 2007. Niche occupation limits adaptive radiation in experimental microcosms. *PLoS One* 2: e193.

Brodie, E. D. III. 1992. Correlational selection for color pattern and antipredator behavior in the garter snake *Thamnophis ordinoides*. *Evolution* 46: 1284–1298.

Brusatte, S. L., M. J. Benton, M. Ruta, and 1 other. 2008. Superiority, competition, and opportunism in the evolutionary radiation of dinosaurs. *Science* 321: 1485–1488.

Carlquist, S. J. 1974. *Island Biology*. Columbia University Press, New York.
Case, T. J. 1979. Character displacement and coevolution in some *Cnemidophorus* lizards. *Forts. Zool.* 25: 235–282.
Chase, J. M. 2007. Drought mediates the importance of stochastic community assembly. *Proc. Natl. Acad. Sci. USA* 104: 17430–17434.
Chase, J. M. and M. A. Leibold. 2003. *Ecological Niches: Linking Classical and Contemporary Approaches*. University of Chicago Press, Chicago.
Cheverud, J. M. 1996. Developmental integration and the evolution of pleiotropy. *Am. Zool.* 36: 44–50.
Ciampaglio, C. N. 2002. Determining the role that ecological and developmental constraints play in controlling disparity: Examples from the crinoid and blastozoan fossil record. *Evol. Dev.* 4: 170–188.
Ciampaglio, C. N., M. Kemp, and D. W. McShea. 2001. Detecting changes in morphospace occupation patterns in the fossil record: Characterization and analysis of measures of disparity. *Paleobiology* 27: 695–715.
Collar, D. C., T. J. Near, and P. C. Wainwright. 2005. Comparative analysis of morphological diversity: Does disparity accumulate at the same rate in two lineages of centrarchid fishes? *Evolution* 59: 1783–1794.
Coyne, J. A., and H. A. Orr. 2004. *Speciation*. Sinauer Associates, Sunderland, MA.
Cracraft, J. 1990. The origin of evolutionary novelties: Pattern and process at different hierarchical levels. In M. Nitecki (ed.), *Evolutionary Innovations*, pp. 21–44. University of Chicago Press, Chicago.
Darwin, C. 1845. *Journal of Researches into the Natural History and Geology of the Countries Visited During the Voyage of H.M.S. Beagle Round the World, under the Command of Capt. FitzRoy, R.N*, 2nd ed. John Murray, London.
Davis, M. A. 2003. Biotic globalization: Does competition from introduced species threaten biodiversity? *BioScience* 53: 481–489.
Dayan, T. and D. Simberloff. 2005. Ecological and community-wide character displacement: The next generation. *Ecol. Lett.* 8: 875–894.
de Queiroz, A. 2002. Contingent predictability in evolution: Key traits and diversification. *Syst. Biol.* 51: 917–929.
Després, L. and M. Cherif. 2004. The role of competition in adaptive radiation: A field study on sequentially ovipositing host-specific seed predators. *J. Anim. Ecol.* 73: 109–116.
Donoghue, M. J. 2005. Key innovations, convergence, and success: Macroevolutionary lessons from plant phylogeny. *Paleobiology* 31(suppl.): 77–93.
Drès, M. and J. Mallet. 2002. Host races in plant-feeding insects and their importance in sympatric speciation. *Phil. Trans. Roy. Soc. Lond. B* 357: 471–492.
Ehrlich, P. R. and P. H. Raven. 1964. Butterflies and plants: A study in coevolution. *Evolution* 18: 586–608.
Elias, D. O., M. M. Kasumovic, D. Punzalan, and 2 others. 2008. Assessment during aggressive contests between male jumping spiders. *Anim. Behav.* 76: 901–910.
Erwin, D. H. 2001. Lessons from the past: Biotic recoveries from mass extinctions. *Proc. Natl. Acad. Sci. USA* 98: 5399–5403.
Erwin, D. H. 2007. Disparity: Morphological pattern and developmental context. *Palaeontology* 50: 57–73.
Erwin, D. H. 2008. Macroevolution of ecosystem engineering, niche construction and diversity. *Trends Ecol. Evol.* 23: 304–310.
Farrell, B. D. 1998. "Inordinate fondness" explained: Why are there so many beetles? *Science* 281: 553–557.

Farrell, B. D. and C. Mitter. 1994. Adaptive radiation in insects and plants: Time and opportunity. *Am. Zool.* 34: 57–69.

Farrell, B. D. and C. Mitter. 1998. The timing of insect/plant diversification: Might *Tetraopes* (Coleoptera: Cerambycidae) and *Asclepias* (Asclepiadaceae) have coevolved? *Biol. J. Linn. Soc.* 63: 553–577.

Fear, K. K. and T. Price. 1998. The adaptive surface in ecology. *Oikos* 82: 440–448.

Foote, M. 1993. Discordance and concordance between morphological and taxonomic diversity. *Paleobiology* 19: 185–204.

Foote, M. 1996. Ecological controls on the evolutionary recovery of post-Paleozoic crinoids. *Science* 274: 1492–1495.

Foote, M. 1997. The evolution of morphological diversity. *Ann. Rev. Ecol. Syst.* 28: 129–152.

Foote, M. 1999. Morphological diversity in the evolutionary radiation of Paleozoic and post-Paleozoic crinoids. *Paleobiology* 25: 1–115.

Forbes, A. A., T. H. Q. Powell, L. L. Stelinski, and 2 others. 2009. Sequential sympatric speciation across trophic levels. *Science* 323: 776–779.

Freckleton, R. P. and P. H. Harvey. 2006. Detecting non-Brownian trait evolution in adaptive radiations. *PLoS Biol.* 4: 2104–2111.

Friedman, M. 2010. Explosive morphological diversification of spiny-finned teleost fishes in the aftermath of the end-Cretaceous extinction. *Proc. Roy. Soc. Lond. B* 277: 1675–1683.

Fukami, T., H. J. E. Beaumont, X. X. Zhang, and 1 other. 2007. Immigration history controls diversification in experimental adaptive radiation. *Nature* 446: 436–439.

Funk, D. J. and P. Nosil. 2008. Comparative analyses of ecological speciation. In K. J. Tilmon (ed.), *Specialization, Speciation, and Radiation: The Evolutionary Biology of Herbivorous Insects*, pp. 117–135. University of California Press, Berkeley.

Funk, D. J., P. Nosil, and W. J. Etges. 2006. Ecological divergence exhibits consistently positive associations with reproductive isolation across disparate taxa. *Proc. Natl. Acad. Sci. USA* 103: 3209–3213.

Fürsich, F. T. and D. Jablonski. 1984. Late Triassic naticid drillholes: Carnivorous gastropods gain a major adaptation but fail to radiate. *Science* 224: 78–80.

Futuyma, D. J. 1998. *Evolutionary Biology*. Sinauer Associates, Sunderland, MA.

Futuyma, D. J. 2005. Progress on the origin of species. *PLoS Biol.* 3: 197–199.

Futuyma, D. J. 2008. Sympatric speciation: Norm or exception? In K. J. Tilmon (ed.), *Specialization, Speciation, and Radiation: The Evolutionary Biology of Herbivorous Insects*, pp. 136–148. University of California Press, Berkeley.

Gavrilets, S. 2004. *Fitness Landscapes and the Origin of Species*. Princeton University Press, Princeton.

Gavrilets, S. and A. Vose. 2009. Dynamic patterns of adaptive radiation: Evolution of mating preferences. In R. K. Butlin, J. R. Bridle, and D. Schluter (eds.), *Speciation and Patterns of Diversity*, pp. 102–126. Cambridge University Press, Cambridge, UK.

Gerhart, M. and J. Kirschner. 1998. *Cells, Embryos, and Evolution: Toward a Cellular and Developmental Understanding of Phenotypic Variation and Evolutionary Adaptability*. Blackwell, Oxford.

Gittenberger, E. 1991. What about nonadaptive radiation? *Biol. J. Linn. Soc.* 43: 263–272.

Givnish, T. J. 1997. Adaptive radiation and molecular systematics: Issues and approaches. In T. J. Givnish and K. J. Sytsma (eds.), *Molecular Evolution and Adaptive Radiation*, pp. 1–54. Cambridge University Press, Cambridge, UK.

Givnish, T. J., K. C. Millam, A. R. Mast, and 7 others. 2009. Origin, adaptive radiation and diversification of the Hawaiian lobeliads (Asterales: Campanulaceae). *Proc. Roy. Soc. Lond. B* 276: 407–416.

Glor, R. E. 2010. Phylogenetic approaches to the study of adaptive radiation. *Ann. Rev. Ecol. Evol. Syst.* in revision.

Gomulkiewicz, R. and R. D. Holt. 1995. When does evolution by natural selection prevent extinction. *Evolution* 49: 201–207.

Gould, S. J. 2002. *The Structure of Evolutionary Theory*. Harvard University Press, Cambridge, MA.

Grant, B. R. and P. R. Grant. 2008b. Fission and fusion of Darwin's finch populations. *Phil. Trans. Roy. Soc. Lond. B* 363: 2821–2829.

Grant, P. R. and B. R. Grant. 2006. Evolution of character displacement in Darwin's finches. *Science* 313: 224–226.

Grant, P. R. and B. R. Grant. 2008a. *How and Why Species Multiply: The Radiation of Darwin's Finches*. Princeton University Press, Princeton.

Grant, P. R. and B. R. Grant. 2009. The secondary contact phase of allopatric speciation in Darwin's finches. *Proc. Natl. Acad. Sci. USA* 106: 20141–20148.

Guyer, C. and J. B. Slowinski. 1993. Adaptive radiation and the topology of large phylogenies. *Evolution* 47: 253–263.

Harmon, L. J., J. B. Losos, T. J. Davies, and 16 others. 2010. Early bursts of body size and shape evolution are rare in comparative data. *Evolution*, in press.

Harmon, L. J., B. Matthews, S. Des Roches, and 3 others. 2009. Evolutionary diversification in stickleback affects ecosystem functioning. *Nature* 458: 1167–1170.

Harmon, L. J., J. Melville, A. Larson, and 1 other. 2008. The role of geography and ecological opportunity in the diversification of day geckos (*Phelsuma*). *Syst. Biol.* 57: 562–573.

Harvey, P. H. and A. Rambaut. 2000. Comparative analyses for adaptive radiations. *Phil. Trans. Roy. Soc. Lond. B* 355: 1599–1605.

Heard, S. B. and D. L. Hauser. 1995. Key evolutionary innovations and their ecological mechanisms. *Hist. Biol.* 10: 151–173.

Hendry, A. P. 2009. Ecological speciation! Or the lack thereof? *Can. J. Fish. Aquat. Sci.* 66: 1383–1398.

Herre, E. A., C. A. Machado, E. Bermingham, and 5 others. 1996. Molecular phylogenies of figs and their pollinator wasps. *J. Biogeog.* 23: 521–530.

Hey, J. and R. Nielsen. 2004. Multilocus methods for estimating population sizes, migration rates and divergence time, with applications to the divergence of *Drosophila pseudoobscura* and *D. persimilis*. *Genetics* 167: 747–760.

Hey, J., Y. J. Won, A. Sivasundar, and 2 others. 2004. Using nuclear haplotypes with microsatellites to study gene flow between recently separated cichlid species. *Mol. Ecol.* 13: 909–919.

Hodges, S. A. and M. L. Arnold. 1995. Spurring plant diversification: Are floral nectar spurs a key innovation? *Proc. Roy. Soc. Lond. B* 262: 343–348.

Holt, R. D. 1977. Predation, apparent competition and the structure of prey communities. *Theor. Pop. Biol.* 12: 197–229.

Hooghiemstra, H., V. M. Wijninga, and A. M. Cleef. 2006. The paleobotanical record of Colombia: Implications for biogeography and biodiversity. *An. Mo. Bot. Gard.* 93: 297–324.

Hughes, C. and R. Eastwood. 2006. Island radiation on a continental scale: Exceptional rates of plant diversification after uplift of the Andes. *Proc. Natl. Acad. Sci. USA* 103: 10334–10339.

Hunter, J. P. 1998. Key innovations and the ecology of macroevolution. *Trends Ecol. Evol.* 13: 31–36.

Ingram, T., L. J. Harmon, and J. B. Shurin. 2009. Niche evolution, trophic structure, and species turnover in model food webs. *Am. Nat.* 174: 56–67.
Janson, E. M., J. O. Stireman III, M. S. Singer, and 1 other. 2008. Phytophagous insect-microbe mutualisms and adaptive evolutionary diversification. *Evolution* 62: 997–1012.
Jeffries, M. J. and J. H. Lawton. 1984. Enemy free space and the structure of ecological communities. *Biol. J. Linn. Soc.* 23: 269–286.
Jones, C. G., J. H. Lawton, and M. Shachak. 1994. Organisms as ecosystem engineers. *Oikos* 69: 373–386.
Jones, C. G., J. H. Lawton, and M. Shachak. 1997. Positive and negative effects of organisms as physical ecosystem engineers. *Ecology* 78: 1946–1957.
Kassen, R. 2009. Toward a general theory of adaptive radiation: Insights from microbial experimental evolution. *Yr. Evol. Biol.* 1168: 3–22.
Kato, M., A. Takimura, and A. Kawakita. 2003. An obligate pollination mutualism and reciprocal diversification in the tree genus *Glochidion* (Euphorbiaceae). *Proc. Natl. Acad. Sci. USA* 100: 5264–5267.
Kozak, K. H., R. A. Blaine, and A. Larson. 2006. Gene lineages and eastern North American palaeodrainage basins: Phylogeography and speciation in salamanders of the *Eurycea bislineata* species complex. *Mol. Ecol.* 15: 191–207.
Kubatko, L. S. 2009. Identifying hybridization events in the presence of coalescence via model selection. *Syst. Biol.* 58: 478–488.
Kubo, T., and Y. Iwasa. 1995. Inferring the rates of branching and extinction from molecular phylogenies. *Evolution* 49: 694–704.
Langerhans, R. B. 2006. Evolutionary consequences of predation: Avoidance, escape, reproduction, and diversification. In A. M. T. Elewa (ed.), *Predation in Organisms: A Distinct Phenomenon*, pp. 177–220. Springer-Verlag, Heidelberg.
Langerhans, R. B., M. E. Gifford, and E. O. Joseph. 2007. Ecological speciation in *Gambusia* fishes. *Evolution* 61: 2056–2074.
Larson, A. and J. B. Losos. 1996. Phylogenetic systematics of adaptation. In M. R. Rose and G. V. Lauder (eds.), *Adaptation*. pp. 187–220. Academic Press, San Diego.
Leigh, E. G. Jr., A. Hladik, C. M. Hladik, and 1 other. 2007. The biogeography of large islands, or how does the size of the ecological theater affect the evolutionary play? *Revue E'cole (Terre Vie)* 62: 105–168.
Lenski, R. E. and M. Travisano. 1994. Dynamics of adaptation and diversification: A 10,000-generation experiment with bacterial populations. *Proc. Natl. Acad. Sci. USA* 91: 6808–6814.
Levinton, J. 1988. *Genetics, Paleontology, and Macroevolution*. Cambridge University Press, Cambridge, UK.
Liem, K. F. 1974. Evolutionary strategies and morphological innovations: Cichlid pharyngeal jaws. *Syst. Zool.* 22: 425–441.
Loeuille, N. and M. Loreau. 2005. Evolutionary emergence of size-structured food webs. *Proc. Natl. Acad. Sci. USA* 102: 5761–5766.
Loeuille, N. and M. Loreau. 2006. Evolution of body size in food webs: Does the energetic equivalence rule hold? *Ecol. Lett.* 9: 171–178.
Losos, J. B. 2009. *Lizards in an Evolutionary Tree: Ecology and Adaptive Radiation of Anoles*. University of California Press, Berkeley.
Losos, J. B. and D. B. Miles. 2002. Testing the hypothesis that a clade has adaptively radiated: Iguanid lizard clades as a case study. *Am. Nat.* 160: 147–157.
Losos, J. B. and R. E. Ricklefs. 2009. Adaptation and diversification on islands. *Nature* 457: 830–836.

Losos, J. B., P. L. N. Mouton, R. Bickel, and 2 others. 2002. The effect of body armature on escape behaviour in cordylid lizards. *Anim. Behav.* 64: 313–321.
Lovette, I. J., E. Bermingham, and R. E. Ricklefs. 2002. Clade-specific morphological diversification and adaptive radiation in Hawaiian songbirds. *Proc. Roy. Soc. Lond. B* 269: 37–42.
Mabuchi, K., M. Miya, Y. Azuma, and 1 other. 2007. Independent evolution of the specialized pharyngeal jaw apparatus in cichlid and labrid fishes. *BMC Evol. Biol.* 7: 10.
MacArthur, R. H. 1972. *Geographical Ecology: Patterns in the Distribution of Species.* Princeton University Press, Princeton.
MacArthur, R. and R. Levins. 1967. The limiting similarity convergence, and divergence of coexisting species. *Am. Nat.* 101: 377–385.
MacFadden, B. J. 1992. *Fossil Horses: Systematics, Paleobiology, and Evolution of the Family Equidae.* Cambridge University Press, Cambridge, UK.
Mahler, D. L., L. J. Revell, R. E. Glor, and 1 other. 2010. Ecological opportunity and the rate of morphological evolution in the diversification of Greater Antillean anoles. *Evolution*, in press.
Mallet, J. 2009. Rapid speciation, hybridization and adaptive radiation in the *Heliconius melpomene* group. In R. Butlin, J. Bridle, and D. Schluter (eds.), *Speciation and Patterns of Diversity*, pp. 177–194. Cambridge University Press, Cambridge, UK.
McGowan, A. J. 2004. Ammonoid taxonomic and morphologic recovery patterns after the Permian-Triassic. *Geology* 32: 665–668.
McPeek, M. A. 2008. The ecological dynamics of clade diversification and community assembly. *Am. Nat.* 172: E270–E284.
McPeek, M. A., A. K. Schrot, and J. M. Brown. 1996. Adaptation to predators in a new community: Swimming performance and predator avoidance in damselflies. *Ecology* 77: 617–629.
Meyer, A., T. D. Kocher, P. Basasibwaki, and 1 other. 1990. Monophyletic origin of Lake Victoria cichlid fishes suggested by mitochondrial-DNA sequences. *Nature* 347: 550–553.
Meyer, J. R. and R. Kassen. 2007. The effects of competition and predation on diversification in a model adaptive radiation. *Nature* 446: 432–435.
Miller, A. H. 1949. Some ecologic and morphologic considerations in the evolution of higher taxonomic categories. In E. Mayr and E. Schüz (eds.), *Ornithologie als Biologische Wissenschaft*, pp. 84–88. Carl Winter, Heidelberg.
Mitter, C., B. Farrell, and B. Wiegmann. 1988. The phylogenetic study of adaptive zones: Has phytophagy promoted insect diversification? *Am. Nat.* 132: 107–128.
Monasterio, M. and L. Sarmiento. 1991. Adaptive radiation of *Espeletia* in the cold Andean tropics. *Trends Ecol. Evol.* 6: 387–391.
Nee, S. 2001. Inferring speciation rates from phylogenies. *Evolution* 55: 661–668.
Nee, S. 2006. Birth-death models in macroevolution. *Ann. Rev. Ecol. Evol. Syst.* 37: 1–17.
Nee, S., E. C. Holmes, R. M. May, and 1 other. 1994. Extinction rates can be estimated from molecular phylogenies. *Phil. Trans. Roy. Soc. Lond. B* 344: 77–82.
Nee, S., A. Ø. Mooers, and P. H. Harvey. 1992. Tempo and mode of evolution revealed from molecular phylogenies. *Proc. Natl. Acad. Sci. USA* 89: 8322–8326.
Nosil, P. and B. J. Crespi. 2006. Experimental evidence that predation promotes divergence in adaptive radiation. *Proc. Natl. Acad. Sci. USA* 103: 9090–9095.

Olson, M. E. and A. Arroyo-Santos. 2009. Thinking in continua: Beyond the "adaptive radiation" metaphor. *BioEssays* 31: 1337–1346.
O'Meara, B. C., C. Ane, M. J. Sanderson, and 1 other. 2006. Testing for different rates of continuous trait evolution using likelihood. *Evolution* 60: 922–933.
Parent, C. E. and B. J. Crespi. 2009. Ecological opportunity in adaptive radiation of Galápagos endemic land snails. *Am. Nat.* 174: 898–905.
Pfenning, K. S. and D. W. Pfennig. 2009. Character displacement: Ecological and reproductive responses to a common evolutionary problem. *Quart. Rev. Biol.* 84: 253–276.
Phillimore, A. B. and T. D. Price. 2008. Density-dependent cladogenesis in birds. *PLoS Biol.* 6: 483–489.
Pinto, G., D. L. Mahler, L. J. Harmon, and 1 other. 2008. Testing the island effect in adaptive radiation: Rates and patterns of morphological diversification in Caribbean and mainland *Anolis* lizards. *Proc. Roy. Soc. Lond. B* 275: 2749–2757.
Pratt, H. D. 2005. *The Hawaiian Honeycreepers: Drepanididae*. Oxford University Press, Oxford.
Price, T. 1997. Correlated evolution and independent contrasts. *Phil. Trans. Roy. Soc. B* 352: 519–529.
Price, T. D. 2008. *Speciation in Birds*. Roberts and Company, Greenwood Village, CO.
Quental, T. B. and C. R. Marshall. 2009. Extinction during evolutionary radiations: Reconciling the fossil record with molecular phylogenies. *Evolution* 63: 3158–3167.
Rabosky, D. L. 2009. Ecological limits on clade diversification in higher taxa. *Am. Nat.* 173: 662–674.
Rabosky, D. L. and I. J. Lovette. 2008. Density-dependent diversification in North American wood warblers. *Proc. Roy. Soc. Lond. B* 275: 2363–2371.
Rainey, P. B. and M. Travisano. 1998. Adaptive radiation in a heterogeneous environment. *Nature* 394: 69–72.
Raup, D. M., S. J. Gould, T. J. M. Schopf, and 1 other. 1973. Stochastic models of phylogeny and the evolution of diversity. *J. Geol.* 81: 525–542.
Rauscher, J. T. 2002. Molecular phylogenetics of the *Espeletia* complex (Asteraceae): Evidence from nrDNA ITS sequences on the closest relatives of an Andean adaptive radiation. *Am. J. Bot.* 89: 1074–1084.
Ree, R. H. 2005. Detecting the historical signature of key innovations using stochastic models of character evolution and cladogenesis. *Evolution* 59: 257–265.
Revell, L. J., L. J. Harmon, and R. E. Glor. 2005. Underparameterized model of sequence evolution leads to bias in the estimation of diversification rates from molecular phylogenies. *Syst. Biol.* 54: 973–983.
Rice, W. R. and E. E. Hostert. 1993. Laboratory experiments on speciation: What have we learned in forty years? *Evolution* 47: 1637–1653.
Ricklefs, R. E. 2009. Aspect diversity in moths revisited. *Am. Nat.* 173: 411–416.
Rieseberg, L. H., M. A. Archer, and R. K. Wayne. 1999. Transgressive segregation, adaptation, and speciation. *Heredity* 83: 363–372.
Roderick, G. K. and D. M. Percy. 2008. Insect-plant interactions, diversification, and coevolution: Insights from remote oceanic islands. In K. J. Tilmon (ed.), *Specialization, Speciation, and Radiation: The Evolutionary Biology of Herbivorous Insects*, pp. 151–161. University of California Press, Berkeley.
Rosenzweig, M. L. and R. D. McCord. 1991. Incumbent replacement: Evidence for long-term evolutionary progress. *Paleobiology* 17: 202–213.

Roy, K. and M. Foote. 1997. Morphological approaches to measuring biodiversity. *Trends Ecol. Evol.* 12: 277–281.
Rundell, R. J. and T. D. Price. 2009. Adaptive radiation, nonadaptive radiation, ecological speciation and nonecological speciation. *Trends Ecol. Evol.* 24: 394–399.
Rundle, H. D. and D. Schluter. 2004. Natural selection and ecological speciation in sticklebacks. In U. Dieckmann, M. Doebeli, J. A. J. Metz, and D. Tautz (eds.), *Adaptive Speciation*, pp. 192–209. Cambridge University Press, Cambridge, UK.
Russell, A. P. 1979. Parallelism and integrated design in the foot structure of gekkonine and diplodactyline geckos. *Copeia* 1979: 1–21.
Rutherford, S. L. and S. Lindquist. 1998. *Hsp90* as a capacitor for morphological evolution. *Nature* 396: 336–342.
Sanderson, M. J. and M. J. Donoghue. 1996. Reconstructing shifts in diversification rates on phylogenetic trees. *Trends Ecol. Evol.* 11: 15–20.
Savolainen, R. and K. Vepsalainen. 2003. Sympatric speciation through intraspecific social parasitism. *Proc. Natl. Acad. Sci. USA* 100: 7169–7174.
Sax, D. F., J. J. Stachowicz, J. H. Brown, and 9 others. 2007. Ecological and evolutionary insights from species invasions. *Trends Ecol. Evol.* 22: 465–471.
Schliewen, U. K., D. Tautz, and S. Pääbo. 1994. Sympatric speciation suggested by monophyly of crater lake cichlids. *Nature* 368: 629–632.
Schluter, D. 1988. Character displacement and the adaptive divergence of finches on islands and continents. *Am. Nat.* 131: 799–824.
Schluter, D. 1994. Experimental evidence that competition promotes divergence in adaptive radiation. *Science* 266: 798–801.
Schluter, D. 1995. Adaptive radiation in sticklebacks: Trade-offs in feeding performance and growth. *Ecology* 76: 82–90.
Schluter, D. 2000. *The Ecology of Adaptive Radiation*. Oxford University Press, Oxford.
Schluter, D. 2003. Frequency dependent natural selection during character displacement in sticklebacks. *Evolution* 57: 1142–1150.
Schluter, D. and P. R. Grant. 1984. Determinants of morphological patterns in communities of Darwin's finches. *Am. Nat.* 123: 175–196.
Seehausen, O. 2004. Hybridization and adaptive radiation. *Trends Ecol. Evol.* 19: 198–207.
Seehausen, O. 2006. African cichlid fish: A model system in adaptive radiation research. *Proc. Roy. Soc. Lond. B* 273: 1987–1998.
Sepkoski, J. J. 1978. A kinetic model of Phanerozoic taxonomic diversity. I. Analysis of marine orders. *Paleobiology* 4: 223–251.
Sepkoski, J. J., F. K. McKinney, and S. Lidgard. 2000. Competitive displacement among post-Paleozoic cyclostome and cheilostome bryozoans. *Paleobiology* 26: 7–18.
Sibley, C. G. and J. E. Ahlquist. 1990. *Phylogeny and Classification of Birds*. Yale University Press, New Haven.
Simberloff, D. 1981. Community effects of introduced species. In T. H. Nitecki (ed.), *Biotic Rises in Ecological and Evolutionary Time*, pp. 53–81. Academic Press, New York.
Simpson, G. G. 1953. *The Major Features of Evolution*. Columbia University Press, New York.
Slatkin, M. 1980. Ecological character displacement. *Ecology* 61: 163–177.
Slowinski, J. B. and C. Guyer. 1989. Testing null models in questions of evolutionary success. *Syst. Zool.* 38: 189–191.

Spencer, C. C., M. Bertrand, M. Travisano, and 1 other. 2007. Adaptive diversification in genes that regulate resource use in *Escherichia coli*. *PLoS Genet.* 3: 0083–0088.
Stanley, G. D. 1981. Early history of scleractinian corals and its geological consequences. *Geology* 9: 507–511.
Stiassny, M. L. J. and J. S. Jensen. 1987. Labroid intrarelationships revisited: Morphological complexity, key innovations, and the study of comparative diversity. *Bull. Mus. Comp. Zool.* 151: 269–319.
Stoks, R., M. A. McPeek, and J. L. Mitchell. 2003. Evolution of prey behavior in response to changes in predation regime: Damselflies in fish and dragonfly lakes. *Evolution* 57: 574–585.
Tokeshi, M. 1999. *Species Coexistence: Ecological and Evolutionary Perspectives*. Blackwell, Oxford.
Tyerman, J. G., M. Bertrand, C. C. Spencer, and 1 other. 2008. Experimental demonstration of ecological character displacement. *BMC Evol. Biol.* 8: 34.
Vamosi, S. M. 2005. On the role of enemies in divergence and diversification of prey: A review and synthesis. *Can. J. Zool.* 83: 894–910.
Vermeij, G. J. 1974. Adaptation, versatility, and evolution. *Syst. Zool.* 22: 466–477.
Vermeij, G. J. 1987. *Evolution and Escalation: An Ecological History of Life*. Princeton University Press, Princeton.
Wagner, G. P. and L. Altenberg. 1996. Perspective: Complex adaptations and the evolution of evolvability. *Evolution* 50: 967–976.
Webb, J. K. and R. Shine. 1994. Feeding habits and reproductive biology of Australian pygopodid lizards of the genus *Aprasia*. *Copeia* 1994: 390–398.
Weiblen, G. D. and G. L. Bush. 2002. Speciation in fig pollinators and parasites. *Mol. Ecol.* 11: 1573–1578.
Whittaker, R. H. 1977. Evolution of species diversity in land communities. *Evol. Biol.* 10: 1–67.
Wilson, E. O. 1992. *The Diversity of Life*. Harvard University Press, Cambridge, MA.
Winkler, I. S. and C. Mitter. 2008. The phylogenetic dimension of insect/plant interactions: A summary of recent evidence. In K. J. Tilmon (ed.), *Specialization, Speciation, and Radiation: The Evolutionary Biology of Herbivorous Insects*, pp. 240–263. University of California Press, Berkeley.
Wright, S. 1932. The roles of mutation, inbreeding, crossbreeding, and selection in evolution. *Proc. 6th Intl. Cong. Genet.* 1: 356–366.
Yamagishi, S., M. Honda, K. Eguchi, and 1 other. 2001. Extreme endemic radiation of the Malagasy vangas (Aves: Passeriformes). *J. Mol. Evol.* 53: 39–46.

Chapter 16

Phylogenetic Progress and Applications of the Tree of Life

David M. Hillis

As buds give rise by growth to fresh buds, and these, if vigorous, branch out and overtop on all sides many a feebler branch, so by generation I believe it has been with the great Tree of Life, which fills with its dead and broken branches the crust of the earth, and covers the surface with its ever branching and beautiful ramifications (Darwin 1859: 130).

The Origin of Species is known, first and foremost, as a formulation of the theory of natural selection: Darwin's explanation of a mechanism to account for biological evolution. But, *The Origin of Species* accomplished much more than providing a mechanism for evolution. It also laid out in great detail the evidence for evolution and popularized the idea that all of life is connected through its evolutionary history. Darwin's image of the Tree of Life captured the attention of biologists and the public alike. As it became increasingly clear that all of life was connected through its evolutionary history, the Tree of Life provided a unifying conceptual umbrella for all of biology.

Darwin clearly understood the concept of evolutionary relatedness across life, and one of his notebooks from 1837 famously included a sketch of a phylogenetic tree, although the term "phylogeny" would not be coined for another three decades (Figure 16.1). This sketch was a precursor to the only figure in *The Origin of Species*, which was a concept diagram of evolutionary relationships presented in the form of a tree (see Bowler, Chapter 2, Figure 2.1). Darwin wrote about the Tree of Life and clearly understood and emphasized the importance of the concept, although he probably never expected the details of the relationships across life to be discovered. Darwin left the job of reconstructing the details of the Tree of Life to others.

Ernst Haeckel took up this job in earnest, using morphological comparisons to reconstruct major branches of the Tree of Life in his 1866 book *Generelle Morphologie der Organismen* (Haeckel 1866). Haeckel's representations

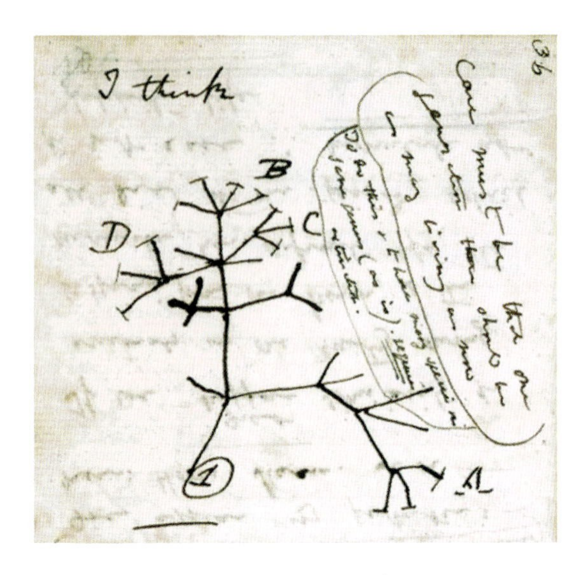

FIGURE 16.1 **Darwin's Sketch of a Phylogenetic Tree from His 1837 Notebook**

of various parts of the Tree of Life were drawn to resemble literal tree branches (Figure 16.2). In addition to summarizing the relationships across major groups, Haeckel (1866) also coined the word *phylogeny* to describe the study of evolutionary relationships.

Despite this early interest in the Tree of Life by Haeckel and many others, reconstructing evolutionary relationships with the data and methods available in the nineteenth century proved difficult. As no explicit techniques were developed for inferring phylogeny during this period, phylogenetics gradually became thought of more as an authoritarian art form than a science. During the first half of the twentieth century, most systematists were concerned more with problems of species, speciation, and geographic variation than with phylogeny. In fact, about the only explicit reference to phylogeny in Julian Huxley's *Evolution: The Modern Synthesis*, published in 1942, is a brief discussion of its possible relevance to taxonomy (Huxley 1942). In *Methods and Principles of Systematic Zoology*, Ernst Mayr et al. (1953; see Futuyma, Chapter 1) divided the history of systematic biology into three periods: (1) a first period: the Study of Local Faunas, which was largely descriptive; (2) a second period: the Acceptance of Evolution, in which evolution was established and phylogeny was emphasized; and (3) a third period: the Study of Populations, in which "...taxonomy is characterized by a study of the evolution *within* species" (Mayr et al. 1953: 9). Phylogeny was relegated primarily to the second period. Although the authors noted that "...it should remain the ultimate aim of the taxonomist to devise a phylogenetic classification" (Mayr et al. 1953: 45), they were not optimistic about the ability of systematists (with the possible exception of paleontologists) to infer phylogeny correctly:

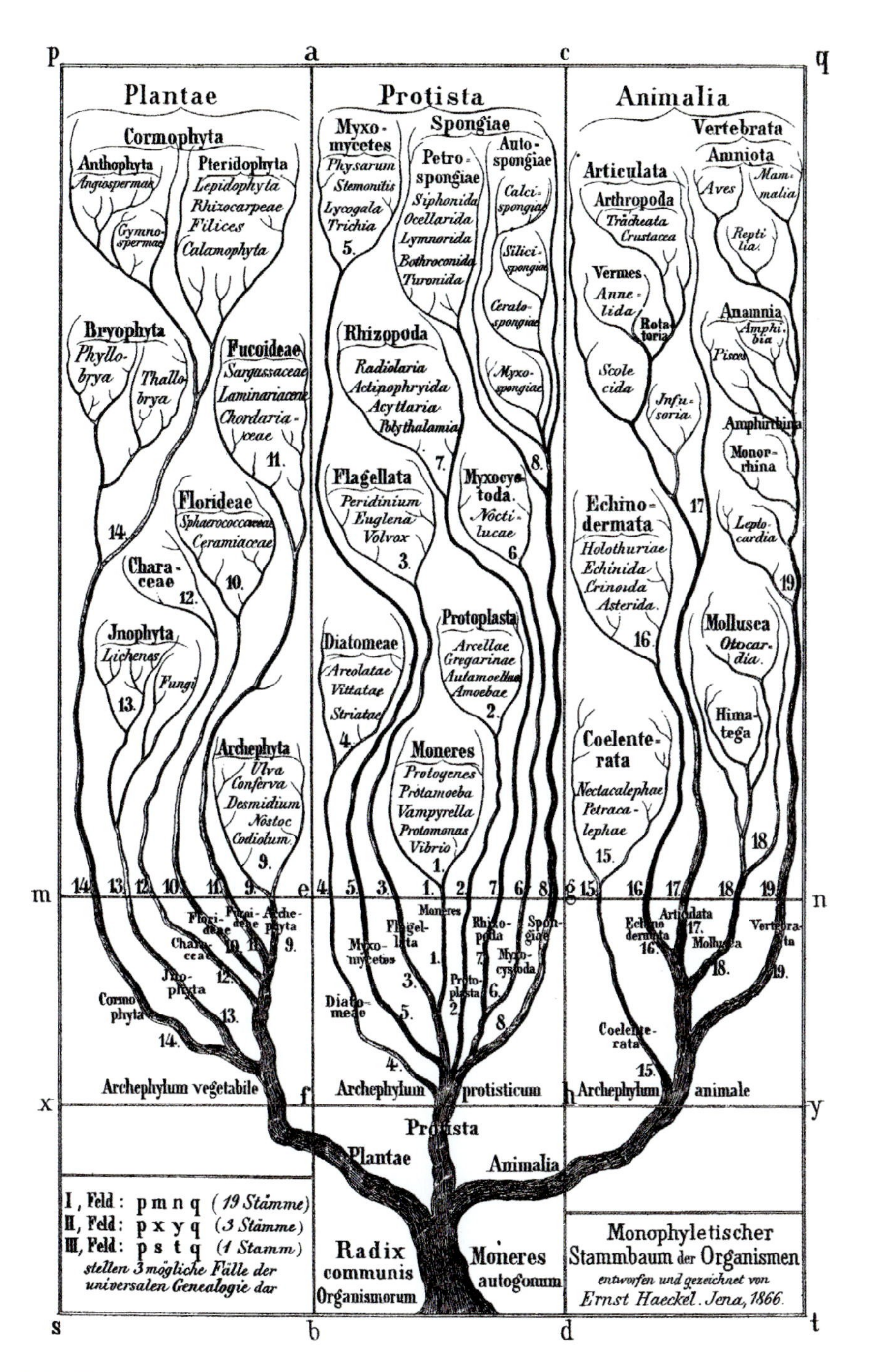

FIGURE 16.2 Ernst Haeckel (1866) Made the First Significant Progress in Reconstructing the Details of the Tree of Life

Every one of the sources of information on phylogeny that has been used in the past has its limitations and pitfalls. This is true for genetics, physiology (including serology), embryology, and zoogeography (Simpson 1945). It is even true for paleontology, because there are several interpretations possible for many fossil remains, particularly if they are incomplete. Still, paleontology (when fossils are available) and comparative morphology are on the whole the most productive sources of phylogenetic information (Mayr et al. 1953: 42).

Other systematists of the first half of the twentieth century did argue for the primacy of phylogeny in systematics, but their influence was limited. For instance, the German botanist Walter Zimmermann (1931) presented a clear discussion of phylogeny and argued strongly for phylogenetic classifications:

Do we want to group phylogenetically, that is, following naturally given relationships? Or do we want to group artificially? Or do we want to group intuitively, following subjective impression? We do not have any other possibilities. Of course, one can do entirely without phylogenetics. One must be aware, however, that then one is forced to group artificially, or 'idealistically;' phylogenetics is the only procedure which groups according to naturally given relationships, the only procedure which, through the act of grouping, directly depicts natural relationships (Zimmerman 1931: 949–950, as translated by Donoghue and Kadereit 1992: 76).

But by the 1950s, although some plant taxonomists were still acknowledging the importance of phylogeny, many were just as skeptical as Mayr was of systematists' abilities to reconstruct phylogeny. The discussion of phylogeny in Earl Core's (1955) textbook *Plant Taxonomy* consists of a single paragraph, other than an historical note on Haeckel's coining of the term:

All modern taxonomic studies are founded on the tendency to make phylogeny (evolutionary development) the basic principle of taxonomy. However, phylogeny is not a tangible subject that may be studied independently and the results applied to taxonomy. Even though it be granted that present-day plants are descended by evolution from plants of the past, we do not in fact know the origin of more than a few species. In many cases we do not know the nearest living relatives. We do not actually know the phylogenetic history of any group of plants, since it lies in the unwritten past; hence our only recourse is to make use of criteria that seem to be pertinent. These criteria are those which have been the subject matter of descriptive and experimental taxonomy—morphology, ecology, cytology, and related sciences. For practical purposes morphology is still the most widely used means of identification and classification. This was true before the days of Darwin and it remains true today. The experimental method throws light on the mechanism of evolution, but requires much time for its operation. Fortunately, it has been discovered that genetic relationships are usually reflected in similarities and differences in morphological structure, which are readily detected by eye, usually on brief examination (Core 1955: 136).

Phylogenies were still something produced by the authority of a given group after much study and consideration, but the explicit methods by which the authorities worked were rarely described. Phylogenetic relationships were simply "detected by eye, usually on brief examination" (Core 1955: 136). As a result, few biologists seemed to take these efforts seriously.

The Rise of Quantitative Methods

In the 1950s and 1960s, two research groups emerged that dealt with the lack of rigor in phylogenetic study, but each in very different ways. Both groups sought to bring explicit, objective, and quantitative methodologies into systematics. One group, whose practitioners initially called themselves numerical taxonomists, decided that phylogeny was unknowable and sought other criteria (primarily overall similarity) as the basis of classification. In the first major textbook of this group, *The Principles of Numerical Taxonomy*, Robert Sokal and Peter Sneath (1963) stated:

Undoubtedly more utter rubbish has been written since the time of Haeckel on supposed phylogenies than on any other biological topic. The fact is that we have a reasonably correct picture of the phylogenies of only a very few taxa and these entirely on the basis of paleontological evidence. Even in paleontology the proportion of fact to speculation is not too high (Sokal and Sneath 1963: 24).

The other research group followed along the lines that Zimmermann and other early phylogeneticists had begun to develop. These systematists retained the conviction that phylogeny should be the central organizing principle in systematics and sought to develop objective, reliable methodologies to infer phylogeny. One of the most influential authors to delimit such methods explicitly was Willi Hennig (1950) in *Grundzüge einer Theorie der Phylogenetischen Systematik* (see Figure 1.11). Because his book was in German and wasn't translated into English until 1966 (as *Phylogenetic Systematics*), his ideas were slow to penetrate the English-speaking world. Initially, the numerical taxonomists were impressed with Hennig's work; although they disagreed on the importance he placed on phylogeny, they agreed with the need for explicit, objective methodologies in systematics. Sokal and Sneath (1963) even lamented the fact that Hennig (1950) had not been read by more English-speaking systematists:

In recent years three comprehensive analytical studies of systematic principles have been published in books by Hennig (1950), Remane (1956), and Simpson (1961). It is especially regrettable that the earlier two books, published in German, have been almost entirely ignored in the English and American literature. Hennig's book presents the issues with particular clarity and objectivity, and there is considerable truth in Kiriakoff's (1959) statement that a number of controversies of the last decade published in the United States are in a sense outdated and could have been guided into more productive channels if Hennig's thoughts had been available to the disputants (Sokal and Sneath 1963: 21).

People who agreed with Hennig that phylogeny should be central to systematics, that phylogenies could be reconstructed from comparisons of extant species, and that classifications should be based directly on phylogeny became known as cladists (after *clade*, a branch in a phylogeny) or phylogenetic systematists. However, Hennig's methodology was not the only one proposed for inferring phylogenies. Lyman Benson, in *Plant Taxonomy: Methods and Principles* (1962), included a "class instruction sheet" from Warren H. Wagner, Jr. on a "graphic method for expressing relationships based upon group correlations of indexes of divergence" (Benson 1962: 273–277). These Wagner trees, as they would become known, were quite similar to the trees produced by Hennig's methods. Other explicit methods for reconstructing phylogenies were published by numerical taxonomists (Camin and Sokal 1965). As the interests of the numerical taxonomists expanded to include phylogeny, the research program that used similarity (rather than phylogeny) as the basis for classification became known as phenetics.

Some systematists from the traditional school of systematics (sometimes called evolutionary systematics), primarily paleontologists such as George Gaylord Simpson (see Figure 1.4), continued to embrace the importance of phylogeny without accepting the new methodologies and theories of classification proposed by Hennig and (later) others. In fact, Simpson (1961) argued that:

With only the rarest exceptions, zoologists now agree that phylogeny is the appropriate theoretical background for taxonomy and that it is essential for understanding and explaining all the associations involved in classification. Important disagreement persists only as to the desirable or practicable relationship between phylogeny and classification, and especially whether phylogeny can and should provide criteria for classification or can and should play its taxonomic role only in interpreting classifications based on other criteria. I hold the former view... (Simpson 1961: 50).

Other traditional systematists continued to ignore everything about "The New Systematics" (the name applied to efforts to combine phylogeny, population biology, genetics, and taxonomy). In his nearly 700-page book *Taxonomy: A Text and Reference Book* (1967), Richard E. Blackwelder noted that there was no room for subjects like phylogeny or evolution:

It must be emphasized that this book is not intended to cover all of systematics. An attempt is made to deal with all aspects of taxonomy, but the other branches of systematics are more appropriately discussed under other headings. It is not intended to cover Evolution, Speciation, Phylogeny, Population Dynamics, or Genetics. Much of what is sometimes called Biosystematics is omitted as belonging to one of the above fields. What has been called The New Systematics by American evolutionists is discussed only to the extent of showing that it has had little effect on taxonomy (Blackwelder 1967: 331).

The Emergence of Schools of Systematic Thought

Authoritarian approaches to systematics and to phylogenetic reconstruction, in particular, were challenged by the two new research groups—both of which argued in favor of more explicit, quantitative methodologies to replace former authoritarianism. However, the cladists defended and expanded the central position of phylogeny in systematics, whereas the numerical taxonomists argued that phylogeny was unknowable, and therefore advanced the concept of similarity as the preferred criterion for the basis of classifications (Hull 1988). Some of the systematists, who were initially in the numerical taxonomy camp because of its early insistence on quantitative methods, later became phylogeneticists as the cladists became more quantitative. The remaining numerical taxonomists who preferred similarity to phylogeny as an organizing principle of classifications became known as pheneticists.

The numerical taxonomists were the first to disrupt the systematic establishment. Their insistence on the necessity of rigorous, mathematical methods in systematics did not receive immediate acceptance. Several early papers by numerical taxonomists were rejected without review from the leading journal of systematics, *Systematic Zoology*, because the editor, Libbie Henrietta Hyman, considered them inappropriate:

One article was rejected because [it was] written in incredibly bad English and another because [it was] too mathematical. Inquiries among subscribers indicate that the journal has had enough for the present of articles about numerical taxonomy. However, further opinions on this matter are desired. An article of this nature is scheduled for the March, 1962, issue but that will be the last of this nature for the present (1961 editor's report, as quoted by Hull 1988: 123).

Numerical taxonomists continued to have problems with Hyman, and appealed to G. G. Simpson, the president-elect of the Society of Systematic Zoology, for help (Hull 1988). Although Simpson was a strong supporter of traditional systematics and an advocate of the importance of phylogeny (see previous discussion), he had also co-authored (with his wife) one of the first books on quantitative methods in biology (Simpson and Roe 1939). Therefore, he had little sympathy for Hyman's reluctance to publish mathematical papers, and when he became president of the Society of Systematic

Zoology, he replaced Hyman with George Byers, an entomologist at the University of Kansas (then the center of numerical taxonomy). The pages of *Systematic Zoology* were thus opened to one of the most vigorous confrontations in modern day biology.

The 1960s, 1970s, and early 1980s witnessed a lively, ongoing debate on the pages of *Systematic Zoology* among the major schools of systematics: the traditional evolutionary systematists, the upstart numerical taxonomists or pheneticists, and the almost evangelical cladists or phylogenetic systematists. Hull (1988) presents an entertaining and insightful account of this debate. Textbooks by evolutionary systematists were published in 1969 (*Principles of Systematic Zoology*, by E. Mayr) and 1974 (*Biological Systematics*, by Herbert H. Ross). An updated textbook on numerical taxonomy (*Numerical Taxonomy*, by Sneath and Sokal) appeared in 1973. The viewpoints of several leading cladists appeared in three textbooks published in 1980 and 1981: *Phylogenetic Patterns and the Evolutionary Process: Method and Theory in Comparative Biology* (Eldredge and Cracraft 1980); *Phylogenetics: The Theory and Practice of Phylogenetic Systematics* (Wiley 1981); and *Systematics and Biogeography: Cladistics and Vicariance* (Nelson and Platnick 1981).

Another group of biologists also appeared in the 1960s: molecular and statistically oriented biologists who had not been directly involved in the debates about systematic philosophy but were interested in inferring phylogenies with molecular data sets. To these molecular systematists, phylogenies were the key to understanding not only the evolutionary history of taxa, but the details of the molecular evolution of genes as well. To them, it was obvious that quantitative techniques (especially statistical approaches) and phylogenetic inference were critical parts of biology, although many molecular systematists had little interest in the connection between phylogeny and classification (Edwards and Cavalli-Sforza 1964; Fitch and Margoliash 1967).

In the 1980s, although the battles continued to be fought among the various research groups or their remnants, the old labels began to lose their meaning. If by cladist one meant someone who considered the inference of phylogeny to be of considerable importance to a modern understanding of biology, accepted that phylogenies could indeed be inferred from biological data, and had adopted at least some cladistic terminology, then the vast majority of biologists were cladists by the end of the decade (Hull 1989). In contrast, virtually all systematists in the 1980s found the use of computers and quantitative techniques essential for their work, so in this sense the numerical taxonomists succeeded. If one used more restrictive definitions, such as requiring that one use only Hennig's original methodology to be called a cladist or that classifications be based strictly on similarity to be called a pheneticist, then the two groups evolved virtually out of existence during the decade.

To add to the confusion, a subgroup of cladists called "pattern cladists" arose briefly during the 1980s. Pattern cladists argued that all theory of processes (including evolutionary theories) should be expunged from systematic analysis and that these studies should concentrate primarily on patterns among organisms. According to this view, any discussion of processes had to follow an analysis of pattern; analysis of pattern and process could not be interwoven. Interested readers can find discussions of pattern cladistics in Hull (1988, 1989), Platnick (1979, 1985), and Ridley (1986).

Recent History

Beginning about 1990, the use of phylogenies and phylogenetic thinking suddenly became much more widespread throughout biology (Hillis and Moritz 1990; Figure 16.3). As phylogenetic methods spread beyond systematics throughout the rest of biology, the labels that had been applied to systematists in the past ceased to be useful. Few pheneticists can be found who still argue for similarity being superior to phylogeny as an organizing principle of biology, just as few nonquantitative systematists have survived to the present. Likewise, most biologists recognize the importance of monophyly in classifications, even though few people now label themselves cladists (that term now tends to be more associated with people who restrict phylogenetic analyses to nonparametric methods). There are still plenty of

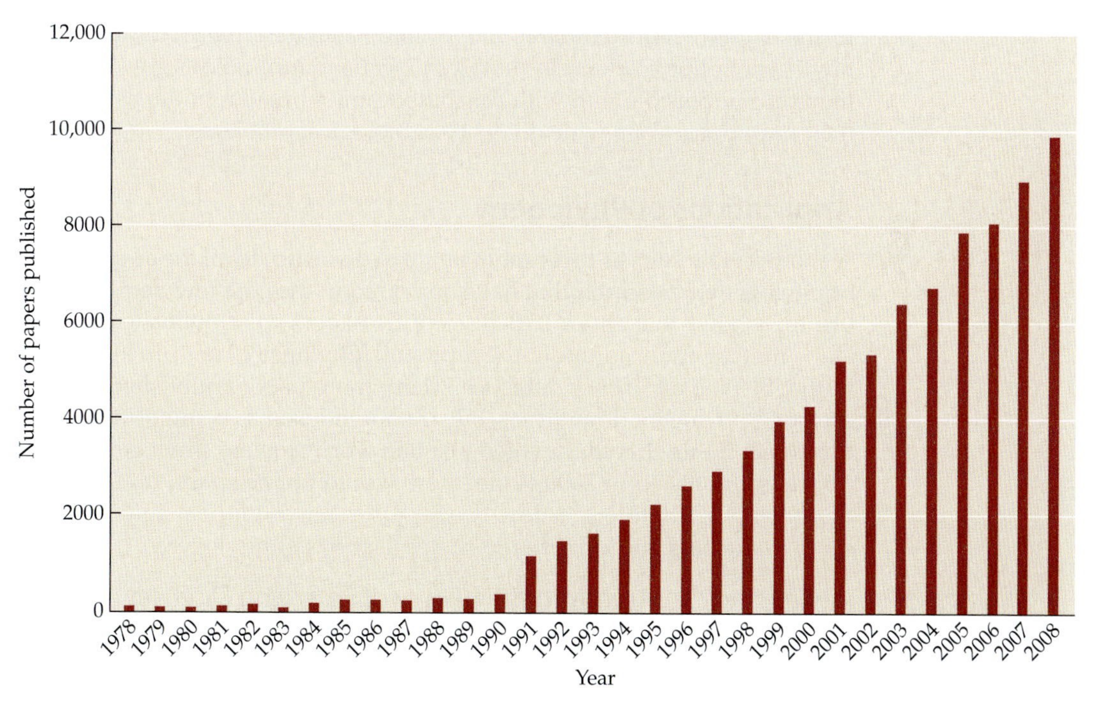

FIGURE 16.3 The Growth of Phylogenetic Research and Applications This graph shows the number of published scientific papers, by year of publication, in a Thomson Web of Science literature search using the terms "phylogeny" and "phylogenetic."

disagreements with regard to systematic theory, methods of phylogenetic inference, the relationship between phylogeny and classification, species concepts, the relationship between pattern and process, and many other topics—systematists just do not fall into discrete groups on these issues anymore. Most systematists no longer worry about the old labels and have moved on to discuss the theories, methodologies, and applications of phylogeny without regard to their sociological origin.

What changes led to the rapid growth of applications of phylogeny over the past two decades (see Figure 16.3)? Four factors were particularly important. The development and implementation of effective phylogenetic algorithms was a critical first step; the book chapter by Swofford and Olsen (1990) summarizing these algorithms had a significant stimulating affect on the field. This development and implementation of algorithms required substantial computational power, so the exponential rise in computer speed and capacity was necessary to make the algorithms practical and useful to a broad audience of biologists. Effective estimation of phylogenetic trees also requires substantial information, and the growth in phylogenetics research (see Figure 16.3) paralleled rapid growth of gene and genomic databases over the same period. Finally, a growing appreciation among biologists that most comparisons in biology require an understanding of underlying phylogenetic relationships (Felsenstein 1985; Harvey and Pagel 1991) fueled the growth in phylogenetic applications. In short, phylogenetic applications grew because they became feasible, and with feasibility came a broader understanding of their importance for a wide diversity of biological problems.

Applications of Phylogeny

A short review such as this cannot begin to do justice to all the applications of phylogenetic research that have appeared in the past few decades. Figure 16.3 shows that about 10,000 scientific papers were published in 2008 alone that are indexed under the topics of "phylogeny" or "phylogenetic" in the ISI Web of Science database. Many more papers published in 2008 included phylogenetic analyses, but are not indexed as being primarily on this topic. If this chapter devoted just one word to every phylogenetic application published in 2008 alone, there would not be space to include all these studies. Therefore, below I cover just a few examples that illustrate some interesting and diverse recent applications of phylogeny.

Phylogenetic Insights into the History of Emerging Diseases

Many emerging diseases in human populations are caused by viruses that have moved from other animal hosts into human populations. Identifying the causative agents of these diseases and deducing their source or sources was once a long and difficult process that required the isolation, culture, and characterization of the pathogen. Such work often required many years of laboratory work, during which time an initial infection could become a global pandemic. Even then, the timing and geographical origin of the zoonotic event was often not discernable. Phylogenetic analysis transformed the field of emerging diseases, and now a phylogenetic analysis can identify the origin and closest relatives of a newly emerging pathogen within days. For example, phylogenetic analyses quickly identified the cause and source of Sin Nombre (Nichol et al. 1993), SARS (Peiris et al. 2003; Ksiazek et al. 2003), and North American West Nile (Lanciotti et al. 2002) epidemics. Without phylogenetic analysis, the origins of these epidemics would have required many years to understand and control, during which time they may have become much more widespread and caused many thousands of deaths. By quickly identifying the causative agents, public health officials were able to institute control measures (such as control of the host organisms, changes in agricultural practices, and appropriate quarantine procedures) that minimized the epidemics. In 1995, the Centers for Disease Control and Prevention of the United States Department of Health and Human Services began publishing a new journal, *Emerging Infectious Diseases*, which now regularly presents phylogenetic analyses of the many emerging pathogens that infect human populations.

Acquired Immune Deficiency Syndrome (AIDS), caused by the Human Immunodeficiency Virus (HIV), provides a good example of the many uses of phylogenetics for understanding an emerging disease. A high-level phylogenetic analysis of isolates of HIV and related Simian Immunodeficiency Viruses (SIVs) shows that these viruses have entered human populations from at least two different host species (Figure 16.4; Hahn et al. 2000). In

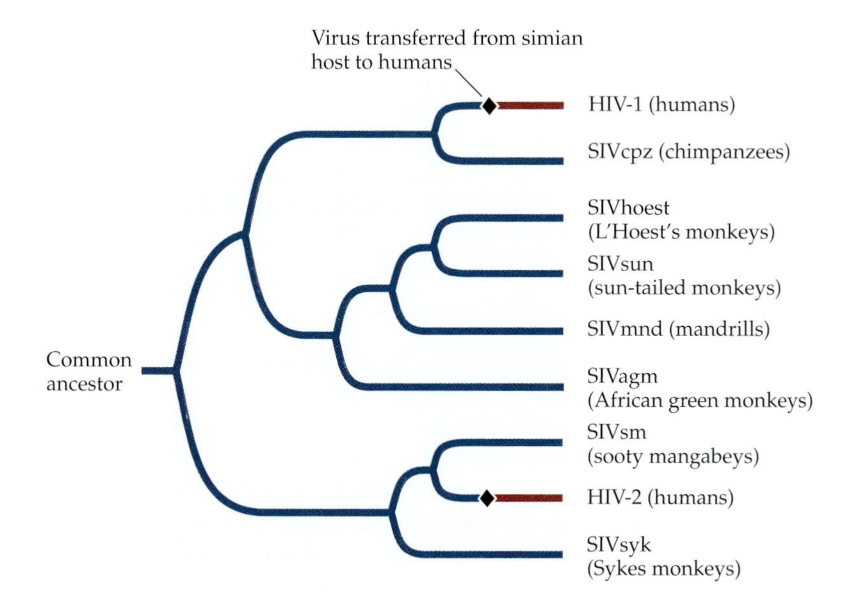

FIGURE 16.4 Origins of Human Immunodeficiency Viruses (HIV-1 and HIV-2) as Revealed by a Phylogenetic Analysis of Relevant Human and Simian Immunodeficiency Viruses Figure from Sadava et al. 2009.

FIGURE 16.5 The Multiple Origins of HIV-1 Strains from SIVcpz (the Immunodeficiency Virus Present in Chimpanzees) and Possibly SIVgor (the Immunodeficiency Virus Present in Gorillas) The numbers above the branches are bootstrap support values. Based on a phylogenetic analysis by Van Heuverswyn et al. 2006.

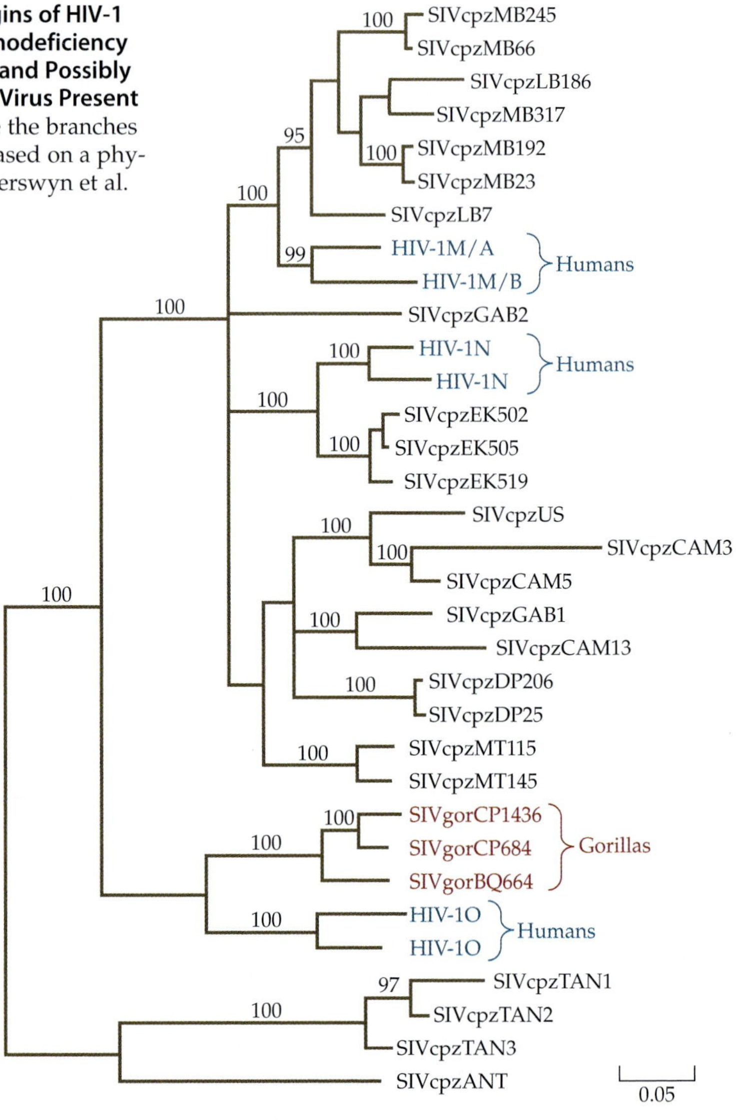

central Africa, chimpanzees were the host of the SIVs (known as SIVcpz strains) that were transferred into human populations (and became known as HIV-1 strains), whereas in western Africa sooty mangabeys were the source of the viruses (SIVsm) that gave rise to HIV-2 strains. If we zoom in on a more detailed view of HIV-1 and SIVcpz phylogeny (Figure 16.5), it is clear that these viruses were transferred to humans from simian hosts on multiple occasions, because the various strains of human viruses are each more closely related to different subsets of chimpanzee viruses (Van Heuverswyn et al. 2006). This finding demonstrates that the major strains of HIV-1 (known as the M, N, and O groups) have moved independently from chimpanzees into humans (either directly, or in the case of the O group, possibly via gorillas (see Figure 16.4; Van Heuverswyn et al. 2006).

If different strains of HIV have moved into humans multiple times, why did these epidemics only become apparent in the 1980s? Did all of these transmissions occur at that time, or did much older infections simply become evident during that decade? To answer that question, it is possible to zoom in further on the tree and examine the variants of the HIV-1 M-group (Figure 16.6; based on data and analyses from Korber et al. 2000). As we

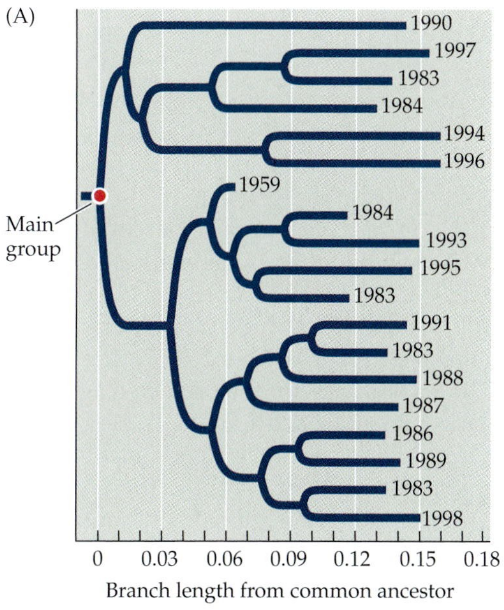

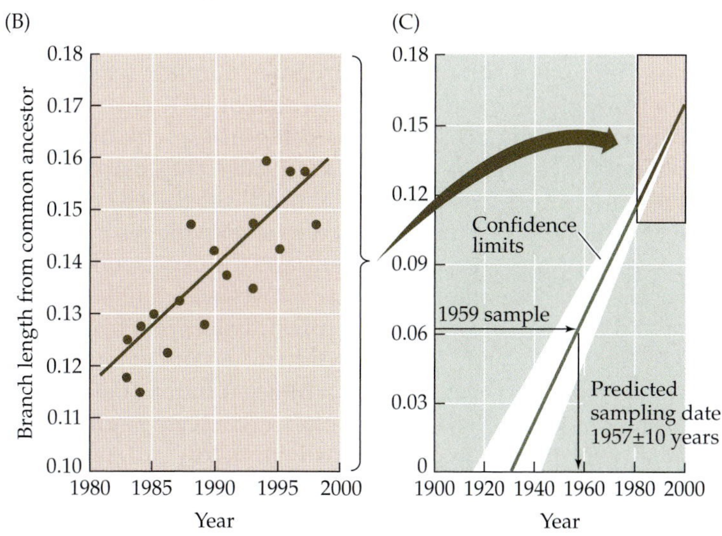

FIGURE 16.6 A Phylogenetic Analysis (A) and Molecular Clock Analysis (B and C) of HIV-1 "Main group" indicates the transmission of the HIV-1 M-group into humans in the 1920s or 1930s. Figure from Sadava et al. 2009, based on more detailed analyses published by Korber et al. 2000.

look in fine detail at the phylogeny of these viruses, we can combine information from the phylogenetic analysis of HIV-1 M-group and the time of isolation of various strains (see Figure 16.6A). The information on the phylogenetic distance of these isolates from their common ancestor can be combined with the information on the dates of their collection; this information allows estimation of the average rate of evolution of HIV-1 (see Figure 16.6B). Projection of this rate back to the base of the HIV-1 M-group radiation allows an estimate of the approximate time of introduction of this strain into human populations (see Figure 16.6C). Bootstrapping the data points provides a method of calculating confidence limits for the time estimates. This analysis suggests an origin of the HIV-1 M-group in the 1920s or 1930s, well before AIDS became known in western medicine in the 1980s.

Is there a way to verify the accuracy of the method of dating this radiation of HIV-1? If HIV-1 has been present in human populations since at least the 1930s, then perhaps the virus could be located in old stored tissue samples taken from humans in Africa in the 1940s or 1950s. A search of such samples has indeed turned up some old samples that are infected with HIV. One of these from 1959 is shown in the phylogenetic analysis in Figure 16.6A. When Korber et al. (2000) used HIV samples collected in the 1980s and 1990s to predict the age of this 1959 sample (see Figure 16.6B), the molecular clock analysis suggested an age of about 1957. The actual age is well within the confidence limits obtained for the predicted age, thus supporting the accuracy of the molecular clock analysis.

The analysis shown in Figure 16.6 (and many other phylogenetic analyses of HIV) have helped epidemiologists understand how these viruses have moved into human populations and become global pandemics. The various analyses are all consistent with the hypothesis that: (1) the transfer of these viruses from various other primates into humans has occurred many times over the centuries and (2) these transfers are associated with the hunting of chimpanzees and sooty mangabeys for food by humans. In the process of killing and skinning the primates, hunters were likely to contract SIVs through small cuts. But, if strains of HIV have been in human populations for many decades or even centuries, why did AIDS only become evident to western medicine in the 1980s? The answer lies in the significant social changes in Africa in the middle of the last century. Until that time, infections of HIV were limited to small villages in Africa. Progression to AIDS typically takes many years, and the expected life span of humans in rural Africa was quite short. The viruses were unlikely to be transmitted beyond a local village, and the transmission rate from HIV-infected individuals was probably low. An epidemic occurs when the average number of transmission events from an infected individual to non-infected individuals exceeds 1.0, so that the number of infected individuals grows exponentially through time. For a virus with a slow progression to disease (such as HIV), the lag time for such an epidemic to become evident can be quite long, especially if the average number of new infections per infected individual is close to 1.0.

Social changes in Africa in the 1950s–1970s provided a perfect storm for local zoonotic viruses to emerge in global pandemics in human populations (Worobey et al. 2008). Wars of independence and civil wars displaced many populations. Africa became increasingly urbanized, with people moving from small villages into the cities. Transportation (both within Africa and between Africa and the rest of the world) became easier and more frequent. Vaccination programs resulted in the widespread use of hypodermic needles, which would sometimes be re-used when supplies were short. Drug use and the sexual revolution also led to increased transmission rates, especially as the virus moved out of Africa. All of these factors resulted in recurrent localized infections becoming global pandemics. Strains of HIV became widespread in western countries in the 1970s, and then began to be evident as cases of AIDS by the early 1980s, at least a half-century after they first entered human populations.

Phylogeny and Forensic Applications

Strains of HIV and many other viruses evolve so quickly that their evolution can be followed over the course of just months and certainly across several years (Hillis 1999b). This phenomenon has led to many forensic applications of phylogeny, in which a particular series of infections among individuals can be traced. When such infections are related to criminal behavior, phylogenetic analyses have proven useful in courts of law.

Forensic uses of phylogenetic analysis include transmission of HIV in rape cases (Leitner et al. 1996, 1997), negligent transmission because of unsanitary dental or medical practices (e.g., Ou et al. 1992; Hillis and Huelsenbeck 1994; Hillis et al. 1994), and purposeful transmission as a means of attempted murder (Metzker et al. 2002). In the latter case, a physician was accused of attempted murder for injecting blood or blood products from one of his HIV-positive patients into his former girlfriend. The prosecution had evidence of means, motive, and opportunity, but needed phylogenetic analysis to demonstrate that the HIV present in the victim was directly related to the HIV present in the physician's patient. A phylogenetic analysis of HIV sequences amplified from the patient, the victim, and other HIV-positive individuals in the local community showed that the HIV sequences present in the victim were imbedded as a subset of the sequences present in the physician's patient (Figure 16.7). Additional analyses of HIV sequences collected at a later date and conducted in an independent laboratory confirmed this finding (see inset, Figure 16.7).

Forensic phylogenetic analyses have also been used to free the innocent, as in a well-publicized case of Bulgarian and Palestinian health-care workers who traveled to Libya on a humanitarian mission. Unfortunately,

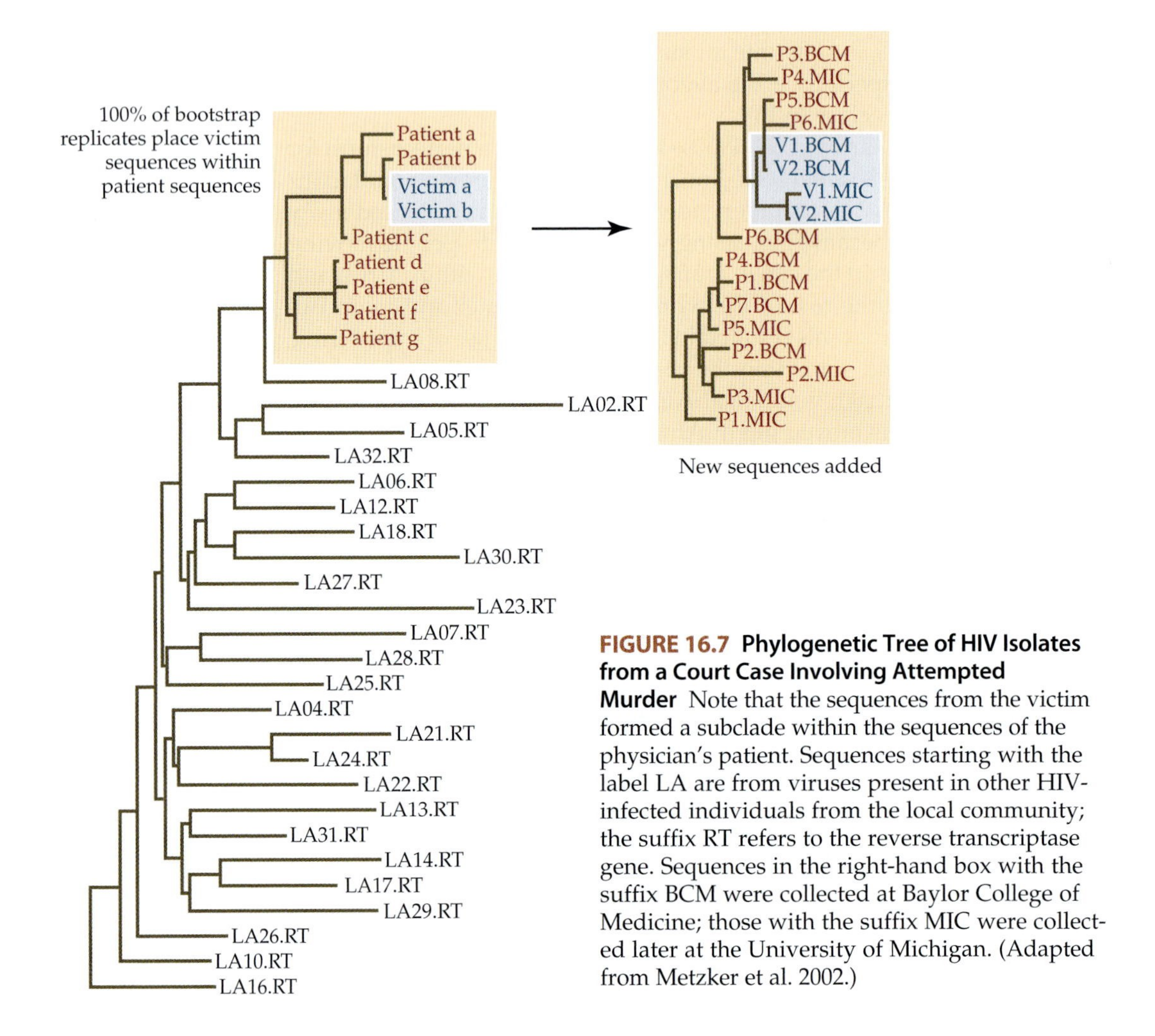

FIGURE 16.7 Phylogenetic Tree of HIV Isolates from a Court Case Involving Attempted Murder Note that the sequences from the victim formed a subclade within the sequences of the physician's patient. Sequences starting with the label LA are from viruses present in other HIV-infected individuals from the local community; the suffix RT refers to the reverse transcriptase gene. Sequences in the right-hand box with the suffix BCM were collected at Baylor College of Medicine; those with the suffix MIC were collected later at the University of Michigan. (Adapted from Metzker et al. 2002.)

shortly after the health-care workers arrived in Libya, several serious viral epidemics became evident at the children's hospital where they were working. They were accused of purposefully introducing the deadly viruses to the hospital patients, and were tried, convicted, and sentenced to death by the Libyan courts. However, phylogenetic analysis of the viruses and molecular clock dating conducted by independent investigators (de Oliveira et al. 2006) clearly showed that the epidemics had begun many years before the arrival of the convicted health-care workers in Libya (Figure 16.8). These analyses were used to pressure the Libyan government to release the innocent health-care workers to Bulgaria, where they were then set free (Butler 2007).

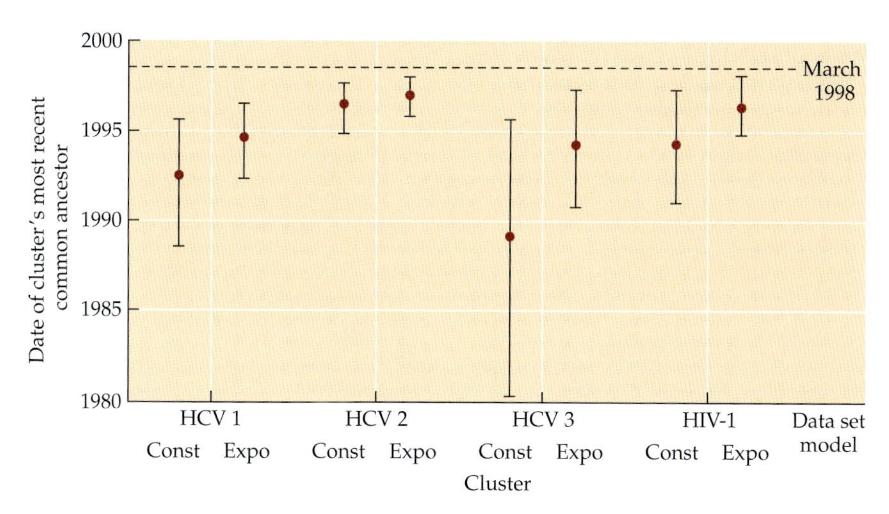

FIGURE 16.8 Bulgarian Health Care Workers Were Freed from a Libyan Death Sentence by a Phylogenetic Analysis and Molecular Clock Analysis of the Viruses that They Had Been Accused of Transmitting to Patients in a Children's Hospital HIV-1 = human immunodeficiency virus-1; Const = constant; Expo = exponential; hepatitis C virus (HCV-1, HCV-2, and HCV-3) = three independent introductions of the virus to the hospital. The heath care workers arrived in Libya in March 1998 (dashed line), well after the epidemics in the hospital had started. (Adapted from de Oliveira et al. 2006.)

Phylogeny and the Development of Improved Vaccines

There are many reasons that phylogenetic analyses are important for vaccine development (Halloran et al. 1998). For example, phylogenetic analyses are critical as a first step in understanding the phylogeographic distribution of variation, as in HIV (McCutchan 1999) and Dengue (Twiddy et al. 2002). When vaccines involve attenuated viruses, as in some polio vaccines, phylogenetic analysis is used to study disease outbreaks and determine if the outbreaks result from reversion of the attenuated virus or from natural reservoirs (Dowdle et al. 2003; Yang et al. 2003; Kew et al. 2004), which helps epidemiologists establish the best strategies for elimination of the disease in a given area (e.g., Korotkova et al. 2003; Alexander et al. 2004).

Phylogenetic analysis of influenza A is used to monitor newly emerging variants, including reassortment among the eight major components of influenza virus through time (e.g., Smith et al. 2009). Phylogenetic analysis of the hemagglutinin protein of influenza (one of the major surface antigens detected by host immunological systems) reveals an asymmetric phylogeny that suggests strong selection for particular variants that are best able to escape immune detection by the virus's hosts (Bush et al. 1999; Figure 16.9). An analysis of synonymous and nonsynonymous nucleotide changes shows that many of the amino acid residues in hemagglutinin are under positive selection for change. These positively selected amino acid residues

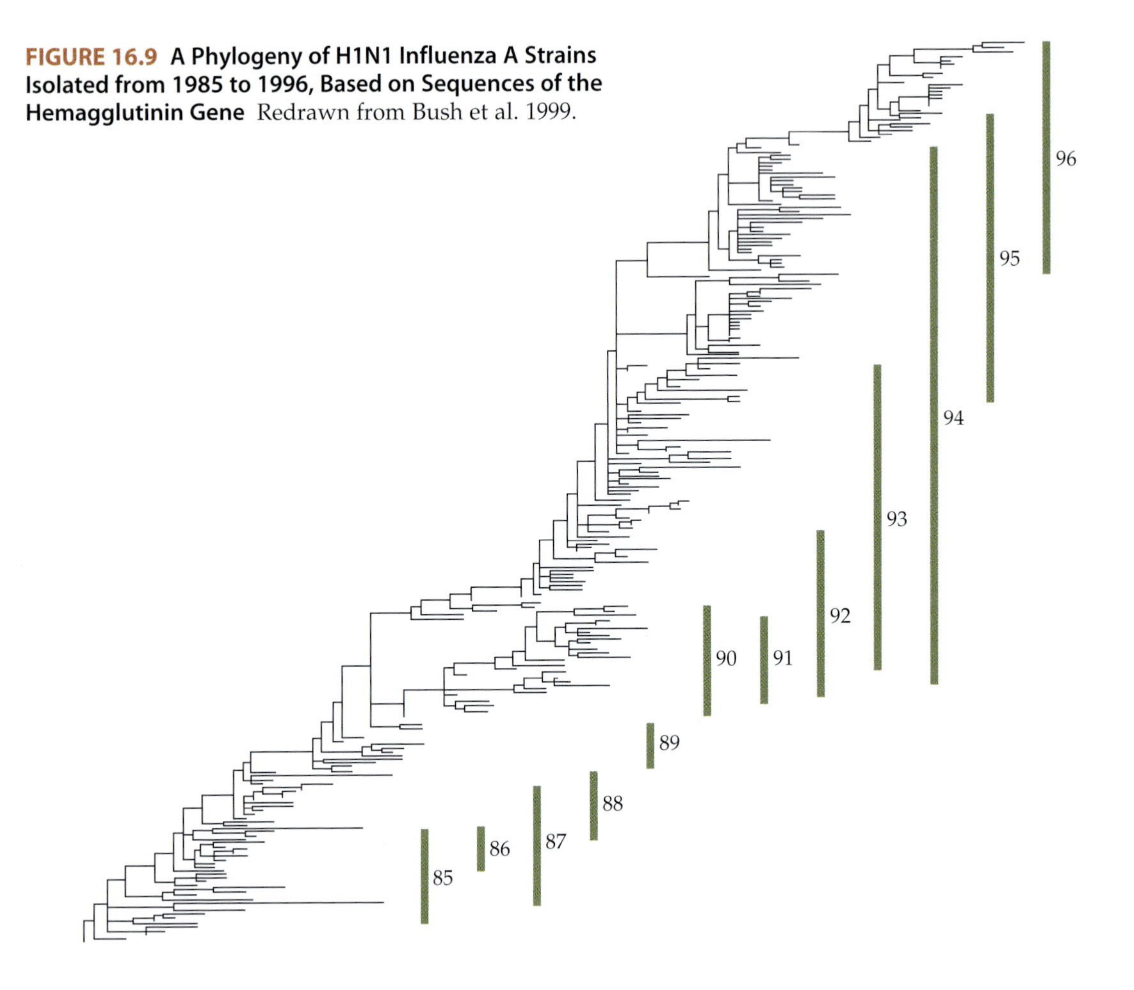

FIGURE 16.9 A Phylogeny of H1N1 Influenza A Strains Isolated from 1985 to 1996, Based on Sequences of the Hemagglutinin Gene Redrawn from Bush et al. 1999.

are concentrated at the distal end and on the external surfaces of the protein, where they are most likely to interact with components of the host's immune system. This distribution of the positively selected amino acid residues is consistent with the explanation that changes in these positions allow for viral escape from detection by the host's immune system (Figure 16.10). This information can then be used to predict (in any given year) which of the currently circulating strains of the virus is most likely to give rise to future successful lineages of influenza (Bush et al. 1999).

Some vaccination programs can even be designed to eradicate pathogens in animal hosts, before the pathogens can be transmitted to human populations. Rabies virus is found in a broad spectrum of mammalian host species, but phylogenetic analyses show that only some of these species serve as a reservoir for the viruses that are transmitted to humans. By identifying the specific sources of the virus, oral vaccination programs can be developed for wild populations of the critical mammalian species (Rupprecht et al. 2004), thus preventing the transmission of the virus into human populations.

Phylogeny and the Function of Extinct Gene Sequences

Analyses of controlled experimental (directly observed) evolutionary histories demonstrated that ancestral sequences of genes could be reconstructed accurately using phylogenetic analysis (Hillis et al. 1992; Bull et al. 1993). Sequences from many long-extinct ancestral species have now been reconstructed using such techniques, and in many cases, these protein sequences have been resurrected, expressed, and studied under controlled laboratory conditions (Thornton 2004). Such analyses require broad understanding of phylogenetic analyses from across the Tree of Life.

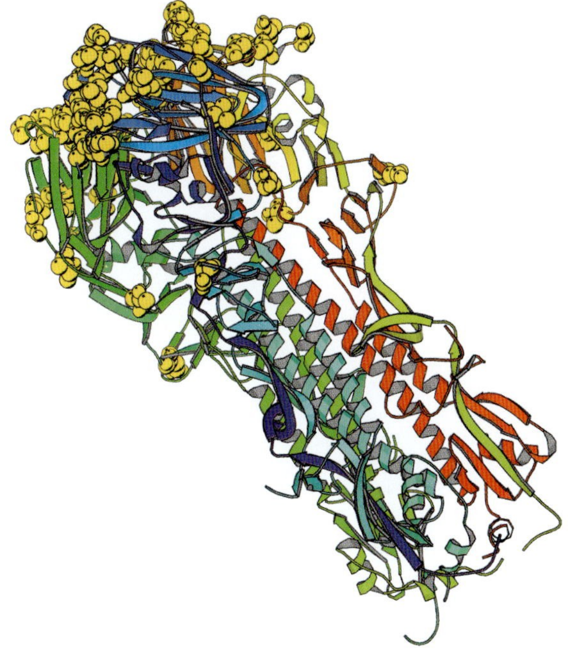

FIGURE 16.10 Model of the Hemagglutinin Protein, Showing Amino Acid Sites that Are Under Positive Selection for Change (Yellow Spheres) From Hillis 1999a.

Shi et al. (2001) and Chang et al. (2002) used phylogenetic analyses and ancestral-state reconstruction methods to resurrect and study the function of visual pigment proteins that existed hundreds of millions of years ago in vertebrate evolution. By reconstructing the inferred protein sequences in the laboratory, they could study the particular sequence of changes that gave rise to important functional shifts in vertebrate vision. In a similar manner, a phylogenetic analysis and reconstruction of bacterial proteins allowed Gaucher et al. (2003) to study the function of these ancient proteins in the laboratory. The protein activity could then be tested under different thermal conditions to infer the paleoenvironment of ancestral bacteria. In another example, phylogenetic analysis and reconstruction of steroid receptors allowed Thornton et al. (2003) to understand the origin of estrogen signaling.

Ugalde et al. (2004) used phylogenetic analyses to reconstruct the genes that produce fluorescent proteins in corrals, expressed these genes in bacteria, and used the analyses to study convergent evolution of fluorescent pigments (Figure 16.11). They were thereby able to demonstrate that the red fluorescent pigment in the corral *Montastraea cavernosa* evolved independently from the red fluorescent pigment in other corrals.

The combination of phylogenetic and functional analyses of genes and proteins is producing a clear understanding of how genes evolve through

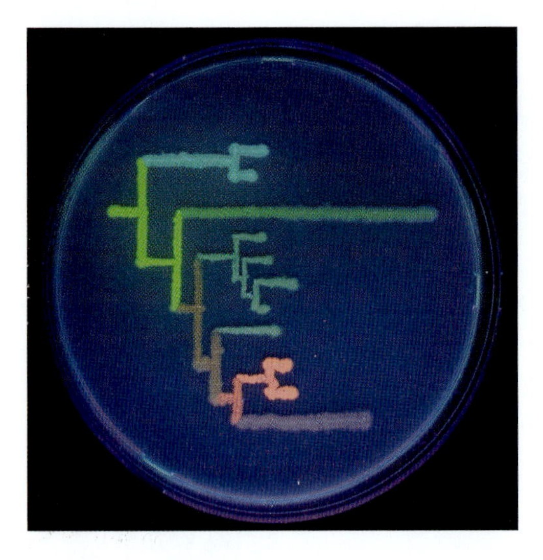

FIGURE 16.11 Extant and Resurrected Corral Pigments, Cloned and Expressed in Bacteria, Plated on an Agar Plate, and Photographed under UV Light From Ugalde et al. 2004; photo courtesy of Mikhail Matz.

time as well as how specific substitutions in genes affect the function of the proteins they encode. These analyses often have applications well beyond the species studied. For example, an analysis of the parallel changes that confer tetrodotoxin resistance within the sodium channel genes of pufferfishes revealed changes that have occurred repeatedly in the evolution of animal sodium channel genes (Jost et al. 2008). *In vitro* expression of these individual changes revealed the mechanisms of specific functional changes in the sodium channels. In this manner, phylogenetic and functional analyses of genes and proteins are allowing biologists to focus on the specific changes (among the many observed) that have given rise to specific attributes and behaviors over evolutionary time.

Phylogeny, Biodiversity Discovery, and the Encyclopedia of Life

Many genes are gained and lost throughout the Tree of Life. However, a few are essential to life, and they can be used to understand the phylogenetic relationships across all living species. For example, the ribosomal RNA genes encode the RNA component of ribosomes, which are required for protein synthesis. Since all organisms need to produce proteins, these essential genes are found in every independently living species. They can be used to reconstruct the evolutionary relationships among the most distantly related branches of the Tree of Life (Figure 16.12), thus producing a visualization of the tree that Darwin (1859) described as covering the surface of the Earth: "with its ever branching and beautiful ramifications" (Darwin 1859: 130). This tree is too large to be easily seen in the pages of this book, but a full poster-sized version can be downloaded from http://www.zo.utexas.edu/faculty/antisense/DownloadfilesToL.html.

A simplified version of the Tree of Life (derived from many studies and adapted from the Tree of Life Appendix in Sadava et al. 2009) is shown in Figure 16.13. The Assembling the Tree of Life program of the U.S. National Science Foundation is coordinating efforts by several groups to fill out the details of the various major branches. Not only will this information database allow the kind of applications described in this chapter, but it will also permit the organization and understanding of biological information that has never been possible before.

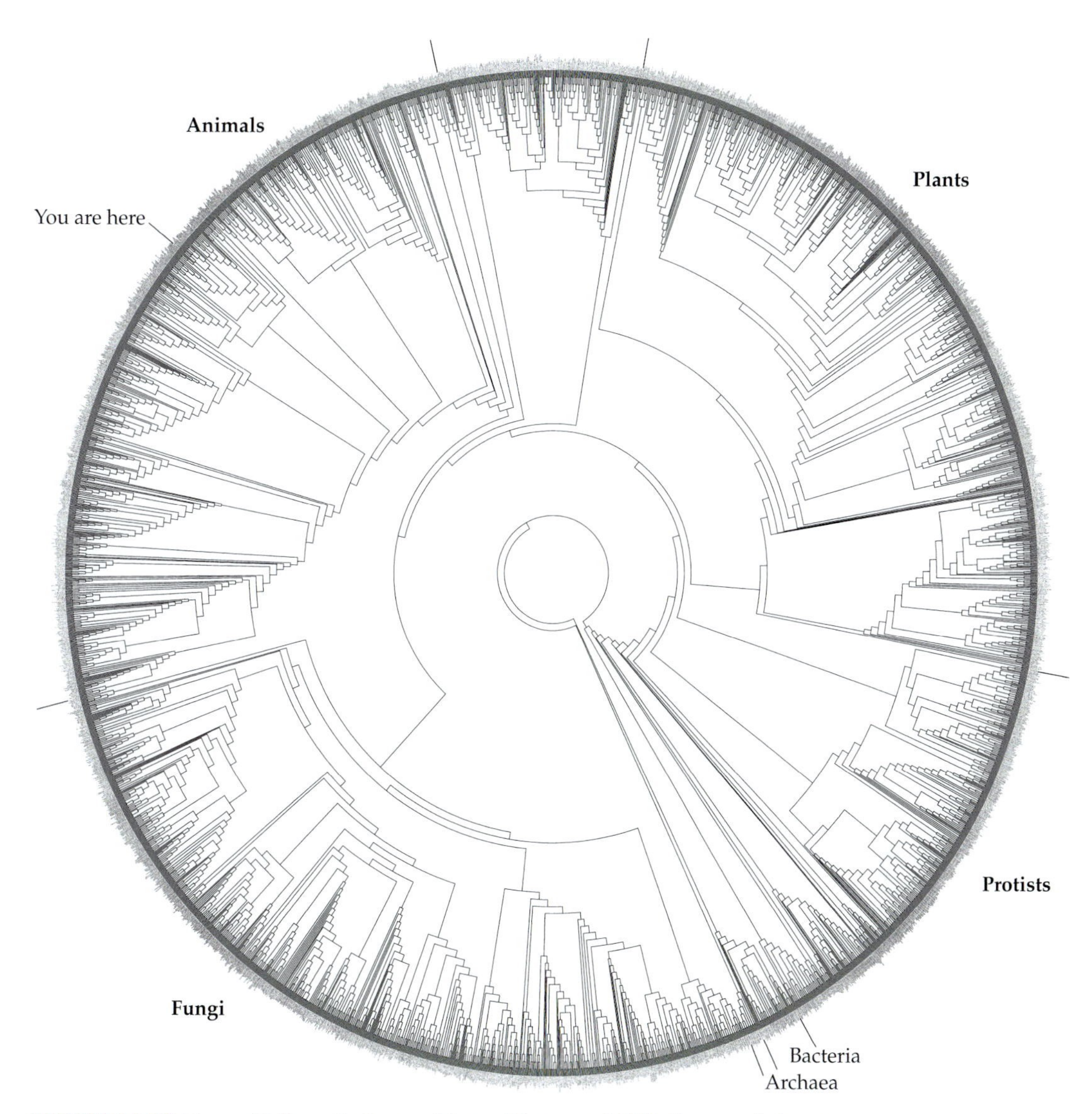

FIGURE 16.12 Tree of Life as Estimated from Ribosomal RNA Genes of about 3000 Species, Sampled from throughout Life A full size (2 meter × 2 meter) poster version of this tree can be downloaded from http://www.zo.utexas.edu/faculty/antisense/DownloadfilesToL.html, where details of the analysis are presented.

The tree depicted in Figure 16.12 shows the relationships among about 3000 species, sampled from throughout life, which represents less than 0.2% of the extant, described species on Earth and roughly the square-root of the number of living species that are estimated to exist on the planet. That

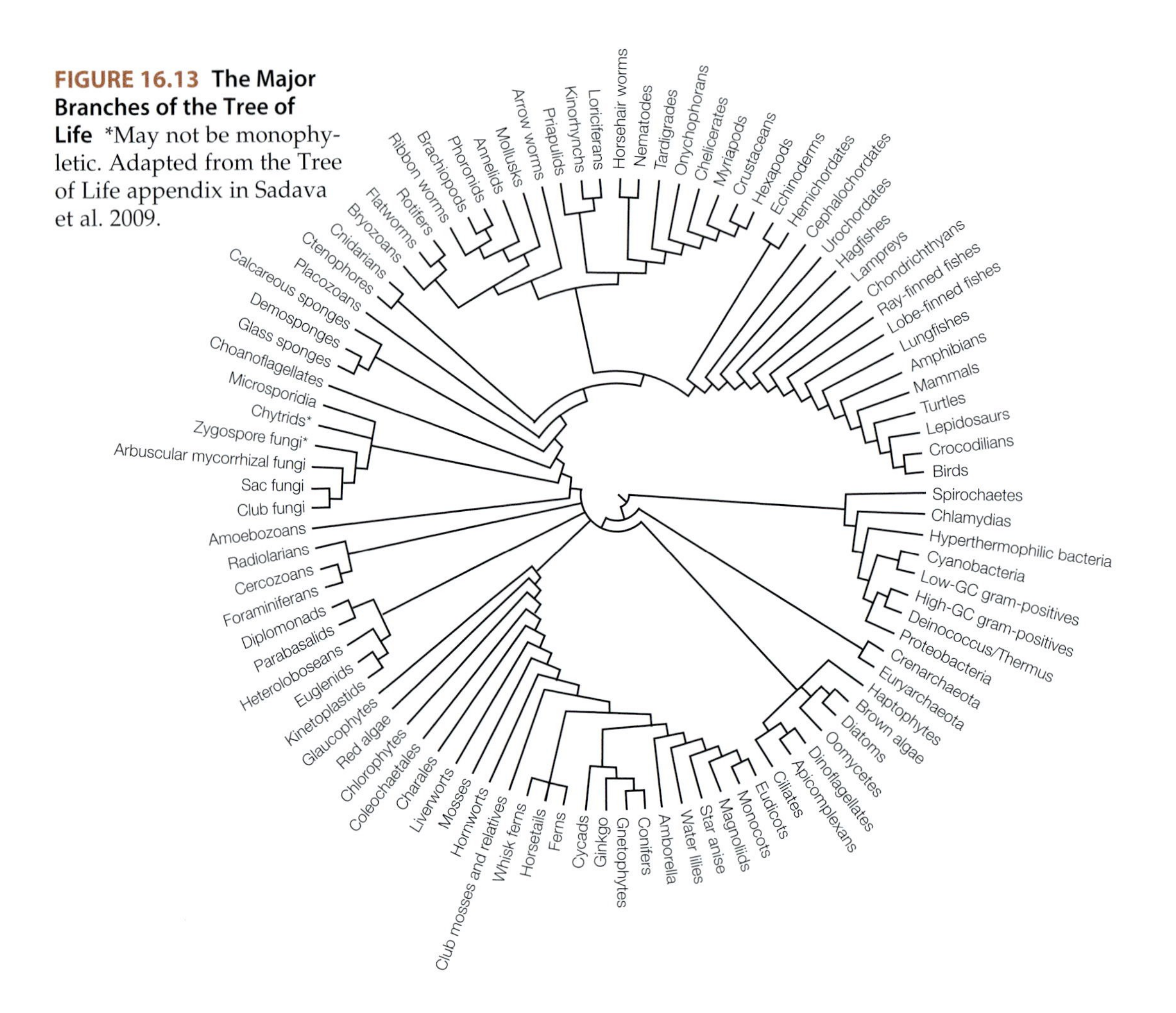

FIGURE 16.13 The Major Branches of the Tree of Life *May not be monophyletic. Adapted from the Tree of Life appendix in Sadava et al. 2009.

means that the complete Tree of Life for living species could be depicted in just two levels of trees shown in the amount of detail depicted in Figure 16.12. Sequences of just a few genes can place any organism into this framework of the Tree of Life. Advances and automation of DNA sequencing now allow any unknown organism to be identified by phylogenetic placement using a few genes. Not only can unknown organisms be identified in this manner, but the trees can also be used as a framework for organizing the entirety of biological literature and information. Already, large databases such as GenBank are organized using such a taxonomic hierarchy, and there are now efforts to build an online reference system and biological database for all species using the Tree of Life as an organizing principle (e.g., see the Encyclopedia of Life project at http://www.eol.org/).

Although many of the undiscovered species on Earth are small (often microscopic) organisms, even some vertebrate groups are surprisingly poorly known. To illustrate the use of the Tree of Life for biological discovery purposes, consider the discovery of a new species of vertebrate at one of the most intensively studied spots in the world: Barton Springs, in the middle of Austin, the capital city of Texas. There is an endemic, endangered species of salamander (*Eurycea sosorum*, the Barton Springs Salamander) known to occur only at Barton Springs. This species was described in 1993 (Chippindale et al. 1993) and was quickly listed as an Endangered Species under the U.S. Endangered Species Act. The city of Austin hired several full-time salamander biologists to monitor the species, improve its habitat, and establish a captive breeding program in case of an environmental accident within the recharge zone of the springs (which is largely within the city limits of Austin). These biologists began a weekly SCUBA survey of the population, in which the habitat for the salamanders was searched extensively. Probably no other species of amphibian in the world has been studied quite as intensively.

After years of weekly surveys, the biologist in charge of the captive breeding program, Dee Ann Chamberlain, noted that some of the juvenile salamanders looked slightly different from others: they were less pigmented and had highly reduced eyes. Upon raising some of these light-colored juveniles, she discovered that the adults were dramatically different from typical salamanders and completely lacked image-forming eyes. Were these salamanders the product of unusual developmental conditions, or could there possibly be another undescribed vertebrate species at such a well-studied and well-known location?

The method used to answer this question is the same one that biologists now routinely use to identify most new species, whether they are bacteria in remote Antarctica (Mikucki et al. 2009) or vertebrates in the middle of a large city. DNA sequences were obtained from the specimens in question, and these sequences were placed within the framework of the Tree of Life (Hillis et al. 2001). There is no doubt, in the case of a salamander, where to start the analysis, but for many organisms the initial identification to a major branch of organisms is not always obvious. In such cases, conserved gene sequences (such as rRNA genes) can quickly place the unknown organism into a major taxonomic group (by placing the sequence into a tree, such as the ones shown in Figure 16.12 and 16.13). Once the relevant taxonomic group is identified, the unknown sample can then be placed in phylogenetic relationship to the most closely known species (in this case, all the described species of salamanders in the world; Figure 16.14). Zooming in on the details of the tree in Figure 16.14 shows that the new sample clusters are from the subclade *Typhlomolge*, rather than with the subclade *Blepsimolge*, to which the Barton Springs salamander belongs (see Figure 16.15). The unknown sample was indeed an undescribed species and was described as *Eurycea waterlooensis*, the Austin Blind salamander. It had been missed during all those years of weekly surveys because the adults are only rarely observed at the surface, and the juveniles had not been examined closely by biologists.

FIGURE 16.14 A Phylogenetic Supertree of All Described Species of Salamanders

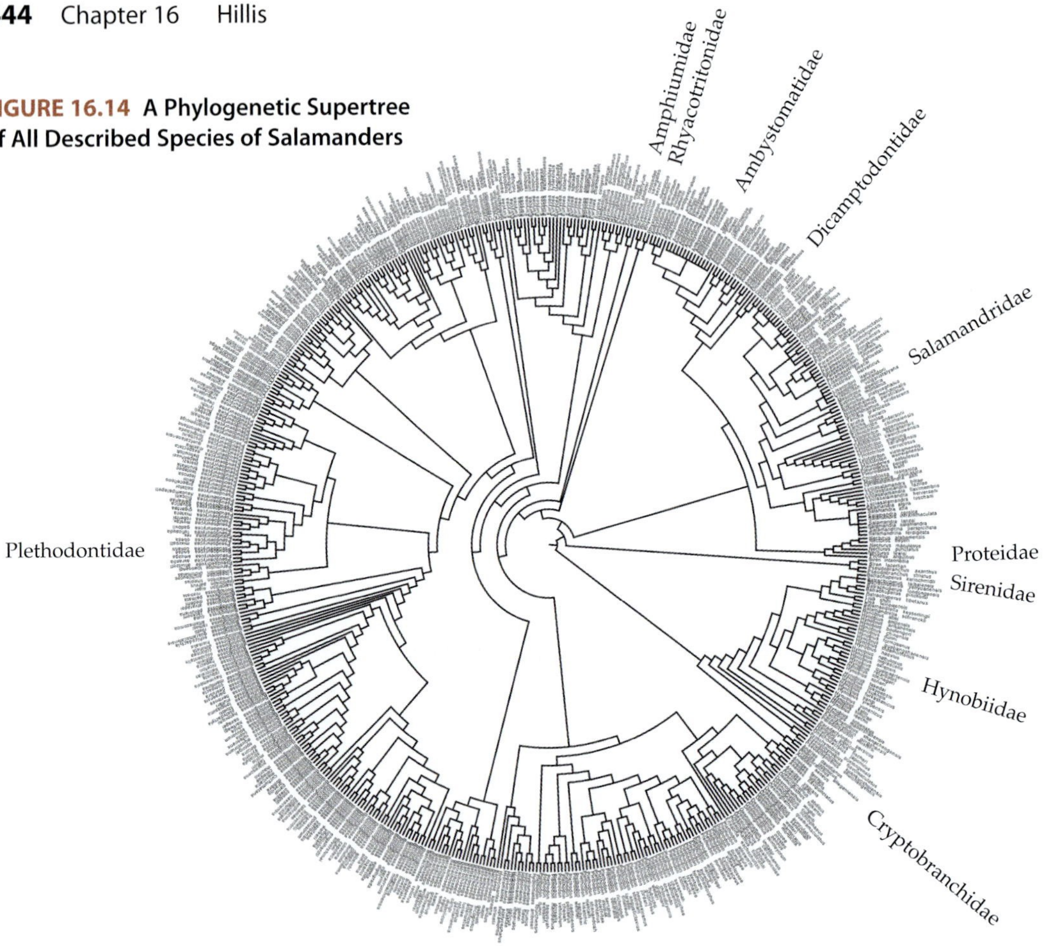

Amphiumidae
Rhyacotritonidae
Ambystomatidae
Dicamptodontidae
Salamandridae
Proteidae
Sirenidae
Hynobiidae
Cryptobranchidae
Plethodontidae

If it is possible to miss a new, spectacular species of vertebrate at an intensively studied site in the middle of an urban area, what are we missing in the rest of the world, and what can we do about it? Fortunately, the development of databases on the Tree of Life offers a solution. All of the methods used to place unknowns into the framework of the Tree of Life (DNA isolation, DNA sequencing, and phylogenetic analysis) have been automated and miniaturized. It now remains to combine these miniaturized methods into a simple, portable device that can be carried anywhere to identify any species. This identification can then be used to access the global database of relevant information. Placement within the Tree of Life gives an automated identification: the sample either falls within the population clustering of known species (which identifies it as that species), or it falls outside the range of variation of known species. In the latter case, its relationship to known species provides an instant classification and facilitates its description as a new species.

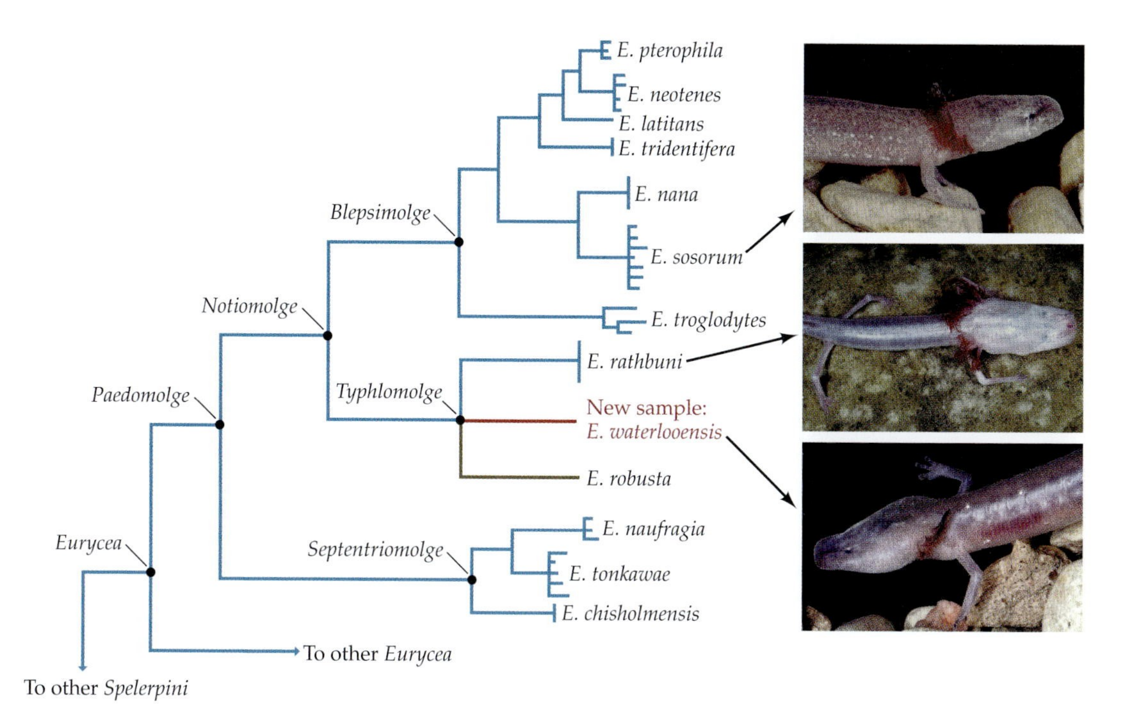

FIGURE 16.15 Details of a Portion of the Salamander Tree, Showing the Relationships among the Central Texas Paedomorphic Salamanders of the Genus Eurycea The three photos show the Barton Springs Salamander (*E. sosorum*, upper), the Texas Blind Salamander (*E. rathbuni*, middle), and the Austin Blind Salamander (*E. waterlooensis*, shown in the tree as "New sample," lower).

Just knowing the identity of a biological sample has little utility without connection to the biological database for the relevant species, which is the purpose of the Encyclopedia of Life project: to provide a connection to the information database of every species on Earth. As these projects (the Tree of Life and Encyclopedia of Life) reach maturity over the coming years, biological research and applications will be revolutionized. With this technology, anyone anywhere will be able to automatically identify biological samples and access all the information that is known about the species.

Although Charles Darwin had amazing vision, he could not have imagined the technology that would allow biologists to reconstruct and utilize the Tree of Life in this manner. Those "ever branching and beautiful ramifications" are now serving to inform us about life on Earth in unprecedented detail.

Literature Cited

Alexander, L. N., J. F. Seward, T. A. Santibanez, and 8 others. 2004. Vaccine policy changes and epidemiology of poliomyelitis in the United States. *J. Am. Med. Assoc.* 292: 1696–1701.

Benson, L. 1962. *Plant Taxonomy: Methods and Principles.* Roland Press, New York.

Blackwelder, R. E. 1967. *Taxonomy: A Text and Reference Book.* John Wiley and Sons, New York.

Bull, J. J., C. W. Cunningham, I. J. Molineux, and 2 others. 1993. Experimental molecular evolution of bacteriophage T7. *Evolution* 47: 993–1007.

Bush, R. M., C. A. Bender, K. Subbarao, and 2 others. 1999. Predicting the evolution of human influenza A. *Science* 286: 1921–1925.

Butler, D. 2007. Libyan ordeal ends: Medics freed. *Nature* 448: 398.

Camin, J. H. and R. R. Sokal. 1965. A method for deducing branching sequences in phylogeny. *Evolution* 19: 311–326.

Chang, B. S., K. Jonsson, M. A. Kazmi, and 2 others. 2002. Recreating a functional ancestral archosaur visual pigment. *Mol. Biol. Evol.* 19: 1483–1489.

Chippindale, P. T., A. H. Price, and D. M. Hillis. 1993. A new species of perennibranchiate salamander (*Eurycea*, Plethodontidae) from Austin, Texas. *Herpetologica* 49: 242–259.

Core, E. L. 1955. *Plant Taxonomy.* Prentice Hall, Englewood Cliffs, NJ.

Darwin, C. 1859. *The Origin of Species.* New American Library of World Literature ed. (1958), New York.

de Oliveira, T., O. G. Pybus, A. Rambaut, and 11 others. 2006. Molecular epidemiology: HIV-1 and HCV sequences from Libyan outbreak. *Nature* 444: 836–837.

Donoghue, M. J. and J. W. Kadereit, 1992. Walter Zimmermann and the growth of phylogenetic theory. *Syst. Biol.* 41: 74–85.

Dowdle, W. R., E. De Gourville, O. M. Kew, and 2 others. 2003. Polio eradication: The OPV paradox. *Rev. Med. Virol.* 13: 277–291.

Edwards, A. W. F. and L. L. Cavalli-Sforza. 1964. Reconstruction of evolutionary trees. In W. H. Heywood and J. McNeill (eds.), *Phenetic and Phylogenetic Classification*, pp. 67–76. Systematics Association Publication 6, London.

Eldredge, N. and J. Cracraft. 1980. *Phylogenetic Patterns and the Evolutionary Process.* Columbia University Press, New York.

Felsenstein, J. 1985. Phylogenies and the comparative method. *Am. Nat.* 125: 1–15.

Fitch, W. M. and E. Margoliash. 1967. Construction of phylogenetic trees. *Science* 155: 279–284.

Gaucher, E. A., J. M. Thomson, M. F. Burgan, and 1 other. 2003. Inferring the palaeoenvironment of ancient bacteria on the basis of resurrected proteins. *Nature* 425: 285–288.

Haeckel, E. 1866. *Generelle Morphologie der Organismen: Allgemeine Grundzüge der organischen Formen-Wissenschaft, mechanisch begründet durch die von C. Darwin reformirte Decendenz-Theorie.* Georg Reimer, Berlin.

Hahn, B. H., G. M. Shaw, K. M. De Cock, and 1 other. 2000. AIDS as a zoonosis: Scientific and public health implications. *Science* 287: 607–614.

Halloran, M. E., R. M. Anderson, R. S. Azevedo-Neto, and 19 others. 1998. Population biology, evolution, and immunology of vaccination and vaccination programs. *Am. J. Med. Sci.* 315: 76–86.

Harvey, P. H. and M. D. Pagel. 1991. *The Comparative Method in Evolutionary Biology.* Oxford University Press, New York.

Hennig, W. 1950. *Grundzüge einer Theorie der Phylogenetischen Systematik.* Deutscher Zentralverlag, Berlin.
Hennig, W. 1966. *Phylogenetic Systematics.* University of Illinois Press, Urbana.
Hillis, D. M. 1999a. Predictive evolution. *Science* 286: 1866–1867.
Hillis, D. M. 1999b. Phylogenetics and the study of HIV. In K. A. Crandall (ed.), *The Evolution of HIV*, pp. 105–121. Johns Hopkins University Press, Baltimore.
Hillis, D. M. and J. P. Huelsenbeck. 1994. Support for dental HIV transmission. *Nature* 369: 24–25.
Hillis, D. M. and C. Moritz. 1990. *Molecular Systematics.* Sinauer Associates, Sunderland, MA.
Hillis, D. M., J. J. Bull, M. E. White, and 2 others. 1992. Experimental phylogenetics: Generation of a known phylogeny. *Science* 255: 589–592.
Hillis, D. M., D. A. Chamberlain, T. P. Wilcox, and 1 other. 2001. A new species of subterranean blind salamander (Plethodontidae: Hemidactyliini: *Eurycea: Typhlomolge*) from Austin, Texas, and a systematic revision of central Texas paedomorphic salamanders. *Herpetologica* 57: 266–280.
Hillis, D. M., J. P. Huelsenbeck, and C. W. Cunningham. 1994. Application and accuracy of molecular phylogenies. *Science* 264: 671–677.
Hull, D. 1988. *Science as a Process: An Evolutionary Account of the Social and Conceptual Development of Science.* University of Chicago Press, Chicago.
Hull, D. L. 1989. The evolution of phylogenetic systematics. In B. Fernholm, K. Bremer, L. Brundin, H. Jörnvall, L. Rutberg, and H.-E. Wanntorp (eds.), *The Hierarchy of Life: Molecules and Morphology in Phylogenetic Analysis*, pp. 3–15. Excerpta Medica, Amsterdam.
Huxley, J. 1942. *Evolution: The Modern Synthesis.* Allen and Unwin, London.
Jost, M. C., D. M. Hillis, Y. Lu, and 3 others. 2008. Toxin-resistant sodium channels: Parallel adaptive evolution across a complete gene family. *Mol. Biol. Evol.* 25: 1016–1024.
Kew, O. M., P. F. Wright, V. I. Agol, and 4 others. 2004. Circulating vaccine-derived polioviruses: Current state of knowledge. *Bull. World Health Organ.* 82: 16–23.
Kiriakoff, S. G. 1959. Phylogenetic systematics versus typology. *Syst. Zool.* 8: 117–118.
Korber, B., M. Maldoon, J. Theiler, and 6 others. 2000. Timing the ancestor of the HIV-1 pandemic strains. *Science* 288: 1789–1796.
Korotkova, E. A., R. Park, E. A. Cherkasova, and 5 others. 2003. Retrospective analysis of a local cessation of vaccination against poliomyelitis: A possible scenario for the future. *J. Virol.* 77: 12460–12465.
Ksiazek, T. G., D. Erdman, C. S. Goldsmith, and 23 others. 2003. A novel coronavirus associated with Severe Acute Respiratory Syndrome. *N.E. J. Med.* 348: 1953–1966.
Lanciotti, R. S., G. D. Ebel, V. Deubel, and 9 others. 2002. Complete genome sequences and phylogenetic analysis of West Nile virus strains isolated from the United States, Europe, and the Middle East. *Virology* 298: 96–105.
Leitner, T., D. Escanilla, C. Franzen, and 2 others. 1996. Accurate reconstruction of a known HIV-1 transmission history by phylogenetic tree analysis. *Proc. Natl. Acad. Sci. USA* 93: 10864–10869.
Leitner, T., S. Kumar, and J. Albert. 1997. Tempo and mode of nucleotide substitutions in *gag* and *env* gene fragments in human immunodeficiency virus type 1 populations with a known transmission history. *J. Virol.* 71: 4761–4770.
Mayr, E. 1969. *Principles of Systematic Zoology.* McGraw-Hill, New York.

Mayr. E., E. G. Linsley, and R. L. Usinger. 1953. *Methods and Principles of Systematic Zoology*. McGraw-Hill, New York.

McCutchan, F. E. 1999. Global diversity in HIV. In K. A. Crandall (ed.), *The Evolution of HIV*, pp. 41–101. Johns Hopkins University Press, Baltimore.

Metzker, M. L., D. P. Mindell, X.-M. Liu, and 3 others. 2002. Molecular evidence of HIV-1 transmission in a criminal case. *Proc. Natl. Acad. Sci. USA* 99: 14292–14297.

Mikucki, J. A., A. Pearson, D. T. Johnston, and 6 others. 2009. A contemporary microbially maintained subglacial ferrous "ocean." *Science* 324: 397–400.

Nelson, G. J. and N. I. Platnick. 1981. *Systematics and Biogeography: Cladistics and Vicariance*. Columbia University Press, New York.

Nichol, S. T., C. F. Spiropoulou, S. Morzunov, and 7 others. 1993. Genetic identification of a hantavirus associated with an outbreak of acute respiratory illness. *Science* 262: 914–917.

Ou, C.-Y., C. A. Ciesielski, G. Myers, and 13 others. 1992. Molecular epidemiology of HIV transmission in a dental practice. *Science* 256: 1165–1171.

Peiris, J. S. M., S. T. Lai, L. L. M. Poon, and 13 others. 2003. Coronavirus as a possible cause of severe acute respiratory syndrome. *Lancet* 361: 1319–1325.

Platnick, N. I. 1979. Philosophy and the transformation of cladistics. *Syst. Zool.* 28: 537–546.

Platnick, N. I. 1985. Philosophy and the transformation of cladistics revisited. *Cladistics* 1: 87–94.

Remane, A. 1956. *Die Grundlagen des natürlichen Systems, der vergleichenden Anatomie und der Phylogenetik*. Geest and Portig, Leipzig.

Ridley, M. 1986. *Evolution and Classification: The Reformation of Cladism*. Longman, New York.

Ross, H. H. 1974. *Biological Systematics*. Addison-Wesley, Reading, MA.

Rupprecht, C. E., C. A. Hanlon, and D. Slate. 2004. Oral vaccination of wildlife against rabies: Opportunities and challenges in prevention and control. *Dev. Biol. (Basel)* 119: 173–184.

Sadava, D., D. M. Hillis, H. C. Heller, and M. R. Barenbaum. 2010. *Life: The Science of Biology*, 9th ed. Sinauer Associates, Sunderland, MA and W. H. Freeman, New York.

Shi, Y., F. B. Radlwimmer, and S. Yokoyama. 2001. Molecular genetics and the evolution of ultraviolet vision in vertebrates. *Proc. Natl. Acad. Sci. USA* 98: 11731–11736.

Simpson, G. G. 1945. The principles of classification and a classification of mammals. *Bull. Amer. Mus. Natl. Hist.* 85: 1–350.

Simpson, G. G. 1961. *Principles of Animal Taxonomy*. Columbia University Press, New York.

Simpson, G. G. and A. Roe. 1939. *Quantitative Zoology*. McGraw-Hill, New York.

Smith, G. J. D., D. Vijaykrishna, J. Bahl, and 10 others. 2009. Origins and evolutionary genomics of the 2009 swine-origin H1N1 influenza A epidemic. *Nature* 459: 1122–1125.

Sneath, P. H. A. and R. R. Sokal. 1973. *Numerical Taxonomy*. W. H. Freeman, San Francisco.

Sokal, R. and P. Sneath. 1963. *The Principles of Numerical Taxonomy*. W. H. Freeman, San Francisco.

Swofford, D. L. and G. J. Olsen. 1990. Phylogeny reconstruction. In D. M. Hillis and C. Mortiz (eds.), *Molecular Systematics*, pp. 441–501. Sinauer Associates, Sunderland, MA.

Thornton, J. W. 2004. Resurrecting ancient genes: Experimental analysis of extinct molecules. *Nat. Rev. Genet.* 5: 366–375.
Thornton, J. W., E. Need, and D. Crews. 2003. Resurrecting the ancestral steroid receptor: Ancient origin of estrogen signaling. *Science* 301: 1714–1717.
Twiddy, S. S., J. J. Farrar, N. V. Chau, and 5 others. 2002. Phylogenetic relationships and differential selection pressures among genotypes of Dengue-2 virus. *Virology* 298: 63–72.
Ugalde, J. A., B. S. Chang, and M. V. Matz. 2004. Evolution of corral pigments recreated. *Science* 305: 1433.
Van Heuverswyn, F., Y. Y. Li, C. Neel, and 13 others. 2006. Human immunodeficiency viruses: SIV infection in wild gorillas. *Nature* 444: 164.
Wiley, E. O. 1981. *Phylogenetics: The Theory and Practice of Phylogenetic Systematics.* John Wiley and Sons, New York.
Worobey, M., M. Gemmel, D. E. Teuwen, and 9 others. 2008. Direct evidence of extensive diversity of HIV-1 in Kinshasa by 1960. *Nature* 455: 661–664.
Yang, C.-F., T. Naguib, S.-J. Yang, and 10 others. 2003. Circulation of endemic type-2 vaccine-derived poliovirus in Egypt from 1983 to 1993. *J. Virology* 77: 8366–8377.
Zimmermann, W. 1931 (1937). Arbeitsweise der botanischen Phylogenetik und anderer Gruppierungswissenschaften. In E. Abderhalden (ed.), *Handbuch der Biologischen Arbeitsmethoden*. Urban und Schwarzenberg, Berlin.

Chapter 17

Paleontological Perspectives on Morphological Evolution

Peter J. Wagner

The fossil record documents over three billion years of evolution. Darwin considered the general succession of morphologies in the fossil record to corroborate the basic idea of evolution by natural selection. However, he also recognized aspects of the fossil record that challenged his ideas. For example, did the sudden appearance of higher taxa reflect a different underlying model of evolution or the imperfections of the fossil record? In the 150 years since the publication of *The Origin of Species*, paleontologists have used fossil data to challenge, corroborate, and augment elements of Darwin's basic model. Here, I will focus particularly on work from the last 40 years. In that time, paleontologists have taken advantage of advances in biological theory and quantitative methods to predictably and quantitatively summarize patterns that Darwin could only describe verbally and to articulate predictions for Darwin's models that Darwin himself was unable to derive. Paleontologists also have devised tools for assessing sampling levels, thus testing whether geology or biology might be responsible for patterns. Fossil data contradict some of Darwin's ideas, such as the idea that differences among higher taxa reflect the unsampled, long-term accumulation of changes. However, fossil data strongly corroborate Darwin's idea that natural selection is the major force for fixing novelties once they do appear.

Morphospace and Related Concepts

Throughout this paper, I frequently discuss concepts and patterns in terms of morphospace. A morphospace describes the distribution and range of observed and potential morphologies. Although the word is relatively new, the basic concept obviously is not: for example, Darwin's (1845: 379–380) summary of the range of forms among Galapagos finches represents a crude verbal description of a morphospace. Similarly, Sewall Wright's (1932) adaptive landscapes tacitly refer to morphospaces as potential selective points for adaptive peaks and valleys (Simpson 1944; Lande 1986).

However, use of the concept of repeatable, quantified morphospaces is relatively new. There are in three basic varieties of morphospaces. Theoretical morphospaces use predetermined characters, such as mathematical descriptions of mollusk shell coiling (Raup 1966; McGhee 1991). Empirical morphospaces use multivariate summaries of morphometric data (Foote 1990). Character-based morphospaces represents a hybrid involving multivariate summaries of predetermined characters (Foote 1992b). All three morphospaces permit repeatable quantitative summaries of what Darwin and other workers of his era could only verbally describe. More importantly, morphospaces enable workers to make calculations and assessments, such as illustrate similarities as well as differences among classes of morphotypes and provide insights into *possible* but unobserved morphotypes that were more difficult to make with simple verbal descriptions.

Morphospaces are a useful device for describing important aspects of morphological evolution that paleontologists measure. Disparity describes the breadth and range of taxon distributions within morphospace (Gould 1991; Foote 1993b; Wills et al. 1994). Intrinsic constraints and/or ecological restrictions can make regions of morphospace difficult to occupy, evolve, or re-evolve, or simply may slow rates of change. Trends reflect shifts in morphospace occupancy over time (McShea 1994). Finally, rates represent the size of "steps" that phylogenies take through morphospace over either time or speciation events. These concepts all are related: disparity, constraint/restrictions, and trends not only all affect one another (Ciampaglio et al. 2001), but they can all be described or modeled in terms of rates (McShea 1994; Foote 1996b). Therefore, I will review disparity, constraints/restrictions, and trends in turn, with particular regard to rates, and emphasize areas in which paleontological research has corroborated, contradicted, or augmented Darwin's ideas.

Morphological Disparity

Overview

Workers throughout evolutionary biology commonly discuss constraints, ecological restrictions, trends, and rates, but until fairly recently only paleontologists studied disparity in great detail. Therefore, I will review disparity in the most detail. The paleontological literature of the 1980s uses the word "disparity" for a variety of related concepts. However, Gould's (1991) definition of disparity as the diversity of morphological types, in contrast to "diversity" as numbers of taxa (i.e., richness) is now the standard definition, and the one that I will use here.

Obviously, the concept of disparity is quite old, and Darwin implicitly referred to what is now called disparity in a number of ways. However, morphological disparity studies require a battery of tools, including numerical taxonomy, morphometrics, and multivariate analysis, which requires computer data to be easily analyzed and expressed. Thus, it was not until the 1990s that Foote (1990, 1991b, 1991c, 1992a, 1992b, 1993a, 1993b, 1994, 1996a, 1996b) and others (Wills et al. 1994; Ciampaglio 2002) built upon frameworks established in the 1960s to advance morphological disparity both methodologically and theoretically to the point where disparity studies offered objective summaries of evolutionary history.

Historical disparity studies based upon extant taxa are becoming common. Such studies use ancestral reconstructions over molecular phylogenies to infer disparity in the past (e.g., Harmon et al. 2003; Ricklefs 2004; Sidlauskas 2007; Adams et al. 2009). Paleontological studies differ fundamentally from these in that they use observed taxa from different time intervals rather than from inferred taxa at inferred time intervals. Thus, paleontological disparity studies do not depend on particular ideas of phylogeny, even if a phylogenetic context is invaluable for corroborating theoretical implications of those studies (see below). More importantly, paleontological disparity studies use extinct taxa that cannot be inferred from modern phylogenies. Arthropods exemplify this issue (Figure 17.1): the great disparity among extant arthropods concentrates largely in regions of morphospace unoccupied during the Cambrian, whereas the nearly equal disparity among Cambrian arthropods concentrates largely in regions unoccupied by extant taxa (Briggs et al. 1992; Wills et al. 1994). Carboniferous arthropods not only link the Cambrian and Recent morphospaces, but also add their own disparity (Stockmeyer Lofgren et al. 2003).

There are a variety of ways in which paleontologists evaluate disparity. Typical disparity metrics summarize the average differences among taxa, such as average pairwise dissimilarity or variance. Ciampaglio et al. (2001) and Wills (2001) provide excellent summaries of different disparity metrics. Consider a simple hypothetical example (Figure 17.2A): in its first stage, a clade has four species distributed evenly in morphospace. The clade then adds new species at a similar underlying rate of morphologic evolution as it did in the first stage. Given either pairwise dissimilarity or variance along the morphological axes 1 and 2, the expansion in morphospace accompanies an increase in disparity (Figure 17.2B).

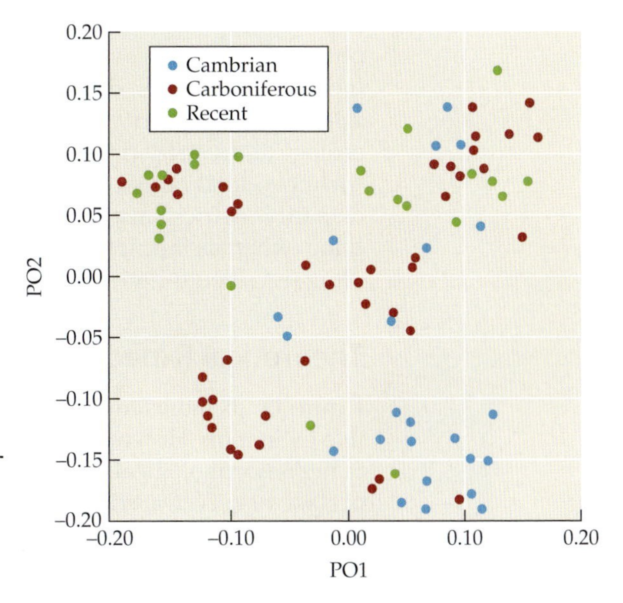

Figure 17.1 Principal Coordinate Summary of Arthropod Morphospace Occupancy within Three Geological Intervals PO1 and PO2 represent the first two principal coordinate axes. (Based on data from Wills et al. 1994 and Stockmeyer Lofgren et al. 2003.)

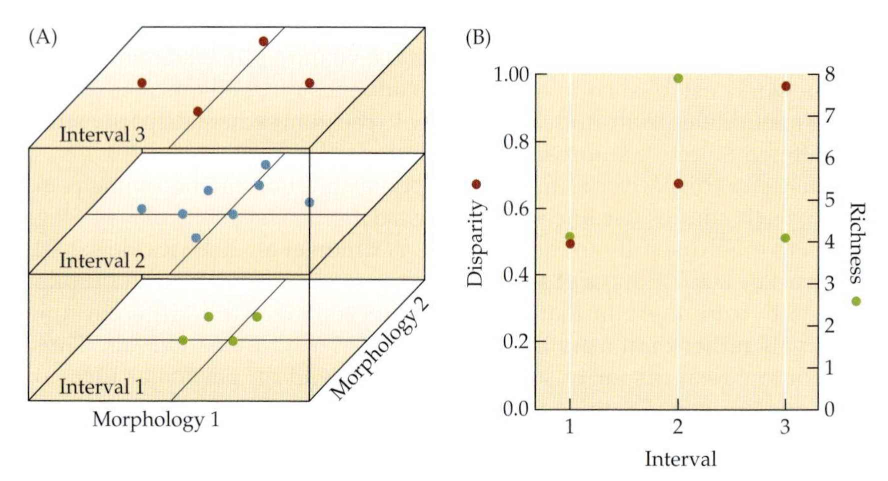

Figure 17.2 Morphologic Disparity: A Simple Example (A) Morphospace distributions of a hypothetical clade. Morphology might represent either measurements of a predefined theoretical character or multivariate summaries of discrete or morphometric characters. (B) Disparity through time (see red circles) is based on the pairwise dissimilarities among the taxa in each interval. Note that summed variances yield the same general pattern. Initially, disparity increases as richness increases (see green circles in B). However, the extinction of the middle of the morphospace strengthens disparity without altering the size of the occupied morphospace and thus increases disparity, while richness declines.

If the primitive morphologies become extinct after Interval 2 with no additions to morphospace (see Interval 3), then the sparser occupation of morphospace increases average pairwise dissimilarities and variances along morphological axes, despite the fact that the size of the morphospace is unchanged. Disparity measured in this manner is conceptually akin to ecological diversity (Hurlbert 1971), that is, it is a product both of the size (richness/total range) and the distributions (evenness/patchiness). These disparity metrics have the added advantage of being fairly robust to sampling, whereas metrics, such as the total range, are very sensitive to sampling (e.g., Foote 1991a; Roy and Foote 1997; Ciampaglio et al. 2001).

Theoretical Expectations of Disparity and Empircal Tests

Figure 17.2B shows high disparity accompanying low richness *after* selective extinction. However, the fossil record seems to offer many examples of high morphological disparity accompanying lower richness at the *outset* of clade history. Darwin describes this phenomenon in general terms when discussing the sudden appearance of a wide variety of metazoans in the Cambrian and also in reference to the sudden appearance of major taxa, such as teleost fishes. Darwin attributes such patterns to geology rather than biology, and I will return to how paleontologists test this sort of idea in the following discussion. Valentine (1969, 1980; Valentine and Campbell 1975; Erwin et al. 1987) as well as Gould (1989, 1991) and others argued that this pattern reflects evolution rather than geology: they invoked models of increasing intrinsic constraints and/or ecological restrictions as representative of slowing rates of morphological evolution.

Foote (1991b, 1993a, 1996b) used simulations to go beyond first-principle arguments and examined the expected relationship between diversification and disparity. He contrasted a wide variety of models, but I will focus on just a few important ones. Given constant step size (i.e., rate of morphological change between ancestors and descendents), Foote showed that we expect disparity to increase linearly as diversity increases exponentially (Figure 17.3A). Given a Valentine–Gould model of elevated early step-size, disparity begins at very high levels despite low richness and does not increase markedly as richness increases (Figure 17.3B). Adding constraints to morphospace will yield somewhat similar patterns; even if step-sizes do not decrease over time (Figure 17.3C), the size of the morphospace cannot continue to increase, resulting in a crowded, less disparate morphospace.

Foote's distinctions are not just theoretical experiments, as the majority of relevant disparity studies show high disparity early in clade histories (Table 17.1). This finding might not be *the* typical pattern, although paleontologists no doubt have been drawn to examples for which simple observation

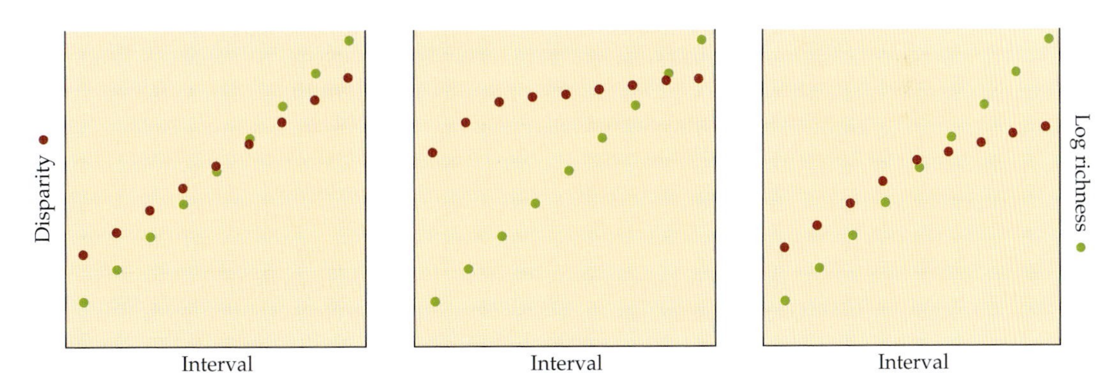

Figure 17.3 Disparity (Red Circles) and Richness (Green Circles) Under Three Models of Morphological Evolution with Exponential Diversification (A) Constant rates of morphologic change over time. (B) Elevated rates of morphologic change early, decreasing markedly after the early intervals. Although peak disparity does not change greatly from constant rates of morphologic change, the clade approaches peak disparity much more rapidly. (C) Constant rates of morphologic change, but with constraints/restrictions on the range of possible designs. Now the same initial start leads to much lower morphologic disparity. (Adapted from Foote 1993a: Figure 1 and Wesley-Hunt 2005: Figure 1.)

TABLE 17.1 DISPARITY PATTERNS

TAXON	STUDY
HIGH EARLY DISPARITY	
Proterozoic–Cambrian acritarchs	Huntley et al. (2006)
Angiosperm pollen	Lupia (1999)
Paleozoic gastropods	Wagner (1995)
Rostroconchs	Wagner (1997)
Ammonites	Saunders et al. (2008)
Ordovician bryozoans	Anstey and Pachut (1995)
Mesozoic articulate brachiopods	Ciampaglio (2004)
Paleozoic crinoids	Foote (1994, 1995)
Mesozoic crinoids	Foote (1996a)
Blastozoans	Foote (1992b)
Stylophorans	Lefebvre et al. (2006)
Holasteroid echinoids	Eble (2000)
Spatangoid echinoids	Eble (2000)
Arthropods	Briggs et al. (1992), Wills et al. (1994)
Crustacea	Wills (1998b)
Insects	Labandeira and Eble (2008)
Tetrapods	Ruta et al. (2006)
Archosaurs	Brusatte et al. (2008)
LOW EARLY DISPARITY	
Trilobites	Foote (1991c)
Olenelloid trilobites	Smith & Lieberman (1999)
Ptychoparioid trilobites	Cotton (2001)
Blastoids	Foote (1991b)
Atelostomate echinoids	Eble (2000)
Aporrhaid gastropods	Roy (1994)
Inarticulate brachiopods	Smith & Bunge (1999)
Paleozoic articulate brachiopods	Ciampaglio (2004)
Priapulids	Wills (1998a)
Carnivore mammals	Wesley-Hunt (2005)
Ungulate mammals	Jernvall (1996)

Source: Adapted from Erwin 2007: Table 1.
Note: Many clades with high disparity retain it late in their histories.

suggests this sort of pattern. Nevertheless, Figure 17.3B is common enough that the issue of whether disparity patterns predict rates of change (and shifts in those rates of change) across the empirical phylogenies, as Foote's simulations indicate that they should, must be addressed. When the necessary phylogenetic context is available, disparity patterns predict rate patterns accurately (Table 17.2). Taxa with relatively low early disparity do

TABLE 17.2 STUDIES EXAMINING RATE SHIFTS ASSOCIATED WITH DISPARITY PATTERN

TAXON	STUDY
HIGH EARLY RATES	
Paleozoic gastropods	Wagner (1995)
Rostroconchs	Wagner (1997)
Blastozoans	Smith (1988) & Foote (1992b)
Bryozoans	Anstey and Pachut (1995)
Coelacanths	Cloutier (1991)
Tetrapods	Ruta et al. (2006)
Archosaurs	Brusatte et al. (2008)
NO SHIFT	
Synapsids	Sidor and Hopson (1998)
Sauropod dinosaurs	Sereno et al. (1999)
Olenelloid trilobites	Smith and Lieberman (1999)
Ptychoparioid trilobites	Cotton (2001)

not show high early rates of change. For example, neither olenelloid nor ptychoparioid trilobites show any major shifts in rates (Smith and Lieberman 1999; Cotton 2001), and trilobites as a whole, show peak disparity reasonably late in clade history (Foote 1991c, 1993a). Among taxa with high early disparity, there are variations on elevated early rates. Tetrapods show a continuous decrease in rates over the middle-late Paleozoic (Ruta et al. 2006), whereas Paleozoic gastropods and Triassic archosaurs show very high rates in the earliest intervals, followed by lower rates in subsequent intervals (Wagner 1995; Brusatte et al. 2008). Within rostroconch mollusks, one derived subclade shows lower rates of change than does the stem-group and other subclades (Wagner 1997). However, that derived subclade is essentially the sole survivor of the end-Ordovician extinction, which thus concentrates the high rates in the first third of rostroconch history.

Elevated Rates of Change as Artifacts of an Imperfect Fossil

Record: Testing Darwin's Geological Hypothesis

Darwin (1859) suggested that the appearance of major groups, such as teleost fishes or trilobites, reflects the poor quality of the fossil record early in clade histories. Darwin (1859) went so far as to posit that the Silurian through the Recent represents half or less of the evolution of metazoans (note that Darwin's definition of Silurian included the Ordovician and workers had only just begun to recognize the Cambrian). Thus, the sediments containing much (if not most) of metazoan evolution are lost or undiscovered. Darwin (1859) also offered a perfectly plausible first-principles argument for how limited biogeographic distribution might hide the early history of teleosts from the fossil record. Therefore, we should ask whether Darwin's non-preservation hypothesis might explain high early disparity and high early measured rates.

For Paleozoic gastropods, Wagner (1995) showed that both the average samples per species and the number of sampling opportunities per time interval are *greater* in the latest Cambrian and Early Ordovician, when rates are high, and that sampling actually is substantially lower in the Middle Ordovician, when rates are moderate. Similarly, Wagner (1997) showed that sampling intensity is greater among Cambro–Ordovician rostroconchs than among post-Ordovician rostroconchs. In both cases, the relationship between sampling and reconstructed rates contradicts Darwin's expectations.

One might counter that Wagner does not actually test Darwin's hypothesis, as Darwin actually proposes long unsampled records *prior to* major taxa first appearing in the fossil record. Ruta et al. (2006) consider this approach for early tetrapods. Although Darwin does not mention tetrapods, they represent a good possible example of his non-preservation hypothesis. The major innovations of tetrapods correspond to high early disparity, and rates of change seem to decrease through their early history (Ruta et al. 2006; Figure 17.4). Although workers scrutinize the early

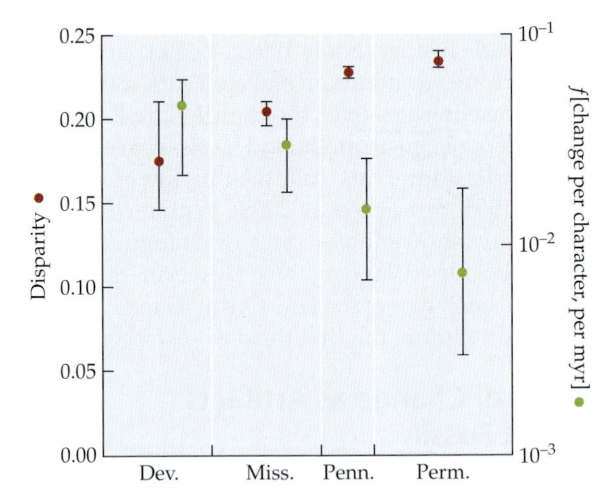

Figure 17.4 Disparity and Average Per-Branch Rates (*f*) of Character Change among Early Tetrapods Red circles denote disparity; green circles denote average per-branch rates of character change. Brackets for disparity represent 95% error bars from 500 bootstrap replications (Foote 1992b). Brackets for rates reflect 25th and 75th percentiles. The Devonian (Dev.)/Mississippian (Miss.) boundary is approximately 359 millions years ago (Ma). The Mississippian / Pennsylvanian (Penn.) boundary is approximately 318 Ma, and the Pennsylvanian/Permian (Perm.) boundary is approximately 299 Ma (Gradstein et al. 2005). "Myr" denotes millions of years. (Adapted from Ruta et al. 2006.)

tetrapod record keenly, that record clearly is not an outstanding record when compared to the marine mollusk record. For example, Niedzwiedzki et al. (2010) document tetrapod tracks from the Eifelian, approximately 12 million years prior to the origin of "tetrapodomorphs" (i.e., the red taxa in Figure 17.5), as assumed by Ruta et al. (2006). Thus, an extrapolation of Darwin's suggestion, that is, that high rates of early evolution in the middle Devonian–Early Carboniferous reflect unsampled evolution from the Early Devonian and earlier, might seem plausible in this case. Here, I will recast the first-principles argument of Ruta et al. quantitatively to illustrate how paleontologists can test and potentially refute Darwin's non-preservation hypothesis.

A phylogenetic context is critical to this argument. Tetrapods are a derived group of sarcopterygian fish, with sister taxa that also appear in the Middle to Late Devonian (see Figure 17.5). Thus, extending the origins of tetrapods to the earliest Devonian (i.e., when lungfish first appear) requires positing range extensions for at least six sarcopterygian fish taxa also (Smith 1988, 1994). If we lump the relatively short Lockhovian and Pragian

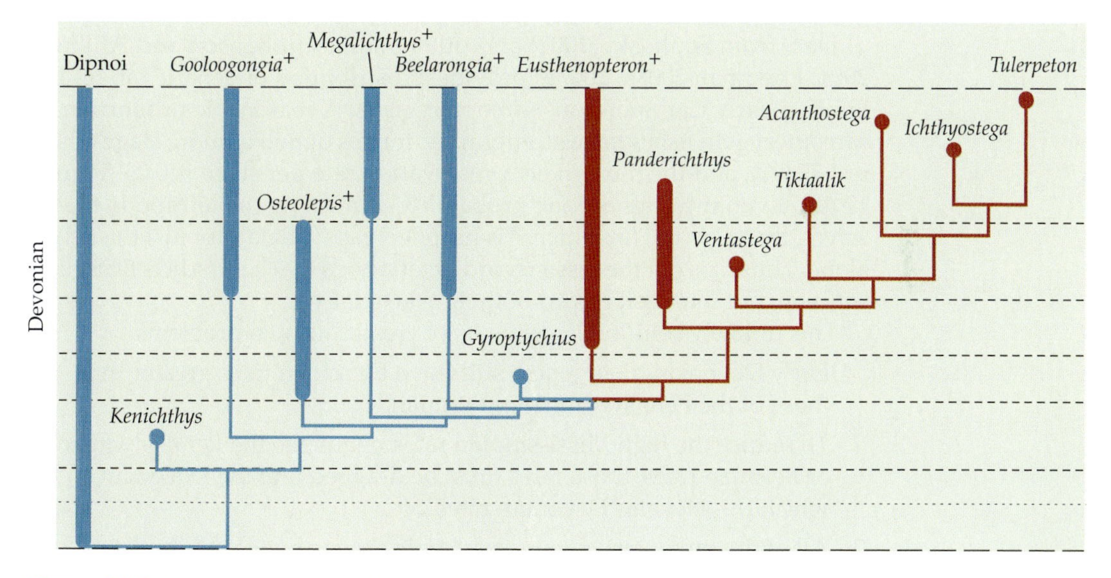

Figure 17.5 Sarcopterygian Phylogeny Divergence times for tetrapods using the analysis by Ruta et al. (2006) that optimizes estimated rates (see red). Divergence times for other sarcopterygians (see blue) reflect general appearance times in the fossil record as given. "+" denotes clade, including named genus. Dashed lines denote stages (see Gradstein et al. 2005). Faint dashed line denotes short stages, lumped with prior stages for preservation rate analyses. Note that a long unsampled record of tetrapods requires a long unsampled record of their closest fish relatives as well. (Graph based on consensus of data from Johanson and Ahlberg 2001, Ruta et al. 2003, Daeschler et al. 2006, Friedman 2007a, and Ruta and Coates 2007.)

stages into a single stage, an earliest Devonian tetrapod origin now requires an additional 11 stages of unsampled sarcopterygian fish record.

An estimate of the probability of these sampling gaps must now be made. The optimal approach involves using numbers of fossiliferous localities and the proportion of times that individual taxa are found in those localities (Strauss and Sadler 1989; Marshall 1990). Such a database is not yet available for early sarcopterygians. An alternative is to estimate R, or the average preservation potential per stage. The exact probability of an early Devonian diversification of tetrapods, given just the sarcopterygian fish record, then, is $(1 - R)$. We can estimate R using distributions of stratigraphic ranges. Although earlier workers use ranges to estimate extinction rates (Simpson 1953; Van Valen 1973), Sepkoski (1975) notes that imperfect preservation truncates stratigraphic ranges and thus affects "survivorship curves." Foote and Raup (1996) further note that as preservation becomes worse, the proportion of taxa sampled from a single time interval increases relative to the true (but unknown) survivorship curve. Numerous taxa known from two or more stages thus reflect (in part) good enough preservation for paleontologists to find taxa in two or more stages. Foote (1997) formalizes this idea to estimate the joint likelihood of average extinction and preservation rates given stratigraphic ranges.

Data from Sepkoski (2002) and other sources (Johanson and Ahlberg 2001; Friedman 2007a, 2007b) provide a distribution of generic ranges for Silurian-early Carboniferous sarcopterygians. A reasonable proportion of sarcopterygian fishes have stratigraphic ranges of two or more stages (Figure 17.6A), and the most likely preservation rate per stage is 0.23 (Figure 17.6B). Even if we ignore the probability of missing the tetrapods themselves, then $P < 0.06$ (in which P is [implied gaps | data]) for just the fishes alone. Thus, it is not the fossil record of tetrapods that contradicts Darwin's solution, but the fossil record of their near-relatives.

This unlikely solution leaves and/or creates its own problems:

1. Early Devonian divergences still leave Devonian rates greater than those of the Pennsylvanian or Permian.
2. To reduce the high Mississippian rates relative to the Pennsylvanian or Permian rates, those rates must be dragged into the Devonian, which (re-)elevates Devonian rates.
3. All of the sarcopterygian morphologic change between lungfishes and tetrapods now must be squeezed into the very beginning of the Devonian.

Accommodating all three problems by pushing the divergences of sarcopterygians still deeper in time simply pushes the problems elsewhere. Moreover, such action requires additional (and increasingly unlikely) gaps in the records not just of sarcopterygians, but also of basal osteichthyans, chondrichthyans, and placoderms. New "oldest evidence of" finds, such as those reported by Niedzwiedzki et al. (2010), still might alter ideas about

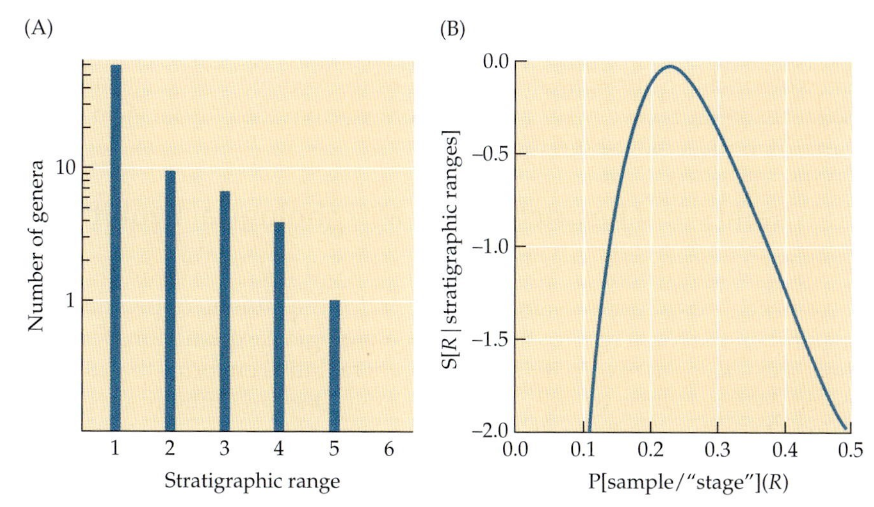

FIGURE 17.6 Stratigraphic Ranges and Preservation Rates (A) Distribution of generic ranges in stages. Note that the Ludlow + Pridoli and the Lockhovian + Pragian both are considered single stages here to make stage lengths roughly comparable in millions of years. (B) Support curves for preservation rates integrating over extinction rates (Foote 1997). Support rescales the log-likelihood to the maximum log-likelihood (Edwards 1992).

ecological correlates of the diversification of major clades. However, much more radical "oldest" finds are necessary to alter ideas about basic rates of morphological evolution in this case.

Obviously, this sort of analysis must be repeated for other taxa. In the case of gastropods, the Paleobiology Database (PBDB; http://pbdb.org as of 1 December 2009) records 1760 occurrences of Cambrian mollusks from 678 localities and 181 formations. These mollusks show shell mineralogies with similar preservation potential as the shell mineralogies of early gastropods (Vendrasco et al. 2010), which means that the absence of gastropods is meaningful (see Bottjer and Jablonski 1988). These mollusks also are phylogenetically very basal to gastropods and are sampled from much of the Cambrian world. Thus, there is no missing record here to explain the apparent elevated rates of change among the earliest known gastropods (Wagner 1995). Even the so-called missing Precambrian record is fairly well known and increasingly well sampled. The PBDB currently is far sparser for Ediacaran data than for Phanerozoic data, but as of December 1 2009, it includes 530 occurrences from 149 localities and 49 formations. These localities preserve both soft-bodied and calcitic fossils and thus should be able to preserve the arthropod and molluscan synapomorphies that Darwin's non-preservation hypothesis predicts existed then. However, although there are possible examples of stem-group members of modern phyla (Fedonkin and Waggoner 1997), none of them clearly nest within subclades of metazoan phyla. Indeed, many Ediacaran taxa are so different from Cambrian metazoans (typically lacking even bilateral symmetry) that workers have questioned the affinity of the Vendobionts to metazoans (Seilacher 1992). So, we now have the sediments: but they do not yield what Darwin predicts they should yield.

Constraints and Restrictions on Morphological Evolution

Constraints Versus Restrictions

If we reject preservation as an explanation for the sudden appearance of disparity in the fossil record, then we need biological explanations for rate shifts. We also need explanations for the general limits on forms that are observed (Wagner 2000). Broadly, two ideas exist. One is essentially a complement to Darwin's explanation for the diversity of forms among Galapagos finches: some clades have great numbers of ecological opportunities early in their history or following mass extinctions (see Losos and Mahler, Chapter 15). The intense resulting selection creates high rates of morphological change (Valentine 1980, 1986; Conway Morris 1989). However, selection afterwards is a *deterrent* to change: incumbent species with adaptations for particular habitats present selection against adaptations for those same habits in other species.

The second idea focuses on changes in intrinsic constraints, such as canalization, increased regulatory networks, refractory gene networks (Valentine and Campbell 1975; Campbell and Marshall 1987; Erwin 1993; Wallace 2000; Davidson and Erwin 2006; see Kirkpatrick, Chapter 7; G. Wagner, Chapter 8). The basic premise is that the evolution of general regulatory systems has made it more difficult for many anatomical features to change, reducing rates of change.

In many cases, paleontologists can document only changes in rates and/or morphospace occupancy. However, in some cases they can distinguish between restrictions and constraints (Table 17.3). Post-extinction rebounds offer good testing grounds, because intrinsic constraint hypotheses predict that it should be difficult for clades to recover disparity, whereas ecological restriction hypotheses predict that newly opened ecospace should encourage high rates of change and thus high levels of disparity. The Permo-Triassic is a particularly good system, because many surviving clades likely bottlenecked at that time (Erwin et al. 1987). Brachiopods and cephalopods suffered major losses at the end-Permian. However, both rapidly re-acquired high disparity in the Triassic (Ciampaglio 2002, 2004; McGowan 2004). The surviving brachiopod clades radiate into morphospace previously occupied by extinct clades, which strongly corroborates the idea that the surviving clades were not constrained in morphospace, but instead restricted by casualties of the end-Permian extinction.

Other clades show evidence of both restrictions and constraints. After the end-Permian extinction, the surviving crinoids do radiate into previously

TABLE 17.3 STUDIES CORROBORATING CHANGES IN ECOLOGICAL RESTRICTIONS AND INTRINSIC CONSTRAINTS

TAXON	STUDY
ECOLOGICAL RESTRICTIONS	
Ordovician–Silurian gastropods	Wagner (1995)
Triassic ammonites	McGowan (2004)
Triassic brachiopods	Ciampaglio (2004)
Carboniferous blastozoans	Ciampaglio (2002)
Triassic crinoids	Foote (1996a, 1999)
Amniotes	Wagner et al. (2006)
INTRINSIC CONSTRAINTS	
Ordovician–Silurian gastropods	Wagner (1995)
Triassic crinoids	Foote (1999)
Lissamphibians	Wagner et al. (2006)
Amniotes	Wagner et al. (2006)

Note: Some studies show evidence of both changing ecological restrictions and intrinsic constraints.

occupied morphospace, but they do not achieve the levels of disparity seen in the early Paleozoic (Foote 1999). Given the absence of ecological analogs to crinoids (e.g., other stalked echinoderms) after the end-Permian, Foote suggests that intrinsic constraints tempered the opportunities offered by ecology. Wagner (1995) partitions general gastropod shell characters into those associated with general differences in internal anatomy from those associated with differences in basic shell functional type. Among modern gastropods, the basic anatomical groups exploit all of the basic gastropod ecological strategies. As ecology does not seem to restrict soft-tissue anatomy among gastropods, increasing restrictions are an unlikely explanation for decreases in step size changes that seem to reflect differences in soft-tissue. The different functional shell types (and thus differences in those types) occur in all major anatomical groups among extant gastropods. As these shell features appear unconstrained among modern gastropods, increasing constraints are an unlikely explanation for decreases in step size among these shell features. Although step size decreases for both types of differences, the decrease is much greater for steps associated with difference in internal anatomy than for steps associated with shell function.

The radiation of tetrapods also corroborates both constraints and restrictions. Wagner et al. (2006) use distributions of changes to examine the number of evolving characters in lissamphibians, amniotes, and stem tetrapods. Lissamphibians occupy the same general habitats as the stem tetrapods and rapidly become the only amphibian-grade tetrapods. If ecological opportunity allowed the appearance of high numbers of characters among amphibian-grade tetrapods, then the same ecological opportunities should have allowed a similar number of characters among lissamphibians. Instead, lissamphibians show a highly restricted character space relative to stem tetrapods. Even if lissamphibians had some sort of "Swiss army knife" character allowing them to exploit habitats that their predecessors exploited with a variety of characters, one still would expect only decreased rates rather than freezing of those other characters. This strongly implicates constraints as the limiting factor. On the other side of the tetrapod tree, amniotes exhibit approximately the same size character space as stem tetrapods. However, both stem tetrapods and amniotes occupy far less than the total tetrapod morphospace, which requires amniotes to gain as well as lose characters. The locking in of old characters for amniotes implicates intrinsic constraints, because amphibian grade tetrapods should not have excluded the increasingly terrestrial amniotes from ecomorphological grades. Conversely, the addition of new characters accompanies the invasion of fully terrestrial habits and thus corroborates the empty ecospace model.

This idea brings us back to Darwin's suggestions that the sudden appearances in the fossil record of teleosts, trilobites, and other major groups require either that his model be incorrect or that the fossil record be highly incomplete. One hundred fifty years later, it is clear that this was predicated on incorrect assumptions, in part, because Darwin himself underestimated natural selection in two ways. One, theoretical (Garcia-Ramos and Kirkpatrick 1997; see Kirkpatrick, Chapter 7) and empirical (Losos et al. 1997) studies indicate that selection operates far more quickly than Darwin seems to have thought possible (Hendry and Kinnison 1999). Two, the idea that ecological restrictions slowed rates of evolution is simply a logical compliment to Darwin's model that he failed to consider explicitly: if ecological opportunity plus natural selection permits radiations, such as those seen in Galapagos finches, then ecological incumbency plus natural selection will restrict morphological change. However, this theory also reflects Amundson's (2005) point that natural selection describes only whether novelties are inherited and not how they are generated. Darwin offers no model for this phenomenon: at best, Darwin's frequent citation of (what is now termed) the Principle of Uniformitarianism (Lyell 1830) indicates that Darwin thought that the same mechanisms existed in the past as exist today—whatever those mechanisms might be. The possibility of changing constraints provides a mechanism for very different amounts of variation over time that is totally lacking from Darwin's model. This possibility also undermines Darwin's supposition that taxa, such as trilobites, teleosts, and others, require extensive time to acquire their distinctive morphologies. Thus, the fossil data contradict Darwin's (at best) vaguely formulated ideas about how morphological variation arises rather than contradicting the idea of natural selection being the primary arbiter by which changes are inherited.

Trends

Overview

Trends offer some of the strongest evidence of natural selection that the fossil record can provide. Curiously, however, Darwin never explicitly refers to trends *among* lineages, as most of his discussion focused on trends *within* lineages. Large-scale trends are a corollary of his theory, only if one supposes that the same (or very similar) selective forces affect numerous related lineages and that the same morphological options would appear for those lineages. The closest Darwin comes to discussing this idea is in his suggestion that Eocene organisms would fare poorly in the modern world (Darwin 1859) or that Paleozoic organisms would fare poorly in the Mesozoic (i.e., Secondary) world. This statement implies that related lineages would undergo parallel improvement over time, leading to morphologies that are better adapted than their predecessors. This idea is essentially a more general version of later paleontological ones, such as Van Valen's (1973) Red Queen hypothesis and Vermeij's (1977, 1987) Predator–Prey Escalation hypothesis. However, Darwin himself does not suggest that the parallel improvements would lead to repeated evolution of the same morphologies: indeed, Darwin confesses that he knows of no way to test this idea! The idea of massive parallel evolution in response to similar general selective pressures, as Vermeij and so many others have suggested, apparently did not occur to Darwin.

Paleontologists were well aware of trends in the nineteenth and early twentieth centuries, which led them to posit mechanisms for deriving the same features repeatedly. Cope (1868) presented early definitions of parallelism that are now equated with general homoplasy. Cope focused particularly on the idea that ontogenetic recapitulation causes homoplasy. Scott (1896) and Osborn (1902, 1935) took this a step further by formally defining parallelism as separate achievements of "latent" homologous features. On one hand, this was a recognition and solution to the problem that Amundson (2005) details: Darwin's model explains what happens to variants, but not what causes variation (see G. Wagner, Chapter 8). Such models, coupled with natural selection, offered the predictions Darwin was unable to derive for features, such as escalation over time. However, some researchers in this school allow only that natural selection accounts for differences among closely related lineages (Osborn 1934, 1935). Otherwise, they largely reject natural selection as an important factor in historical trends and instead invoke orthogenetic models to explain paleontological patterns! Recent works by paleontologists on trends theory (hopefully) have much greater utility and generally augment or complement Darwinian ideas well.

Driven Trends

Let us assume that similar selective forces on closely related species could fix parallel morphological change. Natural selection now offers a mechanism for driven trends (McShea 1994), that is, frequent parallel changes across a phylogeny. It bears noting that driven trends do not necessarily demonstrate selection; for example, driven trends toward decreasing complexity might indicate that transitions from complex to simple character states are easier (and thus more probable) than simple to complex transitions (Strathmann 1978). However, in most cases, similar selective forces represent a good explanation for paleontological trends. Corollary predictions of driven trends include: (1) active displacement of distributions over time, with ancestral regions of morphospace becoming vacant (e.g., loss of minimum sizes; Jablonski 1997) and (2) parallel patterns of skewness in distributions for single characters among subclades of a larger clade (McShea 1994; Wang 2001). Indeed, discredited mechanisms, such as orthogenesis, would generate driven trends if they existed. However, in most cases, similar selective forces represent a good explanation for driven trends.

Numerous paleontological studies document driven trends (Table 17.4). Sophisticated likelihood methods sometimes can identify driven trends with only extant taxa (Hibbet 2006). However, fossil data usually are crucial for easily identifying driven trends. In part, this finding reflects the fact that historical data improve ancestral reconstructions in general (Donoghue et al. 1989; Huelsenbeck 1991). Historical data simply become even more important when driven trends are pervasive (Cunningham 1999). For example, Finarelli and Flynn (2006) show that modern carnivores alone imply larger common ancestors than do modern + fossil carnivores, as methods confound parallel increases in body size for ancestral conditions. Thus, this arena is one in which paleontological data are the most useful for testing hypotheses stemming from Darwin's model, even if Darwin himself did not devise these hypotheses.

Trends as Artifacts

Frequently investigators use the word "trend" simply to describe directionality in distributions without satisfying the criteria listed above. Constraints become important here. Stanley (1973) notes that the classic explanation for increasing body size within clades (termed "Cope's Rule," after Cope 1887) might not be necessary, because clades, such as mammals, begin at small body size and can get much larger but not much smaller, a simple increase in variance will increase the proportion of large animals. Complexity offers a similar example (McShea 1994, 2005) in that organisms cannot have fewer than zero or one parts, but there often is no obvious upper limit on the number of parts. Passive trends (Fisher 1986; Gould 1988) of this sort exist in many systems (McShea 1994; Marcot and McShea 2007; Novack-Gottshall 2008) and augment Darwin's ideas, as they would be expected to exist independently of selection. However, Gould (1988) correctly stresses

TABLE 17.4 CLADES SHOWING EVIDENCE OF DRIVEN TRENDS

TAXON	TRAIT	STUDY
Metazoans*	Decreased complexity	Marcot and McShea (2007)
Crustacea	Increased complexity	Adamowicz et al. (2008)
Paleozoic brachiopods	Muscle geometry	Carlson (1992)
Devonian shelly invertebrates	Increased defensive features	Signor and Brett (1984)
Early gastropods	Decreased shell sinus	Wagner (1996)
Trochonematoid gastropods	Increased shell ornament	Wagner (2001)
Muricid gastropods	Increased defensive features	Vermeij and Carlson (2000)
Plesiosaurs	Size, locomotion	O'Keefe and Carrano (2005)
Pterosaurs*	Increased size	Hone and Benton (2007)
Dinosaurs	Hindlimb dimensions	Carrano (2000)
Dinosaurs*	Increased size	Hone et al. (2005); Carrano (2006)
Synapsids	Decreased skull complexity	Sidor (2001)
Synapsids	Decreased jaw complexity	Sidor (2003)
Mammals*	Increased size	Alroy (1998)
Horses*	Increased size	MacFadden (1986)
Carnivores*	Increased size	Finarelli (2008)

*Denotes studies for which some subclades either showed contrary trends or no trends at all.

that passive trends must be rejected before invoking selective scenarios. McShea (1994) and others (Wagner 1996; Alroy 2000; Wang 2001) put forward several tests for distinguishing passive and driven or active trends based on skewness and related aspects of distributions.

Phylogenetic autocorrelation also can create trends in morphospace. In an early simulation study, Raup and Gould (1974) show that strong clusters in morphospace and temporal shifts in morphospace distributions happen simply because of the chance success and failure of clades. Wagner (1996) reverses Raup and Gould's approach to show that this sort of "phylogenetic hitchhiking" can cause weak but statistically significant active trends without any bias in morphological change. Although this is not a possibility that Darwin considered, it is a necessary corollary to his idea that all taxa are joined by a single phylogeny. Indeed, evolutionary biologists now consider the theoretical removal of phylogenetic autocorrelation to be an important approach to testing hypotheses about adaptations (Felsenstein 1985). Like passive trends, phylogenetic hitchhiking simply augments Darwin's views.

Multimodal Trends

We often describe trends as having single optima. However, there is no reason why this should be true, and it certainly does not follow from Darwin's ideas about natural selection. Although there almost certainly is a passive component to body-size evolution in mammals (Stanley 1973), Alroy (1998) demonstrates that there is also a driven trend towards increasing size. Moreover, Alroy also shows that there appear to be *two* body size optima for mammals, which results in a so-called hole in mammalian size distributions. More importantly, it also indicates a tendency for overlarge or smallish-medium–sized mammals to show driven trends towards *smaller* rather than larger size. Thus, there are minority trends along the same continuum.

Several other studies extend Alroy's point that trends are not homogeneous within clades. For example, although some metazoan clades show driven trends towards decreasing complexity, most simply exhibit passive trends (Marcot and McShea 2007). For both pterosaurs and carnivores, some subclades ignore general trends in body size (Hone and Benton 2007; Finarelli 2008). Among horses, the trend towards large size does not become prominent until the second half of the clade's history (MacFadden 1986). Findings such as these do not contradict Darwin's model; as previously noted, he did not seem to consider driven trends to be an expectation of selection. Instead, such patterns indicate that selective forces and/or the results of selective forces are not homogeneous across time, space, or phylogeny.

Multivariate trends in morphospace might obfuscate patterns along single axes (Cheetham 1987). Wagner and Erwin (2006) demonstrate that several regions of morphospace for Cambro-Devonian gastropods become occupied many more times than expected, given the changes along the individual characters comprising that morphospace and the assumed underlying phylogenetic topology. Notably, the frequently-evolved regions of morphospace match functional models of optimal ecomorphological types (Linsley 1977, 1978). Similarly, marine invertebrates often show trends towards increased protective morphologies (Vermeij 1977; Signor and Brett 1984; Vermeij and Carlson 2000). However, because some morphological adaptations either preclude the necessity or even feasibility of others, trends along any one possible morphological axis are not as strong as are the general trends in morphospace. These examples broaden traditional definitions of trend, but they still corroborate the idea that general selective forces strongly affect morphological evolution.

Species Selection and Punctuated Change: Threats to the Darwinian Paradigm?

Darwin's model of change is anagenetic, envisioning continuous change within lineages. However, Eldredge and Gould (1972) argued that substantial anagenetic change is rare and that change is both pulsed and between lineages. Following this theme, Stanley (1975) and others (Wright 1967; Eldredge and Gould 1972) present a model of differential diversification based on position in morphospace or other characters. Like a driven trend, this would cause active loss of the ancestral condition but (potentially!)

without frequent parallel changes between ancestors and descendants. Species selection has evolved so that current definitions bear little resemblance to Stanley's original idea (Lieberman and Vrba 2005). However, the idea that major trends reflect differential success of species within clades, rather than the differential survival of individuals within species, begs the question of whether traditional depictions of Darwin's model can explain major evolutionary patterns (Gould 1980, 1985).

In assessing whether species selection *sensu* Stanley occurs, one must understand Stanley's challenge of Darwin's model, which requires detailing the assumptions of Stanley's original idea. The first assumption is that change typically is pulsed rather than continuous. The most detailed and sophisticated meta-analysis of trends within lineages to date (Hunt 2007) shows that only a small proportion of published examples best fit directional change models (Figure 17.7). Instead, the vast majority best fit either random walk models (especially for size) or stasis (especially for shape characters).

A second assumption of Stanley's model involves Wright's Rule: evolution in morphospace follows an unbiased, Brownian motion at the species level (Gould and Eldredge 1977; Wright 1967). Contrary to Stanley's

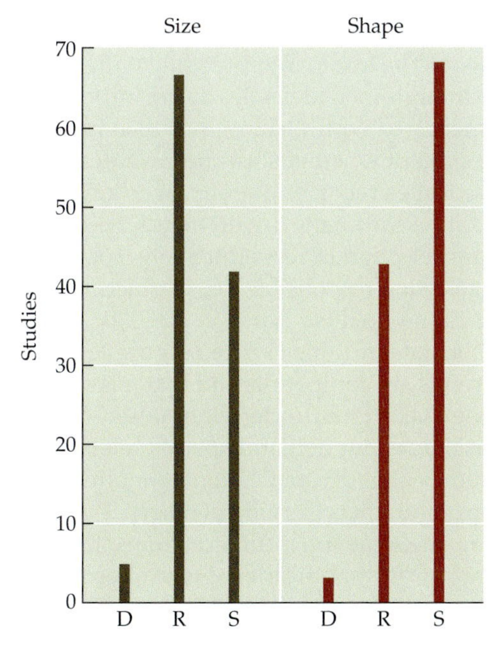

Figure 17.7 Distribution of Most-Likely Models of Character Evolution within Lineages for Size and Shape Characters within Lineages Gould's (2002) definition of stasis includes both random walks and strictly static lineages, but here static lineages show "pull" to a mean morphology, whereas random walks show no demonstrable pull. "D" denotes directional trends, "R," random walks, and "S," static lineages. (Adapted from Hunt 2007: Table 1.)

expectations, clades shown to have driven trends (see Table 17.4) include lineages having random walks and stasis, rather than the individual lineages showing trends. Driven trends also exist in clades for which phylogenetic evidence indicates stasis and pulsed change (Wagner and Erwin 1995). Thus, movement in morphospace can be both punctuated and directional.

A third assumption is that there is too little change within lineages to account for trends. In the same year that Stanley published his model, Lande (1975) made the opposite observation: large scale trends are too *weak* to be products of continuous directional anagenesis. (Haldane (1949) came to a similar conclusion.) Lande estimated that the mortality rates needed for long-term trends in horses amounts to selection culling only one horse in 500,000 per generation. Lande arrived at similar extremely weak selection coefficients for trends in other fossil mammals. Thus, the rare anagenetic trends reported by Hunt (2007) would produce far stronger trends over millions of years than we actually see in the fossil record.

The reconciliation of punctuated equilibrium with natural selection provides explanations for why punctuated change and driven trends are compatible. Ultimately, Gould (2002) favors an extension of Futuyma's (1987) model in which speciation *facilitates* rather than induces change. Here, speciation permits local selective forces to permanently fix novel morphologies that otherwise would be lost to interbreeding with conspecific populations (Lande 1975). Although such ideas clearly are important for reconciling the debate over punctuated equilibrium, they also have two major implications for the question of whether stasis necessitates species selection. First, if some selective forces recur frequently over long periods of time, then even periodic fixation can create driven trends. Second, periodic fixation of change by natural selection in reproductively isolated populations means that the *relevant* selection responsible for, say, increasing body size in horses is much greater than one dead horse in 500,000. These two implications combined with the data in Table 17.4 and Figure 17.7 indicate both that stasis and driven trends are compatible and that it is very possible for natural selection to be the major force underlying pulsed driven trends.

Although stasis does not demand species selection, we cannot dismiss the concept of differential diversification as another major trend that augments rather than contradicts Darwinian theory. Unfortunately, adequately documenting this phenomenon is difficult due to the large amounts of data required to demonstrate significant shifts in speciation and/or extinction rates and because parallel shifts associated with position in morphospace among 2+ subclades are really required to separate Stanley's species selection from classic adaptive radiation models. Liow (2006) demonstrates that deviant trachyleberidid ostracodes have lower extinction rates than do normal trachyleberidids. If origination rates per unit time are the same for deviants and normals, then one would expect the deviants to leave more descendant taxa and thus create trends in trachyleberidid morphospace.

However, Liow (2004) also shows the opposite for crinoids; long-lived genera tend to come from the middle of crinoid morphospace. This presents the intriguing possibility of "stabilizing species selection" (Gilinsky 1981).

Summary

Although workers periodically portray paleontological data as contradicting Darwin's basic ideas, fossil data strongly contradict only the idea that rates of change have been uniform through time. High early disparity in fossil clades and sudden appearances of higher taxa likely do not reflect the imperfections of the geological record. Instead, they probably reflect elevated rates of morphological evolution early in clade histories or following mass extinctions. In some cases, rate shifts apparently reflect changes in intrinsic constraints and thus involve mechanisms that were unknown in Darwin's time. In other cases, they likely reflect changes in ecological restrictions, and thus suggest both that natural selection operates far more quickly than Darwin seems to have supposed and also that selection is a more powerful limiting force than he apparently recognized. Fossil data also provide ample evidence of driven trends or differential acquisition of particular traits, which provide strong evidence of common selective forces frequently affecting related lineages in similar ways. Paleontologists have augmented Darwinian explanations for trends with other ideas that necessarily follow from ideas about constraints (e.g., passive trends) as well as Darwin's basic idea of common ancestry among taxa (i.e., phylogenetic hitchhiking). Ultimately, the most important point to stress is that although fossil data refute Darwin's uniformitarianist notions, these same data build upon, modify, and perhaps even enhance Darwin's most important idea: the importance of natural selection on long-term patterns of morphological evolution.

Acknowledgments

I thank J. Marcot for discussion and references concerning trends and M. Foote and J. S. Levinton for comments. This is PBDB publication #109.

Literature Cited

Adamowicz, S. J., A. Purvis, and M. A. Wills. 2008. Increasing morphological complexity in multiple parallel lineages of the Crustacea. *Proc. Natl. Acad. Sci. USA* 105: 4786–4791.

Adams, D. C., C. M. Berns, K. H. Kozak, and 1 other. 2009. Are rates of species diversification correlated with rates of morphological evolution? *Proc. Roy. Soc. B* 276: 2729–2738.

Alroy, J. 1998. Cope's rule and the dynamics of body mass evolution in North American fossil mammals. *Science* 280: 731–734.

Alroy, J. 2000. Understanding the dynamics of trends within evolving lineages. *Paleobiology* 26: 319–329.
Amundson, R. 2005. *The Changing Role of the Embryo in Evolutionary Thought: Roots of Evo-Devo*. Cambridge University Press, Cambridge, UK.
Anstey, R. L. and J. F. Pachut. 1995. Phylogeny, diversity history and speciation in Paleozoic bryozoans. In D. H. Erwin and R. L. Anstey (eds.), *New Approaches to Studying Speciation in the Fossil Record*, pp. 239–284. Columbia University Press, New York.
Bottjer, D. J. and D. Jablonski. 1988. Paleoenvironmental patterns in the evolution of post-Paleozoic benthic marine invertebrates. *Palaios* 3: 540–560.
Briggs, D. E. G., R. A. Fortey, and M. A. Wills. 1992. Morphological disparity in the Cambrian. *Science* 256: 1670–1673.
Brusatte, S. L., M. J. Benton, M. Ruta, and 1 other. 2008. Superiority, competition, and opportunism in the evolutionary radiation of dinosaurs. *Science* 321: 1485–1488.
Campbell, K. S. W. and C. R. Marshall. 1987. Rates of evolution among Paleozoic echinoderms. In K. S. W. Campbell and M. F. Day (eds.), *Rates of Evolution*, pp. 61–100. Allen and Unwin, London.
Carlson, S. J. 1992. Evolutionary trends in the articulate brachiopod hinge mechanism. *Paleobiology* 18: 344–366.
Carrano, M. T. 2000. Homoplasy and the evolution of dinosaur locomotion. *Paleobiology* 26: 489–512.
Carrano, M. T. 2006. Body-size evolution in the Dinosauria. In M. T. Carrano, T. J. Gaudin, R. W. Blob, and J. R. Wible (eds.), *Amniote Paleobiology: Perspectives on the Evolution of Mammals, Birds, and Reptiles*, pp. 225–268. University of Chicago Press, Chicago.
Cheetham, A. H. 1987. Tempo of evolution in a Neogene bryozoan: Are trends in single morphologic characters misleading? *Paleobiology* 13: 286–296.
Ciampaglio, C. N. 2002. Determining the role that ecological and developmental constraints play in controlling disparity: Examples from the crinoid and blastozoan fossil record. *Evol. Devel.* 4: 170–188.
Ciampaglio, C. N. 2004. Measuring changes in articulate brachiopod morphology before and after the Permian mass extinction event: Do developmental constraints limit morphological innovation? *Evol. Devel.* 6: 260–274.
Ciampaglio, C. N., M. Kemp, and D. W. McShea. 2001. Detecting changes in morphospace occupation patterns in the fossil record: Characterization and analysis of measures of disparity. *Paleobiology* 27: 695–715.
Cloutier, R. 1991. Patterns, trends, and rates of evolution with the Actinistia. *Environ. Biol. Fish.* 32: 23–58.
Conway Morris, S. 1989. Burgess Shale faunas and the Cambrian explosion. *Science* 246: 339–346.
Cope, E. D. 1868. On the origin of genera. *Proc. Acad. Nat. Sci. Phil.* 20: 242–300.
Cope, E. D. 1887. *The Origin of the Fittest*. D. Appleton and Co., New York.
Cotton, T. J. 2001. The phylogeny and systematics of blind Cambrian ptychoparioid trilobites. *Palaeontology* 44: 167–207.
Cunningham, C. W. 1999. Some limitations of ancestral character-state reconstruction when testing evolutionary hypotheses. *Syst. Biol.* 48: 665–674.
Daeschler, E. B., N. H. Shubin, and F. A. Jenkins. 2006. A Devonian tetrapod-like fish and the evolution of the tetrapod body plan. *Nature* 440: 757–763.

Darwin, C. 1845. *Journal of Researches into the Natural History and Geology of the Countries Visited During the Voyage of H.M.S. Beagle round the World, under the Command of Capt. Fitz Roy, R.N.*, 2nd ed. John Murray, London.
Darwin, C. 1859. *On the Origin of Species by Means of Natural Selection, or the Preservation of Favoured Races in the Struggle for Life.* John Murray, London.
Davidson, E. H. and D. H. Erwin. 2006. Gene regulatory networks and the evolution of animal body plans. *Science* 311: 796–800.
Donoghue, M. J., J. A. Doyle, J. Gauthier, and 2 others. 1989. The importance of fossils in phylogeny reconstruction. *Ann. Rev. Ecol. Syst.* 20: 431–460.
Eble, G. J. 2000. Contrasting evolutionary flexibility in sister groups: Disparity and diversity in Mesozoic atelostomate echinoids. *Paleobiology* 26: 56–79.
Edwards, A. W. F. 1992. *Likelihood—Expanded Edition*. Johns Hopkins University Press, Baltimore.
Eldredge, N. and S. J. Gould. 1972. Punctuated equilibria: An alternative to phyletic gradualism. In T. J. M. Schopf (ed.), *Models in Paleobiology*, pp. 82–115. Freeman, Cooper and Co., San Francisco.
Erwin, D. H. 1993. The origin of metazoan development: A palaeobiological perpective. *Biol. J. Linn. Soc.* 50: 255–274.
Erwin, D. H. 2007. Disparity: Morphological pattern and developmental context. *Palaeontology* 50: 57–73.
Erwin, D. H., J. W. Valentine, and J. J. Sepkoski, Jr. 1987. A comparative study of diversification events: The Early Paleozoic versus the Mesozoic. *Evolution* 41: 1177–1186.
Fedonkin, M. A. and B. M. Waggoner. 1997. The Late Precambrian fossil *Kimberella* is a mollusc-like bilaterian organism. *Nature* 388: 868–871.
Felsenstein, J. 1985. Phylogenies and the comparative method. *Am. Nat.* 125: 1–15.
Finarelli, J. A. 2008. Hierarchy and the reconstruction of evolutionary trends: Evidence for constraints on the evolution of body size in terrestrial caniform carnivorans (Mammalia). *Paleobiology* 34: 553–562.
Finarelli, J. A. and J. J. Flynn. 2006. Ancestral state reconstruction of body size in the Caniformia (Carnivora, Mammalia): The effects of incorporating data from the fossil record. *Syst. Biol.* 55: 301–313.
Fisher, D. C. 1986. Progress in organismal design. In D. M. Raup and D. Jablonski (eds.), *Patterns and Processes in the History of Life*, pp. 99–118. Springer-Verlag, Berlin.
Foote, M. 1990. Nearest-neighbor analysis of trilobite morphospace. *Syst. Zool.* 39: 371–382.
Foote, M. 1991a. Analysis of morphological data. In N. L. Gilinsky and P. W. Signor (eds.), *Analytical Paleobiology*, pp. 59–86. Paleontological Society, Knoxville.
Foote, M. 1991b. Morphological and taxonomic diversity in a clade's history: The blastoid record and stochastic simulations. *Contr. Mus. Paleontol., Univ. Mich.* 28: 101–140.
Foote, M. 1991c. Morphologic patterns of diversification: Examples from trilobites. *Palaeontology* 34: 461–485.
Foote, M. 1992a. Rarefaction analysis of morphological and taxonomic diversity. *Paleobiology* 18: 1–16.
Foote, M. 1992b. Paleozoic record of morphological diversity in blastozoan echinoderms. *Proc. Natl. Acad. Sci. USA* 89: 7325–7329.
Foote, M. 1993a. Discordance and concordance between morphological and taxonomic diversity. *Paleobiology* 19: 185–204.

Foote, M. 1993b. Contributions of individual taxa to overall morphological disparity. *Paleobiology* 19: 403–419.
Foote, M. 1994. Morphological disparity in Ordovician-Devonian crinoids and the early saturation of morphological space. *Paleobiology* 20: 320–344.
Foote, M. 1995. Morphological diversification of Paleozoic crinoids. *Paleobiology* 21: 273–299.
Foote, M. 1996a. Ecological controls on the evolutionary recovery of post-Paleozoic crinoids. *Science* 274: 1492–1495.
Foote, M. 1996b. Models of morphologic diversification. In D. Jablonski, D. H. Erwin, and J. H. Lipps (eds.), *Evolutionary Paleobiology: Essays in Honor of James W. Valentine*, pp. 62–86. University of Chicago Press, Chicago.
Foote, M. 1997. Estimating taxonomic durations and preservation probability. *Paleobiology* 23: 278–300.
Foote, M. 1999. Morphological diversity in the evolutionary radiation of Paleozoic and post-Paleozoic crinoids. *Paleobiol. Mem.* 25(Suppl. 2): 1–115.
Foote, M. and D. M. Raup. 1996. Fossil preservation and the stratigraphic ranges of taxa. *Paleobiology* 22: 121–140.
Friedman, M. 2007a. The interrelationships of Devonian lungfishes (Sarcopterygii: Dipnoi) as inferred from neurocranial evidence and new data from the genus *Soederberghia* Lehman, 1959. *Zool. J. Linn. Soc.* 151: 115–171.
Friedman, M. 2007b. *Styloichthys* as the oldest coelacanth: Implications for early osteichthyan interrelationships. *J. Syst. Palaeontol.* 5: 289–343.
Futuyma, D. J. 1987. On the role of species in anagenesis. *Am. Nat.* 130: 465–473.
Garcia-Ramos, G. and M. Kirkpatrick. 1997. Genetic models of adaptation and gene flow in peripheral populations. *Evolution* 51: 21–28.
Gilinsky, N. L. 1981. Stabilizing species selection in the Archaeogastropoda. *Paleobiology* 7: 316–331.
Gould, S. J. 1980. Is a new and general theory of evolution emerging? *Paleobiology* 6: 119–130.
Gould, S. J. 1985. The paradox of the first tier: An agenda for paleobiology. *Paleobiology* 11: 2–12.
Gould, S. J. 1988. Trends as changes in variance: A new slant on progress and directionality in evolution. *J. Paleontol.* 62: 319–329.
Gould, S. J. 1989. *Wonderful Life*. W. W. Norton and Co., New York.
Gould, S. J. 1991. The disparity of the Burgess Shale arthropod fauna and the limits of cladistic analysis: Why we must strive to quantify morphospace. *Paleobiology* 17: 411–423.
Gould, S. J. 2002. *The Structure of Evolutionary Theory*. Belknap Press, Cambridge, MA.
Gould, S. J. and N. Eldredge. 1977. Punctuated equilibria: The tempo and mode of evolution reconsidered. *Paleobiology* 3: 115–151.
Gradstein, F., J. Ogg, and A. Smith. 2005. *A Geologic Time Scale 2004*. Cambridge University Press, Cambridge, UK.
Haldane, J. B. S. 1949. Suggestions as to quantitative measurement of rates of evolution. *Evolution* 3: 51–56.
Harmon, L. J., J. A. Schulte II, A. Larson, and 1 other. 2003. Tempo and mode of evolutionary radiation in iguanian lizards. *Science* 301: 961–964.
Hendry, A. P. and M. T. Kinnison. 1999. Perspective: The pace of modern life: Measuring rates of contemporary microevolution. *Evolution* 53: 1637–1653.
Hibbet, D. S. 2006. Trends in morphological evolution in Homobasidiomycetes inferred using Maximum Likelihood: A comparison of binary and multistate approaches. *Syst. Biol.* 53: 889–903.

Hone, D. W. E. and M. J. Benton. 2007. Cope's Rule in the Pterosauria, and differing perceptions of Cope's Rule at different taxonomic levels. *J. Evol. Biol.* 20: 1164–1170.
Hone, D. W. E., T. M. Keesey, D. Pisani, and 1 other. 2005. Macroevolutionary trends in the Dinosauria: Cope's rule. *J. Evol. Biol.* 18: 587–595.
Huelsenbeck, J. P. 1991. When are fossils better than extant taxa in phylogenetic analysis? *Syst. Zool.* 40: 458–469.
Hunt, G. 2007. The relative importance of directional change, random walks, and stasis in the evolution of fossil lineages. *Proc. Natl. Acad. Sci. USA* 104: 18404–18408.
Huntley, J. W., S. Xiao, and M. Kowalewski. 2006. 1.3 billion years of acritarch history: An empirical morphospace approach. *Precamb. Res.* 144: 52–68.
Hurlbert, S. H. 1971. The nonconcept of species diversity: A critique and alternative parameters. *Ecology* 52: 577–586.
Jablonski, D. 1997. Body-size evolution in Cretaceous molluscs and the status of Cope's Rule. *Nature* 385: 250–252.
Jernvall, J., J. P. Hunter, and M. Fortelius. 1996. Molar tooth diversity, disparity and ecology in Cenozoic ungulate radiations. *Science* 274: 1489–1492.
Johanson, Z. and P. E. Ahlberg. 2001. Devonian rhizodontids and tristichopterids (Sarcopterygii; Tetrapodomorpha) from East Gondwana. *Trans. Roy. Soc. Edin. Earth Sci.* 92: 43–74.
Labandeira, C. C. and G. J. Eble. 2008. The fossil record of insect diversity and disparity. In J. Anderson, F. Thackery, B. Van Wyk, and M. De Wit (eds.), *Gondwana Alive: Biodiversity and the Evolving Biosphere*, pp. Witwatersrand University Press, Johannesburg.
Lande, R. 1975. Natural selection and random genetic drift in phenotypic evolution. *Evolution* 30: 314–334.
Lande, R. 1986. The dynamics of peak shifts and the pattern of morphologic evolution. *Paleobiology* 12: 343–354.
Lefebvre, B., G. J. Eble, N. Navarro, and 1 other. 2006. Diversification of atypical Paleozoic echinoderms: A quantitative survey of patterns of stylophoran disparity, diversity, and geography. *Paleobiology* 32: 483–510.
Lieberman, B. S. and E. S. Vrba. 2005. Stephen Jay Gould on species selection: 30 years of insight. In E. S. Vrba and N. Eldredge (eds.), *Macroevolution: Diversity, Disparity, Contingency*, pp. 113–121. *Paleobiol. Mem.* 31 (Suppl. to No. 2).
Linsley, R. M. 1977. Some laws of gastropod shell form. *Paleobiology* 3:196–206.
Linsley, R. M. 1978. Locomotion rates and shell form in the Gastropoda. *Malacologia* 17: 193–206.
Liow, L. H. 2004. A test of Simpson's "Rule of the survival of the relatively unspecialized" using fossil crinoids. *Am. Nat.* 164: 431–443.
Liow, L. H. 2006. Do deviants live longer? Morphology and longevity in trachyleberidid ostracodes. *Paleobiology* 32: 55–69.
Losos, J. B., K. I. Warheitt, and T. W. Schoener. 1997. Adaptive differentiation following experimental island colonization in *Anolis* lizards. *Nature* 387: 70–73.
Lupia, R. 1999. Discordant morphological disparity and taxonomic diversity during the Cretaceous angiosperm radiation: North American pollen record. *Paleobiology* 25: 1–28.
Lyell, C. 1830. *Principles of Geology*. John Murray, London.
MacFadden, B. J. 1986. Fossil horses from "Eohippus" (*Hyracotherium*) to *Equus*: Scaling, Cope's Law, and the evolution of body size. *Paleobiology* 12: 355–369.

Marcot, J. D. and D. W. McShea. 2007. Increasing hierarchical complexity throughout the history of life: Phylogenetic tests of trend mechanisms. *Paleobiology* 33: 182–200.

Marshall, C. R. 1990. Confidence intervals on stratigraphic ranges. *Paleobiology* 16: 1–10.

McGhee, G. R. Jr. 1991. Theoretical morphology: The concept and its applications. In N. L. Gilinsky and P. W. Signor (eds.), *Analytical Paleobiology*, pp. 87–102. Paleontological Society, Knoxville.

McGowan, A. J. 2004. The effect of the Permo-Triassic bottleneck on Triassic ammonoid morphological evolution. *Paleobiology* 30: 369–395.

McShea, D. W. 1994. Mechanisms of large-scale evolutionary trends. *Evolution* 48: 1747–1763.

McShea, D. W. 2005. The evolution of complexity without natural selection, a possible large-scale trend of the fourth kind. In E. S. Vrba and N. Eldredge (eds.), *Macroevolution: Diversity, Disparity, Contingency*, pp. 146–156. *Paleobiol. Mem.* 31 (suppl. to No. 2).

Niedzwiedzki, G., P. Szrek, K. Narkiewicz, and 2 others. 2010. Tetrapod trackways from the early Middle Devonian period of Poland. *Nature* 463: 43–48.

Novack-Gottshall, P. M. 2008. Ecosystem-wide body-size trends in Cambrian-Devonian marine invertebrate lineages. *Paleobiology* 34: 210–228.

O'Keefe, F. R. and M. T. Carrano. 2005. Correlated trends in the evolution of the plesiosaur locomotor system. *Paleobiology* 31: 656–675.

Osborn, H. F. 1902. Homoplasy as a law of latent or potential homology. *Am. Nat.* 36: 259–271.

Osborn, H. F. 1934. Aristogenesis, the creative principle in the origin of species. *Am. Nat.* 68: 193–235.

Osborn, H. F. 1935. The ancestral tree of the Proboscidea. Discovery, evolution, migration and extinction over a 50,000,000 year period. *Proc. Natl. Acad. Sci. USA* 21: 404–412.

Raup, D. M. 1966. Geometric analysis of shell coiling: General problems. *J. Paleontol.* 40: 1178–1190.

Raup, D. M. and S. J. Gould. 1974. Stochastic simulation and evolution of morphology—towards a nomothetic paleontology. *Syst. Zool.* 23: 305–322.

Ricklefs, R. E. 2004. Cladogenesis and morphological diversification in passerine birds. *Nature* 430: 338–341.

Roy, K. 1994. Effects of the Mesozoic marine revolution on the taxonomic, morphologic, and biogeographic evolution of a group: Aporrhaid gastropods during the Mesozoic. *Paleobiology* 20: 274–296.

Roy, K. and M. Foote. 1997. Morphological approaches to measuring biodiversity. *Trends Ecol. Evol.* 12: 277–281.

Ruta, M. and M. I. Coates. 2007. Dates, nodes and character conflict: Addressing the lissamphibian origin problem. *J. Syst. Palaeontol.* 5: 69–122.

Ruta, M., M. I. Coates, and D. L. J. Quicke. 2003. Early tetrapod relationships revisited. *Biol. Rev.* 78: 251–345.

Ruta, M., P. J. Wagner, and M. I. Coates. 2006. Evolutionary patterns in early tetrapods. I. Rapid initial diversification by decrease in rates of character change. *Proc. Roy. Soc. Lond. B* 273: 2107–2111.

Saunders, W. B., E. Greenfest-Allen, D. M. Work, and 1 other. 2008. Morphologic and taxonomic history of Paleozoic ammonoids in time and morphospace. *Paleobiology* 34: 128–154.

Scott, W. B. 1896. Paleontology as a morphological discipline. *Science* 4: 177–188.

Seilacher, A. 1992. Vendobionta and Psammocorallia: Lost constructions of Precambrian evolution. *J. Geo. Soc. Lond.* 149: 607–613.
Sepkoski, J. J. Jr. 1975. Stratigraphic biases in the analysis of taxonomic survivorship. *Paleobiology* 1: 343–355.
Sepkoski, J. J. Jr. 2002. A compendium of fossil marine animal genera. *Bull. Am. Paleontol.* 363: 1–563.
Sereno, P. C., A. L. Beck, D. B. Dutheil, and 8 others. 1999. Cretaceous sauropods from the Sahara and the uneven rate of skeletal evolution among dinosaurs. *Science* 286: 1342–1347.
Sidlauskas, B. 2007. Testing for unequal rates of morphological diversification in the absence of a detailed phylogeny: A case study from characiform fishes. *Evolution* 61: 299–316.
Sidor, C. A. 2001. Simplification as a trend in synapsid cranial evolution. *Evolution* 55: 1419–1442.
Sidor, C. A. 2003. Evolutionary trends and the origin of the mammalian lower jaw. *Paleobiology* 29: 605–640.
Sidor, C. A. and J. A. Hopson. 1998. Ghost lineages and "mammalness": Assessing the temporal pattern of character acquisition in the Synapsida. *Paleobiology* 24: 254–273.
Signor, P. W. III and C. E. Brett. 1984. The mid-Paleozoic precursor to the Mesozoic marine revolution. *Paleobiology* 10: 229–246.
Simpson, G. G. 1944. *Tempo and Mode in Evolution*. Columbia University Press, New York.
Simpson, G. G. 1953. *The Major Features of Evolution*. Columbia University Press, New York.
Smith, A. B. 1988. Patterns of diversification and extinction in early Palaeozoic echinoderms. *Palaeontology* 31: 799–828.
Smith, A. B. 1994. *Systematics and the Fossil Record—Documenting Evolutionary Patterns*. Blackwell, Oxford.
Smith, L. H. and P. M. Bunje. 1999. Morphologic diversity of inarticulate brachiopods through the Phanerozoic. *Paleobiology* 25: 296–408.
Smith, L. H. and B. S. Lieberman. 1999. Disparity and constraint in olenelloid trilobites and the Cambrian radiation. *Paleobiology* 25: 459–470.
Stanley, S. M. 1973. An explanation for Cope's Rule. *Evolution* 27: 1–26.
Stanley, S. M. 1975. A theory of evolution above the species level. *Proc. Natl. Acad. Sci. USA* 276: 56–76.
Stockmeyer Lofgren, A., R. E. Plotnick, and P. J. Wagner. 2003. Morphological diversity of Carboniferous arthropods and insights on disparity patterns of the Phanerozoic. *Paleobiology* 29: 350–369.
Strathmann, R. R. 1978. The evolution and loss of feeding larval stages of marine invertebrates. *Evolution* 32: 894–906.
Strauss, D. and P. M. Sadler. 1989. Classical confidence intervals and Bayesian probability estimates for ends of local taxon ranges. *Math. Geol.* 21: 411–427.
Valentine, J. W. 1969. Patterns of taxonomic and ecological structure of the shelf benthos during Phanerozoic time. *Palaeontology* 12: 684–709.
Valentine, J. W. 1980. Determinants of diversity in higher taxonomic categories. *Paleobiology* 6: 444–450.
Valentine, J. W. 1986. Fossil record of the origin of Baupläne and its implications. In D. M. Raup and D. Jablonski (eds.), *Patterns and Processes in the History of Life*, pp. 209–222. Springer-Verlag, Berlin, Heidelberg.

Valentine, J. W. and C. A. Campbell. 1975. Genetic regulation and the fossil record. *Am. Sci.* 63: 673–680.
Van Valen, L. 1973. A new evolutionary law. *Evol. Theory* 1: 1–30.
Vendrasco, M. J., S. M. Porter, A. Kouchinsky, and 2 others. 2010. New data on molluscs and their shell microstructures from the Middle Cambrian Gowers Formation, Australia. *Palaeontology* 53: 97–135.
Vermeij, G. J. 1977. The Mesozoic marine revolution: Evidence from snails, predators and grazers. *Paleobiology* 3: 245–258.
Vermeij, G. J. 1987. *Evolution and Escalation—An Ecological History of Life*. Princeton University Press, Princeton.
Vermeij, G. J. and S. J. Carlson. 2000. The muricid gastropod subfamily Rapaninae: Phylogeny and ecological history. *Paleobiology* 26: 19–46.
Wagner, P. J. 1995. Testing evolutionary constraint hypotheses with early Paleozoic gastropods. *Paleobiology* 21: 248–272.
Wagner, P. J. 1996. Contrasting the underlying patterns of active trends in morphologic evolution. *Evolution* 50: 990–1007.
Wagner, P. J. 1997. Patterns of morphologic diversification among the Rostroconchia. *Paleobiology* 23: 115–150.
Wagner, P. J. 2000. Exhaustion of cladistic character states among fossil taxa. *Evolution* 54: 365–386.
Wagner, P. J. 2001. Rate heterogeneity in shell character evolution among lophospiroid gastropods. *Paleobiology* 27: 290–310.
Wagner, P. J. and D. H. Erwin. 1995. Phylogenetic tests of speciation hypotheses. In D. H. Erwin and R. L. Anstey (eds.), *New Approaches to Studying Speciation in the Fossil Record*, pp. 87–122. Columbia University Press, New York.
Wagner, P. J. and D. H. Erwin. 2006. Patterns of convergence in general shell form among Paleozoic gastropods. *Paleobiology* 32: 315–336.
Wagner, P. J., M. Ruta, and M. I. Coates. 2006. Evolutionary patterns in early tetrapods. II. Differing constraints on available character space among clades. *Proc. Roy. Soc. Lond. B* 273: 2113–2118.
Wallace, A. 2000. *The Origin of Animal Body Plans: A Study in Evolutionary Developmental Biology*. Cambridge University Press, Cambridge, UK.
Wang, S. C. 2001. Quantifying passive and driven large-scale evolutionary trends. *Evolution* 55: 849–858.
Wesley-Hunt, G. D. 2005. The morphological diversification of carnivores in North America. *Paleobiology* 31: 35–55.
Wills, M. A. 1998a. Cambrian and Recent disparity: The picture from priapulids. *Paleobiology* 24: 177–199.
Wills, M. A. 1998b. Crustacean disparity through the Phanerozoic: Comparing morphological and stratigraphic data. *Biol. J. Linn. Soc.* 65: 455–500.
Wills, M. A. 2001. Morphological disparity: A primer. In J. M. Adrain, G. D. Edgecombe, and B. S. Lieberman (eds.), *Fossils, Phylogeny and Form: An Analytical Approach*, pp. 55–144. Kluwer Academic/Plenum Publishers, New York.
Wills, M. A., D. E. G. Briggs, and R. A. Fortey. 1994. Disparity as an evolutionary index: A comparison of Cambrian and Recent arthropods. *Paleobiology* 20: 93–131.
Wright, S. 1932. The roles of mutation, inbreeding, crossbreeding and selection in evolution. *Proc. 6th Intl. Cong. Genet.* 1: 356–366.
Wright, S. 1967. Comments on the preliminary working papers of Eden and Waddington. In P. S. Moorehead and M. M. Kaplan (eds.), *Mathematical Challenges to the Neo-Darwinian Interpretation of Evolution*, pp. 117–120. Wistar Institutional Press, Philadelphia.

Chapter 18

The Geological History of Biodiversity

Michael Foote

At least for animals in the marine realm, a number of major features of biodiversity, measured as global genus or family richness, seem firmly established (Sepkoski 1981, 1997; Sepkoski et al. 1981; Raup and Sepkoski 1982; Bambach et al. 2004; Bambach 2006; Alroy et al. 2008):

1. Biodiversity has increased over the past half-billion years.
2. It has fluctuated markedly around its secular rise.
3. A number of drops in diversity are clearly associated with major mass extinction events, although others may be associated instead with reduced origination rates.
4. Superimposed on these secular patterns have been substantial changes in the faunal composition of the biosphere.

With the possible exception of the third point, all of these features were already known in Charles Darwin's day (Phillips 1860).

In attempting to narrow the history of biodiversity into a tractable scope, I will emphasize two topics that would probably have interested Darwin: (1) the imperfection of the geologic record and (2) the proposition that secular trends in biodiversity, at the largest scale, have been shaped by biotic interactions. I will focus on global taxonomic richness of readily fossilized marine invertebrates, which have the most extensively documented fossil record. Compared with other measures, such as morphological and ecological diversity, abundance, and evenness, richness is best understood dynamically, in relation to the underlying processes of origination and extinction.

The Imperfection of the Geological Record

Darwin (1859) devotes considerable space to the problem of geological and paleontological incompleteness, and he paints a picture that might make a would-be paleontologist despair of ever hoping to document the evolution of life with data from the fossil record. One reasonable reading of this part of the text is not so much that Darwin felt he had solid evidence for the dismal state of the record, but rather that the spottiness of the record was required by natural selection, under the assumption that evolutionary change was insensibly gradual:

> *...Why then is not every geological formation and every stratum full of such intermediate links? Geology assuredly does not reveal any such finely graduated organic chain; and this, perhaps, is the most obvious and gravest objection which can be urged against my theory. The explanation lies, as I believe, in the extreme imperfection of the geological record (Darwin 1859: 280).*

In addition to the paucity of intermediate forms, Darwin was troubled by the sudden appearance of major biologic groups early in the Phanerozoic geological record. Although taxonomic richness was not Darwin's main concern, an extremely spotty record would obviously hinder any attempt to document the history of richness.

As early as 1860, Darwin's pessimistic view of the quality of the geologic record had already been questioned. John Phillips, in a volume that documented secular trends in the composition of the global fauna over the course of the Phanerozoic, felt that the paleontological record was rather complete, provided that one looked in the right places. In direct response to Darwin (1859), Phillips wrote:

> *Surely this imperfection of the geological record is overrated. With the exceptions of the two great breaks at the close of the Palaeozoic and Mesozoic periods, the series of strata is nearly if not quite complete, the series of life almost equally so. Not indeed in one small tract or in one section; but on a comparison of different tracts and several sections. For example, the marine series of Devonian life cannot be found in the districts of Wales or Scotland, but must be collected in Devonshire, Bohemia, Russia and America. When so gathered it fills very nearly if not entirely the whole interval between the Upper Silurian and the Carboniferous Fauna. So in England the marine intermediaries of the Oolitic and Cretaceous ages are not given: but the Neocomian Strata supply the want. We have no Meiocene Strata in England, but their place is marked in France and America (Phillips 1860: 207).*

So, we have Darwin seeing a glass perhaps 99% empty and Phillips seeing it more like half full. Rather than trying to resolve the issue based on the data available 150 years ago, we will jump forward to the twenty-first century and consider some of the major advances that allow researchers to cope with paleontological incompleteness. First, geologists simply know more about the fossil record than they did then. Many more regions have been explored, for example, and the record of pre-Cambrian and Cambrian strata has been documented in great detail. Second, geological and paleontological data have been archived in ways that enable systematic retrieval and analysis. Finally, paleontologists and stratigraphers have developed tools for quantifying the degree of completeness of the record and for circumventing many of the biasing effects of incompleteness.

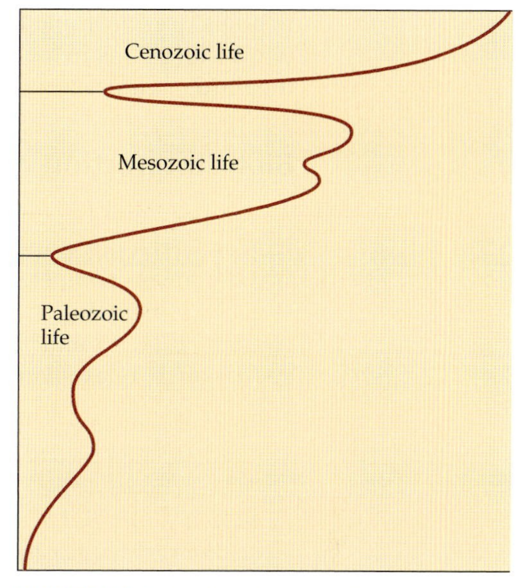

FIGURE 18.1 Phillips's Model of Species Diversity (Abscissa) over Phanerozoic Time (Ordinate) (From Phillips 1860.)

Although Phillips was more optimistic than Darwin about the quality of the fossil record, he recognized that this record could not be read literally, and his *model* of Phanerozoic marine diversity (Figure 18.1) was based in part on his attempts to correct for variation in the spatial and temporal coverage of the record (Miller 2000). His starting point was Morris's (1854) tabulation of British fossil species. Because different periods of geologic time are more or less completely represented in Britain, Phillips (1860) divided the raw count of species by the preserved thickness of strata to obtain an estimate of the number of species per unit time. Although there were earlier and arguably more comprehensive compilations of fossil diversity than the Morris–Phillips combination (Bronn 1849), Phillips's treatment stands out for its clear attempt to deal with the effect that the size of a paleontological sample has on preserved diversity as well as for its quantitative depiction of long-term changes in faunal composition (Phillips 1860: Figure 6). Rudwick (2008) also credits Phillips with recognizing the global, rather than local or regional, significance of his proposed divisions of geological time and with other advances in thinking about geological history.

All depictions of the history of diversity are models, like that of Phillips, each involving different data and, more importantly, having different assumptions about how to convert the raw data to an estimate of diversity. At the one extreme, simply tabulating the number of taxa from an interval of time involves at least three assumptions: (1) Sampling is reasonably complete, so that the observed first and last appearances of taxa are good proxies for their true times of origination and extinction. If this is not the case, then rapid changes in diversity will be poorly resolved and will appear to be spread out over a longer span of time than was in fact the case. (2) Sampling is nearly uniform over time, so that long-term trends and short-term fluctuations are not dominated by changes in the ability to sample fossil taxa. (3) The best-sampled taxa, generally the numerically abundant and well-skeletonized forms, are, in terms of total diversity, representative of the global fauna as a whole. Various attempts to adjust for incomplete and variable sampling generally reject the first two assumptions in favor of alternatives, but they must generally accept the third, *faute de mieux*.

Biotic Interaction

As has been amply pointed out, Darwin (1859: 62) saw the "Struggle for Existence in a large and metaphorical sense," far broader than the idea that organisms compete directly with each other. The principle of natural selection led him to conclude that if environmental conditions are roughly constant, there should be a discernible net improvement over time, so that species alive today should be better suited than their extinct counterparts:

> *There has been much discussion whether recent forms are more highly developed than ancient. I will not here enter on this subject, for naturalists have not as yet defined to each other's satisfaction what is meant by high and low forms. But in one particular sense the more recent forms must, on my theory, be higher than the more ancient; for each new species is formed by having had some advantage in the struggle for life over other and preceding forms. If under a nearly similar climate, the eocene inhabitants of one quarter of the world were put into competition with the existing inhabitants of the same or some other quarter, the eocene fauna or flora would certainly be beaten and exterminated; as would a secondary fauna by an eocene, and a palæozoic fauna by a secondary fauna. I do not doubt that this process of improvement has affected in a marked and sensible manner the organisation of the more recent and victorious forms of life, in comparison with the ancient and beaten forms; but I can see no way of testing this sort of progress (Darwin 1859: 336–337).*

Darwin does not suggest that he has evidence of "this sort of progress," only that it is a logical corollary of the theory of natural selection.

The geologic record of biodiversity provides evidence that large-scale diversity fluctuations (i.e., at the global scale, over tens of millions of years, and involving many major clades) are shaped in part by biotic interaction. Rates of origination and net diversification are negatively correlated with diversity, and rates of extinction may be positively correlated. This diversity dependence is hard to make sense of without biotic interaction. I am not referring, for the most part, to changes in the composition of the biota (e.g., mammals replacing dinosaurs), for which the removal of incumbent competitors by extinction is often an important prerequisite (Jablonski 2008a), but rather to the total level of diversity itself, whose relationship with rates of diversification is consistent with a diffuse biotic interaction that is not strongly clade-specific. The subject of wholesale displacement of one clade by another is a vast one that cannot be treated adequately here (Benton 1987, 1996; Sepkoski 1996; Jablonski 2008b), although I will briefly discuss one particularly interesting example.

Incomplete Sampling

Paleontologists have made great progress in coping with the incompleteness of the fossil record, as reviewed recently by Kidwell and Holland (2002). One of the most powerful approaches has been to predict the expected direction of sampling bias, relative to an opposite effect suggesting little or no bias. For example, Jablonski et al. (2006) documented a strong tendency for fossil bivalve genera to appear first in tropical versus extratropical latitudes. Because there are reasons to think that the tropics are more poorly sampled, Jablonski et al. concluded that the pattern of tropical first appearances is unlikely to result from sampling bias. Similar arguments can be made for the origin of major evolutionary novelties (i.e., orders of marine invertebrates) in the tropics (Jablonski 1993) and in nearshore versus offshore environments (Jablonski and Bottjer 1991). Wagner (1997) showed that morphological transitions decreased in magnitude during the evolutionary radiation of rostroconch mollusks in the early Paleozoic. Spuriously larger transitions between sister taxa would be expected if the earlier representatives were more poorly sampled, since there would be more missing intermediates. But, Wagner was able to show that the earlier forms were in fact better sampled, and he therefore argued that the observed trend was unlikely to reflect sampling bias. Ruta et al. (2006) made a similar case regarding a temporal decline in the size of evolutionary transitions in early tetrapods. Peters (2006) showed that global origination and extinction rates of marine invertebrate genera within time intervals do not correlate well with the durations of stratigraphic gaps flanking these intervals. Because there should be a strong correlation if peaks in origination and extinction were artifacts of gaps in the record (Holland 1995), Peters reasoned that the patterns of origination and extinction were unlikely to be spurious.

Another general approach to assessing whether observed patterns may result from incomplete sampling is the use of *taphonomic control taxa*. If members of a taxon of interest are absent from a set of strata, whereas members of another taxon with similar environmental preferences and preservational properties are present, then the lack of the focal taxon may tentatively be interpreted as a genuine absence. In a phylogenetic analysis that combined morphological character data with information on stratigraphic occurrence, Bodenbender and Fisher (2001) used the presence of stalked crinoid echinoderms as evidence that ecologically similar blastoid echinoderms could have been preserved, if they had been present. Postulated evolutionary trees were regarded as relatively unparsimonious if they require the absence of blastoids in strata where they should have been found if they had in fact existed.

Methods have been developed recently that use explicit models of incomplete sampling, not just to predict the likely consequences of sampling, but also to obtain quantitative estimates of parameters, such as rates of origination and extinction, which are important to an understanding of the dynamics of diversification. For example, Foote and Raup (1996) showed that, if sampling probability and extinction rate are uniform among species, then the frequency distribution of observed stratigraphic ranges can be used to develop a joint estimate of the sampling probability and average extinction rate (for refinements, see Solow and Smith 1997 and Foote 1997). They used this approach to argue that the shorter mean duration of mammal species compared with bivalve species (a classic comparison that goes back to Simpson 1944) is real, not an artifact of less complete sampling of mammals. Wagner (1997) also used this approach in the rostroconch example previously cited. Sampling models have been used to place confidence limits on the times of first and last appearance of individual taxa, which must necessarily be later and earlier, respectively, than their true times of origination and extinction. These models can either assume uniform sampling (Strauss and Sadler 1989; Marshall 1990) or can take empirically calibrated variation in sampling into account (Marshall 1997; Holland 2003). Such models have been applied, for example, to assess the likelihood that the disappearance of certain taxa is an artifact of reduced sampling of their preferred habitat (Holland 2003) and to determine whether gradual patterns of last appearance are consistent with a truly sudden extinction event (Marshall and Ward 1996).

Relaxing the assumption that rates of origination, extinction, and sampling are constant over time, Connolly and Miller (2001) and Foote (2001, 2003) considered the expected probability distribution of taxonomic first and last appearances resulting from a time series of these rates. They then used inverse methods, such as maximum-likelihood, to estimate rates from the observed first and last appearances of marine animal genera. Finally, a number of methods of sampling standardization have been developed, wherein an attempt is made to compensate for temporal variation in the completeness of the fossil record by subsampling a comparable amount of data from each time interval (e.g., Raup 1975; Miller and Foote 1996; Alroy et al. 2001, 2008; Bush et al. 2004). Here, the inverse procedure of Foote (2003) will be combined with sampling standardization to estimate temporal patterns of diversity, origination, and extinction.

Biotic Interaction

The role of biotic interactions in macroevolution has been reviewed recently by Jablonski (2008b), and I will not repeat his efforts. I will instead focus on some new analyses and two exemplary case studies that address the idea that biotic interaction may have influenced long-term changes in global biodiversity. The case for biotic interaction at this large scale derives mainly from the pioneering work of Sepkoski (1978, 1979, 1984, 1991, 1996; Miller and Sepkoski 1988), who extended population biology models, including

Lotka–Volterra dynamics of between-species interaction, to macroevolutionary processes (for reviews, see Miller 1998, 2000).[1]

In Sepkoski's extension of demographic models, species or higher taxa take the place of individuals; rates of taxonomic origination and extinction take the place of birth and death rates; and diversity dependence of taxonomic rates takes the place of density dependence of birth and death rates. Sepkoski often invoked the coupled logistic model, in which there are two or more phases with different parameters, namely, initial origination and extinction rates, strengths of diversity dependence of these rates, and carrying capacities (i.e., fixed, global diversity equilibria). In the model, the diversification rate of each phase at any point in time depends on its parameters as well as the total diversity of all phases. This particular model has been criticized for the fit between model and data, discrepancies between genus- and family-level patterns, and difficulty imagining why the global biota should have a fixed carrying capacity (Benton 1995; Courtillot and Gaudemer 1996; Benton and Emerson 2007; Erwin 2007, 2008; Stanley 2007, 2008). However, the general question of diversity dependent diversification—one of the principal *prima facie* lines of evidence for biotic interaction (Sepkoski 1996; Jablonski and Sepkoski 1996)—should be kept separate from the particular model of diversity dependence. Moreover, although Sepkoski argued that empirical diversity patterns were well described by the coupled logistic model, he arguably saw this model largely as a heuristic tool. Whether or not Sepkoski's particular model holds, there is no denying that he prompted paleobiologists to think about diversification in a new way, and 25 years after he presented a three-phase coupled model, his work is still the starting point for virtually all attempts to model global biodiversity over the Phanerozoic.

Not only does one need to separate the question of diversity dependence from the coupled logistic model, but it also is necessary to separate diversity dependence from the idea that there is an invariant equilibrial diversity—a fixed carrying capacity either for the biota as a whole or for some subset of it (Kitchell and Carr 1985; Benton and Emerson 2007; Erwin 2007). One reasonable interpretation of the results presented here is that the biota tracks a carrying capacity but that this capacity varies considerably over time.

---

[1] Sepkoski's work represents one line of many in which a heavy flux of ideas and approaches from biology into paleontology during the 1960s and 1970s shaped the study of biodiversity. Another important influence came from the theory of island biogeography (MacArthur and Wilson 1967). The paleobiological application of this theory largely focused on possible limits on diversity, such as species-area effects, speciation–extinction balance, and their higher-taxonomic analogues (Webb 1969; Valentine and Moores 1972; Schopf 1974; Flessa 1975; Sepkoski 1976; Mark and Flessa 1977; Flessa and Sepkoski 1978; Rosenzweig 1995).

Overview of Previous Work

History of Diversity

Figure 18.2A shows Sepkoski's classic depiction of family-level diversity of marine animals, along with the principal groups that contribute to his statistically delimited Cambrian, Paleozoic, and Modern evolutionary faunas (Sepkoski 1981), and Figure 18.2B illustrates the three-phase coupled logistic model that he used to represent this history of diversity (Sepkoski 1984). In this model, whose parameters are derived from the observed

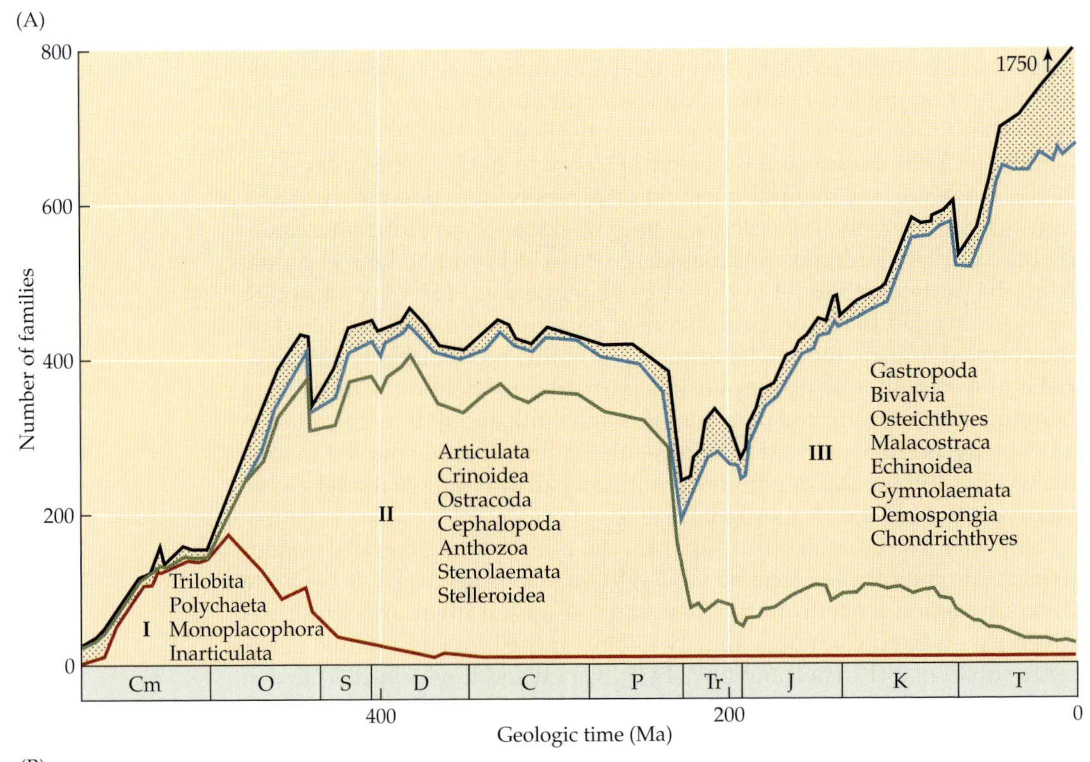

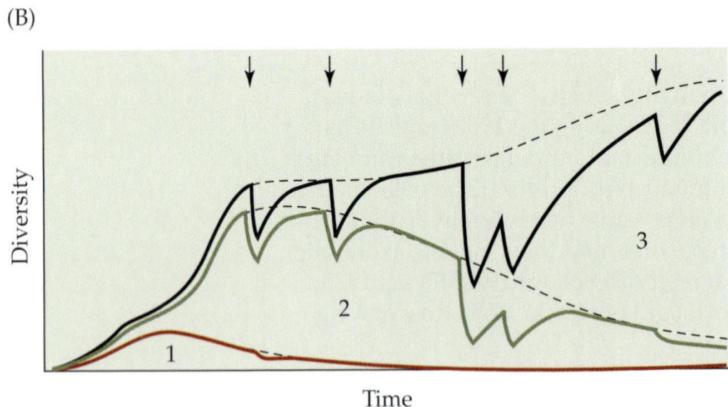

histories of origination, extinction, and diversity of the three evolutionary faunas, there are externally imposed perturbations corresponding to mass extinctions, but the ultimate fate of the three phases, at least in terms of numbers of taxa, is the same with or without such perturbations, as shown by the solid and dotted lines.

Figure 18.3 shows genus diversity over the Phanerozoic, based on the times of first and last appearance of about 31,000 marine animal genera from Sepkoski's (2002) compendium. With even a quick glance at this graph, it is not surprising that a number of workers (Benton 1995; Benton and Emerson 2007; Stanley 2007) have suggested long-term exponential growth in diversity as a better model than simple (Courtillot and Gaudemer 1996) or coupled (Sepkoski 1984) logistic growth. But, much of the apparent growth in diversity is concentrated in the last 10% or so of the Phanerozoic, and this increase may be exaggerated by an increase in the quality of sampling (Raup 1972, 1976; Smith 2001; Peters and Foote 2001) and by the fact that the presence of still-living taxa can be indirectly inferred for longer spans of time than that of extinct taxa, the "Pull of the Recent" (Raup 1979; but see Jablonski et al. 2003).

Sampling issues can be partly circumvented with comprehensive data that include not just the times of first and last appearance of each taxon, but also the intervening occurrences in the stratigraphic record. Miller developed a pioneering database for Ordovician marine invertebrates that compiled thousands of fossil assemblages from around the world, along with their geologic age and various contextual data, such as their paleoenvironment (Miller and Mao 1995; Miller 1997). The database enabled him to ask the question, how would a global diversity trajectory appear if the same amount of data were sampled from each time interval? The result (Miller and Foote 1996) was striking.[2] Whereas the raw data from Sepkoski suggest a rather steady increase in diversity throughout much of the Ordovician

[2] Since I am a co-author of that paper, I should make it clear that my contribution was minor and that the conception of the project and the analyses of taxonomic diversity were entirely Miller's work.

FIGURE 18.2 Sepkoski's Early Representation of Marine Diversity (A) Number of marine animal families over geologic time (Ma, millions of years ago). The taxonomic names within the three fields are the major groups that contribute to the three evolutionary faunas that Sepkoski delimited statistically. The stippled field represents families not attributed to any of these faunas, and the arrow points to present-day diversity. (Cm, Cambrian; O, Ordovician; S, Silurian; D, Devonian; C, Carboniferous; P, Permian; Tr, Triassic; J, Jurassic; K, Cretaceous; T, Tertiary.) (B) Coupled logistic model based on data in (A), with perturbations corresponding to major mass extinctions. Solid and dashed lines indicate diversity trajectories with and without mass extinctions. (A from Sepkoski 1981; B from Sepkoski 1984.)

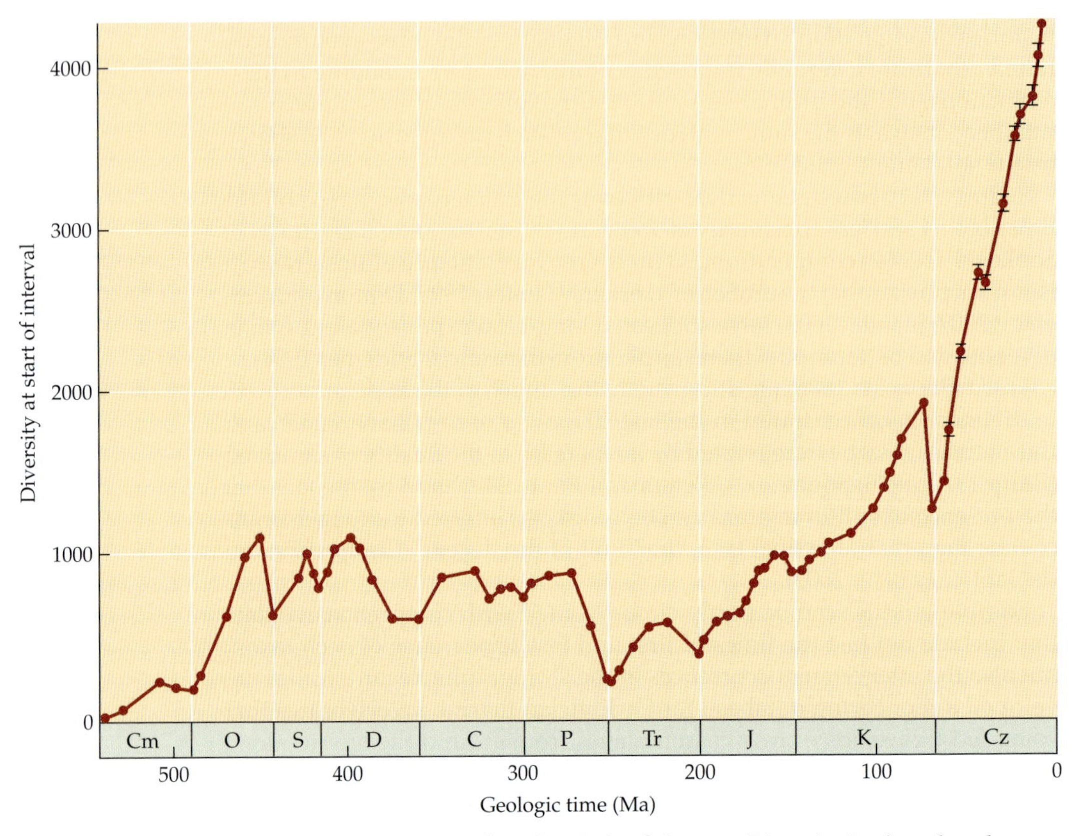

FIGURE 18.3 Diversity of Marine Animal Genera Diversity is plotted as the number of genera known to be extant at the start of each stratigraphic stage. Vertical error bars show ± one standard error, based on bootstrap resampling of the stratigraphic ranges of genera. (Cm, Cambrian; O, Ordovician; S, Silurian; D, Devonian; C, Carboniferous; P, Permian; Tr, Triassic; J, Jurassic; K, Cretaceous; Cz, Cenozoic.) (Based on data in Sepkoski 2002.)

(see Figure 18.3), the "sampling-standardized" diversity trajectory suggested that diversity increased through the middle Ordovician and then leveled off. The obvious question arose: what would Phanerozoic diversity look like if one could do for each time period what Miller had done for the Ordovician? The quest to answer this was in large part responsible for the effort that ultimately led to the Paleobiology Database (http://paleodb.org), a digital archive of the published paleontological record, coupled with diverse analytical tools—effectively a GenBank for paleontology. Although methods of sampling standardization have evolved beyond Miller's initial efforts (Alroy et al. 2001, 2008; Bush et al. 2004; Alroy 2009), his work was seminal.

Dynamics of Diversification

If we are to understand the dynamics of diversification and not just describe its temporal trajectory, it is essential to relate diversity to its underlying evolutionary rates (Sepkoski 1978, 1984; Levinton 1979; Carr and Kitchell 1980; Stanley et al. 1981; Benton 1995; Foote 2000a,b, 2006; Benton and Emerson 2007; Alroy 2008). Some previous approaches to testing for diversity-dependent dynamics are potentially problematic. For example, Stanley (2007: Figure 11) compared average standing diversity in each time interval with the net diversification rate in the same interval. He did not find a significant negative correlation, and therefore concluded that diversity dependence was weak at best. But average diversity and net diversification rate are not logically independent, and their direction of forced dependence will lead, all else being equal, to a positive correlation between diversity and diversification rate, potentially obscuring any true negative correlation that may exist. Alroy (2008) avoided this problem by comparing total diversity in one time interval with rates of origination and extinction in the following interval, and he found a significant positive correlation between diversity and extinction rate. Total diversity in an interval may not be the most relevant figure, however, since much of the diversity generated could become extinct before the following time interval. I therefore suggest here a slight modification of Alroy's approach, namely to compare standing diversity at the start of a time interval with taxonomic rates in that interval. I also use finer temporal resolution than Alroy; average interval lengths are roughly 7 rather than 11 million years. Despite this and other analytical variations described in the following section, I follow the spirit of Alroy's (2008) analysis.

Materials, Methods, and Results

Data for the new results presented here were downloaded from the Paleobiology Database. Details of the download and vetting of data are described in Miller and Foote (2009). What matters for the present discussion is that the data are organized into *collections*, which are lists of co-occurring taxa with contextual information, such as geologic age. The taxonomic level used here is that of the genus, and the presence of a genus in a collection, irrespective of whether it is represented by one or more species, is referred to as an *occurrence*. Given such data, there are many ways to subsample in an effort to even out the temporal variation in the size of the sample (Alroy et al. 2001, 2008; Bush et al. 2004). I have opted for a simple approach known as by-list, *occurrence-weighted* (OW) subsampling (Alroy et al. 2001). In this approach, the basic unit of sampling is the collection, and collections are resampled at random, without replacement, until a given quota of occurrences is obtained for each time interval. This procedure is repeated a number of times (here, 100) and the results are averaged.

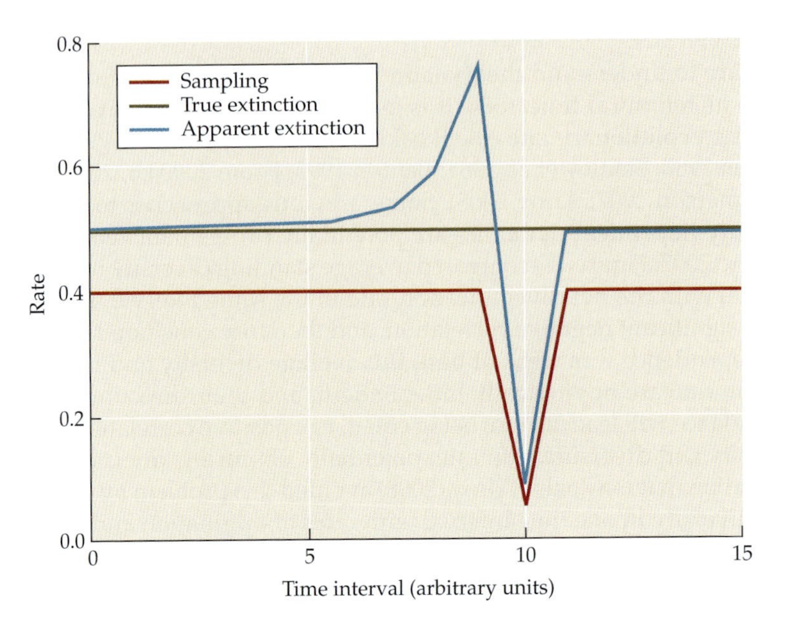

FIGURE 18.4 Model Showing the Per-Taxon Rate of Sampling, True Extinction, and Apparent Extinction A short-lived deficit in sampling in one time interval leads to a spuriously low extinction rate in that interval and spuriously high rates in the intervals leading up to it. Origination rate is affected similarly, with the effect propagating forward in time. (Based on equations in Foote 2000a.)

For each random subsample of the data, the temporal ranges of genera are recorded. The temporal ranges are used to tabulate a matrix **X**, in which the quantity X_{ij} is the number of sampled genera with first appearance in interval *i* and last appearance in interval *j*. Any genus that is sampled at least once before a given time interval and at least once during or after that time interval is credited as extant at the beginning of the interval (Foote 2000a); this count is designated X_b.

Incomplete sampling affects perceived rates of taxonomic origination and extinction as well as standing diversity, regardless of whether sampling is variable or constant. For example, suppose that true extinction rates were constant and there were a dearth of sampling in one time interval (Figure 18.4). Then many of the genera that would have made a last appearance in this interval, if sampling had been constant, will have last appearances in one of the previous time intervals. As a result, apparent extinction rate will be spuriously low in the interval in question and spuriously high in the previous intervals, with the effect decaying exponentially backward in time. If, by contrast, the interval in question is marked by better than average sampling, that interval will show spuriously high extinction and the preceding intervals will show spuriously low extinction. In the case of origination, the effect of incomplete sampling propagates forward in time.

Even if sampling is constant, a true excursion in origination or extinction rate will be smeared out in time because not all of the excess first or last appearances will be captured in the interval of time in which the rate excursion occurs (Signor and Lipps 1982; Foote 2000a) (Figure 18.5). This smearing out of true rates is potentially quite important in assessing diversity dynamics. If, for example, a large decrease in diversity is quickly followed by an increase in the rate of origination, there may be a spurious delay before the acceleration in origination is seen in the fossil record (Foote 2000a, 2003; Lu et al. 2006). This fact may contribute to the apparent lag between extinction and biotic recovery on geologic timescales (Hallam 1991; Erwin 1998, 2001; Sepkoski 1998; Kirchner and Weil 2000).

The effects of incomplete and variable sampling shown schematically in Figures 18.4 and 18.5 can be modeled with some basic assumptions, namely that all taxa extant during an interval of time are characterized by the same rates of origination, extinction, and sampling in that interval, although these rates may vary from one interval to the next (Foote 2003). This model leads to a set of equations relating true rates to expected probabilities P_{ij} that a randomly chosen taxon will have an observed first appearance in interval i and last appearance in interval j (Foote 2003). Given the correspondence between model parameters and probabilities, a likelihood function can be

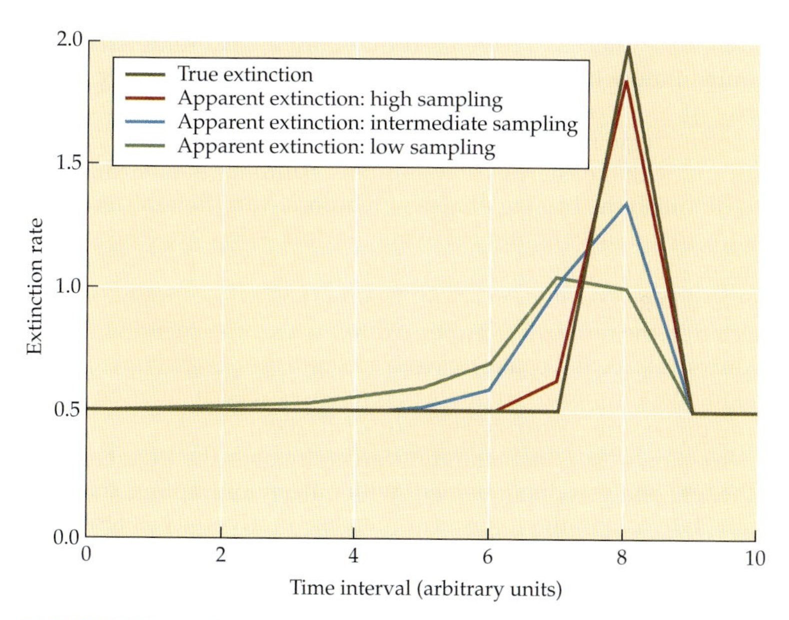

FIGURE 18.5 Model Showing the Effect of Incomplete Sampling on a True Increase in Extinction Rate The true extinction excursion is smeared backward in time, and lower average sampling rates lead to greater smearing. Origination rate is affected similarly, with the effect propagating forward in time. (Based on equations in Foote 2000a; see Signor and Lipps 1982.)

developed that incorporates these probabilities and the observed counts X_{ij}. This function is maximized numerically to obtain the best fitting set of origination, extinction, and sampling rates per interval.

Because the random subsampling of data includes multiple occurrences of genera within their stratigraphic ranges, these occurrences can be used to estimate the sampling probabilities for each time interval. The number of genera inferred to be extant during a time interval is the number sampled at least once before that interval and at least once afterwards. The proportion of these that are also sampled within the interval gives an estimate of the sampling probability for that interval (Paul 1982; Foote and Raup 1996). In the analyses presented, the numerical optimization is constrained to follow these sampling probabilities, hence only the origination and extinction rates are estimated by the inverse procedure. The model used here assumes that originations are clustered at the beginning of each time interval and that extinctions are clustered near the end, rather than being spread throughout the interval. This assumption is supported by previous analysis of data on Phanerozoic marine animals (Foote 2005; cf. Alroy 2008). All rates are converted to instantaneous per-capita rates, per time interval (Foote 2000a, 2003).

In addition to the OW method of subsampling, I have analyzed data using the *lists-unweighted* method (UW; Alroy et al. 2001) and the *shareholder-quorum* method (SQ; Alroy 2009). In addition to total diversity at the lower boundary of a time interval, I have used the count of taxa sampled in the intervals immediately before and immediately after the boundary (two-timers; Alroy 2008). Here I present only those results from the OW analysis of total taxa that are qualitatively consistent with results from the UW and SQ analyses and with the analysis of two-timers. Sampling standardization is meant to diminish the biasing effects of temporal variation in sampling corresponding to the sheer amount of available data. There are other biases that are harder to overcome, such as variation in the range of sampled environments and in geographic coverage. To some extent, geographic coverage is reflected in the number of published references representing a time interval (Alroy et al. 2008). To take this into account, the SQ analysis follows Alroy in taking data from a fixed quota of references at each iteration of the subsampling procedure (Alroy 2008; Alroy et al. 2008).

Figure 18.6 shows estimated standing diversity over the Phanerozoic, based on the OW analysis. In agreement with other subsampling analyses (Alroy et al. 2001, 2008), the long-term increase in diversity is less pronounced than in the raw data (see Figures 18.2 and 18.3), but subsampled diversity shares many features with the raw data, such as drops and rebounds surrounding major extinction events. To minimize possible edge effects (Foote 2000a, 2003) and the effects of sparse Cambrian data, I will focus on the stretch of time between the middle Ordovician Period (Caradoc Epoch, about 460 million years ago) and the Paleogene Period (Oligocene

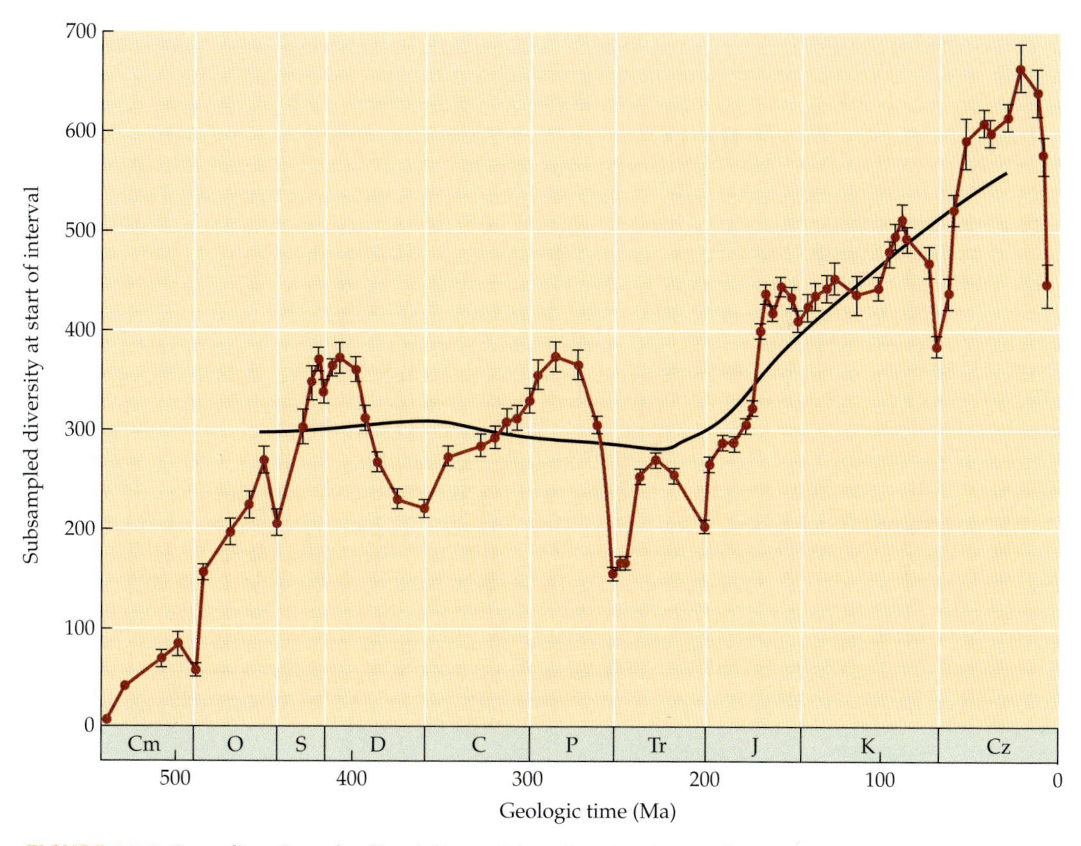

FIGURE 18.6 Sampling-Standardized Genus Diversity Analysis is limited to marine invertebrates. The heavy line in this and subsequent figures shows a locally weighted least squares regression through the focal timespan, from the Late Ordovician through the Paleogene. The method of by-list, occurrence-weighted sampling was used, with a quota of 1012 occurrences per time interval; this quota can be satisfied by all but two stages within the focal timespan (the Induan in the Early Triassic and the Coniacian in the Late Cretaceous). Error bars show ± one standard error (standard deviation of the results of 100 resampling trials). This model of diversity shares certain features with the raw data (see Figure 18.3), such as decreases at most major extinction events and a long-term secular rise, but the increase over the Phanerozoic is more subdued. (Cm, Cambrian; O, Ordovician; S, Silurian; D, Devonian; C, Carboniferous; P, Permian; Tr, Triassic; J, Jurassic; K, Cretaceous; Cz, Cenozoic.) (Based on data from the Paleobiology Database, http://paleodb.org.)

Epoch, about 23 million years ago). Because the time series of diversity and taxonomic rates show temporal trends, it is important to detrend the data if the time series are to be compared to each other (Alroy 2008). Detrending is accomplished by fitting a locally weighted regression (LOWESS), with

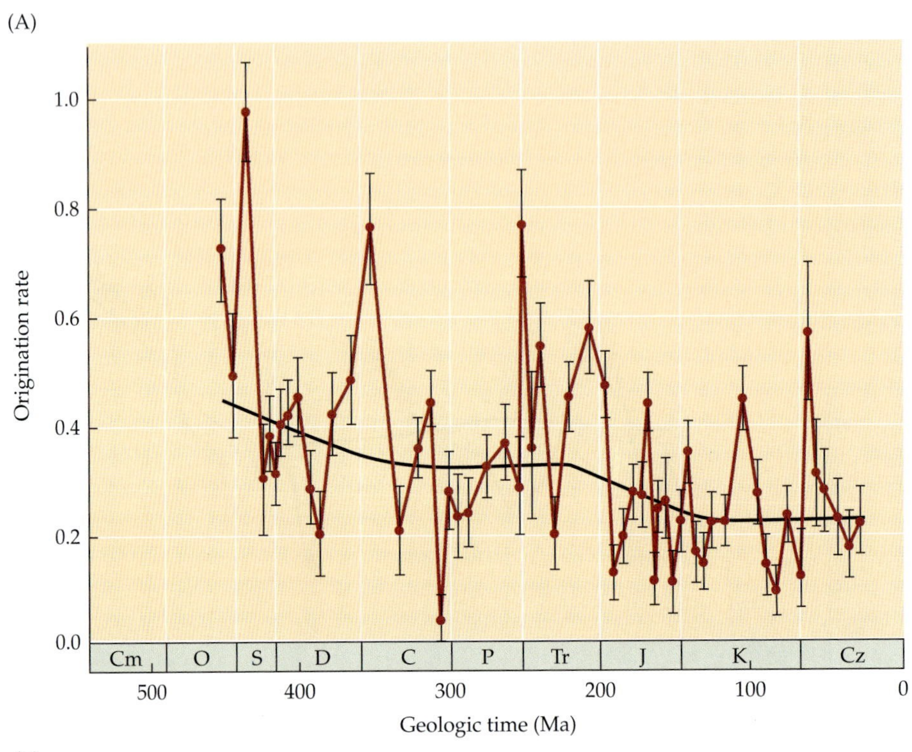
(A)
Origination rate
1.0
0.8
0.6
0.4
0.2
0.0
Cm
O
S
D
C
P
Tr
J
K
Cz
500
400
300
200
100
0
Geologic time (Ma)

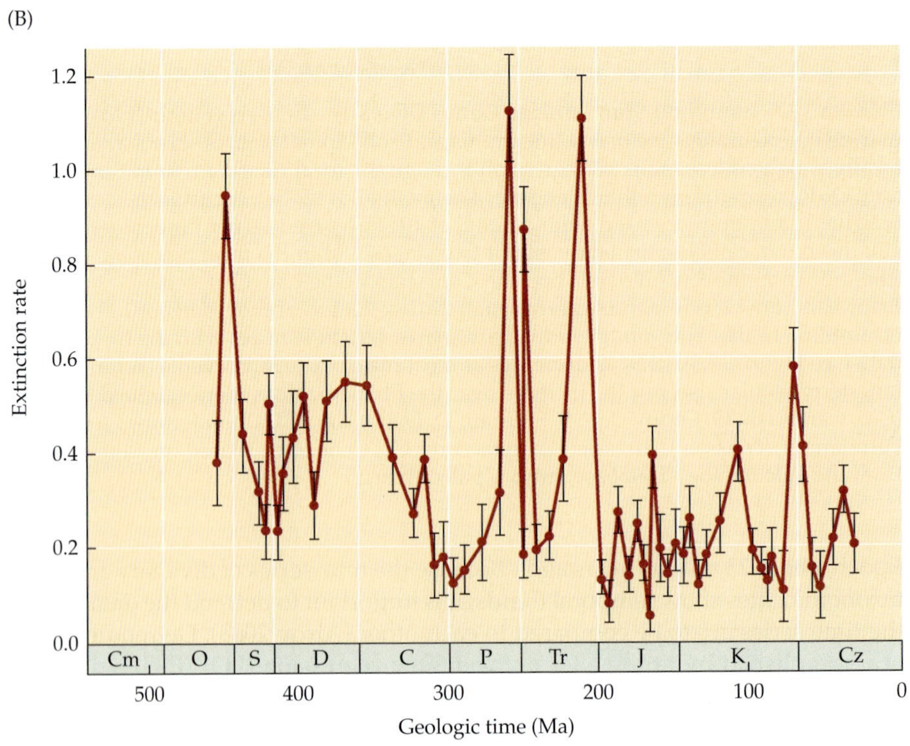
(B)
Extinction rate
1.2
1.0
0.8
0.6
0.4
0.2
0.0
Cm
O
S
D
C
P
Tr
J
K
Cz
500
400
300
200
100
0
Geologic time (Ma)

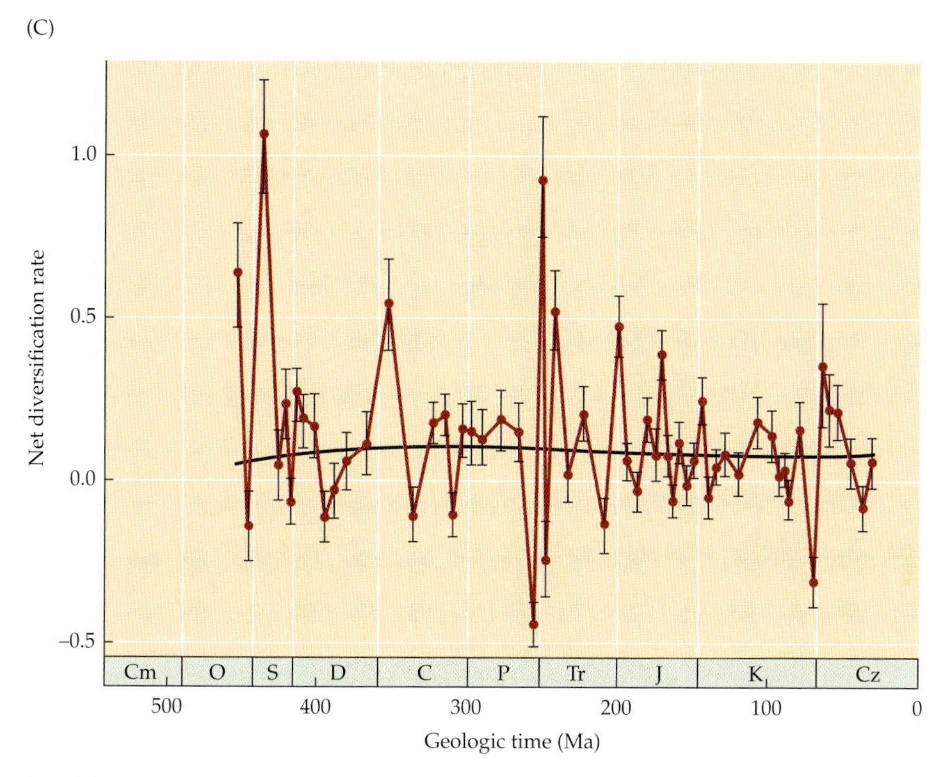

FIGURE 18.7 Taxonomic Rates of Evolution of Marine Invertebrate Genera from the Late Ordovician through the Neogene (A) Origination. (B) Extinction. (C) Net diversification. (See Figure 18.6 for description of heavy lines and error bars.) (Cm, Cambrian; O, Ordovician; S, Silurian; D, Devonian; C, Carboniferous; P, Permian; Tr, Triassic; J, Jurassic; K, Cretaceous; Cz, Cenozoic.)

a smoothing span of 50% of the data points, and analyzing the residuals. Figure 18.7A–C presents the estimated rates of origination, extinction, and net diversification. The black lines in Figures 18.6 and 18.7A–C show the LOWESS regression of each time-series against geologic time. Partly because of the diversity measure used here (Alroy et al. 2008), the time series of diversity is significantly autocorrelated, even if just the residuals from the regression are considered (Spearman rank-order correlation between successive diversity residuals: $r_s = 0.77$, $p < 0.001$). Therefore, I will report the correlations between the first differences in residuals of diversity and taxonomic rates. Results are qualitatively consistent if the raw residuals, rather than their first differences, are analyzed.

The residuals in diversity at the start of each time interval are negatively correlated with the residuals in net diversification rate within the interval (Figure 18.8; Spearman rank-order correlation $r_s = -0.53$, $p < 0.001$), and this correlation holds even if the intervals following the five major mass

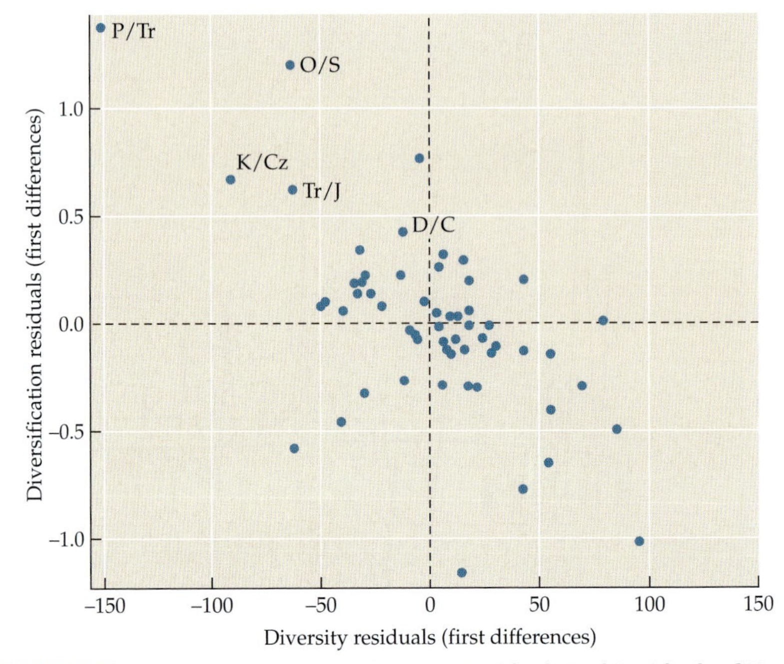

FIGURE 18.8 Comparison between Diversity Residuals and Residuals of Net Diversification Rate Labeled points indicate rebounds after four generally recognized extinction events (O/S, Ordovician/Silurian; P/Tr, Permian/Triassic; Tr/J, Triassic/Jurassic; K/Cz, Cretaceous/Cenozoic) and one additional event (D/C, Devonian/Carboniferous). Extinction rate is elevated in the final two stages of the Devonian: the Frasnian and the Famennian (Bambach 2006; Foote 2007), and the event is generally referred to as the Frasnian–Famennian extinction. For purposes of this paper, the Famennian extinction rate is taken to represent a single Late Devonian event. This usage agrees with some analyses (Foote 2007) that suggest a substantially higher extinction rate in the Famennian than in the preceding stage. Diversification rate is significantly and negatively correlated with diversity whether or not these five recovery intervals are included.

extinction episodes of the Phanerozoic are omitted ($r_s = -0.36$, $p < 0.01$). If the negative and positive diversity residuals are analyzed separately, there is still a pronounced negative correlation for the negative residuals ($r_s = -0.62$, $p < 0.001$), but not for the positive residuals ($r_s = -0.13$, $p = 0.49$). These results suggest that lower than average diversity enhances diversification, but that higher than average diversity does not suppress net diversification (Stanley 2007). This finding would be inconsistent with the canonical logistic model of Sepkoski.

On the whole, analysis of the origination and extinction components of net diversification agrees with previous results (Sepkoski 1978, 1979;

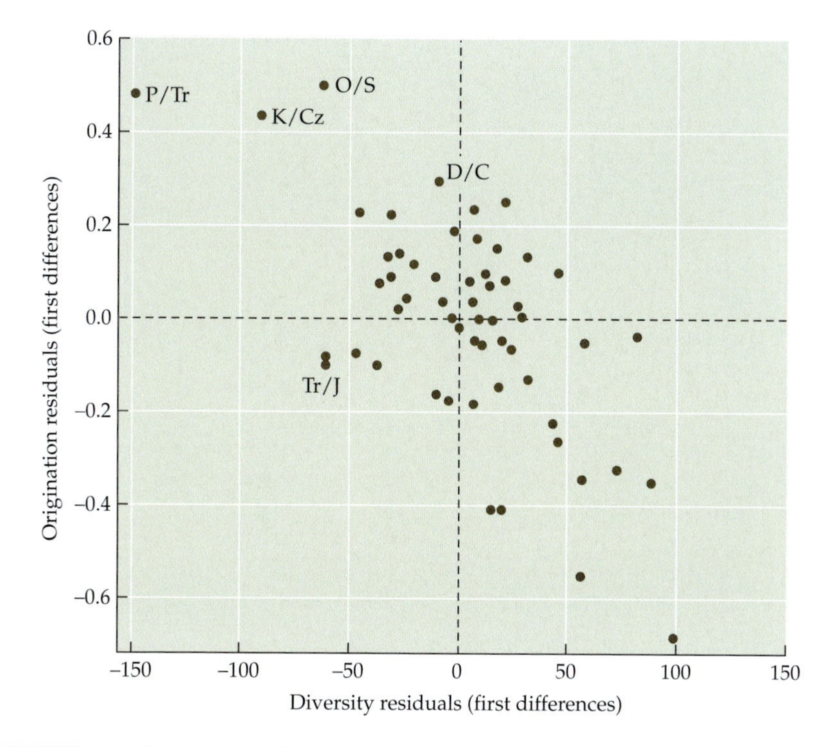

FIGURE 18.9 Comparison between Diversity Residuals and Residuals of Origination Rate Origination rate is significantly and negatively correlated with diversity whether or not the recovery intervals are included. (O/S, Ordovician/Silurian; D/C, Devonian/Carboniferous; P/Tr, Permian/Triassic; Tr/J, Triassic/Jurassic; K/Cz, Cretaceous/Cenozoic.)

Alroy 1998; Connolly and Miller 2002) that suggest substantial diversity dependence of origination (Figure 18.9; $r_s = -0.48$, $p = 0.0014$; $r_s = -0.34$, $p = 0.012$, with recoveries from major mass extinctions omitted), but little net diversity dependence of extinction (Figure 18.10; $r_s = 0.12$, $p = 0.37$). These results stand in contrast to some theoretical suggestions that extinction should be diversity dependent (Levinton 1979) and to the analyses of Alroy (2008), who found diversity in an interval to be significantly and positively correlated with extinction in the following interval but to be virtually uncorrelated with origination in the following interval. Because different approaches yield somewhat different results, the conclusion that can be most safely drawn is that net diversification rate is negatively diversity dependent, irrespective of how this relationship may break down into its origination and extinction components.

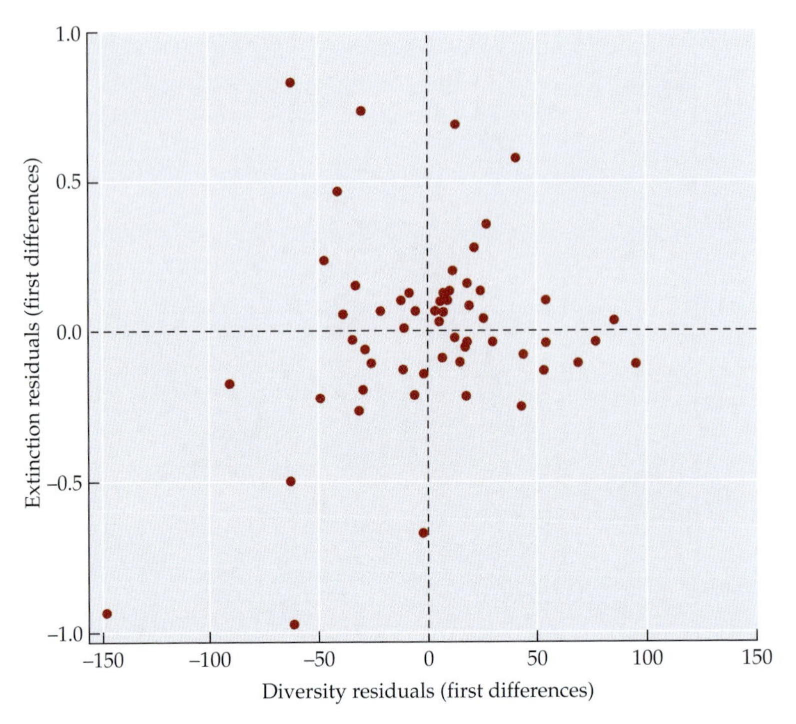

FIGURE 18.10 Comparison between Diversity Residuals and Residuals of Extinction Rate Extinction rate is positively but not significantly correlated with diversity.

Additional Case Studies

Bivalve Genera over the Phanerozoic

An interesting analysis of long-term diversity dynamics for a single major clade is Miller and Sepkoski's (1988) treatment of bivalve diversity at the genus level. After an initial radiation during the Ordovician, bivalve richness increased at a nearly constant average per-capita rate (Figure 18.11). This long-term increase was interrupted by a few negative excursions at extinction events, after which the rate of diversification was accelerated for a short time interval, bringing diversity back to the pre-extinction trajectory. Even though the long-term rate of diversification is roughly constant, the rebounds after perturbations suggest diversity dependence of the net diversification rate rather than simple exponential growth. Indeed, Miller and Sepkoski interpreted this diversity history as one phase of a coupled logistic system with time-homogeneous parameters and superimposed perturbations. The Ordovician increase in diversity, the long-term trend, and the rapid rebounds after extinction events are all accommodated by a single set of initial origination and extinction rates, carrying capacities, and damping coefficients.

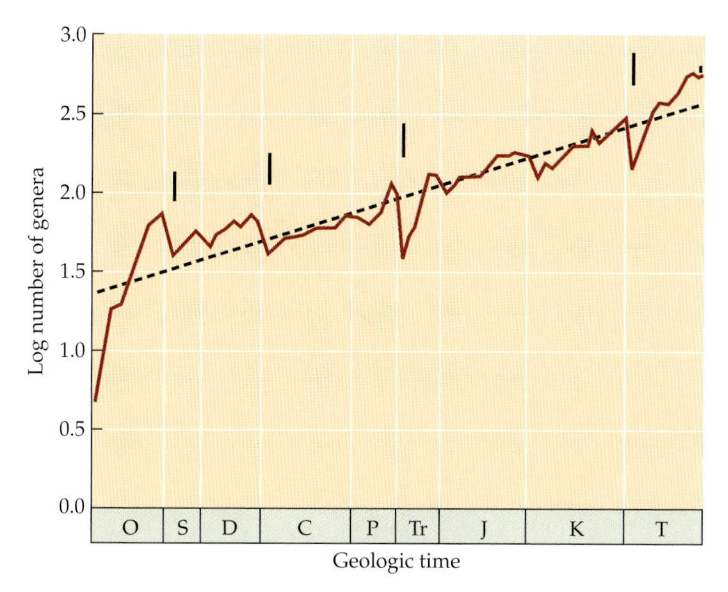

FIGURE 18.11 Miller and Sepkoski's Depiction of Bivalve Genus Diversity Starting in the Ordovician Period Dashed line shows log-linear fit to the data. The entire diversity trajectory can be modeled as one phase of a coupled logistic system with external perturbations (vertical lines) corresponding to major extinction events. (O, Ordovician; S, Silurian; D, Devonian; C, Carboniferous; P, Permian; Tr, Triassic; J, Jurassic; K, Cretaceous; T, Tertiary.) (From Miller and Sepkoski 1988.)

Given the concerns with the quality of the fossil record, can the long-term pattern in bivalves be trusted? After all, analysis of the marine fauna as a whole suggests that attempts to standardize sampling have a considerable impact on the shape of the diversity curve (Alroy et al. 2001, 2008). Jablonski et al. (2003) documented that nearly all fossil bivalve genera in Sepkoski's compendium with living representatives also have fossil representatives in the Plio-Pleistocene (roughly the past 5 million years). They therefore concluded that the observed increase in bivalve diversity is not an artifact of the Pull of the Recent. Because the analysis by Jablonski et al. still leaves open the possibility of long-term changes in the quality of sampling, it is worth trying to correct for this directly with sampling standardization, as depicted in Figure 18.12. In this figure, a LOWESS regression is rather similar to a linear regression—in other words, the temporal pattern is reasonably close to that depicted by Miller and Sepkoski, although there is clearly more deviance than in their representation. In agreement with the raw data, however, there is little net change in the average rate of diversification over some 400 million years, and diversification accelerates greatly following many major drops in diversity, just as Miller and Sepkoski argued.

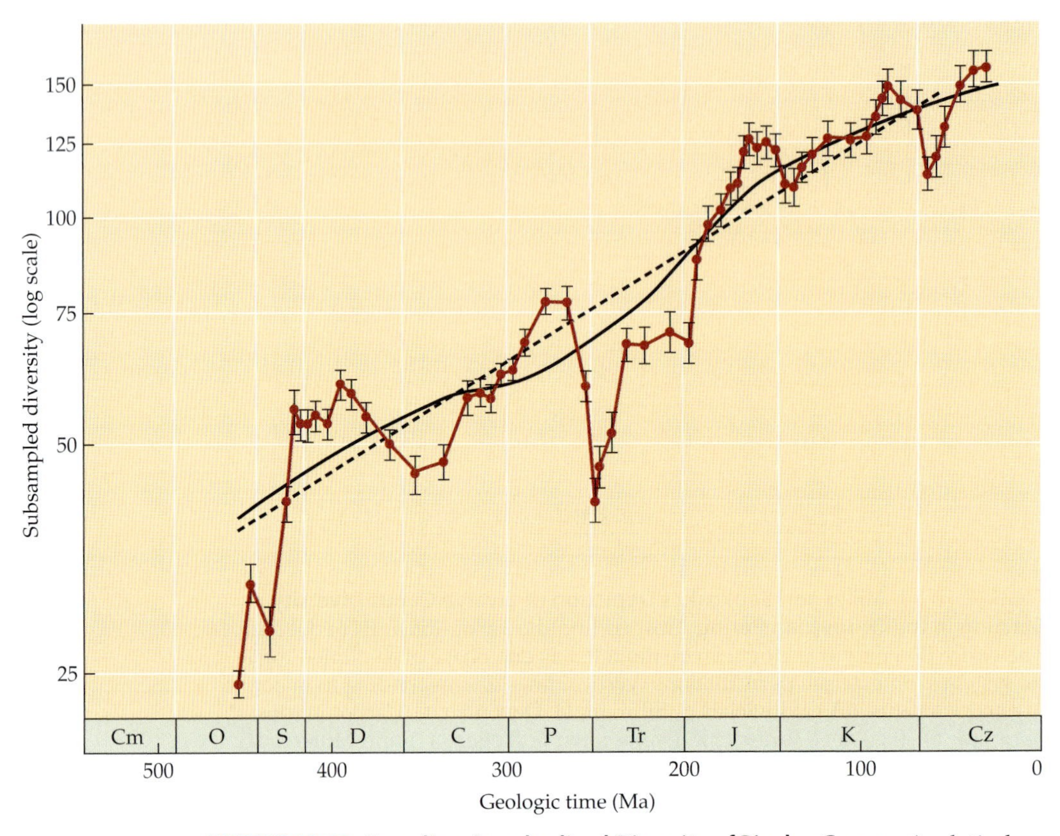

FIGURE 18.12 Sampling-Standardized Diversity of Bivalve Genera Analytical methods, heavy line, and error bars are the same as in Figure 18.6. Dashed line shows log-linear regression of the data points. A quota of 190 occurrences per interval was used; this quota can be satisfied by all but four intervals: Pragian in the Early Devonian; Eifelian in the Middle Devonian; Tournaisian in the Early Carboniferous; and Bashkirian in the Late Carboniferous. Nearly log-linear, long-term trend is consistent with Miller and Sepkoski's (1988) interpretation of the raw data (see Figure 18.11). (Cm, Cambrian; O, Ordovician; S, Silurian; D, Devonian; C, Carboniferous; P, Permian; Tr, Triassic; J, Jurassic; K, Cretaceous; Cz, Cenozoic.) (Based on Data from the Paleobiology Database http://paleodb.org.)

Cheilostome and Cyclostome Bryozoans

While it is possible to establish a reasonable *prima facie* case for biotic interactions shaping large-scale biodiversity patterns (Sepkoski 1996; and results previously presented), there has been substantially less progress in establishing the details of how such patterns emerge. Here I will briefly summarize a case study that demonstrates a possible contributing factor at the finest level of organismic interactions, but still leaves questions unanswered.

McKinney (1992, 1995) has demonstrated that two major groups of bryozoans, the Cyclostomata and Cheilostomata, are not equal competitors for

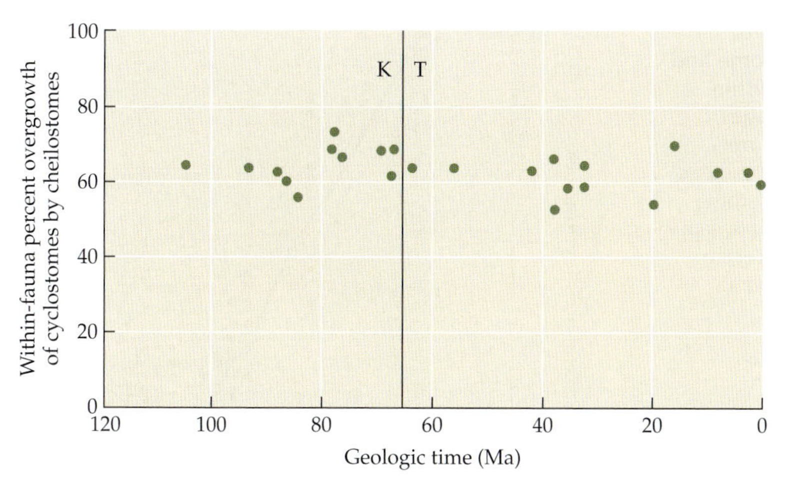

FIGURE 18.13 Geologic History of Overgrowth Relationships Involving Competition between Colonies of Cyclostome and Cheilostome Bryozoans Within individual assemblages, cheilostome colonies tend to overgrow cyclostome colonies about two-thirds of the time, with no long-term trend. K/T marks the boundary between the Cretaceous and Tertiary periods. (From McKinney 1995.)

space on the marine floor. When colonies of the two groups come into contact on a hard substrate, cheilostomes manage to overgrow cyclostomes about two-thirds of the time, and this competitive edge has been roughly constant for the past 100 million years (Figure 18.13). Over the same span of time, the net rate of diversification of cheilostomes has exceeded that of cyclostomes; both clades diversified through the Cretaceous period, but cyclostomes never recovered from the loss of diversity at the end-Cretaceous extinction event (Figure 18.14). Local communities have also become increasingly dominated by cheilostomes, whether this is assessed by species richness (Figure 18.15A) or relative abundance (as skeletal mass; Figure 18.15B). The

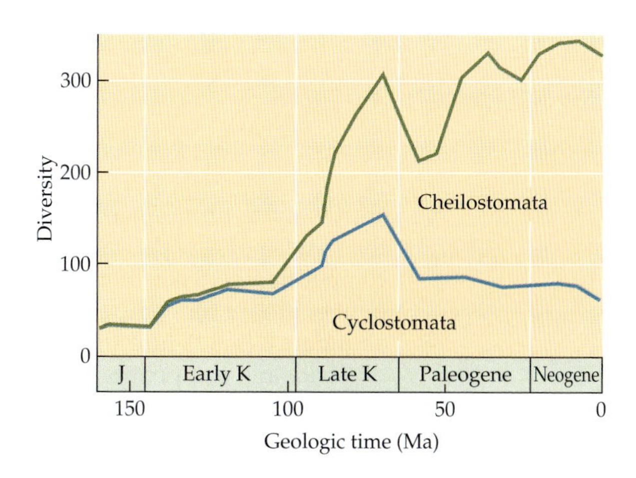

FIGURE 18.14 Global Genus Diversity of Cyclostome and Cheilostome Bryozoans Both groups increased in diversity through the Late Cretaceous, but cyclostomes, in contrast to cheilostomes, failed to diversify following the end-Cretaceous extinction event. (J, Jurassic; K, Cretaceous.) (From Sepkoski et al. 2000.)

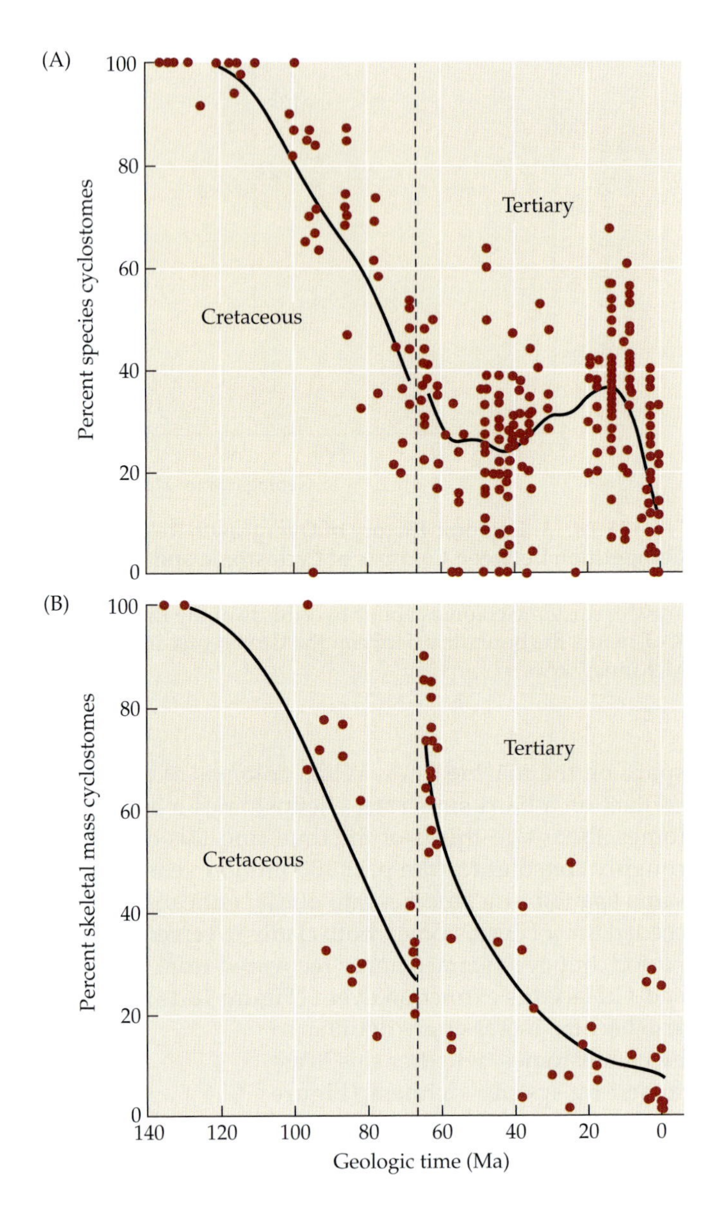

FIGURE 18.15 Comparative Representation of Cyclostome and Cheilostome Bryozoans Within Individual Fossil Assemblages (A) Percent of species that are cyclostomes. (B) Percent of skeletal biomass belonging to cyclostome colonies. Curves fitted to data separately for Cretaceous and Tertiary periods. (From McKinney et al. 1998.)

shift is more pronounced in the latter (McKinney et al. 2001). Based on the pattern of diversification alone, it would be hard to say much about competition between members of the two groups, but given the independent evidence for competitive interaction (see Figure 18.13) and relative abundance (see Figure 18.15B), it is irresistibly tempting to try to connect diversity and competition.

It is not hard to imagine that competitive superiority would lead to greater dominance at the community scale. But how might this translate

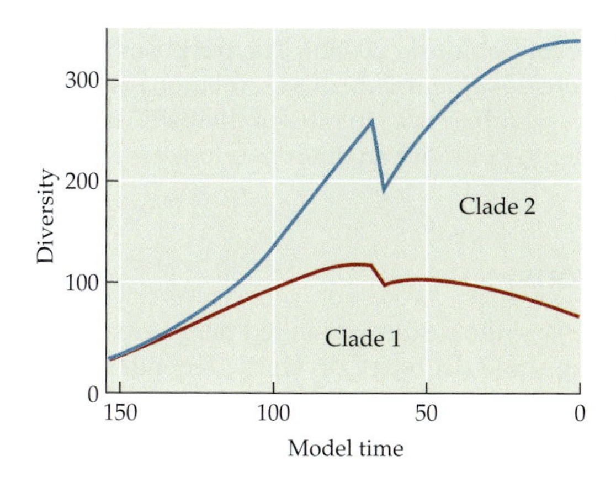

FIGURE 18.16 Two-Phase Coupled Logistic Model with Parameters Intended to Represent Cyclostomes (Clade 1) and Cheilostomes (Clade 2) The perturbation corresponds to the end-Cretaceous extinction event. (From Sepkoski et al. 2000.)

into greater global diversity? One obvious possibility is that it reduces the probability of extinction of the superior competitor. Data on stratigraphic ranges of bryozoan genera suggest that this is not the case, however. Over the past 100 million years (starting in the Cenomanian stage of the Late Cretaceous), the mean per-capita rate of extinction based on Sepkoski's (2002) data was 0.10 (± 0.017) in cheilostomes and 0.098 (± 0.010) in cyclostomes, a trivial difference. Over the same time span, however, cheilostome genera had much higher rates of origination: 0.31 (± 0.031) versus 0.092 (± 0.0095). Indeed, Sepkoski et al. (2000) modeled the cyclostome–cheilostome system with coupled logistic equations (Figure 18.16) and came to the conclusion that suppressed origination is what kept cyclostome diversity from growing following the perturbation of the Cretaceous/Tertiary extinction event.

But what exactly is the link, if any, between competition for space and rate of origination? Or is it just a coincidence? Do cheilostomes compete better for space and have higher rates of origination for other reasons that have nothing to do with competitive ability vis-à-vis cyclostomes? Here we come to one of the major gaps in the understanding of diversification: the causal relationship between organismic traits, on the one hand, and rates of origination and extinction, on the other (Jablonski 2008c). Even many of the clearest examples of a strong statistical correlation between traits and rates stop short of suggesting mechanisms connecting the two. There are important exceptions, of course. For example, the correlation between larval morphology and species-level extinction risk within gastropods evidently exists because larvae that can disperse farther allow species to have broader geographic ranges and therefore greater resistance to extinction (Hansen 1980; Jablonski 1986; Jablonski and Hunt 2006; and Crampton et al. 2010 for an alternative view). Although this particular case is not thought to be underlain by biotic interaction, some preliminary efforts have begun to model the relationship between interaction at the organismic level and dynamics of diversification at the clade level (Jablonski 2008b). The purpose here is not to be negative—even the mere documentation of a correlation between differences in organismic traits and differences in rates of diversification is a major step forward—but rather to point out an often-overlooked avenue for future research.

Discussion and Conclusions

Despite methodological differences, the results presented here agree with many previous studies showing some degree of diversity dependence of taxonomic rates. At the very least, even those studies that are highly skeptical of Sepkoski's particular approach to modeling the diversification of life agree that rates of diversification accelerate after major extinction events (e.g., Benton and Emerson 2007; Erwin 2007, 2008; Stanley 2007, 2008), but it is shown here that diversification is still diversity dependent, if we ignore recoveries from the major mass extinctions.

In a prior study that also used sampling standardization to assess diversity dependence, Alroy (2008) found positive correlations between: (1) total diversity in a time interval and extinction rate in the following interval and (2) extinction rate in one interval and origination rate in the following interval. In contrast to the present study, Alroy did not detect diversity dependence of origination rates. This outcome could well be because, for origination rate in an interval of interest, total diversity in the previous interval—Alroy's measure—is less relevant than standing diversity at the start of the interval of interest. Alroy's correlation between extinction and subsequent origination is consistent with this suggestion, since standing diversity at the start of an interval is affected by origination and extinction rates in the previous interval.

Clearly, different analyses of the data lead to somewhat different interpretations of the details of diversity dynamics, and there is yet no consensus on the particular model that accommodates all these details. Nonetheless, there is clear support for feedback among diversity, origination, and extinction that is consistent with biotic interaction affecting the dynamics of the system. Release from competition is the most commonly invoked reason for the elevation of origination rates following declines in diversity, but Stanley (2007, 2008) instead argues that predation was the key to enhanced origination, as the extinction of predators allowed prey taxa to diversify. At a smaller geographic scale, it is often the case that provinces affected more strongly by an extinction event experience greater invasion in the wake of the extinction (Jablonski 2008a), but this is not always so (Krug and Patzkowsky 2007). Finally, it is important to reiterate that diversity dependence need not imply an equilibrial system. A model in which rates of origination and extinction respond on short time scales to changes in diversity, but in which the global carrying capacity varies, is certainly a live hypothesis (Kitchell and Carr 1985; Hewzulla et al. 1999; Erwin 2007, 2008; Benton and Emerson 2007).

I presume Darwin would have been pleased to see how much progress has been made in circumventing the incompleteness of the fossil record in macroevolutionary studies and to see that the possible influence of biotic interaction on large-scale evolutionary trends has received serious attention. Yet, before we pat ourselves on the back too vigorously, we should keep in mind a few challenges for the coming decades: (1) The details of biotic interaction, for example the relative importance of competition and predation, remain to be worked out. (2) We are still a long way from understanding mechanistically how these interactions, as well as organismic traits more generally, influence rates of speciation and extinction. (3) Although there are a number of candidates for a detailed model of diversity dynamics, we have not determined which is best supported by the available data.

Acknowledgments

I am grateful to Michael Bell and Douglas Futuyma for inviting me to write this paper and to M. Caitlin Fisher-Reid for abundant logistical support. Jeffrey Levinton and Scott Lidgard provided thoughtful reviews. John Alroy generously shared and helped me understand his shareholder quorum subsampling method prior to its publication. David Jablonski, Arnold I. Miller, and Frank M. Richter provided insight on various aspects of this work. Finally, although he has been gone for more than 10 years, Jack Sepkoski continues to inspire and challenge all of us who think about the history of biodiversity. This is Paleobiology Database publication number 113.

Literature Cited

Alroy, J. 1998. Equilibrial diversity dynamics in North American mammals. In M. L. McKinney and J. A. Drake (eds.), *Biodiversity Dynamics: Turnover of Populations, Taxa, and Communities*, pp. 232–287. Columbia University Press, New York.

Alroy, J. 2008. Dynamics of orignation and extinction in the marine fossil record. *Proc. Natl. Acad. Sci. USA* 105(suppl. 1): 11536–11542.

Alroy, J. 2009. A deconstruction of Sepkoski's Phanerozoic marine evolutionary faunas based on new diversity estimates. *Geol. Soc. Am. Abs. Prog.* 41: 507.

Alroy, J., M. Aberhan, D. J. Bottjer, and 32 others. 2008. Phanerozoic trends in the global diversity of marine invertebrates. *Science* 321: 97–100.

Alroy, J., C. R. Marshall, R. K. Bambach, and 22 others. 2001. Effects of sampling standardization on estimates of Phanerozoic marine diversification. *Proc. Natl. Acad. Sci. USA* 98: 6261–6266.

Bambach, R. K. 2006. Phanerozoic biodiversity mass extinctions. *Ann. Rev. Earth Planet. Sci.* 34: 127–155.

Bambach, R. K., A. H. Knoll, and S. C. Wang. 2004. Origination, extinction, and mass depletions of marine diversity. *Paleobiology* 30: 522–542.
Benton, M. J. 1987. Progress and competition in macroevolution. *Biol. Rev.* 62: 305–338.
Benton, M. J. 1995. Diversification and extinction in the history of life. *Science* 258: 52–58.
Benton, M. J. 1996. On the nonprevalence of competitive replacements in the evolution of tetrapods. In D. Jablonski, D. H. Erwin, and J. H. Lipps (eds.), *Evolutionary Paleobiology*, pp. 185–210. University of Chicago Press, Chicago.
Benton, M. J. and B. C. Emerson. 2007. How did life become so diverse? The dynamics of diversification according to the fossil record and molecular phylogenetics. *Palaeontology* 50: 23–40.
Bodenbender, B. E. and D. C. Fisher. 2001. Stratocladistic analysis of blastoid phylogeny. *J. Paleontol.* 75: 351–369.
Bronn, H. G. 1849. *Handbuch einer Geschichte der Natur, Vol. 3, Part 3.* Schweizerbart'sche Verlagshandlung, Stuttgart.
Bush, A. M., M. J. Markey, and C. R. Marshall. 2004. Removing bias from diversity curves: The effects of spatially organized biodiversity on sampling-standardization. *Paleobiology* 30: 666–686.
Carr, T. R. and J. A. Kitchell. 1980. Dynamics of taxonomic diversity. *Paleobiology* 4: 427–443.
Connolly, S. R. and A. I. Miller. 2001. Joint estimation of sampling and turnover rates from fossil databases: Capture-mark-recapture methods revisited. *Paleobiology* 27: 751–767.
Connolly, S. R. and A. I. Miller 2002. Global Ordovician faunal transitions in the marine benthos: Ultimate causes. *Paleobiology* 28: 26–40.
Courtillot, V. and Y. Gaudemer. 1996. Effects of mass extinctions on biodiversity. *Nature* 381: 146–148.
Crampton, J. S., R. A. Cooper, A. G. Beu, and 2 others. 2010. Biotic influences on species duration—interactions between traits in marine molluscs. *Paleobiology* 36: 204–223.
Darwin, C. R. 1859. *On the Origin of Species by Means of Natural Selection.* John Murray, London [facsimile reprint, 1964, Harvard University Press, Cambridge, MA.].
Erwin, D. H. 1998. The end and the beginning: Recoveries from mass extinctions. *Trends Ecol. Evol.* 13: 344–349.
Erwin, D. H. 2001. Lessons from the past: Biotic recoveries from mass extinctions. *Proc. Natl. Acad. Sci. USA* 98: 5399–5403.
Erwin, D. H. 2007. Increasing returns, ecological feedback, and the Early Triassic recovery. *Palaeoworld* 16: 9–15.
Erwin, D. H. 2008. Extinction as the loss of evolutionary history. *Proc. Natl. Acad. Sci. USA* 105(suppl. 1): 11520–11527.
Flessa, K. W. 1975. Area, continental drift, and mammalian diversity. *Paleobiology* 1: 189–194.
Flessa, K. W. and J. J. Sepkoski Jr. 1978. Relationship between Phanerozoic diversity and changes in habitable area. *Paleobiology* 4: 359–366.
Foote, M. 1997. Estimating taxonomic durations and preservation probability. *Paleobiology* 23: 278–300.
Foote, M. 2000a. Origination and extinction components of taxonomic diversity: General problems. In D. H. Erwin and S. L. Wing (eds.), Deep time: Paleobiology's perspective. *Paleobiology* 26(Suppl to No. 4): 74–102.

Foote, M. 2000b. Origination and extinction components of taxonomic diversity: Paleozoic and post-Paleozoic dynamics. *Paleobiology* 26: 578–605.

Foote, M. 2001. Inferring temporal patterns of preservation, origination, and extinction from taxonomic survivorship analysis. *Paleobiology* 27: 602–630.

Foote, M. 2003. Origination and extinction through the Phanerozoic: A new approach. *J. Geol.* 111: 125–148.

Foote, M., 2005. Pulsed origination and extinction in the marine realm. *Paleobiology* 31: 6–20.

Foote, M. 2006. Substrate affinity and diversity dynamics of Paleozoic marine animals. *Paleobiology* 32: 345–366.

Foote, M. 2007. Extinction and quiescence in marine animal genera. *Paleobiology* 33: 261–272.

Foote, M. and D. M. Raup. 1996. Fossil preservation and the stratigraphic ranges of taxa. *Paleobiology* 22: 121–140.

Hallam, A. 1991. Why was there a delayed radiation after the end-Palaeozoic extinctions? *Hist. Biol.* 5: 257–262.

Hansen, T. A. 1980. Influence of larval dispersal and geographic distribution on species longevity in neogastropods. *Paleobiology* 6: 193–207.

Hewzulla, D., M. C. Boulter, M. J. Benton, and 1 other. 1999. Patterns from mass originations and mass extinctions. *Phil. Trans. Roy. Soc. Lond. B* 354: 463–469.

Holland, S. M. 1995. The stratigraphic distribution of fossils. *Paleobiology* 21: 92–109.

Holland, S. M. 2003. Confidence limits on fossil ranges that account for facies changes. *Paleobiology* 29: 468–479.

Jablonski, D. 1986. Background and mass extinctions: The alternation of macroevolutionary regimes. *Science* 231: 129–133.

Jablonski, D. 1993. The tropics as a source of evolutionary novelty: The post-Palaeozoic fossil record of marine invertebrates. *Nature* 364: 142–144.

Jablonski, D. 2008a. Extinction and the spatial dynamics of biodiversity. *Proc. Natl. Acad. Sci. USA* 105(suppl. 1): 11528–11535.

Jablonski, D. 2008b. Biotic interactions and macroevolution: Extensions and mismatches across scales and levels. *Evolution* 62: 715–739.

Jablonski, D. 2008c. Species selection: Theory and data. *Ann. Rev. Ecol. Evol. Syst.* 39: 501–524.

Jablonski, D. and D. J. Bottjer. 1991. Environmental patterns in the origins of higher taxa: The post-Paleozoic fossil record. *Science* 252: 1831–1833.

Jablonski, D. and G. Hunt. 2006. Larval ecology, geographic range, and species survivorship in Cretaceous mollusks: Organismic versus species-level explanations. *Am. Nat.* 168: 556–564.

Jablonski, D. and J. J. Sepkoski Jr. 1996. Paleobiology, community ecology, and scales of ecological pattern. *Ecology* 77: 1367–1378.

Jablonski, D., K. Roy, and J. W. Valentine. 2006. Out of the tropics: Evolutionary dynamics of the latitudinal diversity gradient. *Science* 314: 102–106.

Jablonski, D., K. Roy, J. W. Valentine, and 2 others. 2003. The impact of the pull of the Recent on the history of marine diversity. *Science* 300: 1133–1135.

Kidwell, S. M. and Holland, S. M. 2002. The quality of the fossil record: Implications for evolutionary analyses. *Ann. Rev. Ecol. Syst.* 33: 561–588.

Kirchner, J. W. and A. Weil. 2000. Delayed biological recovery from extinctions throughout the fossil record. *Nature* 404: 177–180.

Kitchell, J. A. and T. R. Carr. 1985. Nonequilibrium model of diversification: Faunal turnover dynamics. In J. W. Valentine (ed.), *Phanerozoic Diversity Patterns: Profiles in Macroevolution*, pp. 277–309. Princeton University Press, Princeton.

Krug, A. Z. and M. E. Patzkowsky. 2007. Geographic variation in turnover and recovery from the Late Ordovician mass extinction. *Paleobiology* 33: 435–454.

Levinton, J. S. 1979. A theory of diversity equilibrium and morphological evolution. *Science* 204: 335–336.

Lu, P. J., M. Yogo, and C. R. Marshall. 2006. Phanerozoic marine biodiversity dynamics in light of the incompleteness of the fossil record. *Proc. Natl. Acad. Sci. USA* 103: 2736–2739.

MacArthur, R. H. and E. O. Wilson. 1967. *The Theory of Island Biogeography*. Princeton University Press, Princeton.

Mark, G. A. and K. W. Flessa. 1977. A test for evolutionary equilibria: Phanerozoic brachiopods and Cenozoic mammals. *Paleobiology* 3: 17–22.

Marshall, C. R. 1990. Confidence intervals on stratigraphic ranges. *Paleobiology* 16: 1–10.

Marshall, C. R. 1997. Confidence intervals on stratigraphic ranges with nonrandom distributions of fossil horizons. *Paleobiology* 23: 165–173.

Marshall, C. R. and P. D. Ward. 1996. Sudden and gradual molluscan extinctions in the latest Cretaceous of western European Tethys. *Science* 274: 1360–1363.

McKinney, F. K. 1992. Competitive interactions between related clades: Evolutionary implications of overgrowth between encrusting cyclostome and cheilostome bryozoans. *Mar. Biol.* 114: 645–652.

McKinney, F. K. 1995. One hundred million years of competitive interactions between bryozoan clades: Asymmetrical but not escalating. *Biol. J. Linn. Soc.* 56: 465–381.

McKinney, F. K., S. Lidgard, J. J. Sepkoski Jr., and 1 other. 1998. Decoupled temporal patterns of evolution and ecology in two post-Paleozoic clades. *Science* 281: 807–809.

McKinney, F. K., S. Lidgard, and P. D. Taylor. 2001. Macroevolutionary trends: Perception depends on the measure used. In J. B. C. Jackson, S. Lidgard, and F. K. McKinney (eds.), *Evolutionary Patterns: Growth, Form, and Tempo in the Fossil Record*, pp. 348–385. University of Chicago Press, Chicago.

Miller, A. I. 1997. Dissecting global diversity patterns: Examples from the Ordovician radiation. *Ann. Rev. Ecol. Syst.* 28: 85–104.

Miller, A. I. 1998. Biotic transitions in global marine diversity. *Science* 281: 1157–1160.

Miller, A. I. 2000. Conversations about Phanerozoic global diversity. In D. H. Erwin and S. L. Wing (eds.), Deep time: Paleobiology's perspective. *Paleobiology* 26 (Suppl. to No. 4): 53–73.

Miller, A. I. and M. Foote. 1996. Calibrating the Ordovician radiation of marine life: Implications for Phanerozoic diversity trends. *Paleobiology* 22: 304–309.

Miller, A. I. and M. Foote. 2009. Epicontinental seas versus open-ocean settings: The kinetics of mass extinction and origination. *Science* 326: 1106–1109.

Miller, A. I. and S. Mao. 1995. Association of orogenic activity with the Ordovician radiation of marine life. *Geology* 23: 305–308.

Miller, A. I. and J. J. Sepkoski Jr. 1988. Modeling bivalve diversification: The effect of interaction on a macroevolutionary system. *Paleobiology* 14: 364–369.

Morris, J. 1854. *A Catalogue of British Fossils*, 2nd ed. The Author, London.

Paul, C. R. C. 1982. The adequacy of the fossil record. In K. A. Joysey and A. E. Friday (eds.), *Problems of Phylogenetic Reconstruction* (Systematics Association Special Vol. No. 21), pp. 75–117. Academic Press, London.
Peters, S. E. 2006. Genus extinction, origination, and the durations of sedimentary hiatuses. *Paleobiology* 32: 387–407.
Peters, S. E. and M. Foote. 2001. Biodiversity in the Phanerozoic: A reinterpretation. *Paleobiology* 27: 583–601.
Phillips, J. 1860. *Life on the Earth: Its Origin and Succession.* Macmillan and Co., Cambridge, UK.
Raup, D. M. 1972. Taxonomic diversity during the Phanerozoic. *Science* 177: 1065–1071.
Raup, D. M. 1975. Taxonomic diversity estimation using rarefaction. *Paleobiology* 1: 333–342.
Raup, D. M. 1976. Species diversity in the Phanerozoic: An interpretation. *Paleobiology* 2: 289–297.
Raup, D. M. 1979. Biases in the fossil record of species and genera. *Bull. Carnegie Mus. Nat. Hist.* 13: 85–91.
Raup, D. M. and J. J. Sepkoski Jr. 1982. Mass extinctions in the marine fossil record. *Science* 215: 1501–1503.
Rosenzweig, M. L. 1995. *Species Diversity in Space and Time.* Cambridge University Press, Cambridge, UK.
Rudwick, M. J. S. 2008. *Worlds Before Adam: The Reconstruction of Geohistory in the Age of Reform.* University of Chicago Press, Chicago.
Ruta, M., P. J. Wagner, and M. I. Coates. 2006. Evolutionary patterns in early tetrapods. I. Rapid initial diversification followed by decrease in rates of character change. *Proc. Roy. Soc. Lond. B* 273: 2107–2111.
Schopf, T. J. M. 1974. Permo-Triassic extinctions: Relation to sea-floor spreading. *J. Geol.* 82: 129–143.
Sepkoski, J. J. Jr. 1976. Species diversity in the Phanerozoic: Species-area effects. *Paleobiology* 4: 298–303.
Sepkoski, J. J. Jr. 1978. A kinetic model of Phanerozoic taxonomic diversity I. Analysis of marine orders. *Paleobiology* 4: 223–251.
Sepkoski, J. J. Jr. 1979. A kinetic model of Phanerozoic taxonomic diversity II. Early Phanerozoic families and multiple equilibria. *Paleobiology* 5: 222–251.
Sepkoski, J. J. Jr. 1981. A factor analytic description of the Phanerozoic marine fossil record. *Paleobiology* 7: 36–53.
Sepkoski, J. J. Jr. 1984. A kinetic model of Phanerozoic taxonomic diversity III. Post-Paleozoic families and mass extinctions. *Paleobiology* 10: 246–267.
Sepkoski, J. J. Jr. 1991. Population biology models in macroevolution. In N. L. Gilinsky and P. W. Signor (eds.), *Analytical Paleobiology* (*Short Courses in Paleontology* 4), pp. 136–156. Paleontological Society, Knoxville.
Sepkoski, J. J. Jr. 1996. Competition in macroevolution: The double wedge revisited. In D. Jablonski, D. H. Erwin, and J. H. Lipps (eds.), *Evolutionary Paleobiology*, pp. 211–255. University of Chicago Press, Chicago.
Sepkoski, J. J. Jr. 1997. Biodiversity: Past, present, and future. *J. Paleontol.* 71: 533–539.
Sepkoski, J. J. Jr. 1998. Rates of speciation in the fossil record. *Phil. Trans. Roy. Soc. Lond. B* 353: 315–326.
Sepkoski, J. J. Jr. 2002. A compendium of fossil marine animal genera. *Bull. Am. Paleontol.* 363: 1–560.

Sepkoski, J. J. Jr., R. K. Bambach, D. M. Raup, and 1 other. 1981. Phanerozoic marine diversity and the fossil record. *Nature* 293: 435–437.

Sepkoski, J. J. Jr., F. K. McKinney, and S. Lidgard. 2000. Competitive displacement between post-Paleozoic cyclostome and cheilostome bryozoans. *Paleobiology* 26: 7–18.

Signor, P. W. III, and J. H. Lipps. 1982. Sampling bias, gradual extinction patterns and catastrophes in the fossil record. *Geol. Soc. Am. Sp. Paper* 190: 291–296.

Simpson, G. G. 1944. *Tempo and Mode in Evolution.* Columbia University Press, New York.

Smith, A. B. 2001. Large-scale heterogeneity of the fossil record: Implications for Phanerozoic biodiversity studies. *Phil. Trans. R. Soc. Lond. B* 356: 351–367.

Solow, A. R. and W. Smith. 1997. On fossil preservation and the stratigraphic ranges of taxa. *Paleobiology* 23: 271–277.

Stanley, S. M. 2007. An analysis of the history of marine animal diversity. *Paleobiology* 33(Suppl. to No. 4): 1–55.

Stanley, S. M. 2008. Predation defeats competition on the seafloor. *Paleobiology* 34: 1–21.

Stanley, S. M., P. W. Signor III, S. Lidgard, and 1 other. 1981. Natural clades differ from "random" clades: Simulations and analyses. *Paleobiology* 7: 115–127.

Strauss, D. and P. M. Sadler. 1989. Classical confidence intervals and Bayesian probability estimates for ends of local taxon ranges. *Math. Geol.* 21: 411–427.

Valentine, J. W. and E. M. Moores. 1972. Global tectonics and the fossil record. *J. Geol.* 80: 167–184.

Wagner, P. J. 1997. Patterns of morphologic diversification among the Rostroconchia. *Paleobiology* 23: 115–150.

Webb, S. D. 1969. Extinction-origination equilibria of late Cenozoic land mammals of North America. *Evolution* 23: 688–702.

Commentary Five

Thinking about Diversity and Diversification: What if Biotic History Is Not Equilibrial?

Joel Cracraft

Contemporary discussions about the evolution of diversity and the processes underlying diversification largely assume that diversity-dependent equilibrial dynamics govern speciation and extinction rate controls. But what if biological evolution is a nonequilibrial process and speciation and extinction are diversity-independent? Contrasts between these two alternative models of diversification are infrequently discussed, which is surprising given the implications of each for understanding the history of life on Earth. This commentary briefly explores some theoretical and empirical underpinnings of the two alternatives and their implications for examining diversification at different spatio-temporal scales. Before beginning, however, the terminology used here should be clarified. Nonequilibrial has two different meanings in the literature of evolution and ecology. The first refers to nonequilibrial thermodynamic systems that, as a consequence of matter and energy flows, results in a far-from-equilibrium organization. Another usage, common in ecology and paleobiology, is to employ nonequilibrium when describing departures or perturbations from expected solutions to models of species diversity that are explicitly equilibrial in their dynamics (Sepkoski 1984; see Kitchell and Carr 1985). As will be discussed, notions of diversity-independent diversification are loosely linked to a nonequilibrial dynamic in biotic history, whereas models of diversity-dependent diversification follow explicitly from equilibrial assumptions.

The goal of diversification theory is to explain patterns of diversity across space and time. There is little, if any, controversy that change in diversity (denoted D, the number of species) is a first-order function of the difference between speciation and extinction rates (S and E, respectively). By themselves, S and E cannot account for all the variance in D within a given area or at every spatiotemporal scale; yet, S and E must play *some* role in explaining all comparisons in D, because species compositions of sampled areas (say, ecosystems) are drawn from more global species pools. This is a restatement of the notion that multiple processes at different levels of organization—from S and E, to events of Earth history, to biotic interactions in the context of local ecology—interact in complex ways to cause changes in diversity. One key to teasing apart causality is to identify the scale at which causes act.

Yet, there are patterns of diversity for which S and E have played a prominent role in explanation, such as the Phanerozoic diversity curve, temporal changes in diversity within and among clades (especially sister-groups), and large-scale spatial gradients in diversity. The behavior of S and E in most of these cases has been causally grounded in theory that was developed to explain the distribution and abundance of *individual organisms* within relatively small-scale ecological systems. In other words, the dynamic causality that accounts for interacting individuals and populations within idealized ecological systems has been extrapolated, usually without modification, to become the mechanism of causation at the level of speciation and extinction rate controls. Accordingly, this approach to explaining diversity can be characterized as being ecologically deterministic.

The Structure of Contemporary Diversification Theory: Ecological Determinism and Equilibrial Thinking

Ecological deterministic causality of large-scale diversification has a long history. Darwin posited the geographic spread of species and clades into empty ecological space with the replacement of competitive inferiors, an idea subsequently amplified by many others, but especially by Simpson (1953) and Darlington (1957). Their expansive vision was subsequently amplified and formalized by the innovative work of Valentine (1973) and Sepkoski (1978, 1979). Sepkoski built his explicitly equilibrial diversification model as an extension of those in population ecology, especially that of MacArthur and Wilson (1969); that is, S and E are conceptual substitutes for birth (B) and death (D) rates, respectively. As Sepkoski noted, equilibrial descriptions of diversification had already gained traction among the players who were transforming the conceptual fabric of paleobiology (Valentine 1973, 1985) as well as among those pushing ecological thinking in new macroecological directions, but arguably it was Sepkoski's work that has made equilibrial diversification the hegemonic view it is today.

As a consequence, contemporary paleobiologists and ecologists concerned with explaining patterns of diversity have seen S and E as being controlled by lower-level ecological processes (e.g., Valentine 1985; McKinney and Drake 1998; Erwin and Wing 2000). In essence, the argument states that S and E, like B and D, are governed by biotic interactions (i.e., competition, predation) coupled to the level of available resources (ecospace) and degree of species packing. As the biotic carrying capacity (K) is approached, these interactions have effects on population structure and distribution, which, in turn, cause S to decrease and E to increase, resulting in an equilibrial diversity: $\hat{D}$; whether or not an equilibrial diversity has ever existed has been debated and is still not resolved (Flessa and Levinton 1975; Cracraft 1985; Benton 1997; Stanley 2007).

A key premise of diversity-dependent diversification is the filling of empty ecospace or new adaptive zones (Simpson 1953), which is postulated to take place via one clade competitively replacing another, through the origin of a key innovation that allows a clade to diversify into previously unoccupied ecospace or perhaps expand it (see Berenbaum and Schuler, Chapter 11; Losos and Mahler, Chapter 15). The actors in this dynamic are species and clades (or surrogate higher taxa), and they are often explicitly the entities that are said to compete, predate, and occupy ecospace. Little, if any, attention is given to the real actors of ecological theory: individual organisms.

Why Biotic History Is Nonequilibrial and Why It Matters

Evolutionary systems are nonequilibrial thermodynamic systems, from astronomical and chemical systems (Frautschi 1988; Hazen 2008), plate tectonics (Anderson 2002), biological evolution (Brooks and Wiley 1988; Brooks 2000), ecosystems (O'Neill et al. 1989; Nielsen 2000), to individual organisms (Peacocke 1983). Open systems are far-from-equilibrium, and each maintains itself by throughputs of matter and energy and develops structural complexity (order) through dissipation; moreover, as irreversible systems, their inherent dynamic is to grow and become more complex (Toussaint and Schneider 1998). Perhaps the clearest example is the growth and development of individual organisms, leading to increased complexity over time. This complexity arises as a result of matter and energy flows that fuel the system, with more energy being dissipated as complexity is increased. Adulthood can be interpreted as a far-from-equilibrium steady state in which the rate of energy dissipation declines as energy is more efficiently utilized (Schneider and Kay 1995; Toussaint and Schneider 1998). This view of the global dynamics of ontogeny, it must be stressed, is not inconsistent with the mechanistic underpinnings of modern developmental biology.

This simple description of how open evolving systems increase in complexity has relevance for understanding diversification in important ways. Because individual organisms use matter and energy fluxes to grow and reproduce, ecological systems composed of those organisms will also have an historical trajectory that is a function of flows through that system (O'Neill et al. 1989; Schneider and Kay 1995). Conceived in this way, the history of the biosphere cannot be described (at least in terms of physical principles) as being equilibrial, because the inherent trajectory of living systems is toward increasing growth and complexity. An important implication of this history is that first-principles alone lead to the expectation that the biosphere will increase in biomass and diversity over time, and arguably, this is what the historical record suggests. But at the same time, there are both external and internal constraints on any system that have the potential to alter fluxes in matter and energy and thus change the history and complexity of the system itself. Consequently, viewing the history of diversification as nonequilibrial carries no implication that ecological processes within ecological communities cannot alter fluxes at local or regional scales.

Why *S* and *E* Are Diversity-Independent

If the biosphere is conceived as a nonequilibrial system, then the expectation of a global *K* disappears (Benton 1997; Benton and Emerson 2007), as does the need to see drivers of *S* and *E* as originating from biotic interactions among individual organisms (for critiques see Cracraft 1992; Stanley 2007). A more plausible alternative is to think of *S* and *E* as being diversity independent (Cracraft 1982, 1985, 1992; Hoffman 1985). A mechanistic explanation for diversification must entail partitioning causation for *S* and *E* and not entangle them by assuming that causation always arises at the within-population level or that taxa have the same dynamics as individual organisms. If, for the sake of argument, one accepts the widely held proposition that most speciation is allopatric, then *S* is a function of (a) the rate of allopatry of populations, (b) the rate of morphogenetic variants that arise in those populations, and (c) the rate of fixation of those variants. Therefore, the first-order determinate of S is the rate of allopatry, which is primarily a function of the rate of barrier formation, which in turn, is a function of Earth history. This causal chain would be true if allopatry occurred either via vicariance or long-distance dispersal across a barrier. Barriers, whether due to geomorphological or climate-driven ecological change, arise as a result of the nonequilibrial evolution of the geosphere (Benton 2009) but can have variable effects at any point in space and time. The remaining two factors (i.e., b and c) function within populations and are necessary but not sufficient for speciation unless allopatry itself is discounted as necessary.

Determinants of the extinction rate, E, in contrast, are sited at different levels of causation because, ontologically, a species does not *go* extinct as an entity, it *becomes* extinct when its last individual dies. Therefore, partitioning the drivers of E in any one instance is complex, as there may be many factors acting simultaneously to cause population decline (Valentine 1973). In terms of large-scale patterns, change in global or regional environmental harshness can be implicated in substantial background extinction and has its mechanistic locus at the level of constraints on matter and energy fluxes of individual organisms as constituents of populations, species, and ecosystems. This process is diversity independent in a global context, but almost always will involve density-dependent causation in a local ecological context.

Literature Cited

Anderson, D. L. 2002. Plate tectonics as a far-from-equilibrium self-organized system. In S. Stein and J. Freymuller (eds.), *Plate Boundary Zone*, pp. 411–425. *Amer. Geophys. Union Monograph, Geodynamics Ser.* 30, Washington, DC.

Benton, M. J. 1997. Models for the diversification of life. *Trends Ecol. Evol.* 12: 490–495.

Benton, M. J. 2009. The red queen and the court jester: Species diversity and the role of biotic and abiotic factors through time. *Science* 323: 728–732.

Benton, M. J. and B. C. Emerson. 2007. How did life become so diverse? The dynamics of diversification according to the fossil record and molecular phylogenetics. *Palaeontology* 50: 23–40.

Brooks, D. R. 2000. The nature of organism: Life has a life of its own. *Ann. New York. Acad. Sci.* 901: 257–265.

Brooks, D. R. and E. O. Wiley. 1988. *Evolution as Entropy: Toward a Unified Theory of Biology*, 2nd ed. University of Chicago Press, Chicago.

Cracraft, J. 1982. A nonequilibrium theory for the rate-control of speciation and extinction and the origin of macroevolutionary patterns. *Syst. Zool.* 31: 348–365.

Cracraft, J. 1985. Biological diversification and its causes. *Ann. Missouri Bot. Gard.* 72: 794–822.

Cracraft, J. 1992. Explaining patterns of biological diversity: Integrating causation at different spatial and temporal scales. In N. Eldredge (ed.), *Systematics, Ecology and the Biodiversity Crisis*, pp. 59–76. Columbia University Press, New York.

Darlington, P. J. 1957. *Zoogeography: The Geographical Distribution of Animals*. John Wiley & Sons, New York.

Erwin, D. H. and S. L. Wing (eds.). 2000. *Deep Time: Paleobiology's Perspective. Paleobiology* 46 (4, supplement): 1–371.

Flessa, K. W. and J. S. Levinton. 1975. Phanerozoic diversity patterns: Tests for randomness. *J. Geology* 83: 239–248.

Frautschi, S. 1998. Entropy in an expanding universe. In B. H. Weber, D. J. Depew, and J. D. Smith (eds.), *Entropy, Information, and Evolution: New Perspectives on Physical and Biological Evolution*. MIT Press, Cambridge, MA.

Hazen, R. M. 2008. The emergence of chemical complexity: An introduction. In L. Zaikowski and J. M. Friedrich (eds.), *Chemical Evolution Across Space and Time*, pp. 2–13. American Chemical Society, Washington, DC.

Hoffman, A. 1985. Biotic diversification in the Phanerozoic: Diversity independence. *Palaeontology* 28: 387–391.

Kitchell, J. A. and T. R. Carr. 1985. Nonequilibrium model of diversification: Faunal turnover dynamics. In J. W. Valentine (ed.), *Phanerozoic Diversity Patterns: Profiles in Macroevolution*, pp. 277–309. Princeton University Press, Princeton.

MacArthur, R. H. and E. O. Wilson. 1969. *The Theory of Island Biogeography*. Princeton University Press, Princeton.

McKinney, M. L. and J. A. Drake (eds.). 1998. *Biodiversity Dynamics: Turnover of Populations, Taxa, and Communities*. Columbia University Press, New York.

Nielsen, S. N. 2000. Thermodynamics of an ecosystem interpreted as a hierarchy of embedded systems. *Ecol. Model.* 135: 279–289.

O'Neill, R. V., A. R. Johnson, and A. W. King. 1989. A hierarchical framework for the analysis of scale. *Landscape Ecol.* 3: 193–205.

Peacocke, A. R. 1983. *The Physical Chemistry of Biological Organization*. Clarendon Press, Oxford.

Schneider, E. D. and J. J. Kay. 1995. Order from disorder: The thermodynamics of complexity in biology. In M. P. Murphy and L. A. J. O'Neill (eds.), *What is Life? Speculations on the Future of Biology*, pp. 161–173. Cambridge University Press, Cambridge, UK.

Sepkoski, J. J. Jr. 1978. A kinetic model of Phanerozoic taxonomic diversity. I. Analysis of marine orders. *Paleobiology* 4: 223–251.

Sepkoski, J. J. Jr. 1979. A kinetic model of Phanerozoic taxonomic diversity. II. Early Phanerozoic families and multiple equilibria. *Paleobiology* 5: 222–251.

Sepkoski, J. J. Jr. 1984. A kinetic model of Phanerozoic taxonomic diversity. III. Post-Paleozoic families and mass extinctions. *Paleobiology* 10: 246–267.

Simpson, G. G. 1953. *The Major Features of Evolution*. Columbia University Press, New York.

Stanley, S. M. 2007. An analysis of the history of marine animal diversity. *Paleobiology* 33 (4, supplement): 1–55.

Toussaint, O. and E. D. Schneider. 1998. The thermodynamics and evolution of complexity in biological systems. *Comp. Biochem. Physiol.* 120A: 3–9.

Valentine, J. W. 1973. *Evolutionary Paleoecology of the Marine Biosphere*. Prentice-Hall, Englewood Cliffs, NJ.

Valentine, J. W. (ed.). 1985. *Phanerozoic Diversity Patterns: Profiles in Macroevolution*. Princeton University Press, Princeton.

Part VI

HUMAN EVOLUTION

Chapter 19

Human Evolution: How Has Darwin Done?

Tim D. White

In *On The Origin of Species*, Charles Darwin's only comment on human origins and evolution (or transformation) was, "...light will be thrown on the origin of man and his history" (Darwin 1859: 488). In his autobiography, Darwin justifies this brevity: "It would have been useless and injurious to the success of the book to have paraded, without giving any evidence, my conviction with respect to his origin" (Darwin 1887: 94). The 50 years between Darwin's birth and the 1859 publication of *On The Origin of Species* had produced little in the acquisition of a hominid fossil record—a trend maintained during the dozen years that followed. Nevertheless, by 1871, with merely a few Neanderthal fossils available, Darwin lavished a huge amount of attention on the topic of human origins. This fact would seem justification enough for an anniversary chapter focused on the tiny limb of the human family within the vast and spreading evolutionary tree that is life on earth.

The 50th and 100th Darwin anniversary celebrations comprised summaries and prognoses (Seward 1909; Tax 1960), but the last half century has far surpassed the two preceding intervals in the astonishing acceleration of the rate at which new molecular, genetic, and fossil data have been acquired. These new data now far more completely document the processes, the pathways, and the patterns of Neogene human evolution. The upcoming 150th anniversary of Darwin's 1871 book on human origins and sexual selection will fail to evoke the celebrations witnessed in 2009, but nonetheless will encourage assessment of how Darwin's 1871 conclusions and predictions stack up against today's knowledge about evolution within the hominid clade. This approach enables evolutionary biologists to engage in both the hindsight and foresight that any significant anniversary should evoke, allowing a broad perspective on what has been learned so far about human evolution and what still needs to be explored two centuries after Darwin's birth.

Here I will not pretend to say anything new about Darwin and his work, nor will I present novel archival research results. This chapter is instead designed to examine what Darwin wrote and to offer a summary of the progress in human evolutionary studies since then. I will first consider what Darwin really did and did not actually write about human origins and evolution, particularly in his 1859 and 1871 books. I will then imagine the kinds of thoughts and questions that Darwin might have had during a hypothetical visit to the Middle Awash study area of Ethiopia's Afar Rift—a location that today offers the Earth's longest succession of hominid occupation.

By comparing Darwin's predictions with evidence of hominid paleobiology currently available, the chapter approaches a progress report on human origins, a topic that Thomas Henry Huxley called: "...the question of questions for mankind—the problem which underlies all others, and is more deeply interesting than any other" (Huxley 1863: 58). Here, Huxley echoes Richard Owen, who had written 3 years earlier that: "The origin of species is the question of questions in Zoology; the supreme problem..." (Owen 1860: 496). As an element of the 2009 progress report, I feel compelled to address the persistent myth that the hominid fossil record is particularly poor. Perhaps, Stephen Jay Gould reified this myth in 2002 when he bemoaned "the spotty data of hominids" (Gould 2002: 833) and asserted: "No true consensus exists in this most contentious of all scientific professions...A field that features more minds at work than bones to study." (Gould 2002: 910). By showcasing how effectively this clade has been probed by paleontologists since Darwin's time, particularly during the last half century, my summary will challenge Gould's assertion. Finally, through evaluation of Darwin's inferences and prognostications about human evolution, one is afforded an opportunity to note some contemporary trends and future directions for this specialized but vital dimension of evolutionary biology.

Principles of Historical and Forensic Sciences

Gould (2002) wrote of "Darwin's struggle to construct a series of procedures offering sufficient confidence to place the sciences of history on a par with the finest experimental work in physics and chemistry" (Gould 2002: 103). Just as Darwin's 1859 book can be thought of as "one long argument," the study of human evolution can be thought of as reading the results of a grand uncontrolled and unreplicable experiment that has left contemporary paleontologists with fragmentary data embedded in an intermittent fossil record.

The data of the neontological world formed during the long experiment of life on Earth are indeed relatively rich, and Darwin put them to unrivalled use by inferring their evolutionary foundations. But, as he lamented the dearth of fossil data in both 1859 and 1871, Darwin also vividly showed how deeply he understood and appreciated the unique value of paleobiology. Today, even with relatively advanced genomics, phylogenetics, and modeling, definitive insights into exactly what happened then and there continue to arise from paleontology. This is because fossil discoveries continue to reveal organisms that cannot be discovered by any other means. No living whale could have illuminated the changes in body proportions and foot elongation in *Rodhocetus*, nor the subsequent hind limb reduction and tail-powered swimming of *Dorudon* (Gingerich et al. 2003). No lungfish is an adequate stand-in for *Tiktaalik* (Shubin et al. 2006) or Middle Devonian tetrapod footprints (Niedzweidzki 2010). No amount of modern chimpanzee behavior, anatomy, or genetics could have accurately predicted, let alone revealed, *Ardipithecus* (White et al. 2009).

Of course, much of what happened in the distant past has vanished without a trace. However, for the traces we can recover, paleobiology serves as a uniquely valuable forensic science with long-term experimental aspects. By providing a time perspective to biology, inaccessible by other means, paleontological science is today firmly seated at the "high table" of science (Sepkoski and Ruse 2009).

A Very Brief History of Hominid Fossil Recovery

Darwin-era scholars could only contemplate a mostly European paleontological record of human evolution, recovered primarily from archaeological contexts (Grayson 1983). Huxley's *Man's Place in Nature* (1863) and Lyell's book on the *Antiquity of Man* (1863) were the prime contemporary contributions of that period. Eighteenth-century, post-Darwinian advances in hominid paleontology first involved the discovery of Neanderthals and their replacements as regular occupants of glacial Europe, via finds such as Cro-Magnon in France (1868), Spy in Belgium (1886), and Krapina in Croatia (1899–1905). Then, when Eugene Dubois found a hominid on Java in 1891, this earlier and more primitive chronospecies of our genus was initially given Haeckel's nomen, *Pithecanthropus.* As the twentieth century opened, debates centered on the place of Neanderthals and the most ancient *Homo*. Scholars, such as Schwalbe and Keith, staked out fossil-sparse positions in a century-long controversy; indeed, the Neanderthal debate has only recently been settled in the face of consilient advances in the paleontological and neontological (biomolecular) realms (Pääbo et al. 2006; Noonan et al. 2006; White et al. 2003).

Hominid paleontology during the twentieth century experienced many new discoveries, progressively from younger to older. The illumination of early hominid speciation and phylogenetic relationships started with Dart's breakthrough discovery of South Africa's Taung child in 1924. However, decades would pass before *Australopithecus* was eventually afforded hominid status. Discoveries at Kromdraai (1938) and Olduvai Gorge (1959) established beyond doubt the presence of a separate, robust hominid clade. Even older fossils from Hadar and Laetoli (1970s) then revealed very primitive *Australopithecus*. The four-million-year barrier was finally broken during the 1990s to reveal the presence of *Ardipithecus* (White et al. 1994, 2009), followed by the recovery of Late Miocene fossils from Ethiopia, Kenya, and Chad (Senut et al. 2001; Brunet et al. 2002; Haile-Selassie 2001).

"A Decidedly Dangerous Book" and beyond: What Darwin Actually Wrote about Human Evolution

Darwin carefully avoided what Gould termed "the mammal that we all love best" (Gould 2002: 833) in *On The Origin of Species* (1859). However, the book's crucial importance regarding what Huxley termed "man's place in nature" occupied center stage even before the book appeared. As Browne (2002) observes, even though unwritten, implications of *On The Origin of Species* for human evolution were immediately and widely appreciated—and hotly debated. Huxley's 1860 review of the work gives a sense of this in a single sentence:

> *Everybody has read Mr. Darwin's book, or, at least, has given an opinion upon its merits or demerits; pietists, whether lay or ecclesiastic, decry it with the mild railing which sounds so charitable; bigots denounce it with ignorant invective; old ladies of both sexes consider it a decidedly dangerous book, and even savants, who have no better mud to throw, quote antiquated writers to show that its author is no better than an ape himself; while every philosophical thinker hails it as a veritable Whitworth gun in the armoury of liberalism; and all competent naturalists and physiologists, whatever their opinions as to the ultimate fate of the doctrines put forth, acknowledge that the work in which they are embodied is a solid contribution to knowledge and inaugurates a new epoch in natural history (Huxley 1860: 541).*

Although the 2009 Darwin 150th meeting and resultant volume celebrate the subsequent impact of *On The Origin of Species*, one must turn to Darwin's other writings to appreciate what he inferred and predicted about human evolution, based on the limited nineteenth century evidence at his command. His major work on the subject was the 1871 two-volume set, *The Descent Of Man, and Selection in Relation to Sex.* In the conclusion to the human evolution volume, Darwin wrote: "...we are not here concerned with hopes or fears, only with the truth as far as our reason allows us to discover it. I have given the evidence to the best of my ability..." (Darwin 1871: 405).

Recognizing the paucity of relevant evidence, Darwin's 1871 writing on human evolution was overwhelmingly cautious—sometimes moving beyond ambiguity to contradiction. Regrettably, subsequent scholarship on human evolution frequently attributes ideas to Darwin that cannot actually be found in his writing, but only exist in secondary, tertiary, or even more derived and less accurate sources. Such misquotation sometimes achieves hilarity, as the case of the creationist website heralding the discovery of *Ardipithecus* as finally disproving Darwin's idea that humans evolved from chimpanzees (White et al. 2009). In the scholarly literature, such misquotations have served to entrench myths and misconceptions about what Darwin really thought and wrote.

Because he had relatively little paleobiological evidence regarding what had actually happened, most of Darwin's writings on human evolution in

1871 constituted extremely well informed inference and speculation, and he carefully labeled it as such. Here I address what Darwin appears to have thought about the limited hominid fossil record available, the degree to which human ancestors resembled living primates, the kinds of environments and selective pressures that were responsible for human origins, and the acquisition sequences involved during the evolution of the human lineage within the hominid clade. Darwin, of course, could not have known that Neanderthals represented a likely second lineage, and it was not until the 1940s that additional, now extinct African hominid lineages were first revealed.

Darwin on the Hominid Fossil Record

The original title chosen by Darwin for his 1871 book was "On the Origin of Man" (Browne 2002: 349), and this title would have proven a much better match for the book's actual content. Since Huxley, in 1863, had already compiled most of the available fossil and anatomical evidence relating to human descent, Darwin had only to cite that work and Lyell's. As Bowler points out, Huxley's was an "excellent survey of the evidence for the fact of human evolution, but said virtually nothing about how the process was supposed to have occurred" (Bowler 1986: 3). In the context of the ongoing battle with Owen, neither Huxley nor Lyell had made much of a contribution regarding the sequence and causes of human evolution. It was to these topics that Darwin turned his attention in the first volume of his oft cited—but regrettably, apparently not often enough read—1871 book.

Darwin had been disappointed by Lyell's *The Geological Evidences of the Antiquity of Man, with Remarks on the Origin of Species by Variation*, describing it in a letter to Hooker as a:

> *...compilation, but of the highest class, for when possible the facts have been verified on the spot, making it almost an original work...It might perhaps be said with truth that he had no business to judge on a subject on which he knows nothing; but compilers must do this to a certain extent. (You know I value and rank high compilers, being one myself!) (Darwin 1887: 8).*

Lyell had painstakingly outlined the Engis, Neanderthal, and other available Pleistocene fossils and their contexts, but Darwin was annoyed by his timidity: "I must say how much disappointed I am that he has not spoken out on species, still less on man" (Darwin 1887: 9). Huxley's 1863 *Man's Place in Nature* had already reviewed the hominid fossil record, concluding that: "...the fossil remains of Man hitherto discovered do not seem to me to take us appreciably nearer to that lower pithecoid form, by the modification of which he has, probably, become what he is" (Huxley 1863: 207–208).

Consequently, there was little point for Darwin to re-review the fossils in his 1871 book. After all, only Neanderthals had surfaced by then, and even as late as 1895 Wallace would remark on the matter: "It is a curious circumstance that, notwithstanding the attention that has been directed to the subject in every part of the world, and the numerous excavations connected with railways and mines, which have offered such facilities for geological discovery, no advance whatever has been made for a considerable number of years in detecting the time or mode of man's origin" (Wallace 1895: 420). In his autobiography, first published posthumously in 1887, Darwin reveals that he spent 3 years writing the human evolution book and identifies the real source of his satisfaction: "it gave me an opportunity of fully discussing sexual selection—a subject which had always greatly interested me" (Darwin 1887: 94). Darwin's work on sexual selection in the second volume indeed reverberates in modern evolutionary biology. Today, the first 1871 volume—Darwin's major work on human evolution—is remembered not as a compilation, but rather as a set of observations and predictions (what today might be considered hypotheses) that have been tested by 140 years of subsequent research results.

Darwin on Hominid Phylogenetics

Haeckel (1868) chose the gibbon and orangutan as the closest living relatives to humans, hinting that he favored the latter. This played a role in Dubois's embarkation for the Indies. Wallace (1889) concluded the same thing, inferring that humans had separated from the rest of the great apes before they diverged from each other. Extending Bowler's (1996) work, Delisle (2006) reviews other nineteenth-century scholarship on this issue and concludes that a:

> *...relatively weak link...apparently existed between taxonomy and phylogeny in the minds of most scholars...For instance, Darwin clearly stated that chimpanzees and gorillas were humankind's closest living allies, although he held, at the same time, that the human line had split from the base of the hominoid branch before the four living hominoid apes had differentiated from one another (Delisle 2006: 56).*

This viewpoint was reflected in classifications wherein even sister species were placed in separate orders, as well as in late nineteenth century debate throughout zoology and paleontology about common ancestry versus parallel evolution in all taxonomic groups (Bowler 1986). By 1895, Wallace was treating hominoid phylogenetics as follows:

> *Yet another important line of evidence as to the extreme antiquity of the human type has been brought prominently forward by Professor Mivart. He shows, by a careful comparison of all parts of the structure of the body, that man is related not to any one, but almost equally to many of the existing apes—to the orang, the chimpanzee, the gorilla, and even to the gibbons, in a variety of ways; and these relations and differences are so numerous and so diverse that, on the theory of evolution, the ancestral form which ultimately developed into man must have diverged from the common stock whence all these various forms and their extinct allies originated (Wallace 1895: 422).*

This position basically followed Darwin's unpublished 1868 phylogenetic sketch (complete with marginalia and topology alterations; Figure 19.1) described by Bowler in 1986 and discussed further by Delisle in 2006. In his unremitting quest for evidence bearing on relationship, Darwin had already generally anticipated how even biochemistry might bear on the issue, and he cited reported similarities between the effects of alcohol on monkeys and humans. As Seward (1909) was to appreciate at the 50th anniversary of *On The Origin of Species*, at a time when the fossils of Java man were new and work on blood serum was moving forward:

> *In quite a different domain from that of morphological relationship, namely in the physiological study of the blood, results have recently been gained which are of the highest importance to the doctrine of descent. Uhlenhuth, Nuttall, and others have established the fact that the blood-serum of a rabbit which has previously had human blood injected into it, forms a precipitate with human blood. This biological reaction was tried with a great variety of mammalian species, and it was found that those far removed from man gave no precipitate under these conditions. But as in other cases among mammals all nearly related forms yield an almost equally marked precipitate, so the serum of a rabbit treated with human blood and then added to the blood of an anthropoid ape gives almost as marked a precipitate as in human...We have in this not only a proof of the literal blood-relationship between man and apes, but the degree of relationship with the different main groups of apes can be determined beyond possibility of mistake (Seward 1909: 129).*

Darwin and his contemporaries were unsuccessful in resolving the phylogenetic relationships among the great apes. His unpublished phylogenetic sketch of 1868 depicts a trifurcation of "Gorilla + Chimp," "orang-utan," and *Hylobates*, following a bifurcation between their "clade" and the lineage leading to *Homo* (twice scribbled out and replaced by "Man"). Darwin added a dotted line between

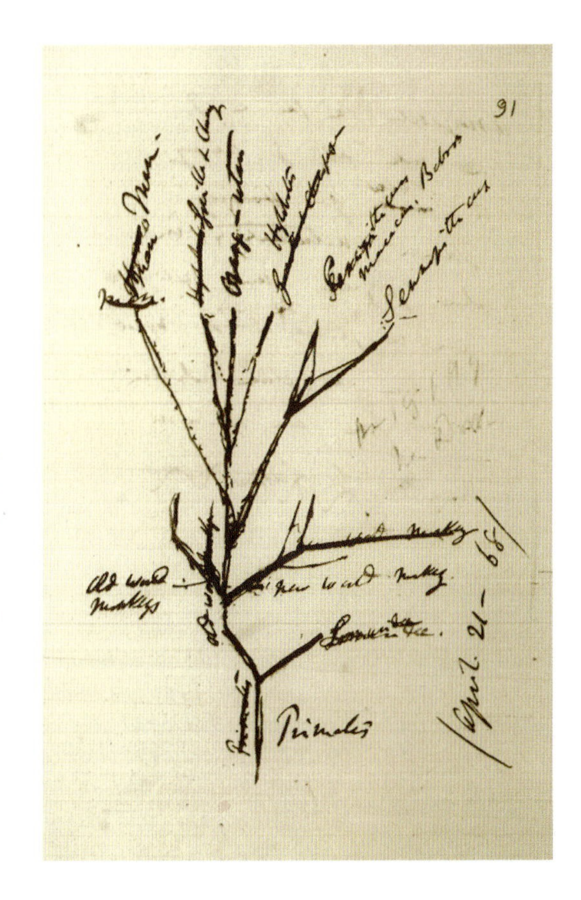

FIGURE 19.1 Darwin's Unpublished Phylogenetic Sketch of 1868 It depicts a trifurcation of "Gorilla + Chimp," "orang-utan," and "*Hylobates*," following a bifurcation between their clade and the lineage leading to *Homo* (scribbled out and replaced by "Man.").

the orangutan and African ape branches as an alternative pathway, but by 1871, he would conclude that African apes were phylogenetically closest to humans.

His use of biogeography and what today would be called careful outgroup analysis (mostly the work of Huxley and others, although this assessment requires further archival research) led Darwin, in 1871, to conclude about the chimpanzee and gorilla that: "...these two species are now man's nearest allies" (Darwin 1871: 199). Even though a considerable amount was known about the anatomy, distribution, and behavior of great apes in Darwin's day and although armed with serum results in the early twentieth century, the phylogenetic relations among the great apes and humans only began to be fully resolved by molecular anthropology during the 1960s when the chimpanzees and humans were shown to be more closely related to each other than either is to a gorilla or, more distantly, the orangutan (Delisle 2006).

Darwin on Hominid Taxonomy

As Hillis (see Chapter 16) has noted, textbook phylogenetic depictions were once rare. It is therefore interesting and significant that discussions of phylogenetic relationships among higher primates, from the time of Haeckel and Darwin forward, involved a plethora of clear phylogenetic depictions in both scholarly and popular work. Darwin was a keen observer both of what is today called phylogenetics and of classification, noting in 1871 that: "If man had not been his own classifier, he would never have thought of founding a separate order for his own reception" (Darwin 1871: 191). Indeed, during mid-nineteenth century, it was common for the great apes to be set apart at the ordinal level as Quadrumana, as traced back to Lamarck and Cuvier by Bowler (1986: 60). However, in a fascinating pre-Hennegian assessment of cladistic versus Simpsonian classification in the human evolution book of 1871, Darwin argued:

> *Nevertheless, under a genealogical point of view it appears that this rank is too high, and that man ought to form merely a Family, or possibly even only a Sub-family. If we imagine three lines of descent proceeding from a common source, it is quite conceivable that two of them might after the lapse of ages be so slightly changed as still to remain as species of the same genus; whilst the third line might become so greatly modified as to deserve to rank as a distinct Sub-family, Family, or even Order. But in this case it is almost certain that the third line would still retain through inheritance numerous small points of resemblance with the other two lines. Here then would occur the difficulty, at present insoluble, how much weight we ought to assign in our classifications to strongly-marked differences in some few points,—that is to the amount of modification undergone; and how much to close resemblance in numerous unimportant points, as indicating the lines of descent or genealogy. The former alternative is the most obvious, and perhaps the safest, though the latter appears the most correct as giving a truly natural classification (Darwin 1871: 195).*

Darwin on the Biogeography of Hominid Origins

Haeckel (1868) proposed "Lemuria" as a sunken continent where the last common ancestor of orangutans and humans lived. He considered these taxa to be phylogenetic sisters. Lyell addressed the paleobiogeography of hominid origins as follows in 1863:

> *The opponents of the theory of transmutation sometimes argue that, if there had been a passage by variation from the lower Primates to Man, the geologist ought ere this to have detected some fossil remains of the intermediate links of the chain. But what we have said respecting the absence of gradational forms between the recent and Pliocene mammalian (p. 436), may serve to show the weakness in the present state of science of any argument based on such negative evidence, especially in the case of Man, since we have not yet searched those pages of the great book of nature, in which alone we have any right to expect to find records of the missing links alluded to. The countries of the anthropomorphous apes are the tropical regions of Africa, and the islands of Borneo and Sumatra, lands which may be said to be quite unknown in reference to their pliocene and post-pliocene mammalia (Lyell 1863: 498).*

He continued:

> *Europe, during the Pliocene period, seems not to have enjoyed a climate fitting it to be the habitation of the quadrumanous mammalian; but we no sooner carry back our researches into miocene times, where plants and insects, like those of Oeninghen, and shells, like those of the faluns of the Loire, would imply a warmer temperature both of sea and land, than we begin to discover fossil apes and monkeys north of the Alps and Pyrenees. Among the few species already detected, two at least belong to the anthropomorphous class. One of these, the Dryopithecus of Lartet, a gibbon or long-armed ape, about equal to man in stature, was obtained in the year 1856 in the upper Miocene strata at Sansan, near the foot of the Pyrenees in the South of France, and one bone of the same ape is reported to have been since procured from a deposit of corresponding age at Eppelsheim near Darmstadt, in a latitude answering to that of the southern counties of England. But according to the doctrine of progression it is not in these miocene strata, but in those of pliocene and post-pliocene date, in more equatorial regions, that there will be the greatest chance of discovering hereafter some species more highly organised than the gorilla and chimpanzee (Lyell 1863: 499–500).*

Darwin was able to draw upon this work, and in 1871 famously concluded:

> On the Birthplace and Antiquity of Man.—*We are naturally led to enquire where was the birthplace of man at that stage of descent when our progenitors diverged from the Catarhine stock. The fact that they belonged to this stock clearly shews that they inhabited the Old World; but not Australia nor any oceanic island, as we may infer from the laws*

> *of geographical distribution. In each great region of the world the living mammals are closely related to the extinct species of the same region. It is therefore probable that Africa was formerly inhabited by extinct apes closely allied to the gorilla and chimpanzee; and as these two species are now man's nearest allies, it is somewhat more probable that our early progenitors lived on the African continent than elsewhere (Darwin 1871: 199).*

Darwin has been widely, but perhaps a little too mythically, credited with *calling the shot* as Africa. Darwin continues this very passage (probably following Lyell's observations previously described) in a decidedly and overtly cautious manner: "But it is useless to speculate on this subject, for an ape nearly as large as a man, namely the *Dryopithecus* of Lartet, which was closely allied to the anthropomorphous *Hylobates*, existed in Europe during the Upper Miocene period; and since so remote a period the earth has certainly undergone many great revolutions, and there has been ample time for migration on the largest scale" (Darwin 1871: 199).

Darwin on the Ecology of Hominid Origins

There is considerable speculation in contemporary paleoanthropology about the degree to which human origins and evolution have involved global, regional, and local environmental changes. In his recent review of this literature, Kingston (2007) follows many authors in attributing the savanna model of hominid origins to Darwin 1871. However, my reading of the original sources has failed to identify any passage in which Darwin makes this argument. The closest comparison I have managed to find is a passage discussing the evolution of bipedality. Darwin clearly inferred a "warm" and "forest clad" arboreal habitat for the pre-bipedal primate that he inferred to have been an ancestor. But, he says nothing more than that this creature came to the ground. He does not state what kind of topography that ground would have shown or what kind of vegetation might have covered it. Darwin certainly did not invoke an arid, savanna, grassland, or open country habitat in the following description:

> *As soon as some ancient member in the great series of the Primates came, owing to a change in its manner of procuring subsistence, or to a change in the conditions of its native country, to live somewhat less on trees and more on the ground, its manner of progression would have been modified; and in this case it would have had to become either more strictly quadrupedal or bipedal (Darwin 1871: 140–141).*

Perhaps, some contemporary authors have confused Charles Darwin with Raymond Dart as the originator of an ill defined and now highly suspect savanna hypothesis of hominid origins, but deeper archival research should be able to illuminate this issue. Given the evidence now available regarding *Ardipithecus's* bipedalism in a woodland habitat (White et al. 2009), it is tempting to think that Darwin intuited bipedalism to have evolved in an environment not entirely dominated by trees, such as we see for *Ardip-*

ithecus. However, I have been able to locate no solid evidence that Darwin predicted this circumstance.

Darwin on Extant Apes, Human Ancestors, "Missing Links," and the Last Common Ancestor (LCA)

The anatomy and behavior of the orangutan, common chimpanzee, and gorilla were fairly well known in the mid-to-late 1800s, and contemporary scholars, such as Owen and Huxley, had paid them so much attention that Darwin was able to easily compile the results, as he forged his inferences about the common ancestors that they once shared and what the earlier phases of human evolution involved.

The physical anthropologist William W. Howells would insightfully write in retrospect for the famous 1950 Cold Spring Harbor Symposium (see White 2009b) that:

> *Darwin and Huxley proceeded of course from comparative anatomy: practically speaking, human or anthropoid fossils were unknown. Their allying of man and the apes was a great victory for the day. At the same time, however, men were men and apes were apes. The two could meet, all right, at a hypothetical crotch where their branches came together in the past. This is the diagram that has fascinated us, and plagued us, ever since. Added to this, the apes outnumbered us four to one which, with other facts, made us look like the aberrant animal, and the apes like the more natural, conservative primates (Howells 1951: 80).*

Darwin was very cautious about employing extant organisms to visualize ancestral ones. After all, in his 1959 *On The Origin of Species*, he had already covered much of this particular intellectual turf. There, he had been crystal clear in his cautions about the dangers of uncritically invoking the richness of the zoological present to conceptualize ancestral forms:

> *In the first place it should always be borne in mind what sort of intermediate forms must, on my theory, have formerly existed. I have found it difficult, when looking at any two species, to avoid picturing to myself, forms directly intermediate between them. But this is a wholly false view; we should always look for forms intermediate between each species and a common but unknown progenitor; and the progenitor will generally have differed in some respects from all its modified descendants (Darwin 1859: 280).*

Nevertheless, contemporary cartoonists seized upon extant chimpanzees and gorillas to ridicule Darwin's 1859 insights and poke fun at the notion that humans were genealogically related to apes. However, Darwin was both clear and cautious on another salient point. Some modern paleoanthropologists have failed to heed his cautions, but Darwin insisted that the last common ancestor shared by extant apes and humans should not be assumed to have been very similar to any living form. His 1871 book was very explicit on this point and well worth citing in the hindsight of the last century-and-a-half of public and professional misconceptions that have too often driven searches for halfway houses and missing links between chimpanzees and humans that were never there in the first place. Darwin famously concluded: "Man still bears in his bodily frame the indelible stamp of his lowly origin" (Darwin 1871: 405), but he was loathe to identify any extant form as an ancestor:

> *But we must not fall into the error of supposing that the early progenitor of the whole Simian stock, including man, was identical with, or even closely resembled, any existing ape or monkey (Darwin 1871: 199).*

> *Some old forms appear to have survived from inhabiting protected sites, where they have not been exposed to very severe competition; and these often aid us in constructing our genealogies, by giving us a fair idea of former and lost populations. But we must not fall into the error of looking at the existing members of any lowly-organised group as perfect representatives of their ancient predecessors (Darwin 1871: 212).*

It was relatively easy for Darwin and his contemporaries to utilize detailed comparative anatomy and embryology in a phylogenetic context. Darwin employed the tools that today are called "commonality" and "parsimony" to identify what are now termed the "primitive" and "derived" conditions of anatomical structures, such as the foot and canine teeth. About these, Darwin wrote in 1871: "The foot, judging from the condition of the great toe in the fœtus, was then prehensile; and our progenitors, no doubt, were arboreal in their habits...The males were provided with great canine teeth, which served them as formidable weapons" (Darwin 1871: 206–207).

Darwin's caution was well founded on the lack of a hominid fossil record, as Owen had emphasized even in the same year of *On The Origin of Species*:

> *Whether, therefore, strata of such high antiquity as the miocene may reveal to us 'forms in any degree intermediate between the chimpanzee and man' awaits an answer from discoveries yet to be made; and the anticipation that the fossil world 'may hereafter supply new osteological links between man and the highest known quadrumana' may be kept in abeyance until that world has furnished us with the proofs that a species did formerly exist which came as near to man as does the orang, the chimpanzee, or the gorilla. Of the nature and habits of the last-named species, which really offers the nearest approach to man of any known ape, recent, or fossil…(Owen 1859: 87).*

In 1871, Darwin wrote about extant African apes as proxies for hominid ancestors:

> *In regard to bodily size or strength, we do not know whether man is descended from some comparatively small species, like the chimpanzee,*

or from one as powerful as the gorilla; and, therefore, we cannot say whether man has become larger and stronger, or smaller and weaker, in comparison with his progenitors. We should, however, bear in mind that an animal possessing great size, strength, and ferocity, and which, like the gorilla, could defend itself from all enemies, would probably, though not necessarily, have failed to become social; and this would most effectually have checked the acquirement by man of his higher mental qualities, such as sympathy and the love of his fellow-creatures. Hence it might have been an immense advantage to man to have sprung from some comparatively weak creature (Darwin 1871: 156–157).

Only from hindsight of new fossil hominids does Darwin's attempt to make the case for the feasibility of a transition from quadrupedalism to bipedalism seem to miss the mark (Lovejoy 2009; Lovejoy et al. 2009):

If the gorilla and a few allied forms had become extinct, it might have been argued with great force and apparent truth, that an animal could not have been gradually converted from a quadruped into a biped; as all the individuals in an intermediate condition would have been miserably ill-fitted for progression. But we know (and this is well worthy of reflection) that several kinds of apes are now actually in this intermediate condition; and no one doubts that they are on the whole well adapted for their conditions of life. Thus the gorilla runs with a sidelong shambling gait, but more commonly progresses by resting on its bent hands. The long-armed apes occasionally use their arms like crutches, swinging their bodies forward between them, and some kinds of Hylobates, without having been taught, can walk or run upright with tolerable quickness; yet they move awkwardly, and much less securely than man. We see, in short, with existing monkeys various gradations between a form of progression strictly like that of a quadruped and that of a biped or man (Darwin 1871: 142–143).

Darwin on the Selective Advantages of Hominid Derivations

Adopting a more theoretical than historical stance in his 1871 book, Darwin contemplated the adaptive significance of human derivations. In some cases, his arguments were decidedly Lamarckian:

Although man may not have been much modified during the latter stages of his existence through the increased or decreased use of parts, the facts now given shew that his liability in this respect has not been lost; and we positively know that the same law holds good with the lower animals. Consequently we may infer, that when at a remote epoch the progenitors of man were in a transitional state, and were changing from quadrupeds into bipeds, natural selection would probably have been greatly aided by the inherited effects of the increased or diminished use of the different parts of the body (Darwin 1871: 120–121).

Furthermore, much of Darwin's thinking was clearly adaptationist, even as it was *Darwinian*:

> *I see no reason to doubt that a more perfectly constructed hand would have been an advantage to them, provided, and it is important to note this, that their hands had not thus been rendered less well adapted for climbing trees. We may suspect that a perfect hand would have been disadvantageous for climbing; as the most arboreal monkeys in the world, namely Ateles in America and Hylobates in Asia, either have their thumbs much reduced in size and even rudimentary, or their fingers partially coherent, so that their hands are converted into mere grasping-hooks (Darwin 1871: 140). If it be an advantage to man to have his hands and arms free and to stand firmly on his feet, of which there can be no doubt from his pre-eminent success in the battle of life, then I can see no reason why it should not have been advantageous to the progenitors of man to have become more and more erect or bipedal. They would thus have been better able to have defended themselves with stones or clubs, or to have attacked their prey, or otherwise obtained food. The best constructed individuals would in the long run have succeeded best, and have survived in larger numbers (Darwin 1871: 142).*

Darwin on How Characters Were Assembled during Human Evolution

Without a paleobiological record of the sequence in which hominid derivations were acquired during human evolution, Darwin merely cited Huxley (1863) and Lyell (1863), and rapidly moved on to write about aspects that were more theoretical than the works of his colleagues. He inferred a great many things about the course of human evolution, but not always transparently. This lack of clarity can be interpreted as a cautious Darwin hedging his bets.

Perhaps the most significant differences in interpreting Darwin involve whether or not he posited an all-encompassing feedback system in which bipedality, brain expansion, canine reduction, and tool use arose gradually and in lock step. Not only has the fossil record subsequently falsified this scenario, but in addition, some scholars have concluded that Darwin never proposed it at all. Did Darwin predict early *Australopithecus* (a small-brained, small-canine biped, with no stone tools) by placing an emphasis on the primacy of bipedality?

Bowler (1986) observes that Darwin's "adaptive scenario" switched attention from the comparison of mental powers to another unique characteristic: upright posture and bipedal locomotion. Bowler calls Darwin's 1871 scenario:

> *...a pioneering effort to grapple with the problem of explaining why the line of human evolution became separated from that of the apes. Later discoveries in the fossil record have at least confirmed that Darwin was*

> *right to suspect that the earliest humans stood upright before the brain began to increase in size. Few of his contemporaries were prepared to admit that the brain did not lead the way in human evolution (Bowler 1986: 188).*

This dilemma may be resolved by examining the following passage:

> *Man alone has become a biped; and we can, I think, partly see how he has come to assume his erect attitude, which forms one of the most conspicuous differences between him and his nearest allies. Man could not have attained his present dominant position in the world without the use of his hands which are so admirably adapted to act in obedience to his will...But the hands and arms could hardly have become perfect enough to have manufactured weapons, or to have hurled stones and spears with a true aim, as long as they were habitually used for locomotion and for supporting the whole weight of the body, or as long as they were especially well adapted, as previously remarked, for climbing trees. Such rough treatment would also have blunted the sense of touch, on which their delicate use largely depends. From these causes alone it would have been an advantage to man to have become a biped; but for many actions it is almost necessary that both arms and the whole upper part of the body should be free; and he must for this end stand firmly on his feet. To gain this great advantage, the feet have been rendered flat, and the great toe peculiarly modified, though this has entailed the loss of the power of prehension. It accords with the principle of the division of physiological labour, which prevails throughout the animal kingdom, that as the hands became perfected for prehension, the feet should have become perfected for support and locomotion (Darwin 1871: 141–142).*

The evolutionary sequence conceived by Darwin is most plainly expressed here:

> *As soon as some ancient member in the great series of the Primates came, owing to a change in its manner of procuring subsistence, or to a change in the conditions of its native country, to live somewhat less on trees and more on the ground, its manner of progression would have been modified; and in this case it would have had to become either more strictly quadrupedal or bipedal (Darwin 1871: 140–141).*

So, Darwin saw subsistence or unspecified environmental change as the most likely causes for the ancestral human to adopt a more terrestrial than savanna existence. He was unwilling to conceptualize the creature as either particularly chimpanzee- or gorilla-like. Discussing the hands and feet, he is obviously invoking a feedback mechanism. He continues with this theme as follows:

> *The free use of the arms and hands, partly the cause and partly the result of man's erect position, appears to have led in an indirect manner to other modifications of structure. The early male progenitors of man were, as*

> *previously stated, probably furnished with great canine teeth; but as they gradually acquired the habit of using stones, clubs, or other weapons, for fighting with their enemies, they would have used their jaws and teeth less and less. In this case, the jaws, together with the teeth, would have become reduced in size... Therefore as the jaws and teeth in the progenitors of man gradually become reduced in size, the adult skull would have presented nearly the same characters which it offers in the young of the anthropomorphous apes, and would thus have come to resemble more nearly that of existing man...A great reduction of the canine teeth in the males would almost certainly, as we shall hereafter see, have affected through inheritance the teeth of the females. As the various mental faculties were gradually developed, the brain would almost certainly have become larger. No one, I presume, doubts that the large size of the brain in man, relatively to his body, in comparison with that of the gorilla or orang, is closely connected with his higher mental powers (Darwin 1871: 144–145).*

One might read these passages to mean that Darwin was suggesting a linear progression, with the brain following the other structures. But on the very next page, he follows that excerpt with the explanation: "The gradually increasing weight of the brain and skull in man must have influenced the development of the supporting spinal column, more especially whilst he was becoming erect" (Darwin 1871: 146). This statement seems more like feedback than an invocation of the primacy of bipedality, particularly given that Darwin concludes his 1871 chapter with: "...but it is quite conceivable that they might have existed, or even flourished, if, whilst they gradually lost their brute-like powers, such as climbing trees, &c., they at the same time advanced in intellect" (Darwin 1871: 157). He goes on to state: "In a series of forms graduating insensibly from some ape-like creature to man as he now exists, it would be impossible to fix on any definite point when the term 'man' ought to be used" (Darwin 1871: 235).

In the second volume of his 1871 treatise, in a summary about sexual selection, Darwin reiterates the theme of gradual acquisition. Here he discusses bipedality and tool use, the latter he promotes as the selective force for brain expansion:

> *It was remarked in a former chapter that as man gradually became erect, and continually used his hands and arms for fighting with sticks and stones, as well as for the other purposes of life, he would have used his jaws and teeth less and less. The jaws, together with their muscles, would then have become reduced through disuse, as would the teeth through the not well understood principles of correlation and the economy of growth; for we everywhere see that parts which are no longer of service are reduced in size (Darwin 1871: 324–325).*

Darwin had already addressed the relationship of these two central characteristics (bipedality and tool use) in his discussion of humans among the anthropoid apes: "No doubt man, in comparison with most of his allies, has undergone an extraordinary amount of modification, chiefly in consequence of his greatly developed brain and erect position; nevertheless we should bear in mind that he 'is but one of several exceptional forms of Primates'" (Darwin 1871: 197).

The persistently gradualistic mode of Darwin's thought and presentation represents an inconvenient fact for those who contend that Darwin diagnosed the sequence of acquisition as bipedality having evolved prior to the brain's expansion. The ambiguity about the sequence of events seems deliberate on Darwin's part, and archival research may shed more light on the matter.

Summarizing Darwin's Predictions about Yet to Be Discovered Fossil Hominids

How, then, should we summarize Darwin's predictions about the hominid fossil record that was still so poor in 1871? It turns out that Darwin's cautious, evidence-based manner makes this question easy to answer, because he predicted so little; but difficult to answer, because he was never very specific and always apt to qualify. He clearly invokes modern apes as bridging the anatomical and behavioral gulfs between humans and Old World Monkeys. However, he just as emphatically cautions against adopting any one of them as a proxy. His writing on the sequence by which anatomical traits were assembled by natural selection is ambiguous, and one gets the sense that if pressed on acquisition order, his response would have been the familiar "it is useless to speculate." Even his prediction that the original hominids would be African was immediately qualified. In short, Darwin fully expected that one day, light would be shed on the issue, but was fully prepared to await the needed paleobiological evidence. Regrettably, he did not live long enough to see very much of it.

Travels through Time in the Afar Rift

Resurrecting Darwin

Darwin lies buried in Westminster Abbey, only feet from Newton, "...by the will of the intelligence of the nation" (Huxley 1882: 597). Bowler's elegant contrafactual assessment of Darwin's impact (Bowler 2008; see Bowler, Chapter 2) shows what might have happened if Darwin had not been born or had died before writing *On The Origin of Species*. If we could somehow resurrect Darwin today, his writings make it clear that he would almost certainly soon ask about the fossil record of human evolution that had accumulated after his death. In his day, he correctly viewed the lack of hominid fossils as resulting from major imperfections in the geological record and from the limited exploration of relevant locations:

> *...the facts given in the previous chapters declare, as it appears to me, in the plainest manner, that man is descended from some lower form,*

> *notwithstanding that connecting-links have not hitherto been discovered (Darwin 1871: 185).*

> *With respect to the absence of fossil remains, serving to connect man with his ape-like progenitors, no one will lay much stress on this fact, who will read Sir C. Lyell's discussion, in which he shews that in all the vertebrate classes the discovery of fossil remains has been an extremely slow and fortuitous process. Nor should it be forgotten that those regions which are the most likely to afford remains connecting man with some extinct ape-like creature, have not as yet been searched by geologists (Darwin 1871: 201).*

Imagining Darwin in the context of our digitally dependent century poses challenges—not the least of which is having Darwin speak in our terms, our dialects, and our jargon. In what follows, I will not try to put words in Darwin's mouth, but rather imagine the kinds of things a resurrected Darwin might ask about, notably, the nature of the evidence now available to understand human evolution. In order to examine comprehensively the accumulated evidence for human evolution, we would have to take resurrected Darwin on another global journey. However, if there were only 1 week for the excursion, there would be no better place to take him than to the study area where my colleagues and I have been working since the 1980s: the Middle Awash Valley of Ethiopia's Afar triangle.

The study area is centered at the base of the Horn of Africa, the continent Darwin identified as the likely birthplace of the hominid clade and where he predicted that those "remains connecting man with some extinct ape-like creature" might someday be found (Darwin 1871: 201). In that remote and desolate corner of Africa, a considerable number of such remains have now been found in eroding sediments comprising the foothills and floor of the Afar rift. On this imaginary, twenty-first century Darwin voyage to the Afar, we would want to take the *young* Darwin, because even today, the region is dangerous and the working conditions physically demanding.

As Bowler (2008) has shown, all biologists should be very glad that Robert FitzRoy did not steer the *Beagle* toward this region of Africa in 1836. Even if the good ship had avoided pirates in the Red Sea and Gulf of Aden rifts, young Darwin would surely have desired to disembark to traverse the Afar. Given that the first European explorers able to return alive from this region did so in the 1920s, it is a good thing for evolutionary biology that the *Beagle* turned to port somewhere east of Madagascar, circumnavigating the Cape of Good Hope and carrying Darwin safely home to England.

The Middle Awash Project and Study Area

The Middle Awash research team has been working in the central axis of the Awash River drainage in Ethiopia investigating a kilometer's thickness of accumulated sediment—sediment that recorded parts of the one-time human evolution experiment within a single depository. Whereas limiting an imaginarily resurrected Darwin to a single geographic area risks compromising the current global understanding of human evolution, given what we know of Darwin's relentless curiosity, we can be confident that he would never lose track of a broader picture.

The integrated primary missions of the Middle Awash project are to conduct scientific research and build Ethiopian national capacity in paleoanthropology (Asfaw and Gilbert 2008). Today, there are approximately 70 PhD-level scientists from 20 nations conducting research in the fields of archaeology, geochemistry, geochronology, invertebrate and vertebrate paleontology, paleobotany, sedimentology, structural geology, taphonomy, and others. Darwin would probably have been most impressed and pleased by the participation of professional researchers from Ethiopia (whom he would have identified as Abyssinian). More than 600 professional scholars, support personnel, and local guides, guards, and laborers have worked together tirelessly on this project. They deserve enormous credit, as the evidence they have managed to compile, in the face of multiple adversities, has not been acquired easily over the last three decades.

The evidence recovered by the Middle Awash research team comprises approximately 19,000 catalogued vertebrate fossils, including 300+ hominids, more than 1800 geological samples, and thousands of stone artifacts. These data are from over approximately 1 km of accumulated sediments, which provide a succession of temporal snapshots across the last 6 Ma (million years ago). These data have been shared via approximately 6500 printed pages (Table 19.1) as well as electronically (http://middleawash.berkeley.edu/middle_awash.php). It can safely be assumed that Darwin would delight in all this evidence, much of it now housed permanently in the National Museum of Ethiopia, in Addis Ababa. Surely, he would wonder how one valley in Africa could have yielded so much evidence.

Darwin wrote in 1859:

> *Lyell has well remarked, the extent and thickness of our sedimentary formations are the result and the measure of the denudation which the earth's crust has elsewhere undergone. Therefore a man should examine for himself the great piles of superimposed strata, and watch the rivulets bringing down mud, and the waves wearing away the sea-cliffs, in order to comprehend something about the duration of past time, the monuments of which we see all around us (Darwin 1859: 349).*

On his imaginary visit to the Afar, Darwin would be astonished to learn of the breakthroughs in geological and geochronological science since his death. He had a firsthand acquaintance with tectonics and other geological forces, so he would surely appreciate the basic elements of Afar geology and find the kilometer of sediments exposed by erosion to be the logical consequence of erosion of the adjacent Ethiopian highlands, the parent rock for this basin's deposits. However, realizing that this kilometer of strata had been chronologically calibrated by interbedded volcanic tephra, lava flows,

TABLE 19.1 MAIN DISCOVERIES IN THE MIDDLE AWASH STUDY AREA, AFAR RIFT, ETHIOPIA

Age (Ma)	Localities	Number of Fauna Specimens	Archaeological Industry	Hominid: Number of Identified Specimens	Initial Publication
0.08	Aduma	21	MSA	4	*AJPA* 03
0.10	Halibee	2646	MSA	16	
0.16	Herto	104	MSA/Late Acheulean	12	*Nature* 03
0.2	Talalak	575	MSA/Late Acheulean	2	
0.5	Bodo	25	Typical Acheulean	3	*Nature* 84/*Science* 04
1.0	Bouri Daka	753	Early Acheulean	11	*Nature* 02
2.0	Guneta	753	Oldowan and later	2	
2.5	Bouri Hata	576	Oldowan	11	*Nature* 99
3.5	Maka	171		12	*Nature* 84, 93
3.9	Belohdelie	54		1	*Nature* 84
4.1	Asa Issie	605		37	*Nature* 06
4.4	Aramis	6432		114	*Nature* 94/*Science* 09
5.3	Amba	550		1	*Nature* 01
5.8	Adu–Asa (Western Margin)	2230		19	*Nature* 01/*Science* 04

Source: Adapted from Asfaw and Gilbert 2008; Haile-Selassie and WoldeGabriel 2009.
Note: Only major hominid announcements listed in publications.
Ma = million years ago.
MSA = Middle Stone Age.

and paleomagnetics across an interval of 6 million years would surely stun Darwin. Similarly, he would be impressed by the unraveling of plate tectonics (accomplished in the 1960s)—these forces ultimately are responsible for forming the rift's modern topography and creating this natural laboratory and repository of African prehistory, all based on separation of the Nubian and Arabian plates at a rate of 17 mm per year and up to 55 mm per year in the Afar rift just north of the Middle Awash.

Aduma and Halibee: Atop the Local Middle Awash Sequence

Darwin's imaginary Middle Awash visit is most easily conceptualized as a trip that progresses back through time, beginning with the modern people who live there today and visiting older and older eroding patches of sediment whose content reveals the sequence by which humans evolved. As we recede deeper in time, fewer and fewer human characters and behaviors are recovered from the fossil record. Today, the nomadic Afar tribe inhabits the Middle Awash study area. Their domestic livestock inadvertently assist erosion of the Afar's emergent sediments. The Afar settlements shift from place to place during a seasonal round and are conditioned by the availability of water.

It was not always this way. Late Stone Age people lived here, and archaeological traces of their hunting and gathering existence are ubiquitous. In the sediments of 80,000 years ago, we have recovered evidence that Middle Stone Age people occupied a riverside habitat (Haile-Selassie et al. 2004a), where they manufactured implements of obsidian (Yellen et al. 2005). Darwin would recognize these as similar to Neanderthal tools found in Europe. He would be surprised to learn that the Aduma hominid crania, although fully anatomically modern, are much older than the few Neanderthal fossils known by 1871. (Explaining to Darwin that something called DNA had been extracted and sequenced would exhaust several sessions around the campfire.)

Of the peculiar Eurasian Neanderthal fossils, the world's first fossil hominids to be found, Darwin wrote apologetically in 1871: "...it must be admitted that some skulls of very high antiquity, such as the famous one of Neanderthal, are well developed and capacious" (Darwin 1871: 146). His admission was predicated on the fact that in 1871, crania with braincases smaller than those of humans but larger than those of apes (i.e., the intermediate crania virtually demanded by his inferences about human evolution) were yet to be found. Further frustrating Darwin and his contemporaries was the fact that they had no way to assess the age of even these geologically recent Ice Age Neanderthals.

Herto: Early Anatomically Modern Homo sapiens

Long before Neanderthals reached their apogee in Europe and long before their apparent demise there, near the end of the Pleistocene, anatomically near-modern people (White et al. 2003) butchered hippopotamus carcasses and practiced mortuary rituals beside a tropical lake (Clark et al. 2003), at what is now Herto in Ethiopia's Middle Awash. These fossils are among Africa's earliest *Homo sapiens*. Dated more than 155,000 years old, they are accompanied by some of the youngest Acheulean handaxes. Darwin would be surprised to learn that these bifacially trimmed implements, very similar to those he knew of from England and France, were found so far afield, and in association with crania of substantial anatomical modernity. He would certainly grasp the phylogenetic consequences of this fact for Europe's younger Neanderthals.

Darwin's views on human race were particularly enlightened for his time (see Richerson and Boyd, Chapter 20). He wrote in 1871:

> *If the races of man were descended, as supposed by some naturalists, from two or more distinct species, which had differed as much, or nearly as much, from each other, as the orang differs from the gorilla, it can hardly be doubted that marked differences in the structure of certain bones would still have been discoverable in man as he now exists. Although the existing races of man differ in many respects, as in colour, hair, shape of skull, proportions of the body, &c., yet if their whole organisation be taken into consideration they are found to resemble each other closely in*

> *a multitude of points. Many of these points are of so unimportant or of so singular a nature, that it is extremely improbable that they should have been independently acquired by aboriginally distinct species or races. The same remark holds good with equal or greater force with respect to the numerous points of mental similarity between the most distinct races of man (Darwin 1871: 231–232).*

A detailed, forensic examination of the 155,000-year-old *Homo sapiens idaltu* crania from Herto places them morphologically closer to living aboriginal Australians than to modern African populations (White et al. 2003). This fossil evidence from Ethiopia is consilient with modern genetic evidence, showing the shallow roots of morphological differentiation among humans, and is also consistent with Darwin's thinking. Following are the words that Darwin wrote on the topic—not exactly a prediction, but nevertheless uncannily matching subsequent evidence: "The spreading of man to regions widely separated by the sea, no doubt, preceded any considerable amount of divergence of character in the several races; for otherwise we should sometimes meet with the same race in distinct continents; and this is never the case" (Darwin 1871: 234).

If Darwin walked about 500 meters to the north of the Herto discoveries, he would cross the outcrop of a 250,000-year-old volcanic ash and observe more stone handaxes directly beneath it. He would wonder what their makers were like. So do we—no hominid remains have yet been found in this particular stratum, only the remains of other, older extinct mammals as well as those thought to be ancestral to the modern African fauna.

Bodo: Not a Human, Not an Ape

Deeper in the Middle Awash sedimentary succession, on the eastern side of today's Awash River, the partial cranium of a large, presumably male, hominid from Bodo was found amongst an even larger proportion of extinct mammals. This fossil would be deeply satisfying to Darwin. Here is a cranium that is patently neither ape nor human—a fossil that today is best classified as *Homo rhodesiensis*. Additionally, it is a virtually perfect ancestral morphotype for the younger hominids from Herto (Figure 19.2) and an obvious descendant from earlier forms in deeper sediments. The Bodo hominids manufactured bifaces on flakes struck from massive basalt cores 500,000 years ago, which is about the time that the Neanderthal clade formed in Europe.

Bouri Daka: Homo erectus

The discovery of *Homo erectus* on Java before the close of the nineteenth century provided the very evidence of hominids that Darwin lacked but did not lived to see. Thus, on a tour of Ethiopia's Middle Awash today, Darwin would be enthralled to visit million-year-old sediments near Bouri village, which yielded the Daka cranium in 1997 (Asfaw et al. 2002; Asfaw and Gilbert 2008). There he would observe less refined handaxes eroding

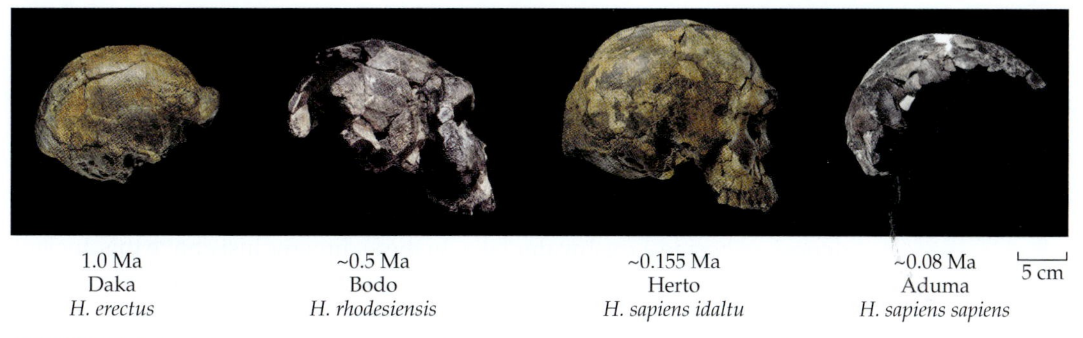

FIGURE 19.2 Middle Awash Pleistocene Crania from a Succession of Middle Awash Strata Note the temporal and phenetic progression. These are the most complete crania for each horizon within the local succession, but similar forms are known from similar time horizons across the continent.

from lake marginal beach sands, along with the fossil remains of a plethora of extinct mammals among which *Homo erectus* lived. Darwin would almost certainly be torn between examining this evidence in detail and asking to be taken to even older outcrops.

Guneta: Expansion from Africa

The Middle Awash succession is more like a series of snapshots through time than a video. Loci of deposition shifted across this basin as tectonic forces influenced volcanism, deposition, and erosion of the accumulated sediments and their contents through time. Only eroding windows of these are today available for exploration.

Two million years ago, sediments were deposited about 25 kilometers to the west of the Daka cranium discovery site, where today the foothills of the Afar rift comprise a series of tilted blocks dissected by erosion. These have now yielded rich assemblages of early hominid technology at localities called Guneta and Kurbi Lefna (unpublished data). An acute observer with extensive geological field experience from his time on the *Beagle*, Darwin would have easily appreciated the anomalous presence of these primitive sharp-edged implements so long ago discarded and suspended in the fine silt of an ancient river's floodplain. This stone tool technology is named "Oldowan" for its original discovery at Olduvai Gorge, Tanzania. The hominids responsible for making these stone tools are not yet well represented in the Middle Awash, and their systematics elsewhere remain in flux even today. They are variously classified as early *Homo erectus* (*ergaster*) and *Homo habilis* and are from southern and eastern Africa. Their 1.8 million-year-old descendants are the first known to have expanded their range beyond Africa. Such remains have been found in Eurasian Georgia as well as in Indonesia.

Bouri Hata: The Earliest Stone Tools and the Origin of Homo

Even as the assembled evolutionary biologists and anthropologists prepared to celebrate the centenary of *On The Origin of Species* at Cold Spring Harbor in 1950, the late physical anthropologist Howells summarized the state of affairs in Pliocene hominid fossil acquisition: "Let us remember that the fossil record is never complete, and that from present indications, the Pliocene still has the principal secrets of human ancestry" (Howells 1951: 81).

About halfway down the kilometer-thick composite sedimentary succession of the Middle Awash is a yellow volcanic ash dated to 2.5 Ma. It was deposited beside a freshwater lake. In the sediments just above this ancient ash are large mammal bones bearing unambiguous traces left by stone tools wielded by hominids foraging on this landscape. Nearby, in the same stratum, are the remains of those who may have fashioned these implements and exhibited these behaviors. The manufacturer of these earliest stone tool kits may be *Australopithecus garhi* (Asfaw et al. 1999; deHeinzelin et al. 1999). In the Gona study area north of the Middle Awash, such stone tools are found in abundance (Semaw et al. 1997).

This stone technology and these meat and marrow-processing behaviors are unique among the primates and would have exposed the bipedal hominids to selective pressures previously not witnessed during human evolution. A threshold in human evolution was crossed as these hominids began to exploit the new niche that stone tools had created for them—the niche of a carnivorous bipedal primate that was able to compete with hyaenids and felids for large mammal carcasses. Their descendants on the human side of that threshold are classified as *Homo*. Darwin would be satisfied that the earliest stone implements involved a change in subsistence and that a novel regime of natural selection would ultimately lead to earth-orbiting satellites. We would explain to him that we use these orbiting tools today in order to target outcrops in the Middle Awash and to pinpoint our discoveries spatially.

Maka, Belohdelie, and Asa Issie: Primitive Australopithecus

The geologist Maurice Taieb was the first scientist to appreciate the paleoanthropological potential of the Afar rift and the first to find fossils in the Middle Awash study area. In the 1970s, his attention was drawn to the Hadar area, north of the Middle Awash and east of Gona. There the Hadar project began to recover fossil hominids much older than those of either early *Homo* or the specialized megadont lineage *Australopithecus boisei* (by that time, representatives of this separate lineage had been known from Olduvai Gorge for more than a decade). The Hadar fossils were also somewhat earlier than fossils of *Australopithecus* that had been collected since the 1920s in South Africa.

Remains of a very small female *Australopithecus* (named "Lucy" by her discoverers) were found at Hadar in 1974 (Johanson and Taieb 1976). Dozens of other fossils, including skulls and associated skeletal parts from males, females, and children from the species *Australopithecus afarensis*, have since been found at Hadar (Johanson and White 1979; Kimbel and Delezene 2009), as well as in Tanzania, where they are accompanied by footprints showing bipedal progression on a freshly fallen volcanic ash (Leakey and Hay 1979; White 1980). An even earlier chronospecies is known from Kenya (Leakey et al. 1995) and from Asa Issie in the Middle Awash, each dated slightly older than 4.0 million years (White et al. 2006).

At this point in his imaginary journey through the Middle Awash study area, Darwin would probably endorse Bowler's (1986) interpretation of Darwin's 1871 text—the one that posited that bipedality preceded brain expansion and/or stone tool use. Darwin's own writing seems ambiguous on the sequence of character assembly. Like the fossils found since the 1920s in southern Africa, these early eastern Africa *Australopithecus* were demonstrably bipedal. Yet, we find no evidence of lithic technology accompanying them, and their brain sizes lie firmly within modern ape ranges. If Darwin ever did favor a feedback model of hominid evolution, he would readily have abandoned it in the face of the wealth of evidence dated to between 3 and 4 million years ago in Africa.

Aramis: Ardipithecus ramidus

Moving down the Middle Awash geological column only 80 meters below the horizon with the area's earliest *Australopithecus* fossils at approximately 4.2 million years ago, Darwin could be not help but be impressed by the contents of a thin but extensively exposed stratum of largely floodplain-deposited, fine-grained sediments. At the end of the twentieth century, from these beds, the Middle Awash research team was able to capture an extraordinary, uniquely high-resolution and high-integrity snapshot of the Afar at 4.4 million years ago (Figure 19.3; White et al. 2009). Some scientists had predicted that hominids between Lucy and the last common ancestor shared with the chimpanzees (i.e., the CLCA) would be like *Australopithecus* all the way back to the fork. Most predicted (or assumed) that when found, such fossils would be increasingly chimpanzee-like.

The hominid fossils from this Pliocene Middle Awash horizon (the same species is also found at the site of Gona to the immediate north) differ substantially from even the earliest and most primitive *Australopithecus*. They are decidedly unlike extant great apes in much of their bony and dental anatomy. These fossils, named *Ardipithecus ramidus*, now include the partial skeleton of a young adult female whose teeth, hands, feet, pelvis, arms, and legs finally provide a paleobiological perspective on the ape-like creature that Darwin predicted would characterize the first phase of human evolution, subsequent to the CLCA.

Ardipithecus ramidus inhabited a largely woodland habitat. The anatomy, context, and geochemical content of its remains provide perspective on the early *Australopithecus* stratigraphically above it in this valley. Because it is older and more primitive than *Australopithecus*, *Ardipithecus* allows investigators to better discern what the last common ancestors with the gorillas and chimpanzees were like, thereby providing fresh perspective on the evolution of these African apes. Darwin would be surprised to find that the entire fossil

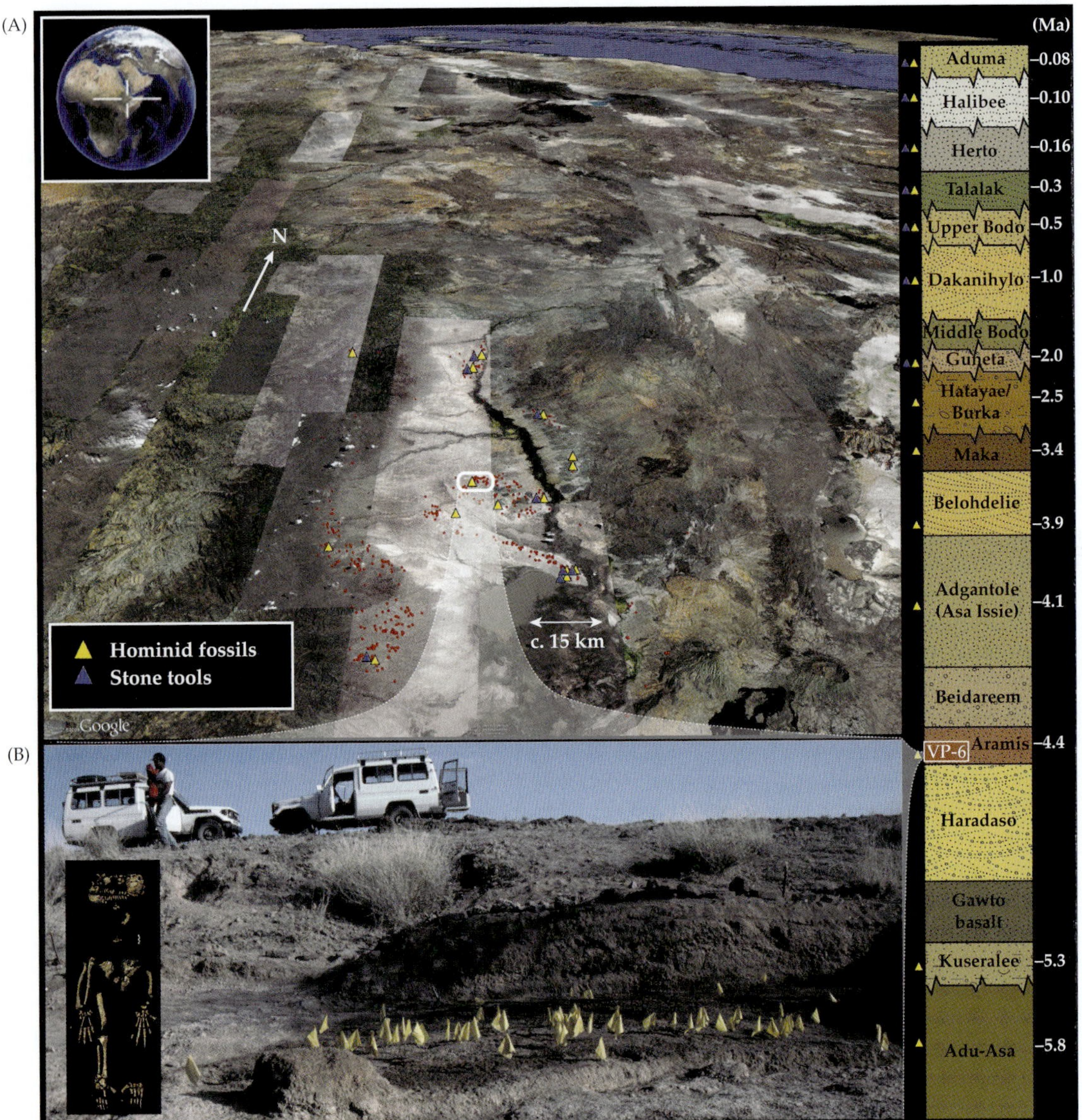

record of these forms is still massively imperfect today, comprising only three teeth from one Pleistocene individual (McBrearty and Jablonski 2005).

Confronted with *Ardipithecus*, Darwin would immediately discern a foot that retained prehensility of the large toe, the significance of which would not have escaped him. Indeed, it was his rival Owen who pointed out 150 years ago that: "The great toe which forms the fulcrum in standing or walking is perhaps the most characteristic peculiarity in the human structure; it

FIGURE 19.3 The Middle Awash Paleoanthropological Study Area, Afar Rift, Ethiopia View is to the north. A composite stratigraphic thickness of more than a kilometer of predominantly fluviatile, lacustrine, and volcaniclastic sediment has produced samples across 14 different time horizons that contain hominid fossils. (A) The small red dots are geological sample points, usually indicating rocks extracted for radioisotopic, geomagnetic, isotopic, and sedimentological studies. Yellow and blue triangles show the location of major hominid-bearing and archaeological occurrences, respectively. (B) The excavation of 4.4 Ma (millions of years ago) deposits containing the ARA VP-6/500 skeleton of *Ardipithecus ramidus* (inset). Yellow flags indicate position and elevation of each fragment recovered from this individual.

is that modification which differentiates the foot from the hand, and gives the character to his order" (i.e., the Bimana) (Owen 1859: 79).

Explaining to Darwin the current work in evolutionary developmental biology that may reveal the genetic substrate of these kinds of anatomical shifts (Prabhakar et al. 2008; Wray and Babbitt 2008) would require more time than available around the campfire on our imaginary field excursion. However, in examining the *Ardipithecus* fossils, Darwin would have noticed that even the largest males had small, non-honing canine teeth and that the pelvis was much broader anteroposteriorly than in any chimpanzee's. Turning to the hand, he would have seen the lack of any metacarpal elongation that characterizes the hands of suspensory orangutans or knuckle-walking African apes. *Ardipithecus* was a mosaic organism, neither ape nor human, with key hominid derivations in both locomotion and social behaviors. It retained primitive (but monkey-like rather than extant ape-like) postcranial characters shared by Miocene apes and a brain no larger than that of a living chimpanzee's. It is impossible to know what Darwin would have made of *Ardipithecus*, but he certainly would have been intrigued. He would have been humble, but probably proud and relieved for having been so explicitly cautious about employing living apes as proxies for hominid ancestors.

Contemporary paleoanthropologists have only begun to digest the significance of *Ardipithecus*. Some primatologists who have justified their studies based on the close genetic affinity of living chimpanzees and humans are dismayed that *Ardipithecus* is not specifically chimpanzee-like. Some geologists who have adopted the savanna model of hominid origins to justify their studies of global climate are dismayed to discover that the earliest hominids were already bipedal in wooded habitats.

Some paleoanthropologists who have previously portrayed *Australopithecus* as a "missing link" between the CLCA and *Homo* point to the narrow time interval between the earliest *Australopithecus* and *Ardipithecus*, arguing that there is insufficient time for this evolution to have occurred. Of course, the rate of evolution depends on the availability of genetic variation translating into phenotypic differences upon which natural selection can act (Bell et al. 2004; Hendry et al. 2008; Bell 2010). The fact that genetic change can occur rapidly is now well supported by genomics (Marques-Bonet et al. 2009; Hughes et al. 2010; see Kolaczkowski and Kern, Chapter 6). Perhaps, it is the residual conditioning of the largely gradualistic Modern Synthesis and/or an under-appreciation of the rapid advances in evolutionary developmental biology and genomics that are responsible for the reluctance of some workers to consider that anatomical evolutionary change can happen rapidly, even among hominids. In considering the relationships between *Ardipithecus* and *Australopithecus*, Darwin would certainly have pointed again to the imperfections of the fossil record (Figure 19.4), the resulting lack of resolution in early hominid phylogeny, and the familiar need for more fossil evidence.

The Western Margin of the Middle Awash: Africa's Earliest Hominids

Curious to see what even earlier hominid fossils might reveal, Darwin would approach the bottom of the Middle Awash succession, appreciating that the overlying succession roughly reveals the sequence by which the specifically human bauplan was assembled during the Neogene, but curious about what lay deeper in time, in the late Miocene. The probable chronospecies *Ardipithecus kadabba* found in the foothills of the western margin of the Middle Awash study area is comparatively poorly known relative to the younger members of its lineage. However, already it shows much the same morphology as similar, approximately 6 million-year-old fossils from Kenya and Chad (Haile-Selassie 2001; Haile-Selassie et al. 2004b; Haile-Selassie and WoldeGabriel 2009)—a morphology which is in keeping with

FIGURE 19.4 Phylogenetic Relationships of Hominids (A) Phylogenetic relationships of extant hominoid primates based on biochemical data and fossil hominid taxa, which are currently time successive in the fossil record and may have evolved phyletically. Resolution of the hominid phylogeny is hampered by gaps in the fossil record. (B) Colored segments (see for taxa) show the temporal ranges of currently available Tanzanian, Kenyan, Chadian, and Ethiopian samples. (C) Phylogenetic hypotheses showing relationships among fossil samples that are currently impossible to fully resolve. Depicted species lineages are the gray bundles that are composed of subspecific strands (populations), with continuity through time and each reticulating with adjacent populations through gene flow. The slice at 6 Ma (millions of years ago) reveals the two known (red) samples of Late Miocene hominids (from Chad and Kenya), schematized here for simplicity within the same bundle, pending additional evidence. *Au. afarensis* is (so far) sampled in the Ethiopian, Kenyan, Tanzanian, and Chadian (hidden behind the bundle) regions. The Ethiopian Afar region has yielded four named, time-successive taxa, including *Ar. ramidus* (see yellow star). The close chronologic and geographic proximity of *Ar. ramidus* and *Au. anamensis* within the Middle Awash stratigraphic succession can be accommodated in discrete stratophenetic arrangements, each with its own predictions about future fossil discoveries. **Hypothesis 1** interprets all known evidence to represent a species lineage evolving phyletically across its entire range. **Hypothesis 2** depicts the same evidence in a relatively rapid *Ardipithecus*-to-*Australopithecus* transition (speciation) occurring between ~4.5 and ~4.2 Ma in a regional (or local) group of populations that might have included either or both the Afar and Turkana rifts. **Hypothesis 3** accommodates the same evidence to an alternative, much earlier peripheral allopatric rectangular speciation model (cladogenesis through microevolution accumulated in a peripheral isolate population, becoming reproductively separated). Other possibilities exist, but at the present time, none can be falsified. To choose among them will require more fossil evidence, including well-documented transitions in multiple geographic locales. (Adapted from White et al. 2009.)

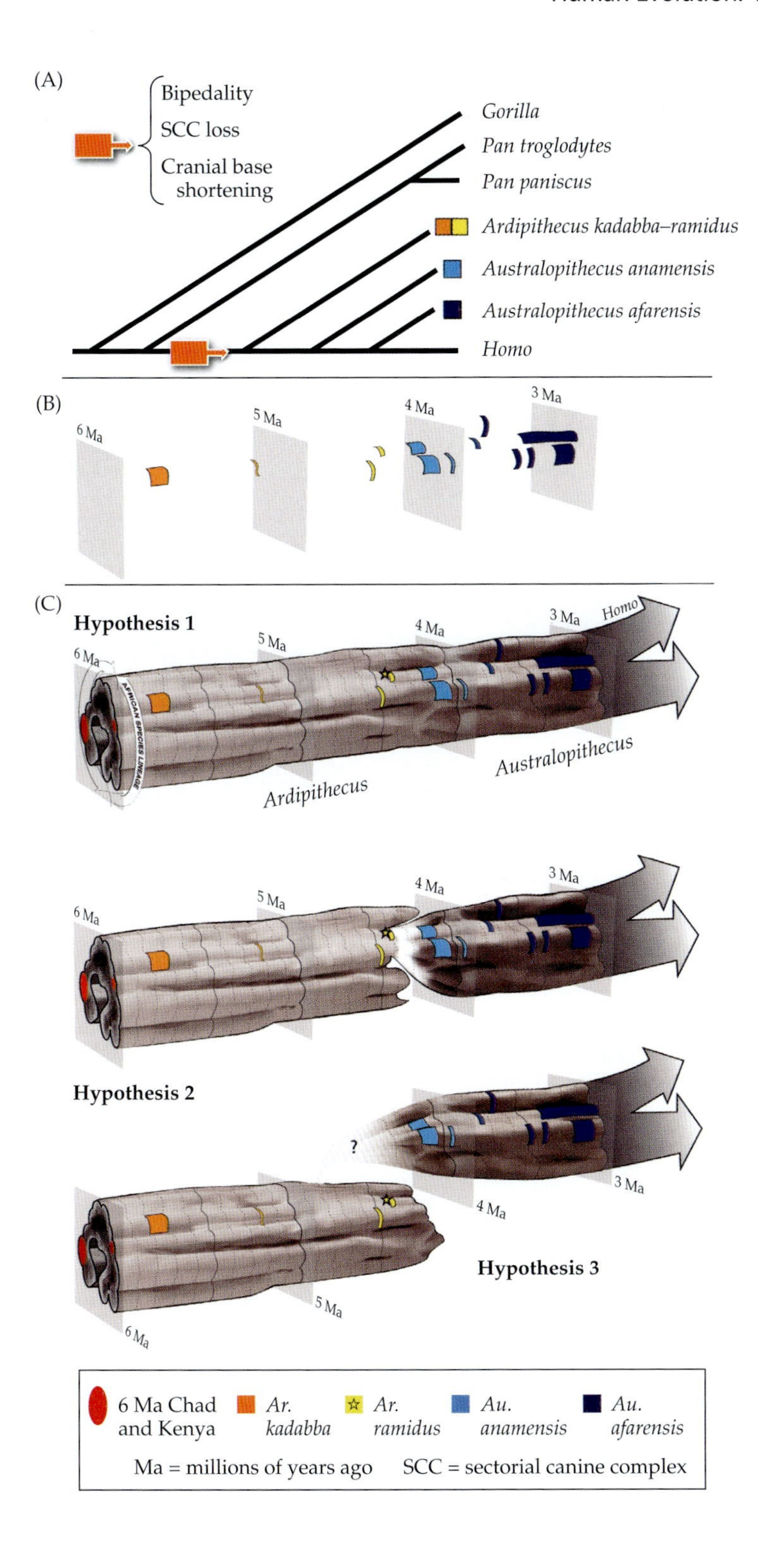
(A)
Bipedality
SCC loss
Cranial base shortening
Gorilla
Pan troglodytes
Pan paniscus
Ardipithecus kadabba–ramidus
Australopithecus anamensis
Australopithecus afarensis
Homo
(B)
6 Ma
5 Ma
4 Ma
3 Ma
(C)
Hypothesis 1
6 Ma
5 Ma
4 Ma
3 Ma
Homo
AFRICAN SPECIES LINEAGE
Ardipithecus
Australopithecus
6 Ma
5 Ma
4 Ma
3 Ma
Hypothesis 2
?
3 Ma
4 Ma
5 Ma
6 Ma
Hypothesis 3
6 Ma Chad and Kenya
Ar. kadabba
Ar. ramidus
Au. anamensis
Au. afarensis
Ma = millions of years ago
SCC = sectorial canine complex

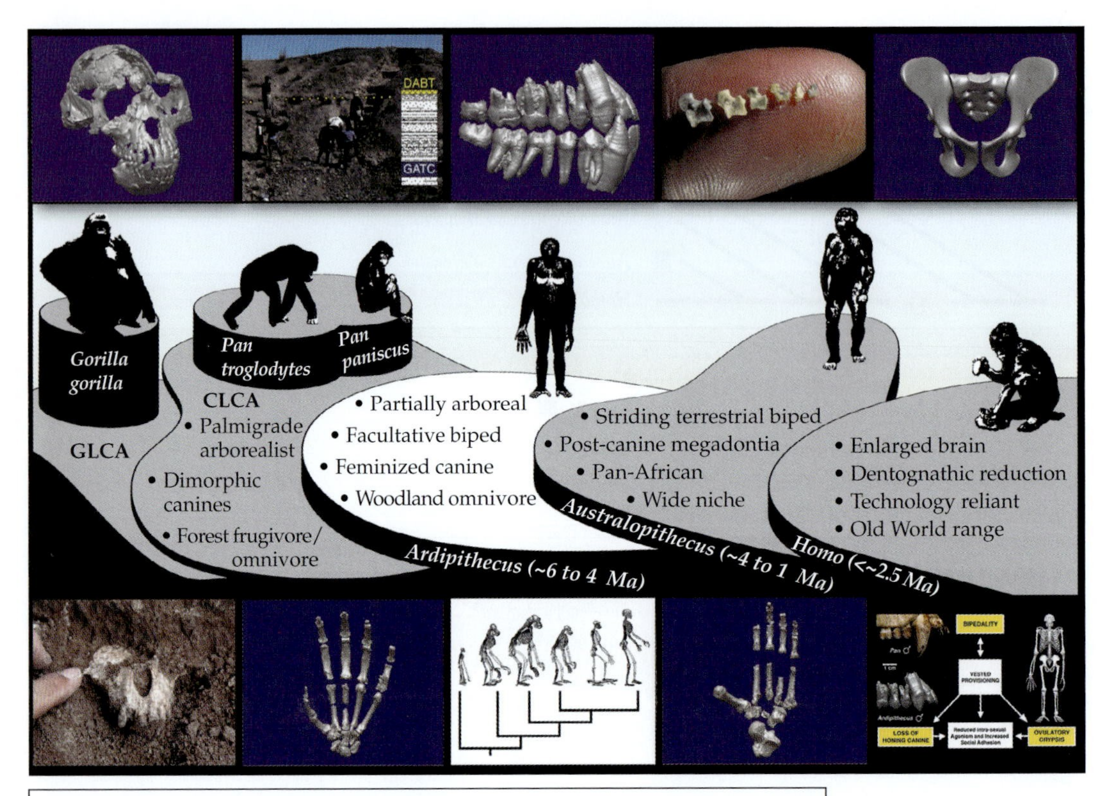

GLCA = Last common ancestor to the gorilla clade and the chimpanzee/human clade
CLCA = Last common ancestor to the chimpanzee and human clades
DABT = Overlying basaltic tuff
GATC = Underlying vitric tuff
Ma = Millions of years ago

FIGURE 19.5 The Evolution of Hominids and African Apes Since the Gorilla/Chimp + Human Last Common Ancestor (GLCA) and Chimp/Human (CLCA) Last Common Ancestor The adaptive plateaus roughly corresponding to three hominid genera are evident, in part, because of the rich contextual and skeletal evidence of *Ardipithecus ramidus*. Smaller frames above and below correspond to aspects of this species elaborated in *Science* papers containing these results. (Adapted from White et al. 2009.)

the genus *Ardipithecus* representing the first phase of hominid evolution on the human side of the CLCA (Figure 19.5).

Summarizing the Middle Awash Succession

Darwin famously wrote in his 1871 book: "Unless we wilfully close our eyes, we may, with our present knowledge, approximately recognise our parentage; nor need we feel ashamed of it" (Darwin 1871: 213). He made this claim based on the wealth of embryological, comparative anatomical, behavioral, and biogeographical evidence available to him and his contemporaries, fully mindful of the missing fossils. Towards the end of his imaginary tour of Ethiopia's Middle Awash, Darwin might have reflected that this African valley's record of hominid evolution documents three successive adaptive plateaus, with major thresholds between them. The last common ancestor that humans shared with chimpanzees in the late Miocene was probably a forest-living, folivorous/omnivorous palmigrade arborealist with dimorphic canines. Its Pliocene descendants on the hominid side (*Ardipithecus*) were partially arboreal, but terrestrially bipedal omnivores, whose males exhibited morphologically and metrically feminized canines. They lived and died in woodland habitats. From this foundation, striding terrestrial bipeds with postcanine megadontia (*Australopithecus*) expanded their niches and ranges, penetrating much of the African continent, and in turn, becoming the substrate for the emergence of lithic-technological primates. These technological hominids (*Homo*) expanded their geographic range to Eurasia as their brains enlarged, their faces shrank, and their penchant for the meat of large mammals intensified.

Human Evolution: Today's Big Picture

It was previously mentioned that Gould (2002) and others have described the hominid fossil record as "spotty." Regrettably, it is still not nearly as complete as any of us would wish. But, to a scholar used to a world in which there were a few undated Neanderthal fossils, twenty-first century paleoanthropology would constitute a paleobiological palette of riches for Darwin. He would be astonished to learn that from a single Spanish cavern there are now literally thousands of fossils from an apparently ancestral segment of the Neanderthal lineage. Precursors have been found from Africa's Cape to the Mediterranean as well as from Java to Beijing. Up and down the geological column, across the Old World, the evidence for hominid evolution is impressive today. It is remarkable how well the spatially restricted Middle Awash record accords with the global evidence of human evolution now available.

Always the generalist and always interested in amassing the complete evidentiary basis for his historical inferences, a resurrected Darwin would stand in the new antiquities research facility in Ethiopia's National Museum in Addis Ababa after visiting the Middle Awash, his mind inevitably wandering beyond that valley in Ethiopia. He would wonder how many other museums held yet more evidence. Of course, a focus on any one depository, no matter how rich its record, does no justice to the currently available wealth of data charting the course of human evolution. Table 19.2 is an attempt to identify the major events revealed so far by global hominid evolutionary studies. Darwin would probably find the wealth of information on which this distillation is based overwhelming, a mere 150 years

TABLE 19.2 MAJOR EVENTS IN HUMAN EVOLUTION

8–15+ Ma	Planet of the Apes: Barcelona to Yunnan, Namibia to Hungary; REAL species diversity; no hominids
6+ Ma	Last common ancestor with chimpanzee; Hominid clade (*Ardipithecus*) established
4.3 Ma	Adaptation to heavily masticated diets: *Australopithecus* established
2.7 Ma	*Homo* clade established Parallel megadontia in *Australopithecus* Oldowan technology and large mammal butchery
>1.8 Ma	First expansion from Africa
600 Ka	Neanderthal clade established
160 Ka	Anatomically modern *Homo sapiens*
30 Ka	Neanderthals go extinct, mini Floresians persist
After that...	Western hemisphere colonized Darwin visits Galápagos *Homo sapiens* threaten global biodiversity

Ma = million years ago.

after publication of his *On The Origin of Species* (and imagine his reaction to the twenty-first century Internet). He would also clearly recognize the dimensions of persistent debates, a few of which are considered below.

The Mode and Tempo of Hominid Evolution

The 1871 book on human evolution reveals Darwin's appreciation of the variable rate of evolutionary change. He wrote: "We are also quite ignorant at how rapid a rate organisms, whether high or low in the scale, may under favourable circumstances be modified: we know, however, that some have retained the same form during an enormous lapse of time" (Darwin 1871: 200). Darwin's ambiguity regarding the sequence of hominid evolution was matched in his considerations of the area of evolutionary mode and tempo, as later in the same book, he went back to his more usual gradualistic perspective: "The great break in the organic chain between man and his nearest allies, which cannot be bridged over by any extinct or living species, has often been advanced as a grave objection to the belief that man is descended from some lower form; but this objection will not appear of much weight... all these breaks depend merely on the number of related forms which have become extinct" (Darwin 1871: 337). However, Huxley's review of *On The Origin of Species* immediately took Darwin to task on this matter:

> *A frequent and a just objection to the Lamarckian hypothesis of the transmutation of species is based upon the absence of transitional forms between many species. But against the Darwinian hypothesis this*

argument has no force. Indeed, one of the most valuable and suggestive parts of Mr. Darwin's work is that in which he proves, that the frequent absence of transitions is a necessary consequence of his doctrine...And Mr. Darwin's position might, we think, have been even stronger than it is if he had not embarrassed himself with the aphorism, 'Natura non facit saltum,' which turns up so often in his pages. We believe, as we have said above, that Nature does make jumps now and then, and a recognition of the fact is of no small importance in disposing of many minor objections to the doctrine of transmutation (Darwin 1859: 569–570).

Well over a century later, Eldredge and Gould (1972) published "Punctuated Equilibria: An Alternative to Phyletic Gradualism." Eldredge, Gould, and Tattersall maintained a loud drumbeat against gradualism during the latter quarter of the twentieth century, and frequently employed hominid paleontology as an exemplar (White 2009a). Cain (2009) accuses Gould of "ritual patricide" against the architects of the Modern Synthesis (particularly George Gaylord Simpson), a contention that is reinforced in noting what Howells wrote for Cold Spring Harbor in 1950:

This reaffirmation rests in turn on what is new. This is largely the realization that the apes of today are just as peculiar as man is; that it is not a case of some aberrant offshoot of the ape stem giving rise to man but more one, as Schultz says, of both branches developing different potentialities inherent in the basic form… It is true that bipedal walking was the more radical line of change. As Washburn says, this was undoubtedly a case of quantum evolution, a conceptual contribution of Simpson's (1944) (Howells 1951: 82).

As I have pointed out elsewhere (White 2009a), hominid evolutionary studies have figured prominently in the debates about evolutionary tempo and mode and seem sure to continue to do so.

Hominid Diversity: Myths and Manias

Given his obsession with phylogeny, Darwin would have been eager to see the latest phylogenetic tree for the Hominidae. Here, he would have entered an arena that remains highly contentious even in the face of the avalanche of paleobiological data of the last 150 years. Given his philosophy, Darwin would have been undeterred; in a letter to Fawcett, 18 September, 1861, he wrote: "How odd it is that anyone should not see that all observation must be for or against some view if it is to be of any service!" (Darwin 1887; http://darwin-online.org.uk/).

The scholarly and economic reasons for continuing battles over the geometry of hominid phylogeny have been reviewed elsewhere (White 2009a), so an anecdote will suffice here. Before World War II, Ralph von Koenigswald collected hominid fossils in central Java by paying local peasants to collect them. Knowing that the scientist would buy the fossils they found and that von Koenigswald was particularly excited by ones that looked most like humans, the local peasants began to break the hominid fossils they found into pieces in order to maximize their profits. Today, are some practicing paleoanthropologists doing the same thing with hominid taxa? The substrate for this practice appears to be an influential popular prediction made by Gould in 1976: "We know about three coexisting branches of the human bush. I will be surprised if twice as many more are not discovered before the end of the century" (Gould 1976: 31).

A decade after Gould penned these words, Bowler (1986) termed the predictable result "taxonomic inflation." The trend has continued to the point that it pervades popular media: "the tree is a complex, tangled bush," the *New York Times* pronounced about hominid phylogeny. Even some Darwin anniversary pundits have uncritically adopted this diversity viewpoint (Pagel, 2009). The assertion of hominid diversity is rarely questioned (Jolly 2001; White 2000, 2009a, 2010).

There exist several potential hominid diversity inflators in hominid phylogenetics. Following Simpson, arbitrary names have regularly been applied to chronospecies along species lineages. Invalid temporal extensions of species extinctions and/or first appearances frequently exaggerate overlap between different hominid taxa. Finally, despite scholars from Darwin to Dawkins valuing it, there remains an under-appreciation of intraspecific variation within hominoid species (White, 2009b).

Of the 25 currently and widely recognized hominid taxa, 12 were named after 1990, and many are mere chronospecies (Figure 19.6). If we used the same approach for the geographically widespread and variable extant mammal known as the tiger, we would obscure the biology of this large Eurasian carnivore. Rather than asking, "how many more hominid taxa can we create," we should be asking why species diversity is so moderate among hominids, particularly compared to other mammals, such as Old World fruit bats (which include 173 extant species in 42 genera). Surely, it is relevant that hominids have been large, mobile, variable transcontinental terrestrial generalists since the Pliocene. At least two hominid species lineages specialized in intelligence and culture. As niche breadth widened through time, opportunities for sympatry were surely reduced, as Dobzhansky (1944) appreciated long before the Darwin centenary.

A fascinating exception that appears to conform to these ecological tenets has recently surfaced in the form of Indonesia's "Hobbits" or *Homo floresiensis* (Brown et al. 2004). With diminutive statures and braincases, the limited available evidence has been interpreted via biological pathways that both Darwin and Wallace would have appreciated (Wallace's Line today separates Flores Island from Java to the immediate west). Some workers have suggested that some kind of familial or other pathology is involved with these subfossil hominids. Others have interpreted the remains (still a single skeleton with some additional elements from other individuals) as

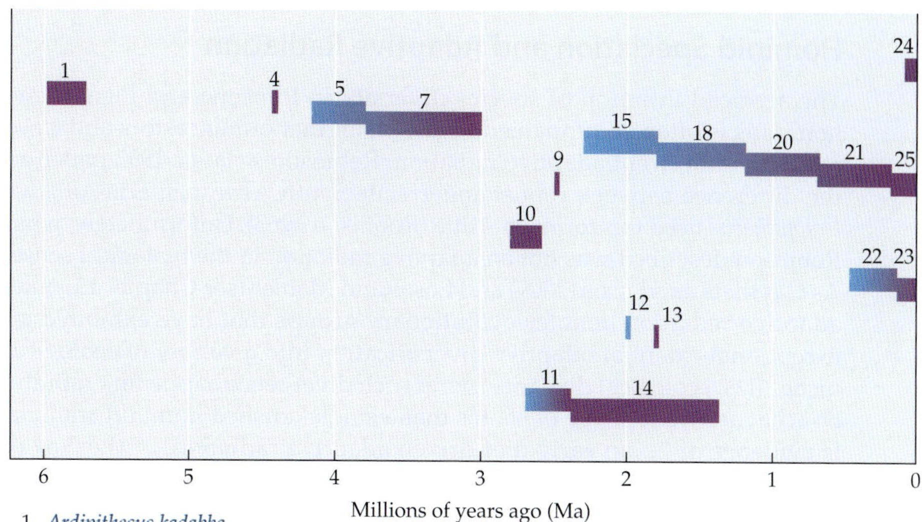

1. ***Ardipithecus kadabba***
2. *Orrorin tugenensis*
3. *Sahelanthropus tchadensis*
4. *Ardipithecus ramidus*
5. ***Australopithecus anamensis***
6. *Kenyanthropus platyops* (= #7)
7. *Australopithecus afarensis*
8. *Australopithecus bahrelghazali* (= #7)
9. ***Australopithecus garhi***
10. *Australopithecus africanus*
11. ***Australopithecus aethiopicus***
12. ***Australopithecus crassidens***
13. *Australopithecus robustus*
14. *Australopithecus boisei*
15. *Homo habilis*
16. *Homo antecessor* (= #20)
17. *Homo georgicus* (= #18)
18. ***Homo ergaster***
19. *Sinanthropus pekinensis* (= #20)
20. ***Homo erectus***
21. ***Homo rhodesiensis***
22. ***Homo heidelbergensis***
23. *Homo neanderthalensis*
24. *Homo floresiensis*
25. *Homo sapiens*

FIGURE 19.6 The Synonomy, Inferred Phylogenetic Relationships, and Temporal Distribution of 25 Hominid Species Currently in Wide Use I consider names in light blue to be junior subjective synonyms and those in dark blue to be subsequently named chronospecies. Apparent phyletic evolution between successive chronospecies is indicated by the abutting, adjacent boxes. Note that the maximum number of separate species at any particular time horizon is only three or four lineages at around 2 Ma. Because speciation is not mere evolutionary change but the multiplication of species, this is not the phylogeny of a speciose group and certainly not an adaptive radiation.

evidence of a very early departure from Africa and colonization of Flores Island by migrant *Australopithecus* or early *Homo*. Another hypothesis is that *Homo erectus* populations were isolated and underwent insular dwarfing on Flores during the Pleistocene. It is also possible that an early *Homo sapiens* immigrant population was confined to this island and—under intense selection and substantial founder effect drift—evolved rapidly into these strange hominids. It is not difficult to imagine Darwin noting the futility of speculation and inevitably calling for more paleobiological evidence.

Hominid Speciation and Adaptive Radiation

The artificial inflation of species diversity in Pliocene and Pleistocene hominids is often accompanied by the claim that our ancestors and close relatives underwent adaptive radiation. Robinson's classic 1963 paper on the difference between robust and gracile South African species of *Australopithecus* used this term in its title (Robinson 1963). Unfortunately, what Robinson described was not an adaptive radiation in the biological sense. As Gavrilets and Losos (2009) and Losos and Mahler (see Chapter 15) note, adaptive radiation denotes evolutionary groups that have exhibited an exceptional extent of adaptive diversification into a variety of ecological niches (i.e., ecological disparity), with such divergence happening rapidly. So what are we to make of claims that a single crushed hominid cranium is evidence of "...an early diet-driven adaptive radiation..." (Leakey et al. 2001: 433)? Contemporary proliferation of hominid species should not be confused with real adaptive radiation via speciation. This practice is at odds with modern thinking in evolutionary biology, and can create confusion for both students and investigators.

The best way to ascertain earlier hominid species lineage diversity is simply to consider single slices through time. At no slice through the known fossil record of this clade is there current evidence of more than four contemporary hominid species, which does not constitute an adaptive radiation, even a small one with relatively low species richness, such as Darwin's finches (Gavrilets and Losos 2009; see Losos and Mahler, Chapter 15).

Building Knowledge by Removing Imperfections: Hominid Paleobiology as a Work in Progress

Darwin's great contribution in 1859 was to identify and bestow natural selection on biology as the mechanism responsible for adaptation. In his meticulously judicious, evidence-based approach, Darwin set the standard in the life sciences for inferring changes through time, appreciating phylogenetic continuity, and evaluating evidence critically and cautiously.

In 1859, Darwin was so frustrated with the general "Incompleteness of the Geological Record" that he devoted a 33-page chapter to the topic. A little over a decade later, he bemoaned the paucity of the hominid fossil record in his 1871 volume on human evolution. Persistent incompleteness of the paleontological record continues today as a fundamental problem plaguing most of contemporary paleobiology. Surely, only an invertebrate paleontologist would pronounce (astonishingly, in a Darwin anniversary issue of *Nature*) that: "...those interested in the history of life must move beyond collecting fossils to creating models" (Erwin 2009: 282).

Although its fossil record has greatly improved, the hominid clade is still no exception to the rule of incompleteness addressed by Darwin in 1871. Indeed, it has only proven possible to test his models when the record has been rendered more complete by fieldwork. On the occasion of the 1909 anniversary of *On The Origin of the Species* publication, Seward wrote:

> *How Darwin would have rejoiced over each of the discoveries here briefly outlined! What use he would have made of the new and precious material, which would have prevented the discouragement from which he suffered when preparing the second edition of* The Descent of Man! *But it was not granted to him to see this progress towards filling up the gaps in his edifice of which he was so painfully conscious (Seward 1909: 130).*

So, how have Darwin's inferences about human evolution fared in the face of an accumulating evidentiary record? In assessing how Darwin did, it is crucial to remember that he was effectively bereft of a hominid fossil record and ignorant of the immensity of deep time. From his vantage point, he certainly saw that light would be shed. From the limited evidence available, he clearly understood that humans had evolved. He accurately cautioned against inference based on the limited data from extant species, like chimpanzees. No matter how closely related to us they might be, such organisms are not living fossils, and often constitute unsuitable proxies for ancestors. Darwin predicted transitional hominid fossils, many of which have now been recovered, particularly in the Pleistocene record. He appreciated that modern human racial diversity does not have deep phylogenetic roots (see also Richerson and Boyd, Chapter 20). He stepped ever so lightly out on a limb to predict that hominid origins were African—just before he wrote to hedge his bet that "it would be useless to speculate."

Perhaps most importantly, Darwin appreciated the unique value of fossils as evidence with which to understand life's history. Neither genomics nor comparative behavioral studies could have given us the windows on the past that provided for the evolution of tetrapods by *Tiktaalik* and of humans by *Ardipithecus*. These kinds of ongoing paleobiological discoveries show the immense power of Darwin's historical, evidence-based inferential approach to what we now call evolutionary biology.

Indeed, Darwin was committed to the use of evidence. To the dismay of those who continue to deny human evolution—and in contradiction to colleagues who have underestimated hominid paleobiology data—a massive amount of new evidence for human evolution has accumulated since the Darwin centenary of 1959. This fossil evidence today provides an enhanced view of human origins; it is the hard (and hard to get) evidence of human evolution for which Darwin longed. As the great American author (and admirer of Darwin) Samuel Clemens expressed it: "I think there is no sense in forming an opinion when there is no evidence to form it on. If you build a person without any bones in him he may look fair enough to the eye, but he will be limber and cannot stand up; and I consider that evidence is the bones of an opinion" (Twain 1896: 10).

Today, unique data from paleobiological records constitute evidence that provide a unique perspective to evolutionary biology. The continuously expanding fossil record of the hominid clade powerfully supports the most basic of Darwin's predictions. As hominid paleobiology probes at the roots and branches of the hominid clade with a wide assortment of tools, ranging from DNA sequencers to micro-CT scanners, much more about human evolution will surely be revealed. But as Lyell, Huxley, and Darwin all appreciated so well, fieldwork is the bedrock for this discipline and still is an essential means by which we understand the global one-time experiment of life.

Given the technologies that we strange bipedal primates have created during our relatively short time on this planet, Darwin's great truth about the interconnectedness of all Earth's life is today a fact with profound implications. So, Mr. Darwin, thank you for revealing to us a view of life that accurately situates our tiny clade in life's tree: it is a perspective on life that is essential to our future.

Acknowledgments

It is an honor, a privilege, and a bit of a mystery to have been invited to Stony Brook University, a campus that over the last 30 years has become a sort of collection point for people involved with questions of primate, specifically human, evolution. There are many local inhabitants of Long Island currently specializing in this topic, and now there are a few transient faculty foreigners, to boot. Any of them could and would have (differently) addressed the broader aspects of hominid evolution considered in this chapter (and might have finished their talk punctually enough to please the resident primatologists). So, I must doubly thank the organizers for presenting a somewhat less standard view, and a decidedly regional perspective from Afar on this Darwin anniversary.

I thank Mike Bell, Doug Futuyma, and all of the participants in the Stony Brook Conference for their patience. Bill Jungers provided valuable review comments. Peter Bowler provided inspiration and insightful comments and corrections. Not a historian of science, I have not ventured far beyond the published record of Darwin's thinking about human origins, but I hope that this limited effort may encourage others to dig deeply into the relevant archives. Thanks to my so-called *Ardipithecus* colleagues, Gen Suwa and Owen Lovejoy, for having provided stimulus and information at many stages. Colleagues on the Middle Awash project did the difficult, dirty, and often dangerous work of extracting the fossil record from the Afar's longest record. Leslea Hlusko endured the long process of articulating this paper and provided important comments and insights. Henry Gilbert and Josh Carlson provided graphics and bibliographic support. Thanks to Mike Bell and Lucy Anderson for their excellent edits of a manuscript left very unpolished when Ethiopian fieldwork led me astray.

Literature Cited

Asfaw, B. and W. H. Gilbert (eds.). 2008. Homo erectus*: Pleistocene Evidence from the Middle Awash, Ethiopia*. University of California Press, Berkeley.

Asfaw, B., W. H. Gilbert, Y. Beyene, and 5 others. 2002. Remains of *Homo erectus* from Bouri, Middle Awash, Ethiopia. *Nature* 416: 317–320.

Asfaw, B., T. D. White, C. O. Lovejoy, and 3 others. 1999. *Australopithecus garhi*: A new species of early hominid from Ethiopia. *Science* 284: 629–635.

Bell, G. 2010. Fluctuating selection: The perpetual renewal of adaptation in variable environments. *Phil. Trans. Roy. Soc. B:* 365: 87–97.

Bell, M. A., W. Aguirre, and N. Buck. 2004. Twelve years of contemporary armor evolution in a threespine stickleback population. *Evolution* 58: 814–824.

Bowler, P. J. 1986. *Theories of Human Evolution*. Basil Blackwell, Cambridge, MA.

Bowler, P. J. 1990. *Charles Darwin: The Man and His Influence*. Basil Blackwell, Cambridge, MA.

Bowler, P. J. 2008. What Darwin disturbed: The biology that might have been. *Isis* 99: 560–567.

Brown, P., T. Sutikna, M. J. Morwood, and 4 others. 2004. A new small-bodied hominin from the Late Pleistocene of Flores, Indonesia. *Nature* 431: 1055–1061.

Browne, J. 2002. *Charles Darwin: The Power of Place*. Knopf, New York.

Brunet, M., G. Franck, D. Pilbeam, and 35 others. 2002. A new hominid from the Upper Miocene of Chad, Central Africa. *Nature* 418: 145–151.

Cain, J. 2009. Ritual patricide: Why Stephen Jay Gould assassinated George Gaylord Simpson. In D. Sepkowski and M. Ruse (eds.), *The Paleobiological Revolution: Essays on the Growth of Modern Paleontology*, pp. 346–363. University of Chicago Press, Chicago.

Clark, J. D., Y. Beyene, G. WoldeGabriel, and 10 others. 2003. Stratigraphic, chronological and behavioural contexts of Pleistocene *Homo sapiens* from Middle Awash, Ethiopia. *Nature* 423: 747–752.

Darwin, C. 1859. *On the Origin of Species by Means of Natural Selection, or the Preservation of Favoured Races in the Struggle for Life*. John Murray, London.

Darwin, C. 1871. *The Descent of Man, and Selection in Relation to Sex*. John Murray, London.

Darwin, F. (ed.). 1887. *The Life and Letters of Charles Darwin, including an Autobiographical Chapter, vol. 3*. John Murray, London.

deHeinzelin, J., J. D. Clark, T. D. White, and 5 others. 1999. Environment and behavior of 2.5-million-year-old Bouri hominid. *Science* 284: 625–629.

Delisle, R. G. 2006. *Debating Humankind's Place in Nature, 1860–2000*. Pearson/Prentice Hall, Upper Saddle River, NJ.

Dobzhansky, T. 1944. On species and races of living and fossil man. *Am. J. Phy. Anthropol.* 2: 3.

Eldredge, N. and S. J. Gould. 1972. Punctuated equilibria: An alternative to phyletic gradualism. In T. J. M. Schopf (ed.), *Models in Paleobiology*, pp. 82–115. Freeman, Cooper and Co., San Francisco.

Erwin, D. 2009. A call to the custodians of deep time. *Nature* 462: 282–283.

Gavrilets, S. and J. B. Losos. 2009. Adaptive radiation: Contrasting theory with data. *Science* 323: 732–737.

Gingerich, P. D. 2003. Land-to-sea transition of early whales: Evolution of Eocene Archaeoceti (Cetacea) in relation to skeletal proportions and locomotion of living semiaquatic mammals. *Paleobiology* 29: 429–454.

Gould, S. J. 1976. Ladders, bushes, and human evolution. *Nat. Hist. Mag.* 85: 24–31.

Gould, S. J. 2002. *The Structure of Evolutionary Theory*. Belknap Press, Cambridge, MA.

Grayson, D. K. 1983. *The Establishment of Human Antiquity*. Academic Press, New York.

Haeckel, E. 1868. *Naturliche Schopfungsgeschichte*. Reimer, Berlin.

Haile-Selassie, Y. 2001. Late Miocene hominids from the Middle Awash, Ethiopia. *Nature* 412: 178–181.

Haile-Selassie, Y. and G. WoldeGabriel (eds.). 2009. *Ardipithecus kadabba: Late Miocene Evidence from the Middle Awash, Ethiopia*. University of California Press, Berkeley.

Haile-Selassie, Y., B. Asfaw, and T. D. White. 2004a. Hominid cranial remains from Upper Pleistocene deposits at Aduma, Middle Awash, Ethiopia. *Am. J. Phy. Anthropol.* 123: 1–10.

Haile-Selassie, Y., G. Suwa, and T. D. White. 2004b. Late Miocene teeth from Middle Awash, Ethiopia, and early hominid dental evolution. *Science* 303: 1503–1505.

Hendry, A. P., T. J. Farrugia, and M. T. Kinnison. 2008. Human influences on rates of phenotypic change in wild animal populations. *Mol. Evol.* 17: 20–29.

Howells, W. W. 1951. Origin of the human stock: Concluding remarks of the chairman. *Cold Spring Harbor Symp. Quant. Biol.* 15: 79–86.

Hughes J. F., H. Skaletsky, T. Pyntikova, and 14 others. 2010. Chimpanzee and human Y chromosomes are remarkably divergent in structure and gene content. *Nature* 463: 536–539.

Huxley, T. H. 1860. Darwin on the origin of species. *Westminster Rev.* 17 (n.s.): 541–570.

Huxley, T. H. 1863. *Man's Place in Nature*. Williams & Norgate, London.

Huxley, T. H. 1882. Charles Darwin. *Nature* 25: 597.

Johanson, D. C. and M. Taieb. 1976. Plio-Pleistocene hominid discoveries in Hadar, Ethiopia. *Nature* 260: 293–297.

Johanson, D. C. and T. D. White. 1979. A systematic assessment of early African hominids. *Science* 203: 321–330.

Jolly, C. J. 2001. A proper study for mankind: Analogies from the papionin monkeys and their implications for human evolution. *Yrbk. Phys. Anthrop.* 44: 177–204.

Kimbel, W. H. and L. K. Delezene. 2009. "Lucy" redux: A review of research on *Australopithecus afarensis*. *Am. J. Phy. Anthropol.* 140: 2–48.

Kingston, J. D. 2007. Shifting adaptive landscapes: Progress and challenges in reconstructing early hominid environments. *Yrbk. Phys. Anthrop.* 134: 20–58.

Leakey, M., C. Feibel, I. McDougall, and 1 other. 1995. New four-million-year-old species from Kanapoi and Allia Bay, Kenya. *Nature* 376: 565–571.

Leakey, M. D. and R. L. Hay. 1979. Pliocene footprints in the Laetolil Beds at Laetoli, northern Tanzania. *Nature* 278: 317–323.

Leakey, M. G., F. Spoor, F. Brown, and 4 others. 2001. New hominin genus from eastern Africa shows diverse middle Pliocene lineages. *Nature* 410: 433–440.

Lovejoy, C. O. 2009. Reexamining human origins in light of *Ardipithecus ramidus*. *Science* 326: 74.

Lovejoy, C .O., G. Suwa, S. W. Simpson, and 2 others. 2009. The great divides: *Ardipithecus ramidus* reveals the postcrania of our last common ancestors with African apes. *Science* 326: 100–106.

Lyell, C. 1863. *The Geological Evidences of the Antiquity of Man with Remarks on the Origin of Species by Variation*, 3rd ed. John Murray, London.

Marques-Bonet, T., J. M. Kidd, M. Ventura, and 17 others. 2009. A burst of segmental duplications in the genome of the African great ape ancestor. *Nature* 457: 877–881.

McBrearty, S. and G. Jablonski. 2005. First fossil chimpanzee. *Nature* 437: 105–108.

Niedzwiedzki, G., P. Szrek, K. Narkiewicz, and 2 others. 2010. Tetrapod trackways from the early Middle Devonian period of Poland. *Nature* 463: 43–48.

Noonan, J. P., G. Coop, S. Kudaravalli, and 8 others. 2006. Sequencing and analysis of Neanderthal genomic mtDNA. *Science* 314: 1113–1118.

Owen, R. 1859. On the classification and geographical distribution of the Mammalia, being the lecture on Sir Robert Reade's foundation, delivered before the University of Cambridge, in the Senate-House, May 10, 1859. To which is added an appendix "on the gorilla," and "on the extinction and transmutation of species." John Parker, London.

Owen, R. 1860. Review of *Origin* and other works. *Edin. Rev.* 111: 487–532.

Pääbo, A., J. K. Pritchard, and E. Rubin 2006. Sequencing and analysis of Neanderthal genomic mtDNA. *Nature* 444: 330–336.

Pagel, M. 2009. Natural selection 150 years on. *Nature* 457: 808–811.

Prabhakar, S., A. Visel, J. A. Akiyama, and 10 others. 2008. Human-specific gain of function in a developmental enhancer. *Science* 321: 1346–1350.

Robinson, J. T. 1963. Adaptive radiation in the australopithecines and the origin of man. In F. C. Howell and F. Bourlière (eds.), *African Ecology and Human Evolution*, pp. 385–416. F. Aldine, Chicago.

Semaw, S., P. Renne, J. Harris, and 4 others. 1997. 2.5-million-year-old stone tools from Gona, Ethiopia. *Nature* 385: 333–336.

Senut, B., M. Pickford, D. Gommery, and 3 others. 2001. First hominid from the Miocene (Lukeino Formation, Kenya). *Comptes Rendus de l'Academie des Sciences, Series IIA—Earth and Planetary Science* 332: 137–144.

Sepkoski, D. and M. Ruse (eds.). 2009. *The Paleobiological Revolution: Essays on the Growth of Modern Paleontology*. University of Chicago Press, Chicago.

Seward, A. C. (ed.). 1909. *Darwin and Modern Science: Essays in Commemoration of the Centenary of the Birth of Charles Darwin and of the Fiftieth Anniversary of the Publication of* The Origin of Species. Cambridge University Press, Cambridge, UK.

Shubin, N. H., E. B. Daeschler, and F. A. Jenkins Jr. 2006. The pectoral fin of *Tiktaalik roseae* and the origin of the tetrapod limb. *Nature* 440: 764–771.

Tax, S. 1960. *Evolution After Darwin, Vol 2: The Evolution of Man*. University of Chicago Press, Chicago.

Twain, M. 1896. *Personal Recollections of Joan of Arc by the Sieur Louis de Conte*. Harper & Brothers, New York.

Wallace, A. R. 1889. *Darwinism*. Macmillan, London and New York.

Wallace, A. R. 1895. *Natural Selection and Tropical Nature: Essays on Descriptive and Theoretical Biology*. Macmillan, London.

White, T. D. 1980. Evolutionary implications of Pliocene hominid footprints. *Science* 208: 175–176.

White, T. D. 2000. A view on the science: Physical anthropology at the millennium. *Am. J. Phy. Anthropol.* 113: 287–292.

White, T. D. 2009a. Ladders, bushes, punctuations, and clades: Hominid paleobiology in the late twentieth century. In D. Sepkoski and M. Ruse (eds.), *The*

Paleobiological Revolution: Essays on the Growth of Modern Paleontology, pp. 122–148. University of Chicago Press, Chicago.
White, T. D. 2009b. Human origins and evolution: Cold Spring Harbor, *déjà vu*. *Cold Spring Harbor Symp. Quant. Biol.* 74: 1–10.
White, T. D. 2010. Q and A: Tim White. *Curr. Biol.* 20: R6–R8.
White, T. D., B. Asfaw, Y. Beyene, and 4 others. 2009. *Ardipithecus ramidus* and the paleobiology of early hominids. *Science* 326: 75–86.
White, T. D., B. Asfaw, D. DeGusta, and 4 others. 2003. Pleistocene *Homo sapiens* from Middle Awash, Ethiopia. *Nature* 423: 742–747.
White, T. D., G. Suwa, and B. Asfaw. 1994. *Australopithecus ramidus*, a new species of early hominid from Aramis, Ethiopia. *Nature* 371: 306–312.
White, T. D., G. WoldeGabriel, B. Asfaw, and 19 others. 2006. Asa Issie, Aramis, and the origin of *Australopithecus*. *Nature* 440: 883–889.
Wray, G. A. and C. C. Babbitt. 2008. Enhancing gene regulation. *Science* 321: 1300–1301.
Yellen, J., A. Brooks, D. Helgren, and 8 others. 2005. The archaeology of Aduma Middle Stone Age Sites in the Awash Valley, Ethiopia. *PaleoAnthropology* 10: 25–100.

Chapter 20

The Darwinian Theory of Human Cultural Evolution and Gene–Culture Coevolution

Peter J. Richerson and Robert Boyd

Darwin realized that his theory could have no principled exception for humans. He put the famous teaser, "Light will be thrown on the origin of man and his history," near the end of *The Origin of Species*. If his evolutionary account made an exception for the human species, the whole edifice might be questioned. As the *Quarterly Review*'s reviewer of *The Descent of Man and Selection in Relation to Sex* (probably the long-hostile and devoutly Catholic St. George Mivart) gloated, the *Descent* "offers a good opportunity for reviewing his whole position"—and rejecting it (Anonymous or St. George Mivart, 1871).

Darwin apparently hoped someone else would apply Darwinism to the origin of humans. Lyell (1863), Huxley (1863), and Wallace (1864, 1869) all wrote on the subject, but their work was unsatisfactory because all three had reservations about a selectionist account of human mental evolution. Darwin's views on the origin of humans did not rely entirely on selection but had a supplementary set of mechanisms consistent with a selectionist account.

Darwin eventually wrote *The Descent of Man*, a rich and sophisticated treatment of evolution, even by contemporary standards. Yet, during his lifetime, Darwin's treatment of evolution generally and of human evolution specifically had many competitors (Bowler 1988). For example, Herbert Spencer and Darwin debated the relative importance of natural selection and inheritance of acquired variation (or acquired characteristics) for evolution of the mind (Richards 1987). Each admitted that both processes were important, but Darwin thought selection was dominant, and Spencer favored inheritance of acquired variation, substantially discounting the importance of selection. Furthermore, Spencer's emphasis of acquired variation reflected his belief that evolution was the "never-ceasing transformation of the homogeneous to the heterogeneous" (Richards 1987), while Darwin was ambivalent about progress.

Despite his prestige, Darwin convinced few of his contemporaries that he had the correct theory for the origin of the human mind. He most strongly influenced the pioneering psychologists Romanes, Morgan, James, and Baldwin, but their importance in psychology waned after the turn of the century (Richards 1987). No twentieth-century social scientist was significantly influenced by *The Descent of Man*, and eminent social scientists are hostile to Darwinism, to this day. How could a theory generate so much controversy, yet for over a century, fail to attract enough critical work to test its worth? Can a satisfactory theory of the evolution of human behavior along Darwinian lines be fleshed out, or is the endeavor fatally flawed?

The first part of this essay is an attempt to understand what sort of theory of human cultural evolution Darwin proposed in *The Descent of Man*, which is difficult for two reasons. Although, Darwin wrote clearly, he lacked important theoretical tools, especially genetics. Believing in the inheritance of acquired variation and of habits, as a special case of it, the modern distinction between genetic and cultural evolution was foreign to him. Yet, he distinguished more conservative traits that had evolved in "primordial times" from those that had influenced more recent evolution of civilizations. For the latter, Darwin often evoked cultural explanations, though he seldom used that word in its modern technical sense. Second, scientific readers of such an iconic scientist are liable to exploit his ambiguities, citing him selectively to favor their own agenda. Any reading of Darwin is certainly influenced by our own theory of gene–culture coevolution, which we sketch in the second part of the essay.

Darwin's Problem with Humans

The Early Notebooks

Darwin's early M and N notebooks on *Man, Mind and Materialism* clarify the importance the human species played in his thinking about evolution (Gruber and Barrett 1974). In 1838, he wrote, "Origin of man now proved.—Metaphysics must flourish.—He who understand baboon would do more toward metaphysics than Locke" (Gruber and Barrett 1974: 281). These words were written during Darwin's most creative period, shortly *before* his first clear statement of natural selection in his notebook on *The Transmutation of Species*. The passage expresses hopeful enthusiasm rather than triumph. He was actively pursuing a materialistic theory of evolution and was convinced that humans would be included. Given the scope of the theory, it could hardly be otherwise. The promise and perils of understanding human origins and behavior remained unavoidable parts of Darwin's agenda. If correct, evolutionary theory could provide powerful tools to understand human behavior, and if humans were not understandable in the terms Darwin set out, perhaps there were deep, general problems with his theory.

Who Would Address Human Evolution?

Darwin knew that most of his contemporaries considered his theory to be dangerously radical, and he long delayed publication of even the biological part of it (Gruber and Barrett 1974). He waited a dozen years after *The Origin of Species* to fulfill his promise to discuss humans. In the Introduction to the first edition of *The Descent of Man*, he discussed his fear that publishing his views on the subject would inflame prejudices against his theory.

Natural selection is a micro-scale process in which local environments favor variants within local populations. It is not an obvious candidate for a process to generate macroevolution. As Darwin confided to his N notebook in 1838:

> *Man's intellect is not become superior to that of the Greeks (which seems opposed to progressive development) on account of the dark ages.—Look at Spain now.—Man's intellect might well deteriorate.-((effects of external circumstances)) ((In my theory there is no absolute tendency to progression, excepting from favorable circumstances!)) (Gruber and Barrett 1974: 339).*

We, along with others, assert that Darwin's skepticism about evolutionary progress and his failure to incorporate it into his theory of human evolution were major reasons why his theory was not popular among his contemporaries (Bowler 1986). Even Huxley favored Spencer's account of acquired inheritance and progress in critical respects (Bowler 1993), though Darwin speaks elsewhere more favorably of progress (Richards 1988). Ambivalence toward progress has concerned evolutionary biologists to the present day (Nitecki 1988) and certainly was not the evolutionary motor for Darwin that it is for Spencer, as the N notebook quotation shows.

Why were Darwin's contemporaries so keen on progressive theories of evolution? Almost all Victorians feared the direction in which a thoroughly Darwinian theory of human origins would lead. As the *Edinburgh Review's* commentator on *The Descent of Man* remarked:

> *If our humanity be merely the natural product of the modified faculties of brutes, most earnest-minded men will be compelled to give up those motives by which they have attempted to live noble and virtuous lives, as founded on a mistake… (Anonymous, or W. B. Dawkins, 1871: 195).*

According to Burrow (1966), a significant segment of Victorian society was skeptical about traditional religion and enthusiastic about evolution. Even the idea that humans were descended from apes did not bother these secular, Christian intellectuals. However, they did believe that human morality required natural laws. If God's Law were dismissed by the scientific as superstition, then it was crucial to find a substitute in natural laws. Spencer's law of progress included the moral sphere, and he willingly drew moral norms from his theory (Richards 1987: 203–213). His theory filled the bill, while Darwin's was ambiguous.

Darwin's Argument

In many respects, Darwin's *The Descent of Man* is more typical of the late twentieth century than of Victorian times. Because progress was not the centerpiece, he did not rank human minds or their moral intuitions on a primitive–advanced scale. The extent to which Darwin subscribed to what is now called the doctrine of psychic unity is often overlooked. Even otherwise knowledgeable scholars believe that Darwin shared the widespread Victorian belief that the living races could be ranked on a primitive–advanced scale (Ingold 1986). Bowler (1993: 70) remarks "*The Descent of Man* takes racial hierarchy for granted and cites the conventional view that whites have a larger cranial capacity than other races." But Darwin's (1874: 81) discussion of the cranial data, some probably influenced by racist preconceptions, is tempered by a footnote, citing Paul Broca's hypothesis that civilization should select for smaller brain size due to the preservation of weak minded individuals who ought otherwise to have been eliminated by the hard conditions of uncivilized life. Darwin also refers in the same passage to the then-recent Neandertal find and to another archaeological sample showing that some ancients had very big brains. Darwin certainly draws no conclusions about racial hierarchy from these data. Alexander Alland (1985) cited Stephen Jay Gould to claim that Darwin shared the typical Victorian idea that the dark races are primitive. This reading of Darwin is mistaken! Even such knowledgeable authors as Desmond and Moore (2009) misunderstand *The Descent of Man* in this regard.

Darwin's first published views on humans appeared in his *Journal of Researches (Voyage of the Beagle)* several years after he had first formulated natural selection but more than a decade before *The Origin of Species* and 25 years before *The Descent of Man* were published. His descriptions of the Fuegans in the *Journal* are often cited as evidence of his typical Victorian views on racial hierarchy. He did use purple Victorian prose to describe the wretched state of the Fuegans, whom he had observed on the *Beagle*:

> *These poor wretches were stunted in their growth, their hideous faces bedaubed with white paint, their skins filthy and greasy, their hair entangled, their voices discordant, and their gestures violent. Viewing such men, one can hardly make one's self believe that they are fellow-creatures, and inhabitants of the same world (Darwin 1845: 243).*

He goes on at length in this fashion, but this is the bait, not the hook of his argument. The passage on the Fuegans begins by describing the environmental rigors of Tierra del Fuego and ends by attributing the low nature of the people to their poor surroundings rather than to inherently primitive qualities:

> *We were detained here for several days by the bad weather. The climate is certainly wretched: the summer solstice was now past (passage is dated December 25) yet every day snow fell on the hills, and in the valleys there*

was rain. The thermometer generally stood at 45° but in the nights fell to 38° or 40° (Darwin 1845: 242).

Darwin continued:

While beholding these savages, one asks, whence could they have come? What could have tempted, or what change compelled a tribe of men, to leave the fine regions of the North, to travel down the Cordillera or backbone of America . . . and then to enter on one of the most inhospitable countries within the limits of the globe?. . . (W)e must suppose that they enjoy a sufficient share of happiness, of whatever kind it may be, to render life worth living. Nature by making habit omnipotent, and its effects hereditary, has fitted the Fuegans to the climate and the productions of this miserable country (Darwin 1845: 246–247).

The argument is consistent with his idea that progress could come only from favorable circumstances. Thus, he is saying that any humans forced to live under such conditions, with such limited technology, would soon behave similarly. Note the reference to hereditary habits; this concept figures large in his mature ideas on human evolution.

Darwin routinely condemned White Christians' morals, for example, when he discussed slavery and the genocidal Argentinean war against the Patagonian natives in the *Journal* (Darwin 1845). He ends the story of an Indian's daring escape:

What a fine picture one can form in one's mind—the naked, bronze-like figure of the old man with his little boy, riding like a Mazeppa [Ukrainian Cossack hero] on a white horse, thus leaving far behind him the host of his pursuers! (Darwin 1845: 124).

His paean against slavery begins:

On the 19th of August, we finally left the shores of Brazil. I thank God I shall never again visit a slave country. To this day, if I hear a distant scream, it recalls with vivid painfulness my feelings when, passing a house near Pernambuco I heard the most pitiable moans, and could not but suspect that some poor slave was being tortured, yet knew that I was as powerless as a child even to remonstrate (Darwin 1845: 561–563).

And ends:

It makes one's blood boil, yet heart tremble, to think that Englishmen and our American descendants, with their boastful cry of liberty, have been and are so guilty: but it is a consolation to reflect that we have made a greater sacrifice than ever made by any nation to expiate our sin (Darwin 1845: 561–563).

Gruber and Barrett (1974) note that Darwin and his entire family shared a deep antipathy to slavery, which was not widely accepted by his contemporaries. This led, for example, to a furious argument with Captain Fitzroy on the *Beagle*. Darwin certainly thought moral progress was possible and that Europeans had achieved some notable advances, counting among them the rule of law and the enactment of just laws, including the end of slavery in the British Empire (Britain freed all colonial slaves in 1838). Darwin's view of progress does not imply a racial hierarchy that would justify extermination or enslavement of the so-called lower races by the higher! We find it odd that contemporary social scientists fail to recognize that Darwin's politics, while not often as explicit as in his views on slavery, were far to the left for his day and not so different from those of today's academic left (Sulloway 1996; Desmond 1989; Richards 1987).

Of course, Darwin's (1874) best efforts on human evolution and behavioral diversity appear in his mature work, *The Descent of Man*. In Chapters 3 and 4, both entitled "Comparison of the Mental Powers of Man and the Lower Animals," he summarizes the issue: "There can be no doubt that the difference between the mind of the lowest man and the highest animal is immense" (Darwin 1874: 170). In these chapters, he struggles with the problem posed by this gap for the gradual emergence of humans from apes. His task would have been easier if he had filled the gap with the living human races, as so many of his contemporaries did. He solved the problem by proposing hypothetical continuities across the gap (as in the *Expression of Emotion in Man and Animals*) rather than bridging it with living savages.

Darwin's argument is clear in Chapter 5 of *The Descent of Man*, "On the Development of the Intellectual and Moral Faculties During Primeval and Civilized Time." In contrast to Wallace, Darwin believed that natural selection produced human mental and social capacities in *primeval* times. In particular, he posits that the foundation of human morals arose by selection on group differences:

> *It must not be forgotten that although a high standard of morality gives but a slight or no advantage to each individual man and his children over other men of the same tribe, yet that an increase in the number of well-endowed men and an advancement in the standard of morality will certainly give an immense advantage to one tribe over another. A tribe including many members who, from possessing in a high degree the spirit of patriotism, fidelity, obedience, courage, and sympathy, were always ready to aid one another, and to sacrifice themselves for the common good, would be victorious over most other tribes; and this would be natural selection (Darwin 1874: 178–179).*

He credits "savages" with sufficient loyalty to their tribes to motivate self-sacrifice to the point of death and with the "instinct" of sympathy. His objective was to ensure that the reader understands that these moral "sentiments and faculties" are primeval and shared by the living savage and the civilized alike. The importance of sympathy in Darwin's evolutionary ethics is noteworthy, as is its role in his detestation of slavery. He cites Adam Smith's (1790) *Theory of Moral Sentiments* on the importance of sympathy. He emphasizes customs acquired by imitation to explain further moral advances. He spends several pages reviewing the tendency of advancing civilization, if anything, to weaken natural selection, and the evidence that retrogressions are common. He summarizes the argument regarding morality:

> *I have already said enough, while treating of the lower races, on the causes which lead to the advance of morality, namely the approbations of our fellow men-the strengthening of our sympathies by habit—example and imitation—reason—experience, and even self-interest—instruction during youth, and religious feelings (Darwin 1874: 185–186).*

And, in current circumstances:

> *With highly civilized nations, continued progress depends in a subordinate degree on natural selection. . . . The more efficient causes of progress seem to consist of a good education during youth while the brain is impressible, and of a high standard of excellence, inculcated by the ablest and best men, embodied in the laws, customs, and traditions of the nation, and enforced by public opinion (Darwin 1874: 192).*

Darwin's list of moral advances by the "lower races" is virtually identical to those of the highest. *Primeval* evolution endowed living savages with the same social instincts as civilized men, enabling the same improvement from "good education" and the rest. Darwin attributed the progress of morality to the strengthening of our sympathies. As we have seen in the *Journal of Researches,* on the basis of sympathy, he indicted Argentineans for killing Indians and Brazilians and Americans for holding slaves, while others defended such acts on the basis of loyalty to one's race (Richards 1987: 599–601).

The climax of Darwin's argument is in *The Descent of Man,* in Chapter 7 "On the Races of Man." He considers the two hypotheses that the races are sufficiently distinct to be different species or are alike in all-important organic respects. First, he outlines all the evidence in favor of the different species hypothesis (Darwin, 1874: 224–231)—his dispassionate tone allows the careless reader to believe that Darwin favors this alternative. However, he immediately demolishes the separate species argument in favor of the trivial differences alternative:

> *Although the existing races differ in many respects, as in color, hair, shape of the skull, proportions of the body, etc., yet, if their whole structure be taken into consideration, they are found to resemble each other closely on a multitude of points. Many of these are so unimportant or of so singular a nature that it is extremely improbable that they should have been independently acquired by aboriginally distinct species or races. The same remark holds good with equal or greater force with respect to the numerous points of mental similarity between the most distinct races of man. The*

American aborigines, Negroes, and Europeans are as different from each other in mind as any three races that can be named; yet I was constantly struck, while living with the Fuegans on board the "Beagle," with the many little traits of character showing how similar their minds were to ours; and so it was with a full-blooded Negro with whom I happened once to be intimate (Darwin 1874: 231–240).

The contrast between Darwin and others, like Ernst Haeckel, who really believed that "natural men are closer to the higher vertebrates than to highly civilized Europeans," was stark (Richards 1987: 596). The behavior of people in various places does differ substantially. Darwin's explanation of these differences in Chapter 7 of *The Descent of Man* downplays differences that people today tend to attribute to genes in favor of those attributed to culture:

He who will read Mr. Tylor's and Sir J. Lubbock's interesting works can hardly fail to be deeply impressed with the close similarity between the men of all races in tastes, dispositions and habits (Darwin 1874: 238).

Darwin's favorable citations of Tylor, the founder of cultural anthropology and an important nineteenth century defender of the Enlightenment doctrine of the psychic unity of all humans, are surely significant and consistent with his sympathy for savages and slaves. Tylor's (1871) postulate of the organic similarity but difference in customs is clear: "For the present purpose it appears both possible and desirable…to treat mankind as homogeneous in nature, though placed in different grades of civilization" (Tylor 1871: 7).

Darwin sometimes uses identical contrast:

As it is improbable that the numerous and unimportant points of resemblance between the several races of man in bodily structure and mental faculties (I do not here refer to similar customs) should all have been independently acquired… (Darwin 1874: 239).

Here Darwin suggests that the main explanation for differences between races is customs, not organic differences; though seeing heritable variation everywhere, Darwin would not treat humans as homogeneous. Darwin's position is difficult to understand because of his concept of inherited habits (see previous quote about the Fuegans). In the preface to *The Descent of Man's* second edition, Darwin reiterated his commitment to the inheritance of acquired variation:

I may take this opportunity of remarking that my critics frequently assume that I attribute all changes of corporeal structure and mental power exclusively to natural selection of such variation as are often called spontaneous; whereas, even in the first edition of the "Origin of Species," I distinctly stated that great weight must be attributed to the inherited effects of use and disuse, with respect both to the body and the mind (Darwin 1874: 3–4).

Darwin considered "inherited habits" to be among the most important forms of the inherited effects of use and disuse. Custom, good education, imitation, example of the best men, and his other versions of culture would tend to become hereditary. He divides traits into more and less conservative poles. Anatomy and basic features of the mind would be influenced little by the inheritance of acquired variation and mainly by selection over long spans of time; in modern terminology, these traits are highly heritable. The more labile traits are quite sensitive to environmental and cultural influences. Although they become inherited, they are susceptible to rapid remodeling by acquired habit. This feature of his theory is generally erroneous, though the discovery of transgenerational transmission of epigenetic elements has given this old idea new legs (Griffiths 2003; Jablonka and Raz 2009). Indeed, some cultural traits are relatively conservative (Nisbett 2003) and fit Darwin's concept of inherited habits. For example, historical relationships can be traced back 6000–8000 years in languages (Atkinson and Gray 2006). Similarly, many Holocene stone tool assemblages are dominated by simple expedient flakes struck from cores, as were the earliest known stone tools about 2.6 million years ago (Haines et al. 2004). The genetics of the early twentieth century were more conceptual than practical, as illustrated by its intractable quarrels over nature versus nurture. Contemporary advances in genomics are furnishing the tools to solve the difficult riddle of the interactions between inheritance and development.

Another problem with interpreting Darwin arises from his determination to minimize qualitative differences between humans and other animals. He was wary that his theory would encounter unbridgeable gaps that would imply unique processes applying only to human evolution, inevitably creating problems for his general theory. *The Descent of Man* reports many observations of animal behaviors requiring nearly human moral and intellectual faculties. Modern behavioral research shows that Darwin exaggerated the capacities of animal minds.

In particular, Darwin imagined that the ability of animals to modify their behavior by imitating others is similar to that of humans. Darwin illustrated this point with his own observations of bees imitating each other. When he observed bumble bees cutting holes in the sides of bean flowers to steal nectar, he attributed later observations of honeybees using the same technique to imitation, with the bees using mental faculties analogous to humans (Galef 1988). Galef and other modern students of animal imitation have demonstrated such effects in many vertebrates and even some invertebrates, but nothing that approaches human capabilities for the cumulative cultural evolution of complex traditions (Whiten and van Schaik 2007). Young children are considerably better imitators than adult hand-reared chimpanzees, with a long history of being rewarded for human-like behaviors (Tomasello 1996; Whiten and Custance 1996). Accurate imitation revolutionizes social learning by increasing the number and sophistication of behaviors acquired (Boyd and Richerson 1996). Individual learning is relatively slow, costly, and produces behaviors only as sophisticated as one animal can invent. Accurate imitation permits cultural sophistication to increase through a succession of innovators, each contributing to the gradual cultural evolution of complex behaviors. Even technology as simple as a stone-tipped spear contains numerous design components that evolved step-by-step. Thus, sophisticated imitation is part of the "great gulf" that Darwin admitted separates humans from their living hominid relatives. To bridge the great gulf, Darwin tended to raise up non-humans, not cast down human races. While he is sometimes said to have biologized humans, he is more accurately guilty of having humanized animals.

The Twentieth Century

Darwin's Evolutionary Theory Used in Biology and Neglected in the Social Sciences

After several decades in "eclipse" (Bowler 1983), Darwin's ideas on evolution were incorporated into the foundations of biology during the first half of the twentieth century. By contrast, most leaders of the emerging social sciences virtually ignored ideas in *The Descent of Man* (Ingold 1986). In psychology and economics, for which detailed histories have been written, this was due largely to idiosyncratic events in the careers of the most prominent Darwinians (Hodgson 2004; Richards 1987). The so-called Social Darwinism that influenced turn-of-the-century sociology and anthropology was thoroughly Victorian in its moral naturalism and progressivism, as the confident recommendations for social policy of its followers indicate. Social Darwinism was more in the spirit of Spencer than of Darwin. Most sociologists and anthropologists rejected Social Darwinism, probably because they found its political uses abhorrent (Hofstadter 1945; Campbell 1965), although Bannister (1979) argues that Hofstadter's famous critique of Social Darwinism substantially mythologizes it. Myth or truth, other pioneers in the social sciences were eager to differentiate their disciplines from biology and downplay its significance. For example, the pioneering student of imitation, Tarde (1903), excluded "biological" considerations in his theory and was apparently unaware of the parallels between his ideas and those in *The Descent of Man*. Still, in 1900, the early psychologists William James, Lloyd Morgan, and James Baldwin all espoused Darwinian theories of psychology (Richards 1987).

Baldwin (1895) particularly reconciled Darwinian concepts with new findings in biology. First, he elaborated a complex theory of imitation based on observations of his own children, noting the emergence of powerful capacities for imitation in late infancy. Second, 5 years before the rediscovery of Mendel, Baldwin drew a sharp distinction between the "machinery of heredity" and imitation:

> *(T)here is instinctive tendency to functions of the imitative type and to some direct organic imitations; but those clear conscious imitations which represent new accommodations and acquirements are not as such instinctive, and so come later as individual acquirements (Baldwin 1895: 294).*

Third, Baldwin envisioned a complex interplay between biology and imitation, as the previous quote suggests. The capacity for imitation is inherently biological and emerges late in the child's first year, as more detailed recent studies have shown (Tomasello 1999, 2008). Although the impulse to imitate may override pleasure and pain, these biologically based perceptions typically affect behavior and what others imitate. In contrast, learned or imitated behaviors could lead humans (and other animals with adaptive phenotypic flexibility) to persist in environments to which they are poorly adapted biologically. Subsequent natural selection acting on heritable variation can eventually transform learned or imitated behavior into innate behavior. Baldwin termed this effect "organic selection," and it is generally known today as the "Baldwin Effect." It was actually discovered independently by Baldwin, T. Hunt Morgan, and Henry Fairfield Osborn (Richards 1987).

As will be seen, all of Baldwin's ideas resonated with late twentieth century theories of gene–culture coevolution. However, they had no immediate impact on the rapidly forming social sciences because few social scientists espoused a Darwinian perspective. The sociologist Albert G. Keller taught a version of social evolution that was truer to the Darwinian tradition than to that of his mentor Sumner, but his influence was negligible (Campbell 1965). The theories of some very influential pioneers contained elements of Darwinian processes, but they were not subsequently developed. For example, Turner (1995) argues that Durkheim's (1893) theory of the division of labor was highly Darwinian at its root, but this feature stimulated no subsequent interest.

Specific research agendas developed with the new social sciences after the turn of the century. For example, psychologists strove to sever their roots from philosophy and embrace rigorous experimental methods. Baldwin, who was a good experimentalist and observationalist in his younger days, always had a large philosophical agenda and turned increasingly toward philosophy, while his younger colleagues turned sharply in the opposite direction. His influence waned when his arrest in a Black house of prostitution ruined his career in the United States (Richards 1987). In anthropology, Franz Boas opposed all forms of theorizing, arguing that field workers should have minimal preconceptions when collecting ethnography (Harris 1979). Similarly, institutional economists, particularly Thorstein Veblen, at the turn of the twentieth century were sophisticated evolutionists, but their influence waned and was finally extinguished (Hodgson 2004). Veblen's later career was also plagued by scandal. He had an unhappy and stormy marriage punctuated by several affairs and eventually a divorce. These contretemps cost him faculty positions at the University of Chicago and Stanford University. Physics rather than biology became the science for economists to emulate.

Another problem was that Darwin's view of evolution ebbed in the early twentieth century (Bowler 1983), just as the social sciences were emerging. When the pioneering geneticists at first discovered mutations with large effects, they rejected the Darwinian principle that selection acts mostly on continuous variation. Darwinian theory and genetics were not reconciled until Ronald Fisher's famous paper in 1918 validated the importance of Darwinian concepts in evolutionary biology (Provine 1971).

The social and biological sciences continued to diverge until mid-twentieth century. Their relationship was largely limited to sterile nature–nurture debates (Cravens 1978). Attempts to heal or at least minimize this rift included Dobzhansky and Montagu's (1947) influential argument that biology produced the substratum for human culture and that culture and biology form a unique, transcendent coevolving complex. Dobzhansky's (1962) book *Mankind Evolving* expands on this theme without ever really specifying how the coevolution works or precisely what transcendence means in this context. The position taken by Dobzhansky and Montagu (1947) was sort of a peace treaty that allowed the biological and social sciences to independently pursue their own agendas, ignoring whatever inconsistencies arose. Individuals such as Lorenz (1966) and Jensen (1969), who broke this peace in the 1950s and '60s, were unsophisticated theorists trapped in the nature–nurture debate. Evolutionary thinkers in the social sciences, such as White (1959), Carneiro (1967), and Lenski and Lenski (1982), remained wedded to the progressive evolutionary theories of Spencer. In essence, no one in the mid-twentieth century followed up late nineteenth century ideas implicit in *The Descent of Man* to create a sophisticated theory of cultural evolution that incorporated the Darwinian theory of genetic evolution. Unification of the social sciences with each other and with biology remains a work in progress (Gintis 2007).

DONALD CAMPBELL'S "VICARIOUS FORCES" Donald Campbell, a polymath psychologist, was the first twentieth-century scientist to tackle the problem of cultural evolution seriously. He made three important arguments. First, in several papers (e.g., Campbell 1960), he proposed that all knowledge processes are related to organic evolution, captured by the idea, "blind variation and selective retention." Campbell (1965) introduced the concept of vicarious forces to characterize the relationship between organic evolution by natural selection and individual learning. Assuming inheritance of acquired variation, psychological forces would shape cultural variation, much as Darwin thought sympathy molded moral progress. Campbell contended that these forces vicariously act as surrogates for natural selection, because they arose by natural selection to shape phenotypes adaptively, as Darwin had argued. Second, Campbell (1965) argued that Darwinian theory should apply to any system of inheritance, including culture. Third, his (1975) article carefully distinguished Darwinian from progressive evolution, showing that there is no material basis for a concept of progress. He asserted that progressive evolutionary theory was simply a description of history without any causal process, given that Spencer's homogeneity-to-heterogeneity mechanism had been discarded.

Campbell's approach emphasizes the interplay of forces that drive cultural evolution. In genetic evolution, the most important forces to change gene frequencies are mutation, genetic drift, gene flow, and natural selection, making organic evolution a process based purely on random variation and selective retention. Cultural evolution must be subject to analogs of these forces as well as to several vicarious forces. Not only are humans subject to natural selection, but they also change their culture by making choices as they learn. Some of the rules for making choices are inherited genetically, and then affect cultural evolution. For example, the distribution of sensory receptors in the nose and mouth may influence whether potential food items are considered pleasant or noxious. Choices of food items by individuals will in turn drive the evolution of a society's cuisine. One might expect vicarious selectors for food items to favor nutritious, healthful diets, because they have been shaped by natural selection, but they may be exploited by items like addictive drugs or overridden by cultural preferences (e.g., the consumption of peppers, which stimulate pain sensors). Culture might also drive organic evolution; for instance, adult milk-sugar (lactose) digestion has evolved within the last few thousand years in milk-consuming human populations (Simoons 1978). Simoons apparently discovered the tip of an iceberg (Hawks et al. 2007), as agriculture has greatly changed the diets, disease exposure, and social life of many human populations, driving a major coevolutionary response over the past 10,000 years.

Campbell forcefully reintroduced Darwinian ideas to social scientists after a lapse of some 60 years. He did not trace specific parallels in his scheme to Darwin and was apparently unaware of the late nineteenth century Darwinian social scientists, such as Baldwin and Veblen. Subsequently, several other evolutionary research programs applied to humans followed pioneering contributions by Cavalli-Sforza and Feldman (1973), E. O. Wilson (1975), and Alexander (1979). They became heavily politicized in the famous sociobiology debate (Segerstråle 2000). Today, Darwinian social science is tolerated, if reluctantly in some quarters, and is a very active field spread across the social sciences and human biology (Laland and Brown 2002). The importance of cultural evolution in human organic evolution is an important debate in the field. One view is that selection ultimately falls on genes and thus, that genetic vicarious selectors strongly constrain cultural evolution. Another view, championed most effectively by Susan Blackmore (1999), is that "memes" are cultural parasites that have driven the evolution of human brains and other genetically coded aspects related to these parasites' support. A third view, to which we subscribe, is that in humans, genes and culture play more or less equally important roles in a coevolutionary system.

A Theory of Gene–Culture Coevolution

Darwinian Principles Applied to Cultural Evolution

The way that culture might make us theoretically interesting, as opposed to merely taxonomically unique, is if it affects the evolutionary process in fundamental ways. Many evolutionary social scientists have been keen to apply the main theoretical and empirical *results* of evolutionary biology, such as Hamilton's inclusive fitness rule, to human behavior. Contrariwise, using the formal, mathematical, experimental, and observational *methods* of Darwinian biology to study cultural evolution has turned out to be an effective way to understand the distinctive processes of cultural evolution and the coevolution of genes and culture.

The argument for applying Darwinian methods to culture is that learning by imitation or teaching is analogous to acquiring genes from parents (Baldwin 1895; Campbell 1965, 1975). In both cases, a determinant of behavior is transmitted from one individual to another, and it is important to consider the population as a whole in analyzing either. As individuals acquire genes or culture, they sample a large population of potential parents and cultural models. Then, evolutionary processes operate on individuals, favoring some cultural and genetic variants and disfavoring others. A population's next generation typically differs from the previous one. As many generations pass, changes accumulate and evolution occurs. Population genetic theory is a formal system that scales up effects on individuals in the short run to longer-term changes within populations. Its basic methods are equally applicable to culture, and evolutionary theory should function in the social sciences in a manner similar to that which it does in biology (Cavalli-Sforza and Feldman 1981). This analogy undoubtedly caused Darwin's failure to make as sharp a distinction between genes and culture as twentieth-century biologists did. Both inheritance systems are historical population processes that frequently result in adaptive diversification of behavior.

Basic Processes of Gene–Culture Coevolution

The task implied by Baldwin's and Campbell's argument is not trivial, because there are many differences between genetic and cultural transmission. Substantial modifications are required to make genetic models mimic culture, and cultural models need to be linked with genetic models to understand the coevolution of genes and culture. These tasks have only begun, but fascinating processes have already been uncovered. Consider a few of the main differences between genes and culture and their evolutionary implications (reviewed by Cavalli-Sforza and Feldman 1981; Boyd and Richerson 1985; Richerson and Boyd 2005; Mesoudi 2007; Henrich 2008).

First, an individual is not restricted to sampling just his or her two biological parents to acquire a cultural trait. Dozens of other people in one's social network may be surveyed as well, and the one individual whose behavior seems best by some standard is chosen as the source of a cultural variant. This process can give inordinate weight to teachers, leaders, or celebrities and generates variation among groups much more rapidly than genetic evolution can.

Second, we are not limited to imitating people of our parental generation; peers, grandparents, and even ancient prophets can influence cultural evolution. Imitating peers, in effect, shortens the generation time in cultural evolution. Traits spread among peers may be harmless fads, important skills, or pathological behaviors. Hunt and Chambers (1976) studied heroin addiction, which spreads mostly among close friends. Parents observe that kindergarten children bring home viruses and bad habits alike! These pathways for cultural transmission likely have not been closed by selection on genes or culture, because the cost of catching a cultural pathogen is outweighed by the benefit of adaptive vicarious selectors to acquire useful traits from people other than one's parents.

Third, individuals acquire and discard items of culture throughout life. One is stuck with one's genes, though *expression* of genes can be modified throughout life. In contrast, culture is acquired gradually, traits acquired early can influence those adopted later, and later enthusiasms displace previous traits. The flexibility to pick and choose allows great scope for vicarious selection in cultural evolution and creates what we call "biases" (Richerson and Boyd 2005) and Cavalli-Sforza and Feldman (1981) called "cultural selection."

Fourth, variations acquired during an individual's lifetime are readily passed on to others by coupling the common animal ability to learn to imitate. Animals without imitation lose what the parents have learned, and the young must relearn each generation. With culture, the results of learning in one generation can be passed on to the next and accumulate over generations. Coupling individual learning to the cultural system allows new non-random variation to be introduced into the cultural system.

Exploring these differences has just begun. Culture is about as complex as genes, making it a tremendous task. No one has devised a precise comparison, but the number of words a high school graduate knows may be a few tens of thousands—that is, about the same as the number of protein-coding genes in our genome. One hundred fifty years after publication of *The Origin of Species*, evolutionary biology remains a vibrant field; cultural evolution is perhaps a half-century behind.

Evolution of Human Uniqueness

Formal models of cultural evolution can be used to study the evolution of the cultural system itself. There are three major differences between humans and our close primate relatives that are basic to understanding human evolution, using the Darwinian framework. Humans have (1) a greater capacity for imitation and the associated massive use of culture, (2) much symbolism and stylistic variation (e.g., many languages) of no obvious practical use, and (3) larger to significantly larger social groups, with relatively high levels of cooperation, coordination, and division of labor. How and why have these differences arisen? Theoretical models can provide some interesting tentative answers.

ESTIMATES OF THE BASIC BENEFITS AND COSTS OF A MASSIVE CAPACITY FOR CULTURE Why do humans have such a large capacity for culture? The standard answers are strongly flavored by non-Darwinian progressivist evolutionary ideas. Almost everyone assumes that human culture is an intrinsically superior method of acquiring and transmitting non-genetically heritable adaptations. The question is not why humans came to have culture, but how and when humans made the breakthrough to a qualitatively superior mode of cultural adaptation. Landau (1991) showed that all accounts of human origins, even by professional paleo-anthropologists, have the structure of folk hero stories. The human species had to perform tasks and overcome obstacles before reaching fully modern form. Even such deep-dyed Darwinians as Lumsden and Wilson remarked that humankind "overcame the resistance to advanced cognitive evolution by the cosmic good fortune of being in the right place at the right time" and that "the eucultural (complex human culture) threshold could at last be crossed" (Lumsden and Wilson 1981: 330–331). The breakthrough hypothesis is plausible if we assume that special, costly cognitive machinery is necessary to imitate complex traditions (Boyd and Richerson 1996). However, such capacities could not easily increase when they are rare, even if complex traditions had been a great adaptive advantage. In other words, mutations that create the ability to learn complex culture could not be advantageous until such complex traditions already existed!

Given the great span of time available, the absence, until recently, of complex capacities for imitation suggests the hypothesis that the costs of elaborate cultural capacity usually outweigh the benefits. If there are intrinsic barriers to the evolution of such capacity, it is surprising that complex culture has only evolved once in the history of life. Perhaps Darwin's "favorable circumstances" recently create the potential for the benefits of a large culture capacity to outweigh its costs.

Simple evolutionary models that link the capacity for individual learning to a capacity for imitation can be used to model the inheritance of acquired variation. They illustrate how culture can have real advantages in some environments (Boyd and Richerson 1985). Suppose individuals inherit some economically important trait by imitating their parents—for example, knowing how much subsistence to derive from hunting versus gathering plants. Individuals must combine traditional knowledge acquired culturally with that acquired by their own experience. We assumed that they use a weighted average. If tradition and individual learning were equally important, the traditional diet is 50% animals, but experience indicated that 90% was best. Then, individuals might end up collecting enough plants to make up 30% of the diet in the first generation, 20% in the second, 15% in the third, and so on. We also investigated similar models in which genes and learning, but not culture, were used to decide what to do.

Under what circumstances should cultural tradition be emphasized over individual experience plus genetic transmission? The answer depends upon two interacting factors: how the environment is changing and the economics of information acquisition and transmission. We assumed that the genetic system is less prone to random transmission errors (mutation) than is cultural transmission and that individual learning is fairly costly or error prone (closely related variables, since learning could always be improved with more time and effort). If the environment is changing very slowly, a fixed genetic rule is superior to any combination of learning and imitation. Selection acting on a conservative inheritance system can track slow environmental change, and the greater errors of learning and imitation are a fitness burden. In contrast, in very rapidly changing environments, any form of transmission from parents is useless—their world is simply too different from their children's. Instead, individuals do best by depending entirely on experience. In intermediate environments, some mixture of individual and social learning is typically the most adaptive system. Culture's greatest advantage is in environments that are changing substantially over tens of generations but not too rapidly within any one generation. By making individual learning cumulative, a cultural system of inheritance can track changing environments more rapidly than genes while reducing the need for costly individual learning.

Assuming that individual learning is costly compared to imitation, the model's results recover Darwin's intuition: inheritance of acquired variation has distinctive advantages in variable environments. Empirical support exists for this result. The origin of culture in humans is associated with the increasingly fluctuating climates of the last few million years (deMenocal 1995; Potts 1996; Vrba et al. 1995; Richerson et al. 2009). Really sophisticated human culture arose during the last few hundred thousand years under the strongly fluctuating Ice Age climates of the Middle and Late Pleistocene. The last glacial period (70,000–10,000 years ago), for which ice cores from Greenland give an especially good picture, was interrupted by numerous, short warm intervals of about 1000 years' duration. The last glacial was more variable than the Holocene at the 10-year limit of resolution in the ice core record, depending upon depth in the core (Ditlevsen et al. 1996). Similar but less intense variation also occurs in three earlier glacials (Martrat et al. 2007). According to our simple model, individual and social learning might be advantageous in this sort of world. Culture may be as much a means of coping with the deteriorating Pleistocene environment as a cosmic breakthrough of progressive evolution. As more and longer high-resolution cores become available, the climate hypothesis will be tested with greater rigor. For example, it predicts that hominin brain size increases will follow increases in millennial and sub-millennial climate variation. Whether or not this prediction is accurate remains to be seen.

However, if our model is correct, many animal lineages should have become increasingly cultural during the Pleistocene. In fact, many bird and mammal lineages with simple social learning systems manifested trends for increasing brain size when environmental variability increased during the Miocene to the late Pleistocene (Jerison 1973). Brain size in both birds and primates is associated with innovative behavior and social learning. Thus, humans are merely the upper tail of the distribution in terms of cognitive and cultural responses to increasing climate variation (Laland and Reader 2010). Interestingly, the west Eurasian Neandertals and the anatomically modern, tropical African humans apparently both evolved very large brains after diverging from a smaller-brained common ancestor (Klein 2009). A strong possibility exists that Neandertals also independently innovated symbolic artifacts about 50,000 years ago (Zilhão et al. 2010). Richerson et al. (2009) argue that small population size of both species probably limited their cultural complexity. Small, isolated living populations also produce very simple toolkits (Henrich 2004). Perhaps much inheritance of acquired variation in many species is carried by epigenetic inheritance systems (Jablonka and Raz 2009). Suppose that specific capacities are necessary for the transmission for complex culture. If these capacities are rare, no complex culture will exist for individuals to acquire, and thus these capacities will provide no adaptive advantage. If so, some piecemeal innovations might have allowed the human lineage to reach a threshold frequency at which the capacity would be advantageous to the individual. What costs might these piecemeal innovations have incurred? Inferring the causes for any lineage's unique evolutionary history is challenging (Nitecki and Nitecki 1992), but science must explain why only the human species has complex culture, despite frequent occurrence of simple social learning in other species. Washburn (1959) proposed that bipedal locomotion made hands free to specialize for tool-making, setting humans on the path to the capacity for complex culture.

OTHER COMPLEX BENEFITS AND COSTS OF A MASSIVE CAPACITY FOR CULTURE Further clues about the value of high capacity for culture emerge from the features of human culture. The capacity to use many people, in addition to parents, as models is a good example. On the benefit side, surveying multiple models is useful to find a better one to imitate. High frequency of a trait among replicate models often indicates a successful trait. As in the model for simple learning plus imitation, these advantages are most useful in variable environments.

On the cost side, imitating people other than parents exposes individuals to the possibility of acquiring pathological cultural traits. Deleterious cultural traits are unlikely to evolve if cultural transmission follows the conservative parent-to-child pattern. Few heroin addicts, for instance, survive to raise children. Natural selection acts against such cultural variants, but addicts can attract peers before the most harmful consequences emerge. With non-parental transmission, natural selection *on cultural variation* can favor the evolution of cultural fragments that act much like viruses, as in Dawkin's famous selfish meme idea (Goodenough and Dawkins 1994; Blackmore 1999). Much as microbial pathogens invade the body by attaching to receptor molecules on the host cell's surface, pathological cultural variants often subvert biases that normally act to favor adaptive behavior. Addictive drugs subvert the pleasure systems in the brain. Natural selection acting on parentally transmitted culture and genetic variation alone could reduce the chances of acquiring such traits, but only by sacrificing the potential benefits from imitating superior non-parents. Parental advice, formal instruction, and mass media propaganda are cultural defenses against pathological cultural variants, just as medicines augment defenses against pathogens.

A massive, sophisticated system of culture is an excellent adaptation; witness the spread of humans from tropical homeland into the Arctic and New World. But loosely speaking, the coevolutionary complexity of coordinating two inheritance systems means that the cultural system even now is far from perfect. We pay for cultural flexibility with a susceptibility to diverse cultural pathologies. *Humans are built for speed, not for comfort.*

The problems created by a second inheritance system are not necessarily as obviously harmful as heroin addiction. Many otherwise puzzling patterns of human behavior may be a by-product of the cultural system. For example, many wealthy societies have recently experienced sustained reductions in birth rates that are now often below replacement (Coale and Watkins 1986). Borgerhoff Mulder (1987) and Irons (1979) have argued based on case studies in East Africa and Iran, respectively, that wealth is efficiently converted into children in traditional rural societies, as we would expect from fitness considerations. Why do modern and modernizing societies behave so contrary to the expectation that wealth should be translated into greater fitness?

Such societies have expanded non-parental routes to transmit culture, which should multiply the pathways by which pathological cultural variants can spread (Newson et al. 2007). (We mean pathological from the perspective of Darwinian fitness and do not mean to suggest that the demographic transition is bad.) Urbanization brings people into contact with many non-kin, diluting the influence of family members who have a kin-selected interest in family members' reproduction. Specialized non-parental roles, such as teaching, are influential in socializing the young. Competition for these roles is keen, and preparation for them imposes delayed reproduction. Those who sacrifice marriage and a large family for a career they value are likely to succeed and to influence their pupils', subordinates', and employees' values and aspirations. Western society with "careers open to talent" seems to have fostered the spread of low fertility norms through cultural means of transmission (Newson and Richerson 2009).

SYMBOLS Do models of cultural evolution offer insights into the existence of elaborate, apparently functionless symbols such as costumes, artistic creations, and complex supernatural belief systems, which, along with other aspects of culture, distinguish the human species? Social groups are usually symbolically marked. For example, modern research universities, bastions of rationality, have a seal, a motto, elaborate graduation rituals, and sports teams that engage in ritualized combat. Even the faculty exhibits remarkable affection for the symbols and rituals of academia. Campbell (1969) noted the similarity of academic disciplines to ethnic groups.

To investigate this problem, we constructed theoretical models in which individuals use marker traits to assess whom to imitate. (Note the analogy to mate-choice sexual selection.) In the first instance, people might benefit from imitating others who are economically successful and have large families. Prestige and success in survival and reproduction are empirically frequently correlated, as Irons (1979) showed. Models also demonstrate that apparently adaptively neutral, symbolic characters can serve as an adaptive marker (McElreath et al. 2003; Boyd and Richerson 1987). In a spatially variable environment with migration, using symbolic criteria can reduce imitation of individuals with inappropriate cultural adaptations.

THE ORIGIN OF COOPERATION AND COMPLEX SOCIETIES The ethnic unit, like human culture, has no close parallel in other species. Altruism in large sophisticated societies of bees, ants, and termites has classically been attributed to inclusive fitness (Hamilton 1964). The workers in canonical insect colonies are all siblings, and each colony contains only few reproductives. African naked mole rats, the mammal with the most complex social organization aside from humans, form colonies comprising a reproductive queen and numerous highly inbred and closely related workers (Sherman et al. 1991). The discovery of insect colonies with moderate-to-low relatedness within colonies (Pedersen et al. 2006) has given rise to the suggestion that colony-level selection may be important in many species (Hölldobler and Wilson 2009). Among our primate relatives, cooperation appears to be largely restricted to close relatives and to partners engaging in reciprocal altruism.

To judge from simple contemporary societies (Johnson and Earle 2000), Upper Paleolithic societies comprised three levels of social organization: the family, the co-residential band, and sets of bands that routinely intermarried, spoke a common language, and shared a set of myths and rituals. Members of this largest unit, often called a tribe, maintain relatively peaceable relations with each other and routinely cooperate. Again, by analogy with simple contemporary societies, there was probably no overall formal leader or council. Rather, forceful, able men probably acted as semiformal headmen of bands. Inter-band affairs were probably regulated by *ad hoc* negotiations dominated but not controlled by the headmen.

Several hypotheses have been proposed to explain human cooperation. Alexander (1987) supposed that human intelligence allows extensive reciprocal altruism and indirect reciprocity involving assistance to others who may be several steps removed from anyone who can help you in return. However, it is difficult to scale this process up to larger groups. Models show that reputation and punishment can help stabilize indirect reciprocity, but the solution to collective action problems seems to require additional processes like group selection (Panchanathan and Boyd 2004). Something like Darwin's proposal for selection among tribes perhaps accounts for human abilities to cooperate (Richerson and Boyd 1999; Sober and Wilson 1998).

Most evolutionary biologists, including Darwin and Hamilton, are skeptical that selection between groups of unrelated individuals is effective (Williams 1966; but see Wilson and Wilson 2007). As with any form of natural selection, group selection must proceed through the differential survival or reproduction of entities that differ for heritable traits. Reproduction of groups must ordinarily be slower than that of individuals, and group death must be infrequent compared to individual death. Migration among groups can also erode differences among them. Any successful group dominated by altruistic individuals may evolve toward selfish qualities because non-cooperators enjoy the benefits of altruism without paying its costs. Inside the group, non-cooperators will increase. Neighboring human societies have far more cultural than genetic variation (Bell et al. 2009), suggesting that the cultural inheritance system is a likelier target for group selection.

Several properties of cultural inheritance make it more responsive to group selection than genes. First, as already noted, a few influential teachers in each group can create great variation among them. The impact of great ethical teachers, like Moses, Christ, Confucius, and Mohammed, on a whole series of civilizations, is evidence for the power of this effect (Cavalli-Sforza and Feldman 1981).

Second, the tendency of newcomers to conform will minimize the homogenizing effects of migration on variation among groups (Henrich and Boyd 1998). As long as migrants are a minority and are not disproportionately influential, resident culture will limit the cultural impact of minority migrants. The assimilation of many immigrants to the United States and to British–American culture is testimony to the power of this effect.

Third, cultural symbols are a potent source of variation among groups (McElreath et al. 2003). Ritual, religious belief, and language isolate groups. They protect groups from the effects of migration, much as in the case of conformity, because people ordinarily tend to admire, respect, and imitate individuals displaying familiar symbolic traits. Cultural chauvinism is all but universal, and important aspects of culture, such as the ethical norms, are often embedded in rich symbolic belief systems.

Finally, selection on cultural groups can be rapid because cultural death and reproduction do not depend upon biological death and reproduction. Defeated groups are often absorbed by the victors or by other friendly groups. In simpler societies, defeat in war typically produces more captives and refugees than dead. Successful societies also attract imitators and uncoerced immigrants who assimilate into a society and build its population (Boyd and Richerson 2009), so a culture could expand without any overt conflict at all (Boyd and Richerson 2002). Cultural group selection may thus be rapid.

Once such a system begins to evolve, selection on genes will have a difficult time opposing cultural evolution. Practices that penalize deviation from social conventions would tend to exclude genotypes that resist conformity to culturally prescribed behavior. Cultural environments can clearly exert strong coevolutionary forces on genes. The development of agriculture, for example, seems to have launched a wave of strong selection on genes (Hawks et al. 2007). Similarly, human "social instincts," as Darwin called them, plausibly evolved by gene–culture coevolution. Primitive culturally transmitted social customs would select for the innate capability to follow social rules, which would in turn, allow the evolution of more sophisticated customs (Richerson and Boyd 1999).

Discussion

The Descent of Man contains a sophisticated theory of human biocultural evolution. Darwin's ignorance of the basis for inheritance may turn out to be prescient if trans-generational epigenetic inheritance turns out to be important. His treatment of culture inheritance foreshadows late twentieth century developments rather exactly. We think Darwin actually understood cultural evolution well and made it his model for organic inheritance. Failure to appreciate his project in *The Descent of Man* has misled many modern readers, including sympathetic and knowledgeable ones. Historians of science have made some headway in explaining why his theory failed to influence the emerging social sciences in the early twentieth century, leading the social sciences to fall decades behind the biological sciences in evolutionary matters. But, much remains to be told.

Human culture may have originated as an adaptation permitting rapid evolution in a noisy Pleistocene environment. The costs of culture include the complexity and clumsiness of a coevolutionary system in which genes and culture are often collaborators but sometimes antagonists. Human ultra-sociality is a super adaptation that underpins our ecological dominance of Earth, yet it is much less perfect than the ultra-sociality of the ants, bees, and termites. In one of our models of gene–culture interaction (Boyd and Richerson 1985), each system of inheritance tends to pull behavior in the direction that favors its own transmission. As one system obtains a small advantage, the other escalates to correct it. This system comes to rest only when the cost of psychic pain becomes a significant selective disadvantage. This result is reminiscent of Sigmund Freud's model of humans, painfully torn between an animal id and a cultural superego as the price of civilization.

Acknowledgments

We thank Doug Futuyma and Mike Bell for organizing the wonderful conference that resulted in this contribution. Thanks also to our fellow participants for making the conference so stimulating and enjoyable. Two anonymous reviewers and Mike Bell much improved our chapter.

Literature Cited

Alexander, R. D. 1979. *Darwinism and Human Affairs*. University of Washington Press, Seattle.

Alexander, R. D. 1987. *The Biology of Moral Systems*. Aldine de Gruyter, New York.

Alland, A. 1985. *Human Nature, Darwin's View*. Columbia University Press, New York.

Anonymous (W. B. Dawkins, Wellesley Index). 1871. Review of *The Descent of Man and Selection in Relation to Sex* by Charles Darwin, *Contributions to the Theory of Natural Selection* by A.R. Wallace, and *On the Genesis of Species* by St. George Mivart. *Edinburgh Rev.* or *Critical J.* July, 195–235.

Anonymous (St. George Mivart, Wellesley Index) 1871. Review of *The Descent of Man and Selection in Relation to Sex* by Charles Darwin. *Quart. Rev.* 131: 47–90.

Atkinson, Q. D. and R. D. Gray. 2006. How old is the Indo-European language family? Progress or more moths to the flame? In P. Forster and C. Renfrew (eds.), *Phylogenetic Methods and the Prehistory of Languages*, pp. 19–31. McDonald Institute for Archaeological Research, Cambridge, UK.

Baldwin, J. M. 1895. *Mental Development in the Child and the Race: Methods and Processes*. Macmillan, New York.

Bannister, R. C. 1979. *Social Darwinism: Science and Myth in Anglo-American Social Thought*. Temple University Press, Philadelphia.

Bell, A. V., P. J. Richerson, and R. McElreath. 2009. Culture rather than genes provides greater scope for the evolution of large-scale human prosociality. *Proc. Natl. Acad. Sci. USA* 106: 17671–17674.

Blackmore, S. 1999. *The Meme Machine*. Oxford University Press, Oxford.

Borgerhoff Mulder, M. 1987. On cultural and reproductive success: Kipsigis evidence. *Am. Anthropol.* 89: 617–634.

Bowler, P. J. 1983. *The Eclipse of Darwinism: Anti-Darwinian Evolution Theories in the Decades Around 1900*. John Hopkins University Press, Baltimore.

Bowler, P. J. 1986. *Theories of Human Evolution: A Century of Debate 1844–1944*. Johns Hopkins University Press, Baltimore.

Bowler, P. J. 1988. *The Non-Darwinian Revolution: Reinterpreting a Historical Myth*. Johns Hopkins University Press, Baltimore.

Bowler, P. J. 1993. *Biology and Social Thought, 1850–1914*. Office for History of Science and Technology, University of California, Berkeley.

Boyd, R. and P. J. Richerson. 1985. *Culture and the Evolutionary Process*. University of Chicago Press, Chicago.

Boyd, R. and P. J. Richerson 1987. The evolution of ethnic markers. *Cult. Anthropol.* 2: 65–79.
Boyd, R. and P. J. Richerson 1996. Why culture is common but cultural evolution is rare. *Proc. Brit. Acad.* 88: 73–93.
Boyd, R. and P. J. Richerson 2002. Group beneficial norms can spread rapidly in a structured population. *J. Theor. Biol.* 215: 287–296.
Boyd, R. and P. J. Richerson 2009. Voting with your feet: Payoff biased migration and the evolution of group beneficial behavior. *J. Theor. Biol.* 257: 331–339.
Burrow, J. W. 1966. *Evolution and Society: A Study in Victorian Social Theory*. Cambridge University Press, Cambridge, UK.
Campbell, D. T. 1960. Blind variation and selective retention in creative thought as in other knowledge processes. *Psychol. Rev.* 67: 380–400.
Campbell, D. T. 1965. Variation and selective retention in socio-cultural evolution. In H. R. Barringer, G. I. Blanksten, and R. W. Mack (eds.), *Social Change in Developing Areas: A Reinterpretation of Evolutionary Theory*, pp. 19–49. Schenkman Publishing, Cambridge, MA.
Campbell, D. T. 1969. Ethnocentrism of disciplines and the fish-scale model of omniscience. In M. Sherif and C. W. Sherif (eds.), *Interdisciplinary Relationships in the Social Sciences*. Aldine, Chicago.
Campbell, D. T. 1975. On the conflicts between biological and social evolution and between psychology and moral tradition. *Am. Psychol.* 30: 1103–1126.
Carneiro, R. L. 1967. Editor's introduction. In R. L. Carneiro (ed.), *The Evolution of Society: Selections From Herbert Spencer's Principles of Sociology*, pp. i–lvii. University of Chicago Press, Chicago.
Cavalli-Sforza, L. L. and M. W. Feldman. 1973. Cultural versus biological inheritance: Phenotypic transmission from parents to children (A theory of the effect of parental phenotypes on children's phenotypes). *Am. J. Hum. Genet.* 25: 618–637.
Cavalli-Sforza, L. L. and M. W. Feldman 1981. *Cultural Transmission and Evolution: A Quantitative Approach*, Monographs in Population Biology 16. Princeton University Press, Princeton.
Coale, A. J. and S. C.Watkins. 1986. *The Decline of Fertility in Europe*. Princeton University Press, Princeton.
Cravens, H. 1978. *The Triumph of Evolution: American Scientists and the Heredity–Environment Controversy*. University of Pennsylvania Press, Philadelphia.
Darwin, C. 1845. *Journal of Researches*. Collier, New York.
Darwin, C. 1874. *The Descent of Man and Selection in Relation to Sex*, 2nd ed. 2 vols. American Home Library, New York.
deMenocal, P. B. 1995. Plio-Pleistocene African climate. *Science* 270: 53–59.
Desmond, A. J. 1989. *The Politics of Evolution: Morphology, Medicine, and Reform in Radical London, Science and its Conceptual Foundations*. University of Chicago Press, Chicago.
Desmond, A. J. and J. Moore. 2009. *Darwin's Sacred Cause: Race, Slavery and the Quest for Human Origins*. Allen Lane, London.
Ditlevsen, P. D., H. Svensmark, and S. Johnsen. 1996. Contrasting atmospheric and climate dynamics of the last-glacial and Holocene periods. *Nature* 379: 810–812.
Dobzhansky, T. 1962. *Mankind Evolving: The Evolution of the Human Species*. Yale University Press, New Haven.
Dobzhansky, T. and M. F. A. Montagu. 1947. Natural selection and the mental capacities of mankind. *Science* 105: 587–590.

Durkheim, E. 1964 [1893]. *The Division of Labor in Society*. Free Press, Glencoe.
Fisher, R. A. 1918. The correlation between relatives on the supposition of Mendelian inheritance. *Trans. Roy. Soc. Edin.* 52: 399–433.
Galef, B. G. Jr. 1988. Imitation in animals: History, definition, and interpretation of data from the psychological laboratory. In T. R. Zentall and B. G. Galef, Jr. (eds.), *Social Learning: Psychological and Biological Perspectives*, pp. 3–28. Lawrence Erlbaum, Hillsdale.
Gintis, H. 2007. A framework for the unification of the behavioral sciences. *Behav. Brain Sci.* 30: 1–61.
Goodenough, O. R. and R. Dawkins. 1994. The "St Jude" mind virus. *Nature* 371: 23–24.
Griffiths, P. E. 2003. Beyond the Baldwin Effect: James Mark Baldwin's "social heredity," epigenetic inheritance, and niche construction. In B. H. Weber and D. J. Depew (eds.), *Evolution and Learning: The Baldwin Effect Reconsidered*, pp. 193–216. MIT Press, Cambridge, MA.
Gruber, H. E. and P. H. Barrett. 1974. *Darwin on Man: A Psychological Study of Scientific Creativity Together With Darwin's Early and Unpublished Notebooks*. Dutton, New York.
Haines, H. R., G. M. Feinman, and L. M. Nicholas. 2004. Household economic specialization and social differentiation. *Ancient Mesoam.* 15: 251–266.
Hamilton, W. D. 1964. Genetic evolution of social behavior I, II. *J. Theor. Biol.* 7: 1–52.
Harris, M. 1979. *Cultural Materialism: The Struggle for a Science of Culture*. Random House, New York.
Hawks, J., E. T. Wang, G. M. Cochran, and 2 others. 2007. Recent acceleration of human adaptive evolution. *Proc. Natl. Acad. Sci. USA* 104: 20753–20758.
Henrich, J. 2004. Demography and cultural evolution: Why adaptive cultural processes produced maladaptive losses in Tasmania. *Am. Antiquity* 69: 197–221.
Henrich, J. and R. Boyd. 1998. The evolution of conformist transmission and the emergence of between-group differences. *Evol. Hum. Behav.* 19: 215–241.
Henrich, J. and R. McElreath. 2008. Dual inheritance theory: The evolution of human cultural capacities and cultural evolution. In R. Dunbar and L. Barrett (eds.), *Oxford Handbook of Evolutionary Psychology*, pp. 571–585. Oxford University Press, Oxford.
Hodgson, G. M. 2004. *The Evolution of Institutional Economics: Agency, Structure and Darwinism in American Institutionalism*. Routledge, London.
Hofstadter, R. 1945. *Social Darwinism in American Thought, 1860–1915*. University of Pennsylvania Press, Philadelphia.
Hölldobler, B. and E. O. Wilson. 2009. *The Super-Organism: The Beauty, Elegance, and Strangeness of Insect Societies*. W. W. Norton, New York.
Hunt, L. G. and C. D. Chambers. 1976. *The Heroin Epidemics. A Study of Heroin Use in the United States, 1965–1975*. Spectrum Publications, New York.
Huxley, T. H. 1863. *Evidence as to Man's Place in Nature*. Williams and Norwood, London.
Ingold, T. 1986. *Evolution and Social Life*. Cambridge University Press, Cambridge, UK.
Irons, W. 1979. Cultural and biological success. In N. A. Chagnon and W. Irons (eds.), *Evolutionary Biology and Human Social Behavior*, pp. 257–272. Duxbury Press, North Scituate.

Jablonka, E. and G. Raz. 2009. Transgenerational epigenetic inheritance: Prevalence, mechanisms, and implications for the study of heredity and evolution. *Quart. Rev. Biol.* 84: 131–176.
Jensen, A. S. 1969. How much can we boost IQ and scholastic achievement? *Harvard Educ. Rev.* 39: 1–123.
Jerison, H. J. 1973. *Evolution of the Brain and Intelligence.* Academic Press, New York.
Johnson, A. W. and T. K. Earle. 2000. *The Evolution of Human Societies: From Foraging Group to Agrarian State*, 2nd ed. Stanford University Press, Stanford.
Klein, R. G. 2009. *The Human Career: Human Biological and Cultural Origins*, 3rd ed. University of Chicago Press, Chicago.
Laland, K. N. and G. R. Brown. 2002. *Sense and Nonsense: Evolutionary Perspectives on Human Behaviour*. Oxford University Press, Oxford.
Laland, K. N. and S. M. Reader. 2010. Comparative perspectives on human innovation. In M. J. O'Brien and S. J. Shennan (eds.), *Innovation in Cultural Systems: Contributions from Evolutionary Anthropology*, pp. 37–51. MIT Press, Cambridge, MA.
Landau, M. 1991. *Narratives of Human Evolution*. Yale University Press, New Haven.
Lenski, G. E. and J. Lenski. 1982. *Human Societies: An Introduction to Macrosociology*, 4th ed. McGraw-Hill, New York.
Lorenz, K. 1966. *On Agression*. Harcourt, Brace and World, New York.
Lumsden, C. J. and E. O. Wilson. 1981. *Genes, Mind, and Culture: The Coevolutionary Process*. Harvard University Press, Cambridge, MA.
Lyell, C. 1863. *Geological Evidences for the Antiquity of Man*. John Murray, London.
Martrat, B., J. O. Grimalt, N. J. Shackleton, and 3 others. 2007. Four climate cycles of recurring deep and surface water destabilizations on the Iberian margin. *Science* 317: 502–507.
McElreath, R, R. Boyd, and P. J. Richerson. 2003. Shared norms and the evolution of ethnic markers. *Curr. Anthropol.* 44: 122–129.
Mesoudi, A. 2007. Using the methods of experimental social psychology to study cultural evolution. *J. Soc. Evol. Cult. Psychol.*1: 35–58.
Newson, L. and P. J. Richerson. 2009. Why do people become modern: A Darwinian mechanism. *Pop. Dev. Rev.* 35: 117–158.
Newson, L, T. S. E. G. Postmes, P. M. Lea, and 3 others. 2007. Influences on communication about reproduction: The cultural evolution of low fertility. *Evol. Hum. Behav.* 28: 199–210.
Nisbett, R. E. 2003. *The Geography of Thought: How Asians and Westerners Think Differently—and Why*. Free Press, New York.
Nitecki, M. H. (ed.). 1988. *Evolutionary Progress*. University of Chicago Press, Chicago.
Nitecki, M. H. and D. V. Nitecki (eds.). 1992. *History and Evolution*. State University of New York Press, Albany.
Odling-Smee, F. J., K. N. Laland, and M. W. Feldman. 2003. *Niche Construction: The Neglected Process in Evolution*. Princeton University Press, Princeton.
Panchanathan, K. and R. Boyd. 2004. Indirect reciprocity can stabilize cooperation without the second-order free rider problem. *Nature* 432: 499–502.
Pedersen, J. S., M. J. B. Krieger, V. Vogel, and 2 others. 2006. Native supercolonies of unrelated individuals in the invasive argentine ant. *Evolution* 60: 782–791.
Potts, R. 1996. *Humanity's Descent: The Consequences of Ecological Instability*. Avon Books, New York.

Provine, W. B. 1971. *The Origins of Theoretical Population Genetics*. University of Chicago Press, Chicago.
Richards, R. J. 1987. *Darwin and the Emergence of Evolutionary Theories of Mind and Behavior*. University of Chicago Press, Chicago.
Richards, R. J. 1988. The moral foundation of the idea of evolutionary progress: Darwin, Spencer, and the Neo-Darwinians. In M. H. Nitecki (ed.), *Evolutionary Progress*, pp. 129–148. University of Chicago Press, Chicago.
Richerson, P. J. and R. Boyd. 1999. Complex societies—the evolutionary origins of a crude superorganism. *Hum. Nat.* 10: 253–289.
Richerson, P. J. and R. Boyd. 2005. *Not By Genes Alone: How Culture Transformed Human Evolution*. University of Chicago Press, Chicago.
Richerson, P. J., R. Boyd, and R. L. Bettinger. 2009. Cultural innovations and demographic change. *Hum. Biol.* 81: 211–235.
Segerstråle, U. 2000. *Defenders of the Truth: The Battle for Science in the Sociobiology Debate and Beyond*. Oxford University Press, New York.
Sherman, P. W., J. U. M. Jarvis, and R. D. Alexander (eds.). 1991. *The Biology of the Naked Mole Rat*. Princeton University Press, Princeton.
Simoons, F. J. 1978. The geographic hypothesis and lactose malabsorption: A weighing of the evidence. *Digest. Dis. Sci.* 23: 963–980.
Smith, A. 1790. *The Theory of Moral Sentiments*, 6th ed. A. Millar, London.
Sober, E. and D. S. Wilson. 1998. *Unto Others: The Evolution and Psychology of Unselfish Behavior*. Harvard University Press, Cambridge, MA.
Sulloway, F. J. 1996. *Born to Rebel: Birth Order, Family Dynamics, and Creative Lives*, 1st ed. Pantheon Books, New York.
Tarde, G. 1903. *The Laws of Imitation*. Holt, New York.
Tomasello, M. 1996. Do apes ape? In C. M. Heyes and B. G. Galef, Jr. (eds.), *Social Learning in Animals: The Roots of Culture*, pp. 319–346. Academic Press, New York.
Tomasello, M. 1999. *The Cultural Origins of Human Cognition*. Harvard University Press, Cambridge, MA.
Tomasello, M. 2008. *The Origins of Human Communication*. MIT Press, Cambridge, MA.
Turner, J. H. 1995. *Macrodynamics: Toward a Theory on the Organization of Human Populations*. Rutgers University Press, New Brunswick.
Tylor, E. B. 1871. *Primitive Culture: Research Into the Development of Mythology, Philosophy, Religion, Art and Custom*. Murray, London.
Vrba, E., G. H. Denton, T. C. Partridge, and 1 other. 1995. *Paleoclimate and Evolution, with Emphasis on Human Origins*. Yale University Press, New Haven.
Wallace, A. R. 1864. The origin of human races and the antiquity of man deduced from the theory of natural selection. *Anthropol. Rev.* 2: 158–187.
Wallace, A. R. 1869. Sir Charles Lyell on geological climates and the origin of species. *Quart. Rev.* 126: 359–394.
Washburn, S. L. 1959. Speculations on the interrelations of the history of tools and biological evolution. *Hum. Biol.* 31: 21–31.
White, L. A. 1959. *The Evolution of Culture: The Development of Civilization to the Fall of Rome*. McGraw-Hill, New York.
Whiten, A. and D. Custance. 1996. Studies of imitation in chimpanzees and children. In B. G. Galef, C. M. Heyes (eds.), *Social Learning in Animals: The Roots of Culture*, pp. 291–318. Academic Press, San Diego.
Whiten, A. and C. P. van Schaik. 2007. The evolution of animal "cultures" and social intelligence. *Phil. Trans. Roy. Soc. Lond. B* 362: 603–620.

Williams, G. C. 1966. *Adaptation and Natural Selection: A Critique of Some Current Evolutionary Thought*. Princeton University Press, Princeton.
Wilson, D. S. and E. O. Wilson. 2007. Rethinking the theoretical foundation of sociobiology. *Quart. Rev. Biol.* 82: 327–348.
Wilson, E. O. 1975. *Sociobiology: The New Synthesis*. Harvard University Press, Cambridge MA.
Zilhão, J., D. E. Angelucci, E. Badal-García, and 14 others. 2010. Symbolic use of marine shells and mineral pigments by Iberian Neandertals. *Proc. Natl. Acad. Sci. USA* 107: 1023–1028.

Part VII

APPLICATIONS OF EVOLUTIONARY BIOLOGY

Chapter 21

Applying Evolutionary Biology: From Retrospective Analysis to Direct Manipulation

Fred Gould

Since the time of Darwin, knowledge about evolutionary processes and genetics has contributed to improving the quality of human life. For at least 90 years, agricultural scientists have used evolutionary theory to search for useful germ plasm (Vavilov 1926; National Academy of Sciences 1978; Doebley et al. 2006; Pickersgill 2009) to improve artificial selection procedures in crop and livestock breeding (Servin et al. 2004; Shi et al. 2009; Zhao et al. 2009), and to develop transgenic crops (Agarwal et al. 2008; Zaidi et al. 2009). Fishery biologists have become aware that selective pressures exerted by specific fishing practices can result in the evolution of fish that mature at smaller sizes, and their more recent investigations are examining how fishing practices can select for changes in fish physiology and behavior (Dunlop et al. 2009). As efforts to conserve biodiversity have increased in the past few decades, insights from evolutionary biology have been used to improve the chances of saving selected, endangered species (Moscarella et al. 2003; Armstrong and Seddon 2008). More recently, evolutionary biology is offering approaches for choosing the most important species to save (Forest et al. 2007; Faith 2008) and methods to predict the response of biological communities to global climate change (Hoffmann and Willi 2008; Aitken et al. 2008; see Davis et al., Commentary 6). In Chapter 16 of this volume, David Hillis discusses how phylogenetic analyses have improved public health as well as the accuracy and efficiency of criminal investigations. Other evolutionary methods are also being applied in the health sector. For example, there has been an increasing interest in use of genetic epidemiology models to understand and manage the evolution of antibiotic resistance (Temime et al. 2008). The field of applied evolutionary biology has grown dramatically in the past 20 years, and all indications are that this growth and diversification will continue.

It would be impossible to adequately review the contributions of applied evolutionary biology to all of the problems just listed in a single chapter. Instead, I will focus on the contributions to the specific area with which I am most familiar, sustainable pest management. The references in the previous paragraph should provide an excellent entre into the literature on other applied problems addressed by evolutionary biologists.

I will use the contributions of evolutionary biology to the improvement of pest management as a means to illustrate three different levels at which evolutionary biology can generally be applied to the needs of society: retrospective analysis, environmental manipulation, and genomic manipulation. An example of retrospective analysis would be determining the evolutionary events that resulted in global resistance to a specific insecticide by a disease-vectoring mosquito. Environmental manipulation might involve using evolutionary theory combined with the retrospective analysis to assess which of several ways to use a novel insecticide (along with other control measures) would most effectively retard adaptation of the same disease-vectoring mosquito to the new insecticide. Genomic manipulation, which has only recently become broadly feasible, would involve alteration of the mosquito's genome in ways that will guide its future evolution away from vectoring the disease. At the end of this chapter, I will briefly describe some of the complex societal issues encountered in applying evolutionary biology to societal problems.

Retrospective Analysis of Insect Adaptation to Control Measures

Early Evolutionary Insights

Insecticide resistance has plagued agricultural productivity for over a century. The first well documented case of insecticide resistance dates back to 1914, when Axel Melander reported an experiment in which the San Jose scale, a fruit tree insect pest, had up to 17.0% survival in one farming location after being sprayed with sulfur-lime, while in two other locations survival was never more than 2.4% (Melander 1914). Melander attributed the difference to variation among locations in the intensity and history of selection by spraying with sulfur-lime. He ruled out inheritance of acquired maternal resistance, because the scale insects are sprayed only about once every 10 generations. In the recorded discussion after the delivery of Melander's paper, there was no opposition to his Darwinian explanation of the results, although many entomologists of his day believed that species characteristics were static (Ceccatti 2009).

In the early 1900s, few biologists were concerned with insecticide resistance, but Melander's colleague, Henry Quayle, concluded after 7 years of observation and experimentation that the California red scale had become resistant to another insecticide, hydrogen cyanide (Quayle 1922). As in the case of the San Jose scale, the time between treatments, two to three generations, ruled out short-term inheritance of maternally acquired resistance. The pesticide industry attempted to explain away resistance as the cause of these failures at chemical control (Moore 1933), but by the early 1940s Quayle could cite seven cases in which there was strong evidence that genetically-based resistance had evolved in response to selection with insecticides (Quayle 1943). In *Genetics and the Origin of Species*, Dobzhansky (1937) concluded the "the spread of the resistant strains constitutes probably the best proof of the effectiveness of natural selection yet obtained" (Dobzhansky 1937: 161). See also Ceccatti (2009) for a more formal historical discussion of this topic.

As use of DDT and other synthetic insecticides became predominant after World War II, concern increased regarding the impact of insecticide resistance on agricultural productivity, and a number of evolutionary biologists began studying the genetics of resistance. Weiner and Crow (1951) conducted artificial selection experiments on *Drosophila* to investigate cross-resistance to multiple compounds. Sokal and Hunter (1954) reported what seems to have been the first case of behavioral resistance to an insecticide. One important question that emerged was whether alleles that conferred high levels of insecticide resistance were from new mutations arising after the use of insecticides or simply alleles that were already present in populations at low frequencies (Crow 1957). Definitive answers to this question were not found until genomic analysis became feasible (see the following discussion).

Using Patterns of Resistance to Test Evolutionary Hypotheses

From the early 1950s onward, reported cases of insecticide resistance increased rapidly (Brown 1971), permitting inference about whether the ability to evolve pesticide resistance varies among insect groups (Figure 21.1). Gordon (1961) noted that insects that utilized a broad range of dietary material and presumably encountered many natural toxicants during their

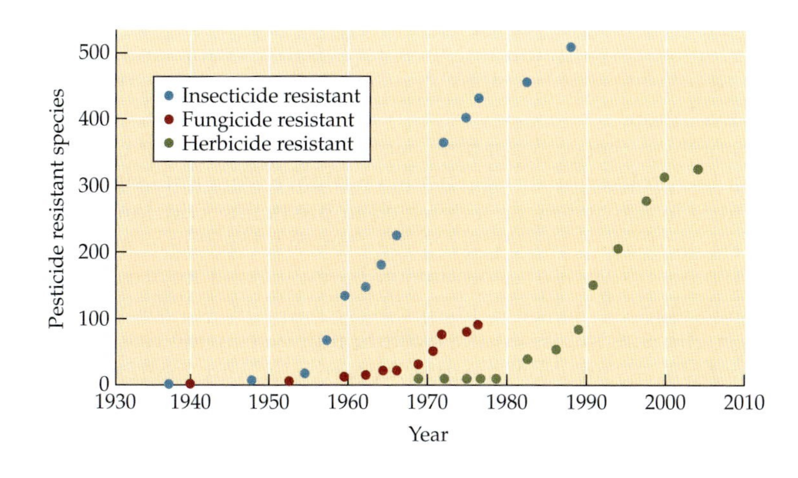

FIGURE 21.1 Time-Line for Cases of Pesticide Resistance in Insects, Weeds, and Pathogens Number of insect, fungus, and weed species that were demonstrated to be resistant to at least one pesticide by a given year. Records for fungi and insects could not be found for recent years. (Redrawn from Gould 1991, Heap 2009.)

evolutionary history were more likely to develop resistance to insecticides. Experimental tests of this hypothesis produced compelling support in some cases (Gould et al. 1982; Lindroth 1989; Li et al. 2000). Related to this hypothesis was the observation that herbivorous pests seemed to be evolving resistance faster than the predators and parasitoids of these pests. A commonly discussed explanation for this was that herbivores were exposed to a diverse array of toxins during their coevolution with plants, while their natural enemies had much less contact with toxins. Therefore, herbivores could be considered pre-adapted to evolve resistance to synthetic insecticides (Georghiou 1972; Tabashnik 1990). A non-exclusive, alternative hypothesis was that resistance evolution in natural enemies could occur only after it had evolved in their prey. If a natural enemy had a mutation for insecticide resistance, but its prey had been exterminated by application of an insecticide, the natural enemy would die of starvation (Huffaker 1971; Georghiou 1972; Croft and Strickler 1983). Tabashnik and Croft (1985) used a population genetic computer simulation model and case studies to distinguish between the two hypotheses and found more support for the prey limitation hypothesis.

Another obvious question was whether species with many generations per year would evolve resistance faster than those with fewer. An initial test of this hypothesis produced a positive correlation between the number of generations per year of fruit tree pest species and the rate of resistance evolution (Tabashnik and Croft 1985). Another study of root-feeding arthropod pests also indicated a positive association between number of generations per year and rate of evolution (Georgiou and Taylor 1986). However, a more detailed analysis of over 600 pests in the empirical literature by Rosenheim and Tabashnik (1990, 1991) failed to find a relationship between generation time and the rate at which insecticide resistance evolves. Instead, it found a strong relationship between the level of pest severity and rate of adaptation (pests that cause more severe damage are generally more intensively sprayed). They concluded that overall selection intensity per year could be just as great for a species with one generation per season as for species with multiple generations in a season, and it was the intensity of selection per unit time that should correlate with the rate of resistance evolution.

These general hypotheses about how evolutionary history and current ecology affect the probability of insecticide resistance were important for economic entomologists who were interested in predicting how long new insecticides would last in specific cropping systems, but these findings also informed basic research in evolutionary biology. Another area of interest from both applied and basic perspectives was whether there were substantial genetic constraints on the evolution of insecticide resistance in pest populations. As mentioned earlier, researchers were interested in whether alleles conferring insecticide resistance were initially at low frequency in most populations or arose from rare mutation events in one or a few populations and thus, acted as constraints on the rate of resistance evolution. They were also interested in how much of a fitness cost these novel, rare alleles would carry.

The first, and perhaps the most striking, molecular genetic investigation of this question focused on organophosphate resistance in the mosquito *Culex pipiens,* which has been sprayed with organophosphate insecticides since the 1960s. Raymond et al. (1991) found that a mosquito strain with multiple copies of an esterase gene had high resistance to organophosphates, and more importantly, that strains of resistant *C. pipiens* from Africa, Asia, and North America that had these specific, multiple esterase genes also had identical flanking DNA sequences around the genes. These findings provided strong evidence that all of these esterase alleles worldwide were derived from a single mutant resistant mosquito. This was not the only allele for organophosphate resistance in *C. pipiens* (Labbe et al. 2007), but its worldwide distribution and high frequency demonstrated that there were constraints on appearance of highly effective mutations. If alleles for effective resistance had been present at low frequencies in many populations prior to use of the insecticide, one would have expected independent evolution on the three continents and a high diversity in the flanking sequences.

C. pipiens has been the subject of long-term monitoring efforts that have shown that ancestral resistance alleles have been replaced over time by other alleles with lower fitness costs and in at least one case, a resistance gene duplication event has resulted in overdominance and maintenance of a polymorphism (Labbe et al. 2007). In total, it is thought that there are only three loci in *C. pipiens* that control resistance to organophosphates (Chevillon et al. 1999). Clearly, there have been historical constraints on how worldwide populations of *C. pipiens* have been able to adapt to organophosphate insecticides.

A different type of constraint is found in insect adaptation to cyclodiene insecticides (e.g., aldrin, dieldrin). In this case, multiple pest taxa (Diptera, Coleoptera, Blattodea) have all become resistant to this class of insecticides based on a precisely parallel evolutionary change in their DNA—a substitution resulting in a single amino acid change at position 302 in the GABA-receptor subunit (Figure 21.2). It is interesting that data from flanking sequences demonstrate that within a species these substitutions have occurred multiple times in different geographic areas (ffrench-Constant et al. 2004).

Can Past Patterns Be Used to Make Predictions?

The studies above certainly provide interesting information on the past dynamics of pest adaptation to insecticides, but the critical question for applied entomologists and farmers is whether there is any way to predict, based on the characteristics of a class of pesticides, if there will be a substantial delay before resistance alleles become prevalent. Unfortunately,

		L	N	R	N	A	T	P/L	A	R	V	S/G (↓A)	L	G	V	T	T
D. melanogaster	*S*	CTC	AAT	CGC	AAT	GCA	ACG	CCG	GCG	CGT	GTG	GCG	CTC	GGT	GTG	ACA	ACC
D. simulans allele 1	*R*	CTC	AAT	CGC	AAT	GCA	ACG	CCG	GCG	CGT	GTG	TCG (↑S)	CTC	GGT	GTG	ACA	ACC
D. simulans allele 2	*S*	CTC	AAT	CGC	AAT	GCA	ACG	CCG	GCG	CGT	GTG	GCG	CTC	GGT	GTG	ACA	ACC
	R	CTC	AAT	CGC	AAT	GCA	ACG	CCG	GCG	CGT	GTG	GGG (↑G)	CTC	GGT	GTG	ACA	ACC
Housefly	*S*	CTT	AAT	CGT	AAT	GCT	ACA	CCA	GCC	CGT	GTA	GCT	TTA	GGT	GTC	ACC	ACT
	R	CTT	AAT	CGT	AAT	GCT	ACA	CCA	GCC	CGT	GTA	TCT (↑S)	TTA	GGT	GTC	ACC	ACC
Red flour beetle	*S*	CTG	AAT	CGT	AAC	GCT	ACT	CTC	GCC	AGA	GTG	GCT	CTG	GGG	GTC	ACC	ACC
	R	CTT	AAT	CGT	AAT	GCT	ACA	CCA	GCC	CGT	GTR	TCT (↑S)	TTA	GGT	GTC	ACC	ACT
German cockroach	*S*	CTG	AAC	CGC	AAY	GCG	ACG	CCC	GCC	CGA	GTC	GCC	CTC	GGG	GTT	ACC	ACT
	R	CTS	AAC	CGC	AAT	GCG	ACG	CCC	GCC	CGA	GTC	TCC (↑S)	CTC	GGG	GTT	ACC	ACT
Yellow fever mosquito	*S*	CTA	AAT	AGA	GAT	GCT	ACA	CCA	GCA	CGT	GTT	GCA	TTA	GGT	GTA	ACC	ACT
	R	CTA	AAT	AGA	GAT	GCT	ACA	CCA	GCA	CGT	GTT	TCA (↑S)	TTA	GGT	GTA	ACC	ACT

FIGURE 21.2 Parallel Evolution of Resistance-Associated Mutations between Species Point mutations within the *dieldrin* gene (*Rdl*) replace the same amino acid across a range of different insect species. The *Rdl* gene encodes a γ-aminobutyric acid (GABA) receptor. The top row (in blue) provides the amino acid sequence. GABA is an important inhibitory neurotransmitter, and blockage of the GABA receptor by cyclodiene or fipronil insecticides results in insect death. Resistance is associated with replacement of alanine (A) at position 302, with either a serine (S) or a glycine (G) residue. Both replacements are encoded by single point mutations, as shown by the resistant (*R*) and susceptible (*S*) nucleotide sequences underneath the predicted amino acids in single letter code. *D. simulans* has two resistance alleles, one of which is identical to the resistance allele in *D. melanogaster*. Alanine, at position 302, is thought to lie in the narrowest part of the integral chloride ion channel, and the replacement of this crucial residue plays a unique dual role in channel resistance, both by reducing insecticide binding and by destabilizing the insecticide-preferred conformation of the receptor. (Adapted from ffrench-Constant et al. 2004.)

even for the same insecticide class, the waiting time can vary dramatically from pest to pest. For the organophosphates described above, Carol Hartley and her colleagues (2006) extracted DNA from museum specimens of the sheep blowfly that had been collected between 1930 and 1949, before organophosphates had been used on this pest. They found that an allele for resistance to one organophosphate already existed before the pesticide was used, and this finding is consistent with the history of rapid adaptation of the blowfly to organophosphates. On the opposite extreme, the boll weevil, a specialized herbivore of cotton that has often been sprayed over a dozen times per season with organophosphates, showed no signs of resistance alleles after 30 years of exposure (Loera-Gallardo et al. 1997).

There was still a belief 40 years ago that insecticidal compounds could be developed that would be impossible for any insects to adapt to because such adaptation would require substantial evolutionary shifts (Williams 1967). Major hopes were pinned on compounds that were analogues of the natural juvenile hormones found in all insects and critical to successful development. Unfortunately, when these analogues were finally developed as pesticides and tested on insects with resistance to other insecticides, cross-resistance was found (Cerf and Georghiou 1972). Furthermore, experiments by Thomas Wilson and his colleagues (Wilson and Ashok 1998; Ashok et al. 1998) identified an allele in *Drosophila* that conferred specific resistance to exogenous juvenile hormone and its analogues.

Today the development of an evolution-proof insecticide is no longer a goal, and general predictions of the rate of evolution of resistance, based on past patterns, is not viewed as reliable enough from an economic perspective. Instead, entomologists use the very pragmatic approach of screening major targeted insect pests for resistance alleles prior to use of any new insecticide. This screening can be an onerous task, but worth the effort when millions of dollars or loss of an environmentally friendly insecticide are at stake. Perhaps the most intensive screening for resistance genes has been for caterpillar pests that are the targets of transgenic crops expressing a toxin from *Bacillus thuringiensis* (Bt)—a topic that will be discussed in detail later in this chapter. Analysis of samples of one cotton pest, the tobacco budworm (*Heliothis virescens*), revealed that before the commercialization of the Bt crops, the frequency of a specific resistance gene in the field was approximately 0.003 (Gould et al. 1997). Subsequent screening of other caterpillar pests found initial frequencies as high as 0.16 for recessive alleles (Tabashnik et al. 2000), but in another target pest, major resistance genes could not be detected at all by intensive screening (Andow et al. 2000; Bourguet et al. 2003). These diverse results certainly argue for pest-specific approaches to delaying resistance evolution.

The United States Department of Agriculture (USDA) and similar government organizations in other countries patrol national borders to keep out potential invasive species, but no similar screening has been set up to detect invasive genes. In cases in which alleles for resistance to a specific insecticide in a specific pest are a constraint to evolution of pesticide resistance, such efforts could result in great savings of money and lower environmental pollution from higher rates of insecticide usage. For less mobile pests, it might even be useful to restrict movement of genes across state lines. Indeed, farmers in New York State blame some of their problems with insecticide-resistant diamondback moths on the shipment of infested broccoli seedlings from Florida, where that pest was already resistant (A. Shelton, personal communication). With the advent of molecular diagnostic tools, detection of such genes has become more feasible.

Environmental Manipulation for Minimizing Pest Evolution

Conventional Insecticides

Estimates of initial frequencies of genes for resistance to novel insecticides can be broadly useful in predicting the risk of rapid evolution of resistance to insecticides, but farmers, governments, and the insecticide industry have always wanted more precise risk predictions as well as economically feasible approaches for delaying evolution of resistance. From these needs arose a field called "resistance management," with the underlying assumption that an understanding of evolutionary mechanisms could be used to manipulate the rate of insect adaptation. The goal of this research is to find ways to reduce the additive genetic variance for effective resistance (i.e., fitness) in a pest population by altering the way it is exposed to the toxicants.

Resistance management research relies heavily on general population genetics models and on detailed hybrid models that incorporate both population genetics and population dynamics parameters (Tabashnik 1990; Gould 1998; Onstad 2008). The general models have been useful in providing insights into the kinds of strategies that might lower additive genetic variance, while the more detailed models often examine one pest in one agricultural system and make specific recommendations for pest control. For the most part, resistance management research has been linked to the field of Integrated Pest Management (IPM), which aims to decrease pesticide use by combining diverse biological, behavioral, and agronomic practices in managing pests.

The first general models of resistance management used analytic, simulation, or optimization techniques (Heuth and Regev 1974; Comins 1977a,b; Georgiou and Taylor 1977). The general focus was on single- and two-locus models (Taylor and Georgiou 1979; Tabashnik and Croft 1985; Uyenoyama 1986; Roush 1989), although some multi-locus and quantitative genetic models were also developed (Plapp et al. 1979; Via 1986). For most of the biological systems examined, the results from the various models were qualitatively similar. These models examined whether and to what extent the rate of resistance evolution was affected by: (1) the pesticide concentration that was sprayed and number of sprays per season, (2) the movement of insect pests between sprayed and unsprayed environments, (3) the decay rate of the insecticide, (4) the fitness cost to pest individuals with one or more resistance alleles, and (5) the interaction of these factors. Whole volumes have been devoted to reviewing the literature on resistance management (National Research Council 1986; Roush and Tabashnik 1990; McKenzie 1996; Onstad 2008). Here I only point out a few interesting findings from the models and how they relate to commercial application.

In the 1950s and 1960s, there was much discussion of whether insects would evolve resistance more rapidly when high doses or low doses of sprays were used (high doses would kill almost every pest in a field, while low doses would kill only enough insects to substantially reduce yield losses). The results from the general deterministic models were that as the dose increased in a closed system (all pest exposed) the rate of evolution would increase but that in an open system where pests moved between sprayed and unsprayed environments, resistance would evolve more slowly if the dose used was high enough to kill pests that were heterozygous for resistance alleles (i.e., made resistance functionally recessive) (Taylor and Georghiou 1979; Tabashnik et al. 2004).

It is fascinating that Melander (1914) had anticipated the results of these models with his simple statement that:

> *What is the economic importance of the appearance in a locality of a resistant strain of the San Jose scale? An alarmist might say that a few such scales would soon result in a totally immune insect, brought about by annual spraying. But viewed from a Mendelian standpoint, the consequences are less direful. If only the resistant individuals survived to reproduce then a pure line might result after repeated sprayings. But always there are some scales missed by the spraying, and these, during the ten generations between sprayings, will produce a population in part, at least, non-resistant. If resistance were a dominant characteristic there would already be a larger proportion of immune individuals than the data show. If it is recessive the crossing with scales missed by the spray would, by the end of each year, produce a majority of susceptible individuals (Melander 1914: 171).*

A more detailed stochastic simulation model based on the biological attributes of the diamondback moth drew a somewhat different conclusion (Caprio and Tabashnik 1992). When the investigators set the mutation rate of the single resistance gene in their model to 10^{-6} and had semi-recessive resistance and a cost of resistance, they found that moderate levels of movement increased the rate of resistance evolution in a landscape with sprayed and unsprayed habitats. In general, there was a tradeoff between movement lowering additive genetic variance but also decreasing the local extinction of resistance alleles.

There remained an overall theoretical result that spraying high doses of pesticides in a heterogeneous landscape could decrease the rate of resistance evolution compared to use of lower doses. However, there were three major practical complaints about this approach: (1) it is expensive, (2) it causes off-target environmental effects, and (3) after spraying, the pesticide decays and at some point, reaches a level at which it differentially affected heterozygotes and totally susceptible individuals. If the pest population was still in the field at that time, the strategy could backfire. Follett et al. (1995) developed a model that included details of the life history of Colorado potato beetle and showed that the timing of insecticide application could affect pest exposure due to decay. Additionally, growth of new leaves would dilute the insecticide, and the emergence of adults from pupae that live below ground would cause delayed contact. The finding from this model and others pointed to the fact that any theory needs to be carefully tied to application and phenology of the target species.

Bruce Tabashnik, a leading researcher in the field of resistance management, concluded that "soft-fail" approaches that did not promise major delays in resistance evolution but were robust to errors about biological and genetic parameter were the best path to take (Tabashnik 1990). Many other researchers and extension personnel agreed with this approach, emphasizing that IPM, which would help decrease the number of sprays per year, could be the best resistance management tool.

Insecticides in Transgenic Plants

In the mid-1990s, the introduction of transgenic insecticidal crops added a new dimension to the field of resistance management because they changed the rules of engagement between pests and toxins. The tools of plant molecular genetics provided interesting new flexibility in toxin characteristics and delivery that were not available with conventional insecticides. Using tobacco as a model system, molecular biologists found that they could transgenically express a variety of toxic proteins, ranging from insect-specific scorpion toxins to the caterpillar-specific toxins from a bacterium that had long been used in pest control by organic farmers. While the scorpion toxins wound up on a back shelf because of their name and origin, the toxins from *Bacillus thuringiensis* (Bt) rose to priority status once it was found that, with some codon optimization, they could be expressed at highly toxic concentrations in a variety of crop plants (Perlak et al. 1991). The discovery of promoter sequences that limit gene expression to specific times, places, and environments gave genetic engineers a tool box enabling them to express toxin genes more-or-less where and when they wanted them expressed.

The evolutionary questions of past decades were back again. Could we predict how long it would take for insects to adapt to Bt toxin-producing crops? A few researchers predicted that adaptation to Bt toxins would be more of an evolutionary hurdle for insect pests than adaptation to conventional insecticides, because insects had long been exposed to these bacterial toxins and if they could have adapted to them, this would have occurred in the past (Bowman 1981). Absent from this prediction was the fact that *B. thuringiensis* was mostly found in the soil, and the insect targets of transgenic crops rarely fed in the soil environment. Furthermore, the diamondback moth, which was famous for adapting to synthetic insecticides, had already become locally resistant to Bt used as an organic insecticide in Hawaii (Tabashnik et al. 1990).

The serious question at the time was whether there was any way to commercially deploy Bt crops on a wide scale in a way that would delay evolution of resistance for even 10 years. A number of possible approaches were proposed. All had the common goal of reducing additive genetic variance for pest fitness in an environment with transgenic crops. Two approaches aimed at lowering exposure to the Bt toxin by using time- and tissue-specific expression of the Bt genes. Economic entomologists have long been aware that much of the feeding by crop pests does not substantially lower yield or quality because of crop tolerance to damage and the fact that feeding on most leaves does not affect quality of the reproductive structures that are often the commercially valuable structures. Laboratory experiments with the tobacco budworm demonstrated that when caterpillars had a choice of food with and without the Bt toxin, they fed much more on the food without the toxin (Gould et al. 1991a). If a bud and fruit specific promoter was used to turn on Bt expression, it was plausible that the pests would be deterred from feeding on commercially valuable tissue but would still survive (Gould 1988, 1998). Such crops have yet to be developed, but over a decade ago, Ciba-Geigy developed a corn variety (from insertion event 176) that did not express Bt toxin in corn kernels to assuage consumer concerns about the toxin (Koziel et al. 1993). Therefore, the possibility of tissue-specific expression is feasible.

A related approach was to express the toxin only at times of the season when the pest caused the most damage, or to trigger expression of Bt toxins only in the presence of numerous pests. Here Ciba-Geigy came up with an ingenious system that used salicylic acid pathway promoters to trigger expression when a dilute solution of aspirin was sprayed on the crop (Williams et al. 1992). This method was partially successful but the time lag between noticing the pest outbreak, spraying the crop, and waiting for gene expression was considered unappealing to farmers (and would not get rid of the headache).

Borrowing from the literature on tritrophic interactions, George Kennedy's and my lab theoretically and experimentally assessed the feasibility of using plants that express such low levels of Bt toxin that they did not kill the pests, but only delayed their growth. We thought this might have three advantages: (1) elimination of strong insecticide selection, (2) retardation of pest development to allow the plant to outgrow the pest, and importantly, (3) creation of a larger window of opportunity for natural enemies to attack smaller stages of the pest. Of course, the result was not that simple. The models indicated that the enhanced interaction between the pests and the natural enemies could increase, decrease, or have no effect on the rate of pest adaptation (Gould et al. 1991b). Furthermore, empirical experiments with the targeted pest, *H. virescens,* which was attacked by parasitoids and a fungal disease, indicated that natural enemies would increase the rate of adaptation by the pest to the Bt toxins in the crop (Johnson and Gould 1992; Johnson et al. 1997).

This low-dose approach was abandoned in favor of a high-dose resistance management strategy for Bt crops. Although the high-dose approach was found to be problematic for conventional insecticides, characteristics of transgenic plants, such as a constant concentration of a caterpillar toxin that did not affect predators, made the use of high doses in plants seem feasible. The idea was to develop cultivars that produce very high doses of the Bt toxin but to interplant them with cultivars that had neither transgenic nor conventional pesticide protection (Gould 1998). The high toxin dose would insure that only rare, homozygous resistant individuals could survive on the Bt-producing plants (Tabashnik et al. 2004). By inter-planting the Bt and non-Bt cultivars in the same or adjacent fields, random mating was expected between the rare homozygous Bt-resistant individuals and the much more abundant homozygous-susceptible insects (produced in large numbers on non-Bt hosts). Population genetic models indicated that even with only 5–10% of the acreage planted to the non-Bt cultivars, resistance could be delayed substantially (Gould 1991). Of course, farmers and industry sales staff were not excited about a product that had any susceptible plants in it.

If it were not for public concerns over transgenic crops and the US Environmental Protection Agency's (EPA) response to these concerns, it is unlikely that any credible resistance management program would have been instituted for these crops. The EPA and USDA fostered additional research and deliberations on the feasibility of various resistance management programs (US-EPA 1998a,b, 2001). Detailed simulation models for crop-specific pest systems and quantitative data on biological parameters that affected model outcomes were obtained. For example, a spatially explicit stochastic model predicted that planting the refuge crop in the same place each year would delay resistance in *H. virescens* more than rotating fields between Bt cotton and refuges (Peck et al. 1999). Models and behavioral experiments with larvae demonstrated that for some pests, inter-planting of Bt and non-Bt cultivars could be counterproductive, because the larvae moved between adjacent plants and were likely to ingest an intermediate dose of toxin (e.g., Onstad and Gould 1998; Davis and Onstad 2000; Tang et al. 2001). Because adults of these same species were strong fliers, it was concluded that there would be sufficient inter-mating of resistant and susceptible genotypes as long as Bt and non-Bt fields were in close proximity.

After consultations with industry, a number of public meetings, and assessment of reports from scientific advisory panels, the EPA settled on a set of crop, region, and pest-specific plans to delay resistance evolution. These plans called for crop cultivars with toxin concentrations high enough to kill heterozygotes, and non-Bt crop acreage large enough to produce 500 fully susceptible pest insects for every homozygous resistant insect, assuming a resistance allele frequency of 0.005 (US-EPA 1998a,b). The set of hybrid models of pest population genetics and population dynamics used to arrive at these plans assumed single-locus control of resistance, because quantitative traits were not expected to evolve in the face of high doses of toxins (US-EPA 1998b).

The first widespread implementation of what became known as "the high dose/refuge approach" came in 1996 when transgenic cotton went on the market. As could be expected, the design and implementation was not without flaws. The crop cultivars that went to market caused 100% mortality of two of the main target pests, *H. virescens* and the pink bollworm, *Pectinophora gossypiella*, and therefore appeared to meet the criteria of a "high dose." Unfortunately, these cultivars only caused 75–95% mortality of another important pest, the corn earworm, *Helicoverpa zea*. A second problem was farmer compliance with the planting and maintenance of non-Bt refuges. Although the EPA and the companies involved made assurances that compliance would be monitored and enforced, this was not the reality at the farm level (J. Bacheler, North Carolina State University, personal communication). It has been difficult to get rigorous data on noncompliance, but a few studies make clear that in some regions it has been substantial (Center for Science in the Public Interest 2009).

Fourteen cropping seasons have passed since the first commercialization of Bt cotton and Bt corn, and the first case of widespread resistance to a Bt crop by a pest was just reported by industry in the past month (Bagla 2010). This new report is too preliminary for further comment, but there have been previous isolated cases of failures as well as cases in which heritable increases in survival have been found in *H. zea* (Ali et al. 2006), fall armyworm, *Spodoptera frugiperda* (Matten 2007), and the African maize stalk borer moth, *Busseola fusca* (van Rensburg 2007; reviewed by Tabashnik et al. 2009).

The only cases in which there were measurable changes in resistance to Bt toxins were in pest species for which the crop did not deliver a high dose (Tabashnik et al. 2009). Furthermore, with *H. zea* (which has been screened for resistance over most of the eastern US cotton belt), the local level of Bt toxin resistance seems to be inversely correlated with the size of the population in the local refuge on non-toxic hosts, which appears to vary from 82% in North Carolina and Georgia to 39% in Mississippi and Louisiana (Gustafson et al. 2006; Tabashnik et al. 2009). Given the lack of a high dose for *H. zea* in cotton or corn, its major alternate host, a population genetic model predicted that a 39% non-toxic refuge population is not sufficient for delaying resistance evolution if the resistance alleles do not carry substantial fitness costs (Gustafson et al. 2006).

When Bt crops were first commercialized, they all contained very closely related Bt toxins to which cross-adaptation was expected (Gould 1998). If that were the case today, the prognosis for insecticidal transgenic crops would not be positive. However, from among hundreds of unique toxic proteins produced by the species *B. thuringiensis* and other microbes, companies have found about a half-dozen toxins that are safe and effective enough for development of new crop products (Tabashnik et al. 2009). Furthermore, the engineering of RNAi constructs into plants that silence essential herbivore genes is showing promise as a novel approach to pest suppression (Mao et al. 2007). With these new genomic tools in hand, companies have already engineered cultivars that express multiple toxins (Gore et al. 2008).

The utility of spraying multiple conventional insecticides at the same time to stymie resistance evolution has been discussed and modeled for many decades (Curtis 1985a; Roush 1989; Tabashnik 1990). The models as well as the limited experimental tests of such mixtures have produced inconclusive results. Significant decreases in the rate of resistance evolution through use of mixtures generally require atypical conditions (Roush 1989; Tabashnik 1990). In contrast, modeling of conventionally bred, insect-resistant wheat varieties indicated that the use of cultivars with multiple resistance factors could substantially delay resistance evolution (Gould 1986). The prospects for transgenic crops with multiple protein toxins seems to be more akin to that of conventionally bred crops than of sprayed insecticides (Gould 1991). A recently developed two-locus analysis of crops with two distinct toxins indicates that the complete replacement of one-toxin cultivars in the field with two-toxin cultivars (that include the original toxin) could actually reverse a pest's adaptation to the original toxin if there was at least a 5% fitness cost to resistance alleles (Gould et al. 2006a). Under these conditions, resistance is never expected to evolve, even if only a small, effective refuge is planted.

In this case as well as in all of the evolutionary scenarios for Bt crops discussed thus far, there is an assumption that there will be no single-step mutation that produces an insect with dominantly inherited resistance to many Bt toxins. To date, efforts in many labs have failed to detect dominant mutants that can survive on plants with high-dose toxins. However, strains with low levels of resistance to multiple toxins have been selected in the lab (Gould et al. 1992), so the possibility does exist. Tabashnik's admonition, previously mentioned, about using a "soft-fail" approach with conventional insecticides has certainly not been adhered to in this system. The price for taking this riskier tack is consistent, thorough monitoring, so that any resistance in a pest population could be quickly discovered and crop deployment could be changed. While this is easier said than done, the EPA and USDA are funding research to develop and employ monitoring systems (EPA 1998a,b, 2001).

Herbicides and Pathogen Resistant Cultivars

Up to this point, I have been focusing entirely on applied evolutionary study of one pest taxon: Insecta. Indeed, Figure 21.1 shows that, at least with regard to pesticides, insect adaptation was more common than was weed and pathogen adaptation through 1990. Therefore, this emphasis is at least partially justified. In the years since 1990, entomologists seem to have given up counting the number of species that have evolved insecticide resistance (Georghiou and Lagunes-Tejeda 1991), but in the world of agricultural weeds, the problem is now receiving more attention (Heap 2009; Gressel 2009; Neve et al. 2009). In the 1950s, weed scientists may have just felt lucky when they looked over the fence at the resistance problems faced by their colleagues in entomology. The renowned plant ecologist, John Harper, tried to raise awareness of the potential for herbicide resistance at that time, but his thoughtful evolutionary analysis did not seem to receive much of a response from weed scientists (Harper 1956). Plant pathologists have had concerns about bacteria and fungi adapting to pesticides, but historically, they have been much more concerned with pathogen adaptation to defense genes in classically bred crops (Vanderplank 1968).

Many of the evolutionary questions addressed by weed scientists and plant pathologists are similar to those of their colleagues in entomology. The major difference is in the answers to these questions, because of the distinct life history and reproductive processes in plants and pathogens (Gould 1995). Most, but not all of the major insect crop pests reproduce sexually in every generation (aphids are one exception). In fungal pathogens, parthenogenesis or cyclical parthenogenesis is the rule, which has a major impact on the maintenance of additive genetic variance in the face of refuges from fungicide spraying. Crop weeds have a range of reproductive strategies ranging from parthenogenesis to obligate outcrossing, so resistance management solutions need to be tailored to the target species (Gould 1995).

Two other complicating biological attributes of weeds are their seed banks and polyploidy. Seed banks can provide a refuge from herbicide selection for over a decade. These refuges could certainly slow down adaptation when outcrossing is common. Recent polyploidy can be either a curse or blessing in terms of slowing resistance evolution. In cases in which the herbicide has a specific protein target site, it will not be too useful for a weed to have a mutation at one of the two to four loci where the gene for that protein resides. However, if metabolic resistance to the herbicide is possible, a weed could alter one copy of an enzyme-coding gene to more specifically target the herbicide while the other copies of that gene continue to carry out normal functions. This is not a black and white distinction between weeds and other pests, since not all gene copies in polyploid plants are functional and other taxa also have duplicated genes.

Plant pathologists embraced evolutionary biology many decades ago when it became apparent that crop cultivars, which had been conventionally bred over periods of 5–15 years to have heightened defenses against pathogens, lost their utility in much less than that time. I am confident that if Dobzhansky had known more about the problems of agricultural plant pathologists, they would have been mentioned in the later editions of his books.

Plant pathologists noticed that cultivars with single genes coding for strong pathogen defense tended to last for fewer years than cultivars with quantitatively inherited defensive traits. In 1968, J. E. Vanderplank reviewed the history of pathogen adaptation and coined the terms vertical and horizontal resistance, respectively, for those defense factors for which substantial pathogen genetic variation existed prior to cultivar release and those for which variation was minimal (Vanderplank 1968). It turned out that most cultivars with defense due to single genes had vertical resistance, and those with polygenic-based defense typically showed horizontal resistance and were, therefore, expected to be less vulnerable to pathogen adaptation. Following Vanderplank's initial work, a number of plant pathologists examined the genetics and evolutionary dynamics of pathogen–host plant interactions in much more detail (e.g., Lamberti et al. 1983; Parlevliet 1983; Leonard 1986).

Armed with data on the history of pathogen evolutionary interactions with crop defenses, some plant pathologists concluded that it was best to breed single gene defenses into cultivars, because it was a much faster process than breeding for polygenic resistance. Other plant pathologists used the same historical data to recommend that whenever it became apparent that there was a strong defense locus in a breeding line, that line should be dropped from the breeding program lest it result in pathogen adaptation (Parlevliet 1983; Parlevliet and Kuiper 1985). Most plant pathologists took a more pragmatic middle ground, but they and government farm advisors were always surveying for adaptation by pathogens after commercial release of any cultivar.

In recent years, with the advent of marker-aided selection based on Quantitative Trait Locus (QTL) methods (Shi et al. 2009), plant breeders sometimes combine a number of single gene defense factors into a single cultivar, analogous to the development of transgenic crops with multiple Bt toxins. How effective this technique is as a resistance management method is yet to be determined.

Genome Manipulation for Directing Pest Evolution

In all of the discussion above, the goal of the applied evolutionary biologist has been to manipulate the biotic and abiotic environment of the pest in a manner that reduces additive genetic variance for pest fitness. This approach is clearly rooted in the general principles of evolution by natural selection, as envisioned by Darwin and Fisher. The next step taken by applied evolutionary biologists was to directly manipulate the pest's genome, so that it would evolve in a desired direction.

Classical Manipulation by Fitness Underdominance Methods

In the 1940s, applied biologists began thinking about whether it was possible to combine their understanding of evolution and pest genetics to directly manipulate pest species to evolve lower fitness or to become biologically more benign (Gould and Schliekelman 2004). As an example, population genetics theory indicates that when matings of two individuals that are homozygous for alternate alleles produced offspring with lower fitness than either of the parents (i.e., fitness underdominance), a population that was initially polymorphic for the two alleles and had random mating would eventually become fixed for one of the two alleles. In a large population, which allele became fixed depended on the initial frequency of the two alleles and the fitness of individuals homozygous for the alternate alleles. If the fitness of the homozygotes is equal, then there is an unstable equilibrium at an allelic frequency of 0.5. The important point for applied biologists was that if one allele conferred considerably lower fitness on homozygotes, it could still become fixed in the population if it initially had a high frequency.

In the 1940s, F. L. Vanderplank was working in Africa on tsetse flies (*Glossina* sp.) that transmit human and animal trypanosomiasis. He realized that he had a system of underdominance that he could use to his advantage (Vanderplank 1944, 1947). Vanderplank was working in one local area that had a significant dry season. The dominant tsetse fly in the area, *Glossina swynnertoni*, was adapted to that seasonality. He released a second tsetse species, *Glossina morsitans*, which was not adapted to the dry season, into this area. He knew that the two species mated randomly but that their offspring were unfit (Vanderplank 1944). After the release, Vanderplank monitored the numbers of both species for 8 months. *G. swynnertoni* numbers declined nearly to extinction, followed by a decrease in *G. morsitans* in the dry season. After 8 months, the survey was ended when local people inhabited the area, cut down the brush habitat required by the remaining tsetse, and started grazing cattle. Clearly, this was a small but significant success based on often-unappreciated population genetic theory. For what seems to have been political reasons, these preliminary results of Vanderplank's were not published until 2005 (Klassen and Curtis 2005). No major eradication program followed up on Vanderplank's initial success, and it is not clear whether a larger program could ever have succeeded, even with adequate infrastructure.

A more widely recognized example of underdominance occurs when strains that differ in chromosomal translocations are mated. Aleksandr Serebrovskii, an adherent of Mendelian genetics in the Lysenko-era Soviet Union, saw the potential to use this phenomenon for pest control. He worked out the mathematics and began some empirical work on houseflies, *Musca domestica*, but apparently never completed his experiments (Serebrovskii 1940). In the 1960s, Chris Curtis independently began investigating the potential to use translocations to spread anti-malaria genes into mosquito populations (Curtis 1968). Curtis and his colleagues managed to conduct a large set of laboratory and field experiments to assess the potential of this and other approaches for manipulating the genomes of disease-vectoring mosquito species (Curtis 1977). After years of effort in this area, Curtis concluded that the tools for driving anti-pathogen genes into mosquito populations were too crude for in-field success (Curtis 1985b). Instead, he recommended a more straightforward genomic manipulation, male sterility based on radiation-induced chromosomal breakage. This radiation-based sterile-male technique was very successful in a few cases, but was not feasible for many pest species (Gould and Schliekelman 2004; Klassen and Curtis 2005).

Transgenic Manipulation of Pest Genomes

Enthusiasm for direct genetic manipulation of pest species waned for about a decade, but with the advent of new transgenic techniques in the 1990s, molecular biologists who had been working with *Drosophila* realized that their methods could be applied to pest problems (Ribeiro and Kidwell 1994). The first molecular genetic techniques for driving anti-malaria genes into mosquitoes involved the use of transposons that had been found to spread rapidly into *Drosophila* populations. The basic idea was to insert anti-malaria genes into an active transposon and integrate the altered transposon into a pest genome, where it would replicate and spread as they do under natural conditions. Unfortunately, experiments in which an additional gene was added to a transposon did not perform as had been hoped (Carareto et al. 1997). Instead, the transposon itself spread into the lab population of *Drosophila*, but the additional gene was lost. In retrospect, this outcome was not too surprising because the fidelity of transposon replication is low, so some replicate copies could have lacked the additional gene. Because smaller transposons tend to replicate and move faster than larger transposons, the versions without the additional gene could easily become fixed.

This first disappointment using transposons led to a search for more robust gene drive systems. Fortunately, as researchers began theoretical work in this area, the Bill and Melinda Gates Foundation became interested in the potential for genetic control of malaria and dengue. Funding from the Gates Foundation has already resulted in a transgenic line of the dengue-vectoring mosquito, *Aedes aegypti*, with a conditional female lethality construct (Fu et al. 2010). Plans are to use this transgenic line in a manner similar to the radiation-induced sterile male approach (Schliekelman and Gould 2000). The female lethality construct has now been backcrossed into an *Ae. aegypti* strain from southern Mexico, where it will soon be tested in field cages (T. W. Scott, personal communication). The female lethality approach has some advantages over the simple sterile male technique in that males with the engineered construct survive and pass on the lethality to a portion of the next generation of females. If multiple insertions of the same construct can be engineered into a release-strain, the impact of the release can be extended as linkage disequilibrium between the constructs diminishes (Schliekelman and Gould 2000).

Beyond development of these female-killing strains, the really exciting and challenging work is being done on development of synthetic selfish genetic elements (Burt and Trivers 2006). The most advanced of these was modeled on a naturally occurring selfish genetic element called *Medea*, in reference to the Greek goddess who killed her children to spite her husband. In *Tribolium* beetles, where this selfish element was first discovered (Beeman et al. 1992), approximately 50% of the offspring from heterozygous females do not inherit a copy of the *Medea* genetic element during embryogenesis and die (Figure 21.3). This process results in a continuous increase in *Medea* gene frequency, even from very low initial numbers, if there are

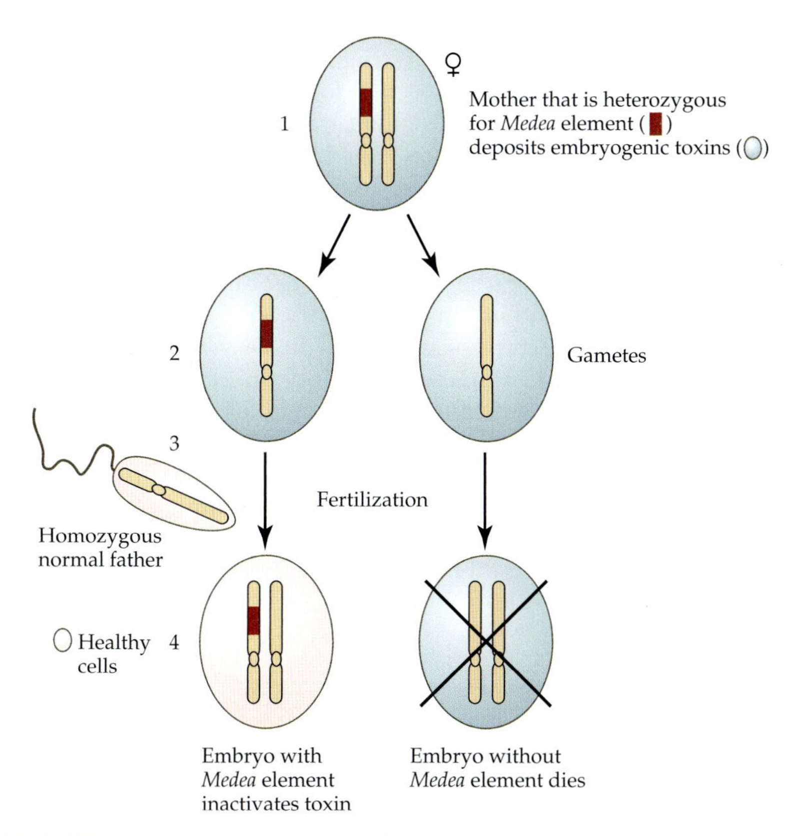

FIGURE 21.3 Diagram of the Action of *Medea* Elements (1) During gametogenesis mothers that are heterozygous or homozygous for *Medea* produce a substance that is toxic during embryogenesis. (2) All haploid eggs contain this substance whether or not they have inherited the *Medea* genetic element. (3) The eggs are typically fertilized by a male lacking the *Medea* element when the frequency of this element is low. (4) In early embryogenesis, those embryos with the *Medea* element destroy or inactivate the toxic substance and survive. The embryos that did not inherit the *Medea* element perish.

no fitness costs associated with the element's presence (Wade and Beeman 1994). If a *Medea* element could be linked with an anti-pathogen gene, it would have ideal properties for converting a mosquito population from a disease vector to a simple nuisance. It is presumed that during gametogenesis, *Tribolium* females with the *Medea* element lay down a substance that is only toxic during embryo development and that offspring inheriting the *Medea* element can detoxify this substance. Although the DNA sequence of the *Tribolium Medea* element is known, the means by which it kills are obscure (Lorenzen et al. 2008).

Bruce Hay, a molecular biologist working with *Drosophila* programmed cell death, recognized that it should be possible to make a synthetic version of the *Medea* element. After 3 years of struggle, he and his colleagues finally succeeded in developing such a construct for *Drosophila* (Chen et al. 2007). Their population cage experiments demonstrated that the synthetic element increased to near fixation in less than 15 generations (Chen et al. 2007). The Hay lab is now working with a team funded by the Gates Foundation to build a modified *Medea* construct for *Ae. aegypti.*

Of equal interest is another class of naturally occurring selfish genetic elements, the homing endonuclease genes (HEGs; Burt 2003). These genetic elements, which are found in fungi, plants, bacteria, and bacteriophages, code for proteins that each recognize a specific DNA sequence in the genome, but only do so when this sequence is not interrupted by the presence of the HEG itself. In a heterozygous individual, the homing endonuclease is produced from the chromosome copy with the HEG, which then cuts the recognition site on the homologous chromosome copy lacking the HEG. DNA repair mechanisms of the cell use the template of DNA from the chromosome with the HEG to repair the cut strand, which results in a cell that is homozygous for the HEG (Figure 21.4). If this process occurs

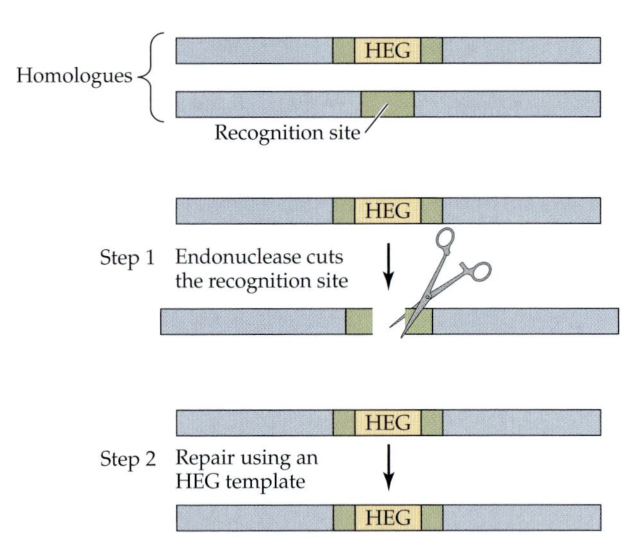

FIGURE 21.4 The Mechanism by which Homing Endonuclease Genes Increase in Frequency within a Population A specific homing endonuclease gene (HEG) is typically found inserted between two specific sequences of DNA within the genome (green). The HEG (yellow) codes for the production of an enzyme that recognizes these two specific coding sequences when they are not interrupted by the presence of an HEG. In individuals that carry the HEG on only one of two homologous chromosomes, the enzyme catalyses a break within the DNA sequence of the chromosome that lacks the HEG (Step 1), which is then naturally repaired using the HEG within the homologue as a template (Step 2). (Adapted from Sinkins and Gould 2006.)

early within gametogenesis, most of that individual's offspring will inherit the HEG. As with the *Medea* element, this super-Mendelian inheritance of HEGs from initially heterozygous individuals results in increasing gene frequency (Burt 2003) that could be subverted to drive a beneficial gene into a pest population. Progress has been made in engineering HEGs into *Anopheles gambiae* mosquitoes that transmit malaria (Papathanos et al. 2009), but many refinements are still needed before a synthetic driving element is perfected.

The Role of Evolutionary Models in Genomic Manipulation of Pests

As with evolutionary analyses of insecticide resistance, general population genetic models are very helpful in determining which genomic manipulation strategies have the potential to succeed, but more detailed models that include population structure and dynamics are critical for analysis of specific pest systems. This is a very new area of endeavor for evolutionary biologists, but a few detailed models have been developed and provide guidance that differs from the simpler models (Yakob et al. 2008). For example, a simple model of *Medea* drive indicates that it is best to release only males carrying *Medea* alleles into the natural pest population, while a model of *Medea* drive that includes age and mating structure suggests that release of only males could be less efficient than releasing both sexes (Huang et al. 2009). Before any in-field releases are conducted, models that specifically take into account the spatial structure of the pest population will be needed in order to assess the best spatial and temporal pattern for releasing the engineered pests. Steps in this direction are progressing (Magori et al. 2009).

Risks and Public Assessment

When evolutionary biologists apply their science to the problems of society, they are almost invariably well meaning. However, having the good of society at heart does not mean that one's contributions always have positive impacts or are viewed as having positive impacts. The public arena is more complex than many scientists expect it to be.

When prominent evolutionary biologists like Pearson (1925), Muller (1935), and R. A. Fisher (1924) threw their support behind the eugenics movement of the early twentieth century, I am sure that they had the betterment of society in mind, and there is no doubt that they gave it careful thought. In this, the early twenty-first century, it would be hard to find a single person at an evolution society meeting anywhere in the world who would defend eugenics.

Today, with renewed interest in applying evolutionary biology, one must be vigilant to ensure that the ideas and projects are appropriate for the problems at hand and that plans are presented in a manner that can be understood by policy makers and the public. When I, or most researchers I know, have a novel idea, there is a tendency to see it in a very optimistic light. I have found that the more critical we are of our own work and the more we listen to criticism from outside our academic circles, the more chance there is of making a useful contribution. Keeping a balance is critical.

The commercialization of transgenic crops has been controversial in many countries across the globe. One main concern in the U.S. was that the widespread use of Bt crops would result in pest adaptation that would render useless the natural formulations of *B. thuringiensis* used by organic farmers. The institution of high-dose/refuge resistance management policies was viewed by some critics as insufficient. They raise questions such as how to estimate the chance that a dominant-acting mutation will negate the utility of all Bt toxins. On the other side were farmers who used conventional pest control and the insecticide industry itself. They brought up different challenges. For example, how can models and lab data assure that the predicted long-term gains from refuges will overcome the very real short-term costs?

Although most of the current opposition to transgenic crops is voiced as concern about health and the environment, there are strong underlying political and ethical concerns that are felt but cannot legally be used as arguments against transgenic technologies that are privatizing ownership of crop seeds built by blending the genes of eukaryotic and prokaryotic organisms (Buttel 2005). Now, as we begin discussing the potential for releasing transgenic pests into the environment, deep-seated ethical and political concerns are again being heard.

Many factors distinguish crop biotechnology from pest biotechnology in both the potential risks and benefits. In some ways, engineered pests seem more benign, but in others, they carry more risk than engineered crops. Transgenic crops could be viewed as a means for capitalism and imperialism to take over a public good, with risks of environmental degradation borne by the wider society. It is hard to put a transgenic mosquito that stops malaria transmission in this category. On one hand, if there is a problem with a transgenic crop, it should generally be possible to recall it, because it cannot survive without human intervention. On the other hand, once a mosquito with a *Medea* or homing endonuclease construct is released, it is expected to spread over large geographic areas, without respect for political borders or private property. Recall is unlikely to be practical, so it is important for researchers to build transgenic strains that work well and last. Perhaps more difficult is the final need to explain to stakeholders why such strains are expected to be safe, effective, and persistent.

From my perspective, the biggest risk entailed in fighting dengue and malaria with transgenic mosquitoes is an evolutionary risk: that is, the chance that a genetic pest management strategy works incredibly well for 5–10 years and then fails due to insect or pathogen evolution. Unlike failure of a Bt crop plant for which the loss can mostly be measured in dollars, the unexpected loss of efficacy of a transgenic mosquito could result in a cohort of children with no herd immunity being exposed to the full onslaught of disease (Gould et al. 2006b). Even worse, after 5–10 years of not worrying about dengue or malaria, health ministries could become less prepared to deal with the epidemic. This risk does not mean that researchers and policy makers should give up on transgenic pest technology, but it does mean that precaution is crucial. Even after the most rigorous assessments of safety, careful monitoring of the efficacy of these transgenic mosquitoes would be critical.

In response to these evolutionary risks, we conceived of a different gene-drive system that is expected to increase the frequency of anti-pathogen genes in a local population for about 50 generations, after which the frequency would decline (Gould et al. 2008). The idea behind this approach is that it would test the effectiveness and evolutionary stability of an anti-pathogen gene for long enough to assess whether more permanent gene-drive mechanisms would be sufficiently safe to use. The many examples in this chapter make it clear that one can never make perfect predictions about the results of evolutionary management of pests and pathogens, but experiments and models are an obvious advancement from anecdotal information and verbal argument.

Transgenic mosquitoes are likely to be less controversial than other engineered pests (Gould 2008). For example, a very difficult decision will need to be made if it becomes possible to engineer tsetse flies not to vector trypanosomes. In the short-term, the spread of a tsetse strain that did not carry trypanosomes could be an immense relief to poor farmers in the central belt of Africa because it would allow them to bring in cattle to help plow new land without the threat of sleeping sickness for themselves or death of their cattle from trypanosomiasis. Unfortunately, the long-term effects of such engineered tsetse flies on this incredibly biodiverse region could be negative. If elimination of tsetse flies resulted in an increase in small, low-input farms, the environmental change might not cause substantial losses to biodiversity (Wilson et al. 1997), but few in-depth studies have been conducted. Furthermore, it is feasible that larger, more environmentally disruptive farms would be developed in those areas. These kinds of problems make many evolutionary biologists decide to head back to academic pursuits, which is understandable. However, these problems do need to be addressed by society, and evolutionary biologists have important insights to contribute.

Looking back over the past 100 years at contributions made by evolutionary biologists to society, it is clear that some mistakes were made (e.g., eugenics), at least through the lens of current societal norms. However, for the most part, insights from evolutionary biology have improved the quality of human life. As the tools of evolutionary biology become more powerful and precise in coming decades, there will likely be many opportunities to contribute further. It is interesting to ponder how society 50 years from now will view the current efforts to apply evolutionary biology.

Acknowledgments

I thank the organizers for giving me the opportunity to participate in the exciting workshop leading to this chapter. M. Bell and B. Tabashnik provided very helpful comments on an earlier draft of the chapter. Interactions fostered by the NIH RAPIDD project on vector-borne diseases improved this work.

Literature Cited

Agrawal, S., R. Singh, I. Sanyal, and 1 other. 2008. Expression of modified gene encoding functional human alpha-1-antitrypsin protein in transgenic tomato plants. *Transgenic Res.* 17: 881–896.

Aitken, S. N., S. Yeaman, J. A. Holliday, and 2 others. 2008. Adaptation, migration or extirpation: Climate change outcomes for tree populations. *Evol. Appl.* 1: 95–111.

Ali, M. I., R. G. Luttrell, and S. Y. Young. 2006. Susceptibilities of *Helicoverpazea* and *Heliothisvirescens* (Lepidoptera: Noctuidae) populations to Cry1Ac insecticidal protein. *J. Econ. Entomol.* 99: 164–175.

Alphey, L. 2002. Re-engineering the sterile insect technique. *Insect Biochem. Mol. Biol.* 32: 1243–47.

Andow, D. A., D. M. Olson, R. L. Hellmich, and 2 others. 2000. Frequency of resistance to *Bacillus thuringiensis* toxin Cry1Ab in an Iowa population of European corn borer (Lepidoptera: Crambidae). *J. Econ. Entomol.* 93: 26–30.

Armstrong, D. P. and P. J. Seddon. 2008. Directions in reintroduction biology. *Trends Ecol. Evol.* 23: 20–25.

Ashok, M., C. Turner, and T. G. Wilson. 1998. Insect juvenile hormone resistance gene homology with the bHLH-PAS family of transcriptional regulators. *Proc. Natl. Acad. Sci. USA* 95: 2761–2766.

Bagla P. 2010. India: Hardy cotton-munching pests are latest blow to GM crops. *Science* 327: 1439–1439.

Beeman, R. W., K. S. Friesen, and R. E. Denell. 1992. Maternal-effect selfish genes in flour beetles. *Science* 256: 89–92.

Bourguet, D., J. Chaufaux, M. Séguin, and 7 others. 2003. Frequency of alleles conferring resistance to Bt maize in French and US corn belt populations of the European corn borer, *Ostrinia nubilalis*. *Theor. Appl. Genet.* 106: 1225–1233.

Bowman, H. G. 1981. Insect response to microbial infections. In H. D. Burges (ed.), *Microbial Control of Pests and Plant Diseases 1970–1980*, pp.769–784. Academic Press, New York.

Brown, A. W. A. 1971. *Insecticide Resistance in Arthropods*. World Health Organization, Geneva.

Burt, A. 2003. Site-specific selfish genes as tools for the control and genetic engineering of natural populations. *Proc. Roy. Soc. Lond. B.* 270: 921–928.

Burt, A. and R. Trivers. 2006. *Genes in Conflict: The Biology of Selfish Elements*. Belknap Press, Cambridge, MA.

Buttel, F. H. 2005.The environmental and post-environmental politics of genetically modified crops and food. *Environ. Pol.* 14: 309–323.

Caprio, M. A. and B. E. Tabashnik. 1992. Gene flow accelerates local adaptation among finite populations: Simulating the evolution of insecticide resistance. *J. Econ. Entomol.* 85: 611–620.
Carareto, C. M. A., W. Kim, M. F. Wojciechowski, and 4 others. 1997. Testing transposable elements as genetic drive mechanisms using *Drosophila* P element constructs as a model system. *Genetica* 101: 13–33.
Ceccatti, J. S. 2009. Natural selection in the field: Insecticide resistance, economic entomology, and the evolutionary synthesis 1914–1951. *Trans. Am. Phil. Soc.* 99: 199–217.
Center for Science in the Public Interest. 2009. *Complacency on the Farm*. CSPI, Washington, DC.
Cerf, D. C. and G. P. Georghiou. 1972. Evidence of cross-resistance to a juvenile hormone analogue in some insecticide-resistant houseflies. *Nature* 239: 401–402.
Chen, C. H., H. Huang, C. M. Ward, and 4 others. 2007. A synthetic maternal-effect selfish genetic element drives population replacement in *Drosophila*. *Science* 316: 597–600.
Chevillon, C., M. Raymond, T. Guillemaud, and 2 others. 1999. Population genetics of insecticide resistance in the mosquito *Culex pipiens*. *Biol. J. Linn. Soc.* 68: 147–157.
Comins, H. N. 1977a. The development of insecticide resistance in the presence of migration. *J. Theor. Biol.* 64: 177–197.
Comins, H. N. 1977b.The management of pesticide resistance. *J. Theor. Biol.* 65: 399–420.
Croft, B. A. and K. Strickler. 1983. Natural enemy resistance to pesticides: Documentation, characterization, theory and application. In G. P. Georghiou and T. Saito (eds.), *Pest Resistance to Pesticides*, pp. 669–702. Plenum, New York.
Crow, J. F. 1957. Genetics of insect resistance to chemicals. *Ann. Rev. Entomol.* 2: 227–246.
Curtis, C. F. 1968. Possible use of translocations to fix desirable genes in insect pest populations. *Nature* 218: 368–369.
Curtis, C. F. 1977. Testing systems for the genetic control of mosquitoes. In D. White (ed.), *XV International Congress of Entomology*, pp. 106–16. Entomological Society of America, College Park.
Curtis, C. F. 1985a. Theoretical models of the use of insecticide mixtures for the management of resistance. *Bull. Entomol. Res.* 75: 259–265.
Curtis, C. F. 1985b. Genetic control of insect pests: Growth industry or lead balloon? *Biol. J. Linn. Soc.* 26: 359–74.
Davis, P. M. and D. W. Onstad. 2000. Seed mixtures as a resistance management strategy for European corn borers (Lepidoptera: Crambidae) infesting transgenic corn expressing Cry1Ab protein. *J. Econ. Entomol.* 93: 937–948.
Dobzhansky, T. 1937. *Genetics and the Origin of Species*, 2nd ed. Columbia University Press, New York.
Doebley, J. F., B. S. Gaut, and B. D. Smith. 2006. The molecular genetics of crop domestication. *Cell* 127: 1309–1321.
Dunlop, E. S., K. Enberg, C. Jorgensen, and 1 other. 2009. Toward Darwinian fisheries management. *Evol. Appl.* 2: 246–259.
Faith, D. P. 2008. Threatened species and the potential loss of phylogenetic diversity: Conservation scenarios based on estimated extinction probabilities and phylogenetic risk analysis. *Cons. Biol.* 22: 1461–1470.

ffrench-Constant, R. H., P. J. Daborn, and G. Le Goff. 2004. The genetics and genomics of insecticide resistance. *Trends Genet.* 20: 163–170.
Fisher, R. A. 1924. The elimination of mental defect. *Eugenics Rev.* 16: 114–116.
Follett, P., F. Gould, and G. G. Kennedy. 1995. High-realism model of Colorado potatobeetle (Coleoptera: Chrysomelidae) adaptation to permethrin. *Environ. Entomol.* 24: 167–178.
Forest, F., R. Grenyer, M. Rouget, and 10 others. 2007. Preserving the evolutionary potential of floras in biodiversity hotspots. *Nature* 445: 757–760.
Fu, G. L., R. S. Lees, D. Nimmo, and 11 others. 2010. Female-specific flightless phenotype for mosquito control. *Proc. Natl. Acad. Sci. USA* 107: 4550–4554.
Georghiou, G. P. 1972. The evolution of resistance to pesticides. *Ann. Rev. Ecol. Syst.* 3: 133–168.
Georghiou, G. P. and A. Lagunes-Tejeda. 1991. *The Occurrence of Resistance to Pesticides in Arthropods.* Food and Agriculture Organization of the United Nations, Rome.
Georghiou, G. P. and C. E. Taylor. 1977. Genetic and biological influences in evolution of insecticide resistance. *J. Econ. Entomol.* 70: 319–323.
Georghiou, G. P. and C. E. Taylor. 1986. Factors influencing the evolution of resistance. In *Pesticide Resistance: Strategies and Tactics for Management*, pp. 157–169. National Academy of Sciences, Washington, DC.
Gordon, H. T. 1961. Nutritional factors in insect resistance to chemicals. *Ann. Rev. Entomol.* 6: 27–54.
Gore, J., J. J. Adamczyk, A. Catchot, and 1 other. 2008. Yield response of dual-toxin Bt cotton to *Helicoverpa zea* infestations. *J. Econ. Entomol.* 101: 1594–1599.
Gould, F. 1986. Simulation models for predicting durability of insect-resistant germ plasm: Hessian fly (Diptera: Cecidomyiidae)-resistant winter wheat. *Environ. Entomol.* 15: 11–23.
Gould, F. 1988. Evolutionary biology and genetically engineered crops. *BioScience* 38: 26–33.
Gould, F. 1991. The evolutionary potential of crop pests. *Am. Sci.* 79: 496–507.
Gould, F. 1995. Comparisons between resistance management strategies for insects and weeds. *Weed Technol.* 9: 830–839.
Gould, F. 1998. Sustainability of transgenic insecticidal cultivars: Integrating pest genetics and ecology. *Ann. Rev. Entomol.* 43: 701–26.
Gould, F. 2008. Broadening the application of evolutionarily based genetic pest management. *Evolution* 62: 500–510.
Gould, F. and P. Schliekelman. 2004. Population genetics of autocidal control and strain replacement. *Ann. Rev. Entomol.* 49: 193–217.
Gould, F., A. Anderson, A. Jones, and 6 others. 1997. Initial frequency of alleles for resistance to *Bacillus thuringiensis* toxins in field populations of *Heliothis virescens*. *Proc. Natl. Acad. Sci. USA* 94: 3519–3523.
Gould, F., A. Anderson, D. Landis, and 1 other. 1991a. Feeding behavior and growth of *Heliothis virescens* larvae on diets containing *Bacillus thuringiensis* formulations or endotoxins. *Entomol. Exp. Appl.* 58: 199–210.
Gould, F., C. R. Carroll, and D. J. Futuyma. 1982. Cross resistance to pesticides and plant defenses: A study of the two-spotted spider mite. *Entomol. Exp. Appl.* 31: 175–180.
Gould, F., M. B. Cohen, J. S. Bentur, and 2 others. 2006a. Impact of small fitness costs on pest adaptation to crop varieties with multiple toxins: A heuristic model. *J. Econ. Entomol.* 99: 2091–2099.

Gould, F., Y. Huang, M. Legros, and 1 other. 2008. A killer-rescue system for self-limiting gene drive of anti-pathogen constructs. *Proc. Roy. Soc. Lond. B.* 275: 2823–2829.

Gould, F., G. G. Kennedy, and M. T. Johnson. 1991b. Effects of natural enemies on the rate of herbivore adaptation to resistant host plants. *Entomol. Exp. Appl.* 58: 1–14.

Gould, F., K. Magori, and Y. Huang. 2006b. Genetic strategies for controlling mosquito-borne diseases. *Am. Sci.* 94: 238–246.

Gould, F., A. Martinez-Ramirez, A. Anderson, and 3 others. 1992. Broad-spectrum resistance to *Bacillus thuringiensis* toxins in *Heliothis virescens*. *Proc. Natl. Acad. Sci. USA* 89: 7986–7990.

Gressel, J. 2009. Evolving understanding of the evolution of herbicide resistance. *Pest Mgmt. Sci.* 65: 1164–1173.

Gustafson, D. I., G. P. Head, and M. A. Caprio. 2006. Modeling the impact of alternative hosts on *Helicoverpa zea* adaptation to Bollgard cotton. *J. Econ. Entomol.* 99: 2116–2124.

Harper, J. L. 1956.The evolution of weeds in relation to resistance to herbicides. *Proc. 3rd Brit. Weed Ctrl. Conf.* 3: 179–188.

Hartley, C. J., R. D. Newcomb, R. J. Russell, and 5 others. 2006. Amplification of DNA from preserved specimens shows blowflies were preadapted for the rapid evolution of insecticide resistance. *Proc. Natl. Acad. Sci. USA* 103: 8757–8762.

Heap, I. M. 2009. International Survey of Herbicide-Resistant Weeds. *http://www.weedscience.org*.

Hoffmann, A. A. and Y. Willi. 2008. Detecting genetic responses to environmental change. *Nat. Rev. Gen.* 9: 421–432.

Huang, Y., A. L. Lloyd, M. Legros, and 1 other. 2009. Gene drive in age-structured insect populations. *Evol. Appl.* 2: 143–159.

Hueth, D. and U. Regev. 1974. Optimal agricultural pest management with increasing resistance. *Am. J. Agric. Econ.* 56: 543–551.

Huffaker, C. B. 1971. The ecology of pesticide interference with insect populations. In J. E. Swift (ed.), *Agricultural Chemicals—Harmony or Discord for Food, People, and the Environment*, pp. 92–107. University of California, Berkeley.

Johnson, M. T. and F. Gould. 1992. Interaction of genetically engineered host plant resistance and natural enemies of *Heliothis virescens* (Lepidoptera: Noctuidae) in tobacco. *Environ. Entomol.* 21: 586–597.

Johnson, M. T., F. Gould, and G. G. Kennedy. 1997. Effect of an entomopathogen on adaptation of *Heliothis virescens* populations selected on resistant host plants. *Entomol. Exp. Appl.* 83: 121–135.

Klassen, W. and C. F. Curtis. 2005. History of the sterile insect technique. In V. A. Dyck, J. Hendrichs, and A. S. Robinson (eds.), *Sterile Insect Techniques: Principles and Practice in Area-Wide Integrated Pest Management*. Springer, Dordrecht.

Koziel, M. G., F. L. Belang, C. Bowman, and 12 others. 1993. Field performance of elite transgenic maize plants expressing an insecticidal protein derived from *Bacillus thuringiensis*. *Biotechnology* 11: 194–200.

Labbe, P., C. Berticat, A. Berthomieu, and 4 others. 2007. Forty years of erratic insecticide resistance evolution in the mosquito *Culex pipiens*. *PLoS Genet.* 3: 2190–2199.

Lamberti, F., J. M. Waller, and N. A. Van derGraaf (eds.). 1983. *Durable Resistance in Crops*. Plenum Press, New York.

Leonard, K. J. 1986. The host population as a selective factor. In M. S. Wolfe and C. E. Caten (eds.), *Populations of Plant Pathogens: Their Dynamics and Genetics*, pp. 163–179. Blackwell, Oxford.

Li, X., A. R. Zangerl, M. A. Schuler, and 1 other. 2000. Cross-resistance to alpha-cypermethrin after xanthotoxin ingestion in *Helicoverpa zea (Lepidoptera: Noctuidae). J. Econ. Entomol.* 93: 18–25.

Lindroth, R. L. 1989. Differential esterase activity in *Papilioglaucus* subspecies: Absence of cross-resistance between allelochemicals and insecticides. *Pestic. Biochem. Physiol.* 35: 185–191.

Loera-Gallardo, J., D. A. Wolfenbarger, and J. W. Norman. 1997. Toxicity of azinphosmethyl, methyl parathion, and oxamyl against the boll weevil (Coleoptera: Curculionidae) in Texas, Mexico, and Guatemala. *J. Agric. Entomol.* 14: 355–361.

Lorenzen, M. D., A. Gnirke, J. Margolis, and 6 others. 2008. The maternal-effect, selfish genetic element *Medea* is associated with a composite Tc1 transposon. *Proc. Natl. Acad. Sci. USA* 105: 10085–10089.

Magori, K., M. Legros, M. Puente, and 4 others. 2009. Skeeter Buster: A stochastic, spatially explicit modeling tool for studying *Aedes aegypti* population replacement and population suppression strategies. *PLoS Negl. Trop. Dis.* 3: e508.

Mao, Y. B., W. J. Cai, J. W. Wang, and 5 others. 2007. Silencing a cotton bollworm P450 monooxygenase gene by plant-mediated RNAi impairs larval tolerance of gossypol. *Nat. Biotech.* 25: 1307–1313.

Matten, S. 2007. *Review of Dow AgroSciences (and Pioneer HiBreds) Submission (dated July 12, 2007) Regarding Fall Armyworm Resistance to the Cry1F Protein Expressed in TC1507 Herculex I Insect Protection Maize in Puerto Rico.* MRID#: 471760D01.

McKenzie, J. A. 1996. *Ecological and Evolutionary Aspects of Insecticide Resistance.* R. G. Landes, Austin.

Melander, A. L. 1914. Can insects become resistant to sprays? *J. Econ. Entomol.* 7: 167–173.

Moore, W. 1933. Studies of the "resistant" California red scale *Aonidiella aurantii* mask in California. *J. Econ. Entomol.* 26: 1140–1161.

Moscarella, R. A., M. Aguilera, and A. A. Escalante. 2003. Phylogeography, population structure, and implications for conservation of white-tailed deer (*Odocoileus virginianus*) in Venezuela. *J. Mammal.* 84: 1300–1315.

Muller, H. J. 1935. *Out of the Night: A Biologist's View of the Future*. Vanguard, New York.

National Academy of Science (NAS). 1978. *Conservation of Germplasm Resources: An Imperative.* National Academy of Sciences, Washington, DC.

National Research Council. 1986. *Pesticide Resistance: Strategies and Tactics for Management.* National Academy Press, Washington, DC.

Neve, P., M. Vila-Aiub, and F. Roux. 2009. Evolutionary thinking in agricultural weed management. *New Phytol.* 184: 783–793.

Onstad, D. W. 2008. *Insect Resistance Management: Biology, Economics, and Prediction.* Academic Press, London.

Onstad, D. W. and F. Gould. 1998. Modeling the dynamics of adaptation to transgenic maize by European corn borer (Lepidoptera: Pyralidae). *J. Econ. Entomol.* 91: 585–593.

Papathanos, P. A., N. Windbichler, M. Menichelli, and 2 others. 2009. The vasa regulatory region mediates germline expression and maternal transmission of proteins in the malaria mosquito *Anopheles gambiae*: A versatile tool for genetic control strategies. *BMC Mol. Biol.* 10: 65.

Parlevliet, J. E. 1983. Models explaining the specificity and durability of host resistance derived from the observations on the barley-*Pucciniahordei* system. In F. Lamberti, J. M. Waller, and N. A. Van derGraaf (eds.), *Durable Resistance in Crops*, pp. 57–80. Plenum Press, New York.

Parlevliet, J. E. and H. J. Kuiper. 1985. Accumulating polygenes for partial resistance in barley to barley leaf rust, *Puccinia hordei*. 1. Selection for increased latent periods. *Euphytica* 34: 7–13.

Pearson, K., and M. Moul. 1925. The problem of alien immigration into Great Britain, illustrated by an examination of Russian and Polish Jewish children. *Eugenics* 1: 5–127.

Peck, S., F. Gould, and S. Ellner. 1999. Spread of resistance in spatially extended regions of transgenic cotton: Implications for management of *Heliothis virescens* (Lepidoptera: Noctuidae). *J. Econ. Entomol.* 92: 1–16.

Perlak, F. J., R. L. Fuchs, D. A. Dean, and 2 others. 1991. Modification of the coding sequence enhances plant expression of insect control protein genes. *Proc. Natl. Acad. Sci. USA* 88: 3324–3328.

Pickersgill, B. 2009. Domestication of plants revisited—Darwin to the present day. *Bot. J. Linn. Soc.* 161: 203–212.

Plapp, F. W., C. R. Browning, and P. J. H. Sharpe. 1979. Analysis of the development of insecticide resistance based on simulation of a genetic model. *Environ. Entomol.* 8: 494–500.

Quayle, H. J. 1922. Resistance of certain scale insects in certain localities to hydrocyanic acid fumigation. *J. Econ. Entomol.* 15: 400–404.

Quayle, H. J. 1943. The increase in resistance in insects to insecticides. *J. Econ. Entomol.* 36: 493–500.

Raymond, M., A. Callaghan, P. Fort, and 1 other. 1991. Worldwide migration of amplified insecticide resistance genes in mosquitoes. *Nature* 350: 151–153.

Ribeiro, J. M. C. and M. G. Kidwell. 1994. Transposable elements as population drive mechanisms: Specification of critical parameter values. *J. Med. Entomol.* 31: 10–16.

Rosenheim, J. A. and B. E. Tabashnik. 1990. Evolution of pesticide resistance—interactions between generation time and genetic, ecological, and operational factors. *J. Econ. Entomol.* 83: 1184–1193.

Rosenheim, J. A. and B. E. Tabashnik. 1991. Influence of generation time on the rate of response to selection. *Am. Nat.* 137: 527–541.

Roush, R. T. 1989. Designing resistance management programs: How can you choose? *Pestic. Sci.* 26: 1989. 423–441.

Roush, R. T. and B. E. Tabashnik (eds.). 1990. *Pesticide Resistance in Arthropods.* Chapman and Hall, New York.

Schliekelman, P. and F. Gould. 2000. Pest control by the release of insects carrying a female-killing allele on multiple loci. *J. Econ. Entomol.* 93: 1566–79.

Serebrovskii, A. S. 1940. On the possibility of a new method for the control of insect pests. *Zool. Zh.* 19: 618–90.

Servin, B., O. C. Martin, M. Mezard, and 1 other. 2004. Toward a theory of marker-assisted gene pyramiding. *Genetics* 168: 513–523.

Shi, A. N., P. Y. Chen, D. X. Li, and 3 others. 2009. Pyramiding multiple genes for resistance to soybean mosaic virus in soybean using molecular markers. *Mol. Breeding* 23: 113–124.

Sinkins, S. P. and F. Gould. 2006. Gene drive systems for insect disease vectors. *Nat. Rev. Genet.* 7: 427–435.

Sokal, R. R. and P. E. Hunter. 1954. Reciprocal selection for correlated quantitative characters in *Drosophila*. *Science* 119: 649–651.
Tabashnik, B. E. 1990. Modeling and evaluation of resistance management tactics. In R. T. Roush and B. E. Tabashnik (eds.), *Pesticide Resistance in Arthropods*, pp. 153–182. Chapman and Hall, New York.
Tabashnik, B. E. and B. A. Croft. 1985. Evolution of pesticide resistance in apple pests and their natural enemies. *Entomophaga* 30: 37–49.
Tabashnik, B. E., N. L. Cushing, N. Finson, and 1 other. 1990. Field development of resistance to *Bacillus thuringiensis* in diamondback moth (Lepidoptera: Plutellidae). *J. Econ. Entomol.* 83: 1671–1676.
Tabashnik, B. E., F. Gould, and Y. Carriere. 2004. Delaying evolution of insect resistance to transgenic crops by decreasing dominance and heritability. *J. Evol. Biol.* 17: 904–912.
Tabashnik, B. E., A. L. Patin, T. J. Dennehy, and 4 others. 2000. Frequency of resistance to *Bacillus thuringiensis* in field populations of pink bollworm. *Proc. Natl. Acad. Sci. USA* 97: 12980–12984.
Tabashnik, B. E., J. B. Van Rensburg, and Y. Carriere. 2009. Field-evolved insect resistance to Bt crops: Definition, theory, and data. *J. Econ. Entomol.* 102: 2011–2025.
Tang, J. D., H. L. Collins, T. D. Metz, and 4 others. 2001. Greenhouse tests on resistance management of Bt transgenic plants using refuge strategies. *J. Econ. Entomol.* 94: 240–247.
Taylor, C. E. and G. P. Georghiou. 1979. Suppression of insecticide resistance by alteration of gene dominance and migration. *J. Econ. Entomol.* 72: 105–109.
Temime, L., G. Hejblum, M. Setbon, and 1 other. 2008. The rising impact of mathematical modeling in epidemiology: Antibiotic resistance research as a case study. *Epid. Infect.* 136: 289–298.
US-EPA. 1998a. The environmental protection agency's white paper on Bt plant-pesticide resistance management. *http://www.epa.gov/EPAPEST/1998/January/Day-4/paper.pdf.*
US-EPA. 1998b. Final report of the subpanel on *Bacillus thuringiensis* (Bt) plant-pesticides and resistance management. *http://www.epa.gov/scipoly/sap/meetings/1998/0298_mtg.htm.*
US-EPA. 2001. Biopesticides registration action document: *Bacillus thuringiensis* plant-incorporated protectants. *http://www.epa.gov/pesticides/biopesticides/pips/bt_brad.htm.*
Uyenoyama, M. K. 1986. Pleiotropy and the evolution of genetic systems conferring resistance to pesticides. *In Pesticide Resistance: Strategies and Tactics for Management*, pp. 207–221. National Academy of Sciences, Washington, DC.
Vanderplank, F. L. 1944. Hybridization between *Glossina* species and suggested new method for control of certain species of tsetse. *Nature* 154: 607–608.
Vanderplank, F. L. 1947. Experiments on the hybridization of tsetse flies and the possibility of a new method of control. *Trans. Roy. Entomol. Soc. Lond.* 98: 1–18.
Vanderplank, J. E. 1968. *Disease Resistance in Plants.* Academic Press, New York.
Van Rensburg, J. B. J. 2007. First report of field resistance by stem borer, *Busseola-fusca* (Fuller) to Bt-transgenic maize. *S. African J. Plant Soil* 24: 147–151.
Vavilov, N. 1926. *Tsentry proizkhozhdeniia kulturnykh rastenii.* Leningrad: Vsesoiuznyi institute prikladnoibotanki. (Contains extensive English translation. *"Studies on the origin of cultivated plants."*)

Via, S. 1986. Quantitative genetic models and the evolution of pesticide resistance. In *Pesticide Resistance: Strategies and Tactics for Management*, pp. 222–235. National Academy of Sciences. Washington, DC.
Wade, M. J. and R. W. Beeman. 1994. The population dynamics of maternal-effect selfish genes. *Genetics* 138: 1309–1314.
Weiner, R. and J. F. Crow. 1951. The resistance of DDT-resistant *Drosophila* to other insecticides. *Science* 113: 403–404.
Williams, C. M. 1967. Third-generation pesticides. *Sci. Am.* 217: 13–17.
Williams, S., L. Friedrich, S. Dincher, and 4 others. 1992. Chemical regulation of *Bacillus thurgingiensis* delta-endotoxin expression in transgenic plants. *Biotechnology* 10: 540–543.
Wilson, C. J., R. S. Reid, N. L. Stanton, and 1 other. 1997. Effects of land use and tsetse fly control on bird species richness in southwestern Ethiopia. *Cons. Biol.* 11: 435–447.
Wilson, T. G. and M. Ashok. 1998 Insecticide resistance resulting from an absence of target-site gene product. *Proc. Natl. Acad. Sci. USA* 95: 14040–14044.
Yakob, L., I. Z. Kiss, and M. B. Bonsall. 2008. A network approach to modeling population aggregation and genetic control of pest insects. *Theor. Pop. Biol.* 74: 324–331.
Zaidi, M. A., G. Y. Ye, H. W. Yao, and 5 others. 2009. Transgenic rice plants expressing a modified Cry1Ca1 gene are resistant to *Spodoptera litura* and *Chilo suppressalis*. *Mol. Biotech.* 43: 232–242.
Zhao, F. P., L. Jiang, H. J. Gao, and 2 others. 2009. Design and comparison of gene-pyramiding schemes in animals. *Animal* 3: 1075–1084.

Commentary Six

A Clade's-Eye View of Global Climate Change

Charles C. Davis, Erika J. Edwards, and Michael J. Donoghue

Recent climate change has had demonstrable effects on plant and animal communities around the world and will pose one of the most significant threats to biodiversity in the coming decades (Walther et al. 2002; Root et al. 2003; Parmesan 2006). Surprisingly, evolutionary biologists have had rather little impact on increasing the understanding of climate change and its consequences for biodiversity (though, there have been some studies of cases of rapid evolution (see Donoghue et al. 2009; Hendry et al. 2010), and phylogenetic approaches to this problem have been limited. But, as in numerous areas of biology (Futuyma 2004; see Hillis, Chapter 16), we see opportunities to use phylogenetic trees to make generalizations of practical importance and to predict responses to climate change. Here, our aim is to briefly highlight, by reference to several recently published examples, some of the ways in which phylogenies might be used to understand and cope with climate change. We focus on plant examples given our expertise, but there are several relevant animal examples that also support our points (e.g., Wiens et al. 2006).

Phylogeny, Historical Climate Change, and Climate Niche Evolution

Phylogenetic trees are commonly used to infer character evolution and historical biogeography and, in turn, to correlate changes with major climatic events in the past or with movements into novel environments. An emerging theme from such studies is that many lineages have persisted in particular biomes for much of their history, despite considerable opportunity (afforded by dispersal and shifting climates) to diversify into other zones (Donoghue 2008; Crisp et al. 2009). This form of phylogenetic niche conservatism is argued to be a major determinant of global biodiversity patterns, for example, the latitudinal species richness gradient (Ricklefs 2004;

Wiens and Donoghue 2004; Mittelbach et al. 2007). The apparent rarity of niche shifts from tropical to temperate biomes may reflect an underlying difficulty in making certain physiological adjustments, such as the evolution of frost tolerance (Donoghue 2008).

However, at the same time, many cases of niche evolution within biomes have been documented (Losos 2008). Plant examples include evening primroses in the arid regions of North America (Evans et al. 2009), honeysuckles in Mediterranean climates (Smith and Donoghue 2010), and subtropical laurel species from Europe and North Africa (Rodríguez-Sánchez and Arroyo 2008). Such studies integrate climate niche models and dated phylogenies to characterize a clade's biogeographic history and are useful in determining the abiotic factors that influence species' ranges. Thus, they can be valuable in assessing responses to future climate change. They are not without their limitations, however, and are unlikely to be particularly helpful in extracting predictions from more ancient evolutionary events for which adequate species occurrence data as well as precise information on paleoclimate and land configurations are often lacking. Examples of studies that have sought to make inferences about a clade's ancestral niche in deep time, independent of a particular geography, include the origin of water-use strategies in Cactaceae (Edwards and Donoghue 2006) and the age and persistence of tropical rain forest clades (Davis et al. 2005).

Interpreting such patterns in relation to current climate change provides some broad generalizations. For instance, based on historical patterns, it seems likely that if the Amazon basin dries (Cox et al. 2004), its newly formed arid community will likely be assembled via migration of pre-adapted desert lineages rather than via significant *in situ* evolution of rainforest species. Thus, the historical importance of habitat tracking in response to climate change should motivate the maintenance of viable corridors for movement. In contrast, many climate shifts may fall within the range of a clade's adaptive potential; shifting precipitation patterns across the already arid American Southwest, for example, may promote rapid adaptive evolution in lineages that already inhabit this zone, as has apparently occurred over the past million years (Evans et al. 2009). Ultimately, by summing over such historical studies, it should be possible to make meaningful generalizations about the relative evolutionary lability of niche-related traits.

Such examples are perhaps the most obvious observations that a clade's-eye view of global climate change affords, but for several reasons, they may not always be very useful in confronting the practical realities of ongoing climate change. Modern landscapes have been so thoroughly modified by humans that it is no longer clear how species will respond. The resulting fragmentation reduces population sizes and impedes habitat tracking, which makes it difficult to draw comparisons with past events. Likewise, the human-induced movement of species around the globe makes it much more difficult to predict how these new, artificial communities will behave (Sax et al. 2007). Thus, although studies of responses to past climate changes provide useful history lessons and a deeper perspective on the relationship between the biosphere and climate, there may often be more powerful ways to incorporate phylogenetic thinking into decision-making and climate change mitigation efforts. These approaches all stem from the basic principle that phylogenies provide the ultimate framework for inferring how different attributes are distributed among organisms, which then can be used to make predictions about the traits of species that have not yet been studied in detail. As we emphasize in the following sections, simple clade-based approaches can help researchers to identify potentially vulnerable lineages and can be used to produce refined global models relevant to understanding the biological impacts of climate change.

Phylogeny and Responses to Current Climate Change

The study by Willis et al. (2008) of rapid changes in species abundances in Thoreau's woods (Concord, Massachusetts, USA) provides a concrete example of the predictive power of phylogeny. They analyzed a data set initiated by the American conservationist Henry David Thoreau. Using statistical methods that incorporate phylogeny, they discovered that entire clades, which apparently have been less able to respond to climate change by adjusting their flowering phenology, have significantly declined in abundance. These results can help identify species and clades that likely face a greater risk of regional extinction as climate change proceeds. For example, we should be particularly concerned about the continued regional loss of species in the Liliaceae and Orchidaceae, but perhaps less so of species in the Brassicaceae and Fabaceae. The latter two clades contain species that are far better able to adjust their phenology to climate change and, thus, contain fewer species that have declined in abundance.

To what extent do these regional results relate to global patterns of decline? For example, should we be concerned about a potential worldwide decline of Orchidaceae due to climate change? If so, the outcome could be severe, and would have especially devastating impacts on regions with low community-wide phylodiversity. For instance, the fynbos of the Cape Floristic Province of southern Africa contains about 9000 plant species, of which approximately 70% are endemic to an area of 90,000 km^2 (Linder and Hardy 2004). This richness is comparable to some Neotropical forests and is significantly greater than other Mediterranean-type ecosystems (Kreft and Jetz 2007). Nevertheless, the fynbos flora is relatively clade-poor with respect to overall plant diversity (Linder and Hardy 2004; Forest et al. 2007). Thus, if the species belonging to any one of the small number of species-rich clades that compose the fynbos flora were negatively affected by climate change, it could disproportionately increase the magnitude of species loss in this system. Along these lines, the Asteraceae, Iridaceae, and Orchidaceae are three species-rich clades that have been shown to be relatively unable to adjust to climate change and are in significant decline in Concord. These three clades compose nearly 25% of the fynbos flora (Kruger and Taylor 1980). If the inability of these clades to respond to climate change extends across communities (i.e., from the temperate region of Concord to the fynbos of South Africa), then climate change-induced losses in the fynbos flora could be far greater than those sustained in phylogenetically more diverse communities. In the end, assessing the likelihood of such scenarios, will require more and better information on the geographic distribution of clades and phylogenetic diversity, a better understanding of the extent to which clade membership predicts climate change response, and knowledge of the regional abiotic factors that influence clade vulnerability across communities and biomes.

A Clade's-Eye View of Ecosystems

The Concord example highlights the utility of phylogenies in predicting how species and clades may respond to climate change. However, climate change research is charged with prediction making at many scales, and some of the most important problems are at the level of biosphere–atmosphere interactions and global biogeochemical cycles. Though less intuitive, a phylogenetic approach can provide important insights into these large-scale problems. Here, we highlight one study that outlined how integrating phylogenetic biology with ecosystem ecology could improve predictions of global carbon cycling under future climate change.

Edwards et al. (2007) explored a dataset of carbonic anhydrase (CA) levels in leaves. CA activity influences leaf oxygen isotope fractionation, which in turn, is used to estimate global primary production (GPP) and the role of terrestrial vegetation in the global carbon cycle (Gillon and Yakir 2001). Gillon and Yakir sampled a number of species for CA activity and found that grasses using the C_4 photosynthetic pathway as a whole had lower CA values than other functional types (e.g., trees, forbs). This result suggested that previous analyses may have grossly underestimated the role of C_4 grasses in GPP and that C_4 grasses may constitute the so-called missing carbon sink (Gillon and Yakir 2001). Reanalyzing these data within a phylogenetic framework, Edwards et al. (2007) found that CA levels were not significantly correlated with C_4 photosynthesis. Instead, one major subclade of grasses, which included a mix of C_3 and C_4 species, contained most of the species with low CA values. On this basis, they suggested the direct use of the low CA clade in calculating GPP. To do so, however, will require better information on the geographic distribution and relative abundance of this clade as well as the development of global carbon models that can take into account new, user-defined vegetation categories (as opposed to only traditional functional or taxonomic categories). More generally, these results nicely clarify the way in which taking phylogeny into consideration can help to refine models that are directly relevant to climate change.

Concluding Thoughts

In all of the cases highlighted here, phylogenetic trees allow investigators to make generalizations that can help in making practical decisions and setting priorities when there is incomplete information. In the end, it is the predictive power of phylogenies that makes them useful in such a wide variety of applications, including understanding and dealing with climate change. Although there have been few concrete practical applications to this problem so far, we see great potential in such approaches and an urgent need for more rapid integration of phylogenetic biology and climate change research.

Literature Cited

Cox, P. M., R. A. Betts, M. Collins, and 3 others. 2004. Amazonian forest dieback under climate-carbon cycle projections for the 21st century. *Theor. Appl. Climatol.* 78: 137–156.

Crisp, M. D., M. T. K. Arroyo, L. G. Cook, and 7 others. 2009. Phylogenetic biome conservatism on a global scale. *Nature* 458: 754–756.

Davis, C. C., C. O. Webb, K. J. Wurdack, and 2 others. 2005. Explosive radiation of Malpighiales supports a mid-Cretaceous origin of tropical rain forests. *Am. Nat.* 165: E36–E65.

Donoghue, M. J. 2008. A phylogenetic perspective on the distribution of plant diversity. *Proc. Natl. Acad. Sci. USA* 105: 11549–11555.

Donoghue, M. J., T. Yahara, E. Conti, and 15 others. 2009. bioGENESIS: Providing an evolutionary framework for biodiversity science. *DIVERSITAS Report* 6: 1–52.

Edwards, E. J. and M. J. Donoghue. 2006. *Pereskia* and the origin of the cactus life-form. *Am. Nat.* 167: 777–793.

Edwards, E. J., C. J. Still, and M. J. Donoghue. 2007. The relevance of phylogeny to studies of global change. *Trends Ecol. Evol.* 22: 243–249.

Evans, M. E. K., S. A. Smith, R. S. Flynn, and 1 other. 2009. Climate, niche evolution, and diversification of the "bird-cage" evening primroses (*Oenothera*, sections *Anogra* and *Kleinia*). *Am. Nat.* 173: 225–240.

Forest, F., R. Grenyer, M. Rouget, and 10 others. 2007. Preserving the evolutionary potential of floras in biodiversity hotspots. *Nature* 445: 757–760.

Futuyma, D. J. 2004. The fruit of the tree of life: Insights into evolution and ecology. In J. Cracraft and M. J. Donoghue (eds.), *Assembling the Tree of Life*, pp. 25–39. Oxford University Press, New York.

Gillon, J. and D. Yakir. 2001. Influence of carbonic anhydrase activity in terrestrial vegetation on the ^{18}O content of atmospheric CO_2. *Science* 291: 2584–2587.

Hendry, A. P., L. G. Lohmann, E. Conti, and 15 others. 2010. Evolutionary biology in biodiversity science, conservation, and policy: A call to action. *Evolution* 64: 1527–1528.

Kreft, H. and W. Jetz. 2007. Global patterns and determinants of vascular plant diversity. *Proc. Natl. Acad. Sci. USA* 104: 5925–5930.

Kruger, F. J. and H. C. Taylor. 1980. Plant species diversity in Cape fynbos: Gamma and delta diversity. *Plant Ecol.* 41: 85–93.

Linder, H. P. and C. R. Hardy. 2004. Evolution of the species-rich Cape flora. *Phil. Trans. Roy. Soc. Lond. B* 359: 1623–1632.

Losos, J. B. 2008. Phylogenetic niche conservatism, phylogenetic signal and the relationship between phylogenetic relatedness and ecological similarity among species. *Ecol. Lett.* 11: 995–1003.

Mittelbach, G. G., D. W. Schemske, H. V. Cornell, and 19 others. 2007. Evolution and the latitudinal diversity gradient: Speciation, extinction and biogeography. *Ecol. Lett.* 10: 315–331.

Parmesan, C. 2006. Ecological and evolutionary responses to recent climate change. *Ann. Rev. Ecol. Evol. Syst.* 37: 637–669.

Ricklefs, R. E. 2004. A comprehensive framework for global patterns in biodiversity. *Ecol. Lett.* 7: 1–15.

Rodríguez-Sánchez, F. and J. Arroyo. 2008. Reconstructing the demise of Tethyan plants: Climate-driven range dynamics of *Laurus* since the Pliocene. *Global Ecol. Biogeogr.* 17: 685–695.

Root, T. L., J. T. Price, K. R. Hall, and 3 others. 2003. Fingerprints of global warming on wild animals and plants. *Nature* 421: 57–60.

Sax, D. F., J. J. Stachowicz, J. H. Brown, and 9 others. 2007. Ecological and evolutionary insights from species invasions. *Trends Ecol. Evol.* 22: 465–471.

Smith, S. A. and M. J. Donoghue. 2010. Combining historical biogeography with niche modeling in the *Caprifolium* clade of *Lonicera* (Caprifoliaceae, Dipsacales). *Syst. Biol.* 59: 322–341.

Walther, G. R., E. Post, P. Convey, and 6 others. 2002. Ecological responses to recent climate change. *Nature* 416: 389–395.

Wiens, J. J. and M. J. Donoghue. 2004. Historical biogeography, ecology, and species richness. *Trends Ecol. Evol.* 19: 639–644.

Wiens, J. J., C. H. Graham, D. S. Moen, and 2 others. 2006. Evolutionary and ecological causes of the latitudinal diversity gradient in hylid frogs: Treefrog trees unearth the roots of high tropical diversity. *Am. Nat.* 168: 579–596.

Willis, C. G., B. Ruhfel, R. B. Primack, and 2 others. 2008. Phylogenetic patterns of species loss in Thoreau's woods are driven by climate change. *Proc. Natl. Acad. Sci. USA* 105: 17029–17033.

Part VIII

PROSPECTS

Chapter 22

Evolutionary Biology: The Next 150 Years

Hopi E. Hoekstra

Darwin was arguably the most prescient thinker that biology has ever witnessed. But, if someone had asked him in 1859 where evolutionary biology would be in 150 years, would he have guessed correctly? He might have predicted that we would have a better understanding of how traits are inherited—a prediction borne out almost 50 years later with the rediscovery of Mendel's laws in 1900. Darwin considered the lack of understanding for how traits are inherited to be the missing link in his argument for evolution by natural selection, and when pushed, he devised his own theory (i.e., pangenesis), which was one of his few major errors. Yet, the ramifications of Mendel's experiments or of subsequent discoveries, like that of the three-dimensional DNA structure by Watson and Crick (1953) a century after *The Origin of Species* was published, along with the resultant technological advances, such as the ability to sequence a complete genome in another 50 years, were unknowable in his day. Nor could Darwin have anticipated the questions that have dominated the field since, such as the relative role of drift and selection in driving molecular evolution (Kimura 1968). With the acknowledgement that technologies, discoveries, and questions will arise that, likewise, cannot be imagined, it is useful—perhaps even stimulating—to speculate about what the next 150 (or more modestly, 50 or even 20) years will hold for evolutionary biology. The organizers of the Darwin 2009 Workshop asked me to speculate on what may lie ahead. My crystal ball, if I have one, is colored by evolutionary genetics and genomics—my main research area—and so necessarily are my predictions.

To predict the advances in the field of evolutionary biology, we (i.e., evolutionary biologists) must first set a direction. One overarching goal of evolutionary biology is to understand how the diversity of life evolved, and more specifically to understand how this variation, both genetic and phenotypic, is generated and maintained in nature. This aim spans sub-disciplines, ranging from the origin of life (see Lazcano, Chapter 14), paleontology (see P. Wagner, Chapter 17), and phylogenetics (see Hillis, Chapter 16) to theoretical population genetics (see Wakeley, Chapter 5) and evolutionary ecology (see Agrawal et al., Chapter 10). Acquiring this knowledge may also help us address some of the most pressing problems of our time, which include concerns about the future evolution (or extinction) of this biodiversity (see Gould, Chapter 21; Davis et al., Commentary 6).

The question of how diversity evolved is of course not a simple one and therefore, must be attacked from multiple angles and at several levels. At the most proximate level, we would like to know what mutations give rise to variation and how that genetic variability is maintained in populations (see Zhang, Chapter 4). Next, the question arises of *how* genetic variation actually produces variation in organisms, for example, through changes in developmental events, pathways, or processes (see Wray, Chapter 9). Deciphering how evolutionary forces act on this phenotypic variation in a given ecological context (see Agrawal et al., Chapter 10; Berenbaum and Schuler, Chapter 11)—and how different properties may constrain (see Kirkpatrick, Chapter 7) or promote (see G. Wagner, Chapter 8) phenotypic change—remains a major challenge. While the study of morphological variation offers a logical starting point, we also want to uncover the mechanisms responsible for variation and evolution of other characters (e.g., behavior) and understand how and why behavioral evolution may be similar to or different from morphological evolution (see Kokko and Jennions, Chapter 12). Moreover, we are not limited to processes occurring within a lineage; rather, our thinking must be extended to the genetic and ecological changes associated with speciation (see Harrison, Chapter 13) and macroevolutionary diversification (see Losos and Mahler, Chapter 15; Foote, Chapter 18). Finally, a temporal component must be added—changes in allele frequencies at specific loci (e.g., ancient DNA studies), genome composition (e.g., comparative genomics), phenotypes (e.g., fossils and character mapping using phylogenetics) and the environment—to understand change through time. Only when all of these pieces are taken together can we start to formulate a complete picture of evolutionary change.

Integration of these diverse approaches has long been an ideal in evolutionary biology. For example, in the introductory chapter (notably entitled "The Problem") of his 1974 book *The Genetic Basis of Evolutionary Change*, Richard Lewontin advocated a merger between scientists working on the genetic processes (e.g., population geneticists studying the impact of different evolutionary forces on changes in allele frequency) and those focusing on forces acting on phenotypes (e.g., field naturalists studying the role of differential survival and reproduction on phenotypic change across generations).

Before proceeding, however, let us review briefly some major advances that have led to this point. Not long after the rediscovery of Mendel's laws of inheritance, Thomas Hunt Morgan's mutational experiments in

Drosophila demonstrated that genes are carried on chromosomes, providing a material basis for heredity. Together, these discoveries also highlighted how phenotypic variation—whether it be the shape of peas or the eye color of flies—can be studied in the laboratory (and eventually in the field). Around the same time, the role of natural selection acting in the wild was being further documented. For example, using a simple general selection model, J. B. S. Haldane calculated the selective advantage necessary for the observed evolution of industrial melanism in peppered moths (Haldane 1924). However, it was the combination of critical contributions by Haldane, Sewall Wright, and most notably R. A. Fisher who showed how the action of many discrete genetic loci studied in the lab could result in continuous trait variation observed in nature. But, arguably, it was Dobzhansky's 1937 book *Genetics and the Origin of Species* that played a key role in bridging the gap between population geneticists and field naturalists. These works and others culminated in the modern evolutionary synthesis, that is, the union of ideas from scientists across several disciplines (from laboratory genetics to field ecology, systematics, and paleontology) about the way evolution proceeds.

From here, as is often the case, it was a technological advance that pushed the field forward. In the 1960s, Richard Lewontin moved to the University of Chicago, where he met Jack Hubby. Lewontin was an evolutionary geneticist with a question: how much genetic variation exists in natural populations? Hubby was a biochemist with a new technique: protein electrophoresis. It was the perfect union. Together, they surveyed 18 loci in *Drosophila pseudoobscura* and reported that a large fraction was polymorphic (Lewontin and Hubby 1966). This result had a great impact, as the discovery of high levels of molecular polymorphism raised the question of what evolutionary forces maintain this variation—a question that preoccupied population geneticists for decades. The ability to survey allozymes in wild populations also offered an exciting opportunity to connect genetic variation to fitness in nature. One of the earliest examples was reported by Watt (1977), who showed that variation at the phosphoglucoisomerase (PGI) locus is associated with fitness differences in natural populations of *Colias* butterflies. Yet, in the following three decades, there have been only modest steps toward cementing the link between genotype, phenotype, and the environment in any one system, that is, until very recently.

In the last few years, case studies have started to accumulate that provide a near complete picture of adaptive change, that is, the genes and mutations that underlie variation and divergence in traits that have documented fitness consequences in nature have been identified (Figure 22.1). To some, this represents the holy grail of evolutionary biology. In threespine stickleback (*Gasterosteus aculeatus*), changes in armor are associated with the invasion of freshwater lakes following the last glacial cycle, starting about 20,000 years ago, and a reduction in armor has measurable fitness consequences (Barrett et al. 2008; Marchinko 2009). At the genetic level, changes in armor are

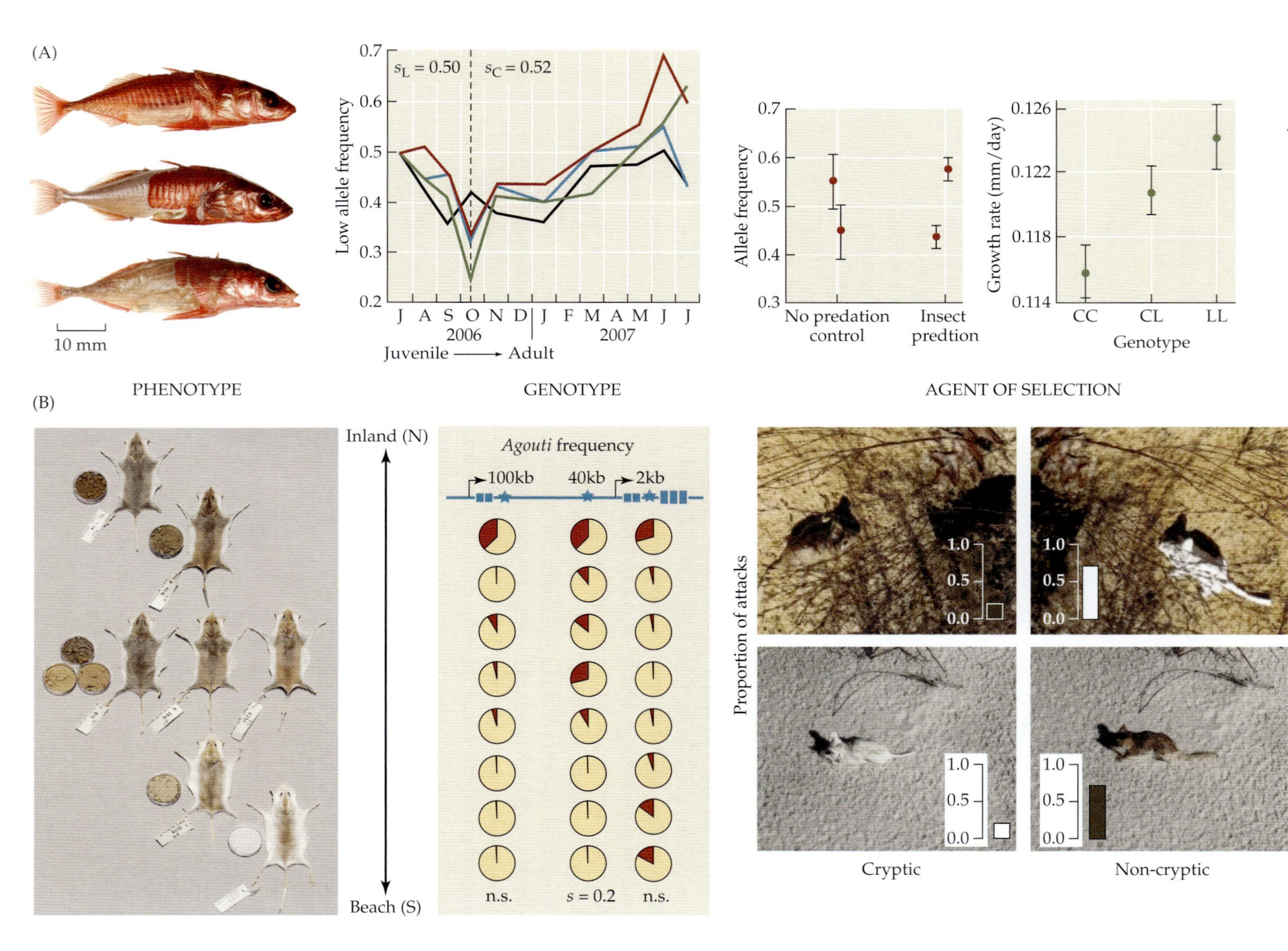
(A)
10 mm
Low allele frequency
0.7
0.6
0.5
0.4
0.3
0.2
$s_L = 0.50$
$s_C = 0.52$
J A S O N D J F M A M J J
2006
2007
Juvenile ⟶ Adult
Allele frequency
0.7
0.6
0.5
0.4
0.3
No predation control
Insect predtion
Growth rate (mm/day)
0.126
0.122
0.118
0.114
CC
CL
LL
Genotype
(B)
PHENOTYPE
Inland (N)
Beach (S)
GENOTYPE
Agouti frequency
100kb
40kb
2kb
n.s.
$s = 0.2$
n.s.
AGENT OF SELECTION
Proportion of attacks
1.0
0.5
0.0
Cryptic
Non-cryptic

FIGURE 22.1 Two Examples for which Both the Targets (Phenotypic and Genotypic) and Agents of Natural Selection Have Been Identified (A) Selection on body armor in the threespine stickleback, *Gasterosteus aculeatus.* Left panel: complete (top), partial (middle), and low (bottom) lateral plate morphs. (From Barrett et al. 2008.) Middle panel: changes in low *Eda* allele frequency within a single generation in four replicate ponds (colored lines). Selection coefficients are given for selection against the low allele from July–October (s_L) and from selection against the complete allele from October–July (s_C). (Adapted from Barrett et al. 2008.) Right panel: relative to the complete *Eda* allele, individuals carrying the low *Eda* allele enjoy decreased predation by insects (left, adapted from Marchinko 2009) and increased growth rates in fresh water (right, adapted from Barrett et al. 2009). (B) Selection on coat color in the oldfield mouse, *Peromyscus polionotus*. Left panel: representative mice and soil sampled from collection sites along a 150-km transect from northwestern Florida (beach) to southeastern Alabama (inland). (From Mullen and Hoekstra 2008.) Middle panel: Allele frequencies at three polymorphic sites (stars) within the pigmentation gene *Agouti* (large boxes: coding exons; small boxes: untranslated exons), sampled from eight populations along the same 150-km transect. Pie charts are arranged North (top) to South (bottom), and frequency of alleles associated with light pelage are indicated in yellow and the dark pelage in red. One of the three *Agouti* single nucleotide polymorphisms (SNP; 40kb), but not the others, varies clinally; the selection coefficient is given for this SNP. (Adapted from Mullen and Hoekstra 2008.) Right panel: Increased attack rates on non-cryptic clay models relative to cryptic clay models on both light (beach) and dark (inland) soils demonstrates that visually hunting predators are an important selective agent targeting color variation within and between *P. polionotus* populations. (Modified from Vignieri and Hoekstra, in press; from Linnen and Hoekstra, in press.)

associated with changes in the *Ectodysplasin* (*Eda*; Colossimo et al. 2005) and the *PitX1* loci (Shapiro et al. 2004). Moreover, deletions of a *cis*-regulatory element in the *PitX1* locus alter pelvis-specific expression of *PitX1* during development (Chan et al. 2010). These phenotypic (and inferred allele) frequencies can be tracked through time, using both museum specimens (Bell et al. 2004; Kitano et al. 2008) and the fossil record (Hunt et al. 2008; Bell 2009). In a second example, oldfield mice (*Peromyscus polionotus*) have dorsal coat colors that closely match their local substrate, and the degree of color matching has a measurable effect on inferred predation rates (Vignieri et al., in press). Differences between a dark mainland and pale beach mouse subspecies have been attributed to three major genes (Steiner et al. 2007), and in one case, to a single amino acid mutation in the *Melanocortin-1 receptor* (*Mc1r*) gene, which affects ligand binding and receptor signaling that is associated with reduced pigmentation (Hoekstra et al. 2006). Variation in pigment allele frequencies also can be traced in space and time (Mullen and Hoekstra 2008; Mullen et al. 2009). Berenbaum and Schuler (see Chapter 11) describe a third such example, in which amino acid substitutions in a cytochrome P450 enzyme enhance the ability of swallowtail butterflies to detoxify defensive compounds in their food plants.

Such studies are now enabling us to answer long-standing questions in evolutionary biology, some first posed by the architects of the modern synthesis. These questions include, but are not limited to: Can adaptation take big leaps or does it proceed through many small steps? Are adaptive alleles generally dominant or recessive? Is evolutionary change limited by mutation? To what extent is evolution constrained? How repeatable is evolution—are the same or different genetic solutions responsible for solving similar ecological problems? And, most recently, how do changes in or regulation of proteins during development produce phenotypic change? Answers to these general questions will come only by studying many traits in myriad species.

While these exemplars are among our most complete stories, it has taken nearly a decade to make the connection between genes, phenotype, and environment for a single trait in a single species. Moreover, at least with the stickleback and mouse examples, these studies are not unbiased, as they represent the best case scenario. In each case, there were *a priori* reasons to think that trait variation impacts fitness, and these morphological traits were easy to measure (and are influenced little, if at all, by the environment). Moreover, their genetic basis proved to be relatively simple—a few genes explain a large proportion of the phenotypic variation. How then can we be hopeful about making more progress in the future studying traits for which the functional consequences are unclear, that are challenging to measure accurately, and for which the genetic basis may be more complex? To do so, we face three major challenges: (1) describe the genomic (and other -omic) variation within and among species; (2) unravel how this genomic information is translated into phenotypic information; and (3) understand how evolutionary forces drive differentiation (and ultimately fitness) in an ecological context, which remains among the hardest tasks, even if (or perhaps because) it is not currently possible to study using high-throughput technology.

The Rise of Genomics

Our ability to comprehensively describe genetic variation by sequencing complete genomes, the basic blueprint of an organism, represents an extraordinary technological advance that is remaking the field of biology. The rate at which whole genome sequences are generated is astonishing. Complete genome sequencing started only 30 years ago when the modest 5368-base pair genome of bacteriophage fX174 was decoded (Sanger et al. 1977). It was quickly followed by several other, larger viral genomes. But, it took almost another 20 years until the first complete genome sequence of a free-living organism, *Haemophilus influenzae* (1.8 megabases), was finished because decoding a genome of this size required both technological and computational advances (Fleischmann et al. 1995).

Less than 6 years after this first complete genome sequence came the first complete *human* genome sequence—2.91 billion base pairs of euchromatic sequence (~5-fold coverage), with 2.1 million identified polymorphisms, at the reported cost of greater than $10 million (International Human Genome Sequencing Consortium [IHGSC] 2001; Venter et al. 2001) and when finished (i.e., the fold-coverage was high enough to provide reliable data),

TABLE 22.1 RAPID CHANGE IN HUMAN GENOME SEQUENCING COSTS

Year	Technology	Reference	Average reported coverage depth (fold)	Reported sequencing consumables cost	Estimated cost per 40-fold coverage
2004	IHGSC 2004	Sanger	5	$300,000,000	—
2007	Levy et al. 2007	Sanger (ABI)	7	$10,000,000	$57,000,000
2008	Wheeler et al. 2008	Roche (454)	7	$1,000,000	$5,700,000
2008	Bentley et al. 2008	Illumina	30	$250,000	$330,000
2009	Pushkarev et al. 2009	Helicos	28	$48,000	$69,000
2010	Drmanac et al. 2010	Nanoarrays	87	$8000	$3700
2010	Drmanac et al. 2010	Nanoarrays	63	$3500	$2200
2010	Drmanac et al. 2010	Nanoarrays	45	$1725	$1500

the cost was estimated to be closer to $300 million (IHGSC 2004). However, as the rate of published complete genome sequences increased, the cost decreased almost exponentially (Table 22.1). To underscore this point, on the first day of the Darwin 2009 Workshop that gave rise to this volume, a publication announced that three human genomes were sequenced at the cost of approximately $4400 (for consumables), with 45- to 87- fold coverage, identification of 3.2 to 4.5 million sequence variants per genome, and a 1-false-variant-per-100-kilobases accuracy (Drmanac et al. 2010)[1]. The rapid fall in sequencing prices is the genomic equivalent of Moore's Law, which describes the long-term trend in which the number of transistors that can be placed on computer chips doubles every 18 months, steadily driving down the cost of computing power.

Although this technology is largely driven by its applications to human disease (i.e., personalized medical genomics), the newest technologies are easily transferable to other species. At the time I write, the complete sequence is known for about 2000 viruses, 600 bacterial species, and roughly 200 eukaryotes including 60 chordates; these numbers will likely be different next week. As sequencing prices continue to fall, project proposals are becoming increasingly ambitious. For example, a recent proposal was put forth to sequence 10,000 vertebrate genomes in 5 years (Genome 10K Community of Scientists 2009). Arguably, all species[2] have the potential to be genome-enabled, just as 20 years ago it became routine to sequence the mitochondrial cytochrome b gene in any eukaryote.

[1] Published in *Science Express* on 5 November 2009.

[2] Genome sequencing of certain species undoubtedly will prove more challenging, such as those with extraordinarily large repetitive genomes or high levels of heterozygosity and those that are recent or ancient polyploids.

But with these new data flowing in at an unprecedented pace, we need tools and infrastructure in place to make them both accessible and usable. First, the need for well-maintained and long-term data repositories has never been greater (Robinson et al., in press). Second, for a genome sequence to be truly useful, it must be assembled and annotated, which is non-trivial. If, for example, the 10K Vertebrate Genome Project goes forward and reaches its goal of completion in 5 years, the processing rate will need to hold steady at five genomes per day (with no weekend or holiday breaks). Is this possible? Luckily, the field of bioinformatics—sometimes referred to as the symbiotic harmony of computer science and biology—is burgeoning. For example, Ensembl, a web-based community resource that provides genome data and analysis tools for a comprehensive set of chordate genomes, just celebrated its 10th anniversary and continues to evolve (Flicek et al. 2010). Moreover, the assembly and annotation of complete genomes likely will become more straightforward as more scaffold genomes (i.e., existing genome sequences that can be used as a reference to aid genome assembly of a new, closely related species) are available and computational algorithms become further automated. In fact, the first genome sequence based primarily on short-read, next-generation sequences, that of the giant panda, is purported to have been assembled in only 2 days, suggesting that the tools are already in place to assemble de novo a genome's worth of small DNA fragments even when a reference genome is unavailable (Li et al. 2010). Thus, the availability of genome sequences and their assembly is unlikely to be a limiting factor as we move forward.

Will the era of genomics end? Not any time soon. Instead I predict that, at least in the near future, it will continue to expand and do so exponentially. Furthermore, just as genome sequencing is becoming faster and cheaper, so too will the next level of -omics: epigenomics, trancriptomics, proteomics, metabolomics, and so forth. But, these inventories of genomic parts are limited in their utility without knowledge of the functional consequences of variation at any of these molecular levels (and we must be open to novel, perhaps unexpected, ways in which DNA sequence can confer function). For example, future studies geared at uncovering the function of noncoding or so-called junk DNA, epigenetic modification (e.g., methylation and chromatin remodeling), and spatio-temporal changes in RNA abundance on phenotype (and fitness) will undoubtedly open many windows into the evolutionary process. But, there is no doubt that evolutionary biology currently is and will continue to be transformed by our ability to sequence the genome of virtually any organism.

How to Build an Organism from Its Genomic Blueprint

Sydney Brenner is a brilliant scientist, probably best known for his contribution to the then emerging field of molecular biology in the 1960s, including his collaboration with Francis Crick to experimentally reveal the triplet nature of the genetic code. He later went on to establish *Caenorhabditis elegans* as a model system for the study of development, for which he received a Nobel Prize. Whether then it was blind enthusiasm, the naiveté of the time, or the shrewd promotion of genetics, that led him to say: "... give me the complete DNA sequence of any organism, and I can reconstruct it"[3] is unclear (R. C. Lewontin, personal communication). At the time (i.e., the 1980s), this sentiment was shared by many—the genome was the key that would unlock the secrets of biological complexity. It has become abundantly clear, however, that this assertion was indeed naïve; without the instruction manual, even with all the parts in hand, it is impossible (even at present) to reconstruct the whole organism.

Historically, molecular research has focused on identifying the parts—individual genes and proteins—and understanding their functions. While the reductionist program has been both successful and enlightening, to understand how whole organisms work, we need to consider the individual components both through developmental time (Carroll et al. 2001) and in the context of their interactions and as part of large networks (Noble 2006). By analogy, knowing all the parts of an airplane lends little to our understanding of how the plane actually functions. This daunting task falls largely under the joint purview of developmental biology and systems biology, which is the study of organisms as an integrated and interacting network of genes, proteins, and biochemical reactions (Sauer et al. 2007). Systems biology is still in its infancy, but is gaining momentum due to a host of new technologies that are high throughput, quantitative, and large-scale (Zhu and Snyder 2002).

At the moment, many of these new systems-level approaches are focused at the cellular level or conducted in relatively simple model organisms. More recently, the evolution of metabolic networks has been investigated, that is, how selection acts to optimize fitness across a landscape of networks (Pfeiffer et al. 2005), which may have implications for the early stages in the origin of life. Eventually, however, new technologies aimed at organisms that are more complex will be required, especially to automate the characterization of expression patterns (e.g., high-throughput *in situ* hybridizations) and the implementation of functional assays (e.g., RNAi, viral vectors, transformations, transgenesis; Kitano 2002). While some tools will necessarily be species- or clade-specific, the most useful advances undoubtedly will be those that are easily transferable across organisms. Thus, with new tools, we may acquire access to a fuller understanding of how genes and genomes produce phenotypes through changes in development, how historical processes have shaped that transformation, and how

---

[3] This is not an exact quote, but has been recounted by many, who were in attendance at Brenner's keynote address at the Cambridge University Symposium (1982) on the occasion of the 100th anniversary of Darwin's death.

constraints and opportunities may limit or promote future evolutionary change (Schwenk et al. 2009).

Thus, just as molecular biology has expanded beyond a focus on single genes or proteins, so too has developmental biology (see Wray, Chapter 9). Establishing a true understanding of how pathways and networks give rise to cellular and organismal phenotypes and how those interactions evolve will require very large experimental data sets. Thus, I predict that the largest advances will come by generating network models, predicting how they affect the phenotype, testing hypotheses derived from these models, and refining the models based on new experimentation. Importantly, understanding the functional interactions of genes, RNA, and proteins will provide a necessary first step in translating the genetic code into phenotypes and, ultimately, into fitness of organisms in nature.

The Genotype–Phenotype Connection in Nature

While systems biology may aim to unravel the details of all gene-network interactions (i.e., to identify the genes and connections necessary to produce a phenotype or whole organism), most evolutionary biologists are primarily concerned with changes in genes or development that contribute to phenotypic variation: in particular, identifying genes that are responsible for local adaptation, phenotypic novelties, and/or the promotion or restriction of diversification. To accomplish this goal, there are at least four major areas in which tremendous growth may be anticipated.

New Model Systems

Over the past several decades, tool and resource development, and hence research effort, have largely (and necessarily) focused on a handful of model organisms. However, these few species are not representative of the vast diversity of life. Thus, a broader array of organisms, replete with genome sequence data and functional tools, is needed to elucidate the genetic and developmental basis of organismal diversity (Jenner and Wills 2007). But at present, we have limitations. While genome sequencing can be done in any and all species of interest, not all species or clades can immediately become new model systems in the traditional sense; there are limited resources and a limited number of research communities to study any particular species. Just as some species are best suited to address specific biological questions,[4] others are better suited, at least initially, to become a model species—those that can be easily obtained, cultured in the lab, genetically crossed, and for which functional tests are feasible (Abzhanov et al. 2008). Yet, as technology, tools, and resources continue to develop, it is feasible to imagine that the term "model organism" eventually and finally will be eliminated from

---

[4] As August Krogh famously said, "for many problems there is an animal on which it can be most conveniently studied" (Krebs 1975).

biology's lexicon. Thus, as a diversity of species with appropriate genetic, genomic, developmental, and functional tools emerge, we will soon be able to ask questions at a number of levels, from variation among individuals to adaptation among populations to the evolution of novel traits between species, and understand macroevolutionary patterns by comparing across a rich array of taxa.

Population Genomics

It is inappropriate to think of "*the*" genome of a single species, because this thinking fails to capitalize on intraspecific variation. In fact, much of evolutionary biology's future promise may rest in the hands of population genomicists, those making large-scale comparisons of genome-scale DNA sequences among individuals sampled from natural populations. Why? While comparisons among phylogenetically dispersed taxa have been extremely useful in understanding genome evolution, such broad comparisons can only go so far in linking genetic to phenotypic variation. It can be argued that the comparison of genomes from collections of variable individuals is the key to understanding variation in phenotypes. Thus, I expect to see a shift in DNA sequencing efforts from single individuals in many species to many individuals in a few species, and in the future, to many, many individuals in many, many species.

The power of population genomic sampling stems from our ability to both apply population genetic tests of non-neutrality and to statistically associate genetic variants with phenotypic variation across the genome. These statistical approaches were initially designed for application to single genes, first to allozyme alleles and then to DNA sequence polymorphisms (e.g., McDonald and Kreitman 1991), but can now be applied to the whole genome (see Kolaczkowski and Kern, Chapter 6). As theoretical population geneticists and statisticians become better at identifying signatures of selection in the genome (Jensen et al. 2005; Nielsen et al. 2005), we will be better poised to identify genomic regions, genes and mutations associated with phenotypes, and the action of natural selection. The success of these statistical approaches also rests on having plenty of data. The Human HapMap project represented a major effort to generate genome-wide population data but was limited in that only single nucleotide variants and those that were most common (>~5% frequency) were considered (The International HapMap Consortium 2005). New efforts aimed at surveying and including rare variants and other forms of variation (e.g., copy-number variants) include the 1000 Genomes Project, which endeavors to create the most detailed compendium of human genetic variation (Kaiser 2008), and a recently proposed 1000 Genomes Project focused on *D. melanogaster*. Additional proposals, such as the 1001 Genomes Project for *Arabidopsis thaliana*, the workhorse of plant genetics, tout the utility of combining quantitative trait locus mapping, population genomic approaches, and genome-wide association studies (Weigel and Mott 2009).

Yet, population genomic descriptions of nucleotide variation, while a tremendously exciting advance, represent a first step. Importantly, methods for the functional validation of candidate genes and mutations identified by genome-wide approaches are needed (see previous discussion), and to assign any adaptive function at all, one must be certain that fitness is carefully measured in nature and the selective agent is established (see the following section).

Phenomics

Compared to our knowledge of genomes, our knowledge of phenotypes remains cursory. Part of the explanation for this imbalance is certainly that phenotype space is vastly more expansive than genotype space (Houle 2010); in other words, genotypes are more easily defined. However, the ability to make the link between genotype and phenotype rests, in part, on our capacity to measure phenotypes accurately and consistently. Ideally, we want to be able to measure objectively the same (homologous) trait in many individuals as well as many traits in a few individuals. While our textbook example of a phenotype is often height or weight, it can be useful to deconstruct these complex phenotypes into more precise traits. At the extreme, functional genomic outputs (e.g., mRNA or protein expression levels) may serve as readily measured quantitative phenotypes (i.e., endo-phenotype) and also can be used to assess genotype–phenotype associations. At the organismal level, high-throughput morphometrics (and accompanying databases) may be another boon, as such approaches allow for the linking of macroevolution-level collections of fossils, microevolution-level data collected from natural populations, and experimental-level altered morphologies of mutants. For example, commercial facilities now specialize in phenomics and are well equipped to do high-throughput screens for rare mutants (e.g., greenhouses that can raise thousands of seedlings) or, alternatively, multi-phenotype screens for a mutant strain (e.g., hundreds of behavioral assays run on a few mutant/transgenic animals). These high-throughput screens are revolutionizing the scale at which geneticists and neurobiologists design and implement experiments (e.g., Tecott and Nestler 2004). While evolutionary biologists have yet to fully capitalize on these large-scale approaches, it is easy to imagine that such an approach can be adapted to address questions about evolutionary diversification and constraint.

Fitness in the Wild

Measuring phenotypes in controlled laboratory environments has many advantages, but what we really want to know is how morphological, physiological, and behavioral variation translates to fitness differences in the wild. As a first step, we would like to carefully characterize changes in the environment at both the micro- and macro- spatial and temporal scales. Long term ecological research (LTER) studies have been set up to study changes in ecological processes over time, yet these studies rarely integrate evolution. Large-scale terrestrial environmental data may soon become available through the National Ecological Observatory Network (NEON; www.neoninc.org). While still in its infancy, NEON holds the promise of providing both environmental information for terrestrial biomes and a model for characterizing additional biomes. However, describing environmental variation through time and space will not alone provide a comprehensive picture because organisms evolve in response to both abiotic and biotic factors, including those that pertain to each species individually (see Davis et al., Commentary 6). Thus, we also want to know how individuals are interacting within an environment, especially with other species.

While developing high-throughput ecological experiments is not straightforward, some initial steps are already being taken to monitor remotely the movements, survival, and reproduction of hundreds of both plant and animal individuals. New miniaturized transmitters exist to record continuously the activity and performance of organisms in nature. For example, the International Cooperation for Animal Research Using Space (ICARUS) initiative aims to establish a remote sensing platform that can track even small animals globally, enabling observations and experiments over large spatial scales (Wikelski et al. 2007). Future instrumentation to measure not only movement, but also physiological state and whole-organism performance and behavior, may capitalize on advances in microfluidics (Whitesides 2006) and imaging (Bimber 2006), which already are proving powerful in laboratory settings. Thus, organismal movement, physiology, and performance may soon be tracked remotely and recorded automatically in large databases. As with increasing amounts of -omic data, a major challenge in generating such organismal and environmental data is an effective cyberinfrastructure. One model, the iPlant Collaborative (iPlant; www.iplantcollaborative.org) has been developed by plant biologists. Done right, such programs will encourage communication among disciplines and the reuse of data models, file formats, application software, and algorithms, while fostering cross-disciplinary exchange of ideas.

However, understanding changes in the environment and even the movement and interactions of organisms within the environment is still not quite enough. In addition, we want to measure fitness, or at least components of fitness, in nature. But how? Measuring fitness in the wild for large numbers of individuals through time represents one of the most important and largest challenges to evolutionary biologists. No matter how advanced technology and tools become, it is difficult to imagine being able to replace fieldwork and the study of natural history, both at a practical and philosophical level.

The Genotype–Phenotype Synthesis

Darwin lived in a very opportune time, a time of exploration, when new specimens were continuously arriving from around the world and new geological data were being amassed. At present, evolutionary biologists are arguably in a comparable position, as new molecular data are being generated at an unimaginable rate, and the possibility of gathering fine-scale environmental data is on the horizon. However, the real question, of course, is how will these data change the way we think about evolutionary biology, if at all? I focused previously on the technological advances that make it possible to imagine (perhaps optimistically) that we will have an ever increasing ability to link changes in the genome to changes in phenotype and/or in the environment. In the following, I will focus on four general areas (from among many), in which Darwin was clearly interested, but for which he could never have anticipated even our current depth of knowledge. These four areas also represent topics in which connecting genotype and phenotype may be especially illuminating in our quest to more fully understand the evolutionary process.

The Mechanistic Basis of Adaptation

Darwin's magnum opus was "one long argument" for the ultimate cause of adaptive change, that is, natural selection. But, even Darwin was curious about the mechanistic basis of phenotypic change—how are traits encoded and passed on from one generation to the next? Perhaps, it was a propitious augury then that Darwin's last publication arose from a collaboration with a young British naturalist by the name of Walter Drawbridge Crick (Ridley 2004), the grandfather of Francis Crick, who would contribute to the discovery of the precise mechanism of inheritance, by deciphering the three-dimensional structure of DNA.

Technological advances have had a fundamental impact on the way we study adaptation. In Lauder and Rose's 1996 book entitled *Adaptation*, only a single chapter was devoted to molecular data, whereas a decade later we are hard pressed to find new advances in the study of phenotypic adaptation without some molecular genetic contribution. Combinations of genomic and transcriptomic approaches are being applied to variation in nature, and will allow us to address fundamental questions about the adaptive process, as previously discussed. Moreover, our increasing knowledge of the genetics of adaptation in wild populations is, in turn, expanding our knowledge about natural selection (Schluter et al., in press). More specifically, genes contain information about the form, strength, timing, history, and (sometimes) the agent of natural selection (e.g., Fitzpatrick et al. 2007; Barrett et al. 2008; Linnen et al. 2009). Thus, I would suggest that some of the most enlightening future studies will combine studies of proximate (i.e., molecular and developmental mechanisms) and ultimate (i.e., selective mechanism) causes of evolutionary change, as both approaches are mutually enlightening.

While some hard-earned progress has been made identifying a handful of genes underlying adaptive variation, new genomic technologies and approaches are increasing the rate at which such genes are being identified and functionally verified. The taxonomic breath of such investigations is also expanding, as is, most importantly, the complexity of traits that are being dissected. As one example, the role of genetic constraint due to pleiotropy and the importance of epistasis in adaptation are difficult to ascertain until we have enough data, both genetic and phenotypic, to test these hypotheses in a statistically rigorous manner (see G. Wagner, Chapter 8). Thus, the lessons we have learned so far by studying a few simple traits in a handful of species may not be fully representative of adaptive change in general, and consequently, a comprehensive view of the adaptive process is yet to emerge.

Therefore, I predict (and hope) that "gene hunting" studies do not end simply with the identification of a gene, but rather continue on to determine the consequences of mutations for gene function (Dean and Thornton 2007), to understand *how* changes in gene function and regulation through development produce phenotypic change (Carroll et al. 2001), and most importantly, perhaps, to truly understand how genetic variation translates to fitness differences in the wild (Nielsen 2009; Linnen and Hoekstra, in press). With this information in hand, we should be well positioned to not only understand what happened in the past, but also start to make predictions about the genetic and phenotypic responses to known environmental change in the future.

From Instinct to Neurobiology

Darwin devoted an entire chapter of *The Origin of Species* to behavior. He concludes that because "no one will dispute that instincts are of the highest importance to each animal" (p. 243) and that behaviors can be heritable, that they can evolve by natural selection in the same way as do morphological characters. Indeed, while much research has focused on morphological adaptation, one classic view posits that change in animals' morphology is often preceded by that in behavior: "a shift into a new niche or adaptive zone is, almost without exception, initiated by a change in behavior" (Mayr 1963: 604). An understanding of the ultimate causes of behavioral variation has been the subject of study first made popular by the classic ethologists (e.g., Niko Tinbergen), who were interested in understanding both the causes and consequences of behavioral variation across many organisms in their natural environments (see Kokko and Jennions, Chapter 12).

However, despite its importance, we still know very little about the proximate evolutionary mechanisms that give rise to behavioral diversity found in nature. Thus, we remain largely ignorant of how behavior evolves and are left with many fundamental questions unanswered. For example: What are the relative contributions of genetic and environmental effects (or learning) to behavioral differences? What are the genetic changes that underlie the differences in behavior found both within and between species? Do these genetic changes act early in development to alter neural circuitry, or does the circuitry remain constant and changes in physiology (e.g., neurotransmitters) underlie behavioral variation? How do such changes happen at a mechanistic level? Are "behavior genes" specific to behavior? How has the extraordinary complexity of human culture evolved from simpler behavior repertoires of our primate ancestors (see Richerson and Boyd, Chapter 20)? The connections among genes, neural circuitry, and the evolution of complex and adaptive behaviors remain a major frontier in biology.

The study of behavior, unfortunately, shares many of the same obstacles as the study of the genetic basis of morphological traits but also introduces a host of new ones. Many behaviors have low heritability, and some are culturally inherited or have a learned component. Behavior is particularly prone to environmental effects; at the extreme, some behaviors require an environmental stimulus (Robinson et al. 2008). Finally, many of the most interesting behaviors (e.g., behaviors that seem to have a clear fitness consequence in the wild) occur in non-model organisms, which have been the focus of field-based behavioral ecologists (see Kokko and Jennions, Chapter 12). However, optimistically, some of these obstacles can be overcome by an increase in experimental scale, automated phenotyping in controlled environments, and the application of genomic technologies to non-model species. In addition, just as studies by developmental systems biologists are helping to elucidate the link between genes and morphology, new large-sale and high-throughput approaches being adopted by neurobiologists are setting the stage to link genes and behavior.

Neurobiologists, too, have been bitten with the genomics bug and have recently spawned a field of study entitled neurogenomics, which is the study of how the genome contributes to the evolution, development, structure, and function of the nervous system (Boguski and Jones 2004). This approach is already being adopted by those studying behavior in diverse species. For example, the songbird neurogenomics initiative (SoNG) aims to develop technical resources to study gene–brain–behavior relationships using song and songbirds as model systems. Already, transcriptome surveys and large-scale neuroanatomical mapping of gene expression are underway, and may define a new functional anatomy of the brain. Neurogenomicists may, in fact, provide the bridge between reductionist (behavioral genetics) and holist (behavioral ecology) approaches to the study of nervous system evolution, analogous to developmental genomics in morphology.

While elucidating the mechanistic details of the behavioral evolution is still in its infancy, I predict we will soon be able to draw a more complete picture of how many behaviors evolve, from the ultimate forces driving behavioral evolution down to its molecular details. This prediction assumes that with improving technologies, genes that underlie behavioral variation will become increasingly easy to identify, and that behavioral evolution (like morphology) may rely on changes to a common genetic toolkit (i.e., a set of genes shared by all animals), so that unraveling the genetic basis of behavior in one species will earn us some predictive power when studying other species.

Genomic Approaches to Studying The Origin of Species

In the opening paragraph of *The Origin of Species*, Darwin suggests his tome will "throw some light on the origin of species—that mystery of mysteries, as it has been called by one of our greatest philosophers" (Darwin 1859: 1). Despite some modest progress in understanding how new species originate (Coyne and Orr 2004; see Harrison, Chapter 13), surprises still abound, and thus, future work in speciation may demand considerable changes in our views, not just minor modifications (Orr et al. 2007).

How species originate remains one of the most fundamental questions in evolutionary biology and thus is likely to be a subject of much growth in the coming years. Perhaps the most exciting progress will be made by using genomic approaches to elucidate the processes and molecular basis underlying speciation. In particular, we would like to know: What are the forces that drive species formation, and in particular, what is the role of natural selection? How often can differentiation occur in the face of continuous or intermittent gene flow? What mechanisms of reproductive isolation are most important and which ones evolve first? What specific genes are involved in preventing gene flow between incipient species? And how do these patterns vary across organisms with different ecology, mating systems, reproductive biology, and sex chromosomes?

Genomic approaches offer some new insights into these questions in two ways. First, technical advances are increasing the rate at which specific mutations and genes responsible for reproductive isolation are being identified. Second, genomic surveys of natural populations have allowed researchers to investigate barriers to gene flow, genomic interactions, and the genetic permeability of species boundaries in the wild (Noor and Feder 2006). While much of the progress involves increases in scale and speed, as these approaches are extended to additional species, general patterns and rules might emerge. In doing so, we are no longer asking questions about whether specific types of genes and types of processes occur, but rather what their relative frequency and importance are for speciation in general (Schluter 2009).

If we are truly to understand how new species originate, two goals must be accomplished. First, the speciation process can be fully explained only if we identify the heritable underpinnings of species formation and the forces responsible for their origins. Thus, understanding genomic patterns of and specific genetic changes driving speciation likely will change the way we understand the fundamentals of diversification. Second, it is equally important to recognize, that with the growing appreciation for the role of selection in species formation, it is impossible to ignore the role of ecology (Sobel et al. 2010; see McPeek, Commentary 3). Thus, a combination of both genomic and ecological approaches will be the key to solving Darwin's "mystery of mysteries."

Molecular Fossils: Reconstructing Evolution's Path

Evolutionary biology is in large part about reconstructing the past. But, in 1859, there was little data from the past; Darwin spends an entire chapter in *The Origin of Species* touting the importance of ancient organisms as evidence for descent with modification (Darwin 1859: Chapter 10) and another (Darwin 1859: Chapter 9) lamenting the imperfection of the fossil record. It was a great boon to Darwin's theory when, just 3 years after its publication, the canonical example of a transitional fossil, *Archaeopteryx*, was unearthed in Germany. Since then, the vastly greater and still rapidly growing knowledge of the fossil record has affected the way we think about morphological change (see P. Wagner, Chapter 17) and diversification (see Foote, Chapter 18) through time, perhaps most spectacularly in hominids (see White, Chapter 19).

With novel phylogenetic approaches used to reconstruct ancestral genes and genomes and an increasing availability of ancient DNA sequences, we now have a complementary approach to deducing past events that will stand alongside discoveries from the fossil record. First and foremost, phylogenetics (i.e., the reconstruction of the relationships among organisms) provides a means of inferring the history and pattern of past events (e.g., phenotypic character evolution). More specifically, the ancestral reconstruction of genes (Liberles 2007) and now even large stretches of genomes (Blanchette et al. 2004) provide an indirect peek into the evolutionary history of molecular function and genome evolution. Second, the combination of successes in identifying mutations that affect phenotype in extant populations and improved technologies that provide direct access to ancient genomes also allow a rare glimpse into the biology of organisms that are now long extinct. Together, these approaches have provided an increasing ability to recreate ancient phenotypes with genetic precision. For example, using phylogenetic methods, ancient opsin gene sequences that provide insight into dinosaur vision have been reconstructed (Chang et al. 2002). Using sequencing of candidate genes from ancient DNA samples, we now know that Neanderthals were lactose-intolerant and some were red headed (Lalueza-Fox et al. 2007), while some mammoths were likely blonde (Römpler et al. 2006). Now imagine reconstructing or sequencing whole genomes and being able to predict the morphology, physiology, or even behavior of many ancient creatures!

Well, we need not wait long, as these data are already rolling in. Applying phylogenetic methods to the complete genome sequences of many extant mammals, researchers have begun to develop tools to reconstruct the genome (at the nucleotide-level) of the *Ur*-mammal (Paten et al. 2008). For example, the first complete ancient human genome of a paleo-Eskimo, some 4000 years old, has already been published (Rasmussen et al. 2010).

(Note the genome sequencing was completed in just two and a half months at a cost of about $500,000.) The authors deduced that this individual was a male member of the Arctic Saqqaq, the first known culture to settle in Greenland. He likely had brown eyes, non-light skin, thick dark hair (although at risk for baldness), shovel-graded front teeth, dry (as opposed to wet) ear wax, and probably had a metabolism and body mass adapted to cold climate. This is just the beginning. Over four billion base pairs of Neanderthal genome have been sequenced (Green et al. 2010). But, even preliminary genomic results have profound implications; for example, Neanderthals might have shared some basic language capabilities with modern humans (Green et al. 2006; Noonan et al. 2008). Such studies clearly have and will continue to shed light on the phenotypic traits, genetic origin, and biological relationship to present-day populations of now extinct individuals, populations, and species.

In addition to reconstructing past organisms, we are now in a position to reconstruct Earth's history with increasing precision. Advances in biogeochemistry and isotope techniques (e.g., Robert and Chaussideon 2006; Shen et al. 2001) have revealed ancient and fine-scale changes in the environment (e.g., temperature and chemical composition)—changes, though subtle relative to large scale fluctuations over Earth's long history, that nonetheless have had dramatic effects on organisms, their physiology, lifestyles, and distributions. But this is a two-way street. Recently, geneticists have been able to provide experimental results to predict the environments of ancient life; for example, by sequencing genes from extant species, then reconstructing ancestral proteins and testing their thermostability, we can infer fluctuations in the Earth's temperature (Gaucher et al. 2008). Thus, convergent predictions from geology and biology together can be used to reliably track changes in the Earth's environment and concomitant change in the organisms living in those environments over time.

Conclusions

The future of evolutionary biology, like that of any science, is difficult to predict. At present, what is clear is that major advances are being made at a breath-taking pace. Arguably, many of these advances are being made by using interdisciplinary thinking, by taking advantage of large-scale discovery-based science, and by working at scales previously unimaginable. While I have focused here on only one slice of evolutionary biology, analogies can certainly be found in other sub-disciplines in the field.

Historically, many of the major fundamental advances in evolutionary biology have been associated with: (1) the unification of fields, such as ecology and evolution (see Agrawal et al., Chapter 10; Berenbaum and Schuler, Chapter 11; McPeek Commentary 3) or microevolution and macroevolution (see P. Wagner, Chapter 17), (2) the reconciliation of points of view, (e.g., Futuyma, Chapter 1), (3) the elaboration of new approaches such as phylogenetics (see Hillis, Chapter 16), coalescent theory (see Wakeley, Chapter 5), genomics (see Kolaczkowski and Kern, Chapter 6), and (4) the incorporation of new fields, such as genetics (see Zhang, Chapter 4) and most recently, developmental biology (see Wray, Chapter 9). Such interdisciplinary research efforts can reap large benefits (Wake 2008). For example, paleontology and developmental genetics, two seemingly disparate fields, can illuminate each other—fossils may suggest developmental processes in basal forms, and developmental studies can aid in interpretation of the characters of extinct taxa, such as the gain of limbs of tetrapod ancestors (Shubin et al. 2009) or the loss of fins in a fish (Chan et al. 2010). Thus, integrative research that crosses traditional boundaries will undoubtedly continue to push our understanding of evolutionary biology further.

Because evolutionary biologists pride themselves on being hypothesis-driven scientists, the practice of systematically collecting data, even before knowing all of the precise ways it may be used, has often been frowned upon. However, it is important to recognize the countless evolutionary studies that already take advantage of existing comparative data, including life history characters, physiological tolerances, behavior, habitat associations, diet, geographic distribution, and morphologies (Futuyma 1998). In almost all cases, these data were collected by systematists or other biologists without any anticipation of their future use in testing new hypotheses. This descriptive natural history at the organismal level has been invaluable and now is being replayed at the genomic level. In both cases, data repositories are built and can be used for purposes not yet envisioned. Thus, researchers pursuing hypothesis-driven versus discovery-driven approaches have (or soon may) come to see each other as allies rather than antagonists (Boguski and Jones 2004). At the very least, -omic data, when viewed in the context of an appropriate evolutionary model for rigorous statistical testing, can be extremely powerful. However, large data sets will never replace imaginative hypotheses—the engines of scientific progress.

What seems increasingly novel as we move forward is the scale at which we can ask questions. We are not likely to be limited by genomic data, and both large-scale phenotypic and environmental data acquisition are not far behind. We may expect that in 50 years major progress in understanding the evolutionary process, both the proximate and ultimate mechanisms of evolutionary change, will come from a complementary (if not collaborative) effort among population genomicists, systems biologists (and/or developmental biologists, neurogenomicists), and organismal biologists. Like the happy marriages of first the Mendelians and biometricians and later of the laboratory geneticists and field naturalists, the union of genomic and organismal biology will continue to advance our understanding of evolution in years to come.

A Note

The responses by the participants of the Darwin 2009 Workshop to my presentation were surprisingly bimodal. Some said that my predictions were too cautious—that we would "know it all" in 10 years, rather than 50 or 100. Others were far less optimistic, claiming that organisms were too complicated to dissect, and thus questioned if we would *ever* be able to know it all. Perhaps not surprisingly, the type of response was highly associated with the scientists' field of study: geneticists (i.e., reductionists) tended to be more optimistic, whereas organismal biologists (i.e., holists) were generally more cautious. I look forward to looking back to see who, if either, was right.

Acknowledgments

I wish to sincerely thank the organizers of the Darwin 2009 Workshop and resulting volume, in particular Mike Bell, Walt Eanes, Doug Futuyma, and Jeff Levinton. Never before have I been given such a flattering, yet utterly daunting, task. I also wish to thank the meeting participants for engaging and stimulating discussions. Additional discussions with and comments from Andrew Berry, David Hillis, and Jonathan Losos were critical to the development of many of the ideas presented here.

Literature Cited

Abzhanov A., C. G. Extavour, A. Groover, and 4 others. 2008. Are we there yet? Tracking the development of new model systems. *Trends Genet.* 24: 353–360.

Barrett, R. D. H., S. M. Rogers, and D. Schluter. 2008. Natural selection on a major armor gene in threespine stickleback. *Science* 322: 255–257.

Barrett, R. D. H., S. M. Rogers, and D. Schluter. 2009. Environment specific pleiotropy facilitates divergence at the *Ectodysplasin* locus in threespine stickleback. *Evolution* 63: 2831–2837.

Bell, M. A. 2009. Implications of fossil threespine stickleback for Darwinian gradualism. *J. Fish Biol.* 75: 1997–1999.

Bell, M. A., W. E. Aguirre, and N. J. Buck. 2004. Twelve years of contemporary armor evolution in a threespine stickleback population. *Evolution* 58: 814–824.

Bentley, D. R., S. Balasubramanian, H. P. Swerdlow, and 193 others. 2008. Accurate whole human genome sequencing using reversible terminator chemistry. *Nature* 456: 53–59.

Bimber, O. 2006. Computational photography—the next big step. *Computer* 39: 28–29.

Blanchette, M., E. D. Green, W. Miller, and 1 other. 2004. Reconstructing large regions of an ancestral mammalian genome in silico. *Genome Res.* 14: 2412–2423.

Boguski, M. S. and A. R. Jones. 2004. Neurogenomics: At the intersection of neurobiology and genome sciences. *Nat. Neurosci.* 7: 429–433.

Carroll, S. B., J. K. Grenier, and S. D. Wetherbee. 2001. *From DNA to Diversity: Molecular Genetics and the Evolution of Animal Design*. Blackwell, Malden.
Chan, F. Y., M. E. Marks, F. C. Jones, and 13 others. 2010. Adaptive evolution of pelvic reduction in sticklebacks by recurrent deletion of a *PitX1* enhancer. *Science* 324: 302–305.
Chang, B. S. W., K. Jönsson, M. A. Kazmi, and 2 others. 2002. Recreating a functional ancestral archosaur visual pigment. *Mol. Biol. Evol.* 19: 1483–1489.
Colosimo, P., K. Hosemann, S. Balabhadra, and 8 others. 2005. Widespread parallel evolution in sticklebacks by repeated fixation of *Ectodysplasin* alleles. *Science* 307: 1928–1933.
Coyne, J. A. and H. A. Orr. 2004. *Speciation*. Sinauer Associates, Sunderland, MA.
Darwin, C. 1859. *On the Origin of Species by Means of Natural Selection, or the Preservation of Favoured Races in the Struggle for Life.* John Murray, London.
Dean, A. M. and J. W. Thornton. 2007. Mechanistic approaches to the study of evolution: The functional synthesis. *Nat. Rev. Genet.* 8: 675–688.
Dobzhanksy, T. 1937. *Genetics and the Origin of Species.* Columbia University Press, New York.
Drmanac, R., A. B. Sparks, M. J. Callow, and 62 others. 2010. Human genome sequencing using unchained base read on self-assembling DNA nanoarrays. *Science* 327: 78–81.
Fitzpatrick, M. J., E. Feder, L. Rowe, and 1 other. 2007. Maintaining a behaviour polymorphism by frequency-dependent selection on a single gene. *Nature* 447: 210–213.
Fleischmann R., M. Adams, O. White, and 7 others. 1995. Whole-genome random sequencing and assembly of *Haemophilus influenzae. Science* 269: 496–512.
Flicek P., B. L. Aken, B. Ballester, and 54 others. 2010. Ensembl's 10th year. *Nucl. Acids Res.* 38: D557–562.
Futuyma, D. J. 1998. Wherefore and whither the naturalist? *Am. Nat.* 151: 1–6.
Gaucher, E. A., S. Govindarajan, and O. K. Ganesh. 2008. Palaeotemperature trend for Precambrian life inferred from resurrected proteins. *Nature* 451: 704–708.
Genome 10K Community of Scientists. 2009. Genome 10K: A proposal to obtain whole-genome sequence for 10,000 vertebrate species. *J. Hered.* 100: 659–674.
Green, R. E., J. Krause, S. E. Ptak, and 8 others. 2006. Analysis of one million base pairs of Neanderthal DNA. *Nature* 444: 330–336.
Green, R. E., J. Krause, A. W. Briggs, and 53 others. 2010. A draft sequence of the Neanderthal genome. *Science* 328: 710–722
Haldane, J. B. S. 1924. A mathematical theory of natural and artificial selection. Part I. *Trans. Camb. Phil. Soc.* 23: 19–41.
Hoekstra, H. E., R. J. Hirschmann, R. J. Bundey, and 2 others. 2006. A single amino acid mutation contributes to adaptive color pattern in beach mice. *Science* 313: 101–104.
Houle, D. 2010. Numbering the hairs on our heads: The shared challenge and promise of phenomics. *Proc. Natl. Acad. Sci. USA* 107: 1793–1799.
Hunt, G., M. A. Bell, and M. P. Travis. 2008. Evolution toward a new adaptive optimum: Phenotypic evolution in a fossil stickleback lineage. *Evolution* 62: 700–710.
International HapMap Consortium. 2005. A haplotype map of the human genome. *Nature* 437: 1299–1320.
International Human Genome Sequencing Consortium (IHGSC). 2001. Initial sequencing and analysis of the human genome. *Nature* 409: 860–921.

International Human Genome Sequencing Consortium (IHGSC). 2004. Finishing the euchromatic sequence of the human genome. *Nature* 431: 931–945.

Jenner, R. A. and M. A. Wills. 2007. The choice of model organisms in evo-devo. *Nat. Rev. Genet.* 8: 311–319.

Jensen, J., Y. Kim, V. DuMont, and 2 others. 2005. Distinguishing between selective sweeps and demography using DNA polymorphism data. *Genetics* 170: 1401–1410.

Kaiser, J. 2008. DNA sequencing: A plan to capture human diversity in 1000 genomes. *Science* 319: 395.

Kimura, M. 1968. Evolutionary rate at the molecular level. *Nature* 217: 624–626.

Kitano, H. 2002. Systems biology: A brief overview. *Science* 295: 1662–1664.

Kitano, J., D. I. Bolnick, D. A. Beauchamp, and 4 others. 2008. Reverse evolution of armor plates in the threespine stickleback. *Curr. Biol.* 18: 769–774.

Krebs, H. A. 1975. The August Krogh Principle: "For many problems there is an animal on which it can be most conveniently studied." *J. Exp. Zool.* 194: 221–226.

Lalueza-Fox, C., H. Rõmpler, D. Caramelli, and 14 others. 2007. A melanocortin 1 receptor allele suggests varying pigmentation among Neanderthals. *Science* 318: 1453–1455.

Lauder, G. and M. Rose. 1996. *Adaptation*. Academic Press, San Diego.

Levy, S., G. Sutton, P. C. Ng, and 30 others. 2007. The diploid genome sequence of an individual human. *PLoS Biol.* 5: e254.

Lewontin, R. C. 1974. *The Genetic Basis of Evolutionary Change*. Columbia University Press, New York.

Lewontin, R. C. and J. L. Hubby. 1966. A molecular approach to the study of genic heterozygosity in natural populations. II. Amount of variation and degree of heterozygosity in natural populations of *Drosophila pseudoobscura*. *Genetics* 54: 595–609.

Li, R., W. Fan, G. Tian, and 122 others. 2010. The sequence and *de novo* assembly of the giant panda genome. *Nature* 463: 311–317.

Liberles, D. A. 2007. *Ancestral Sequence Reconstruction*. Oxford University Press, Oxford.

Linnen, C. R. and H. E. Hoekstra. In press. Measuring natural selection on genotypes and phenotypes in the wild. In *The Molecular Landscape,* Cold Spring Harbor Symposium Volume, Cold Spring Harbor Press, New York.

Linnen, C. R., E. P. Kingsley, J. D. Jensen, and 1 other. 2009. On the origin and spread of an adaptive allele in deer mice. *Science* 325: 1095–1098.

Marchinko, K. B. 2009. Predation's role in repeated phenotypic and genetic divergence of armor in threespine stickleback. *Evolution* 63: 127–138.

Mayr, E. 1963. *Animal Species and Evolution*. Harvard University Press, Cambridge, MA.

McDonald, J. H. and M. Kreitman. 1991. Adaptive protein evolution at the *Adh* locus in *Drosophila*. *Nature* 351: 652–654.

Mullen, L. M. and H. E. Hoekstra. 2008. Natural selection along an environmental gradient: A classic cline in mouse pigmentation. *Cold Spr. Harb. Symp. Evolution* 62: 1555.

Mullen, L. M., S. N. Vignieri, J. A. Gore, and 1 other. 2009. Adaptive basis of geographic variation: Genetic, phenotypic and environmental differentiation among beach mouse populations. *Cold Spr. Harb. Symp. Proc. Roy. Soc. B* 276: 3809.

Nielsen, R. 2009. Adaptationism—30 years after Gould and Lewontin. *Evolution* 63: 2487–2490.
Nielsen, R., C. Bustamante, A. G. Clark, and 10 others. 2005. A scan for positively selected genes in the genomes of humans and chimpanzees. *PLoS Biol.* 3: e170.
Noble, D. 2006. *The Music of Life: Biology Beyond the Genome*. Oxford University Press, Oxford.
Noonan, J. P., G. Coop, S. Kudaravalli, and 8 others. 2008. Sequencing and analysis of Neanderthal genomic DNA. *Science* 314: 1113–1118.
Noor, M. A. F. and J. L. Feder. 2006. Speciation genomics: Evolving approaches. *Nat. Rev. Genet.* 7: 851–861.
Orr, H. A., J. P. Masly, and N. Phadnis. 2007. Speciation in *Drosophila*: From phenotypes to molecules. *J. Hered.* 98: 103–110.
Paten, B., J. Herrero, S. Fitzgerald, and 4 others. 2008. Genome-wide nucleotide-level mammalian ancestor reconstruction. *Genome Res.* 18: 1829–1843.
Pfeiffer, T., O. S. Soyer, and S. Bonhoeffer. 2005. The evolution of connectivity in metabolic networks. *PLoS Biol.* 3: e228.
Pushkarev, D., N. F. Neff, and S. R. Quake. 2009. Single-molecule sequencing of an individual human genome. *Nat. Biotechnol.* 27: 847–850.
Rasmussen, M., Y. Li, S. Lindgreen, and 50 others. 2010. Ancient human genome sequence of an extinct Palaeo-eskimo. *Nature* 463: 757–762.
Ridley, M. 2004. Crick and Darwin's shared publication in *Nature*. *Nature* 431: 244.
Robert, F. and M. Chaussidon. 2006. A palaeotemperature curve for the Precambrian oceans based on silicon isotopes in cherts. *Nature* 443: 969–972.
Robinson, G. E., J. A. Banks, D. K. Padilla, and 18 others. In press. Empowering 21st century biology. *BioScience*.
Robinson, G. E., R. D. Fernald, and D. F. Clayton. 2008. Genes and social behavior. *Science* 322: 896–900.
Römpler, H., N. Rohland, C. Lalueza-Fox, and 6 others. 2006. Nuclear gene indicates coat-color polymorphism in mammoths. *Science* 313: 62.
Sanger, F., G. M. Air, B. G. Barrell, and 6 others. 1977. Nucleotide sequence of bacteriophage phi X174 DNA. *Nature* 265: 687–695.
Sauer, U., M. Heinemann, and N. Zamboni. 2007. Getting closer to the whole picture. *Science* 316: 550–551.
Schluter, D. 2009. Evidence for ecological speciation and its alternative. *Science* 323: 737–741.
Schluter, D., K. B. Marchinko, R. D. H. Barrett, and 1 other. In press. Natural selection and the genetics of adaptation in threespine stickleback. *Phil. Trans. Roy. Soc. B.*
Schwenk, K., D. K. Padilla, G. S. Bakken, and 1 other. 2009. Grand challenges in organismal biology. *Integr. Comp. Biol.* 49: 7–14.
Shapiro, M. D., M. E. Marks, C. L. Peichel, and 5 others. 2004. Genetic and developmental basis of evolutionary pelvic reduction in threespine sticklebacks. *Nature* 428: 717–723.
Shen, Y., R. Buick, and D. E. Canfield. 2001. Isotopic evidence for microbial sulphate reduction in the early Archaean era. *Nature* 410: 77–81.
Shubin, N., C. Tabin, and S. B. Carroll. 2009. Deep homology and the origins of evolutionary novelty. *Nature* 457: 818–823.
Sobel, J. M., G. F. Chen, L. R. Watt, and 1 other. 2010. The biology of speciation. *Evolution* 64: 295–315.
Steiner, C. C., J. N. Weber, and H. E. Hoekstra. 2007. Adaptive variation in beach mice caused by two interacting pigmentation genes. *PLoS Biol.* 5: e219.

Tecott, L. H. and E. J. Nestler. 2004. Neurobehavioral assessment in the information age. *Nat. Neurosci.* 7: 462–466.
Venter, J. C., M. D. Adams, E. W. Myers, and 122 others 2001. The sequence of the human genome. *Science* 291: 1304–1351.
Vignieri, S. N. and H. E. Hoekstra. In press. The selective advantage of cryptic coloration in mice. *Evolution*.
Wake, M. H. 2008. Integrative biology: Science for the 21st century. *BioScience* 58: 349–353.
Watson, J. D. and F. H. C. Crick. 1953. A structure for deoxyribose nucleic acid. *Nature* 171: 737–738.
Watt, W. B. 1977. Adaptation at specific loci. I. Natural selection on phosphoglucose isomerase of *Colias* butterflies: Biochemical and population aspects. *Genetics* 87: 177–194.
Weigel, D. and R. Mott. 2009. The 1001 Genome Project for *Arabidopsis thaliana*. *Genome Biol.* 10: 107–111.
Wheeler, D. A., M. Srinivasan, M. Egholm, and 24 others. 2008. The complete genome of an individual by massively parallel DNA sequencing. *Nature* 452: 872–876.
Whitesides, G. M. 2006. The origins and the future of microfluidics. *Nature* 442: 368–373.
Wikelski, M., R. W. Kays, N. J. Kasdin, and 3 others. 2007. Going wild: What a global small-animal tracking system could do for experimental biologists. *J. Exp. Biol.* 210: 181–186.
Zhu, H. and M. Snyder. 2002. "Omic" approaches for unraveling signaling networks. *Curr. Opin. Cell Biol.* 14: 173–179.

Commentary Seven

The Next 150 Years: Toward a Richer Theoretical Biology

Charles R. Marshall

The last 150 years have seen stunning advances in many areas of biology (Futuyma 2009; see Smocovitis, Commentary 1). As we look forward to the next 150 years, there is great potential for increased cross-fertilization among these areas, and consequently, increased opportunity for a richer theoretical biology. In this essay, I outline some conceptual approaches that might facilitate the realization of such opportunities. However, the development of an integrated theoretical biology is challenging, in part because:

1. Biological structures and processes are organized hierarchically: DNA and RNA lie within cells, which lie within organisms, that are part of populations, that constitute species, that are parts of ecosystems, and so forth.
2. Biological processes operate over many orders of magnitude, both temporally and spatially.
3. Biological systems are generally open and thus biologists can only make limited use of the conservation laws, such as the conservation of energy, that are so important in physics and chemistry (see Cracraft, Commentary 5).
4. Unlike chemistry, with its vast numbers of identical components (e.g., carbon atoms, phosphate ions, amino acids), biological components are typically heterogeneous (e.g., each gene differs from the next, each individual from the next, each species from the next) and the components occur in much smaller numbers.
5. Biological processes operate within an ever-changing Earth system and biota. What has happened in the past plays a major role in what happens in the future, and historical contingency influences evolutionary outcome.

In light of these challenges, I offer four considerations that should facilitate the task of developing a more richly integrated theoretical biology. Given how far evolutionary biology has come since 1859, we have every reason to aim high and set ambitious goals for the next 150 years.

Building Bridges between Hierarchical Levels

Biology is replete with theory, but much of that theory is not formulated in a way that enables a fertile interchange across hierarchical levels. For example, while evolution may be viewed as the filtering of development by ecology (Marshall 1995), there is virtually no theoretical contact between developmental biology and ecology. One of the grand challenges for evolutionary biologists is to bridge such divides. While I do not suppose to prescribe how this might be achieved, a key component of bridging theories is that they actively engage the language and concepts of the sub-disciplines they are trying to bridge.

As a simple example, consider how the gas laws can be used to understand why liquids cool as they evaporate. At a macroscopic level, gases can be understood through the ideal gas equation: $PV = nRT$. Statistical mechanics is the bridging theory that enables an appreciation of this macroscopic formulation from a microscopic standpoint. Thus, pressure is the macroscopic manifestation of the collisions of the gas particles within the enclosing walls of the container, and the macroscopic concept of temperature is related to the velocities of the molecules via the relationship: $1/2mv^2 = 3/2nkT$. Armed with this theory, one gains insight into evaporative cooling using the molecular perspective. The energies (velocities) of the molecules vary within the fluid. During evaporation, the most energetic molecules are the first to escape, leaving behind the less energetic,which results in a lowering of the temperature of the remaining liquid. The cooling that is experienced during evaporation is a consequence of the variation in the energies of the particles in a liquid. By meaningfully crossing the hierarchical level between fluids and their constituent molecules, we gain a sense of insight—perhaps the best criterion for determining whether a hierarchical divide has been crossed successfully.

While the microscopic foundation of the macroscopic gas laws are now understood, there is no succinct theory that bridges the gas laws to an understanding of more encompassing hierarchical levels, such as the weather—for instance, where the wind is going to come from or how hard it will blow. Bridging theories between hierarchical levels may be hard to come by. Can they be found? If so, what will they look like; if not, why not?

Biology has many sub-disciplines, and some are more easily bridged than others. Here, to explain what I mean by bridging theory, I consider just one case for which I feel such theory is needed: the relationship between developmental constraint and ecology.

The Ecological Meaning of Developmental Constraint

While a sophisticated understanding of the nature of developmental constraint is being developed (see G. Wagner, Chapter 8; Wray, Chapter 9), bridging theory is needed to help explain how developmental constraint translates into creating or limiting ecological opportunity (see Agrawal, Chapter 10; Berenbaum, Chapter 11; Losos and Mahler, Chapter 15; McPeek, Commentary 3). How might the concept of developmental constraint be couched in ecological language? As a step towards answering this question, consider the following example: most living mammals have seven neck vertebrae, whereas there is no stringent constraint on the number of neck vertebrae in their sister group, the diapsids. What are the ecological implications of this mammalian constraint? At the Darwin 2009 workshop, Michael Bell suggested that one way to understand the meaning of developmental constraint in an ecological context is to recognize that it may change the arena in which selection will occur. Yet, it may have only minor consequences from a functional point of view; for instance, despite their constraint, mammals have evolved a wide range of neck lengths, from short-necked whales to the long-necked giraffe. Unlike their diapsid cousins, which often respond to selection for neck length by changing the number of vertebrae, mammals have achieved these changes by varying the lengths of their vertebral elements. Thus, while the developmental constraint in mammals appears to have shifted the target of selection at the morphological level from the number of cervical vertebrae to their lengths, it does not appear to have severely constrained the solutions mammals have found to solve ecological problems (e.g., develop long or short necks). However, the limited number of neck vertebrae is likely to impose some constraints; for example, it might be hard to bend the neck into a tight curve or make it exceptionally long, as seen in the longest necked plesiosaurs. Thus, at the morphological extremes, the developmental constraints may indeed have important ecological implications. The questions that need to be answered are, Under what conditions and in what ways does developmental constraint translate into ecological constraint? An intermediate step in answering this question is to determine how developmental constraint limits the range of possible morphologies before the operation of selection—a step that is now within reach with the development of in silico generative morphospaces, whether based on simple geometric properties (Niklas 1994, 2004) or on those informed by our knowledge of the genes involved in morphogenesis (Salazar-Ciudad and Jernvall 2002; Kavanagh et al. 2007).

The Compressibility of Biological Knowledge

How important are the intricate differences between biological entities, such as individuals, species, and ecosystems, when trying to understand the outcomes of ecological and evolutionary processes? To what extent can general theories or conceptual frameworks be developed that provide reliable answers or insights to questions without recourse to so much raw data? In the following, I borrow the concept of image compression for digital images as a conceptual framework to think about this problem. One way to store a digital image is to describe all the pixels, but this approach requires an enormous amount of memory. Alternatively, one can employ a compression algorithm that takes advantage of repetitive patterns in the image to reduce the file size. By introducing a level of abstraction, a file can often be compressed without loss of information (e.g., if the image is stored as a TIFF file). One can further reduce the file size by employing algorithms that cause a loss of information, but in such a way that the overall perception of the image is not seriously compromised (e.g., when making a JPEG file). Thus, compression can be "lossless," in the case of TIFF files, or it can be "lossy," in the case of JPEGs.

One of the most interesting theoretical questions in biology is how compressible biological knowledge can be made. In general, the degree of compressibility will depend on the nature of the study, or the question being asked. On one hand, it is hard to see how biodiversity surveys can be made compressible. On the other hand, the mathematical theory of population ecology or population genetics, with generalized population parameters, fitness coefficients, and other variables, already have a high degree of compressibility. This compressibility carries over into empirical work as well. Thus, one can estimate selection coefficients by measuring survival rates or other factors, without the need to document the cause of death of every individual. The degree of compressibility also depends on what one wants to know. For example, in ecology a researcher might want an exhaustive list of all interactions between species in an ecosystem (no compression at the species level), but insight can be gained by lumping the species into trophic levels, significantly reducing the amount of data required to understand the ecosystem. This is lossy, but comprehensive compression.

Deep Theoretical Questions

In physics, the most difficult theory relates to questions of why the universe is structured the way it is, why gravity is so much weaker than the other forces, and so forth. In biology, there are similar questions: Why are there some one to ten million living species and not hundreds or thousands of times greater or fewer (Hutchison 1959)? Why are average species durations in the order of a few million years? Why did it take approximately 4.5 billion years before the emergence of sentient life after the formation of the Earth? When in human history did the first hints of sentience emerge (see Richerson and Boyd, Chapter 20)? What controls the rates of major phenotypic innovation? What controls the number of trophic levels in ecosystems? These are hard questions, made harder because it is difficult to incorporate historical contingency into their formulations. In fact, what is the overall importance of contingency in shaping the history of life? In the context of the Cambrian explosion of animals, some scholars think contingency is enormously important (Gould 1989), others hardly at all (Conway Morris 1998, 2003), while I have suggested an intermediate position (Marshall 2006). While one might argue that without replicate histories hypotheses about the cause of unique events cannot be tested, these questions need not be resistant to testing. For example, experimentation can, in principle, be achieved in silico (Niklas 1994, 2004). Finally, answering such questions is also made difficult by the way science is (appropriately) funded, with the emphasis on shorter-term explicitly achievable goals. But, these questions are begging for answers, and they may yield to scientific methods if they were given more attention.

Expecting the Unexpected

New data and new theoretical advances can undo the ideas that we hold most dear. For example, in 1964 George Gaylord Simpson concluded that "there is a strong consensus that completely neutral genes or alleles must be very rare if they exist at all" (Simpson 1964: 1537). This statement was challenged with new data and no such consensus exists today. Thus, future advances may render our most cherished models and theories simply wrong. If evolutionary biologists are lucky, discoveries will be made that reshape basic conceptual tenets, just as the discoveries of neutral change or of the importance of mass extinctions in shaping the living biota have done. I say lucky, because revolutionary discoveries will show us, retrospectively, that we had been unknowingly trapped in an impoverished worldview. Thus, as we look to the future, it is best to do more than just view new data as a way to refine existing theory. We should keep our minds open to the possibility that fundamental new discoveries will demand abandonment or major refinement of old theory or the development of new theory. Thus, for example, as whole genome data come pouring in (see Kolaczkowski and Kern, Chapter 6), evolutionary biologists might look to them as more than a way of increasing statistical power to test current theory—who knows what dynamics might be discovered if we have the wit to search for them. The future of evolutionary biology will likely be enriched if we expect the unexpected.

Literature Cited

Conway Morris, S. 1998. The *Crucible of Creation: The Burgess Shale and the Rise of Animals*. Oxford University Press, Oxford.

Conway Morris, S. 2003. *Life's Solution: Inevitable Humans in a Lonely Universe*. Cambridge University Press, Cambridge, UK.

Futuyma, D. J. 2009. *Evolution*, 2nd ed. Sinauer Associates, Sunderland, MA.

Gould, S. J. 1989. *Wonderful Life: The Burgess Shale and the Nature of History*. Cambridge University Press, Cambridge, UK.

Hutchinson, G. E. 1959. Homage to Santa Rosalina, or why are there so many kinds of animals? *Am. Nat*. 93: 145–159.

Kavanagh, K. D., A. E. Evans, and J. Jurnvall. 2007. Predicting evolutionary patterns of mammalian teeth from development. *Nature* 449: 427–432.

Marshall, C. R. 1995. Darwinism in an age of molecular revolution. In C. R. Marshall and J. W. Schopf (eds.), *Evolution and the Molecular Revolution*, pp. 1–30. Jones and Bartlett, Sudbury, MA.

Marshall, C. R. 2006. Explaining the Cambrian "explosion" of animals. *Ann. Rev. Earth Planet. Sci*. 34: 355–384.

Niklas, K. J. 1994. Morphological evolution through complex domains of fitness. *Proc. Natl. Acad. Sci. USA* 91: 6772–6779.

Niklas, K. J. 2004. Computer models of early land plant evolution. *Ann. Rev. Earth Planet. Sci*. 32: 47–66.

Salazar–Ciudad, I. and J. Jernvall. 2002. A gene network model accounting for development and evolution of mammalian teeth. *Proc. Natl. Acad. Sci. USA* 99: 8116–8120.

Simpson, G. G. 1964. Organisms and molecules in evolution. *Science* 146: 1535–1538.

Commentary Eight

The Expansion of Molecular Data in Evolutionary Biology

Joshua S. Rest

Near the end of his voyage on the *Beagle*, Darwin chose to study barnacles because of their peculiar characteristics and because of his instinct that their geological distribution held important insights (Love 2002). Darwin had a fantastic range of potential study subjects that he might have instead chosen to pursue. His observations spanned great biological variety, from howler monkeys, butterflies, parasitic orchids, and toucans to ceiba trees. He probably felt overwhelmed by the conflicting desires to travel and gather more data, versus the need to analyze, theorize, test, and write about his corpus of observations. When Darwin discovered barnacles possessing unexpected features, it may have seemed logical to collaborate and send them to a cirripedologist so that he might instead focus his efforts on the species problem. But there was no such specialist, so he devoted 8 years to becoming a barnacle specialist (Stott 2003). Eight years of meticulous dissection, classification, and correspondence with a network of other naturalists passed, while his essay about the origin of species sat locked in a drawer and stewed in his mind.

Today, the amount of molecular data available to evolutionary biology may seem as overwhelming as the data that Darwin had collected. Like the new organisms and habitats that Darwin observed on the *Beagle*, molecular data offers a seemingly unlimited source of characters for study. However, a single biologist cannot hope to master the diverse types of data now available. A cursory list of recently developed molecular data types includes next-generation sequencing based on emulsion PCR or bridge PCR, pyrosequencing, and chromatin immunoprecipitation followed by sequencing to assay acetylation, methylation, nucleosome positioning, or DNA–protein interactions. GenBank, the repository of sequencing data, has grown exponentially since its inception (Benson et al. 2009), and as of April 2010, contained over 279 billion base pairs. Other types of data have also grown rapidly; for example, the number of known protein–protein interactions in yeast has increased exponentially over time (He and Zhang

2009). The number of studies, the number of interactions detected per study, and the number of authors per study have all commensurately increased. The quantity of data will continue to grow, as it is now proposed to sequence 10,000 vertebrate genomes (Genome 10k 2009), and a systematic cataloging of genome-wide epigenetic traits in worms and flies is well underway (Celniker et al. 2009). This enormous increase in data production and analysis has been catalyzed by, and indeed relies upon, the escalation in processing power described by Moore's law and its corollaries related to data storage and memory (Bell et al. 2009).

All of these data types hold great promise for the future of evolutionary biology. Perhaps the most immediate benefit has been the quantity and quality of allelic polymorphism data from increased sequencing coverage of individuals, populations, and genes, allowing high-resolution determination of population structures (see Zhang, Chapter 4; Kolaczkowski and Kern, Chapter 6). Several of the high-throughput sequencing approaches are being used to identify potentially adaptive variation in wild populations and non-model organisms (Ellegren 2008; Hudson 2008). RNA sequencing allows identification of variation that is focused on genes and gene transcript levels in non-model organisms, because a reference genome is not required (Novaes et al. 2008). Sequencing technology is also used to assess genome-wide organizational, epigenetic, and protein-binding characteristics of DNA. Although such epigenetic and protein-binding analysis is currently practical only in model organisms, the number of species that can be considered to be models continues to increase, and the data generated in these model organisms can be mapped onto the genomes of non-model organisms to assess evolution of these characters. Using this approach across fungal species, for example, transcription factor binding sites involved in ribosomal regulation were shown to have been gained and lost via an intermediate stage during which redundant elements were present (Tanay et al. 2005). This approach has also been used to study the evolution of metabolic- and protein-interaction networks.

These molecular approaches are clearly powerful for evolutionary biologists, but an individual who wants to master a particular area outside her or his field usually cannot afford to spend 8 years becoming an expert in a new discipline. A first solution, of course, is to collaborate with a specialist. The degree of expertise available is demonstrated by the dramatic increase in both scientists and peer-reviewed journals: the number of journals has increased from only hundreds at the end of the nineteenth century to over 10,000 today (Mabe and Amin 2001). It is not hard to recall discoveries in which close collaboration between molecular and evolutionary biologists played a role. One recent example is the result from large-scale sequencing efforts that showed widespread purifying and positive selection acting on the *Drosophila* genome. Nearly two-thirds of mutations identified in noncoding regions are deleterious, and up to half of the amino acid substitutions and one-in-five noncoding substitutions are adaptive (Andolfatto 2005; Begun 2007). These observations challenge the prominence of the neutral theory for describing evolutionary patterns in *Drosophila*. In contrast, one can also think of scientific questions that have remained refractory and situations in which conflict between specializations or disciplines has not helped matters. For example, phylogenetic relationships and the timing of divergences within Neoaves have defied resolution, despite increasing amounts of sequence data, because there is conflict between molecular and paleontological data (e.g., Brown et al. 2008). A lack of interaction between molecular evolutionists and paleontologists is partly to blame; for example, these two communities often interpret the meaning of paleontological dates differently (Brochu et al. 2004). It is clear that scientists in different fields will have to collaborate to find creative solutions to these difficult problems.

A second solution for evolutionary biologists who want to take advantage of publicly available molecular data is to become familiar with a few of the most essential computational tools. As the number of complete genomes, transcriptomes, and associated data increases, so do the analytical tools and aptitude required for analysis. A decade or two ago, with the advent of phylogenetics, many evolutionary biologists became amateur computational biologists. Today, the genomic era requires a high level of analytical sophistication. In particular, large-scale collection of data comes along with a commensurate proportion of errors and requires careful statistical analysis to account for multiple comparisons. Storing, parsing, tabulating, and visualizing genome-scale datasets can be a significant technical challenge (Bell et al. 2009), and the data often require substantial refinement before comparative analysis can even be considered. Novel data sets and hard problems require dedicated research by experts in statistics, bioinformatics, and computer science. Through time, these tools will be integrated into packages for more general use. At present, processing, statistical analysis, and visualization of biological and genomic data are being centralized in the R software environment, and learning this language is currently an efficient way for all biologists to access shared libraries of data-oriented tools. R and its associated libraries are especially useful for visualizing large and complex datasets in ways that can lead to meaningful understanding of the nature of the data.

As magical as this data revolution may seem, however, evolutionary biology will be ill served by merely increasing the amount of data collected. Let us not forget that Linnaeus and Lamarck had gathered copious observations on natural history, but only Darwin actually derived a sound scientific explanation for the data. Even in the jungle of molecular genomics data, careful thinking, good hypotheses, and broad knowledge of organisms will remain central to actual discovery. Data collection and analytical details are only important to the extent that they complement these skills. In this way, how evolutionary biologists *ask* questions will remain fundamentally the same.

Literature Cited

Andolfatto, P. 2005. Adaptive evolution of non-coding DNA in *Drosophila*. *Nature* 437: 1149–1152.

Begun, D. J., A. K. Holloway, K. Stevens, and 10 others. 2007. Population genomics: Whole-genome analysis of polymorphism and divergence in *Drosophila simulans*. *PLoS Biol.* 5: e310.

Bell, G., T. Hey, and A. Szalay. 2009. Beyond the data deluge. *Science* 323: 1297–1298.

Benson, D. A., I. Karsch-Mizrachi, D. J. Lipman, and 2 others. 2009. Genbank. *Nucl. Acids Res.* 37: D26–31.

Brochu, C. A., C. D. Sumrall, and J. M. Theodor. 2004. When clocks (and communities) collide: Estimating divergence time from molecules and the fossil record. *J. Paleontol.* 78: 1–6.

Brown, J. W., J. S. Rest, J. Garcìa-Moreno, and 2 others. 2008. Strong mitochondrial DNA support for a cretaceous origin of modern avian lineages. *BMC Biol.* 6: 6.

Celniker, S. E., L. A. L. Dillon, M. B. Gerstein, and 15 others. 2009. Unlocking the secrets of the genome. *Nature* 459: 927–930.

Ellegren, H. 2008. Sequencing goes 454 and takes large-scale genomics into the wild. *Mol. Ecol.* 17: 1629–1631.

Genome 10k Community of Scientists. 2009. Genome 10k: A proposal to obtain whole-genome sequence for 10,000 vertebrate species. *J. Hered.* 100: 659–674.

He, X. and J. Zhang. 2009. On the growth of scientific knowledge: Yeast biology as a case study. *PLoS Comp. Biol.* 5: e1000320.

Hudson, M. E. 2008. Sequencing breakthroughs for genomic ecology and evolutionary biology. *Mol. Ecol. Resour.* 8: 3–17.

Love, A. C. 2002. Darwin and *Cirripedia* prior to 1846: Exploring the origins of the barnacle research. *J. Hist. Biol.* 35: 251–289.

Mabe, M. and M. Amin. 2001. Growth dynamics of scholarly and scientific journals. *Scientometrics* 51: 147–162.

Novaes, E., D. R. Drost, W. G. Farmerie, and 4 others. 2008. High-throughput gene and SNP discovery in *Eucalyptus grandis*, an uncharacterized genome. *BMC Genomics* 9: 312.

Stott, R. 2003. *Darwin and the Barnacle.* Faber and Faber, London.

Tanay, A., A. Regev, and R. Shamir. 2005. Conservation and evolvability in regulatory networks: The evolution of ribosomal regulation in yeast. *Proc. Natl. Acad. Sci. USA* 102: 7203–7208.

Appendix

Darwin 2009 Workshop Participants

The following is a list of participants (on at least one day) in the Darwin 2009 workshop at Stony Brook University, Stony Brook, NY, USA, November 5–7, 2009. This list is based on the preregistration list and signatures on the attendance roster. A small number of names could not be read or verified and do not appear below.

Sarah K. Abruzzi
Anurag Agrawal
Matthew Aiello-Lemmens
Resit Akçakaya
Mary Alldred
Stephen Baines
Edwin Battley
Michael A. Bell
Paul Bingham
Stephanie Blatch
Carola Borries
Doug Boyer
Peter J. Bowler
Robert Brandon
Leone Brown
Jonathan Bun
Rodrigo Cogni
Jackie Collier
Joel L. Cracraft
Charles Davis
Brigitte Demes
Donna DiGiovanni
Diane Doran
Daniel E. Dykhuizen
Walter F. Eanes
Wendy Erb
M. Caitlin Fisher-Reid
Megan Flenniken
Michael Foote
Nicholas A. Friedenberg
Douglas J. Futuyma
Jin Gao
Owen Gilbert
George Gilchrist
Aman Gill
Lev Ginzburg
Jan Gogarten
Fred Gould
Patrick Goymer
Catherine Graham
Sarah Gray
Rebecca Grella
Jessica Gurevitch
Mark Hall
Will Harcourt-Smith
Richard G. Harrison
Erin Hill-Burns
David M. Hillis
Hopi E. Hoekstra
Sandra Hoffberg
Xia Hua
Emranul Huq
Jukka Jernvall
Mark Jonas
William L. Jungers
Andrew D. Kern
Mark Kirkpatrick
Andreas Koenig
Hanna Kokko
Spencer Koury

Joseph Lachance
Christopher E. Lane
Adam Laybourn
Antonio Lazcano
Justin Ledogar
Peter Lessing
Jeffrey S. Levinton
Jessica Lodwick
Jonathan B. Losos
Bobbi Low
Patrick Lyons
Melissa Mark
Gary Marker
Charles R. Marshall
Michael McCann
Mark A. McPeek
Daniel Moen
Antonia Monteiro
Seungjae Moon
Christopher Morales
Rocio Ng
Kelly L. O'Donnell
Niamh O'Hara
Alison Onstein
Dana Opulente
Kerry M. Ossi
Dianna K. Padilla
Peter J. Park
Juan Luis Parra
Mihaela Pavlicev
Kestrel Perez
Laurie Perino
David Queller
Joshua S. Rest
Peter J. Richerson
F. James Rohlf
Jennifer Rollins
Emily Rollinson
Nicolas Seltzer
Mark Siegal
Natalia Silva
Vassiliki Betty Smocovitis
K. James Soda
Robert R. Sokal
Gina Sorrentino
Joanne Souza
Samuel L. Stanley, Jr.
Jessica Stanton
Elizabeth St. Clair
Randall Susman
Stacey Tecot
Boris Tinco
Frank Turano
Carolina Ulloa
Fredric Vencl
Vivek Venkataraman
Kyle Marian Viterbo
Günter P. Wagner
Peter J. Wagner
John Wakeley
John Waldron
Ramona Walls
Norah Warchola
Omar Warsi
Tim D. White
Gregory A. Wray
Jianzhi Zhang

Index

The letter *n* following a page number indicates that the entry is derived from a footnote.

A

D

E

Q

R

S